BORATE GLASSES, CRYSTALS, & MELTS:

Eighth International Conference

and

PHOSPHATE GLASSES:

First International Conference

held at University of Pardubice, Czech Republic,
on 30 June-4 July 2014

Edited by:
Alex C. Hannon & Ladislav Koudelka

Society of Glass Technology
Sheffield, 2016

Borate 8 – Phosphate 1
Proceedings of the Eighth International Conference on Borate Glasses, Crystals and Melts and *First International Conference on Phosphate Glasses* held at University of Pardubice, Czech Republic, on 30 June–4 July 2014. A collected volume of papers from the Borate and Phosphate Conferences, the papers were originally published in *Physics and Chemistry of Glasses: European Journal of Glass Science and Technology Part B.*

ISBN 978-0-900682-83-4

The objects of the Society of Glass Technology are to encourage and advance the study of the history, art, science, design, manufacture, after treatment, distribution and end use of glass of any and every kind. These aims are furthered by meetings, publications, the maintenance of a library and the promotion of association with other interested persons and organisations.

Society of Glass Technology
9 Churchill Way
Chapeltown
Sheffield S35 2PY, UK
Tel +44(0)114 263 4455
Email info@sgt.org
Web http://www.sgt.org
The Society of Glass Technology is a registered charity no. 237438.

Foreword

This volume contains the proceedings of two conferences, Borate8 and Phosphate1. Or, to name them in full, the *Eighth International Conference on Borate Glasses, Crystals and Melts: Borate8* and *The International Conference on Phosphate Glasses: Phosphate1*. There are 36 papers in total, with 18 from the Borate8 conference and 18 from the Phosphate1 conference. The papers have previously been published individually in the journal *Physics and Chemistry of Glasses: European Journal of Glass Science and Technology Part B*, for which a full peer review process was used. Each paper was published in the journal as soon as it had had been reviewed and edited, and this volume represents the first time that the papers from the conferences have been published together.

The *Eighth International Conference on Borate Glasses, Crystals and Melts: Borate8* and *The International Conference on Phosphate Glasses: Phosphate1* were held during the period 30th June–4th July, 2014 at The University of Pardubice, in Pardubice, The Czech Republic. This followed previous Borate Conferences held at Alfred, USA (1977); Abingdon, United Kingdom (1996); Sofia, Bulgaria (1999); Cedar Rapids, USA (2002); Trento, Italy (2005); Himeji, Japan (2008); and Halifax, Canada (2011), and it was the fourth time that the Conference was held in Europe. The Eighth Conference was dedicated to Professor Stanislav Filatov of Saint Petersburg, Russia, to honour his achievements in crystallography, and in particular the structure of Borate Crystals.

The community of scientists who study phosphate glasses had occasionally held individual meetings over the years, but never with a regular schedule. Noting the long term success of the Borate Conference, it had been decided to hold an International Conference on Phosphates together with the Borate Conference. The meeting in Pardubice was the first time that this had been done, and, as these proceedings show, it was a successful venture that will be repeated in the future. Professor Doris Ehrt of the Otto Schott Institute at the Friedrich Schiller University in Jena, Germany was chosen as the honoree of the first Phosphate Conference, in recognition of her achievements in the glass chemistry of phosphate and fluorophosphate optical glasses.

In total, 158 scientists from 27 countries gathered in Pardubice; 92 participants attended both conferences, whilst 41 attended only the Borate8 meeting, and 25 attended only the Phosphate1 meeting. The participants came from : Austria (3), Belgium (3), Brazil (2), Bulgaria (4), Canada (3), China (1), Croatia (2), Czech Republic (10), Denmark (2), Egypt (1), France (19), Germany (15), Greece (2), India (5), Italy (2), Japan (18), Republic of Korea (1), Liechtenstein (1), Poland (9), Romania (2), Russia (18), Slovakia (4), Spain (3), Switzerland (2), Taiwan (2), United Kingdom (8), and United States of America (16).

The Borate8 Conference started with a welcome from the conference chairman, Ladislav Koudelka, and the presentation of a special piece of glassware to the honoree, Stanislav Filatov. The lecture in honour of Professor Filatov, entitled "Stanislav Konstantinovich Filatov: Life, Crystal Chemistry and Crystal Chemistry of Borates", was presented by Sergey Krivovichev, who described the major role played by Professor Filatov for many years in using crystallography to study many various aspects of crystalline borate phases. This was followed by a plenary talk entitled "High-temperature borate crystal chemistry", given by Professor Filatov. The other invited talks on borates were by: Josef Zwanziger (Dalhousie University, Canada), Hubert Huppertz (University of Innsbruck, Austria), Thibault Charpentier (CEA, Gif-sur-Yvette France), Philip Salmon (University of Bath, United Kingdom), Yuanzheng Yue (Aalborg University, Denmark), Doris Möncke (Friedrich Schiller University, Germany), Giovanna D´Angelo (University of Messina, Italy), and Takayuki Komatsu (Nagaoka University of Technology, Japan).

On the morning of the third day of the week, there were joint sessions on borophosphates, involving participants of both the Borate8 and the Phosphate1 Conferences. These were followed by an excursion to Opočno Castle, and then a conference dinner at Pardubice Castle was enjoyed by the conference participants.

Ladislav Koudelka, the conference chairman, started the Phosphate1 Conference with a welcome address, and the presentation of a special piece of glassware to the honoree, Doris Ehrt. Doris Möncke gave a lecture in honour of Professor Ehrt, entitled "Doris Ehrt - Glass Chemical Research in the Spirit of Otto Schott", in which she described Doris Ehrt's major role over many years in the study of the glass chemistry of phosphate and fluorophosphate optical glasses. This was followed by a plenary talk by Professor Ehrt, entitled "Phosphate and fluoride phosphate optical glasses - properties, structure and applications". The other invited talks on phosphates were by: Seiji Inaba (Tokyo Institute of Technology, Japan), Franck Fayon (CNRS Orléans, France), David Sidebottom (Creighton University, USA), Jonathan Knowles (UCL Eastman Dental Institute, United Kingdom), Gavin Mountjoy (University of Kent), and Antonella Rossi (ETH Zurich, Switzerland and University of Cagliari, Italy).

In total the oral presentations consisted of 40 talks on borates, 10 talks on borophosphates, and 48 talks on phosphates. There were also two poster sessions, with 42 borate posters, and 31 phosphate posters. The Borate8 Conference structure involved the following sessions and chairmen: Borate Opening Session (L. Koudelka), Borate Crystals (A. C. Wright), Borate Structure – NMR (J. W. Zwanziger), Borate Structure – Neutrons and X-rays (A. C. Hannon), Glass Transition, Relaxation, in Borates (E. I. Kamitsos), Ions in Borate Glass (S. A.

Feller), Fundamental Borate Science (A. Takada), Dynamics of Borates (R. E. Youngman), and Borate Applications (H. Huppertz). The joint sessions on borophosphates were as follows: Borophosphates Session 1 (S. Kroeker), Borophosphates Session 2 (S. W. Martin). The Phosphate1 Conference structure involved the following sessions and chairmen: Phosphate Opening Session (Koudelka), Phosphate Structure and Properties (L. Montagne), Phosphate Melts and Relaxation (R. K. Brow), Crystallized Phosphate Glasses (P. Mošner), Phosphate Optical Properties (D. Möncke), Nano/Micro-Structures and Defects in Phosphate Glasses (D. Ehrt), Phosphate Bioglasses and Surfaces (H. Takebe), Phosphate Modelling and Optical properties (F. Muñoz), Phosphate Conductivity (D. L. Sidebottom), and Phosphate Surfaces and Corrosion (A. Moguš-Milanković).

Thanks are due to the chairman of the Borate International Organising Committee, Adrian Wright (University of Reading, UK), and the members of the committee: Alexis Clare (Alfred University, USA), Laurent Cormier (Université Pierre & Marie Curie, France), Guiseppe Dalba (University of Trento, Italy), Yanko Dimitriev (University of Chemical Technology and Metallurgy, Bulgaria), Doris Ehrt (University of Jena, Germany), Steve Feller (Coe College, USA), Alex Hannon (ISIS Facility, UK), Hubert Huppertz (University of Innsbruck, Austria), Stratos Kamitsos (National Hellenic Research Foundation, Greece), Ladislav Koudelka (University of Pardubice, Czech Republic), Scott Kroeker (University of Manitoba, Canada), Akira Takada (Asahi Glass Co. Ltd, Japan), Norimasa Umesaki (SPRING-8, Japan), Natalia Vedishcheva (Institute of Silicate Chemistry, Russia), Randy Youngman (Corning Inc., USA), and Josef Zwanziger (Dalhousie University, Canada).

We are most grateful to the chairman of the Phosphate International Organising Committee, Dick Brow (Missouri University of Science and Technology, USA), and the members of the committee: Ifty Ahmed (University of Nottingham, UK), Hellmut Eckert (W W University, Münster, Germany and University of São Paolo, Brazil), Doris Ehrt (University of Jena, Germany), Jerzy Garbarczyk (Warsaw University of Technology, Poland), Aswini Ghosh (Indian Association for the Cultivation of Science, India), Uwe Hoppe (University of Rostock, Germany), Ladislav Koudelka (University of Pardubice, Czech Republic), Steve Martin (Iowa State University of Science & Technology, USA), Younès Messaddeq (Université Laval, Canada), Lionel Montagne (University of Lille, France), Francisco Muñoz (Ceramics and Glass Institute, Spain), Hiromichi Takebe (Ehime University, Japan), and Hsi-Wen Yang (National United University, Taiwan).

We are most grateful to the members of the Local Organising Committee: Ladislav Koudelka (chairman), Petr Mošner, Jana Holubova, Zdeněk Černošek, Ivana Rösslerova. We appreciated very successful organization by the company ICARIS Ltd, Conference Management, Prague with two most helpful persons Dr Ladislav Červinka and Romana Kočova. The main organizer of the conferences was the Czech Glass Society.

Generous financial contributions in support of the conferences were received from the Faculty of Chemical Technology of the University of Pardubice, and the Town Council of Pardubice City for free transportation tickets for conference participants. Both supports were very important in ensuring the success of the conferences, and are gratefully acknowledged. Pardubice has a pleasant, relaxed atmosphere, and this provided a delightful environment in which the conference delegates could focus on the science of borates and phosphates, and then enjoy their leisure.

The generous sponsorship of Preciosa A.S. and Corning Inc. was important for the success of the conferences, and is gratefully acknowledged. The support of the Society of Glass Technology was essential for the publication of the Conference Proceedings.

The location of the next Conference was considered by the International Organising Committee, and it was decided to hold the next conference in the United Kingdom. Dr Emma Barney of the University of Nottingham will chair the *Ninth International Conference on Borate Glasses, Crystals and Melts: Borate9* and the *Second International Conference on Phosphate Materials: Phosphate2*. The conferences will be held in succession at St Anne's College, Oxford, from 24th to 28th July, 2017. The Borate9 Conference will be held in honour of Professor Steve Feller, of Coe College, USA, recognising his many achievements in the study of the properties and structure of borate glasses. The Phosphate2 Conference will be held in honour of Dr Uwe Hoppe, of the University of Rostock, Germany, in recognition of his huge contribution to the structural studies of phosphate glasses by neutron and X-ray diffraction. All of the participants present in Pardubice are cordially invited to attend.

Ladislav Koudelka (University of Pardubice), Borate8 and Phosphate1 Conference Chairman
Alex Hannon (ISIS Facility), Borate8 Conference Co-Chairman

The Borate8 Conference and Proceedings are dedicated to Professor Stanislav Konstantinovich Filatov to honour his achievements in Crystallography, especially in the study of Borate compounds.

The Phosphate1 Conference and Proceedings are dedicated to Professor Doris Ehrt to honour her achievements in Glass Science, and in particular Fluorophosphate Glasses.

On the left is the Conference Chairman, Ladislav Koudelka, together with the Borate8 Honoree, Stanislav Filatov, enjoying the art glass exhibition at the conference dinner

The Phosphate1 Honoree, Doris Ehrt, addressing the conference

CONTENTS

The journal references for the papers are given using this journal abbreviation:
PC=*Physics and Chemistry of Glasses: European Journal of Glass Science and Technology Part B*

Proceedings of the Eighth International Conference on Borate Glasses, Crystals and Melts: Borate8

Proceedings of the First International Conference on Phosphate Glasses: Phosphate1

EIGHTH INTERNATIONAL CONFERENCE ON BORATE GLASSES, CRYSTALS AND MELTS: BORATE8

University of Pardubice, Czech Republic,
30 June–2 July 2014

PROGRAMME

Monday 30th June

Borate Opening Session *Chair: L. Koudelka*

8:30 Borate Welcome
L. Koudelka

8:40 Stanislav Konstantinovich Filatov: Life, Crystal Chemistry and Crystal Chemistry of Borates
S. V. Krivovichev

9:10 High-Temperature Borate Crystal Chemistry
S. K. Filatov, R. S. Bubnova

9:40 Advances in Photoelastic Studies of Borate Glasses
J. W. Zwanziger

Borate Crystals *Chair: A. C. Wright*

10:40 Recent Advances in the High-Pressure Chemistry of Borates
H. Huppertz

11:10 Recent Advances in the High-Pressure Chemistry of Alkali Metal and Alkali Metal Equivalent Borates
G. Sohr, S. C. Neumair, H. Huppertz

11:30 High Pressure Borate Apatites Exhibiting Boron in a Fourfold Coordination
M. Glätzle, H. Huppertz

11:50 The High Viscosity of Borate Glasses and Crystals
R. S. Bubnova, S. K. Filatov

12:10 Morphology of Phase Separation and Coarsening of Ba–O–SiO_2–B_2O_3 by X-Ray Microtomography
D. Bouttes, D. Vandembroucq, E. Gouillart, D. Dalmas, E. Boller

Borate Structure – NMR *Chair: S. Kroeker*

14:00 The MD-GIPAW Method: Applications to Borate and Borophosphate Glasses
T. Charpentier

14:30 Structure and Speciation in Borogallate, Boroaluminate and Borovanadate Glasses: The View from Multinuclear Magnetic Resonance
S. Kroeker, J. E. C. Wren, V. K. Michaelis, P. M. Aguiar, S. A. Feller

14:50 Structure-Property Relationships in Pyrex® and Related Boroaluminosilicate Glasses
R. E. Youngman, J. C. Mauro, M. M. Smedskjaer

15:10 Intermediate Range Structures of Alkali Borate Glasses as Determined by ^{10}Boron Solid-State Nuclear Magnetic Resonance
R. Rice, J. Brown, S. Feller, D. Holland, M. Smith, J. Berkowitz, K . Tholen, M. McConnell, V. Khristinko, E. Troendle, N. Barnes, K. Goranson, M. Faaborg , M. Affatigato

15:30 A New Insight on Rare-Earth (RE) Metaborate glass Composition REB_3O_6 through a Multi-Spectroscopic Approach
H. Trégouët, D. Caurant, O. Majérus, L. Cormier, T. Charpentier, H. Vezin, D. Pytalev

Borate Structure – Neutrons and X-Rays *Chair: A. C. Hannon*

16:20 Density-Driven Structural Transformations in B_2O_3 Glass
P. S. Salmon, A. Zeidler

16:50 Densified Liquid B_2O_3: Dynamic and Structural Properties
A. Baroni, G. Ferlat, M. Salanne, M. Micoulaut

17:10 A Neutron Diffraction Study of Six $M_2O.M'_2O.5B_2O_3$ Mixed-Modifier Di-Pentaborate Glasses
A. C. Wright, R. N. Sinclair, C. E. Stone, J. L. Shaw, S. A. Feller, R. B. Williams, H. E. Fischer, N. M. Vedishcheva

17:30 Probing the Oxygen Environment of Lithium Borate Glasses/Crystals Using Inelastic X-Ray Scattering at the Oxygen K-Edge
G. Lelong, L. Cormier, G. Radtke, G. Rousse, B. Baptiste, J.-P. Rueff, J. Ablett

17:50 Multi-Edge Spectroscopic Study in the Lithium Borate System
L. Cormier, G. Lelong, L. Hennet, G. Radtke, J.-P. Rueff, J. Ablett

Tuesday 1st July

Glass Transition and Relaxation in Borates *Chair: E. I. Kamitsos*

8:30 Glass Transition and Relaxation in Pressure-Quenched Borate and Borosilicate Glasses
Y. Z. Yue

9:00 Crystallization of $LaBGeO_5$ in La_2O_3–B_2O_3–GeO_2 Boro-Germanate Glass; Second Harmonic Generation in the Transparent Glass Ceramics
E. Fargin, H. Vigourou, L. N. Truong, D. De Ligny, B. Champagnon, M. Dussauze, F. Adamietz, V. Rodriguez, A. Corcoran, Y. Messaddeq, L. F. Santos

9:20 Ionic Conductivity of Binary Lithium Borate Glasses and Melts
H. Fan, L. Del Campo, S. Ory, D. De Sousa Meneses, M. Malki, P. Echegut

9:40 Dissolution Kinetics of Borate Glasses in Aqueous Solutions
R. K. Brow, J. L. George, K. L. Goetschius

Ions in Borate Glass *Chair: S. A. Feller*

10:30 Metal Ions in Borate and Borosilicate Glasses: Cluster Formation and Borate Ligand Speciation
D. Möncke

11:00 Structural Investigation of Highly Modified Eu^{2+}-Sr^{2+}-Borate Glasses
A. Winterstein, D. Möncke, D. Palles, E. I. Kamitsos, L. Wondraczek

11:20 Absorption and Luminescent Thermochromism of Copper- and Chlorine-Containing Potassium-Alumina-Borate Glasses
A. N. Babkina, N. V. Nikonorov, P. S. Shirshnev, A. I. Sidorov

11:40 Microstructure of Porous Glasses in System of SiO_2–B_2O_3–Na_2O Melted in Electric Heated Mini-Melter
A. Saberi, A. Rosin, K. Kyrgyzbaev, T. Gerdes, M. Willert-Porada

12:00 Structural Role of Titanium Ions on the Enhancement of Bioactivity of B_2O_3–SiO_2–Na_2O–CaO Glass System
G. Sahaya Baskaran, G. Jagan Mohini, N. Veeraiah

Fundamental Borate Science (Parallel Session) *Chair: A. Takada*

11:00 First-Principles Simulations of B_2O_3: From Crystals to Liquid Phases
G. Ferlat, H. Hay, A.-P. Seitsonen, T. Charpentier, M. Lazzeri, F. Mauri

11:20 Short- and Intermediate-Range Order in Sodium Borosilicate Glasses: A Quantitative Thermodynamic Approach
N. M. Vedishcheva, A.C. Wright

11:40 O 2p Partial Density of States and Bond Angles Around O Atoms in Borate Glass: Soft X-Ray Emission and *Ab Initio* Molecular Dynamics Studies
S. Hosokawa, H. Sato, K. Mimura, Y. Tezuka, D. Fukunaga, F. Shimojo

12:00 Phase Separation and Magnetic Particles in Borate Glasses
I. Edelman, O. Ivanova, Y. Zubavichus, V. Zaikovskiy, J. Curély, J. Kliava

Dynamics of Borates *Chair: R. E. Youngman*

13:50 Structural, Vibrational And Thermal Properties of Borate Glasses: Influence of Densification and Modifying Role of Alkali Metal Cations
G. D'Angelo

14:20 Structural Units of the Alkali Borate Glasses and Melts
A. A. Osipov, L.M. Osipova, R. T. Zainullina

14:40 Network Dimensionality and Alkali Modification Driven Elastic Phases in Alkali Borates
K. Vignarooban, P. Boolchand, M. Micoulaut, M. Malki, R. Kerner

15:00 Comparison Between RMC Structural Modelling of a Six-Oxide Borosilicate Glass of Nuclear Interest and Experimental Data
O. Bouty, J. M. Delaye, S. Peuget, T. Charpentier, B. Beuneu

Borate Applications *Chair: H. Huppertz*

15:50 Recent Progress in Laser Patterning in Borate Glasses
T. Komatsu

16:20 Non-isothermal Crystallization Analysis of Lithium Borate Glasses
I. Kleman, S. Feller, M. Affatigato

16:40 ZnO and Bi_2O_3 Effect on Borate Glass Structure and Properties
N. Lönnroth, R. Youngman

17:00 Investigation of Zinc Borate Glasses
F. Spadaro, A. Rossi, C. Ricci, E. Laine, J. Hartle, N. D. Spencer

17:20 High Quantum Yield and Low Concentration Quenching of Eu^{3+} Emission in Oxyfluoride Glass with High BaF_2 and Al_2O_3 Contents and Their Glass Structure
K. Shinozaki, M. Affatigato, T. Honma, T. Komatsu

Borate Poster Session *Chair: L. Koudelka*

JOINT BORATE–PHOSPHATE SESSIONS

Wednesday 2nd July

Borophosphate Session I *Chair: S. Kroeker*

8:30 On the Role of Boron on the Structure and Properties of Mixed Glass Former Na^+ Ion Conducting $Na_2O+B_2O_3+P_2O_5$ Glassy Solid Electrolytes
S. W. Martin, P. Mass, R. Christensen, G. Olson, M. Schuch, C. Trott

8:50 The Structure of Boro-Phosphate Glasses Revisited by $^{11}B/^{31}P$ Correlation NMR
G. Tricot, B. Raguenet, A. Pradel, G. Silly, M. Ribes

9:10 Properties and Trends in Na-Containing Boro- and Phospho-Aluminosilicate Glasses in Peralkaline, Peraluminous and Metaluminous Compositional Fields – The Mixed-Network Formers Effect
M. Potuzak, E.A. King

9:30 Low Photoelastic Property and Structure in Water Durable $ZnO–SnO–P_2O_5–B_2O_3$ Glasses
A. Saitoh, T. Grégory, H. Takebe

Borophosphate Session II *Chair: S. Martin*

10:20 Comparison of Structural Orders Between B_2O_3, P_2O_5 and SiO_2 Systems by Computer Simulation
A. Takada

10:40 Transformations in B_2O_3 and P_2O_5 Melts Under Compression and Structure of Densified B_2O_3 and P_2O_5 Glasses
V. V. Brazhkin, Y. Katayama, A. G. Lyapin

11:00 De-Clustering Influence of Al^{3+} Ions on Up-Conversion Efficiency in Yb^{3+}-Tm^{3+}, $Tm^{3+}Er^{3+}$ and Er^{3+}-Ho^{3+} Codoped $CaF_2–B_2O_3–P_2O_5$ Glass System
Y. Gandhi, M. Piasecki, N. Veeraiah

11:20 Thermoluminescence Dose Response of Calcium Fluoro-boro-phosphate Glasses Doped With Some Transition Metal Ions
N. Veeraiah, Y. Gandhi, Sanyal Bhaskar, B. J. R. S. Swamy

FIRST INTERNATIONAL CONFERENCE ON PHOSPHATE GLASSES

University of Pardubice, Czech Republic,
2–4 July 2014
PROGRAMME

Thursday 3rd July

Phosphate Opening Session *Chair: L. Koudelka*

8:30 Phosphate Welcome
L. Koudelka

8:40 Doris Ehrt – Glass Chemical Research in the Spirit of Otto Schott
D. Möncke, M. Müller

9:10 Phosphate and Fluoride Phosphate Optical Glasses – Properties, Structure and Applications
D. Ehrt

Phosphate Structure and Properties *Chair: L. Montagne*

10:10 Structure and Properties of Anisotropic Phosphate Glasses
S. Inaba, H. Hosono, S. Ito

10:40 Chemical and Geometrical Disorder in Phosphate and Silicophosphate Glasses by Solid-State NMR
F. Fayon

11:10 Structure–Properties Relationships in Lithium Oxynitride Phosphate Glasses
F. Muñoz, N. Mascaraque, A. Durán, G. Tricot, L. Montagne, A. C. M. Rodrigues

11:30 Atomic Structure of Ternary Phosphate Glasses: The Compositional Behavior of the Oxygen Coordination Number of a Second Network-Forming Oxide
U. Hoppe

11:50 The Structure of Molybdenum- and Tungsten-Doped Lead Phosphate Glasses
A. C. Hannon, L. Koudelka, I. Rösslerova

12:10 Structure–Property Relations in Antimony Molybdophosphate Glasses
R. E. Youngman, B. G. Aitken

Phosphate Melts and Relaxation *Chair: R. K. Brow*

14:00 Dynamic Light Scattering in Alkali Phosphate Melts and the Role of Intermediate Range Order in the Fragility of Network-Forming Oxides
D. L. Sidebottom, T. Tran, S. E. Schnell

14:30 The Structure of Glasses and Its Evolution Above T_g – Crystallisation, Phase Separation and Species Exchange: Lessons from *In Situ* MAS-NMR
L. van Wüllen, S. Venkatachalam, M. Engelmayer

14:50 Influence of the Structure on Viscosity in Phosphate Glasses
L. Muñoz-Senovilla, S. Venkatachalam, F. Muñoz, L. van Wüllen

15:10 Viscosity Profiles of Phosphate Glasses Through Combined Quasi-Static and Bob-in-Cup Methods
A. J. Parsons, N. Sharmin, S. I. S. Shaharuddin, C. D. Rudd

15:30 Melting Conditions Impact on the Properties of Copper Containing Phosphate Glasses for High-Power and High-Energy Amplifiers
V. I. Arbuzov, Yu. K. Fyodorov, S. I. Nikitina, R. V. Smirnov, V. M. Volynkin, M. V. Voroshilova

15:50 Structure and Glass Relaxation Studies of Melt Quenched and Mechanically Milled Na_2S+P_2S_5 Glasses
S. W. Martin, M. Marple, C. Bischoff, K. Schuller, S. Berbano

Crystallized Phosphate Glasses (Parallel Session) *Chair: P. Mošner*

14:30 Glasses and Glass Ceramics in $Li_{1+x}Cr_xGe_yTi_{2-x-y}(PO_4)_3$ with Li-Conducting NZP Phase
N. Lönnroth, B. Abel, B.G. Aitken

14:50 Phosphate Based Glass-Ceramics for the Sodium Ion Batteries
T. Honma, T. Komatsu

15:10 Correlation between Structural Changes and Electrical Properties of Crystallized Iron Phosphate Glass
A. Moduš-Milanković, L. Pavić, Ž. Skoko, M. P. F. Graca, B. F. O. Costa, M. A. Valente

15:30 Effect of Crystallization on the Magnetic Properties of $40Fe_2O_3–60P_2O_5$ Glass
L. Pavić, M. P. F. Graca, Ž. Skoko, M. A. Valente, *A. Moduš-Milanković*

15:50 Synthesis and Characterization of a New NZP Material Prepared From a Phosphate Based Glass Reactive Sintering
S. Chenu, P. Bénard-Rocherullé, R. Lebullenger, J. Rocherullé

Phosphate Optical Properties *Chair: D. Möncke*

16:40 Structural Approach on Optical Properties of $BaO–Nb_2O_5–P_2O_5$ Glass
N. Kitamura, Y. Hisano, K. Fukumi, S. Kohara, H. Ofuchi, T. Honma, H. Kozuka

17:00 Development of an Ion Conducting Glass for Electro-Optical Applications
M. Rioux, Y. Ledemi, J. Viens, Y. Messaddeq

17:20 Phosphorous Incorporation in Silica During Modified Chemical Vapor Deposition (MCVD) Combined With Solution Doping
F. Lindner, S. Unger, A. Kriltz, A. Scheffel, A. Delith, J. Delith, H. Bartelt

Phosphate Poster Session *Chair: L. Koudelka*

Friday 4th July

Nano/Micro-Structures and Defects in Phosphate Glasses *Chair: D. Ehrt*

8:30 Femtosecond Laser Structuring of Zinc Phosphate Glasses
D.M. Krol, N.W. Troy, L.B. Fletcher, C. Smith, R.K. Brow

9:00 Silver Containing Phosphate Photosensitive Glass: From Dosimetry to Femtosecond Direct Laser Writing
J. C. Desmoulin, S. Thomas, T. Cardinal, S. Danto, N. Marquestaut, P. Hee, M. Vangheluwe, Y. Petit, E. Fargin, L. Canioni, Y. Messaddeq, R. Vallée, M. Dussauze

9:20 Formation of Nanostructured Metallic Films from Smart Active Phosphate Glasses by Bottom-up Process: Ag^0 and Ni^0 Synthesis
R. Schneider, J. F. Felix, P. Santa-Cruz

9:40 Paramagnetic and Diamagnetic P-related Point Defects in 2.5 MeV Electron Irradiated Yb-Doped Phosphate Glasses
V. Pukhkaya, N. Ollier, F. Trompier

Phosphate Bioglasses and Surfaces *Chair: H. Takebe*

10:30 Processing and Utilisation of Phosphate Glass Fibres for Nerve Repair
J. C. Knowles, R. H. Hyun

11:00 Boron Doped Phosphate-Based Glasses: Development of Bio-Active Formulations for Fibre Drawing
N. Sharmin, E. Barney, T. Kemp, A. Parsons, D. Furniss, I. Ahmed, C. Rudd

11:20 Novel Phosphate-Based Glass Fibres for Biomedical Applications: Core-Clad Quaternary Formulations
I. Ahmed, S. I. S. Shaharuddin, D. Furniss, C. D. Rudd

11:40 Dissolution of Sodium Calcium Phosphate Glasses
F. Döhler, A. Mandlule, L. van Wüllen, M. Friedrich, D. S. Brauer

12:00 Surface Modification of Nd:Phosphate Laser Glass
He Dongbing, Hu Lili,Chen Huiyu, Xu Yongchun, Feng Suya

Phosphate Modelling and Optical Properties *Chair: F. Muñoz*

14:00 Progress in Atomistic Models and Simulations of Phosphate Glasses
G. Mountjoy

14:30 Topological Modeling of Calcium Borophosphate Glasses
C. Hermansen, Y. Z. Yue

14:50 Thermodynamic Model and Structure of $ZnO–MoO_3–P_2O_5$ Glasses
M. Liška, J. Macháček, M. Chromčíková, O. Gedeon

15:10 Chemical Activity of AgI and Ionic Conductivity in xAgI(1−x)$AgPO_3$ Glasses
C. B. Bragatto, A. C. M. Rodrigues

15:30 Charge Carrier Mobility and Concentration as a Function of Composition in $AgPO_3AgI$ Glasses
A. C. M. Rodrigues, C. B. Bragatto, M. L. F. Nascimento, J. L. Souquet

15:50 Electro-Chemical Phosphate Sensor
C. O. Normandeau, W. Olivier-Caron, A. K. Landry, J. F. Viens, Y. Messaddeq

Phosphate Surfaces and Corrosion *Chair: A. Moduš-Milankovič*

16:30 Surface Chemistry and Tribological Properties of Polyphosphate Glasses
A. Rossi

17:00 Composition Optimization of Iron Phosphate Glasses for Radioactive Sludge
H. Takebe, N. Kitamura, I. Amamoto, H. Kobayashi, N. Mitamura, T. Tsuzuki

17:20 Immobilization of Radioactive Iodine in Phosphate Glasses
L. Montagne, T. Lemesle, F.O. Mear, L. Campayo, O. Pinet

17:40 Novel Nanomaterials Based on $Li_2O–FeO–V_2O_5–P_2O_5$ Glasses
T. K. Pietrzak, J. E. Garbarczyk, M. Wasiucionek, J. L. Nowiński

Borate Poster Session

Synthesis and Spectral-Luminescent Properties of Huntite-Like Glasses Co-Doped With Ce, Tb and Sb
M. Ziyatdinova, N. Golubev, G. Malashkevich, E. Mamadzhanova, V. Sigaev

Property Comparison of Ternary Borovanadate and Silicovanadate Glasses with Caesium and Lithium as Modifiers
P. J. Rasmussen, D. Delgado, J. Maldonis, C. Parish, M. Bailey, J. McCoy, S. Kroeker, V. Michaelis, N. Morgan, C. Drapes, J. North, A. Ramm, D. Starkenburg, C. Hopkins, M. Affatigato, S. A. Feller

Anomalous Glass Transition Temperature Widths and Structure in Low Modifier Content Borate Glasses
B. Perez, D. Starkenburg, C. Drapes, M. Franke, N. Barnes, E. Troendle, M. Affatigato, S. A. Feller, E. D. Zanotto, H. Eckert

Investigation of $(1-x)(50PbO-20B_2O_3-30P_2O_5)-xMoO_3$ Glasses
I. Rösslerová, L. Koudelka, P. Mošner, L. Montagne, B. Revel

Synthesis and Structure of Eu^{3+}/Dy^{3+} Doped $50ZnO-10WO_3-40B_2O_3$ Glass for Optical Application
L. Aleksandrov, R. Iordanova, L Wondraczek, A. Herrmann, G. Gao, Y. Dimitriev

Scintillation and Dosimeter Properties of $40Li_2O-40B_2O_3-20SiO_2$ Glass with Different Sn Concentrations
T. Yanagida, Y. Fujimoto, H. Masai

NIR Emission of Bi-doped $TiO_2-ZnO-B_2O_3-Al_2O_3$ Glass.
H. Masai, T. Suzuki, Y. Takahashi, T. Fujiwara, Y. Ohishi

Structure of Sodium Niobato-Borophosphate Glasses: Diffraction Results
U. Hoppe, R.K. Brow, A.C. Hannon, M. von Zimmermann

Modeling of Interionic Interaction in Borate Crystals Based on First-Principles Calculation
Y. Ishii, K. Kasahara, N. Ohtori, K. Shiraki, N. Umesaki, M. Salanne, P. A. Madden

Sol-Gel Materials and Glasses in the $TiO_2-B_2O_3-TeO_2$ System
R. Gegova, A. Bachvarova-Nedelcheva, R. Iordanova, Y. Dimitriev

Thermal Transformation Bakerite–Datolite–Okayamalite
L.A. Gorelova, M.G. Krzhizhanovskaya, R.S. Bubnova

Valence State and Structural Geometry of Polyvalent Fe and Ti Ions in Sodium Borosilicate Glasses by X-Ray Absorption Spectroscopy
M. Dubiel, D. Ehrt

Excitation Energy Transfer Processes From Tb^{3+} to Ln^{3+} (Ln=Eu, Sm) in Lead Borate Glasses
J. Pisarska, A. Kos, L. Żur, M. Sołtys, W. A. Pisarski

Pr^{3+} and Er^{3+} Ions in Lead-Free Oxyfluoride Borate Glasses
W. A. Pisarski, J. Pisarska, D. Dorosz, J. Dorosz

Refractive Index Modification in Fluoro-Borate Glasses Containing WO_3 Induced by Femtosecond Laser
Y. Ledemi, J.-P. Bérubé, R. Vallée, Y. Messaddeq

Fluorescence XAFS Measurements and Luminescence Properties of Sm^{3+}-Doped Borate Glasses
S. Fuchi, R. Kimra, M. Tabuchi, Y. Takeda

High-Energy X-Ray Diffraction Measurements and Luminescence Properties of Sm^{3+}-Doped Borate Glasses
K. Watanabe, S. Fuchi, S. Kohara

Crystal Structures and Glass-Ceramics of Solid Solutions $Sr_{1-x}Ba_xBi_2B_2O_7$
A. P. Shablinskii, R. S. Bubnova, S. N. Volkov, S. K. Filatov, M. G. Krzhizhanovskaya, I. A. Drozdova

Thermal Decomposition of $K_{1-x}Cs_xBSi_2O_6$
E. S. Derkacheva, M. G. Krzhizhanovskaya, R. S. Bubnova, S. K. Filatov, V. L. Ugolkov

SMART FRIT: A Concept for Introducing Redox Sensitive Colorants in Glass Melts
A. Saberi, M. Willet-Porada

New Glasses and Glass-Ceramics-Glazes in the $Li_2O-K_2O-CaO-Al_2O_3-SiO_2-B_2O_3$ Complex System From Pegmatite Wastes
B. Rincón-Mora, M. Jordán, F. Pardo, T. Sanfeliú, M. Rincón

Immiscibility and Glass Formation in the System $B_2O_3-MoO_3-CoO-ZnO$
Y. Dimitriev, D. Ilieva, R. Iordanova, E. Kashchieva, I. Petrov

Glass and Glass Ceramics in the $La_2O_3-Gd_2O_3-PbO-MnO-B_2O_3$ System
R. Raykov, A. Staneva, Y. Dimitriev, E. Kashchieva, B. Blagoev

Microstructure and Dielectric Properties of Strontium-Borate Glasses Containing Bismuth and Vanadium Oxides
N.A. Szreder, P. Kupracz, J. Karczewski, M. Gazda, R.J. Barczyński

Micrography and Electrical Properties of Borosilicate Glasses Containing Manganese Ions Prepared by Sol-Gel Method
P. Kupracz, N. A. Szreder, J. Karczewski, M. Gazda, L. Wicikowski, R. J. Barczyński

Phosphate Poster Session

DOI: 10.13036/17533562.56.1.024 Phys. Chem. Glasses: Eur. J. Glass Sci. Technol. B, February 2015, 56 (1), 24–35

Atomic nature of the high anisotropy of borate thermal expansion

Stanislav K. Filatov[1]

Department of Crystallography, St. Petersburg State University, University Emb., 7/9, St. Petersburg, 199034, Russia

Rimma S. Bubnova

Institute of Silicate Chemistry of the Russian Academy of Sciences, Makarova Emb., 2, St. Petersburg, 199034, Russia
Department of Crystallography, St. Petersburg State University, University Emb., 7/9, St. Petersburg, 199034, Russia

Manuscript received 29 September 2014
Revised version received 20 November 2014
Accepted 29 November 2014

One of the unique features of borate crystals is that, on heating, a majority of them exhibit a strong anisotropy of thermal deformations along some directions in the crystal structure. There may even be negative thermal expansion (i.e. contraction on heating) in some directions. An explanation of this phenomenon is suggested in the present work. The contribution of the three main components of such anisotropy is analysed. These three components are shear deformations, hinge deformations and anisotropic thermal vibrations of atoms, which are well known in the theory of the thermal behaviour of crystalline compounds. The co-occurrence of all three components in borates is what makes their anisotropic thermal deformations record breaking. Oxygen compounds of other classes, e.g. silicates, carbonates, etc., are considered from the same perspectives.

1. Introduction

In the last few decades, new topics in borate crystal chemistry have been developed in the field of high pressure/high temperature synthesis of denser borates and their crystal structures determinations by Huppertz *et al*[(1–3)] and *in situ* temperature-dependent[(4–6)] and pressure-dependent[(7)] structural studies. Due to the increased interest in borates, the thermal behaviour of these materials is currently being intensively investigated. The knowledge of the thermal behaviour of matter is required for the development of crystal chemistry, solid state physics and chemistry, for the productive synthesis of materials and their technical applications.

The anomalously frequent occurrence of strongly anisotropic thermal deformations of borate crystals has been noted by Bubnova & Filatov:[(4–6)] among 70 borate crystals, for which thermal deformations were studied mainly by our group, two thirds exhibit strongly anisotropic deformations ($\alpha_{max}/\alpha_{min}\geq5$, where α is the linear coefficient of thermal expansion); notably anisotropy leads to a negative thermal expansion (contraction on heating) along some particular directions in one third of cases. Previously, high anisotropy of thermal expansion has been found in oblique angle (monoclinic and triclinic) inorganic and organic compounds as a result of *shears*, i.e. deformations, that result from changes of unit cell angles which are not constrained by symmetry (one angle for monoclinic crystals, and three angles for triclinic crystals).[(8–10)] For borates, this property is also widely exhibited by crystalline compounds, for which the unit cell angles are all constrained (equal to 90° or 120°), in particular orthorhombic crystals. Evidently, such deformations have another atomic–molecular nature rather than shears. The analysis of this nature is one of the topics reviewed here.

When thermodynamic (T, p, X) parameters change, the crystal structure of a substance adapts to these changes. Atoms can shift relative to each other, tending to reach the most energy efficient system condition, which is usually characterised by a minimal volume for given conditions. For this "fitting" of a crystal structure to a permanently altered environment, a structure should have degrees of freedom. The more degrees of freedom a structure has, and the more versatile these freedoms are, the more adequately the compound adjusts to the altered environment. Monoclinic crystals can serve as examples. They have an additional degree of freedom in comparison to orthogonal crystal systems. This new freedom is the possibility to select a monoclinicity angle which is not constrained by symmetry (β for cell

[1] Corresponding author. Email filatov.stanislav@gmail.com
Original version presented at VIII Int. Conf. on Borate Glasses, Crystals and Melts, Pardubice, Czech Republic, 30 June–2 July 2014. The conference was held in honour of Professor Filatov's contribution to the science of borates.

choice II, and γ for cell choice I). Specifically, because of this possibility to change the angle, the monoclinic system is the most abundant amongst the different crystal systems.[8,9]

When speaking about rigid boron–oxygen groups, first found by Krogh-Moe,[11,12] it makes us wonder which degrees of freedom does a borate crystal structure possess as a combination of rigid groups? An answer to this question was found when studying the crystal structures of a series of borates at various temperatures in the range of about −200–600°C,[13–22] and even under pressure.[7] First studies have shown[13–15] not only that rigid boron–oxygen groups are reproduced nearly uniformly in the structures of different crystalline and vitreous borates, as observed by Krogh-Moe.[11,12] They are also invariable with temperature[4–6] and, presumably, with pressure. In this respect, the rigidity of a group means that the group cannot change its configurations well as the temperature varies. It can be said that a rigid group is devoid of internal degrees of freedom. Therefore, a borate structure constructed of rigid groups can react to temperature changes only by means of external degrees of freedom. A degree of freedom external to a rigid group is its ability to rotate as a unit relative to other groups around mutual oxygen atoms as *a hinge*. Hinge deformations are sharply anisotropic in nature. In essence, a shear (see above) can be considered as a special case of hinge. When the hinge angle increases ($\beta_1 \rightarrow \beta_2$), the adjacent angle should be decreased (Figure 1). Thermal expansion partially compensates for this anisotropy. Nevertheless, thermal hinges are often accompanied by negative thermal expansion (contraction on heating).[5,8–10] Mainly due to hinges, the occurrence of thermal deformations with an anisotropic character, and even negative linear thermal expansion along some certain directions, is inherent to borates, and these effects may be large in comparison to the other classes of oxide compounds.

Hinge deformations are characteristic to one-dimensional (1D), two-dimensional (2D) and three-dimensional (3D) borate crystal structures. Along with this, many borates are constructed from isolated triangular BO_3 anionic groups, or from isolated rings of three corner-sharing triangles, or from other finite B–O groups. Although these 0D-structures do not have the prerequisites for hinge deformations, they nevertheless demonstrate strongly anisotropic deformations of the crystal structure due to a preference for thermal vibrations of O and B atoms perpendicular to the plane of the triangles. Taking into account the wide occurrence of rigid B–O groups in borate crystal structures, the analysis of the atomic nature of strongly anisotropic thermal expansion in borates begins with the thermal behaviour of the rigid groups.

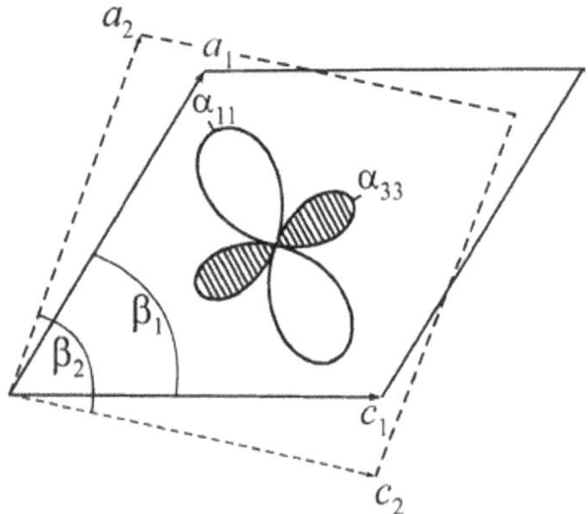

Figure 1. Schematic pattern of shear deformation in comparison with the depiction of the thermal expansion coefficients for the given shear

2. Thermal invariability of rigid boron–oxygen groups

A lot of terms like rigid boron–oxygen group,[11,12] cluster, fundamental building block (FBB),[23–26] superstructural unit,[27,28] repeat anion unit,[29] etc. are used to describe the borate anion hierarchy with little difference in meaning. These terms, nomenclature and borate systematics are considered in recent reviews.[5,6,25–30] Wright has discussed not only geometry, but also the stabilisation energy for the formation of these groups in his excellent review of crystalline and vitreous borate structures.[27] The term "rigid boron–oxygen group" was first introduced by Krogh-Moe.[11,12] It has been widely used for half a century, but the definition of the term was only given recently.[5,6] That is why we mention below some general features of a rigid group according to the references. *Rigid groups* are single triborate B–O rings condensed from three BO_3 and/or BO_4 polyhedra *via* common oxygen atoms (corners) and combinations of the single 3B-rings *by sharing tetrahedra* or, presumably, tetrahedral edges (mainly at high pressure).[5,6] Hence, these rigid groups could be considered as multiple (single, double, triple and so on) triborate groups, i.e. they are formed by condensation of single triborate groups *via* sharing one or two common tetrahedra, independently of the multiplicity of the group.

It was assumed by Filatov & Bubnova[30] that not only BO_3 and BO_4 polyhedra, but also rigid B–O groups built up from these polyhedra maintain their configuration and size with temperature. The assumption was based on the strength of the highly covalent B^{3+}–O^{2-} bonds in BO_3 triangles (1 bond-valence unit (*v.u.*) in each B–O bond, average bond length 1·37 Å) and BO_4 tetrahedra (¾ *v.u.*, 1·47 Å). It was also based upon the invariability of rigid boron–oxygen groups in different borate crystal structures, as first found by Krogh-Moe.[11,12] Until now, more than ten borate crystal structures have been studied using high and low temperature single crystal x-ray experimental diffraction data in the range −250–900°C.[5,6,13–22] The results of the studies confirmed the hypothesis about the thermal invariability of rigid groups.

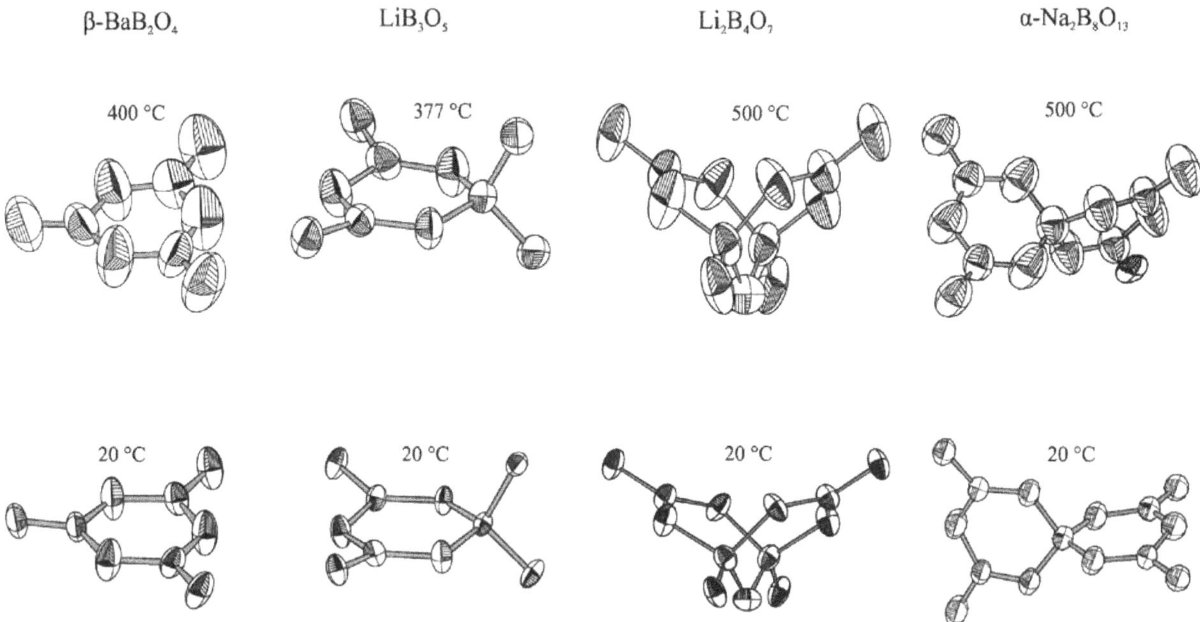

Figure 2. Tri- (β-BaB_2O_4[46] and LiB_3O_5[15]), tetra- ($Li_2B_4O_7$[16]) and pentaborate (α-$Na_2B_8O_{13}$[13]) groups in the structures at different temperatures

Structurally, the borates under study contain mainly triborate, tetraborate, and pentaborate groups. Here we consider typical examples of the thermal structural behaviour of the main rigid groups (Figure 2).

2.1 Triborate (single cyclic) and pentaborate (double cyclic) groups

The fundamental building block of the α-$Na_2B_8O_{13}$ (20, 300, 500°C)[13] structure is a triborate ring, <2Δ□>, composed of two corner-sharing triangles and a tetrahedron, and a pentaborate group, <2Δ□>–<2Δ□>, consisting of two single triborate rings sharing a common tetrahedron. Triborate and pentaborate groups polymerise by sharing terminal oxygen atoms of the groups to form two interpenetrating frameworks.

Both individual B–O and average <B–O> bond lengths (Figure 3A, squares) show a tendency for a weak contraction of BO_3 triangles. However, the B–O bonds elongate after taking into account atomic thermal vibrations (Figure 3A, circles and triangles). A tendency for weak change also holds for tetrahedra, and the average <B–O> bond length practically does not change on heating (Figure 3B, squares), although it increases after corrections (Figure 3B, all signs except squares).

The thermal invariability of the <2Δ□> triborate ring has also been studied (Figures 2, 3) in the well known LiB_3O_5 borate (20, 227, 377°C),[15] which has nonlinear properties. Similarly the thermal invariabil-

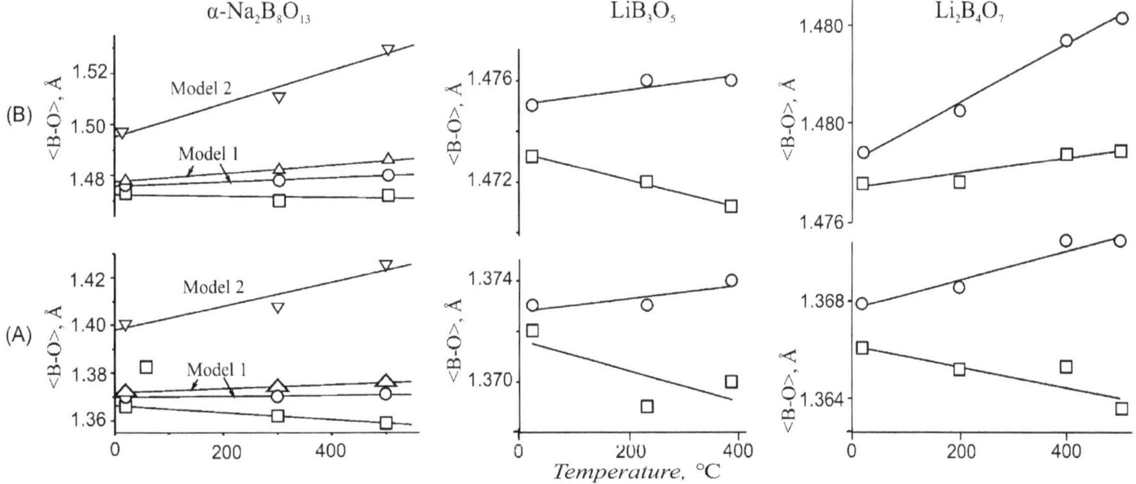

Figure 3. The mean B–O bond lengths in α-$Na_2B_8O_{13}$,[13] LiB_3O_5,[15] and $Li_2B_4O_7$[16] crystal structures as a function of temperature for BO_3 triangles (A) and BO_4 tetrahedra (B): □ – calculated from the atomic coordinates; ○ – calculated using model 1 (rigid-body motion according to Downs[32]); Δ and ∇ – calculated using model 1 (rigid-body motion) and model 2 (non-correlated motion) according to Busing & Levi[31])

ity of the pentaborate group has been studied in the layered structure of α-CsB_5O_8 (20, 300 and 500°C).[14]

2.2 Tetraborate (double cyclic) groups

The tetraborate group, <Δ2□>=<Δ2□>, consists of two <Δ2□> triborate rings condensed to a double ring *via* two common tetrahedra (Figure 2). The groups polymerise by corner sharing (oxygen atoms) to form a double interpenetrating framework $Li_2B_4O_7$ (−150, −100, −50, 20, 200, 400, 500°C) – another well known borate with nonlinear properties. Lithium cations are located in framework channels forming screw 4_1-chains of corner-sharing LiO_4 tetrahedra. Until now, several structural studies using single crystal x-ray and neutron diffraction over a wide temperature range from −150°C up to almost the melting temperature are available in the literature.[16,19,20]

2.3 Pentaborate (double cyclic) groups

A pentaborate, <2Δ□>–<2Δ□>, group consists of two single triborate rings sharing one common tetrahedron. These groups polymerise *via* terminal oxygen atoms to form layers in α-CsB_5O_8 (20, 300 and 500°C).[14] The triborate and pentaborate groups sharing terminal oxygen atoms form two interpenetrating frameworks in α-$Na_2B_8O_{13}$ (20, 300, 500°C)[13] (Figure 2).

2.4 Non-rigid boron–oxygen groups

Non-branched *single cyclic groups* seldom contain more than three corner-shared polyhedra. A few structures are known that contain a ring of four polyhedra (n=4), mainly tetrahedra; for example, finite ring clusters in $Ca_4Mg(CO_3)_2[B_4O_6(OH)_6]$ borcarite (0D), infinite chains in $La(BO_2)_3$ (1D), 4B : <4□>–$^{\infty 2}$ layers in $CaAlB_3O_7$ johadachidolite and $CuTm_2(B_2O_5)_2$ (2D). There are six-membered rings of tetrahedra (n=6) in the $MAl_2(B_4O_{10})O_{0\cdot5}$ family (2D), for example $NdAl_{2\cdot07}(B_4O_{10})O_{0\cdot6}$, and eight-membered rings of alternate triangles and tetrahedra (n=8, 2D) in α-BiB_3O_6 (Figure 4),[33] and so on. On the basis of the definition of rigid groups, it is expected that these groups must be flexible. The temperature dependence of the structural data for α-BiB_3O_6 reported by Stein *et al*[17] demonstrated the thermal flexibility of the non-rigid groups composed of eight corner-sharing polyhedra in this structure.

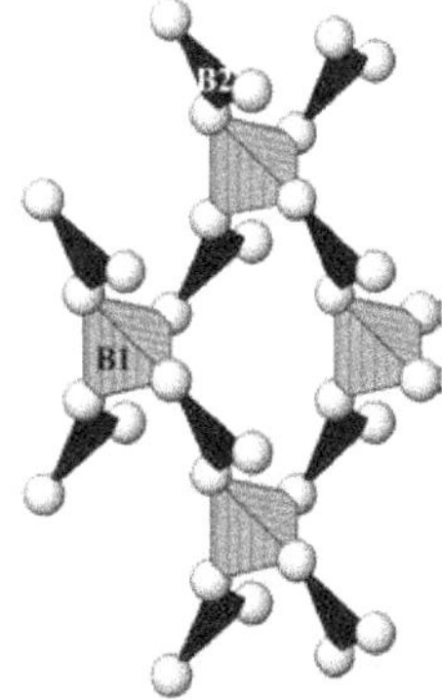

Figure 4. The flexible group of eight polyhedra in the structure of α-BiB_3O_6[17]

2.5 Summary

The average B–O bond lengths calculated from x-ray diffraction data (Figure 3A, squares) show a tendency for a weak contraction of BO_3 triangles in the different groups of borates under study. However, B–O bonds start to elongate after taking into account atomic thermal vibrations (Figure 3A, circles and triangles) using the corrections given by Busing & Levi[31] or Downs.[32] A tendency for a weak bond shortening also holds for tetrahedra (Figure 3B, squares). The average B–O bond length calculated directly from x-ray diffraction data decreases on heating (Figure 3B, squares), although it increases after corrections (Figure 3B, all signs except squares).

The tetraborate, <Δ2□>=<Δ2□>, group, like the triborate group, maintains its configuration and size, and a tendency for weak contraction of bonds is also observed. The average B–O bond lengths in the BO_3 triangles decrease on heating, as is evidenced by uncorrected x-ray diffraction data (Figure 3A, squares), although they increase after taking into account thermal vibrations of atoms (Figure 3A, circles). Tetrahedra are also essentially stable at high temperature, but do not contract (Figure 3B).

The performed studies show that the average B–O bond length changes in triangles and tetrahedra usually do not exceed the error bars ±0·003 Å in the temperature range 20–500°C. Variations of B–Ô–B angles within rigid boron–oxygen groups are usually 0·5° or less, in the temperature range 20–500°C. This value is comparable with the error bars. Angles between rigid groups can vary in a wider range (about 2–3°) over the same temperature range. Within groups considered as non-rigid, the changes in bond angles are similar to those between rigid groups. These results demonstrate that the configuration and size of rigid B–O groups is almost invariant in crystals under temperature change. Based on Krogh-Moe's assumption,[11,12] the structure of borate glasses consists of the same rigid groups as in crystals. The modern representation of borate glass structure has been reviewed by Wright,[27] and in particular it is based on thermodynamic modelling[34,35] that leads to the concept that glasses contain chemical groupings that have the stoichiometry of the crystalline phases in the system under study. Thus, different rigid groups can exist in borate glasses. Because glasses are formed by atoms of the same chemical elements with the same chemical bonding as in crystalline phases, it is assumed that rigid boron–oxygen groups maintain thermal invariability in glass as well.

3. Anisotropy of thermal vibrations of atoms in rigid B–O groups

Temperature can affect the electronic configuration of atoms, affecting the directions of the covalent bonds in the crystal. Thermal vibrations of atoms are generally anisotropic, and generally the maximum amplitudes of atomic vibrations increase intensively with temperature. As follows from the principles of high temperature borate crystal chemistry (rule 1),[5,6] oxygen and boron atoms oscillate perpendicular to a strong, short B–O bond in a boron–oxygen polyhedron: (1) in the case of BO_3 triangles, these atoms oscillate perpendicular to the three strong B–O bonds that are in the plane of the triangle, and as a result of this there is a preference for the maximum amplitude of vibration to be perpendicular to the plane of the triangle (Figure 5), especially when triangles are condensed into groups, because internal atoms have no degree of freedom (cf. Figure 10); (2) in the case of a BO_4 tetrahedron, O and B atoms also oscillate perpendicular to the strong B–O bonds, whilst atoms vibrate maximally randomly due to these bonds are distributed regularly in 3D-space (Figure 5).

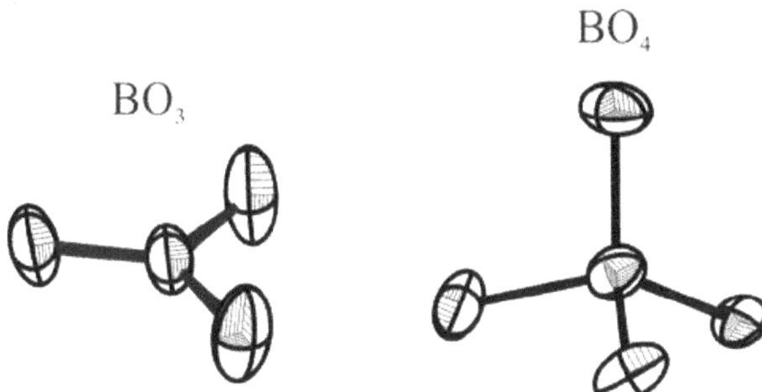

Figure 5. Ellipsoids of atomic thermal vibrations in a BO_3 triangle and a BO_4 tetrahedron

The rigidity of boron–oxygen groups (strength of B–O chemical bonds) gives rise to high anisotropy of atomic thermal displacements, and leads to anisotropy of thermal expansion of the compound. Here we consider the relation between the anisotropic thermal expansion of borates of different dimensionality and the anisotropic atomic thermal vibrations in the various rigid groups of which the borates are composed.

3.1 BO_3 triangle

Nearly half of all borate crystals[36] are built up from isolated BO_3 triangles with a preference for them to be oriented parallel to each other. The majority of such borates exhibit *largely anisotropic thermal expansion.* This anisotropy, as noted above, is primarily ruled by the distribution of the sharply anisotropic thermal atomic oscillations. The triangle as a whole vibrates in the direction nearly perpendicular to the plane of the triangle (Figure 6). In such structures, triangles are depicted in preferred orientation parallel to each

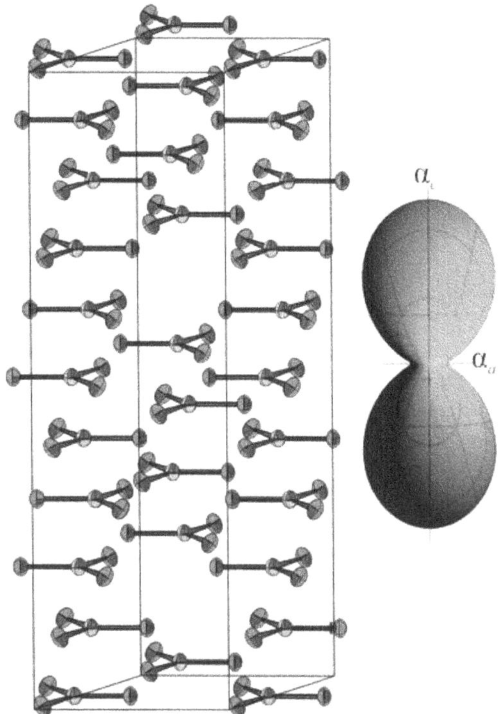

Figure 6. The parallel arrangement of isolated BO_3 triangles in the crystal structure of $LuBO_3$, calcite modification, compared with the thermal expansion pole figure (α_a=1·5×10^{-6} K^{-1}, α_c =13×10^{-6} K^{-1}[38]) [Colour available online]

other, and the maximum expansion of the structure is perpendicular to the planes of triangles (Figure 6). Most *RE*-orthoborates are similar to calcite, aragonite, vaterite and other borates built up from isolated BO_3 triangles.[37] Their maximum expansion, like calcite and aragonite, is perpendicular to the planes of the triangles; for example, see the calcite polymorph of $LuBO_3$,[38] and the aragonite modification of λ-$NdBO_3$.[3,39] A preference for parallel orientation of the planes of the triangles also occurs in the metastable borate $Ba_3Bi_2(BO_3)_4$,[40] although the planes of the triangles are inclined relative to each other, as seen in the *ab* plane (Figure 7). The correlation between the thermal expansion and the orientation of the triangles is also clear.

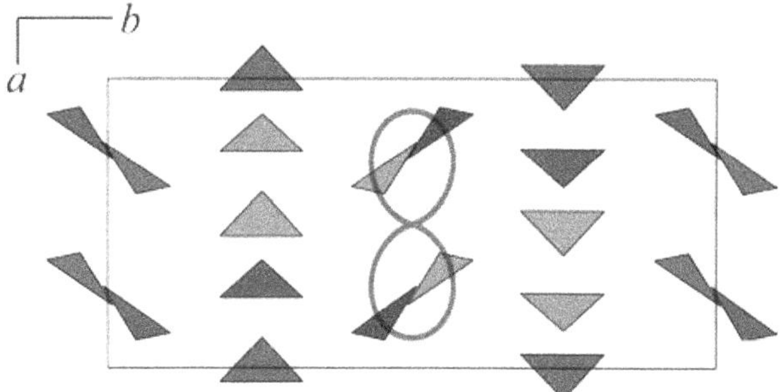

Figure 7. Isolated BO_3 triangles, inclined relative to each other, in $Ba_3Bi_2(BO_3)_4$, shown in comparison with the thermal expansion pole figure (α_a=32×10^{-6} K^{-1}, α_b=−2×10^{-6} K^{-1}, α_c=7×10^{-6} K^{-1} at 700°C[40])

A notable correlation between the thermal expansion and the orientation of the triangles is observed for $LuBO_3$, pseudovaterite modification.[38] This crystal structure is based upon statistically disordered, isolated BO_3 triangles.[41] The triangles are located around the 6_3 axis in such a manner that the planes of BO_3 triangles are parallel to the c axis. Thus we expect that the structure has maximum expansion perpendicular to the c axis. One can see exactly this anisotropy of thermal expansion in Figure 8.[38] This correlation gives the preference to the $LuBO_3$ pseudovaterite modification[41] built up from BO_3 triangles, over the structure built up from BO_4 tetrahedra,[42,43] although powder diffraction patterns of both models could not be distinguished. Moreover synthesised under high pressure/high temperature the borate π-$ErBO_3$[44] is built up from tetrahedra also.

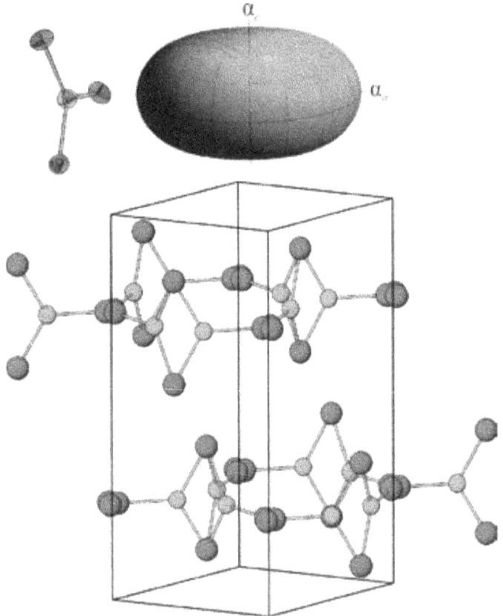

Figure 8. The preferred orientation of isolated BO_3 triangles in $LuBO_3$, pseudovaterite polymorph, compared with the thermal expansion pole figure (α_a=9×10^{-6} K^{-1}, α_c=2×10^{-6} K^{-1} [38]) [Colour available online]

When BO_3 triangles polymerise to form chains, as in SrB_2O_4, the atoms vibrate most perpendicular to the planes of the triangles, and thus it is expected that the structure has maximum expansion perpendicular to the planes of the triangles; this is what is observed in Figure 9. It is notable that the planes of the triangles are arranged in the *ab* plane, and the thermal expansion in this plane is practically isotropic, with α_a=4×10^{-6} K^{-1} and α_b=3×10^{-6} K^{-1}, whilst it increases dramatically almost perpendicular to the BO_3 planes, with α_c=33×10^{-6} K^{-1}.[45]

3.2 Triborate group

By analogy, crystal structures with isolated triborate groups, <3Δ>, expand considerably perpendicular to the plane of the groups, and expand weakly in the plane of the group. As an example, we refer to the

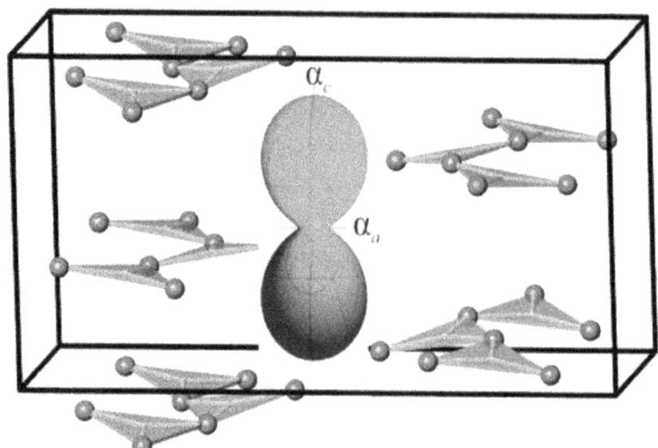

Figure 9. The preferred orientation of chains of BO_3 triangles in SrB_2O_4, compared with the thermal expansion pole figure[45] [Colour available online]

triborate group of β-BaB_2O_4, which is built up from three BO_3 triangles.[22] In response to an increase in temperature, the borate groups vibrate in the same direction more intensively as a whole, and the thermal expansion of the structure is sharply anisotropic, with α_{11}=3×10^{-6} K^{-1} and α_{33}=45×10^{-6} K^{-1} [46] (Figure 10). The internal oxygen atoms, bonded to two B atoms, oscillate more in the direction perpendicular to the plane of a single triborate group than is the case for terminal O atoms, and the group as a whole vibrates in the same direction (see Figure 2).

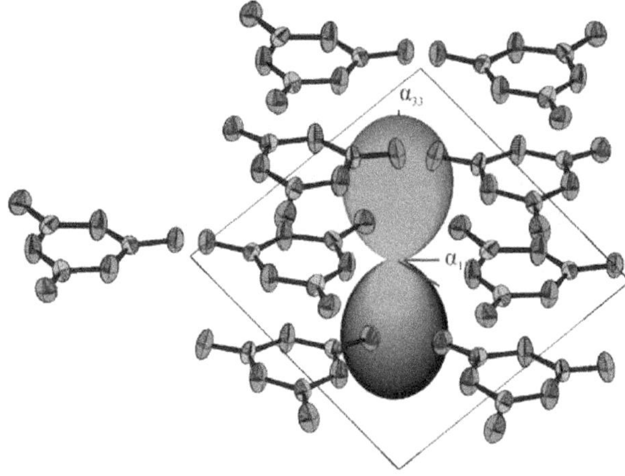

Figure 10. The parallel arrangement of isolated <3Δ> triborate groups in the structure of β-BaB_2O_4,[46] compared with the thermal expansion pole figure. Ellipsoids of thermal vibrations are given with 90% probability [Colour available online]

A situation similar to that for $LuBO_3$, pseudovaterite polymorph (Figure 8), occurs when the thermal expansion of layered $Ba_2Bi_3B_{25}O_{44}$[47] is considered (Figure 11). This latter compound contains a complex layered anion, consisting of three sublayers built from <Δ2□> and <3Δ> triborate groups. These triborate groups are arranged parallel to [001] around the 3_1 axis and perpendicular to the (001) layer plane. Therefore, the B–O bonds are located mainly perpendicular to (001) and the thermal expansion in this direction is minimal, whilst within the plane of the layer the expansion should be maximal. This explanation is in good agreement with the experimental data[47] (Figure 11).

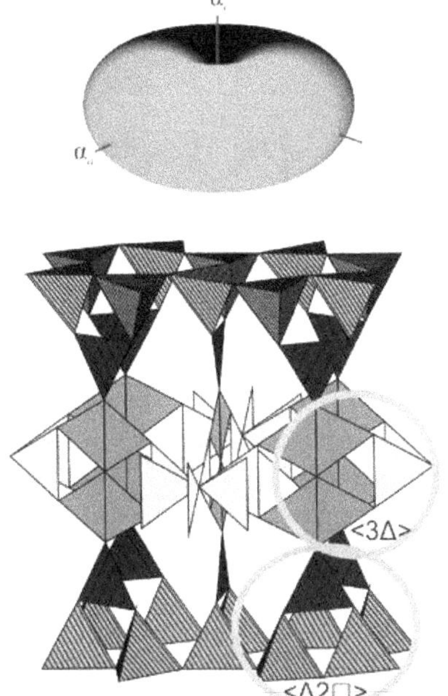

Figure 11. The preferred arrangement of <3Δ> and <Δ2□> triborate groups (parallel to the 3_1 axis) in a layer of $Ba_2Bi_3B_{25}O_{44}$,[47] compared with the thermal expansion pole figure at 700°C [Colour available online]

3.3 Pentaborate group

In the case of <3B>–<3B> pentaborate groups, the internal oxygen and boron atoms of both single rings vibrate perpendicular to the planes of these rings, while the group as a whole oscillates relative to the axis of the <2Δ□>–<Δ2□> group – the dashed line drawn parallel to the plane of both rings in Figure 12. The maximum expansion (α_{max}) occurs perpendicular to the axis of the group and the minimal expansion (α_{min}) is along the axis of group as shown for $Bi_3B_5O_{12}$[48] in Figure 12. Similar character of thermal expansion is typical for pentaborate group in the structures of 0D-, 1D-, 2D- and 3D-dimensionality as noted by Bubnova & Filatov.[5,6] Eleven borates under study expand greatly in directions perpendicular to an axis a group and contract or expand weakly along the group axis independent of the dimensionality.

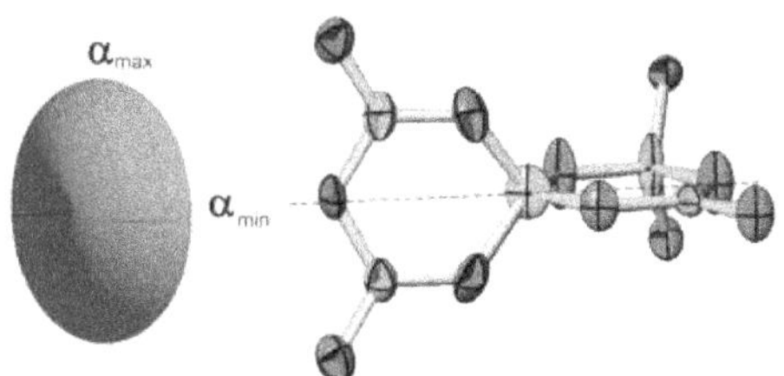

Figure 12. An isolated pentaborate anion in $Bi_3B_5O_{12}$,[48] compared with the thermal expansion pole figure [Colour available online

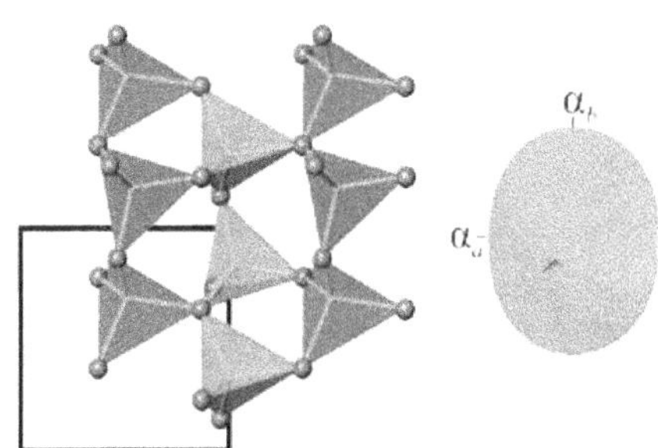

Figure 13. The arrangement of BO_4 tetrahedra in the framework of SrB_4O_7,[45] compared with the thermal expansion pole figure

3.4 BO_4 tetrahedron

In contrast to borates consisting of triangles, crystal structures based mainly upon tetrahedra expand more isotropically. The thermal expansion of SrB_4O_7 borate,[45] consisting solely of BO_4 tetrahedra, serves as a dramatic example of practically isotropic expansion. The main coefficients of the thermal expansion tensor are α_a=7×10^{-6} K^{-1}, α_b=9×10^{-6} K^{-1}, and α_c=8×10^{-6} K^{-1} [45] (Figure 13).

Other examples are given by Bubnova & Filatov.[5,6] Similar tendencies are observed for other rigid boron–oxygen (tetraborate, hexaborate, etc.) groups.

4. Hinge and shear mechanisms of high anisotropy of borate thermal expansion

4.1 Hinges

The term "hinge structure" was introduced by Sleight.[49] As was shown previously for CuO (tenorite) by Domnina *et al*,[50] the thermal expansion of such structures usually has a highly anisotropic character, including negative expansion along certain directions (cf. Figure 1). Hence, if the hinge structure expands in one direction on heating (Figure 14), simultaneously it has to contract slightly in the perpendicular direction.

The hinge mechanism of thermal expansion usually takes place in condensed anions (either 1D

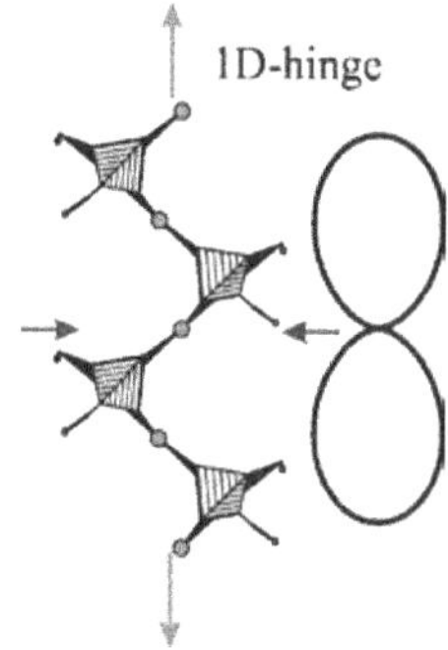

Figure 14. Schematic pattern of hinge deformation, compared with the depiction of the pole figure for the thermal expansion coefficients for a given hinge [Colour available online]

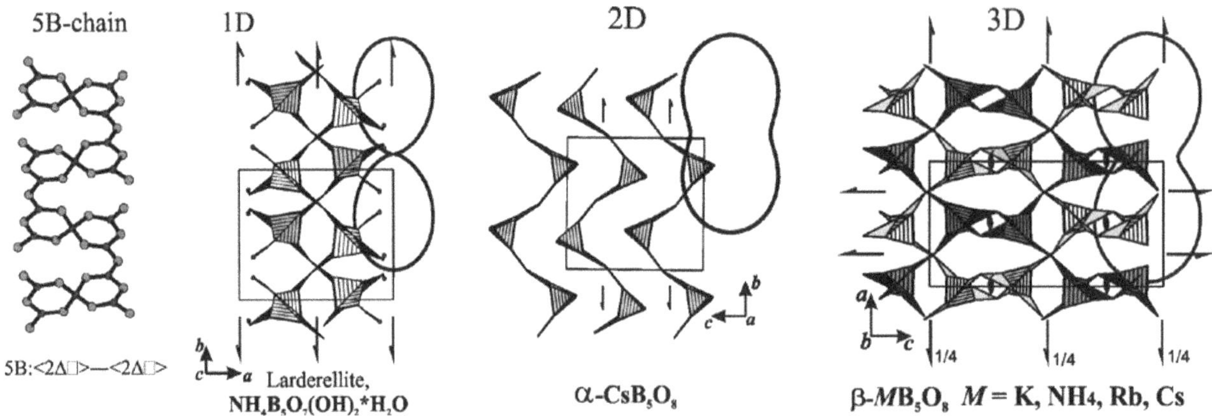

Figure 15. Schematic view of the 5B-chain (left), and the 1D-, 2D- and 3D-hinge structures of pentaborates, together with their thermal expansion pole figures (from left to right side) [Colour available online]

(infinite chains), 2D (layers) or 3D (frameworks)) where the groups turn around the bridging oxygen atoms. As an example, we show the hinge structures composed of <2Δ□>–<2Δ□> pentaborate groups sharing common oxygen atoms to form boron–oxygen chains (Figure 15, *left*); these 5B-chains condense *via* external layers, and form frameworks of different dimensionality from 1D to 3D (double frameworks). The deformations of larderellite,[51] α-CsB_5O_8,[14] and β-MB_5O_8 (M=K, NH_4, Rb, Cs)[52–54] are examples of 1D-, 2D- and 3D-hinges, respectively (Figure 15). These crystal structures contain screw 2_1 chains of the <2Δ□>–<2Δ□> groups (Figure 15, *left side*). On heating, this one-dimensional hinge expands significantly, with a coefficient of thermal expansion of about 60×10^{-6} K^{-1} along the chain. For this reason, the chain slightly contracts or expands weakly in the perpendicular direction. Hence, this hinge mechanism leads to a dramatic thermal expansion along the chains, as the thermal expansion pole figures indicate in Figure 15.

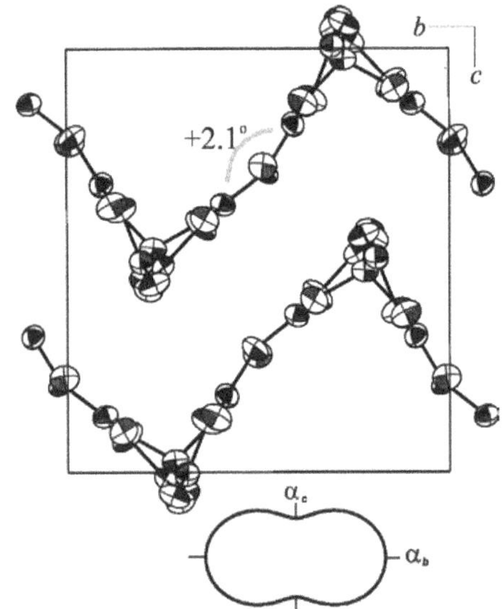

Figure 16. The correlation between the pole figure for the coefficients of thermal expansion and the crystal structure of α-CsB_5O_8 at 500°C (after Filatov et al[14])

The atomic–molecular nature of hinge deformations may be understood from high temperature crystal structure investigations. In particular, untypically for layered structures, caesium pentaborate α-CsB_5O_8[14] exhibits anisotropy of thermal deformations – the maximum thermal expansion occurs in the plane of the layer. The two independent B–Ô–B angles between pentaborate groups show maximal changes (Figure 16, +2·1°) with rising temperature (the angles are 150·7, 151·7, 152·8° and 123·0, 123·1, 125·1° at temperatures of 20, 300, and 500°C, respectively). Thus temperature changes are accompanied by a considerable rotation of the large pentaborate groups (~1 nm) around each other. The corrugated zig-zag layer partially straightens, and causes the maximal thermal expansion α_b in the plane of the layer.

4.2 Shears

Shear deformation is a special case of hinge deformation, when an angle between coordinate axes changes, as first noted by Filatov.[8] The occurrence of angles between the coordinate axes which are not fixed by symmetry, in monoclinic and triclinic crystals, allows these angles to change. Similar to hinges in general, a shear deformation is sharply anisotropic in nature. If the shear angle increases, the adjacent angle should be decreased (cf. Figure 1). The thermal expansion may partially compensate for this anisotropy. Nevertheless, thermal shears are often accompanied by negative linear thermal expansion (see Figure 1, α_{33} tensor axis). The jump of abundance for negative linear thermal expansion when the symmetry is reduced from orthorhombic (a few percent of structures; α, β, γ angles are fixed) to monoclinic (~one third of structures; β angle not fixed) renders the shears an excellent and instructive example of hinges.

Strong anisotropy of thermal expansion of borax, $Na_2B_4O_6(OH)_2.8H_2O$,[55] and $Na_2B_4O_6(OH)_2.3H_2O$, kernite,[56] can be mentioned as examples of shear deformations. On heating, the β angle in the monoclinic

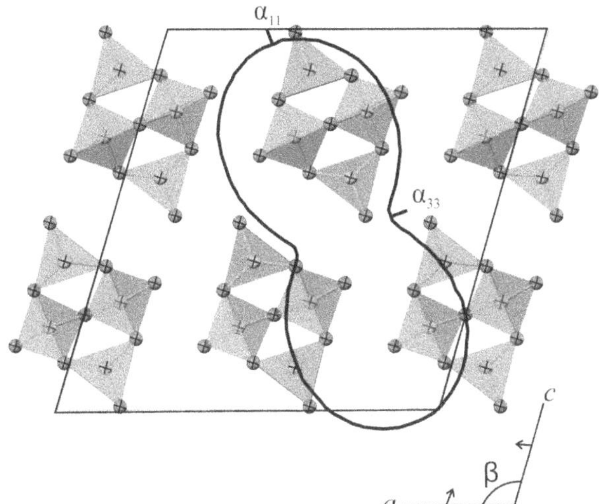

Figure 17. The crystal structure and the thermal expansion pole figure of borax[55] *[Colour available online]*

borax decreases, as shown by the arrows in Figure 17, and the structure expands maximally along the short diagonal of the *ac* plane (bisectors of the β angle) as a shear. As a result, the isolated <Δ2□>=<Δ2□> groups, as well as the chains of NaO_6 octahedra, shift with respect to each other in the monoclinic *ac* plane.

5. Principles of self-organisation of rigid boron–oxygen groups

Information about the anisotropy of thermal expansion for borate crystal structures can be used for understanding the formation process of borates. The parallel alignment of thermal vibration ellipsoids (anisotropic displacements) of B and O atoms in the crystal structures of borates (see Figures 6, 10, 12, 16) is evidence of a collective character of the thermal atomic motion and the motion of the rigid B–O groups, with the ensuing preferred orientation of the rigid B–O groups in many borates. As a result, pentaborate groups are arranged essentially parallel to each other (Figure 18) in borate structures of different dimensionality from 0D- ($Bi_3B_5O_{12}$[48]) to 1D- ($NH_4B_5O_7(OH)_2.H_2O$[51]) and 3D-borates (β-MB_5O_8, M=K, NH_4, Rb, Cs[53,54,57]). The condensation of different rigid groups to form chains, layers and frameworks reveals a similar tendency in the arrangement of these groups. The recent solution of the crystal structure of $Li_3B_{11}O_{18}$ borate,[58,59] known incorrectly for a long time as $Li_2B_8O_{13}$, serves as an example (Figure 19): it consists of 3B- and 5B-groups forming a double framework. As seen in Figure 19, pentaborate groups are also arranged in parallel, connected *via* triborate groups.

This allows the formulation of the following principles of self-organisation of BO_3 triangles, BO_4 tetrahedra, and rigid B–O groups comprised of these polyhedra, with a preference for parallel mutual orientation in the crystal structures of borates:[6]

(1) Thermal motion is the inherent driving force of the self-organisation of atoms and molecules during the crystallisation process. In borates, such an "organising" force is the sharp anisotropy of thermal vibrations of the atoms and the composite rigid B–O groups.

(2) During the crystallisation of borates, the maximal axes of thermal ellipsoids of the atoms in the rigid boron–oxygen groups and elongation directions of the groups tend to keep them away from each other, in order to minimise the mutual interference. If the groups tend to occupy minimal volume, then this can be realised by the parallel orientation (or a preference for parallel orientation) of the rigid groups.

6. Examples of practical applications of matter with highly anisotropic thermal expansion

In technology, sharply anisotropic thermal expansion can have some practical applications. For instance, objects exhibiting almost any possible coefficient of

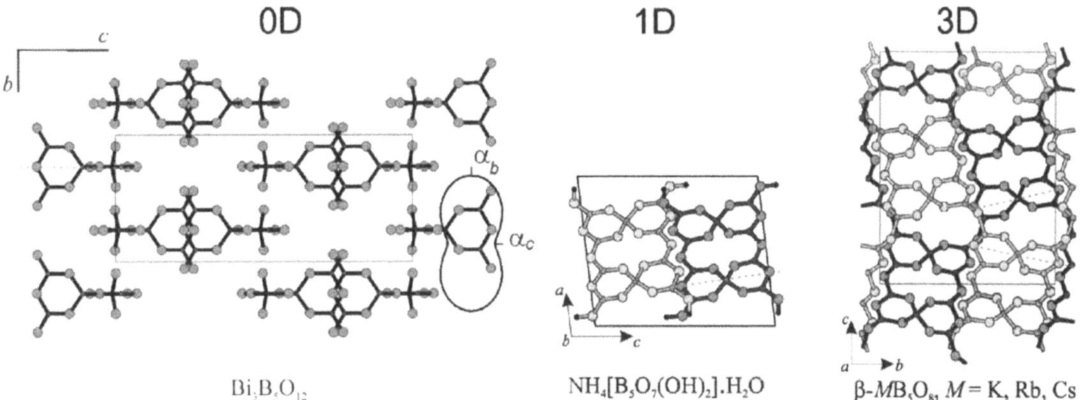

Figure 18. The essentially parallel arrangement of pentaborate groups in the structure of borates: $Bi_3B_5O_{12}$[48] *(0D), $NH_4B_5O_7(OH)_2.H_2O$*[51] *(1D) and β-MB_5O_8 (M =K, NH_4, Rb, Cs)*[53,54,57] *family (3D, two interpenetrating frameworks). Two different chains of $NH_4B_5O_7(OH)_2.H_2O$ and the double B–O framework of β-MB_5O_8 are shown, coloured in light and dark. Dashed lines are noted axes of 5B-groups [Colour available online]*

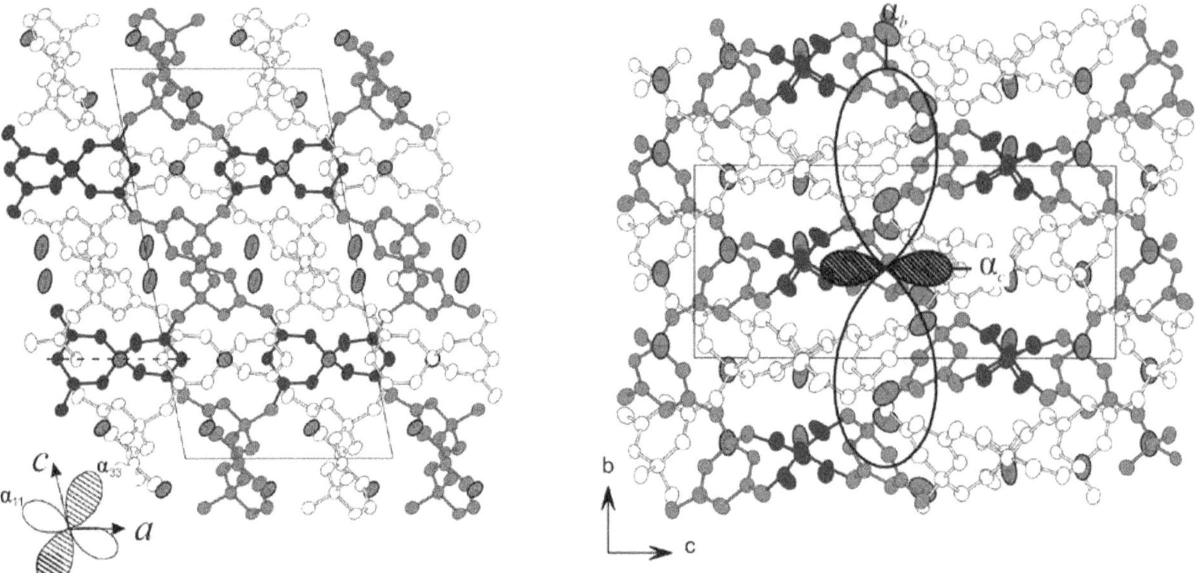

Figure 19. The crystal structure of $Li_3B_{11}O_{18}$, and the pole figure for its thermal expansion coefficients. The double B–O framework of $Li_3B_{11}O_{18}$ is coloured in white and grey, and is projected onto the ac and bc planes. Li atoms are shown by isolated ellipsoids. For the grey framework, $[B_5O_8]^{2-}$ groups connecting the chains of the triborate groups are coloured dark grey (after Sennova et al[59])

linear thermal expansion may be cut from one single crystal of a nonlinear optical LiB_3O_5 borate, which shows extreme anisotropy of thermal expansion, i.e. α_{min}=−71 and α_{max}=101×10^{-6}°C^{-1}.[15]

In nature, due to highly anisotropic thermal deformations, borates may be responsible for the decompression of the rocks which contain them, as well as for increasing permeability by fluids in rock, and in some cases for the accumulation of ores under metamorphic conditions. Similarly, sharply anisotropic thermal expansion of calcite $CaCO_3$ is responsible for the thermal decompression of marbles and for the location of some ore deposits in the marbles.

7. Conclusions

Borate crystals exhibit strongly anisotropic thermal deformations, and often these are record breaking, because they contain rigid boron–oxygen groups, triangular BO_3 anionic groups (and for both of these types of structural group, the thermal deformations can be highly anisotropic for crystals of any symmetry), and unconstrained lattice angles (monoclinic and triclinic crystals).

Rigid groups almost do not change their size and configuration with temperature, and therefore only "adjust themselves" to variation in temperature by mutual rotation around shared oxygen atoms by means of a large hinge mechanism, which is strongly anisotropic by virtue of its nature.

A question can arise, why such anisotropy of structural (in particular thermal) deformations is characteristic for borates, but is not characteristic to the same extent, for example, for silicates? The answer is simple – there are no unalterable atomic groups in silicates, such as rigid boron–oxygen groups; SiO_4 tetrahedra are almost unalterable with temperature, whilst larger polyanions (pyroxene chains, amphibole ribbons, tetrahedral nets, etc.) are not rigid, i.e. have internal degrees of freedom for reconstruction towards reaching minimal energy.

However, silicates exhibit strong anisotropy of thermal expansion in another case – for oblique angle (monoclinic and triclinic) crystals by way of shears connected to changes of unit cell angles.

In their turn, planar anionic groups, like TO_3 triangles, form the basis of the crystal structures of carbonates and nitrates, and thereby compounds of these classes exhibit a strong anisotropy of thermal vibrations of oxygen atoms O and central atoms *T* (C, N) and, as a consequence, a strong anisotropy of thermal deformations is generally characteristic for them.

Thus, the manifestation of strong anisotropy of thermal deformation of crystals has at least three main possible causes: (1) shears, or shear deformations, which are characterised by changes of unit cell angles; (2) strong anisotropy of atomic thermal vibrations in planar anionic groups (TO_3 triangles); (3) hinges, or deformations of an assembly of corner-sharing rigid groups, having no own (internal) degrees of freedom to adjust themselves to varying thermodynamic conditions (T, p, X). The first of the listed reasons can be realised in silicates, the first and the second in carbonates, and all three reasons are realised in borates. That is why borates demonstrate strongly anisotropic thermal expansion most often. It can be assumed that this is characteristic not only for thermal deformations, but also for pressure, composition (chemical) and other types of crystal structure deformations.

Acknowledgements

Stanislav Filatov is grateful to the International Organising Committee, chaired by Adrian Wright and Alex Hannon, and to the Overall Conference Chairman Ladislav Koudelka for honouring him at the Eighth International Conference on Borate Glasses, Crystals and Melts. The authors gratefully acknowledge our borate team – researchers, postdocs, PhD students and students. We are most especially grateful to Maria Krzhizhanovskaya and Natalia Sennova for long-term, productive cooperation in the borate area. The authors would like to thank Ludmila Gorelova and Andrey Shablinskii for help in preparation of figures for the manuscript.

The study was supported by the Russian Foundation of Basic Research (project No. 15-03-05845).

References

1. Huppertz, H. & Eltz, B. Multianvil high–pressure synthesis of $Dy_4B_6O_{15}$: The first oxoborate with edge sharing BO_4 tetrahedra. *J. Am. Chem. Soc.*, 2002, **124**, 9376–9377.
2. Huppertz, H. High–pressure preparation, crystal structure, and properties of $RE_4B_6O_{15}$ (*RE* = Dy, Ho) with an extension of the "fundamental building block"–descriptors. *Z. Naturforsch. B*, 2003, **58**, 278–290.
3. Huppertz, H. New synthetic discoveries *via* high-pressure solid-state chemistry. *Chem. Comm.*, 2011, **47**, 131–140.
4. Bubnova, R. S. & Filatov, S. K. Strong anisotropic thermal expansion in borates. *Phys. Status Solidi B*, 2008, **245**, 2469–2476.
5. Bubnova R. S. & Filatov, S. K. *High-temperature crystal chemistry borates and borosilicates*. Nauka, St.Petersburg, 2008. (In Russian.)
6. Bubnova, R. S. & Filatov, S. K. High-temperature borate crystal chemistry. *Z. Kristallogr.*, 2013, **228** (9), 395–428.
7. Dinnebier, R. E., Hinrichsen, B., Lennie, A. & Jansen, M. High-pressure crystal structure of the non-linear optical compound BiB_3O_6 from two-dimensional powder diffraction data. *Acta Cryst. B*, 2009, **65** (1), 1–10.
8. Filatov, S. K. *Visokotemperaturnaia Kristallohimia (High-Temperature Crystal Chemistry)*. Nedra, Leningrad. 1990. (In Russian.)
9. Filatov, S. K. Negative linear thermal expansion of oblique-angle (monoclinic and triclinic) crystals as a common case. *Phys. Status Solidi B*, 2008, **245**, 2490–2496.
10. Filatov, S. K. General concept of increasing crystal symmetry with an increase in temperature. *Crystallogr. Rep.*, 2011, **56** (6), 953–961.
11. Krogh-Moe, J. Structural interpretation of melting point depression in the sodium borate system. *Phys. Chem. Glasses*, 1962, **3** (4), 101–110.
12. Krogh-Moe, J. Interpretation of the infra-red spectra of boron oxide and alkali borate glasses. *Phys. Chem. Glasses*, 1965, **6** (2), 46–54.
13. Bubnova, R. S., Shepelev, Ju. F., Sennova, N. A. & Filatov, S. K. Thermal behaviour of the rigid boron-oxygen groups in the α-$Na_2B_8O_{13}$ crystal structure. *Z. Kristallogr.*, 2002, **217**, 444–450.
14. Filatov, S., Bubnova, R., Shepelev, Yu., Anderson, J. & Smolin, Yu. The crystal structure of high-temperature α-CsB_5O_8 modification at 20, 300 and 500°C. *Cryst. Res. Technol.*, 2005, **40**, 65–72.
15. Shepelev, Yu. F., Bubnova, R. S., Filatov, S. K., Sennova, N. A. & Pilneva, N. A. LiB_3O_5 crystal structure at 20, 227 and 377°C. *J. Solid State Chem.*, 2005, **178**, 2987–2997.
16. Sennova, N. A., Bubnova, R. S., Shepelev, Ju. F., Filatov, S. K. & Yakovleva, O. I. $Li_2B_4O_7$ Crystal structure in anharmonic approximation at 20, 200, 400 and 500°C. *J. Alloys Comp.*, 2007, **428**, 290–296.
17. Stein, W. D., Cousson, A., Becker, P., Bohatý L. & Braden, M. Temperature-dependent X-ray and neutron diffraction study of BiB_3O_6. *Z. Kristallogr.*, 2007, **222**, 680–689.
18. Filatov, S. K., Shepelev, Yu. F., Alexandrova, Ju. V. & Bubnova, R. S. Structure of bismuth oxoborate $Bi_4B_2O_9$ at 20, 200, and 450°C. *Russ. J. Inorg. Chem.*, 2007, **52** (1), 21–28.
19. Sennova, N. A., Bubnova, R. S., Cordier, G., Albert, B., Filatov, S. K. & Isaenko, L. Temperature dependent changes of the crystal structure of $Li_2B_4O_7$. *Z. Anorg. Allg. Chem.*, 2008, **634**, 2601–2607.
20. Senyshyn, A., Boysen, H., Niewa, R., Banys, J., Kinka, M., Burak, Ya., Adamiv, V., Izumi, F., Chumak, I. & Fuess, H. High-temperature properties of lithium tetraborate $Li_2B_4O_7$. *J. Phys. D*, 2012, **45** (17), 175305.
21. Fofanova, M., Bubnova, R., Albert, B., Filatov, S., Cordier, G. & Egorysheva, A. Structural changes in metastable γ-$Na_2B_4O_7$ between −150°C and 720°C. *Z. Kristallogr.*, 2013, **228** (10), 520–525.
22. Sennova, N. A., Shepelev, Yu. F., Bubnova, R. S. & Filatov, S. K., *private communication*
23. Christ, C. L. & Clark, J. R. A crystal-chemical classification of borate structures with emphasis on hydrated borates. *Phys. Chem. Min.*, 1977, **2**, 59–87.
24. Burns, P. S., Grice, J. D. & Hawthorne, F. C. Borate minerals. I. Polyhedral clusters and fundamental building blocks. *Can. Mineral.*, 1995, **33**, 1131–1151.
25. Hawthorne, F. C., Burns, P. C. & Grice, J. D. The crystal chemistry of boron. *Rev. Miner.*, 1996, **33**, 41–116.
26. Yuan, G. & Xue, D. Crystal chemistry of borates: The classification and algebraic description by topological type of fundamental building blocks. *Acta Cryst.*, 2007, **B63**, 353–362.
27. Wright, A. C. Borate structures: Crystalline and vitreous. *Phys. Chem. Glasses: Eur. J. Glass Sci. Technol. B*, 2010, **51** (1), 1–39.
28. Wright, A. C., Dalba, G., Rocca, F. & Vedishcheva N. M. Review: Borate versus silicate glasses: Why are they so different? *Phys. Chem. Glasses: Eur. J. Glass Sci. Technol. B*, 2010, **51**, 233–265.
29. Touboul, M., Penin, N. & Nowogrocki, G. Borates: A survey of main trends concerning crystal-chemistry, polymorphism and dehydration process of alkaline and pseudo-alkaline borates. *Solid State Sci.*, 2003, **5**, 1327–1342.
30. Filatov, S. K. & Bubnova, R. S. Borate crystal chemistry. *Phys. Chem. Glasses*, 2000, **41**, 216–224.
31. Busing, W. R. & Levy H. A. The effect of thermal motion on the estimation of bond lengths from diffraction measurements. *Acta Crystallogr.*, 1964, **17**, 142–146.
32. Downs, R. T. Analysis of harmonic displacement factors. *Rev. Miner.*, 2000, **41**, 61–87.
33. Fröhlich, R., Bohaty, L. & Liebertz, J. Die kristallstruktur von wismutborat, BiB_3O_6. *Acta Crystallogr. C*, 1984, **40**, 343–344.
34. Shakhmatkin, B. A., Vedishcheva, N. M., Shultz, M. M. & Wright, A. C. The thermodynamic properties of oxide glasses and glass-forming liquids and their chemical structure. *J. Non-Cryst. Solids*, 1994, **177**, 249–256.
35. Vedishcheva, N. M., Shakhmatkin, B. A. & Wright, A. C. Thermodynamic modelling of the structure of glasses and melts: Single-component, binary and ternary systems. *J. Non-Cryst. Solids*, 2001, **293**, 312–317.
36. Becker, P. A contribution to borate crystal chemistry: Rules for the occurrence of polyborate anion types. *Z. Kristallogr.*, 2001, **216**, 523–533.
37. Inorganic Crystal Structure Database (ICSD). Fachinformationzentrum, Karlsruhe, National Institute of Standarts, Gaithersburg. 2012/2.
38. Bubnova, R. S., Krzhizhanovskaya, M. G. & Filatov, S. K., *private communication*
39. Siidra, E. N., Bubnova, R. S. & Krzhizhanovskaya, M. G. Synthesis and anisotropy of thermal expansion of polymorphic aragonite modification of λ-$NdBO_3$. *Fiz. Khim. Stekla*, 2012, **38** (6), 881–884. (In Russian.)
40. Volkov, S. N., Bubnova, R. S., Filatov, S. K. & Krivovichev, S. V. Synthesis, crystal structure and thermal expansion of a novel borate, $Ba_3Bi_2(BO_3)_4$. *Z. Kristallogr.*, 2013, **228** (9), 436–443.
41. Newnham, R. E., Redman, M. J. & Santoro, R. P. Crystal structure of yttrium and other rare-earth borates. *J. Am. Ceram. Soc.*, 1963, **46** (6), 253–256.
42. Lin J. S., Sheptyakov, D., Wang, Y. & Allenspach, P. Structures and phase transitions of vaterite-type rare earth orthoborates: A neutron diffraction study. *Chem. Mater.*, 2004, **16**, 2418–2424.
43. Chadeyron, G., El-Ghozzi, M., Mahiou, R., Arbus, A. & Cousseins, J. C. Revised structure of the orthoborate YBO_3. *J. Solid State Chem.*, 1997, **128**, 261–266.
44. Pitscheider, A., Kaindl, R., Oeckler, O. & Huppertz, H. The crystal structure of π-$ErBO_3$: New single crystal data for an old problem. *J. Solid State Chem.*, 2011, **184**, 149–153.
45. Bubnova, R. S., Firsova, V. A., Krzhizhanovskaya, M. G., Belousova, O. L. & Filatov, S. K., *private communication*

46. Filatov, S. K., Nikolaeva, N. V., Bubnova, R. S. & Polyakova, I. G. Thermal expansion of β-BaB_2O_4 and BaB_4O_7 borates. *Glass Phys. Chem.*, 2006, **32** (4), 471–478.
47. Krivovichev, S. V., Bubnova, R. S., Volkov, S. N., Krzhizhanovskaya, M. G., Egorysheva, A. V. & Filatov, S. K. Preparation, crystal structure and thermal expansion of a novel layered borate, $Ba_2Bi_3B_{25}O_{44}$. *J. Solid State Chem.*, 2012, **196**, 11–16.
48. Filatov, S., Shepelev, Y., Bubnova, R., Sennova, N., Egorysheva, A. V. & Kargin, Y. F. The study of $Bi_3B_5O_{12}$: synthesis, crystal structure and thermal expansion of oxoborate $Bi_3B_5O_{12}$. *J. Solid State Chem.*, 2004, **177**, 515–522.
49. Sleight, W. Compounds that contract on heating. *Inorg. Chem.*, 1998, **37**, 2854–2860.
50. Domnina, M. I., Filatov, S. K., Zuzukina, I. I. & Vergasova, L. P. Thermal deformations of CuO cuprum oxide. *Inorg. Mater.*, 1986, **22**, 1992–1996. (In Russian.)
51. Anderson, J. E., Bubnova, R. S., Filatov, S. K., Polyakova, I. G. & Krzhizhanovskaya, M. G. Thermal behaviour of larderellite, $NH_4[B_5O_7(OH)_2].H_2O$. *Proc. Russ. Miner. Soc.*, 2005, **134** (1), 103–109. (In Russian.)
52. Krjijanovskaya, M. G., Bubnova, R. S., Filatov, S. K., Belger, A. & Paufler, P. Crystal structure and thermal expansion of *b*-RbB_5O_8 from powder diffraction data. *Z. Kristallogr.*, 2000, **215**, 740–743.
53. Bubnova R., Dinnebier, R. E., Filatov, S. & Anderson, J. Crystal structure, thermal and compositional deformations of β-CsB_5O_8. *Cryst. Res. Technol.*, 2007, **42** (2), 143–150.
54. Bubnova, R. S., Anderson, J. E., Krzhizhanovskaya, M. G. & Filatov, S. K. Crystal structure and thermal expansion of ammonium pentaborate $NH_4B_5O_8$. *Glass Phys. Chem.*, 2010, **36** (3), 369–375.
55. Krzhizhanovskaya, M. G., Sennova, N. A., Bubnova, R. S. & Filatov, S. K. Thermal transformations of minerals: Borax – tincalconite – kernite. *Proc. Russ. Miner. Soc.*, 1999, **128** (1), 115–122. (In Russian.)
56. Sennova, N. A., Bubnova, R. S., Filatov, S. K., Paufler, P., Meyer, D. C., Levin, A. A. & Polyakova, I. G. Room, low, and high temperature dehydration and phase transitions of kernite in vacuum and in air. *Cryst. Res. Technol.*, 2005, **40 (6)**, 563–572.
57. Bubnova, R. S., Polyakova, I. G., Anderson, Y. E. & Filatov, S. K. Polymorphism and thermal expansion of the MB_5O_8 crystalline modifications (M = K, Rb) in relation to glass transition of their melts, *Glass Phys. Chem.*, 1999, **25**, 183–194.
58. Rousse, G., Baptiste, B. & Lelong G. Crystal structures of $Li_6B_4O_9$ and $Li_3B_{11}O_{18}$ and Application of the dimensional reduction formalism to lithium borates. *Inorg. Chem.*, 2014, **53**, 6034–6041.
59. Sennova, N., Albert, B., Bubnova, R., Krzhizhanovskaya, M. & Filatov, S. Anhydrous lithium borate, $Li_3B_{11}O_{18}$, crystal structure, phase transition and thermal expansion. *Z. Kristallogr.*, 2014, **229** (7), 497–504.

DOI: 10.13036/17533562.56.1.036 Phys. Chem. Glasses: Eur. J. Glass Sci. Technol. B, February 2015, 56 (1), 36–45

Stanislav Konstantinovich Filatov: Life, Crystal Chemistry and Crystal Chemistry of Borates

S. V. Krivovichev

Department of Crystallography, Institute of Earth Sciences, St. Petersburg State University, 199034 St. Petersburg, Russia
Institute of Silicate Chemistry, Russian Academy of Sciences, 199034 St. Petersburg, Russia

Manuscript received 30 September 2014
Revised version received 18 November 2014
Accepted 19 November 2014

This paper provides a brief overview of the life and professional career of Professor Stanislav Konstantinovich Filatov, St. Petersburg State University, with particular emphasis on his contributions to the crystal chemistry of borates. He has contributed significantly to the development of high temperature crystal chemistry by extensive experimental and theoretical work. In the 1980s, Prof. Filatov established a series of general principles describing the high temperature behaviour of crystalline solids, which explain a range of unique phenomena such as negative thermal expansion. Together with Prof. R. S. Bubnova, he contributed significantly to the crystal chemistry of borates and borosilicates, in particular the understanding of their behaviour at high temperatures. His long-time interest in mineralogy led him to the discovery of more than thirty new minerals, one of which, filatovite, has been named in his honour. In addition to being an excellent scientist, Prof. Filatov has always been an outstanding teacher, who has inspired several generations of young Russian scientists.

1. Introduction

The Eighth International Conference on Borate Glasses, Crystals and Melts, held in July 2014 in Pardubice, Czech Republic, was dedicated to Stanislav Konstantinovich Filatov, Professor of the Department of Crystallography, St. Petersburg State University. Over fifty-five years, his professional and personal life was connected to St. Petersburg (Leningrad) State University, where he went through an impressive career, starting as a student in 1958, and still working now as a Full Professor and former Chairman of the Department of Crystallography. The author of this contribution enjoyed working with him first as a student in 1990, and then as a PhD student, Assistant and close collaborator, taking over the Department chairmanship in 2005. It is therefore my great privilege and honour to devote these pages to the story of Prof. Filatov's life, in both its personal and scientific aspects. The paper is based partially upon the invited talk given at the opening of the Eighth Borate Conference in the beautiful city of Pardubice in early July 2014, and I wish to express my gratitude to Prof. Adrian Wright and Prof. Alex Hannon for the invitation and to Prof. Ladislav Koudelka for the traditional Czech hospitality.

2. Early Life and University Studies

Stanislav Konstantinovich Filatov, known to his students and collaborators also as SK (pronounced: 'æs-ka'), was born on August 6, 1940, in the Dobrinka village of the Voronezh district, Central Russia, to the family of school teacher Konstantin Dmitrievich Filatov and his wife Antonina Mikhailovna Chernyatieva (Figure 1). In SK's postwar childhood, he and his brother Eduard (Figure 2) entertained themselves playing with military relics left in the Russian fields and forests after the fights of the Second World War. In particular, Stanislav was interested in extracting speedometers and other

Figure 1. Parents: mother Antonina Mikhailovna Chernyaeva (left) and father Konstantin Dmitrievich Filatov (right)

Figure 2. SK (right) together with his older brother Eduard. Around 1944

* Corresponding author. Email s.krivovichev@spbu.ru
Original version presented at VIII Int. Conf. on Borate Glasses, Crystals and Melts, Pardubice, Czech Republic, 30 June–2 July 2014

Figure 3. SK's older brother Eduard, military officer

measuring devices from old military vehicles, which is how he became interested in scientists and their instruments. In contrast, his older brother Eduard was more interested in guns. Therefore, it was not a surprise that Eduard became a military officer in the Russian space rocket forces (Figure 3), whereas Stanislav became engaged with scientific activities, preferring to deal with scientific instruments rather than with nice, shiny military equipment. It is, however, noteworthy that the two brothers have been very close to each other, with the elder one always

Figure 4. Memory list given to SK in 1958 by his school. The text is as follows: 'In the day of the last bell ring, before the release from the walls of the school, where you spent ten light years, we, teachers and pupils, wish you a successfull pass of all exams and a happy way on the our life's road to the completion of your desires and duties'

Figure 5. E. S. Fedorov (left) and O. M. Ansheles (right) - the founders of the St. Petersburg crystallographic school

taking care of the younger. In particular, Eduard, being a military officer with a solid military income, provided considerable financial help to Stanislav during his student years in Leningrad (see below), thus taking serious responsibility for the education of his little brother. This was typical for Russian (Soviet) families of the post-war time: parents living in a province were unable to fully support their kids looking for comprehensive education in major cities, and part of the job was usually taken on by their older brothers and sisters.

Soon after the war, the Filatov family moved to the city of Kirsanov, in the Tambov area, where Stanislav finished his school education in 1958 (Figure 4). In the same year he entered Leningrad State University, then and now one of the most prestigious Russian universities. Stanislav started his studies in the Faculty of Geology, choosing crystallography as a major subject. It is worth noting that the St. Petersburg school of crystallography has a great history, being founded by Evgraph Stepanovich Fedorov (1853–1919) (Figure 5), the great Russian scientist who, along with Arthur Schoenflies of Goettingen, derived the 230 space groups that constitute the basis of modern crystallography and crystal chemistry. The Department of Crystallography in Leningrad State University was founded in 1924 by the immediate student of Fedorov, Professor Ossip Markovich Ansheles (1885–1957) (Figure 5).

Figure 6. Student years. During an expedition to Lake Baikal

Figure 7. SK during military service year

Figure 9. SK in the late 1960s during an expedition to the Kola peninsula

Student years in Leningrad, a great time for Stanislav (Figure 6), were interrupted by military service at the Soviet border (1964–1965; Figure 7). Nevertheless, it was a great time too as, according to SK's memories, '... it was a great and rare excuse to do nothing and to simply enjoy life.' This shows that university studies were taken very seriously by Stanislav, and it is of no surprise that he was selected to continue his time in the 'LGU' (abbreviation of Leningradskiy gosudarstvenniy Universitet =Leningrad State University) as a PhD student (=aspirant). At that time, it was already a sign of distinction: only the very few, best students were given this possibility and the competition was extremely hard. SK did his PhD under the supervision of Prof. Viktor Al'bertovich Frank-Kamenetskii (1914–1995) (Figure 8) on the crystal chemistry and thermal stability of zirconium dioxide.[(1)] At the same time SK enjoyed working at the Faculty of Geology, which made possible his participation in geological expeditions, visiting great places like the Kola peninsula in the Russian North (Figure 9).

Figure 8. V. A. Frank-Kamenetskii - PhD Supervisor of SK

3. Professional Career

After finishing his PhD studies and defending his Candidate-of-Sciences dissertation in 1969, SK got the position of Lecturer at the Department of Crystallography, which he held until the 1980s, when he finally became Professor, Full Professor and Chairman of the same Department (in 1990) (Figure 10). The 1990s were extremely difficult years for Russian science. The financial resources supplied to the scientific and educational institutes by the government decreased dramatically. Before that, scientists were treated very

Figure 10. Department of Crystallography, Leningrad State University, in 1974. SK is sitting second from the left

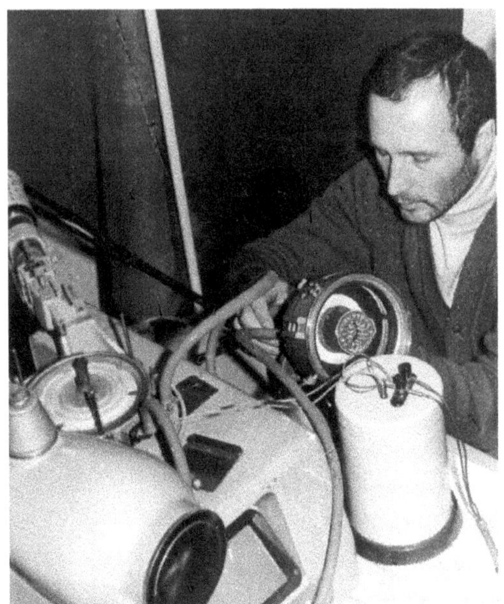

Figure 11. In the 1970s, installing a high temperature camera on the DRON-3 x-ray diffractometer

well in the Soviet Union, and nobody believed that someday such a respectable and valuable activity as science would become 'a job for the poor.' Boris Yeltsin's time was a disaster for Russian science, which had a serious and profound impact upon its future fate and development. However, another aspect of the time was more rewarding: the Iron Curtain (as described by Sir Winston Churchill in his 1946 Fulton speech) disappeared, and Russian science came out of its isolation from the wider world. Publishing in foreign journals and travelling abroad became possible and easy. However, the long years of relative isolation and working for the Soviet military and industrial complex resulted in some very unfortunate circumstances for Russian scientists. Many interesting results, concepts and theories developed in the Soviet Union were published as books and papers in Russian-language journals and publishing companies with no translation into English. This led to the fact that a number of interesting and essential advances made by Soviet scientists remained unknown to their Western colleagues, who sometimes re-invented them many years afterwards (e.g. the concept of mineral evolution[(2)]). All this has to be taken into account when dealing with the history of Russian science in the 20th century. Speaking about the scientific achievements of Professor Filatov, I will describe some of his innovative results that were unfortunately 'buried' in Soviet journals of the 1970s and 1980s.

To be the Department Chairman in the 1990s was a stressful and unpleasant job. SK had to work hard to support the older personnel, at the same making arrangements for future development. Getting new equipment for the university in St. Petersburg was almost impossible, as most of the very modest government money was distributed between academic institutes in Moscow. Without modern infrastructure, research facilities and good salaries, scientific careers became unattractive and the supply of good students was very scarce. SK, using his exceptional lecture talent and scientific enthusiasm, was able to attract some young scientists, who are indebted to him for their professional lives and careers. However, stress never helps, and the time of the 1990s led to many health problems being experienced by such responsible and sensitive people as SK. Thank God, these times are now over.

4. High Temperature Crystal Chemistry

The major theme of SK's research since his PhD studies was the high temperature behaviour of crystalline matter, for which he developed experimental techniques as well as novel theoretical concepts and approaches. He always points out that his main ideas on the mechanisms of thermal expansion of crystals were already understood and roughly formulated in his PhD dissertation.

The experimental basis of SK's research was the use of a high temperature camera installed on the Soviet-made DRON diffractometers (Figure 11). Nowadays high temperature x-ray powder diffraction is a routine technique, and all the necessary attachments and software are usually supplied or can be ordered in a package together with a new x-ray diffractometer. Temperature-controlling equipment and methods to calculate unit cell parameters and their temperature dependence are well elaborated and usually work automatically. In the 1970s, the situation was not that easy: for instance, the angular characteristics of diffraction maxima had to be measured manually using simple rulers on diffractograms printed on extended tapes of a graph paper. Each high temperature x-ray diffraction experiment needed several days of hard work and laborious hand calculations. Yet, the results were so interesting that SK and his students performed more than one hundred and fifty such experiments over approximately twenty five years. Published one-by-one in Russian journals, they constituted a solid database of experimental observations and numerical data that required understanding and generalisation.

The following general topics concerning the temperature dependent behaviour of crystalline solids were analysed and discussed by SK:

(a) thermal deformations as a combination of thermal expansion, shear and hinge deformations;[(3)]
(b) negative thermal expansion and its mechanisms;[(4)]
(c) changes in the symmetry of crystals with changing temperature;[(5)]
(d) plastic crystals, i.e. crystals with rotating molecules (e.g. paraffins);[(6)]
(e) correlation of thermal expansion parameters

Figure 12. Cover of SK's Magnum Opus, 'High-Temperature Crystal Chemistry', published in 1990 in Russian

with other properties of crystals (e.g. optical anisotropy);[7]

(f) relations between high temperature, high pressure and chemistry induced behaviour of crystals.[8]

These and other topics have been discussed in details in SK's book 'High-Temperature Crystal Chemistry. Theory, Methods and Research Results' published in 1990 by the Leningrad-based section of the Soviet geological publishing company 'Nedra' (Figure 12).[9] If translated into English, this book would represent a great addition and independent contribution to the famous book on 'Comparative

Figure 13. After the Agricola Medal award ceremony (from left to right): Prof. Peter Paufler, Rimma S. Bubnova, SK, Prof. Wulf Depmeier [Colour available online]

Figure 14. During the Agricola Medal award ceremony, together with Prof. Falko Langenhorst (on the left) [Colour available online]

Crystal Chemistry' published in 1982 by Robert Hazen & Larry Finger.[10] Unfortunately, this did not happen and the book by SK remains unknown to the Western reader. Later SK published some portions of it as separate papers in English,[11,12] but their impact was much less strong than it could have been in 1990.

Fortunately, not all Western scientists do not read Russian. Professor Peter Paufler of the Technical University of Dresden (former DDR) speaks Russian very well and was acquainted with the research results of SK in the field of high temperature crystal chemistry. Peter Paufler (Dresden) and Wulf Depmeier (Kiel) (Figure 13), being two important figures in German crystallography and mineralogy, nominated SK for the Georg-Agricola medal in Applied Mineralogy awarded by the German Mineralogical Society. The medal was given to SK in 2009 in Halle, Germany (Figure 14). The citation read as following:

> *The Agricola Medal in Bronze is given to Stanislav Konstantinovich Filatov from the State University of St. Petersburg. Stanislav is recognized for his fundamental and sustained research on negative thermal expansion of glasses and minerals. His work has contributed to a basic understanding of the structural causes of this anomalous effect, which in turn has stimulated the development of technical applications.*[13]

5. Crystal Chemistry of Borates: Cherchez la Femme!

Among the minerals and inorganic compounds studied by SK by his favourite high temperature x-ray diffraction techniques, there were several borates. The first one was colemanite, $Ca[B_3O_4(OH)_3].H_2O$, an important ore of boron that occurs in evaporite boron deposits. SK investigated the high temperature behaviour of its structure in 1980,[14] together with

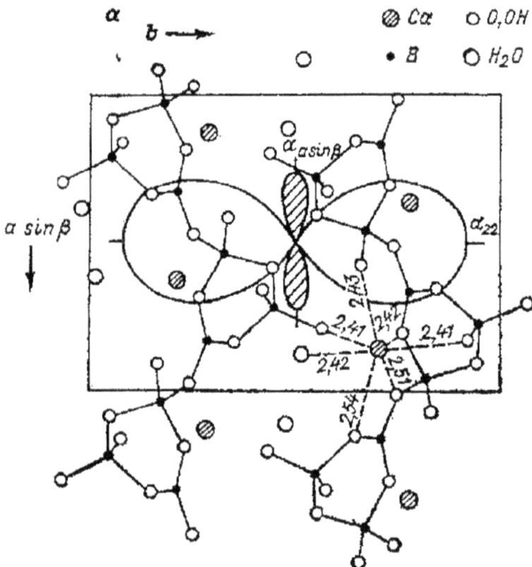

Figure 15. Figure illustrating the thermal expansion behaviour of colemanite. The negative thermal expansion direction is indicated by shaded regions of the section of the figure of thermal expansion coefficients[14]

Viktoria Viktorovna Kondratieva, then a colleague of SK at the Department of Crystallography, LGU. The results were intriguing: despite its chain-like structure, colemanite demonstrated highly anisotropic thermal expansion, with the direction of negative expansion (=contraction) parallel to the extension of the chain (Figure 15)! SK proposed[9] that the reason for the negative thermal expansion is the temperature induced changes in the conformation of the chains, which possess a sufficient flexibility to contract. The presence of relatively large Ca^{2+}

Figure 16. Professor Rimma Sergeevna Bubnova - long-term collaborator of SK and, by chance, his wife

cations in distorted nine-fold coordination does not prevent chain contraction, whereas in hydroboracite, $CaMg[B_3O_4(OH)_3].H_2O$, small octahedral Mg^{2+} cations organized into one-dimensional columns keep borate chains away from contraction.[15]

However, these earlier studies of borates were rather occasional until SK started his fruitful (in all senses of this word) collaboration with Dr Rimma Sergeevna Bubnova, a researcher at the Institute of Silicate Chemistry, USSR Academy of Sciences. She was deeply impressed by SK, both personally and scientifically, and finally became his most close collaborator and, after a while, his wife (Figure 16). Probably, the most valuable result of their collaboration is their beautiful daughter Elena Stanislavovna Filatova (Figure 17). No less beautiful are the results of their scientific collaboration on the crystal chemistry of borates and borosilicates, which I briefly review below.

Figure 17. Elena Stanislavovna Filatova, SK's daugther

The most respected database of inorganic crystal structures at the present time is the Inorganic Crystal Structure Database (ICSD).[16] Its 2012 version contains more than 60 structure reports on borates authored jointly by S. K. Filatov & R. S. Bubnova, and this number is frequently updated by new crystal structure determinations.[17–23] This subset of structures contains data on borates of Li and alkali metals, Bi, Ba, Sr, and other cations. The advances in the crystal chemistry of borates up to 2000 were summarised by the above mentioned authors in their paper 'Borate crystal chemistry'[24] published in *Physics and Chemistry of Glasses*, the predecessor of the *European Journal of Glass Science and Technology Part B*. Since 2000, this paper was cited 64 times, i.e. 4·27 times per year, which contributed significantly to the impact of the journal.

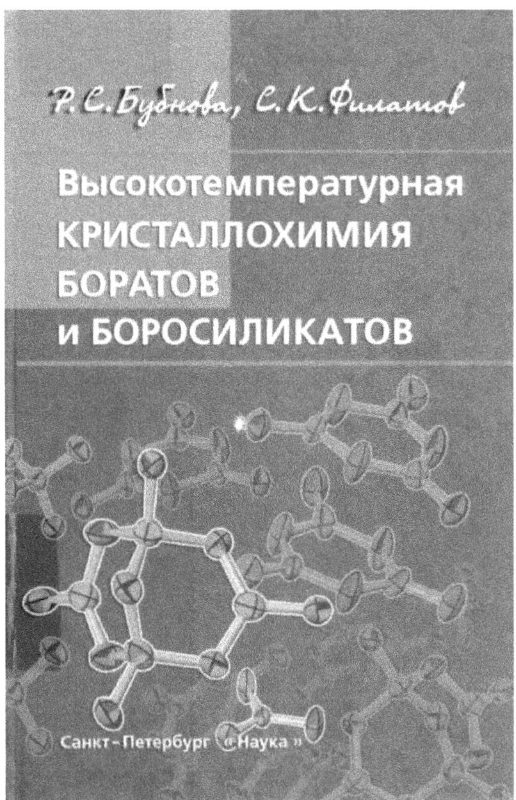

Figure 18. Cover of Rimma S. Bubnova and Stanislav K. Filatov's 760-pages book 'High-Temperature Crystal Chemistry of Borates and Borosilicates' published in 2008 in Russian. Its English translation is forthcoming

However, the family team does not restrict themselves to exploring the crystal chemistry of borates. An important topic for their attention is the high temperature behaviour of crystalline borates, including temperature induced phase transitions (of interest are isosymmetric crystal–glass–crystal transformations), thermal stability, and anisotropy of thermal expansion. The last field is of particular interest for SK, due to his long-time involvement in the development of high temperature crystal chemistry. The remarkable review paper 'High-temperature borate crystal chemistry'[25] was published by SK and Rimma Sergeevna in 2013 in the thematic 'borate' issue of *Zeitschrift für Kristallographie* edited by Hubert Huppertz. The paper is 33 pages long and reviews the current knowledge on the crystal chemistry of borates at high temperatures. The great part of the data reviewed has been obtained by the scientific group headed by the authors. The main idea of the review (along with numerous numerical data) is that of rigid fundamental building blocks (FBBs) that remain intact during thermal transformations. What is really changing is the relations between the FBBs at their links, which behave as atomic scale hinges. This makes the high temperature structural mechanics of borates a really amazing phenomenon, with very frequent occurrence of highly anisotropic behaviour, and even negative thermal expansion (considered in detail in another paper[26]).

The extensive work on borate and borosilicate crystal chemistry culminated in the publication in 2008 of the 760-page book 'High-Temperature Crystal Chemistry of Borates and Borosilicates'[27] (Figure 18), which contains the most comprehensive and detailed treatment of the subject in the world scientific literature. Its updated English translation is forthcoming, thanks to an invitation from the De Gruyter publishing company.

Being a close observer of the hard and systematic work on borates by the scientific group of SK and Rimma Sergeevna, I know that many more new interesting observations and discoveries are ahead, taking into account the many motivated students involved. And with seventeen brand new x-ray diffraction machines acquired by our Department in 2011, the future looks bright and entertaining.

6. Kamchatka Minerals

The story of SK's scientific achievements would be incomplete without mentioning his long-time interest in mineralogy and, in particular, in minerals from the Kamchatka volcanoes. In 1975–76, the Kamchatka peninsula, which SK calls his 'favourite place on Earth,' witnessed the largest basaltic eruption in the history of humankind. It was a fissure eruption, which occurred on the border of the Eurasian continental plate, at the place of its junction with the Atlantic oceanic plate. The eruption itself was absolutely spectacular, but no less spectacular were its consequences. The eruption was followed by intensive fumarolic activities: cooling of magma in a magmatic chamber in the Earth's depths is associated with its degassing. Volcanic gases, reaching the Earth's surface, formed fumaroles with the temperatures reaching ca. 700°C. The gases are rich in metals (in particular, Cu, but also some exotic elements, such as Se and Tl) and their clash with the atmosphere re-

Figure 19. SK (inside) and Dr. Lidiya P. Vergasova (outside) near a fumarole of the Tolbachik volcano, Kamchatka, Russia [Colour available online]

Figure 20. Tolbachik, Kamchatka: SK is drilling a hole to establish a new fumarole [Colour available online]

sults in the crystallisation of an amazing suite of new minerals. In late 1970s and early 1980s, a number of these minerals were collected by mineralogists from the Institute of Volcanology and Seismology of the Russian Academy of Sciences, which, unfortunately, had no proper tools for their characterisation. It was very fortunate, however, that one of them, Dr Lidiya Pavlovna Vergasova, visited Leningrad State University to study x-ray diffraction, and SK served as her supervisor. The beauty and diversity of Kamchatka minerals surprised SK, and another story of his scientific life began. Together with L. P. Vergasova and his numerous students (including the author of this paper), they characterised more than thirty new mineral species. SK especially appreciates fieldwork in Kamchatka, where he usually performs as a digger inside hot and poisonous fumaroles (Figure 19) or a worker drilling the Earth's surface in order to get new fumaroles and their minerals (Figure 20). His contributions to the mineralogy of volcanic exhalative minerals were honoured by a new mineral species, filatovite, $K[(Al,Zn)_2(As,Si)_2O_8]$, that forms as a result of interactions of volcanic gases with host silicate rocks in Tolbachik fumaroles[28] (Figure 21). Filatovite is the first As-bearing feldspar, and is unique among the minerals, in that it contains di- (Zn^{2+}), tri- (Al^{3+}), tetra- (Si^{4+}) and penta- (As^{5+}) valent elements present in the same tetrahedral framework.[29]

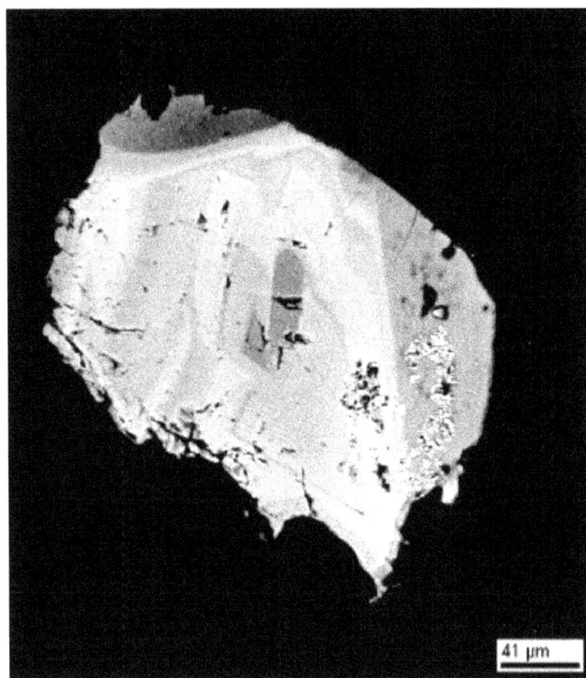

Figure 21. Scanning electron microscopy image of a grain of filatovite, an As-bearing feldspar-group mineral from Kamchatka

The exploration of new minerals from Kamchatka is still an ongoing project led by SK, and his novel discoveries continue to appear in the mineralogical literature.[30–35]

Figure 22. Enjoying traditional Russian music with Prof. E.V. Chuprunov (Nizhnii Novgorod State University). Around 1995 [Colour available online]

7. Anion-centred Tetrahedra

Amongst the most structurally interesting minerals found in the Tolbachik fumaroles are copper minerals that contain so-called 'additional' oxygen atoms. A crystal chemical peculiarity of these atoms is that they are invariably coordinated by four Cu atoms, thus forming oxo-centred OCu_4 tetrahedra. These tetrahedra may link via corners and edges to produce a multitude of structural arrangements, comparable

Figure 23. Two co-chairs of the Commission on x-ray powder diffraction and crystal chemistry of minerals of the Russian Mineralogical Society: SK and Academician Vadim S. Urusov. Miass, Ural, 2007 [Colour available online]

Figure 24. On the Pacific Ocean holding a large piece of algae [Colour available online]

in diversity to those observed in borates. A whole suite of minerals based upon oxo-centred tetrahedra was discovered by SK and co-workers in Kamchatka, which led them to propose a new branch of inorganic crystal chemistry that focuses on structures,[36] whose backbones are formed by polymerisation of anion-centred tetrahedral units. Similar ideas have been expressed in the chemical literature since 1968,[37–40] and our 2001 book[41] and recent review[42] describe the current state-of-the-art of the field.

Figure 25. Listening to gypsy music in Budapest, during the General Meeting of the International Mineralogical Association (2010) [Colour available online]

The possibility for anion-centred tetrahedra to share common edges raised an important problem of systematics for edge-sharing tetrahedral structures that had not been considered in detail in the classical literature on silicates, such as Liebau's book.[43] Therefore it was necessary to elaborate the basic principles of these systematics,[44] which may prove to be useful in providing more information about edge linkages in high pressure borates.[45,46]

8. Epilogue

I hope very much that, in this short paper, I have been able to provide a brief overview of the extremely successful and prolific scientific career of Stanislav Konstantinovich Filatov. Of course, there many things in life beyond writing scientific papers and exploring new research horizons. These include enjoying good friends and colleagues (Figures 22, 23), the Pacific Ocean (Figure 24), gypsy music in Budapest (Figure 25), culinary delights (Figure 26), family (Figure 27) and, of course, the beautiful city of Saint Petersburg (Figure 28), which is truly Russian whilst having a European flavour at the same time. It is its harmony and artistic atmosphere that helped SK in his achievements in scientific and personal life, together with hard work, deep intelligence and generosity with regard to the new generations of scientists.

Figure 26. Preparing meat for the family dinner [Colour available online]

Figure 27. Together with daughter Elena and wife Rimma [Colour available online]

Acknowledgements

In 2014, St. Petersburg scientific crystallographic school headed by S. K. Filatov was awarded the President of Russian Federation grant for leading scientific schools (NSh-1583.2014.5), which is gratefully acknowledged.

References

1. Filatov, S. K. *Crystal Chemistry and Thermal Stability of Zirconium Dioxide*. Candidate of Sciences Dissertation, Leningrad, 1968.
2. Krivovichev, S. V. *Mineral. Mag.*, 2013, **77**, 275.
3. Filatov, S. K. *Dokl. Akad. Nauk SSSR*, 1985, **280**, 369.
4. Filatov, S. K. *Zap. VMO (Proc. Russ. Mineral. Soc.)*, 1982, **111**, 674.
5. Filatov, S. K. *Zap. VMO (Proc. Russ. Mineral. Soc.)*, 1985, **114**, 14.
6. Filatov, S. K., Kotelnikova, E. N. & Aleksandrova, E. A. *Z. Kristallogr.*, 1985, **172**, 35.
7. Domnina, M. I. & Filatov, S. K. *Izv. Akad. Nauk SSSR, Neorg. Mater.*, 1986, **22**, 984.
8. Filatov, S. K. *Zap. VMO (Proc. Russ. Mineral. Soc.)*, 1984, **113**, 172.

Figure 28. Rimma Sergeevna and SK in St. Petersburg, 2006. In the background: 'Standard' yacht, historical reconstruction [Colour available online]

9. Filatov, S. K. *High-Temperature Crystal Chemistry. Theory, Methods and Research Results*, Leningrad, Nedra, 1990. (In Russian.)
10. Hazen, R. M. & Finger, L. W. *Comparative Crystal Chemistry*, London, 1982.
11. Filatov, S. K. *Phys. Status Solidi B*, 2008, **245**, 2490.
12. Filatov, S. K. *Crystallogr. Rep.*, 2011, **56**, 953.
13. German Mineralogical Society News, *Elements*, 2009, **5**, 397.
14. Filatov, S. K. & Kondratieva, V. V. *Izv. Akad. Nauk SSSR, Neorg. Mater.*, 1980, **16**, 475.
15. Kondratieva, V. V. & Filatov, S. K. *Izv. Akad. Nauk SSSR, Neorg. Mater.*, 1986, **22**, 273.
16. *Inorganic Crystal Structure Database (ICSD)*. Fachinformationszentrum, Karlsruhe; National Institute of Standards, Gatesburg.
17. Krzhizhanovskaya, M., Gorelova, L., Bubnova, R. & Filatov, S. *Z. Kristallogr.*, 2013, **288**, 544.
18. Volkov, S. N., Bubnova, R. S., Filatov, S. K. & Krivovichev, S.V. *Z. Kristallogr.*, 2013, **288**, 436.
19. Krivovichev, S. V., Bubnova, R. S., Volkov, S. N., Krzhizhanovskaya, M. G., Egorysheva, A. V. & Filatov, S. K. *J. Solid State Chem.*, 2012, **196**, 11.
20. Krzhizhanovskaya, M. G., Bubnova, R. S., Depmeier, W., Rahmoun, N. S., Filatov, S. K. & Ugolkov, V. L. *Z. Kristallogr.*, 2012, **227**, 446.
21. Volkov, S. N., Filatov, S. K., Bubnova, R. S., Ugolkov, V. L., Svetlyakova, T. N. & Kokh, A. E. *Glass Phys. Chem.*, 2012, **38**, 162.
22. Filatov, S. K., Paufler, P., Georgievskaya, M. I., Levin, A. A., Meyer, D. C. & Bubnova, R. S. *Z. Kristallogr.*, 2011, **226**, 602.
23. Sennova, N., Albert, B., Bubnova, R., Krzhizhanovskaya, M. & Filatov, S. *Z. Kristallogr.*, 2014, **229**, 497–504.
24. Filatov, S. K. & Bubnova, R. S. *Phys. Chem. Glasses*, 2000, **41**, 216.
25. Bubnova, R. S. & Filatov, S. K. *Z. Kristallogr.*, 2013, **228**, 395.
26. Bubnova, R. S. & Filatov, S. K. *Phys. Status Solidi B*, 2008, **245**, 2469.
27. Bubnova, R. S. & Filatov, S. K. *High-Temperature Crystal Chemistry of Borates and Borosilicates*, Nauka, St. Petersburg, 2008.
28. Vergasova, L. P., Krivovichev, S. V., Britvin, S. N., Burns, P. C. & Ananiev, V.V. *Eur. J. Mineral.*, 2004, **16**, 533.
29. Filatov, S. K., Krivovichev, S. V., Burns, P. C. & Vergasova, L. P. *Eur. J. Mineral.*, 2004, **16**, 537.
30. Vergasova, L. P. & Filatov, S. K. *J. Volcanol. Seismol.*, 2012, **6**, 281.
31. Starova, G. L., Vergasova, L. P., Filatov, S. K., Britvin, S. N. & Anan'ev, V. V. *Geol. Ore Dep.*, 2012, **54**, 565.
32. Krivovichev, S. V., Vergasova, L. P., Filatov, S. K., Rybin, D. S., Britvin, S. N. & Ananiev, V. V. *Eur. J. Mineral.*, 2013, **25**, 683.
33. Shuvalov, R. R., Vergasova, L. P., Semenova, T. F., Filatov, S. K., Krivovichev, S. V., Siidra, O. I. & Rudashevsky, N. S. *Am. Mineral.*, 2013, **98**, 463.
34. Vergasova, L. P., Semenova, T. F., Krivovichev, S. V., Filatov, S. K., Zolotarev Jr., A. A. & Ananiev, V. V. *Eur. J. Mineral.*, 2014, **26**, 439.
35. Paufler, P., Filatov, S., Bubnova, R. & Krzhizhanovskaya, M. *Z. Kristallogr.*, 2014, **229**, 725–730.
36. Filatov, S. K., Semenova, T. F. & Vergasova, L. P. *Dokl. Akad. Nauk SSSR*, 1992, **322**, 536.
37. Bergerhoff, G. & Paeslack, J. *Z. Kristallogr.*, 1968, **126**, 112.
38. Caro, P. E. *J. Less-Common. Met.*, 1968, **16**, 367.
39. Carré, D., Guittard, M., Jaulmes, S., Mazurier, A., Palazzi, M., Pardo, M. P., Laurelle, P. & Flahaut, J. *J. Solid State Chem.*, 1984, **55**, 287.
40. Schleid, Th. *Eur. J. Sol. State Inor.*, 1996, **33**, 227.
41. Krivovichev, S. V. & Filatov, S. K. *Crystal Chemistry of Minerals and Inorganic Compounds Containing Anion-Centered Tetrahedra*. St. Petersburg University Press, St. Petersburg, 2001.
42. Krivovichev, S. V., Mentré, O., Siidra, O. I., Colmont, M. & Filatov, S. K. *Chem. Rev.*, 2013, **113**, 6459.
43. Liebau, F. *Structural Chemistry of Silicates. Structure, Bonding and Classification*, Springer, Heidelberg, 1985.
44. Krivovichev, S. V., Filatov, S. K. & Semenova, T. F. *Z. Kristallogr.*, 1997, **212**, 411–417.
45. Huppertz, H. & von der Eltz, B. *J. Am. Chem. Soc.*, 2002, **124**, 9376–9377.
46. Neumair, S. C., Kaindl, R. & Huppertz, H. *J. Solid State Chem.*, 2012, **185**, 1–9.

Phys. Chem. Glasses: Eur. J. Glass Sci. Technol. B, December 2014, **55** (6), 225–236

Short and intermediate range order in sodium borosilicate glasses: a quantitative thermodynamic approach

Natalia M. Vedishcheva, Irina G. Polyakova*

Institute of Silicate Chemistry of the Russian Academy of Sciences, Nab. Makarova 2, St. Petersburg, 199034, Russia

Adrian C. Wright

J.J. Thomson Physical Laboratory, University of Reading, Whiteknights, Reading, RG6 6AF, UK

Manuscript received 7 September 2014
Accepted 11 September 2014

It is shown that the short range and intermediate range order in the structure of sodium borosilicate glasses can be quantitatively described using the thermodynamic concept of the chemical structure. This approach enables the distribution of the basic structural units ($BØ_3$, $BØ_4^-$, $BØ_2O^-$, $BØO_2^{2-}$, Q^4, Q^3 and Q^2), superstructural units (boroxol, pentaborate, triborate, diborate rings and cyclic metaborate anions) and 4-membered borosilicate rings (danburite and reedmergnerite units) to be calculated, without use of adjustable parameters, over the entire glass forming region. The calculations have been made considering structural changes in terms of the minimum Gibbs free energy of the system. A critical analysis has been performed of the fractions of 4-fold co-ordinated boron atoms, $BØ_4^-$, and the glass transition temperatures, T_g, calculated by Mauro and co-workers on the basis of the topological principles of borosilicate glass chemistry.

1. Introduction

It is generally known that properties of glasses are determined by their structure. The structural study of borosilicate glasses is an extremely complex task, due to the large variety of structural units that describe the short range and intermediate range order in the glass structure. The short range order is characterised by presence of borate tetrahedra, $BØ_4^-$, and triangles, $BØ_3$, $BØ_2O^-$, $BØO_2^{2-}$ and BO_3^{3-}, with the number of nonbridging oxygen atoms varying from 0 to 3 (the symbol Ø presents bridging oxygen atoms), and silicon–oxygen tetrahedra, $SiØ_4$ (denoted Q^4), $SiØ_3O^-$ (Q^3), $SiØ_2O_2^{2-}$ (Q^2), $SiØO_3^{3-}$ (Q^1) and SiO_4^{4-} (Q^0), where the number of nonbridging oxygen atoms varies from 0 to 4. The intermediate range order implies the structure considered at the level of superstructural units, such as boroxol, pentaborate, triborate and diborate rings, and cyclic metaborate anions[1], together with 4-membered borosilicate rings, danburite and reedmergnerite, which are also present in sodium borosilicate glasses.[2] The latter units are shown in Figure 1. Structural studies of glasses in the system $Na_2O–B_2O_3–SiO_2$ were carried out since the 1970s by NMR, neutron scattering, Raman and IR spectroscopy methods. Researchers' attention was mainly paid to the short range order in the structure of the borate sub-network, the largest contribution being made, in 1978–1983, by Bray and co-workers.[3,4] The silicate sub-network was studied to a much lesser degree, and results of the most comprehensive study performed by Feller and co-authors are reported in Ref. 5. As to the intermediate range order in sodium borosilicate glasses, to the best of the present authors' knowledge, no information of this sort is currently available in the literature. In this paper, it will be shown that the method of thermodynamic modelling, which is valid for a description of the short range order in the structure of these glasses,[6] can also be used for calculating, over the whole glass formation region, the distribution of borate superstructural units and 4-membered borosilicate rings, together with that of the silicate basic structural units, Q^n, which are also present in borosilicate glasses.

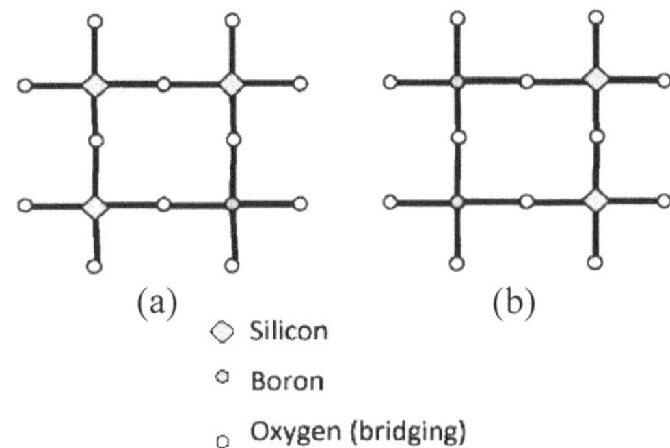

Figure 1. Four-membered borosilicate rings: (a) reedmergnerite and (b) danburite

* Corresponding author. Email ionatali386@gmail.com
Original version presented at VIII Int. Conf. on Borate Glasses, Crystals and Melts, Pardubice, Czech Republic, 30 June–2 July 2014

2. Basic principles of the thermodynamic modelling

This approach is based on the rigorous thermodynamic model of associated solutions. It considers glasses formed from components with different chemical natures as solutions, whose components are the unreacted oxides and the products of their interaction. It is assumed that:

(i) These products, also called chemical groupings, are similar in stoichiometry to the crystalline compounds existing in the phase diagram of the system in question. Therefore, the set of groupings expected to be present in a given glass forming system is chosen on the basis of the set of crystalline compounds;

(ii) A structural similarity between the groupings and crystals also exists, at least, in terms of the ratio of the basic structural units that characterise the short range order in the glass structure. Note that both assumptions yield an observance of the principle of the minimum Gibbs energy of the system;

(iii) The groupings and the unreacted oxides form an ideal solution. The correctness of these assumptions is considered in detail in Ref. 7.

An interaction between oxide components of the system $Na_2O–B_2O_3–SiO_2$ proceeds according to reactions (1)–(3) given below and results in the formation of the products (borates, silicates and borosilicates) whose crystalline analogues can be found in the phase diagram of the system:

$$k\mathrm{Na_2O}+m\mathrm{B_2O_3}=k\mathrm{Na_2O}.m\mathrm{B_2O_3} \quad (1)$$

$$a\mathrm{Na_2O}+b\mathrm{SiO_2}=a\mathrm{Na_2O}.b\mathrm{SiO_2} \quad (2)$$

$$s\mathrm{Na_2O}+p\mathrm{B_2O_3}+r\mathrm{SiO_2}=s\mathrm{Na_2O}.p\mathrm{B_2O_3}.r\mathrm{SiO_2} \quad (3)$$

The mathematical formalism of the model consists in solving the set of equations of the law of mass action written, in an ideal form, for reactions (1)–(3)

$$K_1=\frac{X_{k\mathrm{Na_2O}.m\mathrm{B_2O_3}}}{X^k_{\mathrm{Na_2O}}X^m_{\mathrm{B_2O_3}}} \quad (4)$$

$$K_2=\frac{X_{a\mathrm{Na_2O}.b\mathrm{SiO_2}}}{X^a_{\mathrm{Na_2O}}\cdot X^b_{\mathrm{SiO_2}}} \quad (5)$$

$$K_3=\frac{X_{s\mathrm{Na_2O}.p\mathrm{B_2O_3}.r\mathrm{SiO_2}}}{X^s_{\mathrm{Na_2O}}X^p_{\mathrm{B_2O_3}}X^r_{\mathrm{SiO_2}}} \quad (6)$$

and the equations of the law of mass balance of the oxide components

$$X^*_{\mathrm{Na_2O}}=n_{\mathrm{Na_2O}}+kn_{k\mathrm{Na_2O}.m\mathrm{B_2O_3}}+an_{a\mathrm{Na_2O}\cdot b\mathrm{SiO_2}}+sn_{s\mathrm{Na_2O}.p\mathrm{B_2O_3}.r\mathrm{SiO_2}} \quad (7)$$

$$X^*_{\mathrm{B_2O_3}}=n_{\mathrm{B_2O_3}}+mn_{k\mathrm{Na_2O}\cdot m\mathrm{B_2O_3}}+pn_{s\mathrm{Na_2O}\times p\mathrm{B_2O_3}\cdot r\mathrm{SiO_2}} \quad (8)$$

$$X^*_{\mathrm{SiO_2}}=n_{\mathrm{SiO_2}}+bn_{a\mathrm{Na_2O}\cdot b\mathrm{SiO_2}}+rn_{s\mathrm{Na_2O}\cdot p\mathrm{B_2O_3}\cdot r\mathrm{SiO_2}} \quad (9)$$

In the above equations, the symbol K_i denotes the equilibrium constants of reactions (1)–(3), the symbols $X_{k\mathrm{Na_2O}.m\mathrm{B_2O_3}}$, $X_{a\mathrm{Na_2O}.b\mathrm{SiO_2}}$, $X_{s\mathrm{Na_2O}.p\mathrm{B_2O_3}.r\mathrm{SiO_2}}$, $X_{\mathrm{Na_2O}}$, $X_{\mathrm{B_2O_3}}$ and $X_{\mathrm{SiO_2}}$ are the equilibrium concentrations of various chemical groupings and the unreacted oxides, the symbols $n_{k\mathrm{Na_2O}.m\mathrm{B_2O_3}}$, $n_{a\mathrm{Na_2O}.b\mathrm{SiO_2}}$, $n_{s\mathrm{Na_2O}.p\mathrm{B_2O_3}.r\mathrm{SiO_2}}$, $n_{\mathrm{Na_2O}}$, $n_{\mathrm{B_2O_3}}$ and $n_{\mathrm{SiO_2}}$ are the numbers of moles of these groupings and oxides, and $X^*_{\mathrm{Na_2O}}$, $X^*_{\mathrm{B_2O_3}}$ and $X^*_{\mathrm{SiO_2}}$ represent the analytical content of Na_2O, B_2O_3 and SiO_2, in mole fractions, in a given glass. The equilibrium constants are determined by Equation (10)

$$K=\exp\left(-\frac{\Delta G^0_f}{RT}\right) \quad (10)$$

Where ΔG^0_f is the standard Gibbs free energy of formation from oxides of the compounds existing in the system. These potentials are obtained using the data from reference books,[(8–10)] hence avoiding the use of adjustable parameters. The relationship between the mole fractions, X_i, and the numbers of moles, n_i, of the species i in a given glass is described by Equation (11)

$$X_i=\frac{n_i}{\Sigma n} \quad (11)$$

Here the index i refers both to the chemical groupings and the unreacted oxides. The symbol $\sum n$ denotes the total number of their moles.

As is shown in Ref. 11, at constant temperature, pressure and the fixed numbers of moles of the initial substances, the set of nonlinear Equations (4)–(9) has the only solution, which yields information on the equilibrium concentrations of the chemical groupings and the unreacted oxides. These data enable the chemical structure of glasses to be determined. In case of the system $Na_2O–B_2O_3–SiO_2$, the concept of the chemical structure implies the relative content in the glasses of borate, silicate and borosilicate groupings with different stoichiometry, together with the unreacted oxides, Na_2O, B_2O_3 and SiO_2. The types of the chemical groupings present in these glasses are shown in Table 1.

Recall that the chemical groupings considered in this approach are formed from the basic structural units, the ratio of these units being similar to that in the crystalline compounds with the same stoichiometry. The observance of the principle of the minimum Gibbs energy of a system in the vitreous state requires that, in the groupings, the basic structural units are combined together in the same manner as that in the relevant crystalline compounds. In other words, it can be expected that the chemical groupings present in glasses comprise associates similar to the superstructural units existing in the crystals that form in the given system. Hence, the concept of the chemical structure enables both the short range order and the intermediate range order in the glass structure to be described, at the levels of the basic structural units and superstructural units, respectively. The (super) structural unit species and the quantities in which they are present in the borate, silicate and borosilicate chemical groupings are shown Table 1. Note that, according to the model formalism, the structure of the unreacted oxides B_2O_3 and SiO_2 is considered as similar to that in the vitreous rather than crystalline state. Therefore, the B_2O_3 grouping introduces into glasses both $BØ_3$ triangles and boroxol groups,[12] and the SiO_2 grouping introduces silicon–oxygen tetrahedra with all bridging oxygen atoms, Q^4. Information about the superstructural unit species present in the groupings $Na_2O.5B_2O_3$, $Na_2O.3B_2O_3$ and $Na_2O.2B_2O_3$ is based on the results of Ref. 13, and that for the groupings $Na_2O.4B_2O_3$, $Na_2O.B_2O_3$ and $2Na_2O.B_2O_3$ follows from x-ray diffraction data for the corresponding crystals.[14] The conclusion concerning the presence of danburite and reedmergnerite rings in the groupings $Na_2O.B_2O_3.2SiO_2$ and $Na_2O.B_2O_3.6SiO_2$ is based on information about the structure of the corresponding crystalline borosilicates[14] and the results of an NMR study of Pyrex.[2] The types and numbers of Q^n species present in the groupings $Na_2O.2SiO_2$ and $Na_2O.SiO_2$ are determined on the basis of information about the structures of crystalline compounds with the similar stoichiometries.[15,16] The structure of the crystal $3Na_2O.8SiO_2$ is unknown, therefore it is assumed that the ratio between Na and Si equal to 6:8 implies the ratio between the basic structural units equal to $6Q^3$:$2Q^4$. The validity of this assumption is confirmed by good agreement between the calculated and experimental Q^n values in sodium silicate glasses, over the extended composition region.[7] The information on (super)structural units given in Table 1, together with the calculated equilibrium concentrations, X_i, of the chemical groupings, which bring these units into glasses, form the basis for modelling the glass structure at the two levels.

Table 1. The chemical groupings present in sodium borosilicate glasses, and their relation to the short range and intermediate range order in the glass structure

Chemical groupings	*Types and numbers of the basic structural units and superstructural units introduced into the glasses by 1 mole of each chemical grouping*	
	Basic structural units	*Superstructural units*
B_2O_3	$2BØ_3$	Boroxol ring (½), $BØ_3$ (½)
$Na_2O.5B_2O_3$	$2[BØ_4]^-$, $8BØ_3$	Pentaborate rings (2)
$Na_2O.4B_2O_3$	$2[BØ_4]^-$, $6BØ_3$	Pentaborate ring (1), Triborate ring (1)
$Na_2O.3B_2O_3$	$2[BØ_4]^-$, $4BØ_3$	Triborate rings (2)
$Na_2O.2B_2O_3$	$2[BØ_4]^-$, $2BØ_3$	Diborate ring (1)
$Na_2O.B_2O_3$	$2BØ_2O^-$	Cyclic metaborate anion (⅔)
$2Na_2O.B_2O_3$	$2BØO_2^{2-}$	None
$Na_2O.B_2O_3.2SiO_2$	$2[BØ_4]^-$, $2Q^4$	Danburite ring (1)
$Na_2O.B_2O_3.6SiO_2$	$2[BØ_4]^-$, $6Q^4$	Reedmergnerite rings (2)
SiO_2	Q^4	None
$3Na_2O.8SiO_2$	$2Q^4$, $6Q^3$	
$Na_2O.2SiO_2$	$2Q^3$	
$Na_2O.SiO_2$	Q^2	

2.1. Equations for calculating the structure of sodium borosilicate glasses

This section presents equations used for calculating the content of (super)structural units all over the glass-formation region. For all of the units, the error of calculation does not exceed, on average, ±2%.

2.2. Intermediate range order

In the following, the main attention is given to the structure at the level of superstructural units. In the absence of experimental data, the results obtained in this work can yield a general idea as to the intermediate range order in sodium borosilicate glasses. They can also be considered as guidelines in future structural studies of these glasses. Information about the superstructural units in glasses is of importance for the following reasons. First, as is demonstrated in Ref. 7, the short range order does not reflect all of the complexity of the structure of B_2O_3-containing glasses, whilst superstructural units yield a considerably more detailed picture. Second, a high temperature Raman spectroscopy study[17,18] has revealed that, on quenching sodium borate melts, the re-arrangements at the level of the short range order are "frozen-in" at temperatures noticeably higher than T_g, whilst structural changes at the level of the superstructural units become "frozen-in" at temperatures very close to T_g. This observation is very important, since it points to the fact that the structure of B_2O_3-containing glasses, and hence their properties, are mainly determined by the superstructural units present.

Calculations of the structure at this level have been performed using Equations (12)–(23). In Equations (12)–(18), the numerators represent the numbers of the corresponding superstructural units, and in Equations (19)–(22) the numerators are the numbers of the various independent basic structural units that are not included into superstructural units. In Equations (12)–(22), the denominator described by Equation (23) denotes the sum of the numbers of all of the superstructural units (denoted SSU), together with the independent basic structural units (denoted B^n and Q^n). Due to this, the values calculated using Equations (12)–(22) are normalized to 100% of all of the (super)structural units present in glasses.

$$[\text{Boroxol rings}]=[\text{BO}_3]=\frac{0{\cdot}5X_{B_2O_3}}{\Sigma(SSU+B_n+Q^n)} \quad (12)$$

$$\left[\text{Pentaborate rings}\right]=\frac{2X_{Na_2O.5B_2O_3}+X_{Na_2O.4B_2O_3}}{\Sigma(SSU+B_n+Q^n)} \quad (13)$$

$$\left[\text{Triborate rings}\right]=\frac{2X_{Na_2O.3B_2O_3}+X_{Na_2O.4B_2O_3}}{\Sigma(SSU+B_n+Q^n)} \quad (14)$$

$$\left[\text{Diborate rings}\right]=\frac{X_{Na_2O.2B_2O_3}}{\Sigma(SSU+B_n+Q^n)} \quad (15)$$

$$[\text{Danburite rings}] = \frac{X_{Na_2O.B_2O_3.2SiO_2}}{\Sigma(SSU + B_n + Q^n)} \quad (16)$$

$$[\text{Reedmergnerite rings}] = \frac{2X_{Na_2O.B_2O_3.6SiO_2}}{\Sigma(SSU + B_n + Q^n)} \quad (17)$$

$$[\text{Cyclic metaborate anions}] = \frac{0{\cdot}667\,X_{Na_2O.B_2O_3}}{\Sigma(SSU + B_n + Q^n)} \quad (18)$$

$$[BØO_2^{2-}] = \frac{2X_{2Na_2O.B_2O_3}}{\Sigma(SSU + B_n + Q^n)} \quad (19)$$

$$[Q^4] = \frac{X_{SiO_2} + 2X_{3Na_2O.8SiO_2}}{\Sigma(SSU + B_n + Q^n)} \quad (20)$$

$$[Q^3] = \frac{6X_{3Na_2O.8SiO_2} + 2X_{Na_2O.2SiO_2}}{\Sigma(SSU + B_n + Q^n)} \quad (21)$$

$$[Q^2] = \frac{X_{Na_2O.SiO_2}}{\Sigma(SSU + B_n + Q^n)} \quad (22)$$

$$\begin{aligned}\Sigma(SSU + B_n + Q\ ) = {} & X_{B_2O_3} + 2X_{Na_2O.5B_2O_3} + 2X_{Na_2O.4B_2O_3} \\ & + X_{Na_2O.3B_2O_3} + X_{Na_2O.2B_2O_3} \\ & + 0{\cdot}667X_{Na_2O.B_2O_3} + 2X_{2Na_2O.B_2O_3} \\ & + X_{Na_2O.B_2O_3.2SiO_2} + X_{Na_2O.B_2O_3.6SiO_2} \\ & + X_{SiO_2} + 8X_{3Na_2O.8SiO_2} + 2X_{Na_2O.2SiO_2} \\ & \quad {}_{Na_2O.SiO_2}\end{aligned} \quad (23)$$

2.3. Short range order

Structural studies of the short range order in sodium borosilicate glasses present a problem, which is not reflected in the literature. Although, in recent decades, the accuracy and sensitivity of traditional structural methods keeps increasing, the accumulated quantitative data on the fractions of basic structural units in borosilicate glasses are of limited value. This is due to the fact that NMR and Raman experiments are performed separately on each of the sub-networks, which corresponds to the hypothetical case of the independent existence of the borate and silicate sub-networks. This is illustrated by the data from Ref. 19, where both sub-networks have been studied quantitatively in several glasses by NMR and Raman spectroscopy. It has been found that, in the glass $30Na_2O.10B_2O_3.60SiO_2$ (mol%), the content of tetrahedra, $BØ_4^-$, and symmetric triangles, $BØ_3$, in the borate sub-network is 75 and 25%, respectively. The silicate sub-network of the same glass is built of tetrahedra Q^4 and Q^3, whose respective content is 20 and 80%. Thus, the entire content of the borate and silicate structural units in this glass is equal to 200%. The same is true for other compositions studied in Ref. 19. It is clear that this result does not enable changes in the structure to be described quantitatively as a function of the glass composition or the structure–property relationship to be established.

The concept of the chemical structure allows the short range order in the glass structure to be calculated for both independent and co-existing borate and silicate sub-networks. Here, the calculations are limited to the example of 4-fold co-ordinated boron atoms. As this is the most studied structural unit, extensive experimental data are available for it in the literature, which allows a comparison of the calculated and experimental fractions of $BØ_4^-$ tetrahedra. In case of the sub-networks considered separately, the content of $BØ_4^-$ units is calculated using Equations (24) and (25). The numerator of Equation (24) presents the number of $BØ_4^-$ tetrahedra in a given glass, and the denominator described by Equation (25) is the total number of all borate structural units, tetrahedra and triangles. In this case, the sum of fractions of all borate units in the borate sub-network is equal to 100%, i.e. as in experimental studies.

$$[BØ_4^-] = \frac{\begin{aligned}2(&X_{Na_2O.5B_2O_3} + X_{Na_2O.4B_2O_3} + X_{Na_2O.3B_2O_3} \\ &+ X_{Na_2O.2B_2O_3} + X_{Na_2O.B_2O_3.2SiO_2} + X_{Na_2O.B_2O_3.6SiO_2})\end{aligned}}{\Sigma B_n} \quad (24)$$

$$\begin{aligned}\Sigma B_n = 2\big(&X_{B_2O_3} + 5X_{Na_2O.5B_2O_3} + 4X_{Na_2O.4B_2O_3} \\ &+ 3X_{Na_2O.3B_2O_3} + 2X_{Na_2O.2B_2O_3} + X_{Na_2O.B_2O_3} \\ &+ X_{2Na_2O.B_2O_3} + X_{Na_2O.B_2O_3.2SiO_2} + X_{Na_2O.B_2O_3.6SiO_2}\big)\end{aligned} \quad (25)$$

In case of the borate and silicate sub-networks considered jointly, the fraction of boron–oxygen tetrahedra is calculated using Equation (26), whose numerator is similar to that in Equation (24). The difference is in the denominator, which presents, as shown by Equation (27), the total number of all borate *and* silicate structural units, B_n, and Q^n, respectively. In this way, the calculated content of $BØ_4^-$ tetrahedra is normalised to 100% of all structural units present in a given glass.

$$[BØ_4^-] = \frac{\begin{aligned}2\big(&X_{Na_2O.5B_2O_3} + X_{Na_2O.4B_2O_3} + X_{Na_2O.3B_2O_3} \\ &+ X_{Na_2O.2B_2O_3} + X_{Na_2O.B_2O_3.2SiO_2} + X_{Na_2O.B_2O_3.6SiO_2}\big)\end{aligned}}{\Sigma(B_n + Q^n)} \quad (26)$$

$$\begin{aligned}\Sigma(B_n + Q^n) = {} & 2X_{B_2O_3} + 10X_{Na_2O.5B_2O_3} + 8X_{Na_2O.4B_2O_3} \\ & + 6X_{Na_2O.3B_2O_3} + 4X_{Na_2O.2B_2O_3} + 2X_{Na_2O.B_2O_3} \\ & + 2X_{2Na_2O.B_2O_3} + 4X_{Na_2O.B_2O_3.2SiO_2} \\ & + 8X_{Na_2O.B_2O_3.6SiO_2} + X_{SiO_2} + 8X_{3Na_2O.8SiO_2} \\ & + 2X_{Na_2O.2SiO_2} + X_{Na_2O.SiO_2}\end{aligned} \quad (27)$$

3. Results and discussion

3.1. Intermediate range order

At this level, the glass structure presents a combination of borate superstructural units and 4-membered

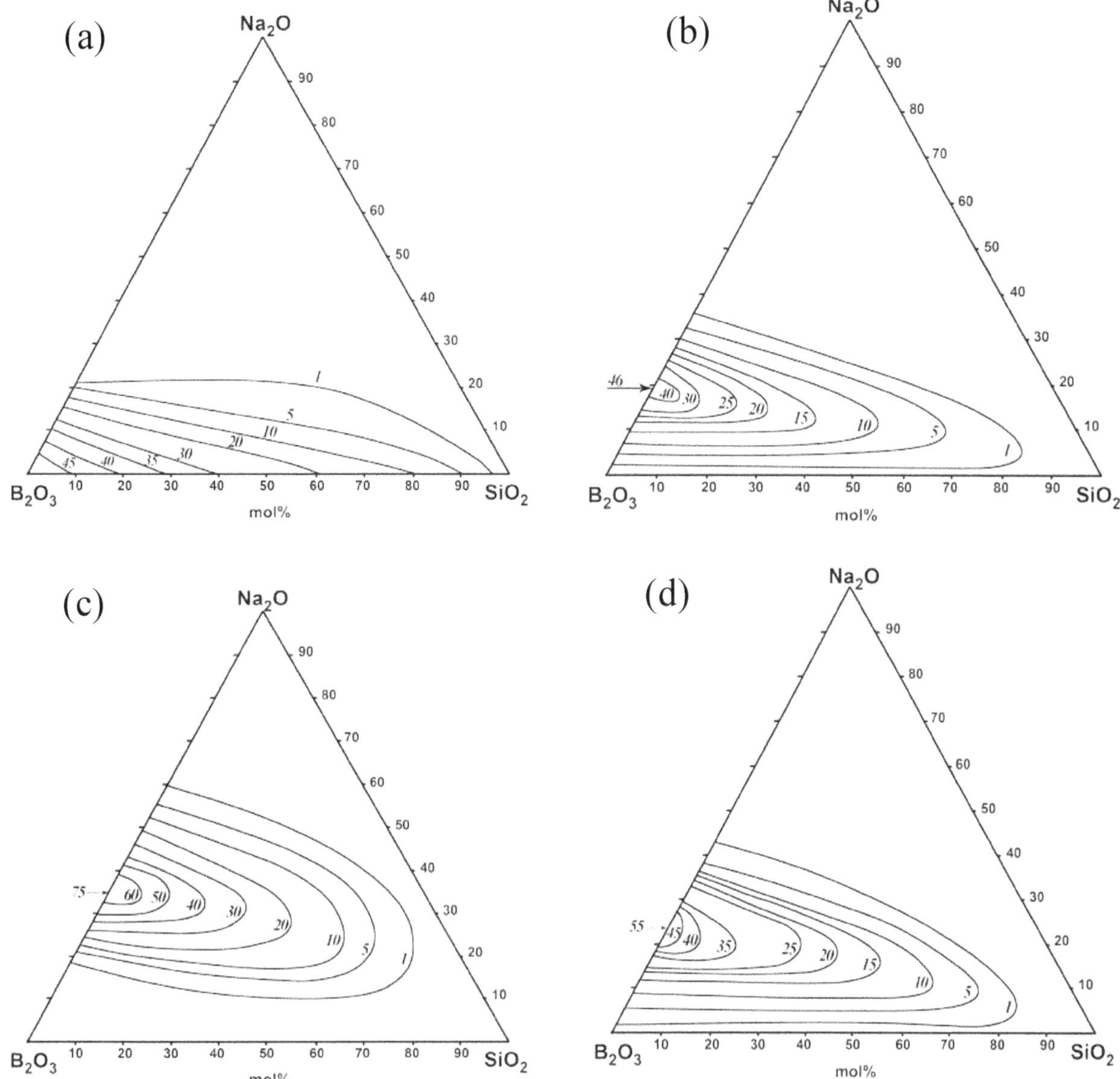

Figure 2. The content in sodium borosilicate glasses of (super)structural units with all bridging oxygen atoms: (a) boroxol and BØ$_3$, (b) pentaborate, (c) diborate and (d) triborate

borosilicate rings, together with independent borate and silicate basic structural units. The fractions of these structural elements have been calculated using Equations (12)–(23), at 400–600°C depending on the glass composition, and the results for borate superstructural units with all bridging oxygen atoms are shown in Figure 2, for cyclic metaborate anions and BØO$_2^{2-}$ units in Figure 3, and for silicon–oxygen tetrahedra, Q^n, and borosilicate rings in Figure 4. In these graphs, the contours represent a constant content (in %) of the corresponding units. Note that, in Figure 2(a), every line refers to both boroxol rings and triangles BØ$_3$, which are present in glasses in equal amounts.[12] Figure 4(d) shows the total content of borosilicate rings, danburite and reedmergnerite, present in glasses. In all of the cases shown in Figures 2–4, the observed trends in the content of (super)structural units can be explained in terms of chemical interactions proceeding in the system Na_2O–B_2O_3–SiO_2. In particular, on the introduction of even the smallest amounts of the oxide Na_2O into the system B_2O_3–SiO_2, all possible reactions of the types (1)–(3) immediately start to proceed, although to a different extent, depending on the glass composition. This results in a gradual decrease in the equilibrium concentrations of the unreacted oxides B_2O_3 and SiO_2 in the glasses as the Na_2O content grows, which is responsible for a decrease in the content of boroxol rings and BØ$_3$ triangles (Figure 2(a)) and silicon–oxygen tetrahedra, Q^4 (Figure 4(a)). It is seen that both borate units are present over the composition region up to ~20 mol% Na_2O, whilst Q^4 species can be found in glasses with a noticeably higher Na_2O content approaching ~50 mol%. This is explained by the fact that, due to its alkaline nature, Na_2O tends to primarily interact with B_2O_3 as a more acidic oxide

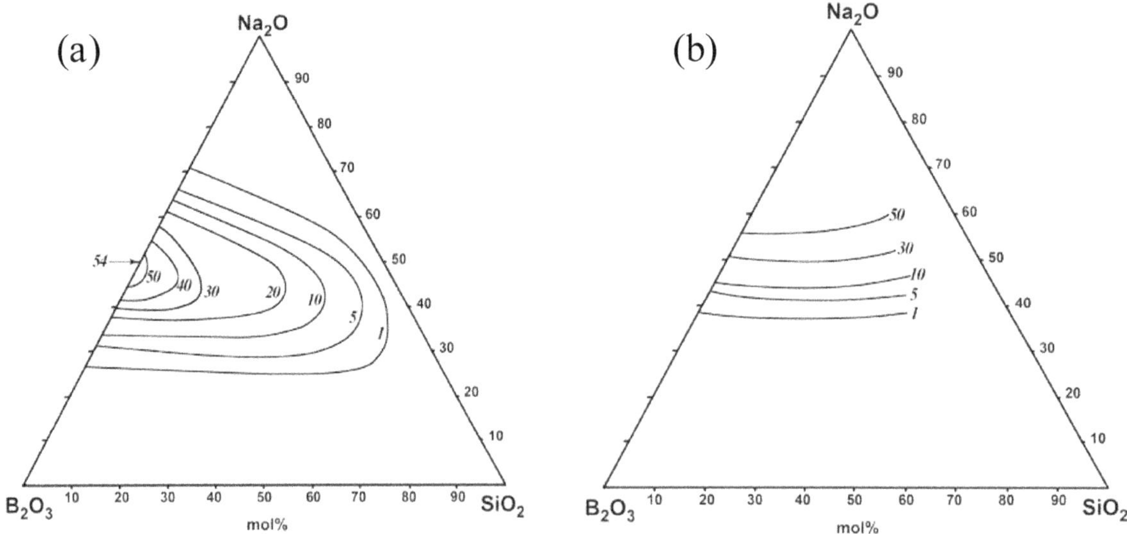

Figure 3. The content in sodium borosilicate glasses of borate anions: (a) cyclic metaborate and (b) $BØO_2^{2-}$

than SiO_2. As a result, SiO_2 remains unreacted over a larger composition region than B_2O_3.

Figures 2 and 3 show that, as the Na_2O content gradually increases, boroxol rings and $BØ_3$ triangles are replaced by pentaborate and triborate superstructural units, which are followed by diborate units and, then, by cyclic metaborate rings and $BØO_2^{2-}$ units. This sequence, as well as the composition regions over which the above units are present in glasses, adequately depict the features observed in the intermediate range order of sodium borate glasses shown in Figure 5. This is due to the fact that, as follows from the calculations of the chemical structure of the systems Na_2O–B_2O_3[(7)] and Na_2O–B_2O_3–SiO_2, the formation of the chemical groupings $Na_2O.5B_2O_3$, $Na_2O.4B_2O_3$ and $Na_2O.3B_2O_3$, which introduce pentaborate and triborate superstructural units, prevails over the low alkali composition region, whilst the groupings $Na_2O.B_2O_3$ and $2Na_2O.B_2O_3$, which introduce cyclic metaborate rings and $BØO_2^{2}$ units, predominantly form over the high alkali composition region. The difference between the glasses in the two systems is only in the amounts of each unit: the further from the Na_2O–B_2O_3 side of the composition triangle (i.e. the smaller the B_2O_3 content in ternary glasses), the lower is the content of all of the borate structural units, as compared to that in binary glasses.

Similar tendencies are observed in case of silicate structural units, Q^n. Both in ternary glasses (Figure 4(a)–(c)) and in binary sodium silicate glasses (Figure 6), an increase in the Na_2O content results in a gradual replacement of tetrahedra with all bridging oxygen atoms, Q^4, by tetrahedra with one nonbridging oxygen atom, Q^3, and then by those with two nonbridging oxygen atoms, Q^2. According to the chemical structure of the systems Na_2O–SiO_2[(7)] and Na_2O–B_2O_3–SiO_2, the chemical groupings $3Na_2O.8SiO_2$ and $Na_2O.2SiO_2$, which introduce Q^3 units, predominantly form at a lower Na_2O content than that at which the grouping $Na_2O.SiO_2$ (Q^2 tetrahedra) prevails. The specific feature of ternary glasses is the presence of 4-membered borosilicate rings, danburite and reedmergnerite. Their total content is shown in Figure 4(d), and it is seen that the region of the maximum amount of these units is located in the SiO_2-rich corner of the composition triangle.

Structural changes observed on going from the Na_2O–B_2O_3 to Na_2O–SiO_2 side of the composition triangle can similarly be explained in terms of chemical interactions. According to information on the chemical structure of the system Na_2O–B_2O_3–SiO_2, a gradual replacement of borate superstructural units and borate anions by silicate Q^n species and 4-membered borosilicate rings, which is shown in Figures 2–4, is due to the fact that the formation of borate chemical groupings according to reaction (1) is a dominating process over the B_2O_3-enriched region, and the formation of sodium silicate and sodium borosilicate chemical groupings according to reactions (2) and (3), respectively, prevails over the SiO_2-rich region. In the middle of the composition triangle, which can be considered as a transition

Table 2. The content (%) of superstructural and basic structural units in the glasses $10Na_2O.20B_2O_3.70SiO_2$ (10–20–70), $20Na_2O.60B_2O_3.20SiO_2$ (20–60–20) and $33Na_2O.33B_2O_3.34SiO_2$ (33–33–34)

(Super)structural units	*Glass composition (mol%)*		
	10–20–70	*20–60–20*	*33–33–34*
Boroxol and	1	3	0
$BØ_3$	1	3	0
Pentaborate	3	17	0
Triborate	6	32	2
Diborate	2	12	24
Cyclic metaborate	0	0	8
$BØO_2^{2-}$	0	0	0
Danburite	3	4	10
Reedmergnerite	12	1	2
Q^4	71	28	11
Q^3	1	0	39
Q^2	0	0	4

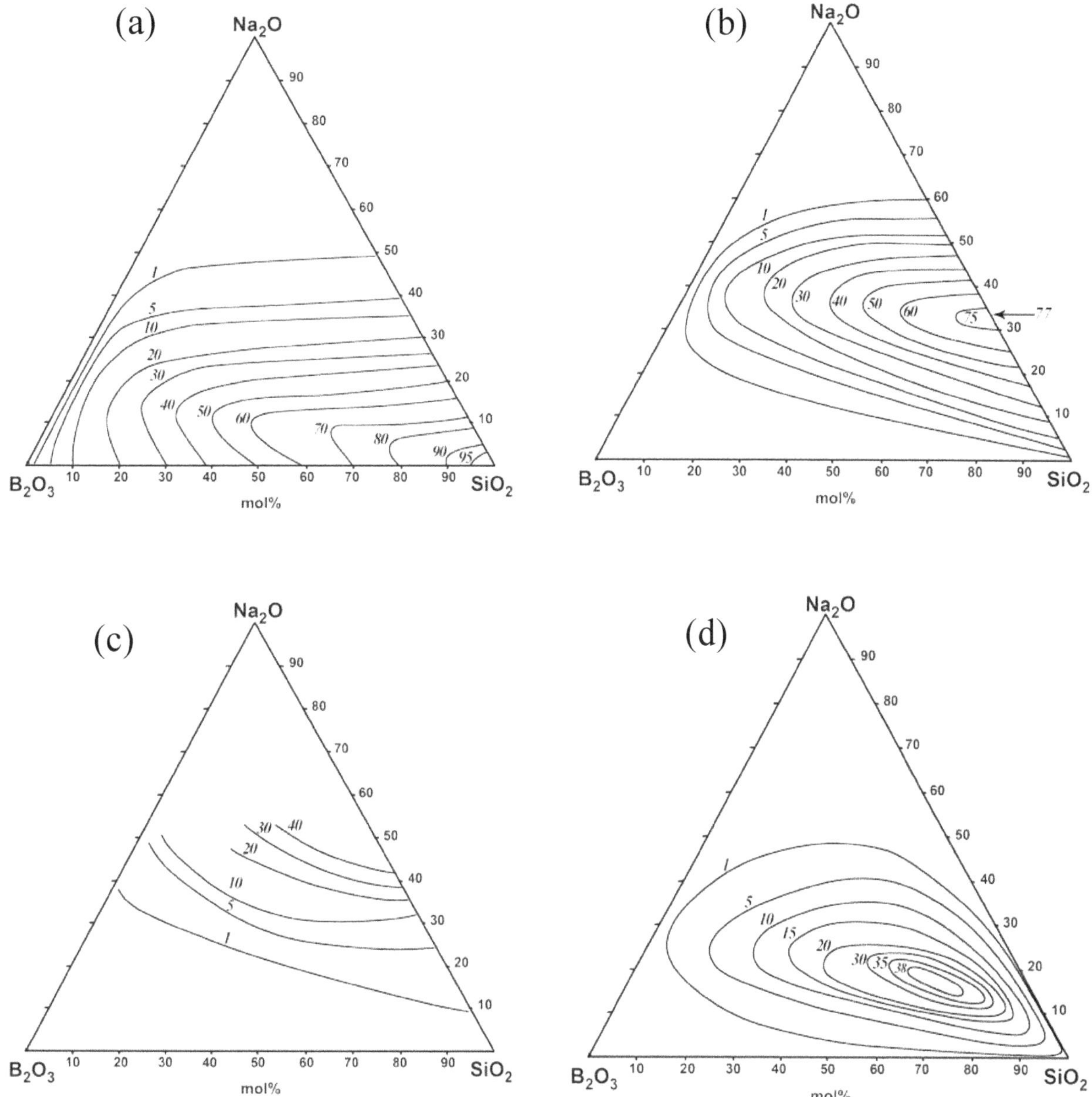

Figure 4. The content in sodium borosilicate glasses of silicate basic structural units: (a) Q^4, (b) Q^3, (c) Q^2, and (d) the total content of danburite and reedmergnerite rings

region, borate superstructural units and 4-membered borosilicate rings co-exist with silicate Q^n tetrahedra in comparable amounts.

To get an impression of the structure of sodium borosilicate glasses over the whole glass forming region, all ten triangles shown in Figures 2–4 should be superimposed. The influence of the glass composition on the intermediate range order in their structure is illustrated in Table 2, by the example of three glasses whose compositions are scattered over the composition triangle, *viz.* $10Na_2O.20B_2O_3.70SiO_2$, $20Na_2O.60B_2O_3.20SiO_2$ and $33Na_2O.33B_2O_3.34SiO_2$. An analysis of the data in Table 2 reveals the following. First, in these glasses, the content of boroxol rings and $BØ_3$ triangles is almost zero. This points to the fact that the B_2O_3 is completely bound to the products of interaction between the constituent oxides. It is seen that none of the glasses contain $BØO_2^{2-}$ triangles, which is due to the fact that these units appear in glasses containing more than 33 mol% Na_2O (see Figure 3(b)). It is also seen that, when going from vitreous $10Na_2O.20B_2O_3.70SiO_2$ to vitreous $33Na_2O.33B_2O_3.34SiO_2$, the content of Q^4 units keeps decreasing, which points to a more and more pronounced interaction between sodium oxide and silica as the analytical content of Na_2O becomes larger. In vitreous $10Na_2O.20B_2O_3.70SiO_2$, the presence of various borate superstructural units, together with 4-membered borosilicate rings (danburite and reedmergnerite) in comparable amounts, means that Na_2O is equally involved into reactions (1) and (3), which leads to the formation of borate chemical groupings, $Na_2O.5B_2O_3$, $Na_2O.4B_2O_3$ $Na_2O.3B_2O_3$ and $Na_2O.2B_2O_3$, and of borosilicate groupings, Na_2O.

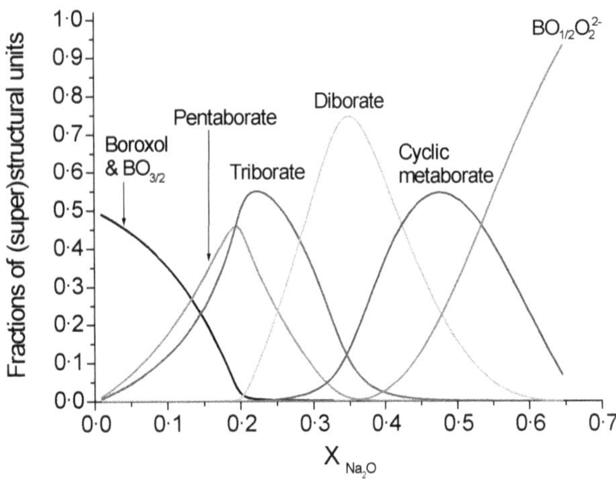

Figure 5. The content of (super)structural units in glasses of the system Na_2O–B_2O_3 [Colour available online]

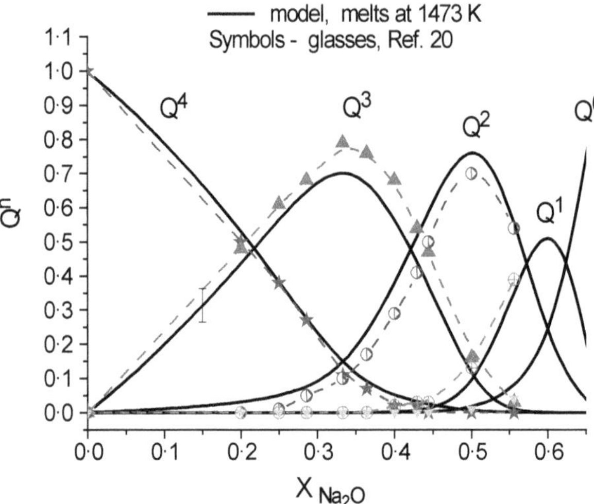

Figure 6. The content of Q^n units in glasses and melts of the system Na_2O–SiO_2 [Colour available online]

$B_2O_3.2SiO_2$ and $Na_2O.B_2O_3.6SiO_2$. The negligibly small amount of Q^3 units indicates that Na_2O does not tend to interact with SiO_2 to form the binary silicate groupings, $3Na_2O.8SiO_2$ and $Na_2O.2SiO_2$, according to reaction (2). Since the formation of ternary compounds is energetically more profitable than the formation of binary compounds,[(8–10)] in low alkali borosilicate glasses reaction (3) is the preferable way for the oxides Na_2O and SiO_2 to interact. This is confirmed by a noticeable amount of 4-membered borosilicate rings in this glass. The absence of Q^2 units is explained by the fact that the high alkali chemical grouping $Na_2O.SiO_2$, which introduces this species into glasses, does not form in any noticeable amounts in the composition with a low analytical content of Na_2O.

In vitreous $20Na_2O.60B_2O_3.20SiO_2$, the content of 4-membered borosilicate rings is the lowest among all three compositions considered, which is due to the lowest analytical content of the oxide SiO_2. The overall content of borate superstructural units is the largest, which is explained by the highest analytical content of B_2O_3 in this glass. The absence of Q^3 and Q^2 units is due to the fact that, as in the previous case, in the composition with 20 mol% Na_2O, a more preferable way for the oxides Na_2O and SiO_2 to interact is to form borosilicate chemical groupings (reaction (3)) rather than binary silicate ones (reaction (2)). The formation of the latter products requires a higher analytical content of Na_2O, which is demonstrated below.

In vitreous $33Na_2O.33B_2O_3.34SiO_2$, due to the balanced amounts of the oxide components, Na_2O is equally involved into reactions (1), (2) and (3). As follows from the chemical structure of this glass, the total content of sodium borate and sodium borosilicate chemical groupings is comparable to the total content of the various sodium silicate groupings. This results in a similar balance between the overall content of borate superstructural units and 4-membered borosilicate rings (46%) and the total content of various Q^n units (54%). Among other specific structural features of this glass are the largest variety of Q^n species and the presence of cyclic metaborate anions. Both are responsible for the presence of a noticeable number of nonbridging oxygen atoms in this glass.

3.2. 4-membered borosilicate rings and phase separation

The specific nature of danburite and reedmergnerite rings, which include both $BØ_4^-$ and Q^4 tetrahedra, suggests a unique role of these units in the structure of borosilicate glasses. It can be assumed that 4-membered borosilicate rings serve as a link that binds together the borate and silicate sub-networks. This assumption has convincingly been confirmed by the results of an NMR study of Pyrex,[(2)] which reveal that the borate sub-network formed from borate superstructural units is connected to the silicate sub-network built from Q^4 tetrahedra through danburite and reedmergnerite rings, as is shown in Figure 7. This fact leads to the conclusion that the

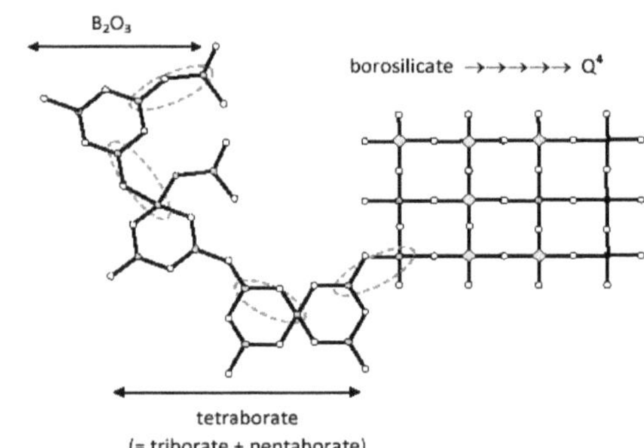

Figure 7. The schematic of the Pyrex structure proposed in Ref. 2. Charge balancing Na^+ cations have been omitted for clarity but will be in close proximity to $BØ_4^-$ units [Colour available online]

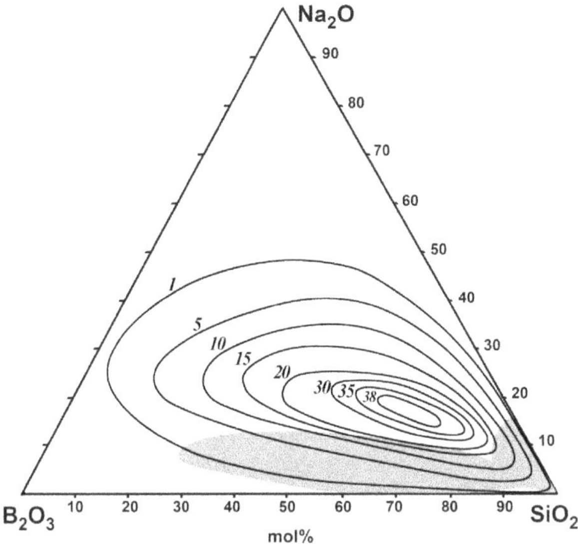

Figure 8. The total content of danburite and reedmergnerite rings, and the phase separation region[21] in sodium borosilicate glasses [Colour available online]

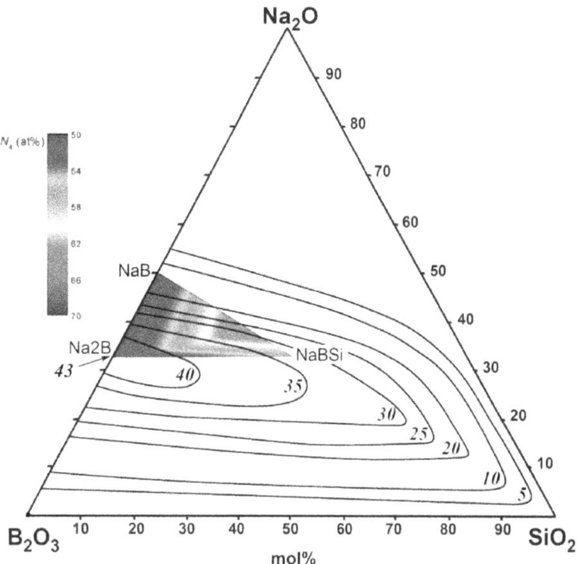

Figure 9. The calculated fraction of $BØ_4^-$ units in sodium borosilicate glasses over the composition triangle: the thermodynamic model used in this work and the topological model (coloured triangle)[22] [Colour available online]

tendency of sodium borosilicate glasses towards phase separation can be related to the amount of borosilicate rings present. In Figure 8, the region of phase separation in the system Na_2O–B_2O_3–SiO_2, at 600°C, as determined in Ref. 21, is superimposed on the total content of danburite and reedmergnerite rings. A combination of the two regions allows the conditions to be quantitatively determined under which a given glass remains single-phase. As is seen from Figure 8, over the composition region where the analytical content of SiO_2 exceeds 50 mol%, glasses are single phase if the total content of 4-membered borosilicate units is larger than 15–20%. Glasses with a lower content of danburite and reedmergnerite units are phase separated. At the analytical content of SiO_2 below 50 mol%, glasses remain single phase at a noticeably lower concentration of borosilicate rings, e.g. for glasses with 25–30 mol% SiO_2, the required content of these rings is only 1–5%. Such a difference between high silica and low silica glasses is due to the difference in the ratio in their networks of Q^4 tetrahedra, borate superstructural units and 4-membered borosilicate rings, and will be considered in detail elsewhere.

3.3. Short range order

This level in the glass structure is considered in terms of the fraction of 4-fold co-ordinated boron atoms. The fractions of $BØ_4^-$ units have been calculated using Equations (26) and (27), hence the results obtained over the entire glass forming region are normalised to 100% of all borate and silicate units present. These results are compared with those calculated on the basis of the topological model.[22] The triangle considered in Figure 17(a) of Ref. 22 is shown in Figure 9, where it is superimposed on the composition triangle for the Na_2O–B_2O_3–SiO_2 system. Since the corners of the literature triangle correspond to the compositions $Na_2O.2B_2O_3$ (denoted Na2B), $Na_2O.B_2O_3$ (NaB) and $Na_2O.B_2O_3.SiO_2$ (NaBSi), it is clear that the area considered[22] represents only a small part of the glass forming region. Figure 9 also shows the results obtained using the concept of the chemical structure. It is seen that the content of 4-fold co-ordinated boron in glasses systematically decreases on going from the Na_2O–B_2O_3 side to Na_2O–SiO_2 side of the composition triangle, and tends to zero on approaching the latter side. This completely agrees with the nature of chemical processes proceeding in the Na_2O–B_2O_3–SiO_2 system, *viz.* at a high B_2O_3 content, the formation of sodium borate chemical groupings, which bring 4-fold co-ordinated boron atoms into glasses (reaction (1)), prevails over the formation of sodium silicate and sodium borosilicate chemical groupings (reactions (2) and (3), respectively). On approaching the Na_2O–SiO_2 side of the triangle, the proceeding of reactions of the type (1) slows down, which results in a gradual decrease in the content of $BØ_4^-$ tetrahedra.

A comparison of the results obtained using the concept of the chemical structure with the results of the topological model[22] reveals a serious disagreement between them. From Figure 9, it is seen that, when going from the composition $Na_2O.B_2O_3$ to $Na_2O.B_2O_3.SiO_2$ in the triangle,[22] the fraction of $BØ_4^-$ units first increases from ~50% to ~64% and then decreases to ~56%. These estimations have been made on the basis of the colour scale shown by the triangle. Figure 9 shows that the fraction of $BØ_4^-$ tetrahedra

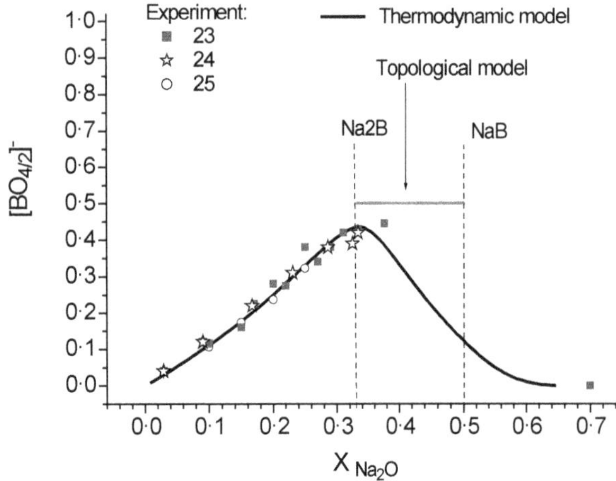

Figure 10. The calculated and experimental fractions of $BØ_4^-$ units in sodium borate glasses [Colour available online]

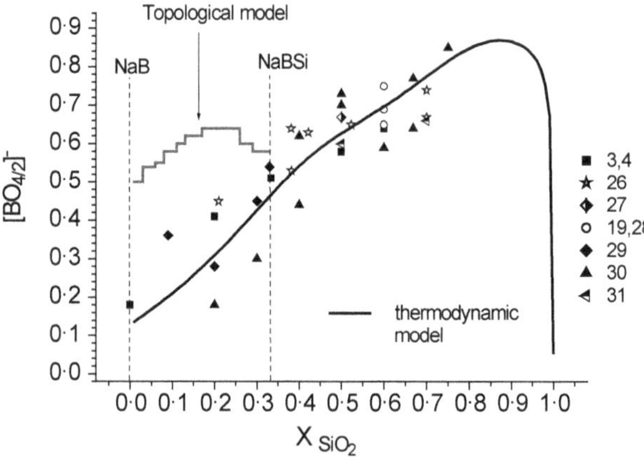

Figure 11. The calculated and experimental fractions of $BØ_4^-$ units in sodium borosilicate glasses in the cut with $R=[Na_2O]/[B_2O_3]=1$ [Colour available online]

obtained in this work systematically increases from a value exceeding 10% at the composition $Na_2O.B_2O_3$ to the value in excess of 30% at the composition $Na_2O.B_2O_3.SiO_2$. In view of this difference, it has been decided to compare the results of both models with the experimental data available in the literature for glasses in the systems $Na_2O–B_2O_3$ and $Na_2O–B_2O_3–SiO_2$. This choice is explained by the fact that one of the sides of the triangle from Ref. 22 presents only a part of the binary system, and the other side is just a part of the ternary system. Note that, for a comparison with experimental data, the fractions of $BØ_4^-$ units modelled in this paper have been calculated using Equations (24) and (25). Hence, the values in question are normalised to 100% of the fractions of all borate units in the borate sub-network, as is done in structural studies. A comparison between the model and experimental[23–25] results for sodium borate glasses is shown in Figure 10. It is seen that the topological model neither correctly describes the concentration dependence of the fractions of $BØ_4^-$ units in binary glasses nor reflects the difference in the structure of glasses and crystals. The latter follows from the overestimated value of the fraction of $BØ_4^-$ units, which is equal to 0·5,[22] whilst it is known that, in glasses, it is never that high.[23–25] The value of 0·5 can only be found in crystalline compounds of the diborate composition.[14] This difference between the vitreous and crystalline states is clearly explained using the concept of the chemical structure.[7]

Figure 11 shows the model and experimental[3,4,19,26–31] fractions of 4-fold co-ordinated boron atoms in sodium borosilicate glasses, in the cut with $R=[Na_2O]/[B_2O_3]=1$. As is seen, the values obtained using the topological model are noticeably higher than the experimental data, and the shape of the curve depicting the concentration dependence of the model values is wrong. It should be noted that, even if the topological model could yield an accurate description of the experimental data, the limitation to such a small part of the glass forming region would not allow a comprehensive understanding of specific features of the glass structure. Note that the concept of the chemical structure correctly accounts for experimental data shown in Figure 11.

3.4. Glass transition temperature as described by the topological model

A reliable prediction of glass transition temperatures, T_g, is claimed to be one of the main advantages of the topological model.[22] Therefore, it is interesting to compare the predictions of this temperature made for sodium borosilicate glasses, as shown in Figures 16(a) and 17(b) of Ref. 22 with the available experimental data.[32,33] This comparison is made in Figure 12, where solid and dashed black lines depict the isotherms that describe changes in the experimental temperatures T_g

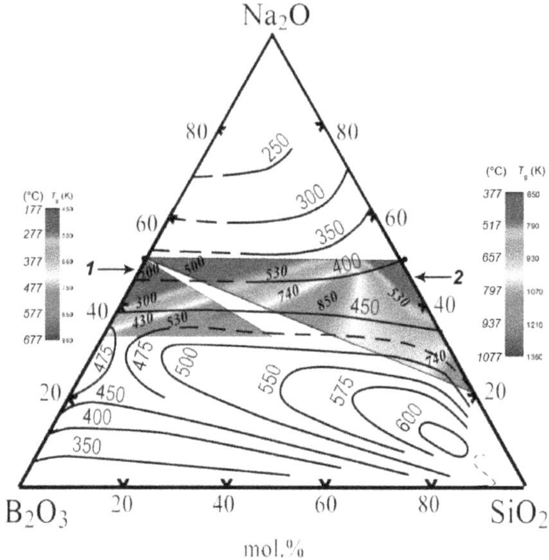

Figure 12. The calculated[22] and experimental[32,33] glass transition temperatures, T_g, for sodium borosilicate glasses [Colour available online]

(°C) as a function of composition. Two triangles with the calculated temperatures[22] are superimposed on that with the experimental data. The triangle from Figure 17(b), whose corners are the compositions $Na_2O.2B_2O_3$, $Na_2O.B_2O_3$ and $Na_2O.B_2O_3.SiO_2$ (further denoted as Triangle 1) is located on the left hand side of the triangle with the experimental data. On the right hand side, the triangle from Figure 16(a), whose corners are the compositions $Na_2O.B_2O_3$, $Na_2O.SiO_2$ and $20Na_2O.80SiO_2$ (Triangle 2) is located. Next to the triangles,[22] there are the colour scales (K[22] and °C) that enable the values of the calculated temperatures to be estimated. The temperatures estimated by the present authors are given in different colour fields: 200, 300, 430 and 530°C (Triangle 1) and 500, 530, 740 and 850°C (Triangle 2). An analysis of the data shown in Figure 12 reveals the following.

1. **Comparison of Triangles 1 and 2.** The calculated temperatures presented in these triangles do not agree, although they are a continuation of each other, since both occupy the area limited by the cuts with the constant content of Na_2O equal to 33 and 50 mol%. When going from Triangle 1 to Triangle 2, the temperatures change abruptly, rather than gradually, and the value of these jumps is close to 300 degrees. The most illustrative is the difference in the temperatures over the area adjacent to the metaborate composition, $Na_2O.B_2O_3$ (~200°C in Triangle 1 and ~500°C in Triangle 2).
2. **Comparison of the calculated and experimental temperatures.** In Triangle 1, over the composition region with 40–50 mol% Na_2O, the calculated values of T_g are significantly lower than the experimental ones. However, over the region with the Na_2O content varying from 33·3 to 40 mol%, this difference noticeably decreases, and the calculated values become rather close to the experimental ones exceeding them by no more than 50–60°C. This is the best of all of the results yielded by the topological model over both triangles. However the composition region, to which this result refers, is very narrow being limited by the compositions containing approximately 33–35 mol% Na_2O. Over Triangle 2, all of the calculated temperatures are much higher than the experimental values: the largest difference is as large as ~400 degrees, being observed in the vicinity of the maximum calculated value of T_g (~850°C – calculated and 450°C – experimental).
3. **Character of the changes in T_g as a function of composition.** As is seen from Figure 12, over the whole triangle from the Na_2O–B_2O_3 side to the Na_2O–SiO_2 side, the experimental values of T_g systematically increase as the Na_2O content changes from ~75 to ~30 mol%. It is also seen that, within the area occupied by Triangles 1 and 2, the experimental isotherms are practically horizontal. This means that, at a fixed content of Na_2O, the values of T_g remain constant irrespective of the $[SiO_2]/[B_2O_3]$ ratio, and they vary only if the Na_2O content changes. In contrast to this, the topological model shows, in the central part of Triangle 2, the region of maximum T_g (estimated by the present authors as ~850°C), which is separated by a region with lower temperatures from another region of maximum T_g, the latter being located in the vicinity of the Na_2O–SiO_2 side, over the narrow region with compositions with 20–25 mol% Na_2O.
4. **Position of the region with the maximum T_g values.** The calculated region of the maximum temperatures refers to the compositions containing over 33 mol% Na_2O, whilst the experimental region is located in the high silica corner of the composition triangle, where the Na_2O content is approximately 10–15 mol%. It should be noted that, within the limits of the calculation uncertainty, a similar position is occupied by region

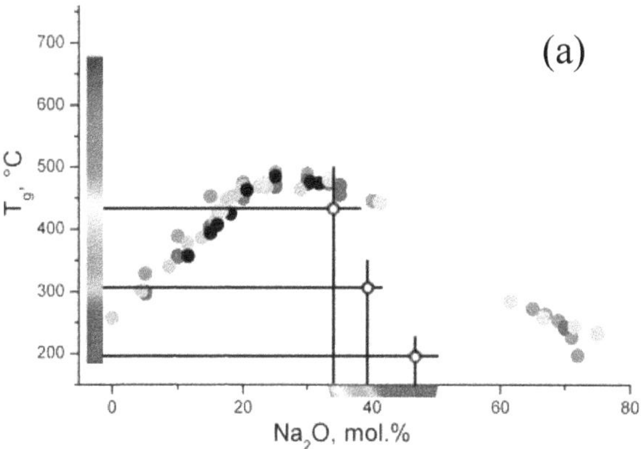

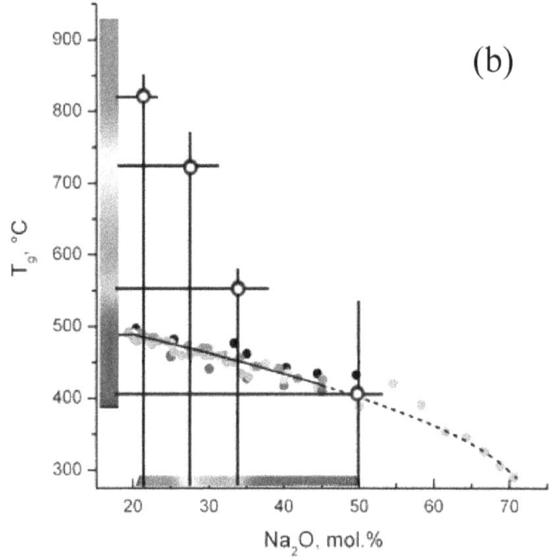

Figure 13. The calculated[22] (open circles) and experimental[32] (coloured circles) glass transition temperatures, T_g, for glasses in the systems (a) Na_2O–B_2O_3 and (b) Na_2O–SiO_2 [Colour available online]

of the maximum total content of danburite and reedmergnerite rings (See Figure 4(d)). Due to this, it appears that the experimentally observed variation of T_g with glass composition should be interpreted in terms of the total content of 4-membered borosilicate rings rather than the fraction of $BØ_4^-$ units.

5. **Glass transition temperatures in the systems $Na_2O–B_2O_3$ and $Na_2O–SiO_2$.** A comparison of the calculated and experimental[32] T_g in the binary systems is shown in Figure 13. The calculated[22] values are taken from Figure 16(a) for the system $Na_2O–SiO_2$ and from Figure 17(b) for the system $Na_2O–B_2O_3$. These temperatures were estimated from narrow strips immediately adjacent to the sides $Na_2O.2B_2O_3–Na_2O.B_2O_3$ (Triangle 1) and $Na_2O.SiO_2–20Na_2O.80SiO_2$ (Triangle 2), which were placed along the respective abscissas. The colour scales determining the temperatures[22] and shown in this work in Figure 12, on both sides of the triangle, were located along the ordinates, the temperatures being converted to degrees centigrade. From the points corresponding to the same colour on both axes, perpendiculars were erected, their crossing points giving the required values of T_g. In Figure 13, these are shown by open circles. The experimental values are indicated by various symbols, and it is seen that, in both binary systems, the calculated temperatures also noticeably disagree with the experimental data.

4. Conclusions

The model of associated solutions allows the distribution of the basic structural units and superstructural units to be calculated over the entire glass-forming region, with the observance of the principle of the minimum Gibbs energy of the system $Na_2O–B_2O_3–SiO_2$. On the basis of these results, quantitative criteria have been developed that characterise the tendency for sodium borosilicate glasses to phase separate. It has been shown that the concentration dependence of T_g over the entire glass forming region closely follows the content of 4-membered borosilicate rings. In addition, it must be concluded that the topological model of Mauro and co-workers[22] fails to account both for the structure of sodium borosilicate glasses and for the composition dependence of the glass transition temperature, T_g.

Acknowledgement

Armenak A. Osipov (Institute of Mineralogy of the Urals Branch of the Russian Academy of Sciences, Miass) is thanked for helpful discussions.

References

1. Wright, A. C. *Phys. Chem. Glasses: Eur. J. Glass Sci. Technol. B*, 2010, **51**, 1.
2. Howes, A. P., Vedishcheva, N. M., Samoson, A., Hanna, J. V., Smith, M. E., Holland D. & Dupree, R. *Phys. Chem. Chem. Phys.*, 2011, **13**, 11919.
3. Yun, Y. H. & Bray, P. J. *J. Non-Cryst. Solids*, 1978, **27**, 363.
4. Dell, W. J., Bray, P. J. & Xiao, S. Z. *J. Non-Cryst. Solid*, 1983, **58**, 1.
5. Bhasin, G., Bhatnagar, A., Bhowmik, S., Stehle, C., Affatigato, M. & Feller, S. A. *Phys. Chem. Glasses*, 1998, **39**, 269.
6. Vedishcheva, N. M., Shakhmatkin, B. A. & Wright, A. C. *J. Non-Crystalline Solids*, 2004, **345&346**, 39.
7. Vedishcheva, N. M. & Wright, A. C. In: *Glass. Selected Properties and Crystallization*, Ed. J. W. P. Schmelzer, De Gruyter, Berlin/Boston, 2014, p 269.
8. Barin, I. & Knacke, O. *Thermodynamic Properties of Inorganic Substances*, Springer-Verlag, Heidelberg, 1973.
9. Chase, M. W., Davis, C. A., Downey, J. R., Frurip, D. J. & McDonald, R. A. *JANAF Thermochemical Tables*, Third Edition, Am. Chem. Soc., Am. Inst. Phys., Nat. Bur. Stand., 1985.
10. Barin, I. *Thermochemical Data of Pure Substances*, 2nd Edition, VCH, Weinheim, 1993.
11. Zeldovich, Ya. B. *Zh. Fiz. Khim. (J. Phys. Chem.)*, 1938, **11**, 685. (In Russian.)
12. Zwanziger, J. W., Youngman, R. E. & Braun, M. In: *Borate glasses, crystals and melts*, Eds. A. C. Wright, S. A. Feller & A. C. Hannon, Society of Glass Technology, Sheffield, 1997, p 21.
13. Wright, A. C. and Vedishcheva, N. M. *Phys. Chem. Glasses: Eur. J. Glass Sci. Technol. B*, 2013, **54**, 147.
14. Bubnova, R. S. & Filatov, S. K. *Vysokotemperaturnaya Kristallokhimiya Boratov i Borosilikatov (High-temperature Crystal Chemistry of Borates and Borosilicates)*, Nauka, St. Petersburg, 2008, p 760. (In Russian.)
15. Donnay, G. & Donnay, J. D. *Am. Miner.*, 1953, **38**, 163.
16. Grund, A. & Pizy, M. M. *Acta Crystallogr.*, 1952, **5**, 837.
17. Osipov, A. A. & Osipova, L. M. *Glass Phys. Chem.*, 2009, **35**, 121.
18. Osipov, A. A. & Osipova, L. M. *Glass Phys. Chem.*, 2009, **35**, 132.
19. Bunker, B. C., Tallant, D. R., Headley, T. J., Turner, G. L. & Kirkpatrick, R. J. *Phys. Chem. Glasses*, 1988, **29**, 106.
20. Maekawa, H., Maekawa, T., Kawamura, K. & Yokokawa, T. *J. Non-Cryst. Solids*, 1991, **127**, 53.
21. Polyakova, I. G. *Glass Phys. Chem.*, 1997, **23** (1), 45.
22. Smedskjaer, M. M., Mauro, J. C., Youngman, R. E., Hogue, C. L., Potuzak, M. & Yue, Y. *J. Phys. Chem. B*, 2011, **115**, 12930.
23. Bray, P.J. & O'Keefe, J. G. *Phys. Chem. Glasses*, 1963, **4**, 37.
24. Kroeker, S., Aguiar, P. M., Cerquiera, A., Okoro, J., Clarida, W., Doerr, J., Olesiuk, M., Ongie, G., Affatigato, M. & Feller, S. A. *Phys. Chem. Glasses: Eur. J. Glass Sci. Technol. B*, 2006, **47**, 393.
25. Osipova, L. M. & Osipov, A. A. *Proc. XII Russ. Conf. Structure and Properties of Metal and Slag Melts*, **3**, Ekaterinburg, 2008, p. 41. (In Russian.)
26. Milberg, M. E., O'Keefe, J. G., Verhelst, R. A. & Hooper, H. O. *Phys. Chem. Glasses*, 1972, **13**, 79.
27. Scheerer, J., Müller-Warmuth, W. & Dutz, H. *Glastechn. Ber.*, 1973, **46**, 109.
28. Bunker, B. C., Tallant, D. R., Kirkpatrick, R. J. & Turner, G. L. *Phys. Chem. Glasses*, 1990, **31**, 30.
29. El-Damrawi, G., Müller-Warmuth, W., Doweidar, H. & Gohar, I. A. *J. Non-Crystal. Solids*, 1992, **146**, 137.
30. Martens, R. & Müller-Warmuth, W. *J. Non-Cryst. Solids*, 2000, **265**, 167.
31. Fleet, M. E. & Muthupari, S. *J. Non-Crystalline Solids*, 1999, **255**, 233.
32. SciGlass, Version 6.6, 1998–2006.
33. Polyakova, I. G. *Phys. Chem. Glasses*, 2000, **41**, 247.

Phys. Chem. Glasses: Eur. J. Glass Sci. Technol. B, December 2014, 55 (6), 274–279

Scintillation and dosimeter properties of $40Li_2O.40B_2O_3.20SiO_2$ glass with different Sn concentrations

T. Yanagida,[1] *Y. Fujimoto*

Kyushu Institute of Technology, 2-4 Hibikino, Wakamatsu, Kitakyushu, 808-0196 Japan

H. Masai

Institute for Chemical Research, Kyoto University, Gokasho, Uji, Kyoto 611-0011, Japan

Manuscript received 3 August 2014
Revised version received 26 September 2014
Accepted 3 October 2014

We have investigated the scintillation and dosimeter properties of Sn-doped $40Li_2O.40B_2O_3.20SiO_2$ glasses and the Sn-free glass. The emission peak in the photoluminescence spectrum appeared at 350 nm and the quantum yield of the Sn-doped glasses was 20–50%, monotonically increasing with Sn concentration. The photoluminescence decay constants of the Sn-doped glasses were characterised by fast (a few ns) and slow (5–6 μs) components while, for the Sn-free glass, only the fast component was apparent. For the scintillation spectra under x-ray irradiation, the peak emission wavelength in the photoluminescence spectrum was similar for both Sn-free and Sn-doped glasses, suggesting that the absorbed energy was dominantly transferred to the host emission. In thermally stimulated luminescence, glow peaks of all glasses were observed at 250–300°C. We have also confirmed that the thermally stimulated luminescence for all of the glasses exhibited a proportionality to x-ray dose.

Introduction

Ionising radiation-induced luminescences are classified into two kinds. One kind is a scintillation which converts the absorbed energy of the ionising radiation to UV-visible photons immediately,[1] and the other kind is storage-type luminescence for personal dosimetry. Common storage-type luminescences in dosimeters are thermally stimulated luminescence (TSL),[2] optically stimulated luminescence (OSL),[3] and radiophotoluminescence (RPL).[4] The fields of application for ionising radiation sensing are quite wide, such as medical imaging,[5] security systems in airports,[6] oil logging,[7] astrophysics,[8] and particle physics.[9] Generally, bulk and transparent materials are used for ionising radiation sensors to absorb high energy radiation efficiently and to lead photons to photodetectors with as little loss as possible. To fulfil these requirements, almost all commercially available materials for ionising radiation sensors are in single crystal form, and thus they have been studied well. However, there has been insufficient investigation of glass materials for ionising radiation sensors, and only two materials have become commercially available products. One is Li-glass scintillator (Saint Gobain)[10] for thermal neutron detectors, and the other is Ag-doped $Na_2O–AlPO_4$ for RPL dosimeters[11] (Chiyoda Technol. Corp.). Thus there remains a large scope for the study of optical glass materials for ionising radiation sensing applications.

Recently, our group has reported the emission properties of Sn-doped phosphate[12,13] and borate glasses[14] for phosphor applications. It is notable that Sn-doped glasses showed high quantum yield (QY) in photoluminescence (PL), due to the Sn^{2+} centre possessing a ns^2–nsnp transition. For the phosphate glass system, white light emission with a high QY was also attained by codoping of Sn^{2+} and Mn^{2+}.[13] By examination of these glass systems, we have found that the emission of Sn^{2+} depends on the chemical composition of the glass. Therefore, it is worthwhile to examine the emission properties of Sn^{2+} in other glass systems. Here, we focus on a borosilicate host glass.

The aim of this study is to investigate the scintillation and dosimeter properties of Sn-free and Sn 0·1, 0·2, 0·5, and 1% doped $40Li_2O.40B_2O_3.20SiO_2$ glasses in comparison with the optical properties. The advantage of this system is its light chemical composition. Previous reports have been given for structure analysis in borosilicate glasses,[15–19] for photoluminescence,[20–25] and for cathodoluminescence of tin in oxide glasses.[26] However, no reports were found about the ionising radiation detection capabilities. Particularly for personal dosimetry, tissue equivalent materials are preferable, and so conventional dosimeters are Al_2O_3,[3] LiF,[2] and $Na_2O–AlPO_4$.[4] Therefore, in the present work, the dose response function from

[1] Corresponding author. Email yanagida@lsse.kyutech.ac.jp
Original version presented at VIII Int. Conf. on Borate Glasses, Crystals and Melts, Pardubice, Czech Republic, 30 June–2 July 2014

1 to 1000 mGy exposure was tested, since the response under low dose is a most important property for personal dosimetry.

Experimental

The starting materials for the Sn-doped $40Li_2O.40B_2O_3.20SiO_2$ glass (mol%) were Li_2CO_3 (99·99%), B_2O_3 (99·9%), SiO_2 (99·99%) and SnO (99·5%). These chemicals were mixed and put into a platinum crucible to be melted at 1100°C in an ambient atmosphere. After melting for 30 min, the glass melt was quenched on a steel plate at about 200°C to obtain the glass. The quenched glass was annealed at its glass transition temperature, T_g, for 1 h, and then cut and polished mechanically. T_g and the melting temperature, T_m, were determined by a differential thermal analysis (DTA) system (Rigaku TG-DTA 8120) operating at a heating rate of 10°C/min.

For the basic optical properties, the in-line transmittance was evaluated by using a V670 spectrometer (JASCO) in the range 190–2700 nm. The PL quantum yield was also evaluated using a Hamamatsu Quantaurus-QY. The absolute QY was calculated by the following equation, $QY=N_{emit}/N_{absorb}$, where N_{emit} and N_{absorb} are the numbers of emission and absorption photons, respectively. Under 250 nm excitation, emissions were accumulated from 300 to 750 nm for the QY calculation. PL decay time profiles were recorded by using a Hamamatsu Quantaurus-τ. The excitation and monitoring wavelengths were 280 and 350 nm, respectively. PL decay times were determined by the instrumental response deconvoluted double exponential function approximation.

Scintillation (radioluminescence) spectra were measured by using a CCD-based spectrometer (Andor DU920P CCD and SR163 monochromator) under x-ray exposure.[26] Scintillation decay time profiles were observed by a pulse x-ray equipped afterglow characterisation system in a fast mode.[27] For the dosimeter properties, the TSL was investigated by using a Nanogray TL-2000. The detailed method for TSL evaluation has been described previously.[28,29] The x-ray doses in the TSL evaluations were from 1 to 1000 mGy, calibrated by an ionisation chamber. Except TSL for the glow curves, all experiments were carried out at room temperature.

Results and discussion

Figure 1 shows the glass samples. These were homogenous with a size of 10×10×1 mm³, without apparent cord. The values of T_m for the glasses were about 800°C, and thus the melting temperature is sufficient with a duration of 30 min. Wide surfaces were optically polished for characterisation. In ionising radiation detectors, bulk and transparent materials are attractive for users. To detect ionising radiation efficiently, generally bulk materials are required. In this sense, for example, film scintillators are not used except some special applications like an image intensifier. High transparency in the bulk form is also an important property, since the transportation efficiency of photons to photodetectors largely affects the detection performance. Figure 1 shows that both points were satisfied by these glasses.

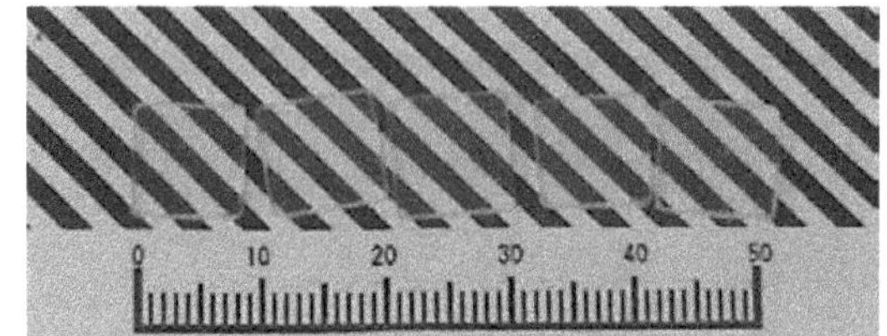

Figure 1. Photograph of $xSnO.40Li_2O.40B_2O_3.20SiO_2$ glasses (x=0, 0·1, 0·2, 0·5, and 1 mol%)

Optical in-line transmittance spectra are shown in Figure 2. In all glasses, 80–90% transmittance was achieved at wavelengths longer than 300 nm. The effect of Sn-doping was not so strong, and when the Sn doping concentration increased, absorption bands due to Sn^{2+} appeared at around 230 nm. The high transmittance for visible wavelengths is consistent with Figure 1. For the nondoped sample, the transmittance did not reach zero at 190 nm. The detection sensitivity of the instrument dropped at this wavelength and the absorption edge of $40Li_2O.40B_2O_3.20SiO_2$ glass had a wavelength in the vacuum ultra violet (VUV) region. SiO_2 is well known as a window material for VUV detectors, and so the presence of SiO_2 would affect the absorption edge. For the nondoped and Sn 1% doped samples, shallow absorption was observed around 270 nm and the origin was unclear.

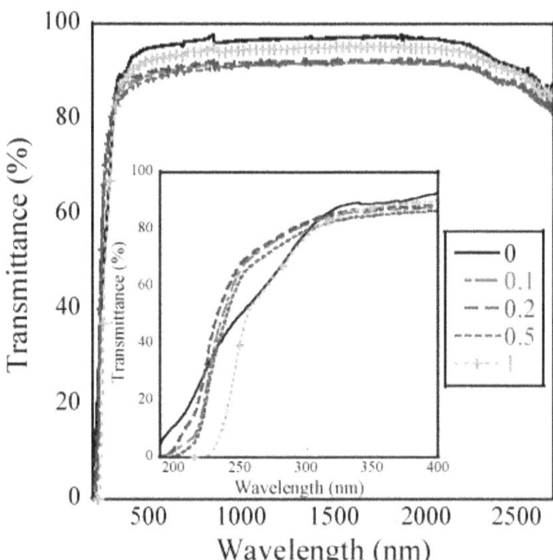

Figure 2. In-line transmittance spectra of Sn-free and Sn-doped $40Li_2O.40B_2O_3.20SiO_2$ glasses. The concentrations of Sn were 0·1, 0·2, 0·5, and 1 mol%. The inset expands spectra from 190 to 400 nm. The thickness of the samples was 1 mm [Colour available online]

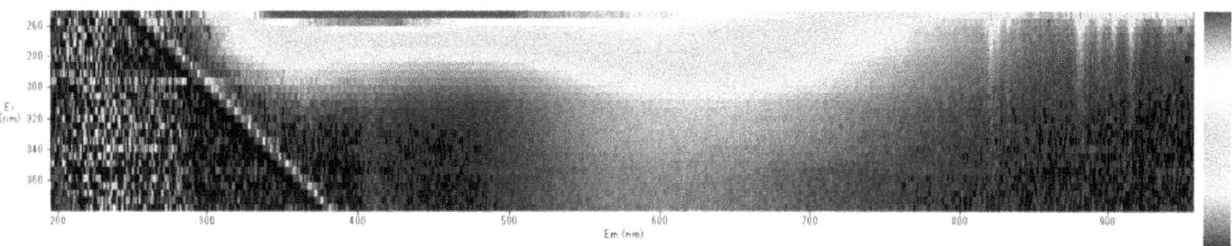

Figure 3. Photoluminesence emission map of Sn 1 mol% doped $40Li_2O.40B_2O_3.20SiO_2$ glass

Figure 3 exemplifies the PL emission map of the Sn 1 mol%-doped glass. Efficient excitation was done at 250 nm, which was the shortest wavelength of the instrument. The emission was very broad from 300 to 750 nm, and the peak wavelength was around 350 nm. The observed spectrum shape was typical for Sn-doped phosphors.[12–14] When compared with Sn-doped phosphate glass,[13] the emission peak shifted to shorter wavelength, due to the difference of the host.

The PL QY as a function of Sn concentration is shown in Figure 4. The PL QY monotonically increased with Sn concentration, and the Sn 1 mol% doped glass showed the highest QY of 50%. Alhough we could expect higher QY for higher Sn concentration, the obtained QY value is lower than that of previous reports.[12–14] It has been reported that the Sn^{2+}/Sn^{4+} ratio of Sn-doped glass prepared in air condition is not so high.[24,30,31] Recently, we have examined the Sn^{2+}/Sn^{4+} ratio of the 1 mol% Sn added glass. Although the details will be reported in another paper, the ratio is below 5%, and most of the Sn^{2+} was oxidised during the preparation process. It is a consensus that, for Sn-doped glasses, the presence of Sn^{2+} is necessary for high luminescence efficiency. Although we have not yet determined the reason, the present preparation scheme or chemical composition is not suitable for high QY.

The PL decay time profiles of the Sn-free and Sn 1% doped glasses are shown in Figure 5. The decay time profiles of all of the Sn-doped glasses were of a similar shape, characterised by fast (a few ns) and slow (5–6 μs) components, after excitation pulse deconvoluted fitting. For the Sn-free glass, very fast (~1 ns), fast (a few ns) and slow decay components were observed, and the fast component can be ascribed to the host emission. The very fast one was due to the excitation pulse. The origin of the slow component can also be ascribed to the host emission, and it was not clear whether this slow emission was luminescence or afterglow in this timing range. In Figure 2, a shallow absorption around 270 nm was observed clearly in nondoped and Sn 1% doped samples and slightly in other samples. The excitation wavelength of PL decay time evaluations coincided with this shallow absorption. Therefore, the host (possibly defects) can be excited by this wavelength.

Figure 6 summarises the PL decay times of all of the glasses. Since the timing resolution of the instrument was quite fast (several tens of ps), the error bars are smaller than the symbols used for the data points. The decay times due to Sn^{2+} became faster monotonically as the doping concentration increased, while the host emission was relatively stable against Sn concentration. The former observation can be interpreted simply; when the number of emission centres increased, the emission efficiency was im-

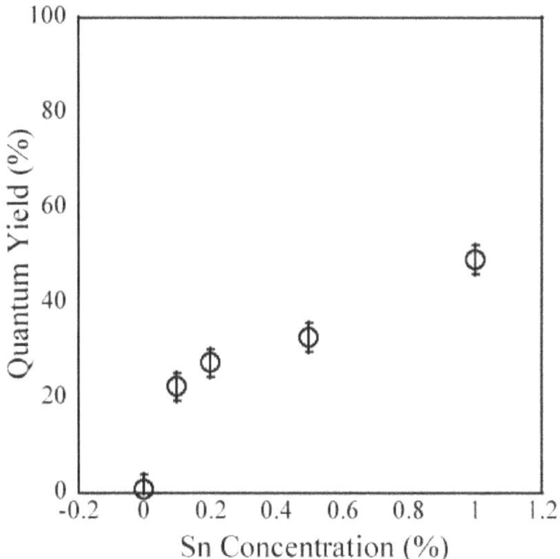

Figure 4. Photoluminescence quantum yield plotted against Sn concentrations

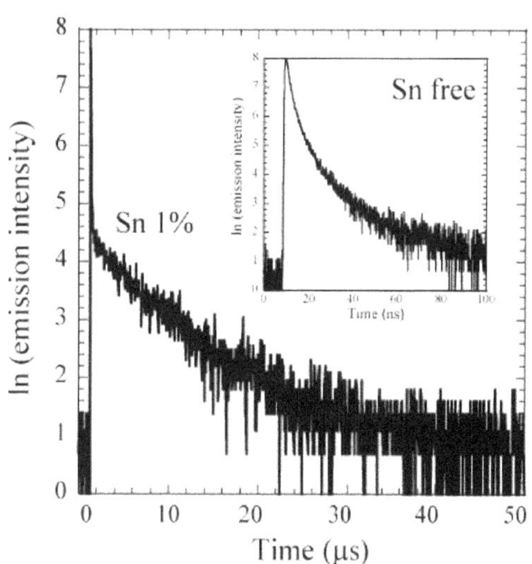

Figure 5. The photoluminescence decay time profile of the Sn-doped glass with a Sn concentration of 1 mol%. The inset shows the photoluminescence decay time profile of the Sn-free glass

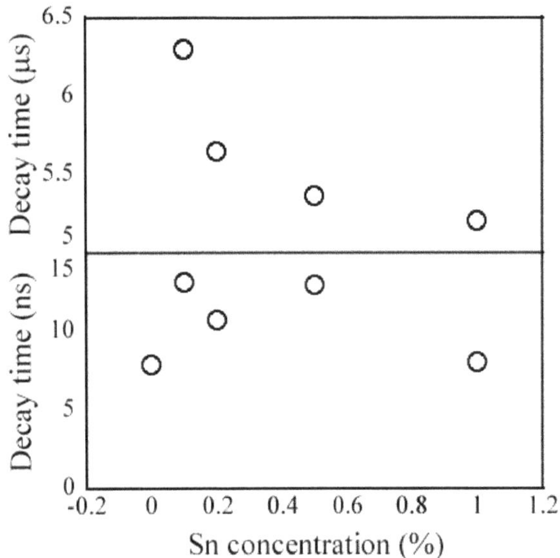

Figure 6. Photoluminescence decay time plotted against Sn concentration. For the Sn-free glass, no ms component was observed

proved. Such kinds of relation can be seen in general for phosphors. On the other hand, the reason for the latter phenomenon was not clear. One possibility was that the number of emission centres in the host (possibly due to defects) was not changed by the introduction of Sn ions.

After optical characterisation, we examined the scintillation properties. Figure 7 shows the x-ray induced radioluminescence (scintillation) spectra. The spectral shapes became narrower than those in PL and the emission peak of Sn-doped glasses appeared at around 340 nm. In addition, no significant difference was observed between the Sn-free and Sn-doped glasses. In scintillation, there are mainly three processes for emission. First, ionising radiation interacts with the host and a primary electron is created. Then, the primary electron ejects and excites many secondary electrons, like avalanche multiplication, and finally these secondary electrons reach emission centres, and emit scintillation photons. In the $40Li_2O.40B_2O_3.20SiO_2$ glass system, most of secondary electrons did not reach emission centres. A possible interpretation is that the number of host emission sites is higher than the number of Sn^{2+} centres, so that most secondary electrons are captured preferentially in host emission centres. In ionising radiation detectors, the most conventional photodetector is a photomultiplier tube (PMT) and they generally have a peak sensitivity to photons from 300 to 500 nm. Thus Sn-doped $40Li_2O.40B_2O_3.20SiO_2$ glasses are suited to the use of a PMT, at least in scintillation.

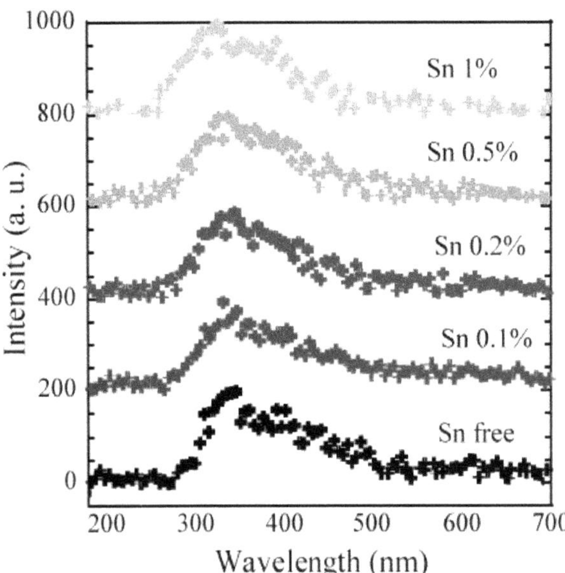

Figure 7. X-ray induced radioluminescence spectra of the glass samples

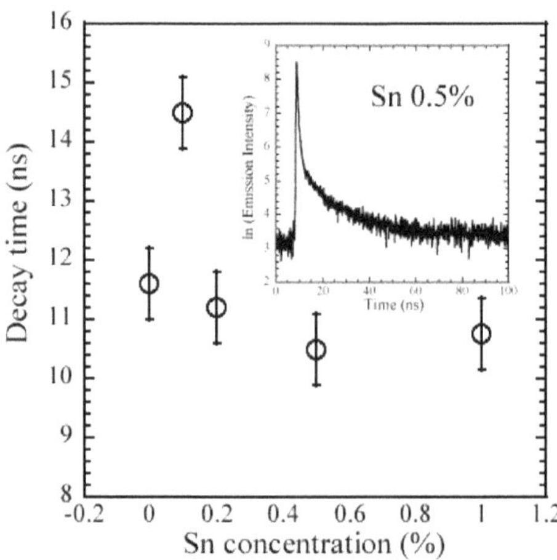

Figure 8. X-ray induced scintillation decay time plotted against Sn concentration. The inset shows the decay time profile of the Sn 0·5 mol% doped sample

In Figure 8, scintillation decay times under pulse x-ray excitation are shown. As for the radioluminescence, no significant difference was observed between Sn-free and Sn-doped glasses. The scintillation decay times were quite fast (11–14 ns) with similar values to the faster component in the PL decay. In PL proper-

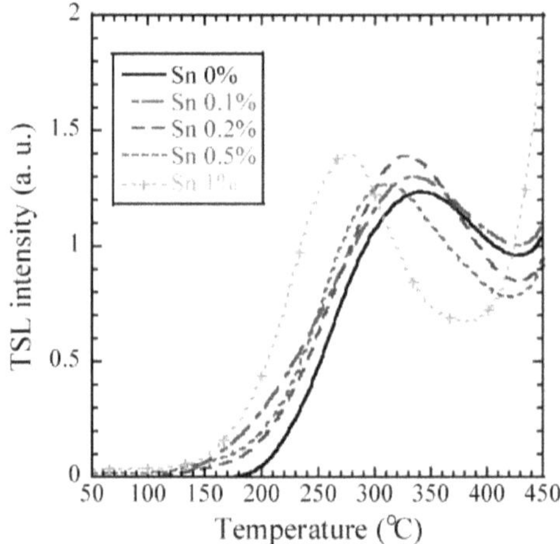

Figure 9. The thermally stimulated luminescence glow curves of the glass samples after 1 Gy exposure. The heating rate was 1°C/s

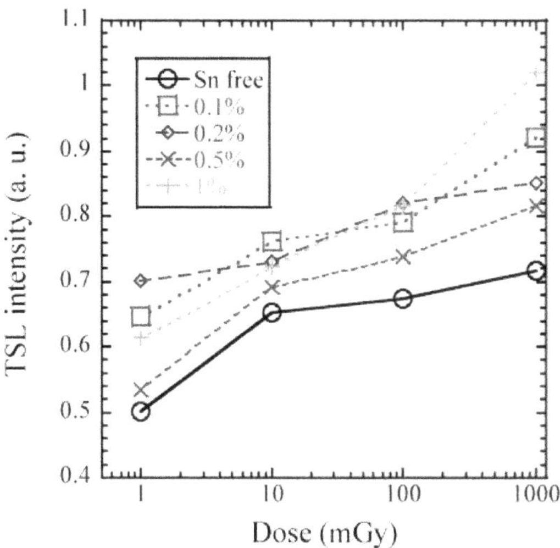

Figure 10. The dose response functions of Sn-free and Sn-doped $40Li_2O.40B_2O_3.20SiO_2$ glasses from 1 to 1000 mGy. The concentrations of Sn were 0·1, 0·2, 0·5, and 1 mol%

ties, Sn^{2+} emission centres surely exist. Thus in the Sn-doped $40Li_2O.40B_2O_3.20SiO_2$ glass system, the energy absorbed from the ionising radiation was not transferred efficiently to the Sn emission centres. A possible explanation is the same as that for the x-ray induced radioluminescence, that most secondary electrons are captured by host emission centres. Taking into account the x-ray induced scintillation spectra as well as the scintillation decay, Sn-doping in the $40Li_2O.40B_2O_3.20SiO_2$ glass system did not affect the scintillation properties much.

As for dosimeter properties, the TSL glow curves of the glasses are exhibited in Figure 9. For all glasses, a glow peak was observed at around 250–320°C. When the Sn concentration increased, the glow peak shifted to lower temperature. Unlike the scintillation responses, the TSL properties were actually affected by Sn-doping. At present, no theoretical model is proposed to predict or explain the glow peak, and so we cannot discuss this point further. Since the glow curve showed a single glow peak, it is convenient to use these glasses for dosimeter applications.

Figure 10 shows the dose response functions of all glasses from 1 mGy to 1000 mGy x-ray exposure assuming a personal dosimetry. The TSL intensity showed a positive proportionality against x-ray dose, and especially the Sn 1 mol% doped sample exhibited the best linearity among the present glasses. At least in this dose range, the present glass system works correctly as a dosimeter. If one were to apply these materials to practical applications, the lower dose response (10–100 μGy) should be examined. To calibrate such a low dose in general laboratories is difficult, and such an experiment is possible in some special facilities. Taking into consideration the luminescence, scintillation, and dosimeter properties, denser Sn doping would improve these properties and technical progress in the synthesis is required.

Conclusions

Sn-free and Sn-doped $40Li_2O.40B_2O_3.20SiO_2$ glasses were prepared by a conventional melt quenching method. In the photoluminescence spectra, Sn-doped glasses exhibited emission from 300 to 750 nm. For the Sn-doped glasses, the photoluminescence decay time exhibited components of a few ms and a few ns, ascribed, respectively, to host and Sn^{2+} emission. In scintillation, the absorbed energy of the ionising radiation was not transferred efficiently to Sn emission centres, and mainly host emission was observed. In the thermally stimulated luminescence glow curves, glow peaks appeared at 250–320°C, and the peak temperature shifted to lower temperature with increasing Sn concentration. The sample doped with 1% Sn exhibited the best performance in the dose response function from 1 to 1000 mGy.

Acknowledgements

This work was mainly supported by a Grant in Aid for Scientific Research (A)-26249147 from the Ministry of Education, Culture, Sports, Science and Technology of the Japanese government (MEXT) and partially by JST A-step. Partial assistance from Nippon Sheet Glass Foundation for Materials Science and Engineering, Tokuyama Science foundation, Iketani Science and Technology Foundation, Hitachi Metals Materials Science Foundation, Mazda Foundation, JFE 21st century Foundation, and The Asahi Glass Foundation, the Cooperative Research Project of Research Institute of Electronics, Shizuoka University, Collaborative Research Program of Institute for Chemical Research, Kyoto University (2014-31), and the SPRITS program, KyotoUniversity are also gratefully acknowledged.

References

1. Yanagida, T. Study of rare-earth-doped scintillators. *Opt. Mater.*, 2012, **35**, 1987–1992.
2. Mckeever, S. W. S. *Thermoluminescence of Solids*, Cambridge University Press, 1985.
3. Yukihara, E. G. & McKeever, S. W. S. *Optically Stimulated Luminescence: Fundamentals and Applications*, Wiley, 2011.
4. Miyamoto, Y., Takei, Y., Nanto, H., Kurobori, T., Konnai, A., Yanagida, T., Yoshikawa, A., Shimotsuma, Y., Sakakura, M., Miura, K., Hirao, K., Nagashima, Y. & Yamamoto, T. Radiophotoluminescence from silver-doped phosphate glass. *Radiat. Meas.*, 2011, **46**, 1480–1483.
5. Yanagida, T., Yoshikawa, A., Yokota, Y., Kamada, K., Usuki, Y., Yamamoto, S., Miyake, M., Baba, M., Sasaki, K. & Ito, M. Development of Pr:LuAG Scintillator Array and Assembly for Positron Emission Mammography. *IEEE. Nucl. Trans. Sci.*, 2010, **57**, 1492–1495.
6. Totsuka, D., Yanagida, T., Fukuda, K., Kawaguchi, N., Fujimoto, Y., Yokota, Y. & Yoshikawa, A. Performance test of Si PIN photodiode line scanner for thermal neutron detection. *Nucl. Instrum. Methods A*, 2011, **659**, 399–402.
7. Yanagida, T., Fujimoto, Y., Kurosawa, S., Kamada, K., Takahashi, H., Fukazawa, Y., Nikl, M. & Chani, V. Temperature dependence of scintillation properties of bright oxide scintillators for well-logging. *Jpn. J. Appl. Phys.*, 2013, **52**, 076401.
8. Yamaoka, K., Ohno, M., Terada, Y., Hong, S., Kotoku, J., Okada, Y., Tsutsui, A., Endo, Y., Abe, K., Fukazawa, Y., Hirakuri, S., Hiruta, T., Itoh, K., Kamae, T., Kawaharada, M., Kawano, N., Kawashima, K.,

Kishishita, T., Kitaguchi, T., Kokubun, M., Madejski, G. M., Makishima, K., Mitani, T., Miyawaki, R., Murakami, T., Murashima, M. M., Nakazawa, K., Niko, H., Nomachi, M., Oonuki, K., Sato, G., Suzuki, M., Takahashi, H., Takahashi, I., Takahashi, T., Takeda, S., Tamura, T., Tanaka, T., Tashiro, M., Watanabe, S., Yanagida, T. & Yonetoku, D. Development of the HXD-II Wideband All-sky Monitor Onboard Astro-E2, *IEEE Trans. Nucl. Sci*, 2005, **52**, 2765–2772.

9. Ito, T., Yanagida, T., Sato, M., Kokubun, M., Takashima, T., Hirakuri, S., Miyawaki, R., Takahashi, H., Makishima K., Tanaka T., Nakazawa K., Takahashi T., Honda T., A 1-Dimensional Gamma-ray Position Sensor based on GSO:Ce Scintillators Coupled to a Si Strip Detector, *Nucl. Instrum. Methods A*, 2007, **579**, 239–242.
10. Saint Gobain, http://www.crystals.saint-gobain.com/
11. Chiyoda Technol Corp., http://www.c-technol.co.jp/technol_eng/first_pages/02personal_dosi.html
12. Masai, H., Takahashi, Y., Fujiwara, T., Matsumoto, S. & Yoko, T. High Photoluminescent Property of Low-Melting Sn-Doped Phosphate Glass. *Appl. Phys. Expr.*, 2010, **3**, 082102.
13. Masai, H., Fujiwara, T., Matsumoto, S., Takahashi, Y., Iwasaki, K., Tokuda, Y. & Yoko, T. White Light Emission of Mn-Doped SnO-ZnO-P_2O_5 Glass Containing No Rare Earth Cation. *Opt. Lett.*, 2011, **36**, 2868–2870.
14. Masai, H., Yamada, Y., Suzuki, Y., Teramura, K., Kanemitsu, Y. & Yoko, T. Narrow Energy Gap between Triplet and Singlet Excited States of Sn^{2+} in Borate Glass. *Sci. Rep.*, 2013, **3**, 3541.
15. Smedskjaer, M. M., Mauro, J. C., Youngman, R. E., Hogue, C. L., Potuzak, M. & Yue, Y. Topological Principles of Borosilicate Glass Chemistry. *J. Phys. Chem. B*, 2011, **115**, 12930–12946.
16. Bunker, B. C. Multinuclear nuclear-magnetic-resonance and Raman investigation of sodium borosilicate glass structures. *Phys. Chem. Glasses*, 1990, **31**, 30–41.
17. Du, L.-S. & Stebbins, J. F. Solid-state NMR study of metastable immiscibility in alkali borosilicate glasses, *J. Non-Cryst. Solids*, 2003, **315**, 239–255.
18. Krohn, M. H., Hellmann, J. R., Mahieu, B. & Pantano, C. G. Effect of tin-oxide on the physical properties of soda-lime-silica glass. *J. Non-Cryst. Solids*, 2005, **351**, 455–465.
19. Maia, L. F. & Rodrigues, A. C. M. Electrical conductivity and relaxation frequency of lithium borosilicate glasses. *Solid State Ionics*, 2004, **168**, 87–92.
20. Ehrt, D. UV-absorption and radiation effects in different glasses doped with iron and tin in the ppm range. *Comptes Rendus Chimie*, 2002, **5**, 679–692.
21. Reisfeld, R., Boehm, L. & Barnett, B. Luminescence and nonradiative relaxation of Pb^{2+}, Sn^{2+}, Sb^{3+}, and Bi^{3+} in oxide glasses. *J. Solid State Chem.*, 1975, **15**, 140–150.
22. Ehrt, D. Photoluminescence in the UV-VIS region of polyvalent ions in glasses. *J. Non-Cryst. Solids*, 2004, **348**, 22–29.
25. Hayakawa, T., Enomoto, T. & Nogami, M. β-band photoluminescence and Sn-E ′ center generation from twofold-coordinated Sn Centers in SiO_2 glasses produced via sol-gel method. *Jpn. J. Appl. Phys.*, 2006, **45**, 5078–5083.
23. Jimenez, J. Emission Properties of Sn^{2+} and Sm^{3+} Co-doped Barium Phosphate Glass. *J. Electron. Mater.*, 2014, **43**, 3588–3592.
24. Segawa, H., Inoue, S. & Nomura, K. Electronic states of SnO-ZnO-P_2O_5 glasses and photoluminescence properties. *J. Non-Cryst. Solids*, 2012, **358**, 1333–1338.
26. Yanagida, T., Kamada, K., Fujimoto, Y., Yagi, H. & Yanagitani, T. Comparative study of ceramic and single crystal Ce:GAGG scintillator, *Opt. Mater.*, 2013, **35**, 2480–2485.
27. Yanagida, T., Fujimoto, Y., Ito, T., Uchiyama, K. & Mori, K. Development of x-ray induced afterglow characterisation system, *Appl. Phys. Expr.*, 2014, **7**, 062401.
28. Yanagida, T., Fujimoto, Y., Kawaguchi, N. & Yanagida, S. Dosimeter properties of AlN, *J. Ceram. Soc. Jpn.*, 2013, **121**, 988–991.
29. Yanagida, T., Fujimoto, Y., Watanabe, K. & Fukuda, K. Dosimeter properties of Ce and Eu doped $LiCaAlF_6$, 2014, in press, DOI: 10.1016/j.radmeas.2014.02.021.
30. Williams, K. F. E., Johnson, C. E., Greengrass, J., Tilley, B. P., Gelder, D. & Johnson, J. A. Tin oxidation state, depth profiles of Sn^{2+} and Sn^{4+} and oxygen diffusivity in float glass by Mössbauer spectroscopy *J. Non-Cryst. Solids*, 1997, **211**, 164–172.
31. Bent, J. F., Hannon, A. C., Holland, D. & Karim, M. M. A. The structure of tin silicate glasses. *J. Non-Cryst. Solids*, 1998, **232**, 300–308.

Investigation of (100–x)[0·5PbO.0·2B_2O_3.0·3P_2O_5].$x$$MoO_3$ glasses

Ivana Rösslerová, Ladislav Koudelka, Jana Holubová, Petr Mošner*

Department of General and Inorganic Chemistry, Faculty of Chemical Technology, University of Pardubice, 532 10 Pardubice, Czech Republic

Lionel Montagne & Bertrand Revel

Université Lille Nord de France – UCCS - UMR CNRS 8181, USTL, 59655, Villeneuve d'Ascq, France

Manuscript received 2 September 2014
Revised version received 4 November 2014
Accepted 6 November 2014

Lead borophosphate glasses of the composition (100–x)[0·5PbO.0·2B_2O_3.0·3P_2O_5].xMoO_3 with 0–70 mol% MoO_3 were prepared and studied. Thermal studies were carried out with DTA and dilatometry. The glass transition temperature in these glass series decreases steadily with increasing MoO_3 content from 467 to 342°C for the range x=0–70 mol% MoO_3. The glass structure was studied by ^{31}P and ^{11}B MAS NMR and by Raman spectroscopy. These studies were devoted to the investigation of changes in boron coordination and their dependence on the MoO_3 content and the B_2O_3/P_2O_5 ratio. When the molybdenum oxide and boron oxide content is low, tetrahedral BO_4 units prevail in the structural network. With higher B_2O_3 content, as well as with higher MoO_3 content, trigonal BO_3 groups are also formed. ^{11}B MAS NMR spectra of BO_4 units in the glasses with a low boron content reveal two symmetrical signals for BO_4 units, ascribed to $B(OP)_3O$ and $B(OP)_2O_2$ units with B–O–P and $B^{(4)}$–O–$B^{(4)}$ bonds. In the ^{11}B spectra of glasses with a higher B_2O_3 or MoO_3 content, the number of $B(OP)_3O$ units decreases, and some $B(OP)O_3$ units appear as well. The formation of mixed BO_4 units with B–O–P, B–O–Mo and B–O–B bonds results in several close resonances from which the relative proportion of borate structural units can be obtained. The effect of changing the B_2O_3/P_2O_5 ratio on the structure of lead borophosphate glasses modified with MoO_3 is also discussed.

1. Introduction

The addition of heavy metal oxides such as Nb_2O_5, MoO_3, WO_3 or TeO_2 to borophosphate glasses is of interest in order to modify their properties,[1–6] e.g. the addition of Nb_2O_5 to potassium borophosphate glasses substantially improves their chemical durability.[2] Similarly, the addition of tungsten oxide to zinc borophosphate glasses[3] increases the glass transition temperature as well as the chemical durability. In contrast, the addition of molybdenum oxide to zinc borophosphate glasses[4,5] results in a decrease in the glass transition temperature, T_g.

Structural studies of zinc borophosphate glasses modified with tungsten oxide or molybdenum oxide have been reported by Šubčík *et al*[3–5] Heavy metal oxide addition also strongly affects the chemical durability of borophosphate glasses since the glass network structure is modified. The formation of mixed phosphate–molybdate units $B(OP)_{4-x}(OMo)_x$ was revealed by ^{11}B NMR spectroscopy, whilst the presence of MoO_4 and MoO_6 units was proposed in these glasses on the basis of the Raman spectra.[4,5] Similar structural units were also identified in lead borophosphate glasses modified with MoO_3[6] in the compositional series (1–x)[50PbO.10B_2O_3.40P_2O_5].$x$$MoO_3$ where glass formation was observed in the range 0–70 mol% MoO_3.

Information on the coordination of boron in borophosphate glasses can be obtained from ^{11}B MAS NMR spectroscopy, because these spectra possess the ability to discriminate between tetrahedral BO_4 coordination and trigonal BO_3 coordination,[7] due to the different ranges of chemical shift values for BO_4 and BO_3 units. Good discrimination of these units can be obtained with the measurement of these spectra at high magnetic field,[8] which gives better resolution of the fine details of the spectra.

We measured ^{11}B MAS NMR spectra at a static magnetic field of 9·4 T,[9] in our former study of PbO–B_2O_3–P_2O_5 glasses, whereas in the study of (100–x)[0·5PbO.0·1B_2O_3.0·4P_2O_5].$x$$MoO_3$ glasses[6] we measured ^{11}B MAS NMR spectra on a spectrometer at 18·8 T. These measurements allowed us to obtain a finer structure for the resonance of BO_4 units, and also to determine the correct ratio between BO_3 and BO_4 units.

In this paper we report the study of PbO–B_2O_3–P_2O_5–MoO_3 glasses in another series, (100–x)[0·5PbO.0·2B_2O_3.0·3P_2O_5].$x$$MoO_3$, with the ratio

* Corresponding author. Email Ladislav.Koudelka@upce.cz
Original version presented at VIII Int. Conf. on Borate Glasses, Crystals and Melts, Pardubice, Czech Republic, 30 June–2 July 2014

B_2O_3/P_2O_5=20/30. We measured ^{11}B MAS NMR spectra with a high resolution in order to be able to sensitively investigate changes in boron coordination. We compared the obtained results with those obtained with the series (100–x)[0·5PbO.0·1B_2O_3.0·4P_2O_5].$x$$MoO_3$ and also studied the dependence of structural changes on changing the B_2O_3/P_2O_5 ratio from 5/45 to 20/30.

2. Experimental

PbO–B_2O_3–P_2O_5–MoO_3 glasses were prepared from analytical grade PbO, MoO_3, H_3BO_3 and H_3PO_4, using a total batch weight of 30 g. The homogenised starting mixtures were slowly calcined up to 600°C for 2 h to remove water. The reaction mixtures were subsequently heated up to 1000–1350°C in a platinum crucible with a lid. The melt was held at a maximum temperature for 20 min, and then poured into a preheated graphite mould. The obtained glasses were then transferred to an annealing furnace for 4 h at a temperature 5°C below the glass transition temperature, T_g, and then cooled to room temperature. The volatilisation losses checked by weighing were not significant, and hence the batch compositions can be considered as reflecting the actual compositions. The amorphous character of the glasses was checked with x-ray diffraction. A structureless spectrum was obtained for all the glass compositions.

The glass density, ρ, was determined for bulk samples using Archimedes' method with toluene as the immersion liquid. The molar volume, V_M, was calculated as $V_M=M/r$, where M is the average molar weight of the glass composition aPbO.$b$$B_2O_3$.$c$$P_2O_5$.$d$$MoO_3$, calculated for $a+b+c+d=1$. The thermal behaviour of the glasses was studied using a Netzcsh DTA 404 PC, operating in DSC mode at a heating rate of 10°C min^{-1}. The measurements were carried out with 100 mg powder samples under an inert atmosphere of N_2. The thermal expansion coefficient, α, the glass transition temperature, T_g, and the dilatometric softening temperature, T_d, were measured on bulk samples with dimensions of 25×5×5 mm, using a dilatometer Netzsch DIL 402 PC. Based on the obtained dilatation curves, the coefficient of thermal expansion, α, was determined as a mean value in the temperature range 100–200°C, the glass transition temperature, T_g^b, was determined from the change in slope of the elongation versus temperature plot, and the dilatation softening temperature, T_d, was determined as the maximum of the expansion trace corresponding to the onset of viscous deformation. The dilatometric measurements were carried out in air at a heating rate of 3°C min^{-1}.

The Raman spectra were measured on bulk samples at room temperature, using a Horiba-Jobin Yvon LaBRam HR spectrometer. The spectra were recorded in back scattering geometry under excitation with Nd:YAG laser radiation (532 nm) at a power of 12 mW on the sample. The spectral slit width was 1·5 cm^{-1} and the total integration time was 50 s.

^{31}P MAS NMR spectra were measured at 9·4 T on a 400 MHz BRUKER Avance spectrometer with a 4 mm probe. The spinning speed was 12·5 kHz and the relaxation (recycling) delay was 180 s. The chemical shifts of ^{31}P nuclei are given relative to H_3PO_4 at 0 ppm. ^{11}B MAS NMR spectra were measured at 18·8 T on a BRUKER Avance 800 spectrometer with a 2·5 mm probe. The spectra were acquired with a single short (0·5 µs) radiofrequency pulse ($\pi/12$) in order to compensate for the different nutation behaviour of the two boron sites and use the obtained spectra for BO_3/BO_4 quantification. The recycling delay was 10 s, and the spinning rate was 20 kHz. The chemical shifts of ^{11}B nuclei are given relative to BPO_4 at −3·6 ppm. The NMR spectra deconvolution was carried out using the Dmfit NMR software.[10] BO_4 lines are known to be subject to a negligible second-order quadrupolar effect; hence the decomposition was performed with a Gaussian-type function assuming that the line shape is dominated by the chemical shift distribution.

3. Results and discussion

We prepared eight glass samples from the compositional series (100–x)[0·5PbO.0·2B_2O_3.0·3P_2O_5].$x$$MoO_3$ in this study. The composition of all the studied glasses is provided in Table 1. X-ray diffraction confirmed that all the samples were amorphous, which was also confirmed by the character of the Raman spectra, showing only broad vibrational bands characteristic of glass. Glasses containing MoO_3 had a blue colour, ascribed to the presence of Mo^{5+} ions,[11] which were detected by ESR spectroscopy. The ratio of Mo^{5+}/Mo^{total} obtained by the double integration of ESR spectra is provided in Table 1. As can be seen in this table, the fraction of molybdenum ions that are Mo^{5+} is lower than 0·7%, and varies within the range 0·41–0·67%.

We have determined the density and molar volume of the prepared glasses and their values are also

Table 1. Density, ρ, molar volume, V_M, glass transition temperature T_g^a (onset on DTA curve), T_g^b (measured by dilatometry), thermal expansion coefficient, α, and the relative amount of Mo^{5+} ions of (100–x)[0·5PbO.0·2B_2O_3.0·3P_2O_5].xMoO_3 glasses

x [mol% MoO_3]	$\rho\pm0·02$ [g cm^{-3}]	$V_m\pm0·5$ [cm^{3}mol^{-1}]	$T_g^a\pm2$ [°C]	$T_g^b\pm2$ [°C]	$\alpha\pm0·3$ (100–200°C) [ppm/°C]	Mo^{5+}/Mo^{total} [%]±0·05
0	5·12	32·9	467	469	12·2	-
10	5·05	33·1	454	457	11·9	0·67
20	5·00	32·7	445	445	11·9	0·62
30	4·91	32·8	424	423	12·4	0·63
40	4·80	33·3	404	402	12·4	0·60
50	4·71	33·1	377	379	12·6	0·47
60	4·55	33·7	359	356	12·2	0·53
70	4·40	34·4	342	343	13·5	0·41

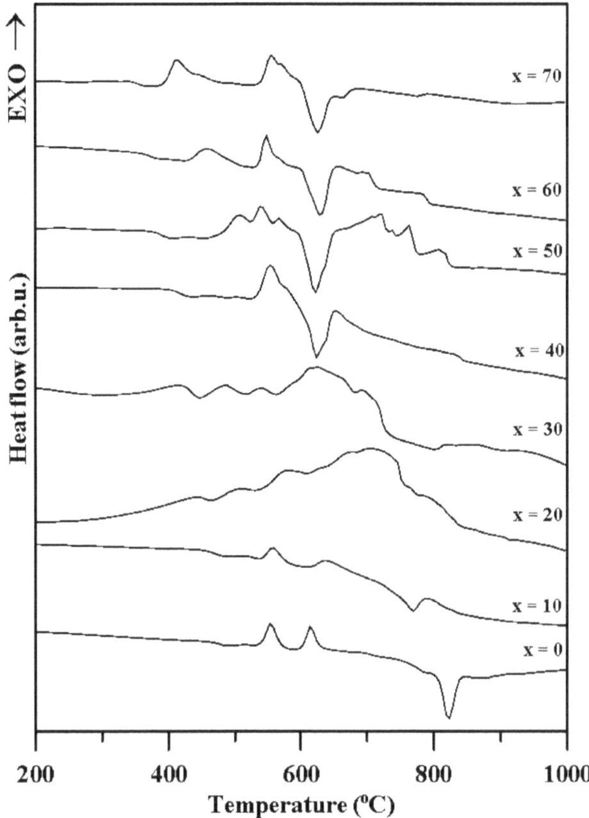

Figure 1. The DSC curves of the glass series (100−x)[0·5PbO.0·2B$_2$O$_3$.0·3P$_2$O$_5$].xMoO$_3$

given in Table 1. The density of glasses in the compositional series (100−x)[0·5PbO.0·2B_2O_3.0·3P_2O_5].$x$$MoO_3$ decreases with increasing MoO_3 content, whereas the molar volume slightly increases.

The thermal properties of the glasses were studied with DTA and dilatometry. The DSC curves obtained for glasses from the series (100−x)[0·5PbO.0·2B_2O_3.0·3P_2O_5].$x$$MoO_3$ are shown in Figure 1. As can be seen in this figure, most glasses crystallise upon heating, some of them in several steps. For the glasses with x=20 and 30 mol% MoO_3, no pronounced crystallisation peaks can be seen, and thus we consider these glasses as relatively stable. The values of the glass transition temperature, T_g^a, from the DSC curves were determined as the onset of the change in the thermal capacity of the samples in the glass transition region (see Table 1). The dilatometric measurements also provide values for the glass transition temperature, T_g^b, (see Table 1) and also the values of the thermal expansion coefficient, a, determined within the temperature range 100–200°C. These values are also given in Table 1.

The compositional dependence of the glass transition temperature in the series (100−x)[0·5PbO.0·2B_2O_3.0·3P_2O_5].$x$$MoO_3$ is shown in Figure 2 along with the compositional dependence of the glass transition temperature obtained previously from the series (100−x)[0·5PbO.0·1B_2O_3.0·4P_2O_5].$x$$MoO_3$.[(6)] The glass transition temperature of the studied glasses (de-

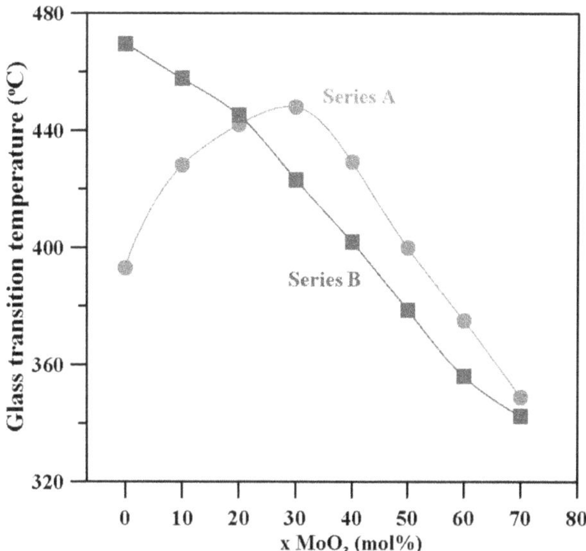

Figure 2. The compositional dependences of the glass transition temperature, T_g, in the glass series (100−x)[0·5PbO.0·1B$_2$O$_3$.0·4P$_2$O$_5$].xMoO$_3$ (Series A) and (100−x)[0·5PbO.0·2B$_2$O$_3$.0·3P$_2$O$_5$].xMoO$_3$ (Series B). The lines are only a guide to the eye [Colour available online]

noted as Series B) decreases monotonically with increasing MoO_3, whereas in the previously studied series (Series A[(6)]) there is a maximum at 30 mol% MoO_3. The effect of the changing B_2O_3/P_2O_5 ratio in the glass series (100−x)[0·5PbO.$y$$B_2O_3$.(0·5−$y$)$P_2O_5$].$x$$MoO_3$ on the evolution of T_g values is shown in Figure 3 for B_2O_3/P_2O_5 ratios from 5/45 to 20/30. The compositional dependence of T_g on the B_2O_3/P_2O_5 ratio only has a non-monotonic character in the glasses with 20 mol%

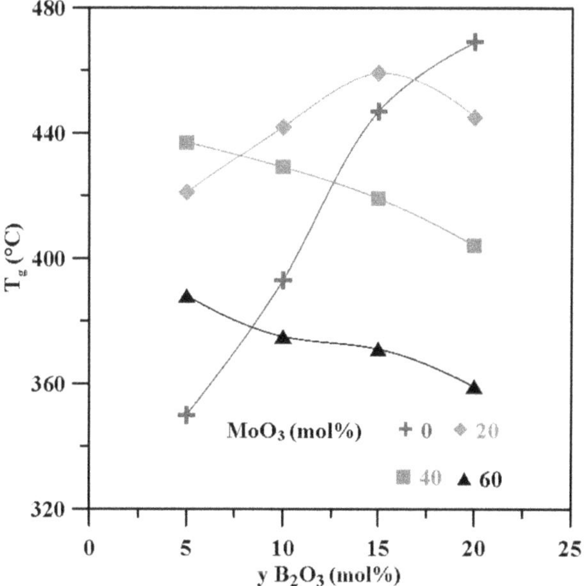

Figure 3. The compositional dependences of the glass transition temperature, T_g, in the glass series (100−x)[0·5PbO.yB$_2$O$_3$.(0·5−y)P$_2$O$_5$].xMoO$_3$ with 0, 20, 40 and 60 mol% MoO$_3$ (b). The lines are only a guide to the eye [Colour available online]

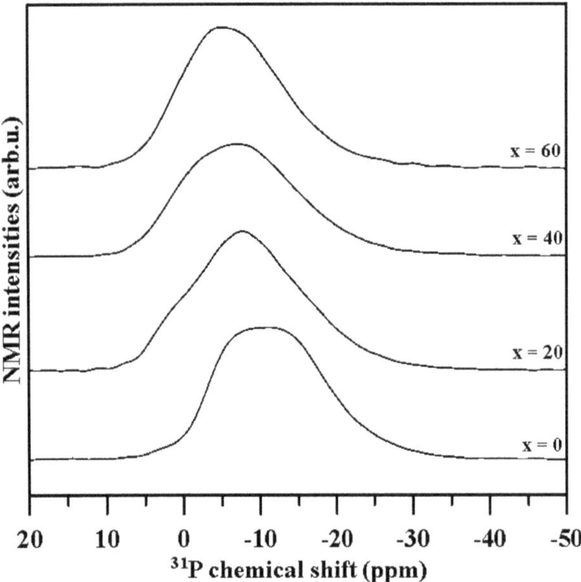

Figure 4. The ^{31}P MAS NMR spectra of (100−x) [0·5PbO.0·2B_2O_3.0·3P_2O_5].xMoO$_3$ glasses

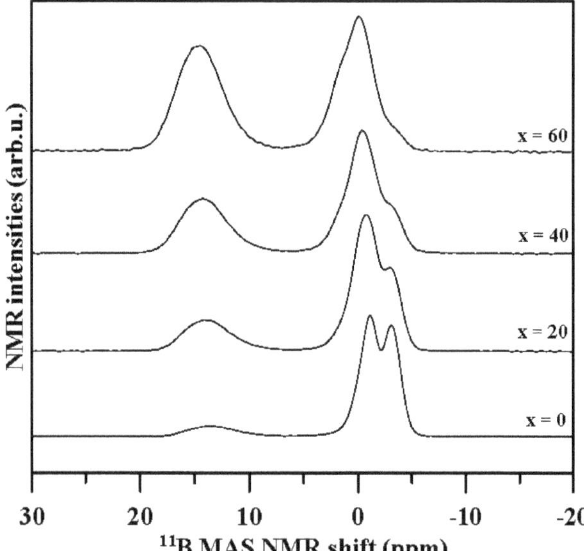

Figure 5. The ^{11}B MAS NMR spectra of (100−x) [0·5PbO.0·2B_2O_3.0·3P_2O_5].xMoO$_3$ glasses

MoO_3, and reveals a maximum of T_g for the B_2O_3/P_2O_5 ratio 15/35. Otherwise, T_g values decrease with increasing B_2O_3/P_2O_5 ratio for the glasses containing 40 and 60 mol% MoO_3. Additions of only B_2O_3 to lead metaphosphate glass result in a steady increase in the glass transition temperature from 255 to 469°C within the region 0–20 mol% B_2O_3[12], whereas a slow decrease in T_g values can be observed in the glass series with 40 and 60 mol% MoO_3 with increasing B_2O_3/P_2O_5 ratio. The reason for these changes in T_g will in all probability be in the reticulation of the glass network, both for B_2O_3 or MoO_3. Additions of MoO_3 to lead metaphosphate glass result in an increase in T_g only within the concentration range 0–50 mol% MoO_3, and a further increase in the MoO_3 content results in a decrease in T_g.[13] The observed increase in T_g at low concentrations of MoO_3 could be ascribed to the transformation of the chain-like metaphosphate glass structure by P–O–Mo linkages to nonbridging oxygen atoms. At higher MoO_3 content, the glass network is reticulated and the formation of weaker Mo–O–Mo linkages (bond strength D^0=560·2±20·9 kJ/mol[14]), replacing P–O–P linkages (bond strength D^0=599±12·6 kJ/mol[14]), or heteropolar P–O–Mo linkages results in a weakening of the chemical bonds inside the structural network. A similar reticulation of the metaphosphate chain structure can be achieved with the addition of B_2O_3 to lead metaphosphate glass.[9,12] The addition of another MoO_3 or B_2O_3 to the reticulated borophosphate network results in a decrease in T_g, as observed in Figure 3.

The ^{31}P MAS NMR spectra of (100−x)[0·5PbO.0·2B_2O_3.0·3P_2O_5].xMoO_3 glass series are shown in Figure 4. The spectrum of the glass without MoO_3 (x=0) contains at least two resonances within the range −5 to −15 ppm. With increasing MoO_3 content, the maximum of the resonances shifts slightly downfield. According to previous studies of borophosphate glasses, the differentiation of Q^n units cannot be achieved[15] due to the reticulation of the borophosphate network by borate units. It can only be stated from the shift of the resonance maxima that the evolution of the NMR spectra is dependent on the depolymerisation of the phosphate chains in the glass structure, due to the incorporation of molybdate units and the formation of P–O–Mo bonds within the network structure.

The ^{11}B MAS NMR spectra of the (100−x)[0·5PbO.

Table 2. NMR shifts and relative amounts of borate structural units in (100−x)[0·5PbO.yB_2O_3.(0·5−y)P_2O_5].xMoO_3 glasses, obtained from the decomposition of their ^{11}B MAS NMR spectra

x [mol% MoO$_3$]	y (B$_2$O$_3$)	BO$_3$ δ [ppm] ±0·2	BO$_3$ [%] ±2	B(OP)$_3$O δ [ppm] ±0·2	B(OP)$_3$O [%] ±2	B(OP)$_2$O$_2$ δ [ppm] ±0·2	B(OP)$_2$O$_2$ [%] ±2	B(OP)O$_3$ δ [ppm] ±0·2	B(OP)O$_3$ [%] ±2
0	0·20	13·5	10·0	−3·3	30·3	−1·0	57·4	1·6	2·3
20	0·20	13·8	22·0	−3·3	18·6	−0·8	55·3	2·0	4·1
40	0·20	14·1	35·4	−3·4	10·9	−0·5	46·0	1·9	7·7
60	0·20	14·5	48·5	−3·5	3·4	−0·3	36·3	1·9	11·8
20	0·05	12·6	0·6	−3·6	96·4	−1·1	3·0	-	-
20	0·15	13·1	6·4	−3·4	50·4	−1·1	43·2	-	-
40	0·05	12·6	2·5	−3·6	71·3	−0·8	25·4	1·9	0·8
40	0·15	13·9	25·4	−3·5	22·5	−0·7	45·6	1·7	6·5
60	0·05	13·7	11·5	−3·5	30·7	−0·9	50·5	1·5	7·3
60	0·15	14·3	40·0	−3·6	6·9	−0·6	45·9	1·7	7·2

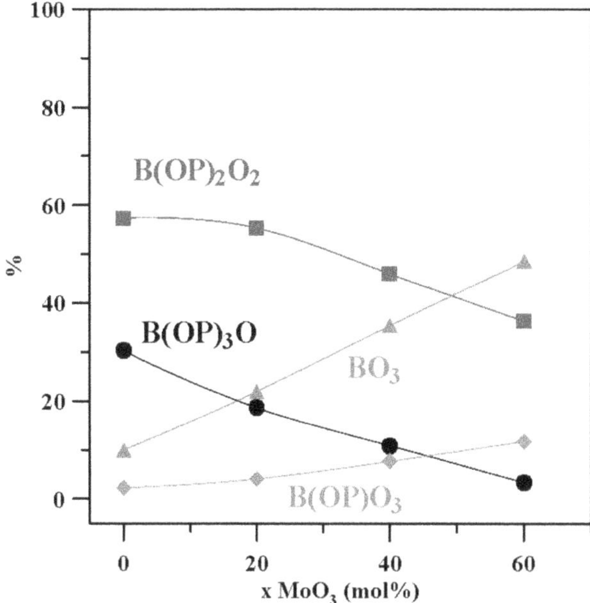

Figure 6. The compositional dependences of the relative amounts of borate structural units in (100−x)[0·5PbO.0·-2B$_2$O$_3$.0·3P$_2$O$_5$].xMoO$_3$ glasses, as obtained from the decomposition of their ^{11}B MAS NMR spectra [Colour available online]

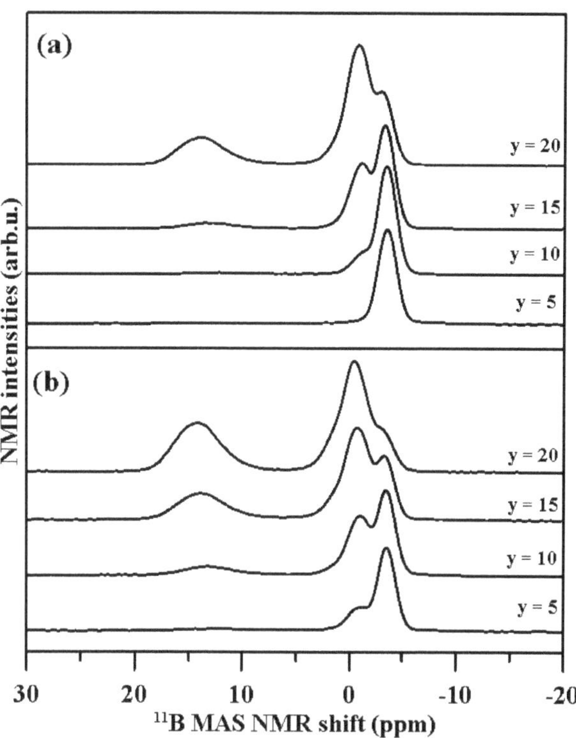

Figure 7. The ^{11}B MAS NMR spectra of (100−x)[0·5PbO. yB$_2$O$_3$.(0·5−y)P$_2$O$_5$].xMoO$_3$ glasses for x=20 mol% MoO$_3$ (a) and x=40 mol% MoO$_3$ (b)

0·2B$_2$O$_3$.0·3P$_2$O$_5$].xMoO$_3$ glass series are shown in Figure 5. Previous ^{11}B MAS NMR studies of borophosphate glasses[7] revealed that BO$_3$ units in borophosphate glasses have NMR shift values ranging from +10 ppm to +15 ppm, whereas for BO$_4$ units the NMR resonance is within the region from −5 ppm to +1 ppm. Lead borophosphate glasses with a low B$_2$O$_3$ content (5–10 mol%) reveal a single narrow signal with a chemical shift value of −3·6 ppm, usually ascribed to tetrahedral B(OP)$_4$ units.[9] A small resonance due to BO$_3$ units, peaking at δ=+13·5 ppm, can be seen in the glass without MoO$_3$ (x=0) (Figure 5). With increasing MoO$_3$ content in these glasses, its intensity increases, reaching 48·5% in the glass with x=60 mol% MoO$_3$. In the chemical shift region between 0 and −4 ppm, two resonances of similar magnitude can be seen peaking at −3·3 and −1·0 ppm for the glass without molybdenum oxide (x=0). The intensity of the first resonance decreases with increasing MoO$_3$ content, while that of the second resonance increases, and another resonance appears on its downfield side at 1·5–2·0 ppm. We have decomposed the ^{11}B MAS NMR spectra into several components and the results are shown in Table 2.

A resonance close to −4 ppm was usually ascribed in previous studies to the presence of B(OP)$_4$ structural units.[16,9] These measurements were carried out with NMR spectrometers operating at 9·4 T, and thus new measurements of the ^{11}B MAS NMR spectra at 18·8 T revealed a better separation of close BO$_4$ resonances in these spectra. From the study of the ^{11}B MAS NMR spectra of silver borophosphate glasses with the rotational echo double resonance (REDOR) technique, Elbers *et al*[17] came to the conclusion that boron atoms in BO$_4$ units can form a maximum of three B–O–P linkages, and that no BPO$_4$ like units (i.e. $B_{4P}^{(4)}$ sites) are formed in silver borophosphate glasses. According to Elbers *et al*,[17] the resonance at −3·3 to −3·5 ppm can therefore be ascribed to tetrahedral boron sites with three B–O–P linkages, i.e. B(OP)$_3$O units, or $B_{3P}^{(4)}$ sites. The second resonance observed at –0·3 to −1·0 ppm in the spectra corresponds to tetrahedral boron sites with two B–O–P links, i.e. B(OP)$_2$O$_2$ units, or $B_{2P}^{(4)}$ sites. As the amount of BO$_3$ units in the glass with x=0 is only 10% (see Table 2) and the resonance ascribed to B(OP)$_2$O$_2$ units amounts to approximately 57% of all boron atoms, the presence of primarily B$^{(4)}$–O–B$^{(4)}$ linkages participating in the formation of B(OP)$_2$O$_2$ units can be assumed. The third resonance at 1·6–2·0 ppm, according to Elbers *et al*,[17] corresponds to tetrahedral boron sites with only one B–O–P bond, i.e. to B(OP)O$_3$ structural units, or $B_{1P}^{(4)}$ sites.

The evolution of borate structural units with the addition of MoO$_3$ (Table 2) is shown in Figure 6. This shows the steady increase in the number of BO$_3$ units participating in the structural network of the studied borophosphate glasses, whereas the number of B(OP)$_2$O$_2$ and B(OP)$_3$O units decreases, and that of B(OP)O$_3$ slightly increases. We are of the opinion that the tetrahedral BO$_4$ units in these glasses with x>0 contain not only B$^{(4)}$–O–P linkages, but also B–O–Mo and B$^{(4)}$–O–B$^{(3)}$ and B$^{(4)}$–O–B$^{(4)}$

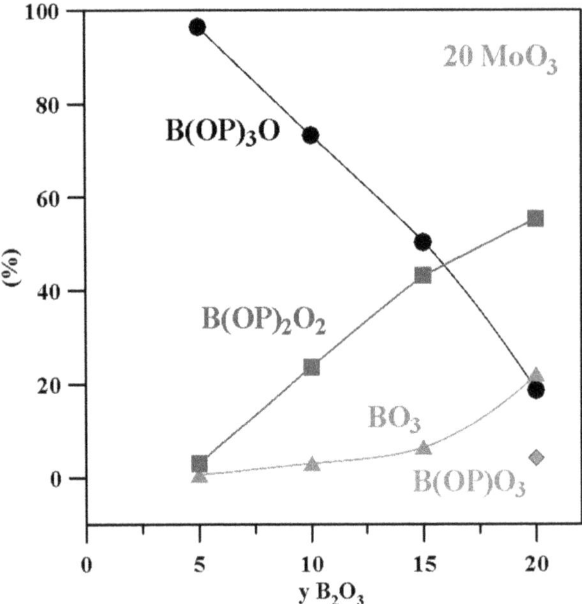

Figure 8. The compositional dependences of the relative amounts of borate structural units in $80[0{\cdot}5PbO.yB_2O_3.(0{\cdot}5-y)P_2O_5].20MoO_3$ glasses, as obtained from the decomposition of their ^{11}B MAS NMR spectra

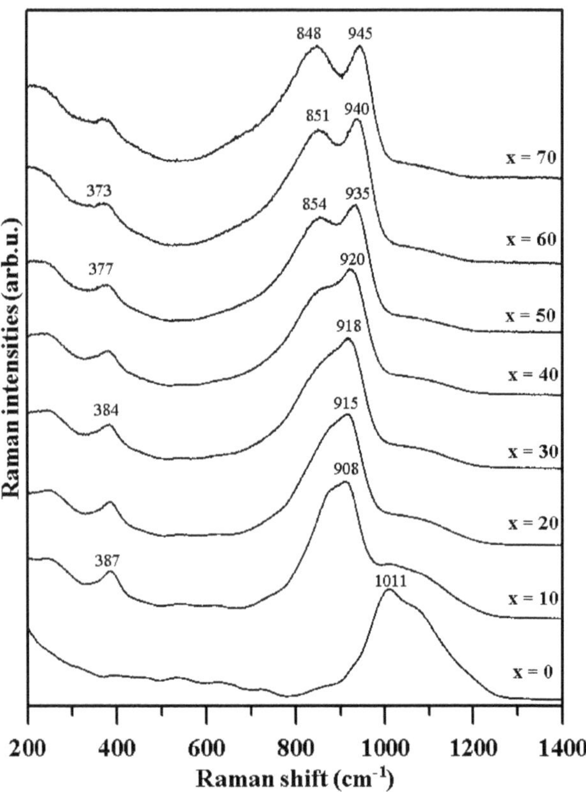

Figure 9. The Raman spectra of $(100-x)[0{\cdot}5PbO.0{\cdot}2B_2O_3.0{\cdot}3P_2O_5].xMoO_3$ glasses

linkages. We also assume that with an increasing MoO_3 content and decreasing boron content the participation of B–O–Mo bonds in the network structure will increase.

In order to evaluate the effect of the changing B_2O_3/P_2O_5 ratio on the ^{11}B MAS NMR spectra, we have plotted the spectra for the $(100-x)[0{\cdot}5PbO.yB_2O_3.(0{\cdot}5-y)P_2O_5].xMoO_3$ glasses with 20 mol% MoO_3 (Figure 7(a)) and 40 mol% MoO_3 (Figure 7(b)). When comparing the evolution of NMR spectra for the series with 20 and 40 mol% MoO_3, several differences can be observed. In the series with 40 mol% MoO_3, the participation of BO_3 units in the glass structure is higher, as well as the participation of the $B(OP)_2O_2$ and $B(OP)_3O$ structural units. Their amounts were obtained by the decomposition of the ^{11}B MAS NMR spectra and are provided in Table 2. We have plotted their dependence on the B_2O_3/P_2O_5 ratio for the series with 20 mol% MoO_3 in Figure 8.

The Raman spectra of the $(100-x)[0{\cdot}5PbO.0{\cdot}2B_2O_3.0{\cdot}3P_2O_5].xMoO_3$ glass series are shown in Figure 9. The spectrum of the parent $50PbO.20B_2O_3.30P_2O_5$ borophosphate glass reveals one dominant broad band in the high frequency region with a maximum at 1011 cm^{-1} ascribed to the symmetric stretching vibration of nonbridging phosphate–oxygen bonds in PO_4 units.[9] A new strong broad band at 800–900 cm^{-1}, composed of at least two bands, appears in the Raman spectrum of the glass with 10 mol% MoO_3. This band can be ascribed to the vibrations of molybdate groups, which are extremely strong Raman scatterers in comparison with PO_4 groups, and their vibrational bands suppress the vibrational bands of phosphate units at 950–1100 cm^{-1}. With increasing MoO_3 content, the shape of the molybdate vibrational bands changes, and their maxima shift in opposite directions. The maximum of the high frequency band shifts from 908 to 945 cm^{-1}, whereas the maximum of the low frequency band shifts from ~870 to 848 cm^{-1} in the glass with 70 mol% MoO_3. The strong band at 908–945 cm^{-1} is usually ascribed to the vibrations of Mo–O terminal bonds in MoO_n polyhedra.[18] Sekiya *et al*[19] assume that the sharp and intense peaks in the region of 800–1000 cm^{-1} can be assigned to the stretching vibration between the molybdenum atom and the unshared oxygen atoms. The frequency of the vibration depends on the mean distance between the molybdenum and the unshared oxygen atoms rather than on the coordination number of the molybdenum atoms.[19] A band at 848–870 cm^{-1} has an asymmetric shape, and there is a shoulder on its low frequency side ascribed to the vibrations of the Mo–O–Mo linkages,[6] the intensity of which increases with increasing MoO_3 content. An increase in the strength of the asymmetric band with increasing MoO_3 content, and its shift to lower wavelengths, is ascribed to the increasing amount of Mo–O–Mo linkages connecting MoO_6 octahedra. The structure of crystalline MoO_3 is composed of MoO_6 octahedra, and we therefore assume that the studied glasses with high MoO_3 content will also contain MoO_6 octahedra, which tend to form clusters via Mo–O–Mo linkages.

Conclusions

A study of a second compositional series of lead borophosphate glasses $(100-x)[0{\cdot}5PbO.0.2B_2O_3.0{\cdot}3P_2O_5].xMoO_3$ has revealed that homogeneous glasses in this series can also be prepared in a wide concentration range of x=0–70 mol% MoO_3. The decreasing glass transition temperature of these glasses with increasing MoO_3 content results from the decreasing bonding forces inside the structural network with the replacement of P–O–P and P–O–B bonds by weaker P–O–Mo and B–O–Mo bonds. The investigation of the ^{11}B MAS NMR spectra of these glasses resulted in the conclusion that the tetrahedral BO_4 units in these glasses contain various bonding linkages. The formation of mixed tetrahedral structural units with B–O–P, B–O–Mo and B–O–B bonds results in several close resonances, which can be differentiated when the NMR spectra are measured with a high resolution. We have accepted the conclusions of previous studies of silver borophosphate glasses with the REDOR technique for the explanation of the NMR spectra, which demonstrated that BO_4 units in borophosphate glasses contain a maximum of three B–O–P bonds. The presence of $B(OP)_4$ units would result in the formation of crystalline BPO_4. We assume that MoO_6 octahedra are the main structural units in the studied borophosphate glasses, primarily in glasses with a high MoO_3 content, even though we cannot evaluate the coordination of molybdenum atoms from the position of vibrational bands of molybdate polyhedra in the Raman spectra. This conclusion is supported by the presence of MoO_6 octahedra in crystalline MoO_3.

Acknowledgements

The Czech authors are grateful for financial support from the Grant Agency of the Czech Republic (Grant No. 13-00355S). The FEDER, Region Nord Pas-de-Calais, Ministère de l'Education Nationale de l'Enseignement Superieur et de la Recherche, CNRS, and USTL are acknowledged for funding of NMR spectrometers.

References

1. Koudelka, L., Mošner, P. & Šubčík, J. Study of structure and properties of modified borophosphate glasses. *IOP Conf. Series: Mater. Sci. Eng.*, 2009, **2**, 012015.
2. Koudelka, L., Pospíšil, J., Mošner, P., Montagne, L. & Delevoye, L. Structure and properties of potassium niobato-borophosphate glasses. *J. Non-Cryst. Solids*, 2008, **354**, 129–133.
3. Šubčík, J., Koudelka, L., Mošner, P., Gregora, I., Montagne, L. & Revel, B. WO_3-doped zinc borophosphate glasses. *Phys. Chem. Glasses: Eur. J. Glass Sci. Technol. B*, 2009, **50**, 243–248.
4. Šubčík, J., Koudelka, L., Mošner, P., Montagne, L., Revel, B. & Gregora, I. Structure and properties of MoO_3-containing zinc borophosphate glasses. *J. Non-Cryst. Solids*, 2009, **355**, 970–975.
5. Šubčík, J., Koudelka, L., Mošner, P. & Černošek, Z. Zinc borophosphate glasses doped with molybdenum oxide. *Adv. Mat. Res.*, 2008, **39–40**, 97–100.
6. Koudelka, L., Rösslerová, I., Černošek, Z., Mošner, P., Montagne, L., Revel, B. & Tricot, G. Structure and properties of lead borophosphate glasses doped by molybdenum oxide. *Phys. Chem. Glasses: Eur. J. Glass Sci. Technol. B*, 2012, **53**, 245–253.
7. Mackenzie, K. J. D. & Smith, M. E. *Multinuclear Solid-State NMR of Inorganic Materials*, Pergamon, Amsterdam 2002.
8. Raguenet, B., Tricot, G., Silly, G., Ribes, M. & Pradel, A. Revisiting the "mixed glass former effect" in ultra-fast quenched borophosphate glasses by advanced 1D/2D solid state NMR. *J. Mater. Chem.*, 2011, **21**, 17693–17704.
9. Koudelka, L., Mošner, P., Zeyer, M. & Jäger, C. Structural study of $PbO–B_2O_3–P_2O_5$ glasses by NMR, Raman and IR spectroscopy. *Phys. Chem. Glasses*, 2002, **43C**, 102–107.
10. Massiot, D., Fayon, F., Capron, M., King, I., LeCalvé, S., Alonso, B., Durand, J. O., Bujoli, B., Gan., Z. & Hoatson G. Modelling one- and two-dimensional solid-state NMR spectra, *Magn. Reson. Chem.*, 2002, **40**, 70–76.
11. Cozar, O., Magdas, D. & Ardelean I. EPR study of molybdenum-lead phosphate glasses. *J. Non-Cryst. Solids*, 2008, **354**, 1032–1035.
12. Mošner, P. & Koudelka, L. Glass-forming ability, thermal stability and chemical durability of lead borophosphate glasses. *Phosph. Res. Bull.*, 2002, **13**, 197–200.
13. Koudelka, L., Rösslerová, I., Holubová, J., Mošner, P., Montagne, L. & Revel, B. Structural study of $PbO-MoO-P_2O_5$ glasses by Raman and NMR spectroscopy. *J. Non-Cryst. Solids*, 2011, **357**, 2816–2821.
14. Lide, D. R. (Ed.) *CRC Handbook of Chemistry and Physics*, CRC Press, Boca Raton 2001, p.9–55.
15. Zeyer-Düsterer, M., Montagne, L., Palavit, G. & Jäger, C. Combined ^{17}O NMR and ^{11}B–^{31}P double resonance NMR studies of sodium borophosphate glasses, *Solid State Nucl. Magn. Reson.*, 2005, **27**, 50–64.
16. Brow, R. K. & Tallant, D. R. Structural design of sealing glasses. *J. Non-Cryst. Solids*, 1997, **222**, 396–406.
17. Elbers, S., Strojek, W., Koudelka, L. & Eckert, H. Site connectivities in silver borophosphate glasses: new results from ^{11}B{^{31}P} and ^{31}P{^{11}B} rotational echo double resonance NMR spectroscopy. *Solid State Nucl. Magn. Res.*, 2005, **27**, 65–76.
18. Chowdari, B. V. R., Tan, K. L. & Chia, W. T. Structural and physical characterization of $Li_2O-P_2O_5-MO_3$ (M = Cr, Mo, W) ion conducting glasses. *Mater. Res. Soc. Symp. Proc.*, 1993, **293**, 325–336.
19. Sekiya, T., Mochida, N. & Ogawa, S. Structural study of MoO_3-TeO_2 glasses. *J. Non-Cryst. Solids*, 1995, **185**, 135–144.

Crystallization kinetics of borosilicate glasses for CHROMPIC nuclear waste vitrification

Mária Chromčíková, Magdaléna Teplanová, Alfonz Plško, Magdaléna Lissová & Marek Liška*

Vitrum Laugaricio – Joint Glass Center of IIC SAS, TnUAD, and FChPT STU, Študentská 2, Trenčín, SK-91150, Slovakia

Manuscript received 25 September 2014
Revised version received 12 November 2014
Accepted 13 November 2014

Borosilicate glasses used for vitrification of CHROMPIC radioactive waste into a boroaluminosilicate glass matrix are considered. The composition dependence of the crystallisation kinetics of glasses with chemical composition close to that currently used for CHROMPIC vitrification was studied. Pairs of glass samples were studied in which the amount (in wt%) of each of the seven components were increased and decreased as follows: (3·5±3·5)Li_2O, (9·0±5·0)Na_2O, (14·5±3·5) B_2O_3, (5·5±5·5)Al_2O_3, (5·5±5·5)TiO_2, (4·5±4·5)Fe_2O_3, (57·5±5·5)SiO_2. A crystallisation peak was observed for only five glasses, and a crystallisation kinetics analysis could only be performed for these glasses. The crystallisation kinetics were studied by DTA at constant heating rates of 2·5, 5·0, 7·5, 10·0, 12·5, and 15·0°C/min. The kinetic parameters of crystallisation were calculated by a model-fitting approach, using the Johnson–Mehl–Avrami–Kolmogorov kinetic model and a nonlinear least squares method. It was found that decreasing the content of network modifying alkaline oxides, Li_2O, and Na_2O, as well as increasing the content of network forming oxides, Al_2O_3, and TiO_2, increases the tendency of the glass to crystallise.

1. Introduction

Vitrification is a high temperature process that can fix liquid radioactive waste into a glass matrix.[1,2] CHROMPIC is a high level radioactive waste (HLW) containing potassium chromate and potassium dichromate, which has been produced by cooling of fuel elements after their removal from reactor A1 of NPP Jaslovské Bohunice. CHROMPIC is fixed into a boroaluminosilicate glass matrix at the vitrification line in Jaslovské Bohunice.[1] The vitrified high level radioactive waste is stored in long term underground deposits. The chemical durability of the vitrified product is a very important parameter. Similarly, the possible crystallisation and mechanical stresses are important, due to possible changes of the mechanical properties of the glass ingots (e.g. fragility). The present contribution deals with the compositional dependence of the crystallisation ability of glasses with composition close to the currently used borosilicate frit. The aim of the present study is not only the characterisation of the crystallisation ability of the currently used glass composition, but also the estimation of changes of crystallisation ability that can be connected with changes of glass composition during the production of the glass frit, and by compositional changes during the vitrification process.

* Corresponding author. Email maria.chromcikova@tnuni.sk
Original version presented at VIII Int. Conf. on Borate Glasses, Crystals and Melts, Pardubice, Czech Republic, 30 June–2 July 2014

2. Method

The experimental data were interpreted by a model-fitting approach in terms of the nucleation growth model formulated by Avrami.[3–5] This model describes the time dependence of the fractional crystallisation as follows:

$$\alpha=1-\exp[-(kt)^{m}] \quad (1)$$

where k and m are constants, and α is the conversion fraction after time t. The constant m may be interpreted in terms of the crystal growth dimensionality. After differentiation with respect to time, the Johnson–Mehl–Avrami–Kolmogorov (JMAK) rate equation is obtained:[6–8]

$$\frac{d\alpha}{dt}=km(1-\alpha)\left[-\ln(1-\alpha)\right]^{\left(1-\frac{1}{m}\right)} \quad (2)$$

The validity of Equation (2) is based on the assumption that homogeneous nucleation or heterogeneous nucleation at randomly dispersed pre-existing nuclei takes place. Furthermore, it is assumed that the growth rate of crystals is independent of time, and is the same in all directions (isotropic crystal growth). The temperature dependence of k is expressed by the Arrhenius equation:

$$k=A\exp[-E/RT] \quad (3)$$

where A is a pre-exponential factor, E is the activation energy, T is the thermodynamic temperature, and R is the molar gas constant. The model-fitting approach assumes that the mechanism of the process is known, and the rate equation is derived with respect to the knowledge of the process mechanism.[8,9]

Experimental DTA curves were processed as described by Plško *et al.*[10] $\alpha(T)$ and $d\alpha(T)/dT$ were obtained from a DTA curve measured with a constant heating rate regime, $T(t)=T_0+\beta t$. The empirical form of the JMAK rate equation, with Arrhenius dependence on temperature, was considered as the basic kinetic equation for crystallisation:

$$\beta\frac{d\alpha}{dT} = A\exp\left[-E/RT\right]m\left(1-\alpha\right)\left[-\ln\left(1-\alpha\right)\right]^{\left(1-\frac{1}{m}\right)} \quad (4)$$

where: α – degree of conversion, t – time, k – rate constant, m – JMAK exponent, A – pre-exponential constant, E – formal activation energy, R – molar gas constant, T – thermodynamic temperature, T_0 – initial temperature, β – constant heating rate.

The values of the kinetic parameters (A and E) and the Avrami coefficient, m, were calculated by a two-step process of regression analysis. Firstly, Equation (4) was linearised, and the linear least squares problem was solved. The next step was connected with solving of a nonlinear least squares problem using Equation (4). The results of linear regression were used as the starting values of the optimised parameters. MATLAB® software was used for these calculations.

3. Experimental

Glasses were prepared from high purity oxides and carbonates by slow heating of the reaction mixture up to 600°C to remove water. Subsequently the mixture was heated up to 1200–1600°C in a platinum crucible. The mixture was held at temperature for 20 min, and hand mixing was performed at this temperature.

DTA measurements were performed using a Netzsch STA 449 F1 Jupiter. Firstly measurements within the temperature interval 30–1000°C were performed on the base glass using four different ranges of glass powder particle size (0·071–0·045 mm, 0·315–0·200 mm, 0·630–0·500 mm, and 0·800–0·710 mm) at a heating rate of 10°C/min in an inert nitrogen atmosphere. DTA curves of all 15 glasses were measured for the fraction 0·800–0·710 mm at a heating rate of 10°C/min. The same fraction was used for all glasses to assure the same extent of surface nucleation. On the basis of the DTA results, the following glasses with well resolved crystallisation peaks were chosen for the crystallisation kinetics study: Li_2O-, Na_2O-, Al_2O_3+, and TiO_2+. DTA curves of these five glasses were recorded with heating rates of 2·5, 5·0, 7·5, 10·0, 12·5 and 15·0°C/min. The glass transition temperature, T_g, was measured by DTA and by thermodilatometry using a TMA Netzsch 402.

Powder diffraction on a Panalytical Empyrean X-ray diffractometer was used to identify the crystallised phase. A rutile phase was identified in all of the crystallised glasses.

4. Results and discussion

The notation and compositions of the studied glasses are summarised in Table 1, together with the T_g values obtained by DTA and thermodilatometry. It can be seen that there is good agreement between the pairs of T_g values. The DTA T_g values are typically a little higher. The glass composition was obtained by optical emission spectroscopy with inductively coupled plasma (ICP OES) after glass decomposition by $HClO_4$ and HF. The glass compositions are designed in pairs by increasing and decreasing the amounts (in wt%) of individual oxides by the same increment, and proportionally changing the content of the other oxides. The exact value of increase and decrease of content of individual components is a compromise between the need to acquire well measurable change of properties on one side, and the requirement of not very different composition from the currently used glass frit on the other. The base glass corresponds to the composition of the glass currently used for CHROMPIC vitrification.

The DTA curves for different ranges of particle size for powdered base glass are compared in Figure 1. It can be seen that the T_g value is not affected by the size of the glass powder particles. The particle size

Table 1. Glass compositions (wt%) and glass transition temperatures obtained by DTA, T_g^{DTA}, and thermodilatometry, T_g^{TMA}

No.	Glass	Li_2O	Na_2O	B_2O_3	Al_2O_3	TiO_2	Fe_2O_3	SiO_2	T_g^{DTA} [°C]	T_g^{TMA} [°C]
1	Base	3·42	9·02	14·50	4·26	5·28	4·33	59·19	494	487
2	Li_2O-	-	9·40	15·00	4·40	5·51	4·56	61·09	520	525
3	Li_2O+	2·77	8·76	14·00	4·27	5·32	4·33	60·55	497	489
4	Na_2O-	3·61	5·07	15·1	4·43	5·43	4·53	61·83	490	486
5	Na_2O+	3·25	12·9	13·9	4·04	5·06	4·17	56·69	492	483
6	B_2O_3-	3·58	9·06	11·00	4·38	5·50	4·46	62·02	494	485
7	B_2O_3+	3·27	8·56	18·00	3·75	4·91	4·09	57·42	492	483
8	Al_2O_3-	3·64	9·59	15·30	-	5·46	4·62	61·38	496	493
9	Al_2O_3+	3·34	8·54	13·70	9·14	4·80	4·10	56·48	492	484
10	TiO_2-	3·66	9·47	15·30	4·64	-	4·67	62·25	490	482
11	TiO_2+	3·26	8·48	13·70	4·01	10·6	4·08	55·83	498	493
12	Fe_2O_3-	3·67	9·59	15·20	4·58	4·45	-	62·49	506	495
13	Fe_2O_3+	3·20	8·54	13·80	4·23	5·09	8·61	56·53	485	478
14	SiO_2-	3·93	10·34	16·40	4·88	5·80	4·86	53·80	488	481
15	SiO_2+	2·94	7·79	12·60	3·84	4·55	3·80	64·50	501	494

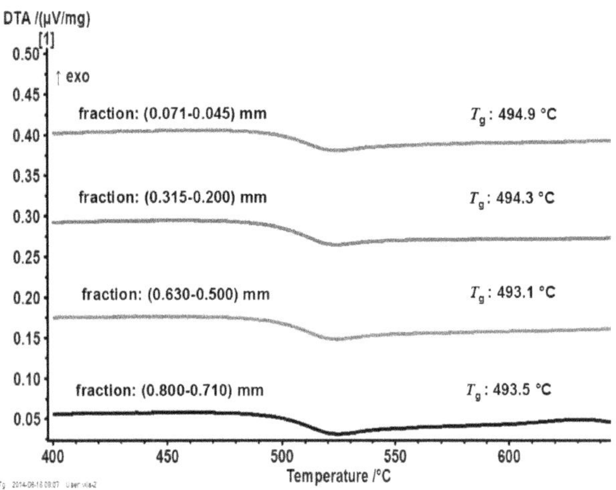

Figure 1. Dependence of the DTA curve of the base glass on selected powder fraction (10°C/min) [Colour available online]

0·80–0·71 mm was then chosen for the DTA study of all other glasses. The DTA curves of glass pairs for which at least one well shaped crystallisation peak was found are compared with the DTA curve of the base glass in Figures 2–5. A crystallisation peak was observed for only five glasses (base glass, Li_2O-, Na_2O-, Al_2O_3+, TiO_2+), and it is only for these glasses that crystallisation kinetic analysis could be performed. For the kinetic analysis, DTA curves recorded with heating rates of 2·5, 5·0, 7·5, 10·0, 12·5, and 15·0°C/min were used. As an example, the dependence of the DTA curve for TiO_2+ glass on heating rate is shown in Figure 6. The parameters of the JMAK kinetic equation, obtained by regression treatment of the experimental data are summarised in Table 2. No simple compositional trends can be seen in the obtained results. When compared with the base glass, the A value of Li_2O- is significantly higher, while for the other glasses this value is significantly lower. The same trend can be seen for the value of formal activation energy, E. The opposite trend is

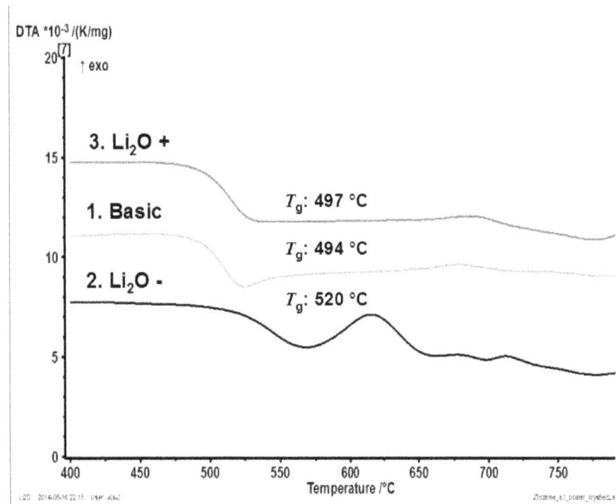

Figure 2. DTA curves of Li_2O+ and Li_2O- glasses compared with the base glass (10°C/min) [Colour available online]

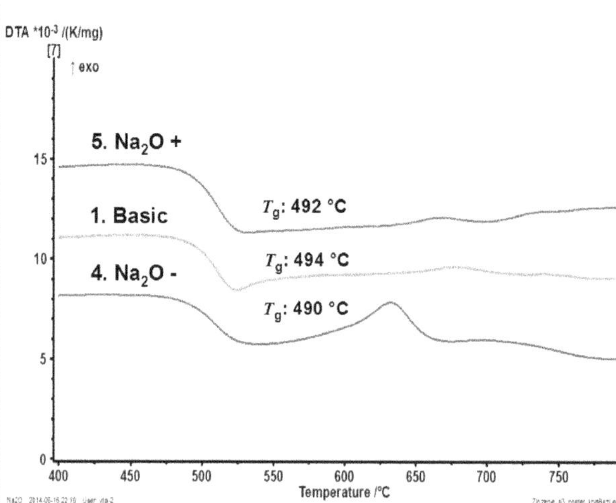

Figure 3. DTA curves of Na_2O+ and Na_2O- glasses compared with the base glass (10°C/min) [Colour available online]

Table 2. Estimates of the parameters of the JMAK kinetic model obtained by nonlinear regression treatment of the experimental data

Glass	A [min⁻¹]	E [kJ/mol]	m
Base	$8·3×10^{12}±7·3×10^5$	238±0·03	1·63±0·01
Li_2O-	$1·1×10^{19}±9·3×10^5$	330±0·02	1·20±0·01
Na_2O-	$1·3×10^6±1·6×10^5$	106±0·91	2·35±0·01
Al_2O_3+	$3·4×10^5±8·2×10^4$	110±2·00	2·19±0·03
TiO_2+	$1·5×10^8±2·3×10^7$	140±1·20	2·22±0·01

found for the shape parameter, m, i.e. a lower value is observed for Li_2O- glass, with higher values for the other glasses.

5. Conclusions

Fifteen borosilicate glasses with compositions based on that used for vitrification of CHROMPIC nuclear waste have been studied. A DTA crystallisation peak was observed for only five compositions, and the crystallisation kinetics were studied for these glasses.

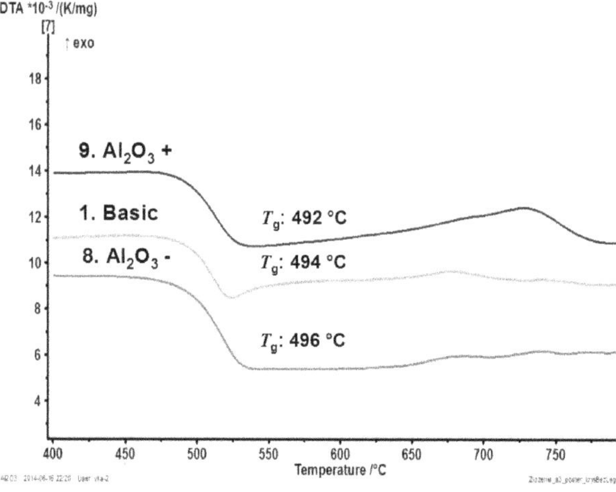

Figure 4. DTA curves of Al_2O_3+ and Al_2O_3- glasses compared with the base glass (10°C/min) [Colour available online]

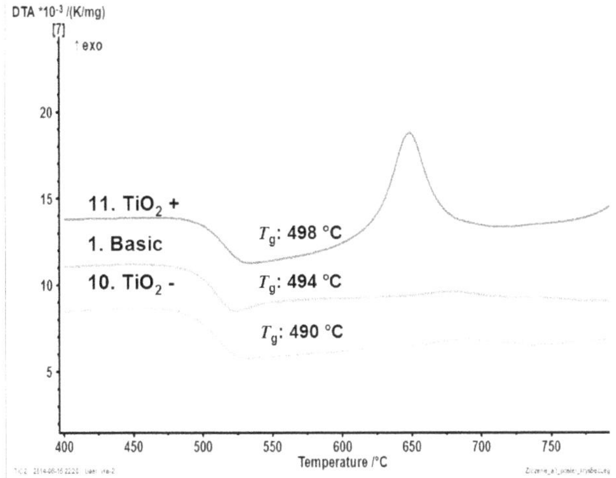

Figure 5. DTA curves of TiO_2+ and TiO_2- glasses compared with the base glass (10°C/min) [Colour available online]

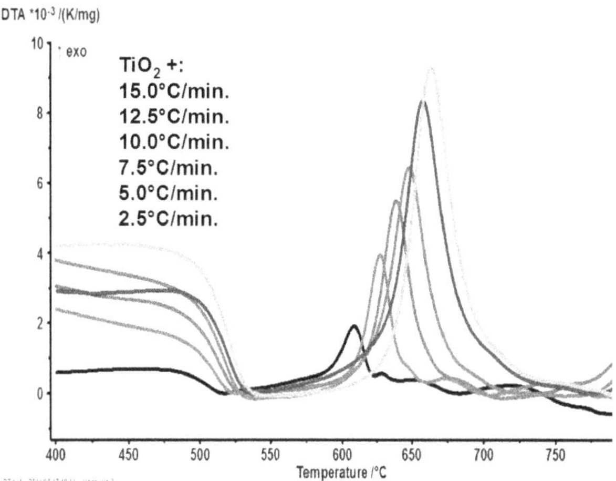

Figure 6. DTA curve of TiO_2+ glass obtained at different heating rates [Colour available online]

No simple trends of the JMAK (Johnson–Mehl–Avrami–Kolmogorov) parameters were obtained with respect to the glass network structure changes caused by increasing/decreasing relative amounts of glass modifying/forming oxides. The highest value of the formal activation energy was obtained for Li_2O- glass. For other modified glasses, the activation energy was lower then for the base glass composition. Similarly the Li_2O- glass provided the lowest value of the exponent m. The obtained spread of values of the JMAK parameters cannot readily be rationalised in terms of glass structure changes. From the point of using the glass in the HLW (high level waste) vitrification process, it can be concluded that decreasing the content of network modifying alkaline oxides, Li_2O and Na_2O, as well as increasing the content of glass network forming oxides, Al_2O_3 and TiO_2, increases the tendency of the glass to crystallise.

Acknowledgement

This work was supported by the Slovak Grant Agency for Science under the grant VEGA 1/0006/12, and by the Slovak Research and Development Agency project APVV-0487-11.

References

1. http://www.javys.sk/sk/jadrove-zariadenia/technologie-spracovania-a-upravy-rao/vitrifikacna-linka, September 2014.
2. Ojovan, M. I. & Lee, W. E. *An introduction to nuclear waste immobilisation.* Elsevier, Oxford 2005.
3. Avrami, M. Kinetics of phase change. I. General theory. *J. Chem. Phys.*, 1939, **7**, 1103–1112.
4. Avrami, M. Kinetics of phase change. II. Transformation-time relations for random distribution of nuclei. *J. Chem. Phys.*, 1940, **8**, 212–224.
5. Avrami, M. Kinetics of phase change. III. Granulation, phase change, and microstructure kinetics of phase change. *J. Chem. Phys.*, 1941, **9**, 177–184.
6. Kolmogorov, A. E. On the statistic theory of metal crystallisation (in Russian). *Izv. Akad. Nauk SSSR Ser. Mater.*, 1937, **1**, 355–359.
7. Johnson, W. A. & Mehl, R. F. Reaction kinetics in processes of nucleation and growth. *Trans. Am. Inst. Min. Metall. Pet. Eng.*, 1939, **135**, 416–458.
8. Šesták, J., Šatava, V. & Wendlandt, W. W. The study of heterogeneous processes by thermal analysis. *Thermochim. Acta*, 1973, **7**, 333–336.
9. Šimon, P. & Thomas, P. S. Application of isoconversional methods for the processes occurring in glassy and amorphous materials. In: Šesták, J., Šimon, P., Editors. *Thermal analysis of micro, nano- and non-crystalline materials: Transformation, crystallisation, kinetics and thermodynamics.* Springer, Dordrecht, 2013. pp. 225–246. ISBN 978-90-481-3149-5.
10. Plško, A., Liška. M. & Pagáčová, J. Crystallisation kinetics of Al_2O_3-Yb_2O_3 glasses. *J. Therm. Anal. Calorim.*, 2012, **108**, 505–509.

DOI: 10.13036/17533562.56.2.053 Phys. Chem. Glasses: Eur. J. Glass Sci. Technol. B, April 2015, 56 (2), 53–58

Low frequency Raman scattering in the $BaO–B_2O_3$ glasses: the effect of temperature

A. A. Osipov & L. M. Osipova

Laboratory of Experimental Mineralogy and Physics of Minerals, Institute of Mineralogy UB RAS, Miass, Russian Federation

Manuscript received 7 August 2014
Revised version received 15 October 2014
Accepted 14 November 2014

The low frequency Raman spectra of a series of barium borate glasses (with 20, 25, 30 and 35 mol% BaO) were measured and analysed from room temperature to 500°C. There are two contributions to the low frequency Raman spectra: vibrations (boson peak and low lying vibrational modes) and relaxations (quasi-elastic scattering (QES)). It was found that increasing the BaO content leads to a small shift of the boson peak to higher frequency. An increase in the temperature does not lead to significant changes in the shape of the low frequency envelope or to a change of the boson peak frequency. Nevertheless, an increase in the temperature leads to a growth in the intensity of the low frequency minimum below the boson peak ($\omega<\omega_{BP}$). This indicates an increase of the QES contribution to the low frequency spectra. The increase in the QES contribution depends on the glass composition, and increases with increasing BaO content.

Introduction

As is well known, the low frequency light scattering spectra of disordered systems (glasses and amorphous solids) are significantly different from those observed for crystalline solids. The Debye model for an elastic continuum predicts that the vibrational density of states (VDOS) is a quadratic function of frequency, $g_{Deb}(\omega)\propto\omega^2$. The VDOS of various crystals is indeed as predicted by the Debye law in the GHz–THz region. Noncrystalline materials, however, exhibit some extra low energy vibrational modes, as compared with $g_{Deb}(\omega)$. The excess density of vibrational states appears as a peak with maximum intensity several times higher than the Debye VDOS. The excess VDOS is manifested in the low frequency light scattering spectra in the form of a broad asymmetric peak (the so-called boson peak), which is characterised by a sharper increase in intensity from low frequency, followed by a more gradual decline for $\omega>\omega_{BP}$ (ω_{BP} is the boson peak position). It should be noted that the excess VDOS is not manifested only in light scattering spectra, but also in neutron scattering spectra, and the low temperature heat capacity.[1–4] The boson peak in the Raman spectra of oxide glasses is usually located in the 20–100 cm^{-1} region.[5–8] Malinovsky & Sokolov[9] have shown that scaled low frequency spectra (in relation to ω_{BP} and $I(\omega_{BP})$) have a universal form: the spectral contour near ω_{BP} is described well by the log-normal distribution. Currently, there is no common opinion about the origin of the boson peak. However, there is a widespread model in which the vibrational excitations responsible for the excess VDOS in glasses are localized on the scale from several tens to hundreds of atoms.[1,8–10]

Another important feature of the low frequency light scattering spectra is so-called quasi-elastic scattering (QES). The QES component appears as a strong intensity line centred at ω=0 cm^{-1}, and with a half-width of about 10–15 cm^{-1}. The distinguishing characteristic of the QES is its anomalous dependence on temperature: vibrational components of the Raman spectra follow $n(\omega,T)+1$ (Stokes components) and $n(\omega,T)$ (anti-Stokes components) dependences, whilst QES exhibits a stronger temperature variation than given by the Bose factor.[3,11] (Here, $n(\omega,T)=1/[\exp(\hbar\omega/kT)-1]$ is the Bose–Einstein statistical factor, whilst $\hbar$ and k are the Planck and Boltzmann constants, respectively.) Usually, QES is attributed to some relaxational process, which in disordered solids can proceed even though the system is macroscopically "frozen" ($T<T_g$).[11] Thus, the total low frequency Raman contour is formed from at least two contributions – the boson peak and QES.

There is now quite a large amount of work regarding the structure of barium borate glasses and melts,[12–15] but only a limited amount of information on the low frequency region of the Raman spectra of $BaO–B_2O_3$ glasses.[16] The behaviour of the boson peak depending on the glass composition was examined in detail by Kojima & Kodama,[16] but the QES has not previously been investigated. Furthermore, it has repeatedly been supposed in the literature that there is a certain intrinsic relationship between the QES and the boson peak.[17,18] Therefore, an aim of this paper is to study the low frequency light scattering as a whole, i.e. both the boson peak and the QES, depending on glass composition and temperature.

* Corresponding author. Email armik@mineralogy.ru
Original version presented at VIII Int. Conf. on Borate Glasses, Crystals and Melts, Pardubice, Czech Republic, 30 June–2 July 2014

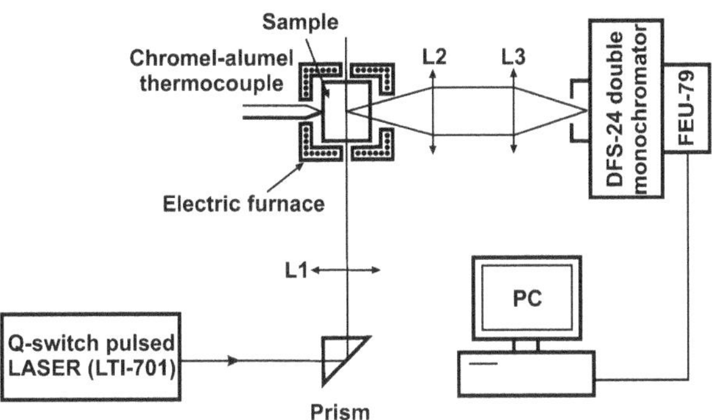

Figure 1. Schematic illustration of the optic arrangement for the low frequency Raman scattering measurements

Experimental

Vitreous xBaO.(100−x)B_2O_3 (x=20, 25, 30 and 35 mol%) samples were prepared by the conventional melt-quench method. Special purity grade boron oxide and reagent grade barium carbonate were used as the starting materials. The initial reagents were carefully dried, weighed and mixed in the required ratios. Next, the batch (15 g) was melted in a platinum crucible at 1100–1200°C for 2 h to obtain bubble-free melts. The homogenised bubble-free melt was poured into a special steel mould to form a glassy parallelepiped (7×7×10 mm, approximately).

Raman scattering measurements were made with 90° scattering geometry, using a double grating DFS-24 monochromator. An uncooled FEU-79 photomultiplier, operating in the photon counting regime, was employed to collect the low frequency Raman spectra. A solid state laser (LTI-701) with wavelength 532 nm and average power 500 mW was used as the excitation source for the light scattering spectra. The spectral width of the slit was 2 cm^{-1} in all experiments. For measurements above ambient temperature, a home made compact electrical furnace was used (Figure 1). The temperature inside the furnace was controlled within ±1°C by a chromel–alumel thermocouple positioned near the sample. Raman scattering measurements were performed over the temperature range from ambient to 500°C.

Results and discussion

The low frequency Raman spectra of the xBaO.(100−x)B_2O_3 (x=20, 25, 30 and 35 mol% BaO) glasses, recorded at ambient temperature, are shown in Figure 2. The spectra are presented in the reduced representation ($I^{red}(\omega)=I^{exp}(\omega)\{\omega[n(\omega,T)+1]\}^{-1}$). The amplitude was normalised to unity at the maximum of the low frequency peak ($I(\omega_{max})$). As Figure 2 demonstrates, an increase in BaO concentration leads to an insignificant shift of the low frequency maximum to the higher frequency side, without changes of the shape of the spectral contour around ω_{max}. However, the contribution from low lying vibrational modes, at approximately $\omega>2\omega_{max}$, is observed to increase with

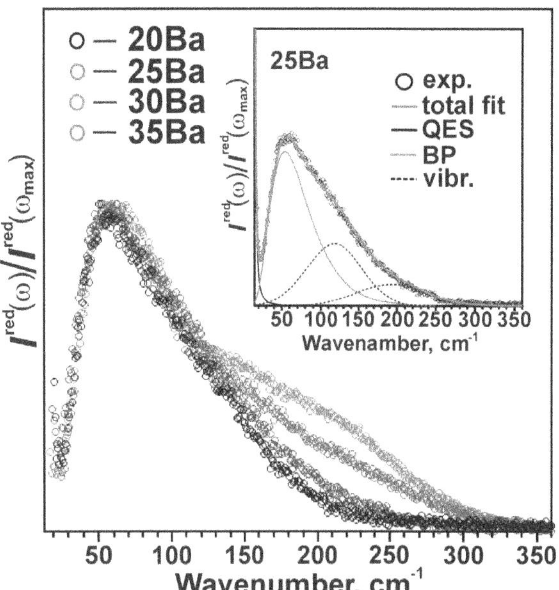

Figure 2. Stokes-side, reduced and scaled low frequency Raman spectra of the xBaO.(100−x)B_2O_3 (x=20, 25, 30 and 35 mol% BaO) glasses. The inset shows a representative fit of the Raman spectrum of 25BaO.75B_2O_3 glass [Colour available online]

increasing x. The increase in intensity of the low lying vibrational modes with x suggests that these modes may be attributed to vibrations of Ba^{2+} ions near their equilibrium positions. Such an assignment is in accordance with far infrared data,[13] according to which the addition of BaO to B_2O_3 leads to the appearance of a new band, the intensity and frequency of which increases with x. Moreover, all spectra exhibit a clear minimum for $\omega<\omega_{max}$. These minima are a result of overlap between the boson peak and the QES. Thus, the total low frequency scattering can be represented as a sum of the three contributions:

$$I^{LF}_{total}(\omega)=I^{QES}(\omega)+I^{BP}(\omega)+I^{vibr}(\omega) \qquad (1)$$

For this reason, each spectrum was represented by a superposition of these three contributions. A Lorentzian line was used to describe the QES. A lognormal function ($I(\omega_{BP})=A\exp\{-[\ln(\omega/\omega_{BP})]^2/2s^2\}$), which takes into consideration the asymmetric shape of the boson peak, was employed to describing this feature, and the contribution of the low lying vibrational modes was described by two Gaussian bands. Because the Ba^{2+} cations can occupy two types of site with differences in the anionic environments,[13] one can expect that at least two components should correspond to their vibrational motions. Fitting with these four components (Lorentzian line, lognormal function and two Gaussian bands) allows us to reproduce the original spectra with a correlation factor ≥0·996. This approach is illustrated in the inset of Figure 2, demonstrating the various contributions to the low frequency Raman spectrum of 25BaO.75B_2O_3 glass. The deconvolution results show that, as x increases, there is an insignificant shift of the

Table 1. Peak parameters of the low lying components in the room temperature low frequency Raman spectra of $BaO–B_2O_3$ glasses. The QES and boson peak region are characterised by ω_{BP}, σ, ω_{min} and R (see text for details). The position, full width at half maximum (FWHM) and relative intensity are given for the two Gaussian components, and the subscripts L and H indicate the low frequency and high frequency Gaussian components respectively. Experimental errors are ±3 cm^{-1}, ±5 cm^{-1} and ±0·01 for the peak position, FWHM and relative intensity of the Gaussian components, respectively

x/mol% BaO	ω_{BP}/cm^{-1}	σ	ω_{min}/cm^{-1}	R	$[\omega/FWHM/I]_L$	$[\omega/FWHM/I]_H$
20	55±2	0·46±0·01	23±2	0·23±0·01	117/83/0·36	192/105/0·12
25	54±2	0·47±0·01	24±2	0·22±0·01	118/83/0·35	192/105/0·16
30	59±2	0·46±0·01	23±2	0·22±0·01	128/83/0·31	200/106/0·26
35	58±2	0·47±0·01	23±2	0·20±0·01	127/83/0·23	202/107/0·39

boson peak position toward higher frequency without change in its width (Table 1). According to Kojima & Kodama,[16] the boson peak position for barium borate glasses does not depend on glass composition. However, our data are evidence of a weak tendency for ω_{BP} to increase (by approximately 3 cm^{-1}) with increase in the BaO content. The peak positions of the low lying vibrational components depends on the glass composition; as can be seen from Table 1, both bands shift slightly toward higher frequencies (by ~10 cm^{-1}) with increasing x and, in addition, their total intensity is increased. Furthermore, significant redistribution of the intensity of Gaussian bands is observed with increasing x; the intensity of the line at 118 cm^{-1} (low frequency component) is three times higher than that located at 192 cm^{-1} (high frequency Gaussian component) in the spectrum of the glass with 20 mol% BaO, while the opposite situation is observed for the sample

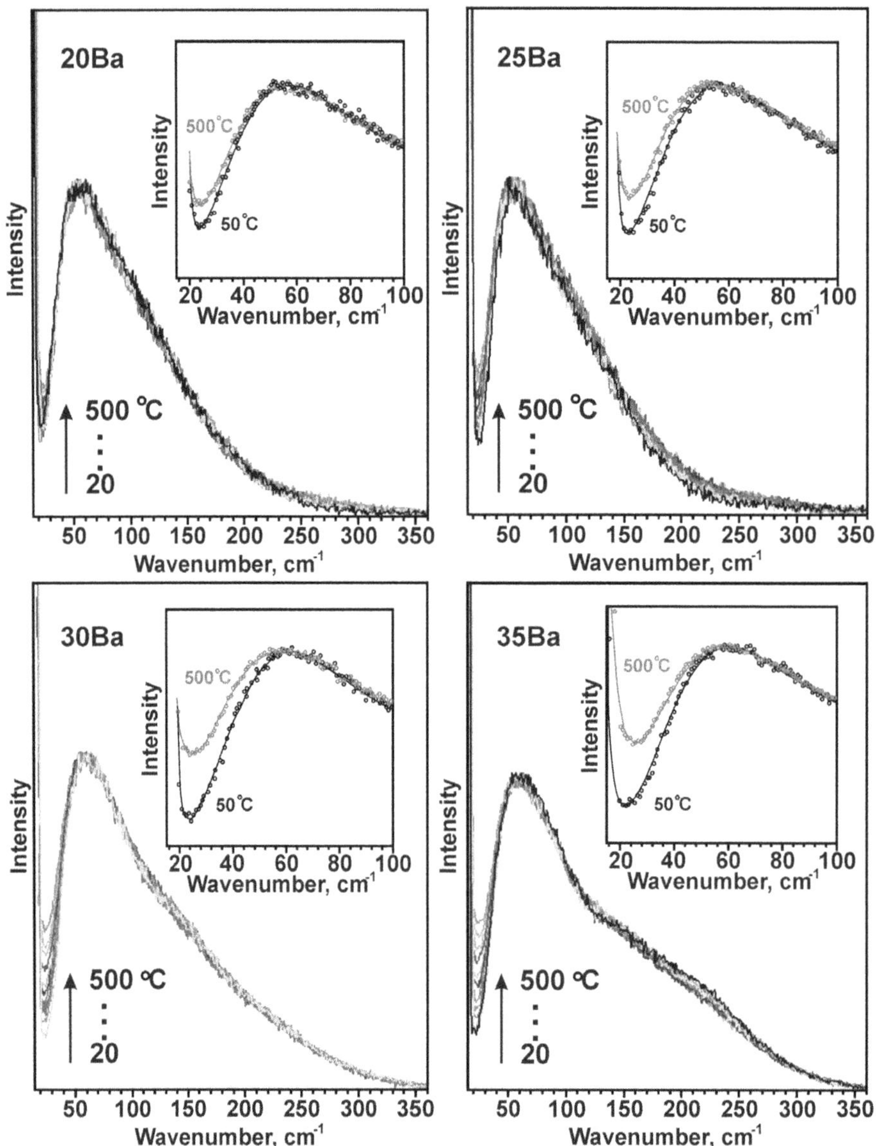

Figure 3. Temperature reduced and scaled low frequency Raman spectra of the barium borate glasses measured at various temperatures [Colour available online]

with 35 mol% BaO. The intensity of the high frequency Gaussian component is higher more than 50% larger than the low frequency Gaussian component in the last case. This indicates that considerable metal ion site redistribution occurs with increasing BaO content.

Another important feature of the low frequency Raman spectra is the minimum located at $\omega<\omega_{BP}$. The formation of this minimum is a result of the overlap of closely located bands, i.e. the QES and the boson peak. It has been shown[17,19,20] that the ratio $R=I(\omega_{min})/I(\omega_{BP})$ ($I(\omega_{min})$ is the intensity of the light scattering at the minimum) can serve as a measure of the QES contribution to the low frequency Raman spectra at fixed temperature. Our data allow us to assert that ω_{min} does not depend on the composition of the glasses studied here, whereas the R values (measured at room temperature) slightly decrease with increasing BaO content (see Table 1).

Temperature reduced and scaled low frequency Raman spectra of barium borate glasses, recorded from room temperature to 500°C, are shown in Figure 3. The shape of the low frequency envelope of all spectra does not change significantly with increasing temperature, with the exception of the frequency range around ω_{min}. An increase of intensity at ω_{min} is observed with increasing temperature for all studied glasses. The magnitude of the increase in $I(\omega_{min})$ depends on the glass composition, and increases with increasing BaO content.

The temperature dependence of the ratio $R(T)$ is a quantitative characteristic of the QES contribution to the low frequency Raman spectra, and in the investigated temperature range $R(T)$ can be described by a straight line within the limits of experimental error. The slope of this line increases with increasing x (see Figure 4).

A more adequate way to estimate the QES contribution to the low frequency spectra is to determine the integrated intensity of the QES (after subtraction of the intensity of the boson peak and the low lying vibrational modes). Ideally, the true contribution of the boson peak and low lying vibrational modes can be obtained by the measurement of the Raman spectra at extremely low temperatures, where all the relaxations responsible for QES are suppressed. However, we were unable to perform low temperature measurements and the spectra obtained at room temperature were accepted as a basis. In addition, the available equipment did not allow us to record spectra from very low frequencies (less than 10–15 cm^{-1}). This fact imposes certain restrictions on the lower limit of the frequency range in which the determination of the integrated QES contribution is possible. Finally, because the shape of the spectral contour at $\omega>\omega_{BP}$ remained practically unchanged when the temperature increases from 20 to 500°C, the upper limit of the frequency range was also limited. Figure 5 illustrates the choice of the lower and upper limits

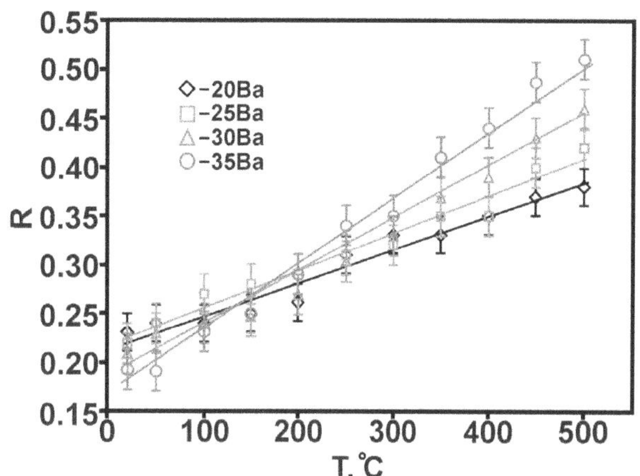

Figure 4. Temperature dependence of $R=I(\omega_{min})/I(\omega_{BP})$ for the barium borate glasses [Colour available online]

to determine the increase in the intensity of the QES with temperature, in accordance with the expression:

$$\Delta S(T) = \int_{\omega_1}^{\omega_2} \left[I(T,\omega) - I(20,\omega) \right] d\omega \tag{2}$$

where $I(T,\omega)$ is the Raman intensity at a certain temperature T, and $I(20,\omega)$ is the Raman intensity at room temperature. For definiteness, the lower limit, ω_1, was chosen to be equal to the average of the values of ω_{min} for the room temperature spectra. The upper limit, ω_2, was chosen as the point of intersection of the fitted curves for two different spectra. (All experimental spectra were fitted with a Lorentzian, a lognormal function and two Gaussian bands). The obtained behaviour of $\Delta S(T)$ is shown in Figure 6. One can see that the $\Delta S(T)$ dependence can be described by straight line. The slope of this line depends on the glass composition, and increases by a factor of more than five as the BaO content increases.

The temperature dependencies of ω_{BP}, σ and ω_{min} for all the studied barium borate glasses are summarised in Table 2. As one can see, ω_{min} does not

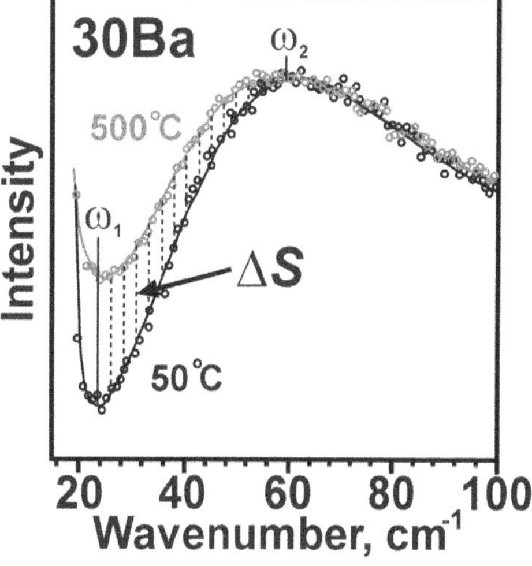

Figure 5. Illustration of the determination of the integral QES contribution to the low frequency Raman spectra of the barium borate glasses [Colour available online]

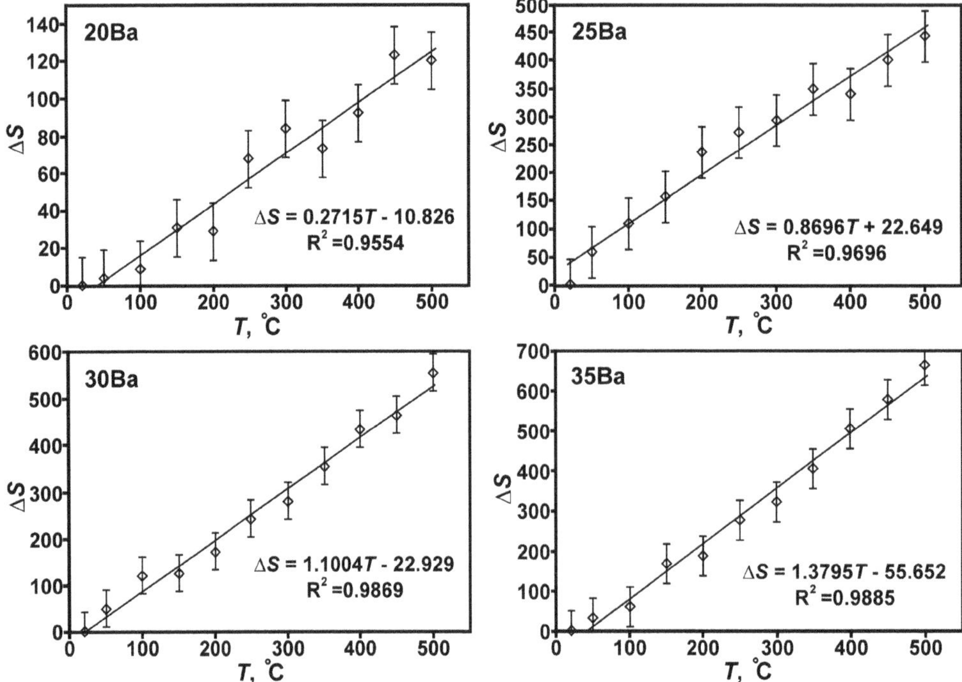

Figure 6. The integrated quasielastic scattering intensity, $\Delta S(T)$, for the barium borate glasses

change with increasing temperature in the 20Ba and 25Ba spectra. On the other hand, there is a shift of ω_{min} toward higher frequency in the Raman spectra of glasses with higher BaO concentration. This shift can be explained by the substantial increase of the QES contribution to the low frequency spectra of the samples with 30 and 35 mol% BaO, in comparison with the glasses containing 20 and 25 mol% BaO. However, no significant changes in the lognormal distribution were found. The σ parameter does not depend on either glass composition or temperature within experimental error. Averaging over all studied spectra gives σ=0·47±0·01. A dependence of the boson peak frequency on temperature was also not found.

According to Sokolov *et al*,[17,20,21] the ratio R, measured at the glass transition temperature, T_g, correlates with fragility $m=\partial(\log_{10}\eta)/\partial(T_g/T)|_{T=Tg}$; more fragile glass formers have higher values of R. Although we did not reach temperatures as high as T_g for the glasses studied here, we approached close enough to the glass transition. (The values of T_g for our glasses are ~555, 590, 604 and 604°C for the 20Ba, 25Ba, 30Ba and 35Ba samples, respectively, according to data from Refs 13 and 22). Thus, it is reasonable to expect that the strong dependence of the QES on glass composition ($R(x)$ and $\Delta S(x)$), which is observed at temperatures closest to T_g (i.e. at 500°C), may be attributed to an increase in the fragility of the studied glasses with BaO content. On the other hand, the temperature dependence of the QES intensity, dI_{QES}/dT, correlates with fragility, rather than $R(T_g)$, as shown by Ribeiro.[23] From this point of view, it is interesting to consider a possible correlation between $d(\Delta S(x))/dT$ values and the fragility of the barium borate glasses. Unfortunately, we do not have any information on the fragility of glasses studied here. However, as shown by Chryssikos *et al*,[24] increasing x results in a sigmoidal increase of the fragility index for xMO(or M_2O).(100−x)B_2O_3 glasses (M=Li, Na, Ca, Sr and Ba), with x values in the range from 0 to 40 mol% (see Figure 1 in Ref. 24). Using this fact and data presented in Figure 6, we plotted $d(\Delta S(x))/dT$ versus

Table 2. The temperature dependence of the peak parameters of the low lying components in the room temperature low frequency Raman spectra of BaO–B_2O_3 glasses, ω_{BP}, σ and ω_{min}. ω_{BP} and ω_{min} are in cm^{-1}

T/°C	20BaO.80B_2O_3			25BaO.75B_2O_3			30BaO.70B_2O_3			35BaO.65B_2O_3		
	ω_{BP}	σ	ω_{min}	ω_{BP}	σ	ω_{min}	ω_{BP}	σ	ω_{min}	ω_{BP}	σ	ω_{min}
20	55	0·45	23	54	0·47	24	59	0·46	23	58	0·46	22
50	55	0·46	23	55	0·46	24	59	0·45	24	59	0·46	22
100	54	0·46	23	55	0·45	24	59	0·46	24	58	0·46	23
150	54	0·46	24	55	0·45	22	60	0·47	24	58	0·46	23
200	53	0·46	23	54	0·47	23	60	0·45	24	58	0·47	23
250	53	0·47	23	54	0·47	23	58	0·46	24	58	0·47	24
300	53	0·47	23	55	0·47	24	58	0·46	25	57	0·47	24
350	53	0·47	24	55	0·47	24	58	0·47	24	56	0·47	25
400	54	0·47	24	54	0·47	24	58	0·47	25	56	0·48	25
450	52	0·47	24	54	0·47	24	58	0·48	25	56	0·48	25
500	53	0·47	23	54	0·47	24	57	0·47	25	57	0·48	26

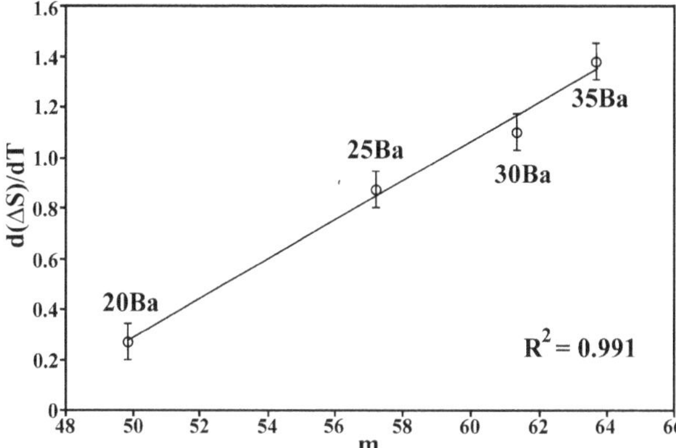

Figure 7. The parameter $d(\Delta S(x))/dT$ against fragility index, m, for the barium borate glasses

m values (based on data from Ref. 24) as is shown in Figure 7. One can see that a linear correlation is indeed observed between these two parameters. This indicates that a strong dependence of the QES intensity on the glass composition (at temperatures close to T_g) can indeed be related to an increase in the fragility of barium borate glasses with increasing x.

Conclusions

Analysis of the low frequency Raman spectra of a series of barium borate glasses (containing 20 to 35 mol% BaO) has revealed the following:

(1) The low frequency Raman spectra of the studied glasses can be represented as a sum of three contributions: quasielastic scattering (QES), boson peak and low lying vibrational modes.

(2) The boson peak position, ω_{BP}, slightly depends on the glass composition: a weak increase in boson peak position (by ~3 cm^{-1}) is observed with increasing BaO content.

(3) The position of the minimum, ω_{min}, which is observed at $\omega<\omega_{BP}$, does not depend on the glass composition at room temperature, whereas the ratio $R=I(\omega_{min})/I(\omega_{BP})$ slightly decreases with increasing BaO content.

(4) Increasing the temperature does not affect ω_{min} for 20Ba and 25Ba glasses, while ω_{min} shifts to higher frequency in the spectra of the glasses with more than 25 mol% BaO.

(5) Both R and the integrated QES intensity, ΔS, increase linearly with increasing temperature. The magnitude of the increase in these quantities depends on the glass composition, and increases with increasing x. The strong composition dependence of these parameters at 500°C can be related to an increase in the fragility of the glasses with increasing BaO content.

(6) The parameter $d(\Delta S(x))/dT$ correlates with the fragility of the glasses.

(7) The shape of the spectral contour near ω_{BP} does not depend on the glass composition and temperature.

Thus, in the investigated temperature range, an increase in temperature leads to an increase of the QES contribution, without significant change of the vibrational contributions, to the low frequency Raman spectra.

References

1. Malinovskiĭ, V. K., Novikov, V. N. & Sokolov, A. P. Nanometre-scale structure of disordered bodies, *Phys.-Usp.*, 1993, **36** (5), 440–444.
2. Fontana, A., Rossi, F., Viliani, G., Caponi, S., Fabiani, E., Baldi, G., Ruocco, G. & Dal Maschio, R. The Raman coupling function in the disordered solids: a light and neutron scattering study on glasses of different fragility. *J. Phys.: Condens. Matter.*, 2007, **19** (20), 205145.
3. Malinovsky, V. K., Novikov, V. N., Surovtsev, N. V. & Shebanin, A. P. Investigation of amorphous states of SiO_2 by Raman scattering spectroscopy. *Phys. Solid State*, 2000, **42** (1), 65–71.
4. Matsuda, Y., Kawaji, H., Atake, T., Yamamura, Y., Yasuzuka, S., Saito, K. & Kojima, S. Non-Debye excess heat capacity and boson peak of binary lithium borate glasses. *J. Non-Cryst. Solids*, 2011, **357** (2), 534–537.
5. McIntosh, C., Toulouse, J. & Tick, P. The Boson peak in alkali silicate glasses. *J. Non-Cryst. Solids*, 1997, **222**, 335–341.
6. Baranov, A. V., Perova, T. S., Petrov, V. I., Vij, J. K. & Neilsen, O. F. Nature of the boson peak in Raman spectra of sodium borate glass systems: influence of structural and chemical fluctuations and intermolecular interactions. *J. Raman Spectrosc.*, 2000, **31** (8–9), 819–825.
7. Kojima, S. & Kodama, M. Boson peak in modified borate glasses. *Phys. B.*, 1999, **263–264**, 336–338.
8. Schroeder, J., Wu, W., Apkarian, J. L., Lee, M., Hwa, L. G. & Moynihan, C. T. Raman scattering and Boson peak in glasses: temperature and pressure effects. *J. Non-Cryst. Solids*, 2004, **349**, 88–97.
9. Malinovsky, V. K. & Sokolov, A. P. The nature of the boson peak in Raman scattering in glasses. *Solid State Commun.*, 1986, **57** (9), 757–761.
10. Buchenau, U., Galperin, Yu. M., Gurevich, V. L. & Schober, H. R. Anharmonic potentials and vibrational localization in glasses. *Phys. Rev. B*, 1991, **43** (6), 5039.
11. Brodin, A. & Torell, L. M. Relaxational and low-wavenumber vibrational dynamics of strong glass formers B_2O_3 and GeO_2 probed by Raman scattering. *J. Raman Spectrosc.*, 1996, **27** (10), 723–730.
12. Solntsev, V. P., Davydov, A. V., Malinovsky, V. K. & Surovtsev, N. V. Raman scattering study of crystalline and melting states of $BaO\times2B_2O_3$. *J. Cryst. Growth*, 2010, **312** (20), 2962–2966.
13. Yiannopoulos, Y. D., Chryssikos, G. D. & Kamitsos, E. I. Structure and properties of alkaline earth borate glasses. *Phys. Chem. Glasses*, 2001, **42** (3), 164–172.
14. Osipov, A. A. & Osipova, L. M. Raman scattering study of barium borate glasses and melts. *Phys. Chem. Solids*, 2013, **74** (7), 971–978.
15. Kabanov, V. O., Sedmale, G. P. & Yanush, O. V. O structure barievoboratnyh stekol po dannym spektroscopii KR. *Fiz. Khim. Stekla*, 1990, **16** (2), 174–177. (*In Russian.*)
16. Kojima, S. & Kodama, M. Raman scattering study of boson peak in barium borate glass. *Jpn. J. Appl. Phys.*, 1996, **35** (5B), 2899–2902.
17. Sokolov, A. P., Kisliuk, A., Quitmann, D., Kudlik, A. & Rosslerk, E. The dynamics of strong and fragile glass formers: vibrational and relaxation contributions. *J. Non-Cryst. Solids*, 1994, **172–174** (1), 138–153.
18. Hong, L., Begen, B., Kisliuk, A., Novikov, V. N. & Sokolov, A. P. Influence of pressure on fast relaxation in glass-forming materials. *Phys. Rev. B*, 2010, **81** (10), 104207.
19. Wang, Y., Nakamura, M., Matsuda, O. & Murase, K. Raman-spectroscopy studies on rigidity percolation and fragility in Ge-(S, Se) glasses. *J. Non-Cryst. Solids*, 2000, **266–269**, 872–875.
20. Sokolov, A. P., Rossler, E., Kisliuk, A. & Quitmann, D. Dynamics of strong and fragile glass formers: Differences and correlation with low-temperature properties. *Phys. Rev. Lett.*, 1993, **71** (13), 2062–2065.
21. Novikov, V. N., Ding, Y. & Sokolov, A. P. Correlation of fragility of supercooled liquids with elastic properties of glasses. *Phys. Rev. E.*, 2005, **71**, 061501.
22. Lower, N. P., McRae, J. L., Feller, H. A., Betzen, A. R., Kapoor, S., Affatigato, M. & Feller, S. A. Physical properties of alkaline-earth and alkali borate glasses prepared over an extended range of composition. *J. Non-Cryst. Solids*, 2001, **293–295**, 669–675.
23. Ribeiro, M. C. C. Low-frequency Raman spectra and fragility of imidazolium ionic liquids. *J. Chem. Phys.*, 2010, **133**, 024503.
24. Chryssikos, G. D., Kamitsos, E. I. & Yiannopoulos, Y. D. Towards a structural interpretation of fragility and decoupling trends in borate systems. *J. Non-Cryst. Solids*, 1996, **196**, 244–248.

DOI: 10.13036/17533562.56.3.085 Phys. Chem. Glasses: Eur. J. Glass Sci. Technol. B, June 2015, 56 (3), 85–97

A neutron diffraction study of six $M_2O.M'_2O.5B_2O_3$ mixed-modifier di-pentaborate glasses

Adrian C. Wright, Roger N. Sinclair,‡ Cora E. Stone, Joanna L. Shaw*

J.J. Thomson Physical Laboratory, University of Reading, Whiteknights, Reading, RG6 6AF, UK

Steven A. Feller, Richard B. Williams

Physics Department, Coe College, Cedar Rapids, IA 52402, USA

Henry E. Fischer

Institut Laue-Langevin, 6 Rue Jules Horowitz, B.P. 156, 38042 Grenoble Cedex 9, France

Manuscript received 18 July 2014
Revised version received 10 December 2014
Accepted 5 February 2015

Twin-axis neutron diffraction experiments have been performed to study structural aspects of the mixed-modifier effect for six ternary di-pentaborate glasses of composition $M_2O.M'_2O.5B_2O_3$ (M/M′=Li/Na, Li/Rb, Na/K, Na/Rb, Na/Cs, and Na/Ag). In reciprocal space, the method of additivity is employed to deduce information concerning the thickness/size of the network cages containing the network-modifying cations, and the zero-Q limit of the diffraction pattern, I(0), to investigate the extent of the nanoheterogeneity in the glasses. The fraction of 4-fold co-ordinated boron atoms, x_4, is obtained from the area under the first (B-O) peak in the real-space total correlation function, T(r) and, with the exception of vitreous $Na_2O.Cs_2O.5B_2O_3$ and $Na_2O.Ag_2O.5B_2O_3$, no evidence is found to indicate the presence of a significant fraction of non-bridging oxygen atoms. An interpretation of the neutron diffraction data, in terms of a thermodynamically-based structural model for the network-modifying cation conductivity and mixed modifier effect in borate glasses, is presented in an accompanying paper (pp. 98–107).

1. Introduction

This paper is the last in a series of three on the structure of binary and ternary alkali, silver and alkaline earth borate glasses, with particular emphasis on the environment of the network-modifying cations and the role of superstructural units in the borate network. The first paper[1] investigated the structure of binary alkali, silver and alkaline earth borate glasses, as a function of the network-modifying cation species, at constant network-modifier fraction, x_M, and involved eight di-pentaborate glasses of composition $2M_2O.5B_2O_3$ (M=Li, Na, K, Rb, Cs or Ag) and $2MO.5B_2O_3$ (M=Ca or Ba).[1] In a complementary study, the second paper[2] considered the structure as a function of x_M for three network-modifying cation species (sodium, rubidium and cæsium). This work has now been extended to six ternary mixed-modifier di-pentaborate glasses of composition $M_2O.M'_2O.5B_2O_3$ (M/M′=Li/Na, Li/Rb, Na/K, Na/Rb, Na/Cs and Na/Ag), corresponding to equimolar combinations of glasses from Ref. 1, to investigate the structural origins of the mixed-modifier effect. The present paper reports the results of a neutron diffraction study, whilst their interpretation in terms of a structural model for the cationic conductivity of borate glasses and the mixed-modifier effect is discussed in Ref. 3.

1.1. Di-pentaborate crystallography

A detailed survey of the structures of the crystalline phases of both single- and mixed-modifier borates has been presented in Ref. 4. As may be seen from Tables A1 and A3 of Ref. 4, structures have not been reported for any of the crystalline single-modifier di-pentaborates that are known to exist ($2Na_2O.5B_2O_3$, $2Rb_2O.5B_2O_3$ and $2BaO.5B_2O_3$), nor for $2Li_2O.5B_2O_3$, whose existence is uncertain. On the other hand, crystal structures are known for three ternary mixed-modifier di-pentaborates, *viz.* $Na_2O.Cs_2O.5B_2O_3$,[5] $Na_2O.Tl_2O.5B_2O_3$[6] and $K_2O.Cs_2O.5B_2O_3$,[5] all of which have borate networks based solely on the di-pentaborate group (Ref. 4, Table 6). This superstructural unit is, however, 5-connected, and so both Zachariasen's criteria,[7] as modified for borate glasses (Ref. 4, Section 9), and constraint theory[8,9] would suggest that, if di-pentaborate groups exist in the corresponding glasses, this will only be in very small numbers, indicating that chemical groupings based on di-pentaborate groups are unlikely to play a major

* Corresponding author. Email a.c.wright@reading.ac.uk
Original version presented at VIII Int. Conf. on Borate Glasses, Crystals and Melts, Pardubice, Czech Republic, 30 June–2 July 2014
‡ Deceased: 18th June 2007

role in the chemical structure of the present glasses.

The number of mixed-modifier crystalline phases reported for compositions other than di-pentaborate is also limited (*cf.* Refs 4 and 10). In addition to $Na_2O.Cs_2O.5B_2O_3$, the following mixed-modifier crystalline phases are known to occur in the six ternary systems considered here: Na/Ag tetraborate ($0{\cdot}4Na_2O.0{\cdot}6Ag_2O.4B_2O_3$), Na/K triborates ($Na_2O.2(Na/K)_2O.9B_2O_3$, $Na_2O.2K_2O.9B_2O_3$ and $(Na/K)_2O.2K_2O.9B_2O_3$), Li/Na and Li/Rb diborates ($Li_2O.Na_2O.4B_2O_3$[11] and $Li_2O.Rb_2O.4B_2O_3$) and Na/Rb and Na/Cs orthoborates ($2Na_2O.Rb_2O.B_2O_3$ and $2Na_2O.Cs_2O.B_2O_3$). (A reference is only given for $Li_2O.Na_2O.4B_2O_3$, since this compound is not included in Table A3 of Ref. 4.) Unfortunately, however, the necessary thermodynamic data are not available for these mixed-modifier compounds, which precludes thermodynamic modelling of the corresponding glasses, especially for those systems where there are ternary triborate or diborate crystalline phases, since these lie on either side of the present di-pentaborate composition, and may therefore contribute significantly to the chemical structure.

1.2. Structural studies of mixed-modifier borate glasses

There have been many structural studies of mixed modifier glasses, and so the discussion here will be restricted to those that are of particular relevance to the present work. Perhaps the most important NMR data are those of Ratai *et al*,[12,13] who have investigated mixed-alkali tri-heptaborate glasses, of composition $3[xNa_2O.(1-x)M_2O].7B_2O_3$ (M=Li, K, Rb and Cs), from four of the systems studied here, their total network-modifier fraction, $x_M=0{\cdot}3$, being very close to that of the present di-pentaborate samples ($x_M=0{\cdot}286$). Ratai *et al*[12,13] find that, within their experimental uncertainty, cation mixing has no measurable influence on the fraction, x_4, of the boron atoms that are 4-fold co-ordinated, as represented by the equilibrium of Equation (1) of Ref. 3. Hence they conclude that, to first order, the structural origins of the mixed-alkali effect appear to be unrelated to the network structure. For their Li/Na, Na/K and Na/Rb glasses, x_4 is consistent with the assumption that all of the network-modifying cations are associated with $BØ_4^-$ tetrahedra, where Ø indicates a bridging oxygen atom, i.e.

$$x_4=x_M/(1-x_M) \tag{1}$$

whereas their experimental results for the Na/Cs glasses are consistently lower than this value, suggesting the presence of nonbridging oxygen atoms ($BOØ_2^-$ triangles).

As to the distributions of the network-modifying cation species, Ratai *et al*[12,13] conclude that, for their Li/Na and Na/K glasses, the model that yields the best agreement with experiment is statistical cation mixing within a homogeneously-distributed total population, there being no evidence for special inter-cation arrangements such as unlike-cation pairing or like-cation segregation. These data thus support the concept of site mismatch as the fundamental principle underlying the mixed-alkali effect. For the Na/Rb glasses, on the other hand, the data suggest some degree of phase segregation. Ratai *et al*[12,13] also find that the replacement of a smaller cation, M, by a larger cation, M′, leads to a slight compression of the M sites and a corresponding expansion of the M′ sites. The modification of the sites occupied by the two network-modifying cation species in mixed modifier borate glasses, as compared to those in the corresponding single-modifier glasses, is supported by the far-infrared spectra of Chryssikos & Kamitsos[14] for sodium-rubidium and sodium-cæsium diborate glasses.

In a further paper on mixed-modifier $xNa_2O.(1-x)Rb_2O.4B_2O_3$ tetraborate and $3[xNa_2O.(1-x)Rb_2O].7B_2O_3$ tri-heptaborate glasses, Epping *et al*[15] confirm the non-statistical mixing of the Na^+ and Rb^+ network-modifying cations, and suggest some preference for like-cation interactions (domain segregation effects), especially for the rubidium-rich compositions and at the lower total cation content. They conclude that their results support the idea that the details of the unlike-cation mixing are influenced by the size difference (site mismatch) between the two network-modifying cation species.

A combined neutron diffraction and reverse Monte Carlo study of a $Li_2O.Rb_2O.4B_2O_3$ mixed-modifier diborate glass, together with the two corresponding end-member diborate glasses, has been reported by Cormier *et al.*[16] These authors conclude that the environments of the two network-modifying cation species in the mixed-modifier glass are slightly modified, relative to those in the single-modifier end-member glasses. A peak in the real-space total correlation function, $T(r)$, that grows with increasing rubidium content, is associated with a Rb–O first neighbour distance (bond length) of ~2·9 Å, in excellent agreement with the value (2·87 Å) found in Ref. 2. Within experimental error, the fraction, x_4, of the boron atoms that are 4-fold co-ordinated in the mixed-modifier glass (0·42) is the same as that in the two end-member glasses.

Vegiri *et al*[17] have performed molecular dynamics simulations of $0{\cdot}3[(1-x)Li_2O.xCs_2O].0{\cdot}7B_2O_3$ and $0{\cdot}3[(1-x)Li_2O.xNa_2O].0{\cdot}7B_2O_3$ mixed-modifier tri-heptaborate glasses, as a function of the mixing fraction, x, and find that x_4 decreases from Li^+ to Cs^+ and exhibits a negative deviation from linearity as Cs^+ cations are substituted for Li^+ cations. No such deviation was found for the $0{\cdot}3[(1-x)Li_2O.xNa_2O].0{\cdot}7B_2O_3$ glasses. For both systems, when there is mixing of modifiers, the first (O) co-ordination shell for the higher

Table 1. Sample data and neutron wavelengths (±0·0001 Å). $\Delta\rho$ is the difference between the number density of the mixed modifier glass and that of the end-member average; i.e. the departure from Vegard's law

Sample	^{11}B enrichment	λ (Å)	$\rho°$ (atom Å^{-3})	$(\rho_M^\circ+\rho_{M'}^\circ)/2$ (atom Å^{-3})	$\Delta\rho$ (atom Å^{-3})
$Li_2O.Na_2O.5B_2O_3$	99·27%	0·5031	0·09820	0·09663	0·00157
$Li_2O.Rb_2O.5B_2O_3$	99·27%	0·5027	0·08363	0·08686	−0·00323
$Na_2O.K_2O.5B_2O_3$	99·27%	0·5031	0·08483	0·08558	−0·00075
$Na_2O.Rb_2O.5B_2O_3$	99·98%	0·5017	0·08010	0·08198	−0·00188
$Na_2O.Cs_2O.5B_2O_3$	99·27%	0·5031	0·07601	0·07868	−0·00267
$Na_2O.Ag_2O.5B_2O_3$	99·27%	0·5031	0·08758	0·09101	−0·00343

field strength cation (Li^+) becomes better defined (decreased Li–O distances), whilst that of the lower field strength cation becomes less well defined (increased Na–O or Cs–O distances). Of particular relevance to the present study is that these effects arise from the fact that the dominant network-modifying cation configurations around the ***nonbridging oxygen atoms*** in the mixed-modifier glasses involve unlike-cation species; i.e. there is a preference for unlike-cation pairing around nonbridging oxygen atoms, which is confirmed by both calculated and experimental IR spectra for the $0{\cdot}3[(1-x)Li_2O.xCs_2O].0{\cdot}7B_2O_3$ system.

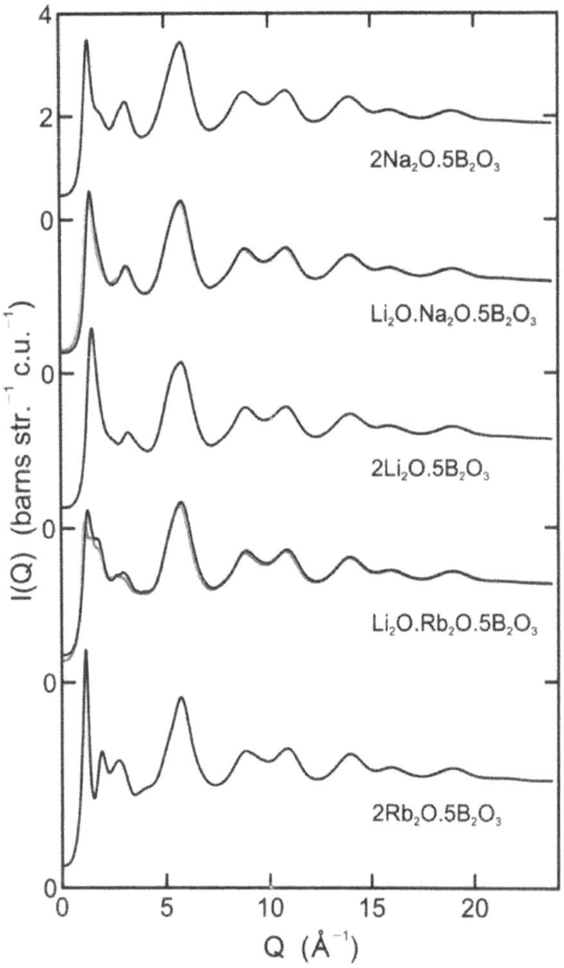

Figure 1. Corrected normalised diffraction patterns for the Li/Na and Li/Rb glasses. The black lies denote the experimental data, and the red lines the average of the two end-member glasses [Colour available online]

2. Experimental procedure

The sample preparation and characterisation, diffraction measurements, experimental corrections and data reduction were the same as those in Ref. 1. The ^{11}B enrichment and average atomic number densities, $\rho°$, for the six samples are given in Table 1, together with the incident neutron wavelengths, λ. The corrected normalised diffraction patterns, $I(Q)$, are compared with those for the relevant end-member $2M_2O.5B_2O_3$ and $2M'_2O.5B_2O_3$ glasses in Figures 1 to 3, and the corresponding real-space correlation functions, $T(r)$, are shown in Figures 4 to 6. In each case, the black lines denote the experimental data, and

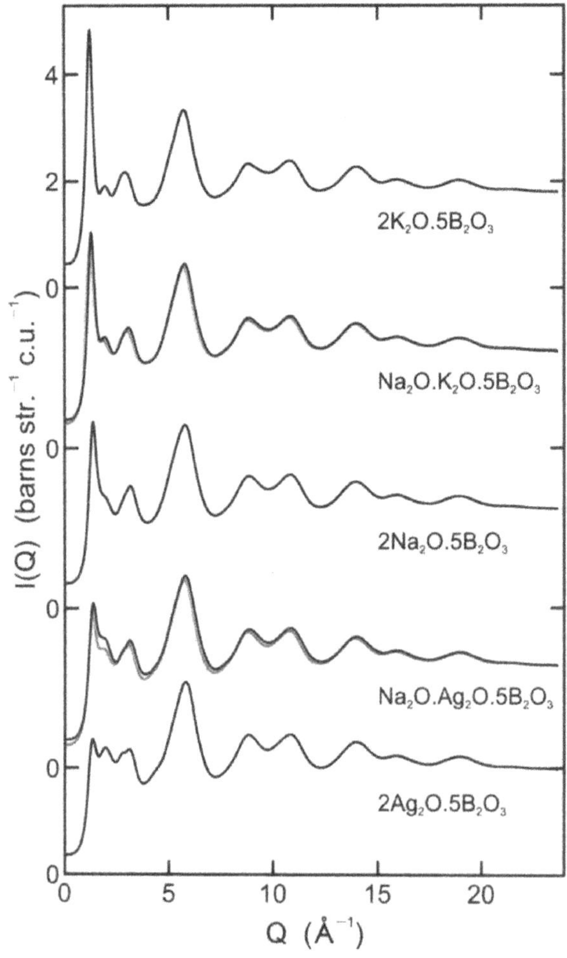

Figure 2. Corrected normalised diffraction patterns for the Na/K and Na/Ag glasses (Key as Figure 1) [Colour available online]

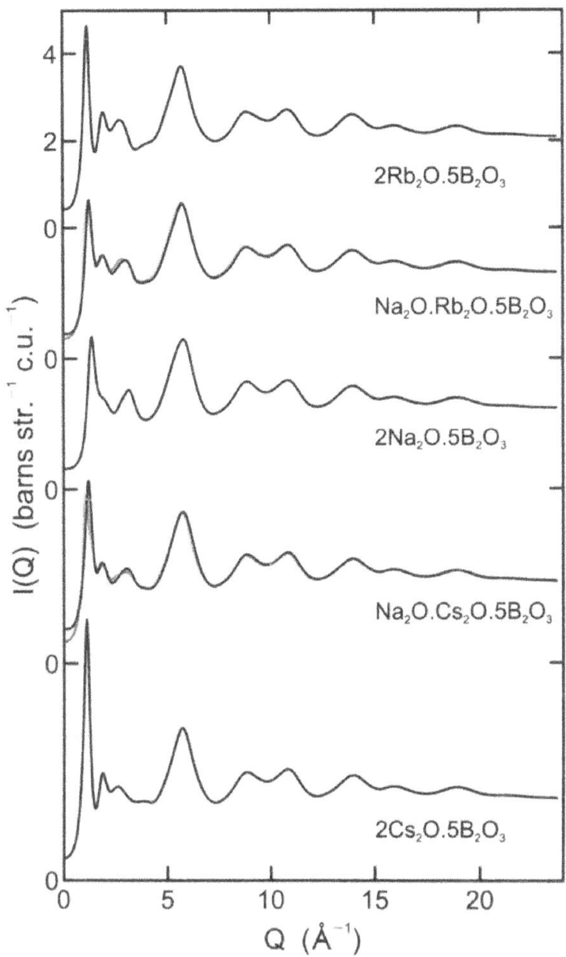

Figure 3. Corrected normalised diffraction patterns for the Na/Rb and Na/Cs glasses (Key as Figure 1) [Colour available online]

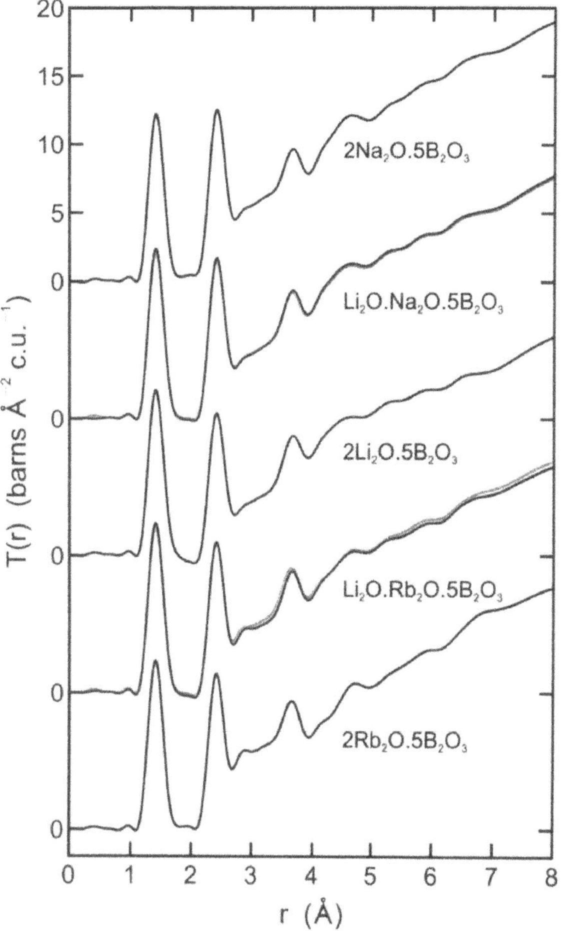

Figure 4. Real-space total correlation functions for the Li/Na and Li/Rb glasses (Key as Figure 1) [Colour available online]

the red lines the average of the curves for vitreous $2M_2O.5B_2O_3$ and $2M'_2O.5B_2O_3$.

3. Results

To avoid constant references to Part 1,[1] parameters for the single-modifier, end-member glasses that are relevant to the ensuing discussion are summarised in Table 2. Although the mixed-modifier effect is confined to transport properties, it is interesting to note from Table 1 that, with the exception of vitreous $Li_2O.Na_2O.5B_2O_3$, which exhibits a positive departure, $\Delta\rho$, from Vegard's law, the number density for all of the mixed modifier glasses is slightly lower than the average of the corresponding end-member glasses, suggesting a slightly more open network structure.

3.1. Diffraction patterns

Above the peak at ~6 Å^{-1}, the structure in the diffraction patterns (i.e. the distinct scattering) for all of the mixed-modifier and end-member glasses is remarkably similar, indicating that there is no obvious variation in the short-range order of their borate networks. In some cases, there are very small differences between the diffraction patterns (Q>6 Å^{-1}) for the mixed-modifier glasses and their corresponding

Table 2. Structural parameters for the binary end-member glasses. Q_1 is the position of the first diffraction peak, d_1 ($=2\pi/Q_1$) is the real-space periodicity, r_1 and r_2 are B–O distances from a two-peak fit to the first peak in T(r), and $n_{B(O)}$ is the total B(O) coordination number derived from the two peaks[1]

Sample	$\rho°$ (atom Å^{-3})	Q_1 (Å^{-1})	d_1 (Å)	r_1 (Å)	r_2 (Å)	$n_{B(O)}$
$2Li_2O.5B_2O_3$	0·10150	1·513±0·020	4·15±0·06	1·366	1·474	3·30
$2Na_2O.5B_2O_3$	0·09176	1·358±0·010	4·63±0·04	1·377	1·468	3·36
$2K_2O.5B_2O_3$	0·07939	1·215±0·005	5·17±0·02	1·383	1·460	3·33
$2Rb_2O.5B_2O_3$	0·07221	1·159±0·005	5·42±0·02	1·368	1·475	3·37
$2Cs_2O.5B_2O_3$	0·06559	1·094±0·005	5·75±0·03	1·369	1·473	3·31
$2Ag_2O.5B_2O_3$	0·09026	1·301±0·030	4·83±0·11	1·364	1·461	3·38

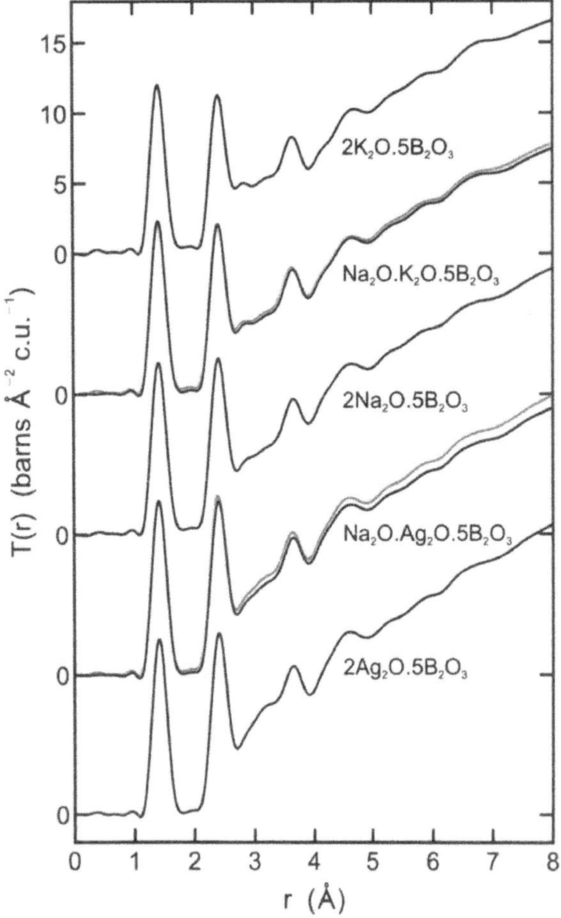

Figure 5. Real-space total correlation functions for the Na/K and Na/Ag glasses. (Key as Figure 1) [Colour available online]

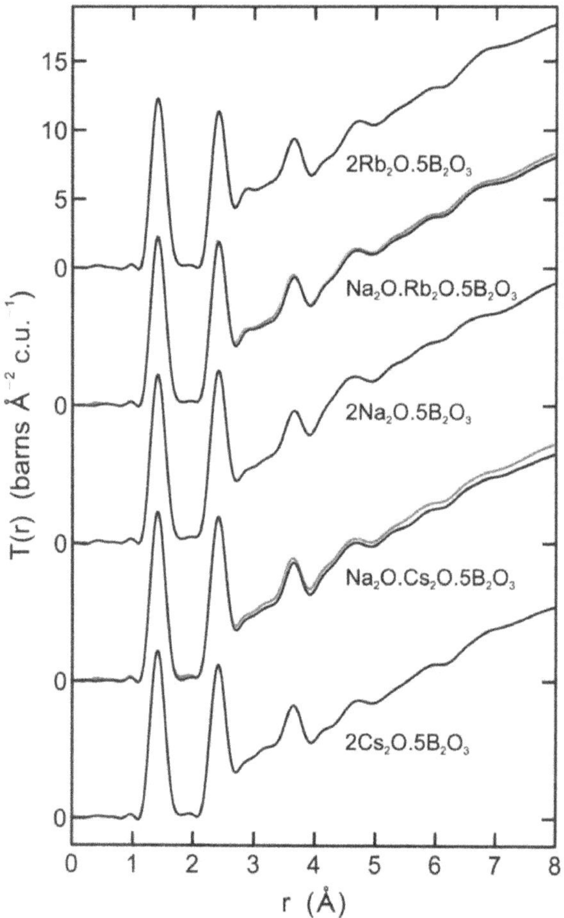

Figure 6. Real-space total correlation functions for the Na/Rb and Na/Cs glasses (Key as Figure 1) [Colour available online]

end-member averages. These are, however, mainly confined to small changes in the magnitude and fall-off of the self-scattering, which are within the experimental uncertainties and the effect of small differences in *"water"* impurity levels.

3.2. First diffraction peak

The major changes in the diffraction patterns for the mixed-modifier glasses, compared to those for the corresponding end-member averages, occur in the region of the first diffraction peak, as may be seen from Figures 7 and 8, in which the low-Q region ($0 \leq Q \leq 5$ Å^{-1}) is replotted on an expanded Q scale. In every case, the maximum of the first diffraction peak for the mixed-modifier glasses is displaced to higher Q with respect to that for the end-member average. For vitreous $Li_2O.Rb_2O.5B_2O_3$, the changes in the diffraction pattern in the region of the first diffraction peak, compared to the end-member average, are very similar to those observed by Cormier *et al*[16] for their $Li_2O.Rb_2O.4B_2O_3$ sample.

The positions, Q_1, of the first diffraction peak for the mixed-modifier glasses were obtained from a Lorentzian fit to the raw diffraction pattern, and are given in Table 3, together with the corresponding real-space periodicity, $d_1(=2\pi/Q_1)$. As an example, the fit for vitreous $Na_2O.Rb_2O.5B_2O_3$ is shown in Figure 9. In every case the peak position for the mixed modifier glass lies between those for the appropriate end-member glasses.

3.3. Zero-Q limit

Although the accuracy of the zero-Q limit, $I(0)$, of a diffraction pattern, obtained by extrapolating $I(Q)$ to zero Q, is significantly lower than that for $I(0)$ extracted from a small-angle neutron scatter-

Table 3. Parameters from fits to the first diffraction peak. Q_1 is the position of the peak and d_1 ($=2\pi/Q_1$) is the real-space periodicity

Sample	*Q_1 (Å^{-1})*	*d_1 (Å)*
$Li_2O.Na_2O.5B_2O_3$	1·439±0·020	4·37±0·06
$Li_2O.Rb_2O.5B_2O_3$	1·287±0·010	4·88±0·04
$Na_2O.K_2O.5B_2O_3$	1·277±0·010	4·92±0·04
$Na_2O.Rb_2O.5B_2O_3$	1·241±0·005	5·06±0·02
$Na_2O.Cs_2O.5B_2O_3$	1·188±0·010	5·29±0·05
$Na_2O.Ag_2O.5B_2O_3$	1·344±0·030	4·68±0·10

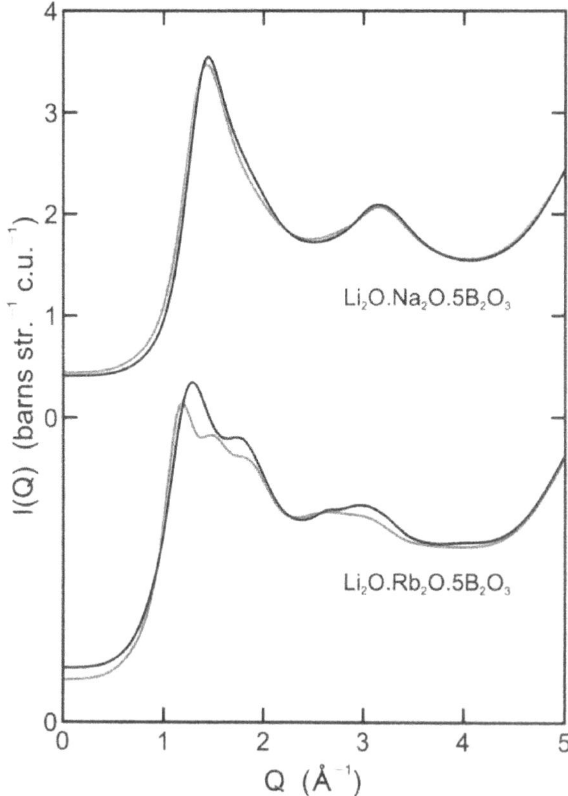

Figure 7. Low-Q region of the corrected normalised diffraction patterns for the Li/Na and Li/Rb glasses (Key as Figure 1) [Colour available online]

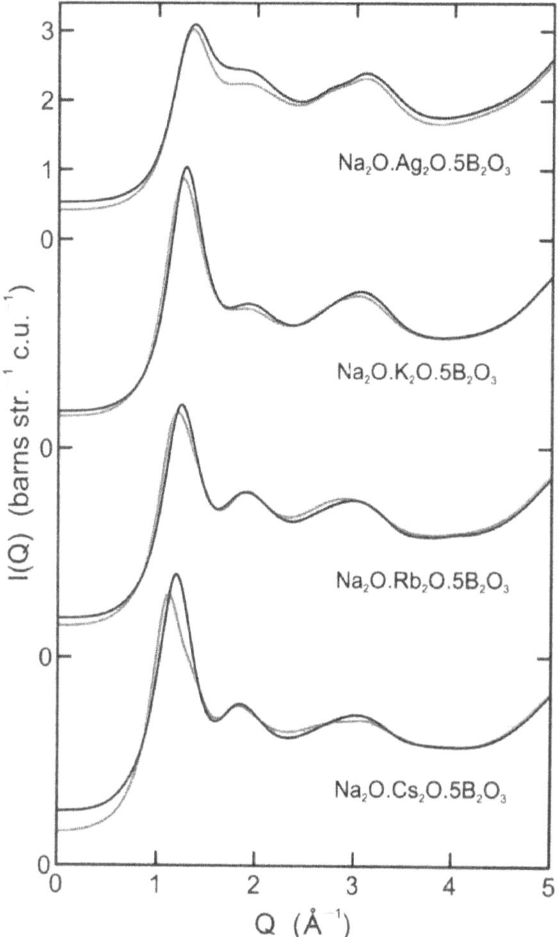

Figure 8. Low-Q region of the corrected normalised diffraction patterns for the Na/Ag, Na/K, NaRb and Na/Cs glasses (Key as Figure 1) [Colour available online]

ing (SANS) experiment, the ***relative*** values for the mixed-modifier and end-member glasses should be more reliable. Hence it is interesting to note from Figures 7 and 8 that, with the exception of vitreous $Li_2O.Na_2O.5B_2O_3$, $I(0)$ is higher for the mixed-modifier glasses than for the end-member averages, indicating increased longer-range fluctuations in scattering density. The greatest increase in these fluctuations occurs in the case of vitreous $Li_2O.Rb_2O.5B_2O_3$ and $Na_2O.Cs_2O.5B_2O_3$, for which there is the greatest difference in the size of the two network-modifying cation species.

3.4. Real-space correlation functions

The similarity between the real-space total correlation functions, $T(r)$, for the mixed modifier glasses and their corresponding end-member averages is again remarkable, the only differences being the result of the displacement of the average number density, $\rho°$, for the mixed-modifier glasses from that for their end-member averages. An attempt was made to compare the mixed-modifier glasses to their end-member averages via the difference, $\Delta T(r)$, between their real-space correlation functions, but the results were inconclusive due to statistical noise and small uncertainties in both composition and data normalisation (including neutron scattering lengths and cross-sections). Although Cormier *et al*[16] do not over-plot the real-space total correlation for their $Li_2O.Rb_2O.4B_2O_3$ sample with the end-member average, it does appear from their Figure 2, that any differences between these two curves would also be small.

Two-peak fits (*cf.* Ref. 1) were performed to the first (B–O) peak in $T(r)$ for each mixed-modifier glass, that for vitreous $Na_2O.K_2O.5B_2O_3$ being given in Figure 10. The total B(O) co-ordination numbers are summarised in Table 4, together with the inter-atomic distances, r_1 and r_2, and the reliability factor, R_χ. As

Table 4. Parameters for a two-peak fit to the first peak in $T(r)$. The estimated uncertainty on the combined coordination number $n_{B(O)}$ is ±0·05, whilst those on r_1 and r_2 are discussed in Ref. 1

Sample	r_1 (Å)	r_2 (Å)	$n_{B(O)}$	R_χ
$Li_2O.Na_2O.5B_2O_3$	1·364	1·475	3·37	0·037
$Li_2O.Rb_2O.5B_2O_3$	1·363	1·476	3·37	0·005
$Na_2O.K_2O.5B_2O_3$	1·368	1·475	3·35	0·006
$Na_2O.Rb_2O.5B_2O_3$	1·368	1·474	3·34	0·024
$Na_2O.Cs_2O.5B_2O_3$	1·367	1·471	3·33	0·005
$Na_2O.Ag_2O.5B_2O_3$	1·361	1·474	3·32	0·003

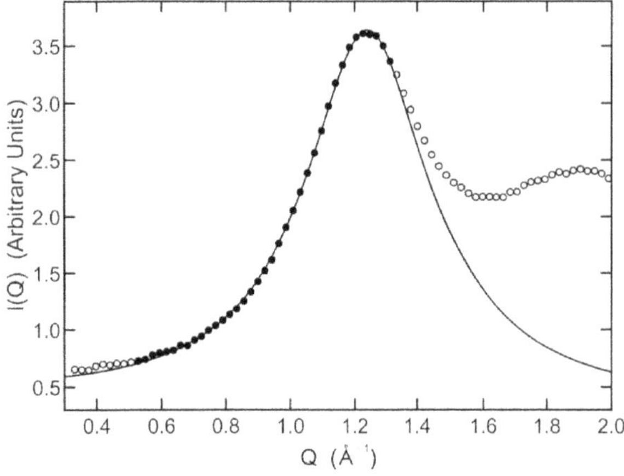

Figure 9. Lorentzian fit to the first diffraction peak for vitreous $Na_2O.Rb_2O.5B_2O_3$. Circles, experimental data and solid line, fit. The fitting range is indicated by the closed circles

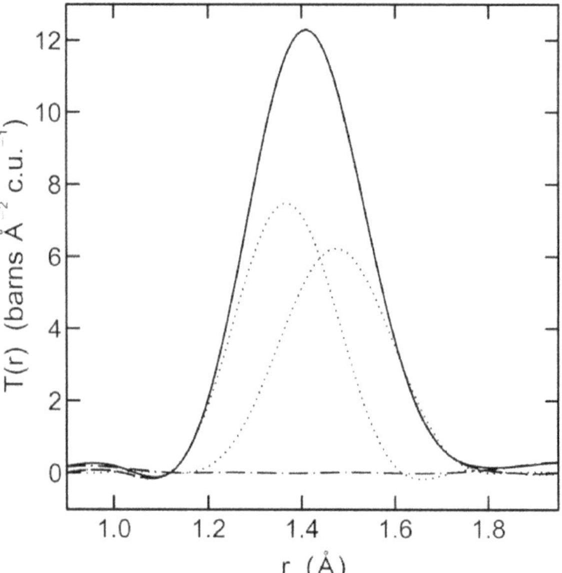

Figure 10. Peak fit to the experimental real space correlation function, T(r), *for vitreous $Na_2O.K_2O.5B_2O_3$. Black solid line, experiment; dashed line, fit; dotted lines, individual peaks and dash-dotted line, residual*

discussed in Part 1,[1] the experimental uncertainty on the overall B(O) co-ordination number is ±0·05, but the average B–O bond lengths are less reliable, and the corresponding peak widths are not meaningful, due to the systematic variation in the trigonal (r_1) and tetrahedral (r_2) bond lengths within individual superstructural units. (See Figure 13 of Ref. 1, and the associated discussion.) It is, however, interesting to note that the spread in each of the three peak fit parameters in Table 4 is less than for the single-modifier end-member glasses.

4. Discussion

There are two important factors that govern the mixed-modifier effect, *viz.* the nature of the conduction pathways and the relative spatial distribution of the two species of network-modifying cations.[3] Even though Ratai *et al*[12,13] were unable to detect any significant change in x_4 as one cation species is substituted for another, and concluded that the structural origin of the mixed-alkali effect in alkali borate glasses appears, to first order, to be unrelated to the framework structure, nevertheless there ***must*** be a close relationship between these two factors and the underlying nature of the borate network. Hence, before discussing the neutron diffraction data, it is instructive to first consider the likely form of this network for borate glasses, as revealed by crystallography and thermodynamic modelling.

4.1. Borate network

As demonstrated by Figure 14 of Ref. 1, it is extremely difficult to detect changes in the distribution of superstructural unit species, even between the five single-alkali di-pentaborate end-member glasses, and so it is even more difficult for the present mixed-modifier glasses. The striking similarity between the structure in $T(r)$ (i.e. in the differential correlation function, $D(r)$) for the six mixed-modifier glasses and the corresponding end-member averages means that their distributions of superstructural unit species are indistinguishable by diffraction within the experimental uncertainty.

The borate *"networks"* in crystalline alkali and related borates up to the diborate composition ($x_M \leq ⅓$) are summarised in Table 2 of Ref. 18. At network-modifier fractions higher than the tetraborate composition ($x_M > ⅕$), the majority of the crystalline phases have structures incorporating a single three-dimensional borate network, with the network-modifying cations located within the cages defined by this network. The situation at lower x_M ($0{\cdot}0 < x_M \leq ⅕$) is very different in that, with the exception of α-$Cs_2O.5B_2O_3$ and $3Cs_2O.13B_2O_3$ (single-layer structures, with the network-modifying cations located between the layers), all of the crystalline structures in this composition region have two independent three-dimensional interpenetrating networks. These arise as a result of the difficulty of packing planar two-dimensional boroxol groups and the large pentaborate groups, which take the form of two planar 3-membered rings at right angles to each other, around the network-modifying cations (see Ref. 8). In this type of structure, there are no *"closed"* network cages and the network-modifying cations are located in the space separating the two networks. Note that crystalline $Li_2O.2B_2O_3$ also has two interpenetrating networks, presumably due to the difficulty of efficiently packing diborate groups around the very small Li^+ cations.

A similar pattern is exhibited by the mixed-modifier crystalline phases mentioned in Section 1.1. The mixed-modifier Na/K triborates and the diborate $Li_2O.Rb_2O.4B_2O_3$ have structures based on a single three-dimensional borate network, whilst $0{\cdot}4Na_2O.0{\cdot}6Ag_2O.4B_2O_3$ (mixed-modifier tetraborate) and $Li_2O.Na_2O.4B_2O_3$[11] (mixed-modifier diborate) have two independent interpenetrating networks. The structures of the mixed-modifier di-pentaborates $Na_2O.Cs_2O.5B_2O_3$,[5] $Na_2O.Tl_2O.5B_2O_3$[6] and $K_2O.Cs_2O.5B_2O_3$[5] all take the form of *"thick"* layers with the larger network-modifying cations (Cs^+ or Tl^+) incorporated within these layers and the smaller network-modifying cations (Na^+ or K^+) situated between the layers.

The likely existence of locally-independent interpenetrating networks in borate glasses was first raised by Krogh-Moe,[19] by comparison with the structure of crystalline $Li_2O.2B_2O_3$. However, with the possible exception of $Li_2O.Na_2O.5B_2O_3$, such a structure is unlikely for the present mixed-modifier di-pentaborate glasses (x_M=0·286). These five glasses almost certainly have a structure predominantly comprising a single three-dimensional network with the network-modifying cations confined within the network cages, since the boroxol group fraction is effectively zero at the di-pentaborate composition, and the pentaborate group fraction is likely to be small, as may be seen from Figure 2 of Ref. 8, which shows the distribution of (super)structural units in sodium borate glasses obtained from thermodynamic modelling. It is thus interesting to conjecture as to the shape of the *"cages"* that might be present in such a borate network, especially since they are likely to be highly non-spherical, and take the form of *"flattened disks"*, due to the planar nature of the 3-membered rings incorporated into boroxol, pentaborate and triborate groups, and the rigid shape of the relatively large diborate group. A thick-layer structure similar to that of the three crystalline mixed-modifier di-pentaborate glasses is highly unlikely for the corresponding glasses, due to the difficulty of incorporating extended layers and the 5-connected di-pentaborate groups (*cf.* Section 1.1) into an isotropic disordered network.

The results obtained for x_4 by Ratai *et al*[12,13] for their Na_2O–Cs_2O–B_2O_3 glasses (~0·39) are consistently less than the fraction (0·43) given by Equation (1), and so it is noteworthy that the B(O) co-ordination number, $n_{B(O)}$, obtained for vitreous $Na_2O.Cs_2O.5B_2O_3$ (Table 4) is also at the lower end of the range for the present mixed-modifier glasses, and that this is also true for vitreous $2Cs_2O.5B_2O_3$[1] (Table 2). Hence, from the combined NMR and neutron diffraction data it may be concluded that there is almost certainly a higher fraction of $BO\emptyset_2^-$ basic structural units (nonbridging oxygen atoms) in the Cs_2O-containing glasses. A possible explanation for this is that the introduction of $B\emptyset_4^-$ basic structural units leads to network compaction, whereas the large Cs^+ cations require greatly expanded network cages, thus leading to the formation of nonbridging oxygen atoms (broken B–Ø–B bridges).

The lowest B(O) co-ordination number in Table 4 is for vitreous $Na_2O.Ag_2O.5B_2O_3$ ($n_{B(O)}$=3·32), suggesting an increased fraction of $BO\emptyset_2^-$ basic structural units (nonbridging oxygen atoms) compared to the $2Na_2O.5B_2O_3$ ($n_{B(O)}$=3·36) and $2Ag_2O.5B_2O_3$ ($n_{B(O)}$=3·38) end-member glasses (Table 2). This is consistent with the results of Varsamis *et al*[20] for a series of $xNa_2O.(1-x)Ag_2O.3B_2O_3$ mixed modifier triborate glasses, who find that network-modifying cation mixing leads to a minimum in x_4 (Ref. 20, Figure 2).

4.2. Network-modifying cation environment

The main differences between the mixed-modifier glasses and their corresponding end-member averages occur in reciprocal space, in the region of the first diffraction peak (Figures 7 and 8). Due to peak overlap, the first peak maximum for the end-member average diffraction patterns, $I(Q)$, lies between those for the individual end members. For vitreous $Li_2O.Na_2O.5B_2O_3$ and $Na_2O.Ag_2O.5B_2O_3$ the maximum is close to the average of the two end members, but for the other mixed modifier glasses, where the difference in cation size is much greater than for $Na_2O.Ag_2O.5B_2O_3$, the maximum for the mixed-modifier glass is much closer to that for the larger cation.

In every case, the first diffraction peak for the mixed-modifier glass moves to slightly higher Q, and its height increases, relative to the end-member average. For vitreous $Na_2O.K_2O.5B_2O_3$, $Na_2O.Rb_2O.5B_2O_3$ and $Na_2O.Cs_2O.5B_2O_3$, the peak shift increases with the size difference between the two network modifying cation species, whereas that for vitreous $Li_2O.Na_2O.5B_2O_3$ and $Na_2O.Ag_2O.5B_2O_3$ remains relatively small (~0·01 Å^{-1}). The greatest effect occurs for the $Li_2O.Rb_2O.5B_2O_3$ glass, for which the real glass has only one shoulder on the high-Q side of the first diffraction peak, whilst the end-member average has two (Figure 7).

First consider the sequence $Na_2O.K_2O.5B_2O_3$, $Na_2O.Rb_2O.5B_2O_3$, $Na_2O.Cs_2O.5B_2O_3$ (Figure 8). As discussed in Refs 1 and 2, the position of the first diffraction peak can be correlated with the periodicity, $d_1(=2\pi/Q_1)$, (Tables 2 and 3) associated with a succession of network cage walls, and hence with the size of the cages containing the network-modifying cations. The fact that the first diffraction peak moves to higher Q, with respect to the end-member average, and has a position that is dominated by the contribution from the larger network-modifying cation, means that the ***average*** width/thickness (perpendicular to the above cage walls) of the network cages containing this larger cation must ***contract***, whilst the increase in peak height indicates that the contributions from the two cation species have moved closer together. This is contrary to the findings of Ratai *et al*[12,13] (*cf.* Section 1.2) who conclude that, on average, there is an ***expansion*** of the sites occupied by the larger cation, accompanied by a contraction of those for the smaller cation; i.e. that the maximum should move to lower Q, and the end-member peaks should move further apart, leading to a reduction in peak height. It might be argued that the contraction of the larger cation sites is due to the fact that there are fewer of the larger cations to fit into the network cages. However, the opposite appears to be true, as may be seen from the data for the Cs_2O–B_2O_3 system in Table 2 of Ref. 2. With decreasing Cs_2O fraction, the periodicity associated with the network sites occupied by the Cs^+ cations systematically increases from 5·75 Å at the cæsium di-pentaborate composition to 6·03 Å for vitreous cæsium enneaborate (x_M=0·1). This can be correlated with the presence at lower x_M of a significant fraction of planar boroxol groups and (relatively) large pentaborate groups, both of which are more difficult to efficiently pack around the Cs^+ cations than are the smaller triborate and non-planar diborate groups.[8]

In comparing the neutron diffraction data with the molecular dynamics simulations of Vegiri *et al*,[16] two important points should be noted. First, the fraction, x_2, of $BO\O_2^-$ basic structural units (i.e. of nonbridging oxygen atoms) in the MD simulations is very much higher (~0·11–0·24), and that of $B\O_4^-$ tetrahedra very much lower (~0·19–0·33) than the corresponding fractions extracted from the neutron and NMR[12,13,15] data. (Equation (1) predicts that x_4 should be equal to 0·43 at a total modifier fraction of 0·3.) The second problem is that, in the real glasses, the $B\O_4^-$ tetrahedra are almost exclusively incorporated into superstructural units, which has a profound effect on their packing around the network-modifying cations. Although Vegiri *et al*[17] do not comment on the superstructural unit fractions in their simulated structures, the general consensus would suggest that they are very much lower than in the real glasses, if not zero. The overall result of the higher x_2, lower x_4 and the lack of superstructural units is that the average network-modifying cation environment will be unduly biased towards those sites involving nonbridging oxygen atoms, which Vegiri *et al*[17] associate with the contraction of the Li(O) and the expansion of the Na(O) and Cs(O) co-ordination shells.

The structure of crystalline $Li_2O.Na_2O.4B_2O_3$[11] is similar to that of $Li_2O.2B_2O_3$, and is based on two independent interpenetrating networks of diborate groups. However, Mączka *et al*[11] note that partial substitution of Na^+ for Li^+ leads to a reduction in the crystalline symmetry from tetragonal to orthorhombic. The two interpenetrating networks produce *"distinct tunnels"* that are selectively occupied by the Li^+ (along [010]) and Na^+ (along [100]) cations. This suggests that the structure of vitreous $Li_2O.Na_2O.5B_2O_3$ may similarly involve locally-independent interpenetrating networks, and hence that the average network-modifying cation sites may differ from those in the other five glasses. In this context, it is interesting to note that vitreous $Li_2O.Na_2O.5B_2O_3$ is the only mixed-modifier glass to exhibit a positive deviation from Vegard's law, and that the difference between $I(Q)$ for this glass and its end-member average is very small (not much larger than the experimental uncertainty), even within the region of the first diffraction peak, and as $Q \rightarrow 0$. This may be related to the fact that both the Li^+ and Na^+ cations are small enough to fit between the locally-independent networks without significant network distortion. It is also noteworthy that this is the only mixed-modifier glass for which $I(0)$ is lower than the end-member average, indicating lower overall spatial fluctuations in scattering density.

4.3. Average network number density

The variation of the network number density (number of borate structural units per unit volume), ρ_N°, with the network-modifier fraction, x_M, has already been discussed for binary alkali borate systems in Refs 4 and 21, where it was shown that, for the larger network-modifying cations, once any suitably-sized available network sites/cages have been occupied, the addition of further cations leads to site expansion and a reduction in ρ_N°. Thus ρ_N° first increases with increasing x_M, passes through a maximum and then decreases. The larger the network-modifying cation, the lower is the value of x_M at which the maximum occurs, as may be seen from the data of Kodama[22,23] plotted in Figure 34(A) of Ref. 4. (The data for the cæsium borate system do not exhibit a maximum, but decrease immediately from the value at zero x_M, suggesting that there are extremely few, if any, cages in the initial borate (i.e. vitreous B_2O_3) network that are naturally large enough to contain a Cs^+ cation.)

In connection with the discussion in Section 4.1, it is interesting to note from Figure 34(A) of Ref. 4 that all of the alkali borate glasses except Li_2O–B_2O_3 exhibit a slight dip in ρ_N° in the neighbourhood of the tetraborate composition; i.e. at the crossover point for the corresponding crystalline borates between two independent interpenetrating networks and a single network (*cf.* Table 2 of Ref. 18). As may be seen from Figure 2 of Ref. 8, for the Na_2O–B_2O_3 system, this is also approximately the point at which the fraction of the most difficult to pack pentaborate groups is a maximum, and also that where the planar boroxol group fraction goes to zero. This association of the dip with the presence of pentaborate groups is consistent with the fact that the chemical structure of lithium borate glasses (Figure 30 of Ref. 4) does not include lithium pentaborate chemical groupings.[18]

With the exception of vitreous $Li_2O.Na_2O.5B_2O_3$, the average network number density is lower than the end-member average (see Table 1), which is contrary to what might be expected for a statistically-random spatial distribution of occupied network cages, especially given the contraction of the cages containing the larger network-modifying cations. Consider the case of a binary random packing of spheres (i.e. of spheres of two different sizes), where it is possible to achieve a higher overall packing density relative to a packing of single-sized spheres. By analogy, therefore, if the distribution of the occupied network cages is statistically random, with respect to the occupying cation species, it would be expected that the overall packing of the network cages would be more efficient, and hence that ρ_N° for the mixed-modifier glass would be higher than the end-member average, and not the reverse. The fact that ρ_N° is lower than the end-member average thus mitigates against a random distribution of occupied cages, and hence a statistically-random spatial distribution of the two network-modifying cation species.

Suppose, on the other hand, that the mixed-modifier glass exhibits nanoheterogeneity in the form of separate regions occupied by each cation species. In this case, at first sight, it might be expected that the average network number density would be much closer to the end-member average. However, this is to neglect the interconnecting regions between the single-modifier regions, which are likely to be more disordered, and hence of lower ρ_N°, thus leading to an average network number density lower than the end-member average, as observed experimentally. Again, therefore, the data are consistent with separate single-modifier regions.

4.4. Nanoheterogeneity

It is now generally accepted that glasses having more than one component are nanoheterogeneous and incorporate fluctuations in both density and composition, leading to corresponding fluctuations in scattering density, as revealed by neutron and x-ray data recorded at low scattering vector magnitudes, Q. According to the cybotactic theory (*cf.* Refs 24 and 25), there are fluctuations in the degree of order with more-ordered regions, in which the structure and composition approach those of related crystalline phases, being interconnected by regions of lower order. An alternative way of viewing these fluctuations in scattering density is in terms of the chemical structure, as revealed by thermodynamic modelling (see Parts 1[1] and 2[2]). In this case, the structure is conceptualised as consisting of an assembly of chemical groupings with the same stoichiometries, and similar short-range structures, to those of the crystalline compounds forming in the given system. Thus, for systems comprising two or more components with different chemical natures (basic and acidic), the cybotactic and chemical groupings are essentially equivalent.

The subject of nanoheterogeneity in binary single-modifier borate glasses has already been discussed in Parts 1[1] and 2.[2] For these glasses, nanoheterogeneity arises from the spatial arrangement of the various chemical groupings that make up their chemical structure. In the case of ternary mixed-modifier glasses, the situation is much more complicated, and difficult to resolve, due to the lack of thermodynamic data for the ternary mixed-modifier crystalline compounds. In principle, there are four possibilities:

(i) Random spatial mixing of individual single-modifier chemical groupings,
(ii) Random mixing of individual single-modifier and mixed-modifier chemical groupings,
(iii) A structure dominated by mixed-modifier chemical groupings,

and

(iv) Clusters of chemical groupings involving a single network-modifying cation species; i.e. larger regions having overall structures characteristic of the individual end-member glasses.

Whilst the discussion of Sections 4.1 and 4.3 specifically addressed the possibility of nanoheterogeneity in the form of single-modifier regions, it is important to realise that these single-modifier regions, if present, will also themselves be nanoheterogeneous, as in the case of the single-modifier end-member glasses (*cf.* Refs 1 and 2). This latter nanoheterogeneity arises as a direct consequence of the distribution of the chemical groupings that make up their chemical structure.

If chemical nanoheterogeneity is present, e.g. in the form of single-modifier regions or chemical groupings, and the volume fraction occupied by the interfacial regions is small, the diffraction pattern, $I(Q)$, and real-space correlation function, $T(r)$, can be approximated by a sum of those for the individual regions/groupings in the correct proportions to make up the overall composition. The method of additivity[26–28] assumes the converse to be true, namely that additivity of the component diffraction patterns or correlation functions is indicative of chemical nanoheterogeneity. However, whilst this converse is not necessarily true, additivity does indicate possible nanoheterogeneity, whereas the absence of additivity would mean that significant nanoheterogeneity of the proposed form is extremely unlikely. The great similarity between the structures in both $I(Q)$ at higher Q (Q>6 Å) and $T(r)$ for the mixed-modifier glasses and corresponding end-member averages, therefore, is ***not inconsistent*** with the presence of nanoheterogeneity in the form of separate end-member regions.

For all of the present mixed-modifier glasses, except $Li_2O.Na_2O.5B_2O_3$, the zero-Q limit, $I(0)$, is higher for the mixed-modifier glass than for the corresponding end-member average; i.e. the presence of two network-modifying cation species leads to an increase in the fluctuations in scattering density, which is greatest for vitreous $Li_2O.Rb_2O.5B_2O_3$ and $Na_2O.Cs_2O.5B_2O_3$. The fact that the neutron scattering length, $\bar{b}$, is larger for rubidium than for cæsium (0·709×10^{-12}, *cf.* 0·542×10^{-12} cm[29]), whilst the increase in $I(0)$ is greater for vitreous $Na_2O.Cs_2O.5B_2O_3$ than for vitreous $Na_2O.Rb_2O.5B_2O_3$, suggests that the magnitude of the increase in scattering density fluctuations mainly results from the difference in the size of the two network-modifying cation species (i.e. from fluctuations in ρ_N°), rather than from the difference in their neutron scattering lengths.

4.5. Spatial distribution of network-modifying cations

As indicated in Section 1.2, the NMR data of Eckert and co-workers[12,13,15] suggest a statistically-random distribution of the two network modifying cation species among the occupied sites for their $Li_2O–Na_2O–B_2O_3$ and $Na_2O–K_2O–B_2O_3$ glasses,[12,13] but their $Na_2O–Rb_2O–B_2O_3$ results[12,13,15] indicate a non-statistical mixing of the Na^+ and Rb^+ network-modifying cations, and some preference for like-cation interactions (domain segregation effects). Given that the NMR data refer to the ***average*** sites for the two cation species, it is interesting to note that, in the $Li_2O–Na_2O–B_2O_3$ and $Na_2O–K_2O–B_2O_3$ systems, mixed modifier crystalline phases exist at relatively low modifier fractions ($x_M \lesssim 1/3$ *cf.* Section 1.1), whereas this is not the case for the $Na_2O–Rb_2O–B_2O_3$ system. Hence it might be expected that mixed-modifier chemical groupings play a significantly larger role in the chemical structure of the $Li_2O–Na_2O–B_2O_3$ and $Na_2O–K_2O–B_2O_3$ glasses than they do in the $Na_2O–Rb_2O–B_2O_3$ glasses, which is consistent with the NMR data.[12,13,15]

Kamiya *et al*[30] have investigated the structure of $Tl_2O–B_2O_3$ and $Na_2O–Tl_2O–B_2O_3$ glasses using the x-ray heavy-element technique, and conclude that the Tl^+ cations in both systems exist as pairs or larger clusters with a shortest $Tl^+–Tl^+$ distance of ~4·0 Å. As discussed in Ref. 4, however, Prins[31] has pointed out that the identification of prominent maxima in the electronic differential radial distribution function with $Tl^+–Tl^+$ interactions is not necessarily correct, as may be seen by comparing the area under the $Tl^+–Tl^+$ peak for a pair of such cations (6400 el.^{2}) with that for $Tl^+–O$ interactions, assuming a $Tl^+(O)$ co-ordination number of 8 (10240 el.^{2}). The Tl^+ cation radius for 8-fold co-ordination (1·60 Å[32]) is identical to that for Rb^+, and the network number density, ρ_N°, for vitreous $2Tl_2O.5B_2O_3$ lies between those for vitreous $2K_2O.5B_2O_3$ and $2Rb_2O.5B_2O_3$, suggesting that the size of the network cages occupied by the Tl^+ cations is similar to that occupied by Rb^+ cations, or between those for K^+ and Rb^+ cations. Hence, if the two Tl^+ cations of a given pair occupy adjacent cages, the position of the first diffraction peaks for vitreous $2K_2O.5B_2O_3$ and $2Rb_2O.5B_2O_3$ would indicate a $Tl^+–Tl^+$ distance of ~5·3 Å, rather than the value of ~4·0 Å obtained by Kamiya *et al.*[30]

A further x-ray study using the heavy element technique has been reported by Hanson & Egami[33] for a sodium-cæsium trisilicate glass ($0·595Na_2O.0·405Cs_2O.3SiO_2$). Comparing with a binary cæsium silicate glass, these authors conclude that the two network-modifying cation species segregate submicroscopically into separate regions in the glass, and that their data do not support theories of the mixed-modifier effect that require relatively immobile unlike-cation pairs.

Based on the relatively large neutron scattering length for rubidium (0·709×10^{-12} cm[29]), Cormier *et al*[16] have employed arguments analogous to the x-ray heavy element technique to identify features at 4·7 Å and 7 Å in their neutron real-space total correlation functions for their $Rb_2O–B_2O_3$ and $Li_2O.Rb_2O.4B_2O_3$ glasses as being due to Rb–Rb interactions. The first of these distances is again significantly shorter than the values obtained in the present work for $Rb_2O–B_2O_3$ glasses up to the diborate composition (5·41–5·48 Å[2]), and for vitreous $Li_2O.Rb_2O.5B_2O_3$ (4·88 Å – *cf.* Table 3), as obtained from the fits to the first diffraction peak. If Cormier *et al*'s assignment[16] is correct, then their Rb–Rb distance may correspond to that between adjacent Rb^+ cations occupying the same flattened/elongated network cage. However, this seems unlikely, because similar features occur at ~4·7 and ~6·7 Å in the real-space correlation functions for most of the present single-modifier[1] and mixed modifier glasses, suggesting that they arise from the borate network, rather than being due to distances separating network modifying cations.

In crystalline $0·4Na_2O.0·6Ag_2O.4B_2O_3$,[34] the Na^+ cations randomly replace the Ag^+ cations in crystalline $Ag_2O.4B_2O_3$ to form a solid solution, but it should be noted that this solid solution does not extend all the way to $Na_2O.4B_2O_3$, since the latter is not isostructural with $Ag_2O.4B_2O_3$. However, the fact that vitreous $Na_2O.Ag_2O.5B_2O_3$ has the largest departure, $\Delta\rho$, from Vegard's law in respect of its number density demonstrates that the distribution of the Na^+ and Ag^+ network-modifying cations among the available sites in this glass is not statistically-random; i.e. the glass cannot be considered as a solid solution.

4.6. Mixed-modifier effect

The initial objective of the present neutron diffraction study was to provide structural information upon which to base a theory of the mixed-modifier effect and, in particular, to probe the environment (Section 4.2) and the spatial distribution (Section

4.5) of the network-modifying cations. However, in practice, complications arise due to the chemical nanoheterogeneity associated with the chemical structures of both the single-modifier[1] and mixed-modifier glasses, as revealed by the zero-Q limit of $I(Q)$; e.g. in Figures 7 and 8. It is therefore necessary to understand that neutron diffraction and NMR spectroscopy[12,13,15] experiments yield only the ***average*** environment and cation-cation separations for each network-modifying cation species. As will be explained in Ref. 3, this is important because only a very small fraction of the network-modifying cations are actually mobile and contribute to the cationic conductivity. These ***mobile*** network-modifying cations are situated in the more highly disordered interfacial/boundary regions between chemical groupings, and hence their environments are more distorted, and less well defined, than the average found experimentally. This leads[3] to a theory of the network-modifying cationic conduction and mixed-modifier effect, based on Ingram's ***cluster-bypass model***[35–38] plus the ***dynamic structure model*** of Bunde *et al*,[39,40] in which Ingram's clusters[35–38] are replaced by the (chemical) groupings that define the chemical structure of any given glass.

5. Conclusions

Neutron diffraction data for the six mixed-modifier di-pentaborate glasses reveal only small differences between these glasses and the corresponding end-member averages, except in the region of the first diffraction peak, and in respect of the zero-Q limit, $I(0)$. The fact that the structure in the diffraction patterns, $I(Q)$, at higher Q, and that in the real-space total correlation functions, $T(r)$, is indistinguishable from those for the end-member averages indicates that, within the experimental uncertainty, the overall distribution of (super)structural unit species for each of the mixed-modifier glasses is closely similar to that for their end-member average, which is ***consistent*** with a chemical structure based primarily on single-modifier chemical groupings.

The values of x_4 obtained for the mixed-modifier glasses are mostly in accord with that expected in the absence of nonbridging oxygen atoms. However, x_4 for vitreous $Na_2O.Cs_2O.5B_2O_3$ is particularly low, as is x_4 for the $2Cs_2O.5B_2O_3$ end-member glass,[1] suggesting the presence of nonbridging oxygen atoms. Similarly, the fact that x_4 for vitreous $Na_2O.Ag_2O.5B_2O_3$ is lower than x_4 for vitreous $2Na_2O.5B_2O_3$ and $2Ag_2O.5B_2O_3$ confirms the conclusion of Varsamis *et al*[20] that cation mixing in the $Na_2O–Ag_2O–B_2O_3$ system leads to a reduction in x_4.

A small shift in the position, Q_1, of the first diffraction peak for all of the mixed-modifier glasses to higher Q, relative to their end-member averages, leads to the conclusion that the ***average*** sites occupied by the larger network-modifying cation species slightly contract in a direction perpendicular to the walls that separate successive network cages. The corresponding real-space periodicity, d_1 ($=2\pi/Q_1$), for vitreous $Li_2O.Rb_2O.5B_2O_3$ and $Rb_2O–B_2O_3$ yields a minimum separation between Rb^+ network-modifying cations occupying adjacent network cages[1] that is significantly larger than the Rb–Rb distance found by Cormier *et al*,[16] using the neutron diffraction equivalent of the x-ray heavy element technique.

An important aspect, common to the structures of the binary single-modifier and ternary mixed-modifier glasses, is that they are nanoheterogeneous, exhibiting fluctuations in both composition and density, as revealed by the zero-Q limit of the diffraction pattern, $I(Q)$. For all of the mixed-modifier glasses except $Li_2O.Na_2O.5B_2O_3$, the magnitude of the resulting fluctuations in scattering density is larger than that for the end-member average, again suggesting a chemical structure based predominantly on single-modifier chemical groupings. As will be explained in an accompanying paper,[3] the nanoheterogeneous chemical structure plays an important role in respect of the network-modifying cationic conductivity of both the single and mixed-modifier glasses, due to the fact that only a very small fraction of the network-modifying cations are actually mobile[3] and occupy sites that are more disordered than the average sites revealed by diffraction and NMR experiments, thus leading to a theory of the cationic conductivity and mixed-modifier effect, based on a reformulated combination of the ***cluster-bypass***[35–38] and ***dynamic structure***[39,40] ***models*** of Ingram and co-workers.

Acknowledgements

This work was supported financially by the UK EPSRC, and CES and JLS would like to thank the EPSRC and BNFL, respectively, for PhD studentships. The sample preparation at Coe College was funded under US National Science Foundation grant DMR 0904615, and the authors are also grateful to the Institut Laue-Langevin for neutron beam time and experimental assistance.

References

1. Wright, A. C., Sinclair, R. N., Stone, C. E., Shaw, J. L., Feller, S. A., Kiczenski, T. J., Williams, R. B., Berger, H. A., Fischer, H. E. & Vedishcheva, N.,M. *Phys. Chem. Glasses: Eur. J. Glass Sci. Technol. B*, 2012, **53**, 191.
2. Wright, A. C., Sinclair, R. N., Stone, C. E., Shaw, J. L., Feller, S. A., Williams, R. B., Fischer, H. E. & Vedishcheva, N. M. *Eur. J. Glass Sci. Technol. B: Phys. Chem. Glasses*, 2014, **55**, 74.
3. Wright, A. C. & Vedishcheva, N. M. *Phys. Chem. Glasses: Eur. J. Glass Sci. Technol. B*, 2015, **56**, 98.
4. Wright, A. C. *Phys. Chem. Glasses: Eur. J. Glass Sci. Technol. B*, 2010, **51**, 1.
5. Tu, J.-M. & Keszler, D. A. *Inorg. Chem.*, 1996, **35**, 463.
6. Penin, N., Touboul M. & Nowogrocki, G. *J. Alloys Compd.*, 2004, **363**, 104.
7. Zachariasen, W. H. *J. Am. Chem. Soc.*, 1932, **54**, 3841.
8. Wright A. C. & Vedishcheva, N.M. *Phys. Chem. Glasses: Eur. J. Glass*

Sci. Technol. B, 2013, **54**, 147.
9. Wright, A. C. *Int. J. Appl. Glass Sci.*, 2013, **4**, 214.
10. Bubnova, R. S. & Filatov, S. K. *Vysokotemperaturnaya Kristallokhimiya Boratov i Borosilikatov (High-temperature Crystal Chemistry of Borates and Borosilicates)*, Nauka, St. Petersburg, 2008.
11. Mączka, M., Waśkowska, A., Majchrowski, A., Żmija, J., Hanuza, J., Peterson, G. A. & Keszler, D. A. *J. Solid State Chem.*, 2007, **180**, 410.
12. Ratai, E., Chan, J. C. C. & Eckert, H. *Phys. Chem. Chem. Phys.*, 2002, **4**, 3198.
13. Ratai, E.-M., Janssen, M., Epping, J. D., Chan, J. C. C. & Eckert, H. *Phys. Chem. Glasses*, 2003, **44**, 45.
14. Chryssikos, G. D. & Kamitsos, E. I. In: *Borate Glasses, Crystals & Melts*, Eds A. C. Wright, S. A. Feller & A. C. Hannon, Soc. Glass Technol., Sheffield, 1997, p. 128.
15. Epping, J. D., Eckert, H., Imre, A. W. & Mehrer, H. *J. Non-Cryst. Solids*, 2005, **351**, 3521.
16. Cormier, L., Calas, G. & Beuneu, B. *J. Non-Cryst. Solids*, 2007, **353**, 1779.
17. Vegiri, A., Varsamis, C.-P. E. & Kamitsos, E. I. *Phys. Rev. B*, 2009, **80**, 184202.
18. Wright, A. C. *Int. J. Appl. Glass Sci.*, 2015, **6**, 45.
19. Krogh-Moe, J. *Phys. Chem. Glasses*, 1965, **6**, 46.
20. Varsamis, C.-P. E. Kamitsos E. I. & Chryssikos, G. D. *Phys. Chem. Glasses*, 2000, **41**, 242.
21. Wright, A. C. Dalba, G. Rocca F. & Vedishcheva, N. M. *Phys. Chem. Glasses: Eur. J. Glass Sci. Technol. B*, 2010, **51**, 233.
22. Feller, S., Affatigato, M., Giri, S. K., Basu, A. & Kodama, M. *Phys. Chem. Glasses*, 2003, **44**, 117.
23. Kodama, M. & Kojima, S. *Phys. Chem. Glasses: Eur. J. Glass Sci. Technol. B*, 2014, **55**, 1.
24. Wright, A. C. *J. Non-Cryst. Solids*, 2014, **401**, 4.
25. Wright, A. C. *Int. J. Appl. Glass Sci.*, 2014, **5**, 31.
26. Porai-Koshits, E.A. *O stekloobraznom sostoyanii (rentgenovskoye issledovaniye) [On the vitreous state (an x-ray diffraction study)]*, PhD thesis, Kazan', 1942.
27. Porai-Koshits, E. A. *Compt. Rend. (Dokl.) Acad. Sci. URSS*, 1943, **40**, 346. [tr. of *Dokl. Akad. Nauk. SSSR*, 1943, **40**, 394.]
28. Ida, M. *Proc. Phys. Math. Soc. Jpn*, 1944, **26**, 55.
29. Sears, V. F. *Neutron News*, 1992, **3**, 26.
30. Kamiya, K., Sakka, S., Mizuno, T. & Matusita, K. *Phys. Chem. Glasses*, 1981, **22**, 1.
31. Prins, J. A. In: *Physics of Non-Crystalline Solids*, Eds J. A. Prins, North-Holland, Amsterdam, 1965, p. 39.
32. Shannon, R. D. & Prewitt, C. T. *Acta Crystallogr. B*, 1969, **25**, 925.
33. Hanson, C. D. & Egami, T. *J. Non-Cryst. Solids*, 1986, **87**, 171.
34. Krogh-Moe, J. *Acta Crystallogr.*, 1965, **18**, 77.
35. Ingram, M. D. *Phys. Chem. Glasses*, 1987, **28**, 215.
36. Ingram, M. D. *Philos. Mag. B*, 1989, **60**, 729.
37. Ingram, M. D., Chryssikos G. D. & Kamitsos, E. I. *J. Non-Cryst. Solids*, 1991, **131–133**, 1089.
38. Ingram, M. D., Mueller W. & Torge, M. In: *The Physics of Non-Crystalline Solids*, Eds L. D. Pye, W. C. LaCourse & H. J. Stevens, Taylor & Francis, London, 1992, p. 441.
39. Bunde, A., Ingram M. D. & Maass, P. *J. Non-Cryst. Solids*, 1994, **172–174**, 1222.
40. Bunde, A., Ingram M. D. & Russ, S. *Phys. Chem. Chem. Phys.*, 2004, **6**, 3663.

DOI: 10.13036/17533562.56.3.098 Phys. Chem. Glasses: Eur. J. Glass Sci. Technol. B, June 2015, 56 (3), 98–107

A thermodynamically-based structural model for the cationic conductivity and mixed-modifier effect in borate glasses

Adrian C. Wright[1]

J.J. Thomson Physical Laboratory, University of Reading, Whiteknights, Reading, RG6 6AF, UK

Natalia M. Vedishcheva

Institute of Silicate Chemistry of the Russian Academy of Sciences, Nab. Makarova 2, St. Petersburg, 199034, Russia

Manuscript received 10 December 2014
Revised version received 5 February 2015
Accepted 5 February 2015

Based on the neutron diffraction data discussed in an accompanying paper (pp 85–97), and on the results of thermodynamic modelling using the model of associated solutions, it is concluded that the (network-modifying) cationic conduction, and the mixed-modifier effect, in alkali and related borate glasses is best described by a reformulated combination of the **cluster-bypass** *and* **dynamic structure models**. *In this revised approach, the clusters are replaced by the (chemical) groupings that characterise the chemical structure of the glasses, and the network-modifying cation transport is via conduction pathways located in the interfacial/boundary regions that separate these groupings. The proposed model can explain the rapid fall-off in the cationic conductivity on the initial introduction of the second network-modifying cation species, and the increasing magnitude of the conductivity depression with increasing total modifier content. A comparison with the case of the corresponding silicate glasses indicates that the spatial distribution of the network-modifying cations in silicate glasses is more complicated than suggested by the well known two-dimensional schematic diagram representing the modified random network theory.*

1. Introduction

"It (the mixed-alkali effect) *may perhaps be the most thought-provoking topic in all of glass science.*"
Arun K. Varshneya[1]

The mixed-modifier effect, also known as the mixed-alkali effect, the mixed-mobile-ion effect[2] or the poly-alkali effect (Russian workers), has been well documented in the literature; e.g. in the reviews by Isard,[3] Day[4] and Ingram,[5] and in the textbook by Varshneya.[1] Various structural theories have been advanced for the network-modifying cationic conductivity and the mixed-modifier effect, but none of them is universally accepted, the problem being that no one theory can account for all of the experimental data.[3–5] The fact that they were mostly developed for silicate glasses means that borate glasses provide an important test of these theories, due to the inherent differences between their bonding and structure as compared to those of the corresponding silicate glasses. In the present paper, the results of a combined neutron diffraction and thermodynamic modelling study of both single-[6,7] and mixed-modifier[8] borate glasses are interpreted in terms of a reformatted combination of the ***cluster-bypass***[5,9–11] and ***dynamic structure***[12,13] ***models*** of Ingram and co-workers. Comments are also included as to the extension of this revised model to the case of silicate glasses.

The mixed-modifier effect involves the extremely nonlinear variation in transport properties as one (guest) network-modifying cation species, M'^{+}, is substituted for another (host), M^{+}, at constant total network-modifying cation content. In the case of the ionic conductivity, the initial introduction of the guest cation species, M'^{+}, leads to an extremely rapid drop in conductivity, whilst the diffusion rate for both cation species systematically increases with their cation fraction, there being no extremum in the mobility of either species. Characteristically, the mobility of the majority (host) cation is greater than that of the minority (guest) cation, up to the crossover point, where the two mobilities are equal.

The underlying basis of eight theories of the mixed-modifier effect are summarised in Table 4 of Ref. 4, where Day[4] divides them roughly into two categories. Four of these theories emphasise structural features of the glass network, and concentrate on either the network-modifying cation distribution or size. The other four theories link the mixed-modifier effect to differences in the bonding and co-ordination environment between the two cation species, but are not based on any specific structural difference between single- and mixed-modifier glasses. Instead,

[1] Corresponding author. Email a.c.wright@reading.ac.uk
Based on a paper presented at VIII Int. Conf. on Borate Glasses, Crystals and Melts, Pardubice, Czech Republic, 30 June–2 July 2014

they assume some form of interaction between unlike-cation species, which lowers their mobility. The concept of unlike cation pairing was introduced by Hendrickson & Bray,[14,15] and is also a component of various defect models.[5,16] The original suggestion was based on the development of an *"interaction energy"* between adjacent cations of differing mass, but this would predict a mixed-cation effect on varying the ^6Li:^7Li ratio in a lithium borate glass, which is not observed.[5] Thus Ingram[5] concludes that the conductivity anomaly is not controlled by a single cationic parameter, such as mass or radius. On the other hand, Varshneya[1] comments that any proposed mechanism for the mixed-modifier effect must somehow involve a strong interaction between different network-modifying cation species, such that a large number of the majority (host) cations is immobilised (or the majority cation mobility is very much reduced) by each minority (guest) cation.

2. Borate *versus* silicate glasses

Before discussing the network-modifying cationic conduction and the mixed-modifier effect in borate glasses, it will be helpful to briefly summarise the way in which their bonding and structure influence their transport properties.[17] Unlike silicates, where the sp^3 hybrid orbitals on the silicon atoms lead to a tetrahedral structural unit and a stable octet configuration, boron atoms have three electrons in their valence shell and hence can only form three covalent bonds (sp^2 hybrid orbitals) with oxygen, thus yielding a planar (triangular) structural unit. This leaves the boron atoms with only six electrons in their valence shell, and a vacant p_z orbital with its axis perpendicular to the plane of the triangle. It is this vacant p_z orbital that is the fundamental reason for the marked differences between the properties of borate and silicate glasses,[17,18] since it can accept a donor bond from an oxygen sp^3 lone-pair orbital, which results in a precarious balance between the co-ordination numbers of 3 and 4, and a very low activation energy for their inter-conversion via the equilibrium[†]

$$BØ_4^- \rightleftharpoons BOØ_2^- \quad (1)$$

where Ø represents a bridging oxygen atom. Hence viscous flow can proceed via a combination of bond switching and the inter-conversion of trigonal and tetrahedral boron atoms, e.g. as shown by the mechanisms in Figures 11 and 12 of Ref. 17. These same mechanisms also allow the transport of negative charge, and/or the modification of the network cage structure during electrical (ionic) conduction.

For alkali silicate glasses, it has been proposed that clustering of the network-modifying cations, together with their associated nonbridging oxygen atoms, leads to their being concentrated into channels, or into two-dimensional boundary regions between silica-rich clusters, as represented by the modified random network model.[19] However, in the case of Na_2O–B_2O_3 glasses, NMR data[20,21] indicate a near homogeneous spatial distribution of the Na^+ cations for $x_M \geq 0·15$, with *"some degree"* of clustering only being found at lower cation concentrations. It has therefore been suggested[17] that the much higher resistivity for the borate glasses is due to their lack of similar cationic conduction channels, especially since they are much closer in temperature to T_g. They also have only half as many network-modifying cations per network structural unit, for a given x_M, as silicate glasses. On the other hand, it is interesting to conjecture whether there is a contribution to their electrical conductivity from the movement of negative charge via the mechanisms of Figures 11 and 12 of Ref. 17.

3. Cationic conduction

There are several important aspects of network-modifying cationic conduction that are common to both single- and mixed-modifier glasses. The first point to note (*cf.* Ref. 16) is that, upon quenching from the supercooled melt, the network-modifying cations, M^+ or M^{2+}, do ***not*** passively occupy sites (cages) formed by the surrounding vitreous network, but that they tailor their environment to suit their own particular requirements. Thus the size (radius) of the network-modifying cations determines that of the sites in which they reside, and their nominal positive charge must be compensated by the appropriate number of negative charges ($BØ_4^-$ tetrahedra or nonbridging oxygen atoms), thus determining the cation co-ordination number, $n_{M(O)}$. The fact that the network-modifying cations configure their environment to suit their size/charge leads to the expectation that there should, on average, be very little difference between the initial (as quenched) distribution of sites for a given network-modifying cation species in the end-member and mixed-modifier glasses; i.e. in mixed-modifier glasses, the average local environment of M^+ cations is largely unaffected by the addition of M'^+ cations,[22] and *vice versa*.

In the case of borate glasses and crystals at low x_M ($x_M \lesssim 1/3$), where the negative charges are almost entirely confined to $BØ_4^-$ tetrahedra incorporated into superstructural units, the charge compensation is complicated by the difficulty of efficiently packing the (relatively) large superstructural units around (especially the smaller) network-modifying cations, such that the positive and negative charges are in close proximity.[23] This leads to a wider distribution of network-modifying cation environments/sites, compared to the corresponding silicates, even in the

[†] Note that the convention used here for (super)structural unit formulæ follows that of Ref. 15.

crystalline state; e.g. in β-$Na_2O.3B_2O_3$, the unit cell includes Na^+ cations with co-ordination numbers of 6, 7 and 8.[24] Note also that similar packing considerations determine the size/shape of the initially vacant sites/cages that participate in cationic diffusion.

On quenching a glass incorporating a single network-modifying cation species, M^+, there are two types of cation site, those occupied by M^+ cations (denoted **M**) and those that are unoccupied (E).‡ The diffusion of the network-modifying cations under the influence of an applied electric field involves their successive movements (jumps) from an occupied to an unoccupied site, followed by the relaxation of both sites, as characterised by a relaxation time, τ:[22]

$\mathbf{M} \rightarrow E$

or

$\mathbf{M} \rightarrow M$

where E and M (in *italics*), respectively, represent a site that was initially vacant (following the initial quenching from the super-cooled melt) and one that has been recently vacated by an M^+ cation. Thus each site retains a *"memory"* of its previous status, until it has completely relaxed. This concept of site relaxation, after being occupied/vacated by a network-modifying cation, as introduced by Maass *et al*,[22] subsequently evolved into the ***dynamic structure model*** of Bunde *et al*.[12,13]

The site relaxation can be of two distinct types; *viz.* displacive and reconstructive (topological). Displacive relaxation does not involve a change in the topology of the borate network, and proceeds via changes in bond length (small), bond angles and bond torsion angles. It is relatively rapid, since it is coupled to the atomic thermal vibrations. The relaxation time for topological relaxation, on the other hand, is significantly longer, because it involves bond switching, as in the case of viscous flow. For this reason, topological relaxation is more important at higher temperatures, and especially in the supercooled melt above T_g, but it may not be totally arrested at lower temperatures. However, as pointed out by Ingram,[25] for glasses where it is thought that the M^+ cations are mainly jumping directly into M sites, the observed ***activation volumes*** for cation transport relate quite closely to the size of the M^+ cations involved, which is exactly what is expected for displacive relaxation. (See, for example, Table 1 of Ref. 26 and Ref. 27.) For borate glasses, bond switching is governed by two conflicting factors. It is facilitated by the low (activation) energy required for the change in the boron atom co-ordination number, from 4 to 3, or *vice versa* (e.g. via Equ. (1)), whereas it is obstructed by the need to break-up and reform superstructural units, a process that requires a much higher activation energy. If the jump time is greater than τ, a previously occupied vacant E site will have relaxed, and so the jump will be more difficult, and relaxation of the newly occupied site will be necessary. The time for this relaxation will depend upon whether topological relaxation is required in addition to displacive relaxation. Maass *et al*[22] comment that, if an original E site is occupied by an M^+ or M'^+ cation for more than a certain time (i.e. long enough for topological relaxation to occur), then on being vacated it will initially relax displacively to an M or M' site. If it occurs, further (topological) relaxation back to its original E configuration will require a longer relaxation time. Bunde *et al*[13] distinguish between two different types of E site, those that were occupied by an M^+ cation immediately following quenching from the supercooled melt, and those that were initially unoccupied. In the former case, the M site formed upon the exit of the M^+ cation will subsequently relax to an E' site. They point out[13] that such an E' site is more favourable in respect of occupation by an M^+ cation than those of the (latter) E sites that do not incorporate an adjacent compensating negative charge. Note that the vacant (E) sites formed during quenching exhibit a wider distribution of sizes/shapes than those that are occupied, and some may be remote from a negative charge, especially at lower network-modifier contents. Some of the E sites may be large enough to take a network-modifying cation, but the majority will not, as may be inferred from the discussion of the average network number density, ρ_N°, in Section 4.3 of Ref. 8, and from the evolution of the first diffraction peak in the neutron diffraction patterns with increasing x_M, for the glasses containing the larger network-modifying cations.[6,7] The smaller the network-modifying cation, the greater will be the number of E sites that are large enough to accept it without requiring topological relaxation.

The cation mobility is determined both by the relative spatial distribution of the occupied and unoccupied sites, which determines the conduction pathways, and by the site relaxation times. The mobility of a specific network-modifying cation will be lower if it moves from an occupied site to one that has been vacant for longer than the relaxation time, τ, and higher if it can jump to a site that has (recently) been vacated within time τ. This facilitates diffusion along well-defined conduction pathways, where the cations can move rapidly into recently-occupied vacant sites within the relaxation time, τ.

As will be discussed further in Section 4, an extremely important fact in relation to the cationic conductivity is that only a very small fraction of the network-modifying cations are actually mobile.[28]

‡ Note that the symbol convention employed here differs from that of Maass *et al*,[22] and in the later papers by Bunde *et al*,[12,13] in that the cations A^+ and B^+ (Section 6) are replaced by M^+ and M'^+, respectively, and the initially empty site $\bar{C}$ has become E. Vacant sites are indicated by italics, rather than a superscript bar, and occupied sites are denoted in bold type.

Ravaine *et al*[29] have pointed out that the important parameter, with respect to the transport properties of the M^+ network-modifying cations in glasses, is the activity of the oxide M_2O, rather than its analytical concentration, which is very important, because the activity differs from the concentration by several orders of magnitude. This idea has been further developed by Shakhmatkin *et al*,[30] who demonstrated that the electrical conductivity is mainly determined by the chemical potential of M_2O and hence, since the systems in question can be considered as ideal solutions, by the equilibrium concentration of ***"unreacted"*** M_2O.

To identify the origin of this *"unreacted"* M_2O, it is necessary to understand that glass formation by a basic network modifier plus an acidic network former is not the result of melting, but of chemical reactions,[31] and that the chemical structure of the ensuing glass-forming (supercooled) melt is characterised by a system of chemical equilibria of the form[23,32]

$$m\mathrm{M_2O} + n\mathrm{B_2O_3} \rightleftharpoons m\mathrm{M_2O}.n\mathrm{B_2O_3} \quad (2)$$

with an equilibrium constant given by

$$K=\exp(-\Delta G_f^\circ/RT) \quad (3)$$

ΔG_f° is the standard Gibbs free energy of formation of $m\mathrm{M_2O}.n\mathrm{B_2O_3}$ from the oxides, and its magnitude is such that equilibrium (2) is heavily biased towards the right hand side, but a very small concentration of M_2O remains *"unreacted"*, together with its associated *"unreacted"* B_2O_3. This *"unreacted"* M_2O is constantly being formed and consumed during the ongoing equilibria. Note also that, at low network-modifier fractions, x_M, the chemical structure is dominated by a much larger concentration of unreacted B_2O_3, as may be seen from Figure 5.5 of Ref. 32.

An alternative way of presenting the above equilibria, which will be helpful in Section 4, is to rewrite the equations in terms of ***basic chemical units***,[17,23] as represented by a superstructural unit plus the necessary number of network-modifying cations. Thus, for the three common superstructural units incorporating $BØ_4^-$ tetrahedra, and found in modified borate glasses up to at least the diborate composition ($x_M \lesssim ⅓$),

$$½(\mathrm{M_2O} + 5\mathrm{B_2O_3}) \rightleftharpoons \mathrm{M^+} + \mathrm{B_5O_6Ø_4^-} \text{ (pentaborate group)} \quad (4)$$

$$½(\mathrm{M_2O} + 3\mathrm{B_2O_3}) \rightleftharpoons \mathrm{M^+} + \mathrm{B_3O_3Ø_4^-} \text{ (triborate group)} \quad (5)$$

and

$$\mathrm{M_2O} + 2\mathrm{B_2O_3} \rightleftharpoons 2\mathrm{M^+} + \mathrm{B_4O_5Ø_4^{2-}} \text{ (diborate group)} \quad (6)$$

These basic chemical units then agglomerate to form the (chemical) groupings that characterise the chemical structure of any given (borate) glass. Thus it is to be expected that the internal structure of the chemical groupings will be based on the relevant superstructural units.[23] Note also that, within the supercooled melt, further equilibria exist between different superstructural unit species, as discussed in Ref. 17.

4. Conduction pathways

Another extremely important aspect of network-modifying cation conduction in glasses concerns the location of the conduction pathways. Clearly, given the fact that only a very small minority of these cations are actually involved in the conduction process, the conduction pathways must be restricted to specific regions of the network structure, which only comprise a small fraction of the total volume. In other words, the conduction pathways must be related to the fluctuations in both density and composition that characterise the longer-range structure of glasses having more than one component, and must reflect the chemical structure of the melt at the fictive temperature, T_f. The situation for borate glasses up to the diborate composition ($x_M \lesssim ⅓$) is more complex than that for silicate glasses, as may be seen from the discussion of Section 4.1 of Ref. 8. At low x_M ($x_M \lesssim 0{\cdot}2$), where the difficulties in packing the large pentaborate and planar boroxol groups suggest a structure based on locally-independent interpenetrating networks, there will be few *"closed"* cages, and the network-modifying cations can move through the space between these networks. At higher x_M, and specifically at the di-pentaborate composition, the structure is much more likely to predominantly involve a single three-dimensional network, such that electrical (cationic) conductivity will proceed via the passage of the network-modifying cations through *"gateways"* linking adjacent *"flattened"* cages.

Two of the theories summarised by Day (Table 4 of Ref. 4) rely on some form of chemical heterogeneity, either involving spatial fluctuations in the network-modifying cation concentration (alkali-rich and alkali-poor regions) or phase separation. However, as Isard[3] points out, observable phase separation is not a prerequisite for the mixed-modifier effect. Isard[3] also rejects a much older explanation of the mixed-modifier effect based on the early crystallite theory.

In his review, Ingram[5] discusses the network-modifying cationic conduction in glasses and the mixed-modifier effect in terms of the strained-mixed-cluster model of Goodman,[33–35] in which glass formation is the result of the crystallisation of the super-cooled melt being frustrated by the presence of crystalline-like clusters (i.e. cybotactic/chemical groupings[8]) corresponding to two or more different crystalline phases with very similar free energies. Ingram[5] argues that the network-modifying cation pathways are located in the more-disordered interconnecting material between the clusters (cybotactic/chemical groupings), as indicated by the

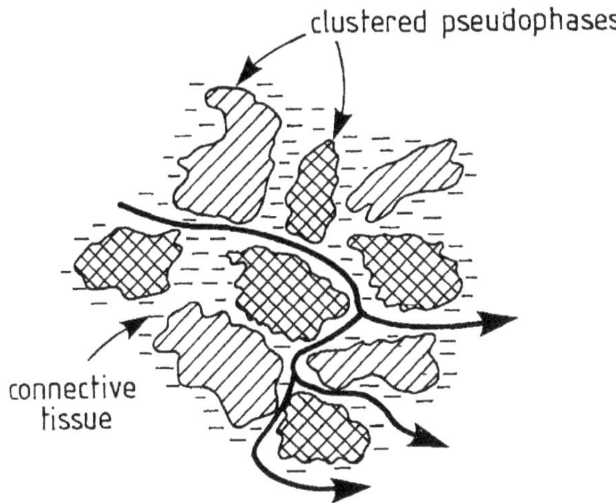

Figure 1. Cationic conduction within the connective tissue between more ordered clusters, which constitute identifiable pseudophases (perhaps 9–15 Å) resembling crystalline phases[10]

two-dimensional schematic diagram in Figure 1.[10] This is consistent with these interconnecting regions having a lower density, and a more open network, leading to more rapid site relaxation and hence higher cationic mobility. In subsequent papers (e.g. Refs 9–11), Ingram has further developed his initial ideas into a generalised ***cluster-bypass model***, with the conduction pathways being confined to the connective tissue between clusters. The exact nature of the clusters themselves is still a matter of conjecture, but it is suggested that they are more highly ordered than the connective tissue, perhaps with structures related to that of corresponding crystalline phases, or having some form of increased non-crystalline order.[9] Ingram's[5,9–11] clusters can thus be identified with the chemical groupings as defined by thermodynamic modelling.

If, as outlined in Section 3, the cationic conductivity is mainly determined by the chemical potential of M_2O, and hence by the equilibrium concentration of ***"unreacted"*** M_2O,[30] the concentration of the ***mobile*** network-modifying M^+ cations will be much too small for them to form M_2O chemical groupings in their own right. By definition, however, they cannot be incorporated into the (chemical) groupings corresponding to the various $M_2O–B_2O_3$ chemical compounds, and so they must be confined to the interfacial/boundary regions between adjacent chemical groupings; i.e. at the point where the various equilibrium reactions that define the chemical structure of the supercooled melt are actually taking place. Hence the number of conduction pathways will be very much lower than if conduction took place through the chemical groupings themselves.

It has already been indicated that the interfacial/boundary regions are likely to be of lower network number density, and more disordered, than the chemical groupings, which will facilitate the relaxation processes that accompany the diffusion of the network-modifying cations. In the case of borate glasses at relatively low x_M ($x_M \lesssim 1/3$), the structures of the chemical groupings will be dominated by superstructural units, which will make the bond switching accompanying topological relaxation much more difficult. On the other hand, the interconnecting borate network between chemical groupings is likely to incorporate a much higher concentration of independent basic structural units, thus greatly reducing the activation energy required for bond switching (e.g. Equation (1)), since it is unnecessary to break-up and reform superstructural units. In addition, the highly-probable existence of locally-independent interpenetrating networks at lower x_M suggests that the number of bridging bonds between adjacent chemical groupings of different type is likely to be reduced, possibly due to the presence of an increased fraction of $BØ_2^-$ basic structural units (nonbridging oxygen atoms), thus resulting in fewer *"gateways"* to impede the motion of the mobile network-modifying cations. Note that the overall content of nonbridging oxygen atoms at low x_M is exceedingly small, and comparable to that of the *"unreacted"* M_2O and that, in this composition region, these nonbridging oxygen atoms are also confined to the interfacial/boundary regions. (It is assumed that the activation energy required to pass through a gateway is significantly larger than that for moving between adjacent cation sites within the same *"cage"*.) The *E* sites are also likely to be larger in the interfacial/boundary regions, thus facilitating their occupation by a network-modifying cation. Similarly, the M(O) co-ordination polyhedra will be much more disordered than those of the corresponding cations located within the chemical groupings. Thus the sites occupied by the ***mobile*** network-modifying cations, and through which they move, will ***not*** correspond to the ***average*** site for that cation species as determined experimentally, for example from diffraction and EXAFS experiments. This leads to the inevitable conclusion that ***any theory of the mixed-modifier effect that is based solely on these average sites must be treated with extreme caution.***

5. Conduction mechanism

As discussed in Section 3, the diffusion of the M^+ network-modifying cations in a single-modifier glass, under the influence of an applied electric field, involves their jumping from an occupied (**M**) site to one that is unoccupied (*E* or *M*), followed by any necessary relaxation of both sites during the ensuing relaxation time, τ. After jumping, the network-modifying cations then have two options in respect of their next jump. They can either return to their original site, unless this has already been occupied by another cation, or jump to another vacant site. Note, however, that, if a cation has jumped to a site that is too small (i.e. a higher energy site), there is an incentive for it to return to its original (if still unoccupied) site, until the relaxation of its new site is complete, thus reducing the probability of a *"successful"* jump. An important factor governing this choice is that the applied electric field is several orders of magnitude lower than that which a positively-charged network-modifying cation experiences from its associated negatively-charged nonbridging oxygen atom or $BØ_4^-$ tetrahedron ($2{\cdot}3\times10^{10}$ V m^{-1} at a distance of 2·5 Å from a single electronic charge). Hence there is only a very small bias towards jumps in the direction consistent with the applied electric field.

A much greater influence on the favourability of any given jump is the equilibrium site energy following relaxation, and this is primarily determined by the proximity of any negative charges to the cation position. If this negative charge is incorporated into the *"gateway"* between the sites, their energies are likely to be comparable. On the other hand, a jump from a site with a nearby negative charge to one without such a charge will be much less favourable, unless an appropriate negative charge can be generated by topological relaxation.

In this connection, it is important to note that the M^+ network-modifying cations in the interfacial/boundary regions between chemical groupings are of two distinct types, those from the *"unreacted"* M_2O, together with their associated O^{2-} anions, and those that form part of the boundaries of the chemical groupings. During the conduction process, these two cation species can exchange identities, which is merely a low-temperature manifestation of equilibria (2) and (4)–(6) in Section 3. Similarly an O^{2-} anion can swap identities with a nonbridging oxygen atom, via the donation of one of its lone pairs into the vacant p_z orbital on a trigonal boron atom, and the retraction of this boron atom's bond to its original nonbridging oxygen atom. Thus, although the O^{2-} anions themselves are too large to migrate under the influence of the applied field, this does provide a possible mechanism for the movement of negative charge. Note that, whereas the exchange of identities between an O^{2-} anion and a nonbridging oxygen atom gives rise to the movement of a negative charge, a simple change of identity between M^+ network-modifying cations does not in itself lead to the transport of a positive charge.

A highly-stylised two-dimensional schematic representation of the conduction mechanism for a single-modifier glass is shown in Figure 2. In reality, the various chemical groupings (blue) will be non-spherical and of varying size, such that they fit together much more closely (*cf.* Figure 1), thus greatly reducing the volume fraction of the interfacial/boundary regions (yellow). The motion of the network-modifying cations, M^+, from left to right, under the influence of the electric field, is denoted by the black arrows, which reveal that conduction occurs along multiply-branched, interconnected pathways within the (yellow) interfacial/boundary regions between the individual (blue) chemical groupings.

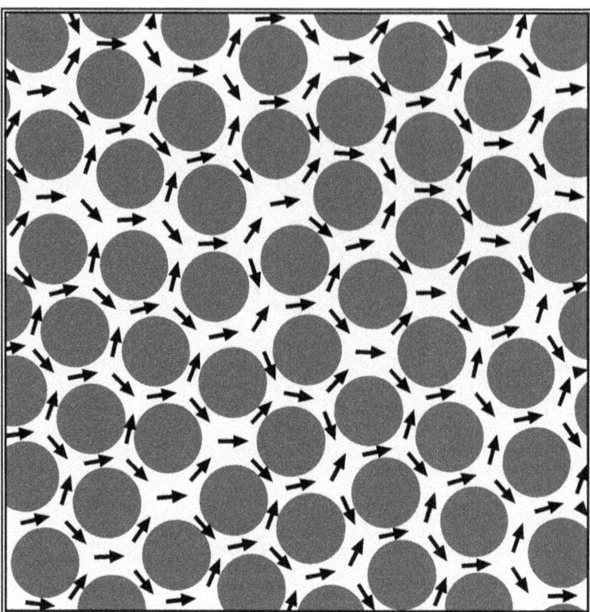

Figure 2. The conduction mechanism for a binary single-modifier glass. The chemical groupings are shown in blue, and the interfacial/boundary regions in yellow. The flow of the network-modifying cations from left to right, under the influence of the applied electric field, is represented by the black arrows [Colour available online]

6. Mixed-modifier glasses

On quenching a ternary mixed-modifier glass, M_2O–M'_2O–B_2O_3, from the supercooled melt, there will be three different cation sites: those occupied by M^+ cations (**M**), those containing M'^+ cations (**M'**), and those that are initially empty (*E*). On the application of an electric field, the cation mobility will be determined by the activation energy required for the cations to jump to a neighbouring empty site and, if necessary, the time for that site to relax once it has been occupied. In the following description of the cation motion, it is assumed that the M'^+ cation is significantly larger than the M^+ cation. In this case, the cation diffusion involves the co-operative jumping of cations from their occupied to vacant sites and, if this occurs within a time that is short with respect to the relaxation time, τ, the following jumps will be (relatively) easy:

(i) $\mathbf{M} \rightarrow M$

(ii) $\mathbf{M} \rightarrow M'$

(iii) $\mathbf{M'} \rightarrow M'$

whereas

(iv) $\mathbf{M'} \rightarrow M$

will be (much) more difficult, due to the longer relaxation time required after jumping. The ease of jumping into a site (*E*) that was initially empty (i.e. immediately following quenching),

(v) $\mathbf{M} \rightarrow E$

and

(vi) $\mathbf{M'} \rightarrow E$

will depend on the size of the M^+ and M'^+ cations relative to that of the initially empty site, *E*, and the extent of the required relaxation. Which of the above jumps, (i)–(vi), predominate will depend upon the nature of the conduction pathways and, in particular, upon whether the two cation species have independent pathways or share the same pathways.

The majority of the structural models and computer simulations of the mixed-modifier effect assume independent cation pathways for the two network-modifying cation species, and that these pathways can be blocked by immobile unlike-cation pairs/dimers.(14,15) For example, the existence of independent pathways is supported by Sakka *et al*,(36) who have demonstrated that the Tl^+ contribution to the conductivity of mixed-modifier Na_2O–Tl_2O–B_2O_3 glasses can be predicted from that of Tl_2O–B_2O_3 glasses with the same Tl^+ content. Similarly, Ratai *et al*(20,21) have concluded that network-modifying cation transport proceeds via diffusion pathways consisting of well-matched target sites that were previously occupied by the same cation species, such that each cation species moves along independent diffusion paths. Hence the overall mobility is determined by the connectivity of these paths. In the case of random mixing, this connectivity decreases with decreasing cation fraction, and so the mobility of the relevant cation species also decreases. Ratai *et al*(21) do, however, admit that there are two aspects of the mixed-alkali effect that are difficult to explain:

(i) the relative magnitude of the mobility depression increases sensitively with total network-modifying cation content (i.e. with decreasing unreacted network former (B_2O_3)),

and

(ii) the cation mobility decreases most rapidly with small substitutions of the second cation.

In this connection, they suggest that a slight compression of the M^+ sites, in the neighbourhood of those accommodating the larger cation species, means that these M^+ sites are *"geometrically distorted"* and thus represent unfavourable target sites for the M^+ cations originating from regular M^+ sites, leading to a marked reduction in the M^+ path connectivity. This, they argue,(21) may at least in part explain the rapid fall-off in conductivity on the initial introduction of the second network-modifying cation species.

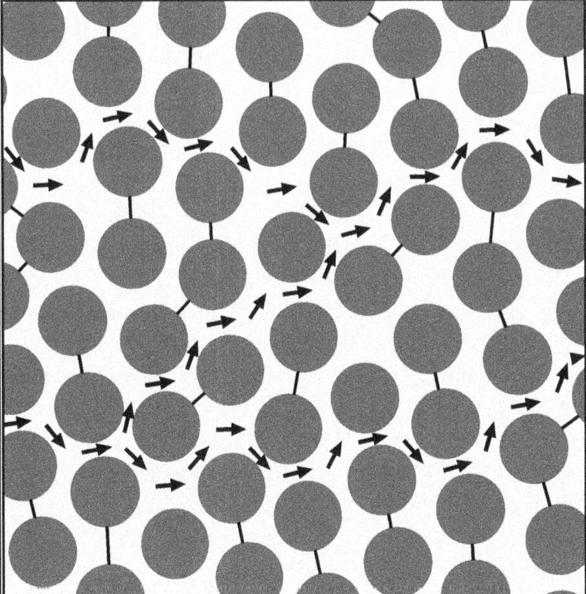

Figure 3. The flow of the M^+ cations (black arrows) through a ternary M_2O–$M_2'O$–B_2O_3 glass, at very low $M_2'O$ fractions, $x_{M'}$. The thin black lines linking the blue M_2O–B_2O_3 chemical groupings denote the impediments formed by unlike-cation pairs/dimers [Colour available online]

7. Low $M_2'O$ fractions

As just indicated, one of the features of the mixed-modifier effect that is most difficult to explain is the extremely rapid fall-off in conductivity on the introduction of only very small amounts of the second modifier, $M_2'O$. This rapid fall-off can be ascribed to the fact that, at very low second modifier fractions, $x_{M'}$, there are insufficient M'^+ cations to form $M_2'O$–B_2O_3 or M_2O–$M_2'O$–B_2O_3 chemical groupings, and so these cations will occur randomly within the interfacial/boundary regions between the M_2O–B_2O_3 chemical groupings, thus forming relatively immobile unlike-cation pairs/dimers and impeding the movement of the M^+ cations, as illustrated in Figure 3, where these impediments are indicated by the thin black lines. Note that, for this impeding action to occur, the concentration of M'^+ cations can be extremely low, and much less than that of the ***mobile*** M^+ cations. In Figure 3, the presence of only 28 *"immobile"* cation pairs/dimers (M'^+ cations), i.e. less than half of the number (61) of chemical groupings that are within (or partially within) the frame, reduces the number of pathways across the figure to only 2, with a single cross-connection. Experimental measurements (e.g. Figure 15 of Ref. 5) indicate that, even at very low $M_2'O$ fractions, the diffusion coefficient of the (impurity) M'^+ cations is not zero, albeit extremely small. Hence the presence of an M'^+ cation in an M^+ cation pathway does not completely block the diffusion of either cation species, but does drastically reduce their combined mobilities.

It is interesting to note that, in Figure 15 of Ref. 5, the diffusion coefficient for minority Cs^+ cations

($x_{Cs}\rightarrow 0$) is much lower than that for minority Na^+ cations ($x_{Na}\rightarrow 0$). First consider the case of a large minority (Cs^+) cation. Every jump of a large Cs^+ cation into a smaller host (Na^+) cation site requires a long (reconstructive) relaxation time, which also holds up the Na^+ cations behind it. On the other hand, it is much easier for a small minority (Na^+) cation to jump into a site that has recently been vacated by a large (Cs^+) host cation. In this case, the overall mobility is restricted by the need for successive Cs^+ cations to squeeze through the original Na^+ cation site. In addition, if an Na^+ cation can pass through a vacated Cs^+ site in a time that is short with respect to τ, the site will not have time to relax reconstructively before being occupied by the next Cs^+ cation.

A related explanation has been proposed by Bunde *et al*,[13] who point out that more of the E and E' sites are available to the smaller (Na^+) cations, without the need for topological relaxation, than is the case for the larger (Cs^+) cations. Ingram and co-workers[25,27,37] have also shown that, in mixed-cation glasses, the activation volumes reach a maximum value close to the sum of the end-member activation volumes, which may point to the existence of a co-operative mechanism, whereby empty M and M' sites can be interchanged to permit the onward flow of network-modifying cations.

8. Higher M'_2O fractions

Once the concentration of the M'^+ cations increases to the point where M'_2O–B_2O_3 and/or M_2O–M'_2O–B_2O_3 chemical groupings can be formed, an alternative mechanism comes into play, in that the occupancy of the conduction pathways between chemical groupings will depend on the cation species present in each grouping. There are two main possibilities for the chemical grouping species formed on quenching from the supercooled melt:

(i) There is clustering of like cations, due to the chemical structure of the glass being dominated by chemical groupings corresponding to single-modifier chemical compounds.
(ii) An association energy exists between unlike cations, leading to unlike-cation dimers[14,15] and/or a predominance of mixed-cation chemical groupings.

Thus, if a cation pathway runs between two single-modifier chemical groupings with the same network-modifying cation species, the conduction pathway is likely to be unique to that cation species. On the other hand, if the two chemical groupings involve different cation species, or one/both of the chemical groupings correspond to a ternary mixed-modifier compound, the conduction pathway will be occupied by cations of both species, and the number of unlike cation pairs/dimers within a given interfacial/boundary region, and hence the impediment of the cationic conduction, will be very much greater than at low $x_{M'}$ (Section 7). Note that, in the case of mixed-cation pathways, the formation of unlike-cation dimers can easily be understood from a stereochemical point of view in that, upon quenching from the supercooled melt, it is easier to fit a large cation into such a conduction pathway if there are smaller cations on either side of it. In the presence of an increased fraction of $BOØ_2^-$ basic structural units in the interfacial/boundary region between chemical groupings, this is consistent with the molecular dynamics simulations of Vegiri *et al*,[38] discussed in Sections 1.2 and 4.2 of Ref. 8, who find that the dominant cation configuration around the nonbridging oxygen atoms in their mixed-modifier glasses involves unlike cation species. If, therefore, it is assumed that cation transport is impeded within mixed-cation pathways, due to the much higher activation energy required for a larger cation to jump into a site previously occupied by a smaller cation, conduction can very much more easily proceed via pathways between like-cation single-modifier chemical groupings. Thus, on adding a second modifier, the number of easy conduction pathways will rapidly diminish with the introduction of its associated chemical groupings, since conduction will be impeded through all of the boundary regions separating these chemical groupings from those of the host cation species.

The effect of the introduction of (red) M'_2O–B_2O_3 and/or M_2O–M'_2O–B_2O_3 chemical groupings on the conduction process is illustrated in Figure 4, in which the interfacial/boundary regions where cation motion is very greatly impeded are again indicated by a (thick) black barrier. As can be seen, the replacement of only 11 (18%) of the 61 chemical groupings within, or partly within, Figure 2 leads to a dramatic reduction in the cationic conductivity, such that only a single unbranched conduction pathway remains across the figure. (Of course, at these higher M'_2O fractions, $x_{M'}$, randomly occurring M'^+ cations will also be found in the interfacial/boundary regions between two M_2O–B_2O_3 chemical groupings, thus further reducing the cationic conductivity.)

Eventually, the M_2O fraction will be reached for which percolation of the M^+ cation pathways through the glass is no longer possible. Similarly, at some point, the M'_2O fraction will become large enough to allow percolation of the M'^+ cation pathways. The order in which these two limits occur is unclear, i.e. whether there is a concentration region where neither of the cation pathways percolate, but the conductivity cross-over must occur between them.

The increasing magnitude of the mobility depression with total network-modifying cation content (Section 6) may, at least in part, be understood from the fact that, at lower M_2O + M'_2O contents, the chemical structure incorporates unreacted B_2O_3 chemical groupings (*cf.* Figure 5.5 of Ref. 32), whose

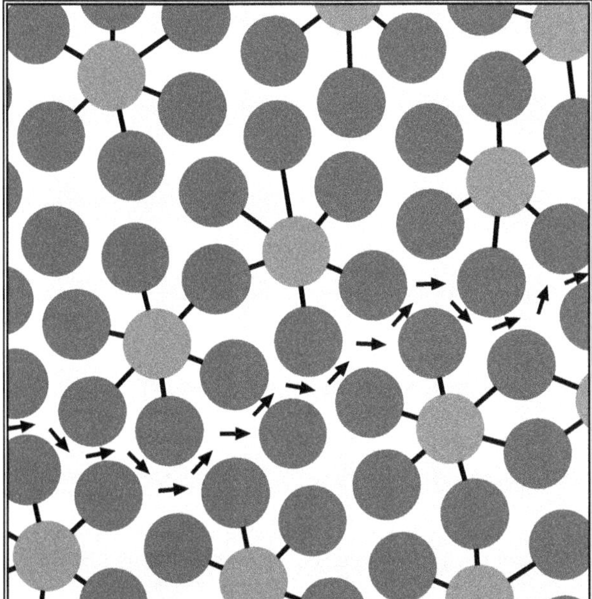

Figure 4. The flow of the M^+ cations (black arrows) through a ternary M_2O–M'_2O–B_2O_3 glass, at higher M'_2O fractions, $x_{M'}$. Chemical groupings that only involve M^+ cations are in blue, and those that include M'^+ cations are in red. The thick black lines between blue and red chemical groupings denote impediments formed by (multiple) unlike-cation pairs/dimers [Colour available online]

fraction decreases with increasing $M_2O + M'_2O$. The only cation species contributing to the interfacial/boundary regions between an M_2O–B_2O_3 (M'_2O–B_2O_3) and a B_2O_3 chemical grouping will be M^+ (M'^+), and so the percolation limit for each network-modifying cation species pathway at a constant total network modifier fraction will be extended to a lower M_2O (M'_2O) fraction, and the mobility depression will not be as large as that at higher total modifier content.

9. Silicate glasses

As in the case of borate glasses, the thermodynamic model of associated solutions[32] reveals a (chemical) structure for modified silicate glasses based on chemical groupings having the stoichiometries of related crystalline phases. The main differences are that, for silicates, there are fewer chemical compounds, and hence possible chemical grouping stoichiometries, within the glass-forming region (*cf.* Figure 5.11 of Ref. 32), and that some of the crystalline phases associated with the modified chemical compounds/groupings (e.g. a-$Na_2O.2SiO_2$) contributing to their chemical structure are based on two-dimensional silicate layers formed from $SiØØ_3^-$ basic structural units. Hence, at low x_M ($x_M \lesssim 1/3$), there is a fundamental difference between the ***average*** network-modifying cation sites in silicate and borate glasses. In silicates, the network-modifying cations are located close to one or more negatively-charged nonbridging oxygen atoms whereas, in borates, those within chemical groupings are almost entirely associated with negatively-charged $BØ_4^-$ tetrahedra, which are themselves mostly incorporated into relatively stable superstructural units.

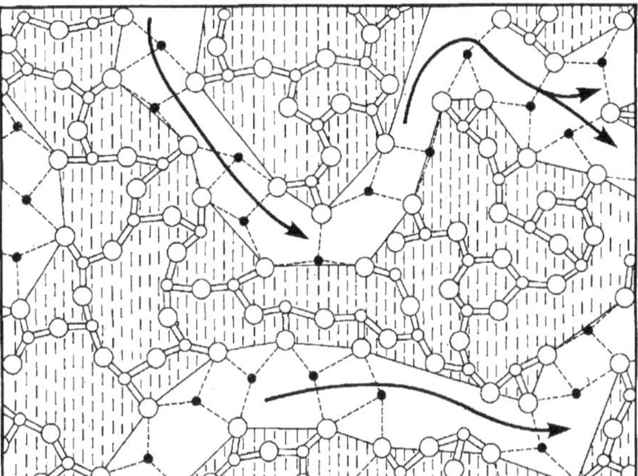

Figure 5. Cationic conduction (arrows) along the "percolation channels" of a modified random network[9]

For silicate glasses, it has been suggested[9,39,40] that cationic conduction takes place along the channels of a modified random network,[19] as represented by the arrows in Figure 5.[9] However, Figure 5,[9] which is based on three-dimensional regions of pure silica separated by two-dimensional interfacial/boundary regions containing the network-modifying cations (or alternatively a silica structure incorporating one-dimensional cation channels), is inconsistent with the fact that only a small (*"unreacted"*) fraction of the network-modifying cations are involved in the conduction process. In reality, there must be two different types of network-modifying cation site, those within the chemical groupings themselves and those located within the interfacial/boundary regions, in which reside the cations from the *"unreacted"* network modifier(s). If there were many fewer network-modifying cations, the modified random network diagram of Figure 5[9] would correspond to the case of very low x_M, where the M_2O fraction is too small for the alkali silicate basic chemical units (e.g. $M^+ + SiØØ_3^-$) to agglomerate to form M_2O–SiO_2 ($M_2O.2SiO_2$) chemical groupings. Thus, although the details of the conduction mechanism may be slightly different, the cationic conduction in silicate glasses is still based on the small fraction of network-modifying cations that is associated with *"unreacted"* modifiers, and takes place along conduction pathways located within the interfacial/boundary regions between chemical groupings.

10. Conclusions

A structural explanation/model has been advanced for the (network-modifying) cationic conduction in both binary and ternary (mixed-modifier) borate glasses, which incorporates the ***cluster-bypass model*** of Ingram and co-workers[5,9–11] and the ***dynamic structure model*** of Bunde *et al*,[12,13] and is consistent with the neutron diffraction data and thermodynamic modelling of Refs 6–8. In this reformulated model, Ingram's clusters[5,9–11] are replaced by the (chemical) groupings that define the chemical structure of the glasses, but the cationic conduction pathways are still located within the interfacial/boundary regions that separate these groupings. The revised model can explain both the extremely rapid fall-off in conductivity upon the initial introduction of a second network-modifying cation species, and the fact that, for any given ternary system, the magnitude of the conductivity depression increases with increasing total modifier content. It is also concluded that, since the network-modifying cationic conductivity only involves ***"unreacted"*** cations, and takes place within the more highly disordered interfacial/boundary regions between chemical groupings, any theory of the cationic conductivity and the mixed-modifier effect that is based on the ***average*** cation sites, e.g. as determined by diffraction and/or EXAFS studies, ***must be regarded as highly dubious.***

Whilst the structural model proposed here for the network-modifying cationic conductivity and the mixed-modifier effect has been specifically presented for borate glasses, many aspects of the model apply equally to silicate glasses. The main differences for silicates are the absence both of superstructural units, and of the change in co-ordination number for the network-forming cations. However, the conduction pathways are still restricted to the interfacial/ boundary regions between chemical groupings, and are much more restricted than implied by the two-dimensional modified random network diagram in Figure 5.[9]

Acknowledgement

The authors would like to thank Malcolm D. Ingram for his helpful comments/suggestions in respect of an earlier version of this manuscript.

References

1. Varshneya, A. K. *Fundamentals of Inorganic Glasses*, Second Edition, Society of Glass Technology, Sheffield, 2006.
2. Moynihan, C. T. In: *Glass ... Current Issues*, Eds A. F. Wright & J. Dupuy, Martinus Nijhoff, Dordrecht, 1985, p. 456.
3. Isard, J. O. *J. Non-Cryst. Solids*, 1969, **1**, 235.
4. Day, D. E. *J. Non-Cryst. Solids*, 1976, **21**, 343.
5. Ingram, M. D. *Phys. Chem. Glasses*, 1987, **28**, 215.
6. Wright, A. C., Sinclair, R. N., Stone, C. E., Shaw, J. L., Feller, S. A., Kiczenski, T. J., Williams, R. B., Berger, H. A., Fischer, H. E. & Vedishcheva, N. M. *Phys. Chem. Glasses: Eur. J. Glass Sci. Technol. B*, 2012, **53**, 191.
7. Wright, A. C., Sinclair, R. N., Stone, C. E., Shaw, J. L., Feller, S. A., Williams, R. B., Fischer, H. E. & N. M. Vedishcheva, *Phys. Chem. Glasses: Eur. J. Glass Sci. Technol. B*, 2014, **55**, 74.
8. Wright, A. C., Sinclair, R. N., Stone, C. E., Shaw, J. L., Feller, S. A., Williams, R. B. & Fischer, H. E. *Phys. Chem. Glasses: Eur. J. Glass Sci. Technol. B*, 2015, **56** (3), 85.
9. Ingram, M. D. *Philos. Mag.*, 1989, **B60**, 729.
10. Ingram, M. D., Chryssikos, G. D. & Kamitsos, E. I. *J. Non-Cryst. Solids*, 1991, **131–133**, 1089.
11. Ingram, M. D., Mueller, W. & Torge, M. In: *The Physics of Non-Crystalline Solids*, Eds L. D. Pye, W. C. LaCourse & H. J. Stevens, Taylor & Francis, London, 1992, p. 441.
12. Bunde, A., Ingram M. D. & Maass, P. *J. Non-Cryst. Solids*, 1994, **172–174**, 1222.
13. Bunde, A., Ingram M. D. & Russ, S. *Phys. Chem. Chem. Phys.*, 2004, **6**, 3663.
14. Hendrickson, J. R. & Bray, P. J. *Phys. Chem. Glasses*, 1972, **13**, 43.
15. Hendrickson ,J. R. & Bray, P. J. *Phys. Chem. Glasses*, 1972, **13**, 107.
16. LaCourse, W. C. *J. Non-Cryst. Solids*, 1987, **95&96**, 905.
17. Wright, A. C., Dalba, G., Rocca, F. & Vedishcheva, N. M. *Phys. Chem. Glasses : Eur. J. Glass Sci. Technol. B*, 2010, **51**, 233.
18. Wright, A. C. *Int. J. Appl. Glass Sci.*, 2015, **6**, 45.
19. Greaves, G. N. *J. Non-Cryst. Solids*, 1985, **71**, 203.
20. Ratai, E., Chan, J. C. C. & Eckert, H. *Phys. Chem. Chem. Phys.*, 2002, **4**, 3198.
21. Ratai, E.-M., Janssen, M., Epping, J. D., Chan, J. C. C. & Eckert, H. *Phys. Chem. Glasses*, 2003, **44**, 45.
22. Maass, P., Bunde, A. & Ingram, M. D. *Phys. Rev. Lett.*, 1992, **68**, 3064.
23. Wright, A. C. & Vedishcheva, N. M. *Phys. Chem. Glasses: Eur. J. Glass Sci. Technol. B*, 2013, **54**, 147.
24. Krogh-Moe, J. *Acta Crystallogr.*, 1972, **B28**, 1571.
25. Ingram, M. D. *Private commun.*, 2015.
26. Ingram, M. D. & Imrie, C. T. *Solid State Ionics*, 2011, **196**, 9.
27. Imre, A. W., Voss, S., Berkemeier, F., Mehrer, H., Konidakis, I. & Ingram, M. D. *Solid State Ionics*, 2006, **177**, 963.
28. Moynihan, C. T. & Lesikar, A. V. *J. Am. Ceram. Soc.*, 1981, **64**, 40.
29. Ravaine, D., Azandegle, E. & Souquet, J. L. *Silic. Ind.*, 1975, **40**, 333.
30. Shakhmatkin, B. A., Vedishcheva, N. M. & Wright, A. C. In: *Borate Glasses, Crystals & Melts*, Eds A. C. Wright, S. A. Feller & A. C. Hannon, Society of Glass Technology, Sheffield, 1997, p. 189.
31. Wright, A. C. *Glass Technol.: Eur. J. Glass Sci. Technol. A*, 2009, **50**, 127.
32. Vedishcheva, N. M. & Wright, A. C. In: *Glass: Selected Properties and Crystallization*, Ed. J. W. P. Schmelzer, de Gruyter, Berlin, 2014, Chapter 5, p. 269.
33. Goodman, C. H. L. *Nature*, 1975, **257**, 370.
34. Goodman, C. H. L. In: *The Structure of Non-Crystalline Materials 1982*, Eds P. H. Gaskell, J. M. Parker & E. A. Davis, Taylor & Francis, London, 1983, p. 151.
35. Goodman, C. H. L. *Phys. Chem. Glasses*, 1985, **26**, 1.
36. Sakka, S., Matusita, K. & Kamiya, K. *Phys. Chem. Glasses*, 1979, **20**, 25.
37. Bandaranayake, P. W. S. K., Imrie, C. T. & Ingram, M. D. *Phys. Chem. Chem. Phys.*, 2002, **4**, 3209.
38. Vegiri, A., Varsamis, C.-P. E. & Kamitsos, E. I. *Phys. Rev.*, 2009, **B80**, 184202.
39. Greaves, G. N., Gurman, S. J., Catlow, C. R. A., Chadwick, A. V., Houde-Walter, S., Henderson, C. M. B. & Dobson, B. R. *Philos. Mag.*, 1991, **A64**, 1059.
40. Greaves, G. N. In: *The Physics of Non-Crystalline Solids*, Eds L. D. Pye, W. C. LaCourse & H. J. Stevens, Taylor & Francis, London, 1992, p. 453.

Phys. Chem. Glasses: Eur. J. Glass Sci. Technol. B, August 2015, 56 (4), 128–138

Sol-gel synthesis of composites in the ternary TiO_2–TeO_2–B_2O_3 system

R. Iordanova,[1] *R. Gegova,*[1] *A. Bachvarova-Nedelcheva*[*,1] & *Y. Dimitriev*[2]

[1] *Institute of General and Inorganic Chemistry, Bulgarian Academy of Sciences, "Acad. G. Bonchev" str., bld. 11, 1113 Sofia, Bulgaria*
[2] *University of Chemical Technology and Metallurgy, 8 Kl. Ohridski, blvd, 1756 Sofia, Bulgaria*

Manuscript received 27 September 2014
Revised manuscript received 5 December 2014
Accepted 8 December 2014

The gel formation region in the ternary TiO_2–TeO_2–B_2O_3 system has been determined. Transparent monolithic gels are obtained for compositions between 30 and 100 mol% TiO_2, up to 65 mol% TeO_2 and 75 mol% B_2O_3. Titanium butoxide, telluric (VI) acid and boric acid were used as precursors. Differences were observed in the degree of decomposition of Ti butoxide in the presence of H_3BO_3 and H_6TeO_6 acids. The phase transformations of the obtained gels in the temperature range 200–700°C were investigated by XRD. Composite materials containing an amorphous phase and different crystalline phases (metallic Te, α-TeO_2, anatase, rutile and $TiTe_3O_8$) were prepared. IR results show that the short range order of the amorphous phases, which are part of the composite materials, consist of TiO_6, BO_3, BO_4, and TeO_4 structural units. Since free B_2O_3 is not detected in the three-component amorphous compositions, it is suggested that they have a better connectivity between the building units compared to binary TiO_2–B_2O_3 compositions. The UV-Vis spectra of the as-prepared gels exhibited a red shift of the cut-off due to the presence of boron and tellurium units.

1. Introduction

Sol-gel derived materials have attracted much attention from materials scientists due to such advantages as good homogeneities, ease of composition control, lowering of the melting or sintering temperature, and better control of the microstructure. The major peculiarities and possibilities of sol-gel processing have been discussed in some excellent overviews.[1–5] The method is immensely suitable for obtaining glasses that are usually difficult to prepare by melt quenching owing to their high melting temperature or tendencies toward crystallisation and phase separation. It is of scientific and technical interest to study the structural evolution at various stages of the sol-gel to glass or ceramic processes, and to investigate the factors that are important for the structure and properties of the obtained materials. This motivated us to perform the sol-gel experiments described in this paper.

Mackenzie *et al*[6] grouped the components in gel-derived amorphous solids into two categories: vitrification oxides and non-glass formers or modifiers. This classification may be considered as an analogy to the classical classification of oxides on the basis of their tendency to form glass from supercooled melts.[7,8] Until now, mainly silicate, titanate, zirconate and niobate gels have been studied.[2,9–12] This is another reason to focus our investigations on a new combination of oxides.

Even though B_2O_3 is a well known classical glass former, the sol-gel method is not often used for the synthesis of amorphous borate materials. Achievements with regard to the synthesis, structure and properties of borate glasses and glass ceramics have been reported in periodically organised, specialised conferences (1977, 1996, 1999, 2002, 2005, 2008, 2011, 2014). The first gel-derived borate materials were reported for the systems B_2O_3–Li_2O[13] and B_2O_3–BaO.[14] Various gel-derived borosilicate glasses with a more homogeneous network compared to the traditional melting route have been prepared.[15–19] Several authors established that the short range order of alkaline borate gels and multicomponent borosilicate gels closely resembles that of the bulk melt quenched glasses.[20,21] Soraru *et al*[22] synthesised hybrid $RSiO_{1.5}$–B_2O_3 (R=Me, Et, Vi) gels in which trigonal BO_3 units are incorporated into the siloxane network via B–O–Si bridges. It was shown that boron ions have a positive role for the bioactivity of sol-gel prepared samples in the SiO_2–P_2O_5–CaO–B_2O_3 system.[23] Using sol-gel technology, many component Ti, Zr, B and Al-modified hybrid materials have been fabricated, and it has been shown that the nature of the organic group plays a major role in the design of new sol-gel materials.[24] Ota *et al*[25] have established that the composition region for glass formation by as-prepared B_2O_3–Na_2O–TiO_2 gel corresponds fairly well to the composition region for glass formation by the twin-roller method. Until now, connectivity between borate and titanate units is questionable, and it is known that boro-titanate glasses are not

* Corresponding author. Email albenadb@svr.igic.bas.bg
Original version presented at VIII Int. Conf. on Borate Glasses, Crystals and Melts, Pardubice, Czech Republic, 30 June–2 July 2014
DOI: 10.13036/17533562.56.4.128

obtained from melts. The incorporation of boron in TiO_2 extends its visible light absorption.[26,27] The photoactivity of TiO_2–B_2O_3 and B_2O_3–SiO_2–TiO_2 catalysts is improved by boron content.[28,29]

TiO_2 is an intermediate oxide, which does not form glass by conventional quenching, but enhances the glass-forming tendency in several systems, such as K_2O–TiO_2, Cs_2O–TiO_2, PbO–TiO_2, and Ln_2O_3–TiO_2.[30] On the other hand, TiO_2 is successfully used to produce amorphous and crystalline titanate materials via the sol-gel process.[2,4,5,12,31] Titanium alkoxides are extremely reactive toward hydrolysis, and many investigations have been performed concerning the modification of titanium precursors to control their reactivity.[4,32–34] Due to the unique properties of titania, its sol-gel derived products are extensively used as catalysts, supports, protective coatings, optical materials, self-cleaning surfaces and as an active ingredient of sunscreen cosmetic products.[2]

TeO_2 is usually classified as a conditional network-former, and does not form glass alone, but many component tellurite glasses are attractive because of their excellent optical properties (high refractive index, high third order nonlinear properties, wide optical window from the ultraviolet to the infrared region, etc).[7,35] The sol-gel method is an alternative way to extend the application of new tellurite compositions. Powders and films in the TiO_2–TeO_2 and TeO_2–PbO–TiO_2 systems have been reported in several papers, irrespective of the anomalously high hydrolysis rate of Te(VI) alkoxides.[36–39] It has been established that sol-gel derived tellurite coatings and glasses exhibit greater resistance to devitrification than melt quenched glasses.

We have focused on the ternary TiO_2–TeO_2–B_2O_3 system, containing one classical network former (B_2O_3), one conditional network former (TeO_2) and one intermediate oxide (TiO_2). Until now, melt quenched glasses in this system have been obtained in a narrow concentration range. For binary TiO_2–TeO_2, glasses have been obtained above 75% TeO_2.[40–42] In the binary system TeO_2–B_2O_3, glasses have been obtained below 25 mol% B_2O_3, due to the wide region of stable and metastable phase separation.[43–46] In the third binary system, TiO_2–B_2O_3, glasses have not been obtained, and over a wide concentration range TiO_2 crystallises and phase separation has been observed.[18] In the ternary system, TiO_2–TeO_2–B_2O_3, melt quenched glasses have been obtained with high TeO_2 content (above 60 mol%), and microheterogeneous structures have been observed over a wide concentration range.[43,47] In this study we selected TiO_2 rich compositions that are difficult to prepare by the melt quenching method. Moreover, there is no previous information on gel formation in this ternary system.

This paper deals with sol-gel processing using one organic (Ti butoxide) and two inorganic (Te(VI) acid and boric acid) precursors. The main aims were to study the gel stability, the phase and structural evolution upon heating in the temperature range 200–700°C, and to verify the compatibility of the structural units in the amorphous phase.

2. Experimental

2.1 Gelling, drying and heat treatment

The scheme for sol-gel synthesis of TiO_2–TeO_2–B_2O_3 compositions is presented in Figure 1. A combination of Te(VI) acid (Aldrich), along with Ti butoxide (Fluka AG) and H_3BO_3 (Merck) precursors dissolved in ethylene glycol ($C_2H_6O_2$) (99% Aldrich), was used. Telluric acid (H_6TeO_6) was selected to overcome the problem with the high hydrolysis rate of tellurium (VI) alkoxide that has been analysed in several pa-

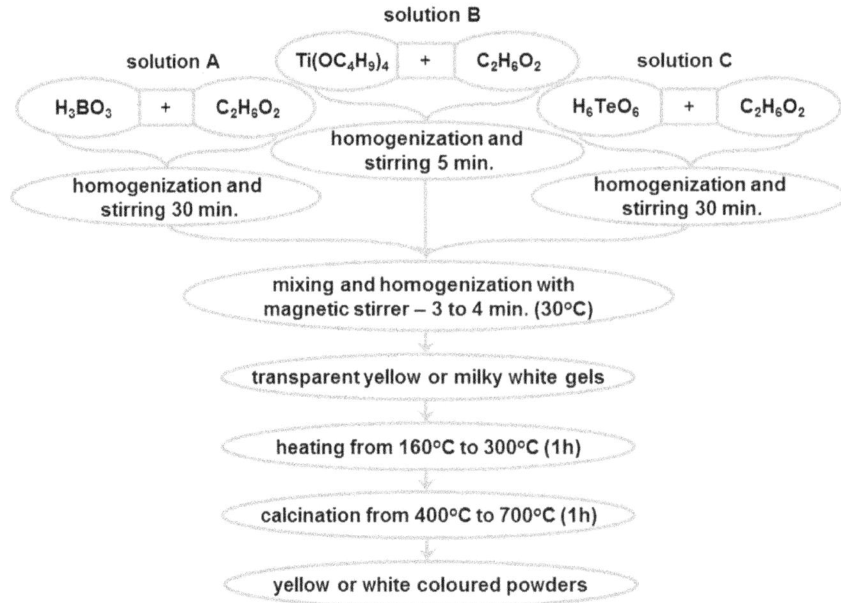

Figure 1. Scheme for the sol-gel synthesis of TiO_2–TeO_2–B_2O_3 gels [Colour available online]

pers.[37,48,49] The precursor solutions were subjected to 5–30 min intensive stirring at room temperature in order to achieve complete dissolution. No direct addition of water was made to the precursor solutions. The sol-gel hydrolysis reaction was acquired from absorbed atmospheric moisture. The measured pH varied in the range 4–5, depending on composition. The gelation time for the investigated compositions was from 1 to 5 min. The aging of gels was performed in air for several days to complete the processes. The obtained gels were subjected to stepwise heating from 200 to 700°C for one hour exposure time in air. Heat treatment at 200–300°C was performed in air in order to hydrolyse any unreacted -OR groups, as well as to decrease their content. Further increase of temperature (300–700°C) was important to verify the phase and structural transformations of the gels. Selected gels, situated in two sections: (i) with 40 mol% TiO_2 and (ii) with 80 mol% TiO_2 were subjected to more detailed phase and structural analysis (Figures 4 and 5, Table 1).

2.2 Sample characterisation

Powder XRD patterns were measured at room temperature with a Bruker D8 Advance diffractometer using Cu-K_α radiation. The thermal stability of selected as-synthesised gels was determined by DTA/TG analysis. A Seteram LABSYSIS™ EVO 1600 apparatus was used for recording of the thermo diagrams over the range from room temperature up to 700°C. A Pt–Pt/Rh thermocouple and Al_2O_3 as a reference material were used. The heating rate was 10°/min in air atmosphere under an air flow of 20 mL/min in order to determine the released gas phases during the increase of temperature, using a mass spectrometer OminStar Preffer vacuum. The infrared spectra were measured in the range 1600–400 cm^{-1} using the KBr pellet technique on a Nicolet-320 FTIR spectrometer with 64 scans and a

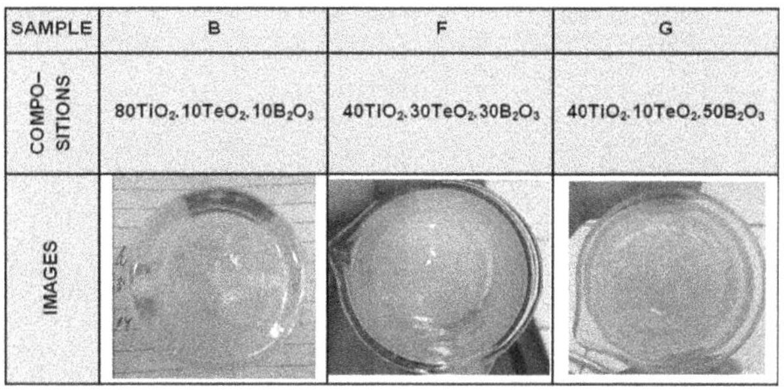

SAMPLE	B	F	G
COMPO-SITIONS	$80TiO_2.10TeO_2.10B_2O_3$	$40TiO_2.30TeO_2.30B_2O_3$	$40TiO_2.10TeO_2.50B_2O_3$
IMAGES			

Figure 2. Images of selected binary and ternary gels [Colour available online]

resolution of ±1 cm^{-1}. The optical absorption spectra of the powdered samples in the wavelength range 200–1000 nm were recorded by a UV–Vis diffused reflectance Spectrophotometer "Evolution 300", using a magnesium oxide reflectance standard as the baseline. The absorption edge and the optical band gap were determined following Dharma *et al.*[50] The bandgap energies, E_g, of the samples were calculated by Planck's equation: $E_g=hc/\lambda=1240/\lambda$, where E_g is the bandgap energy (eV), h is the Planck's constant, c is the velocity of light (m/s), and λ is the wavelength (nm).

3. Results

3.1. Phase transformations and thermal stability of the gels

Transparent and monolithic gels were obtained by applying the scheme shown in Figure 1 and their images are given in Figure 2. The gel formation region determined at room temperature is situated in the composition range 30–100 mol% TiO_2, up to 65 mol% TeO_2 and 75 mol% B_2O_3 (Figure 3). Transparent gels were obtained when the B_2O_3 content is up to 30–40 mol%, while above 50 mol% the obtained gels become opaque. Similar results have been reported by Ota *et al*,[25] who used boron tributoxide and titanium tetra-isopropoxide to synthesise gels containing less

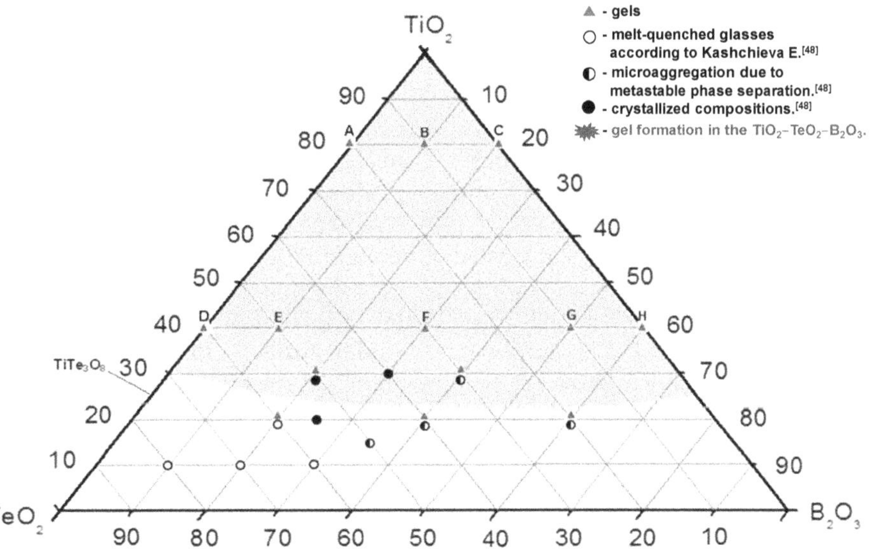

Figure 3. Gel formation region in the investigated TiO_2–TeO_2–B_2O_3 system [Colour available online]

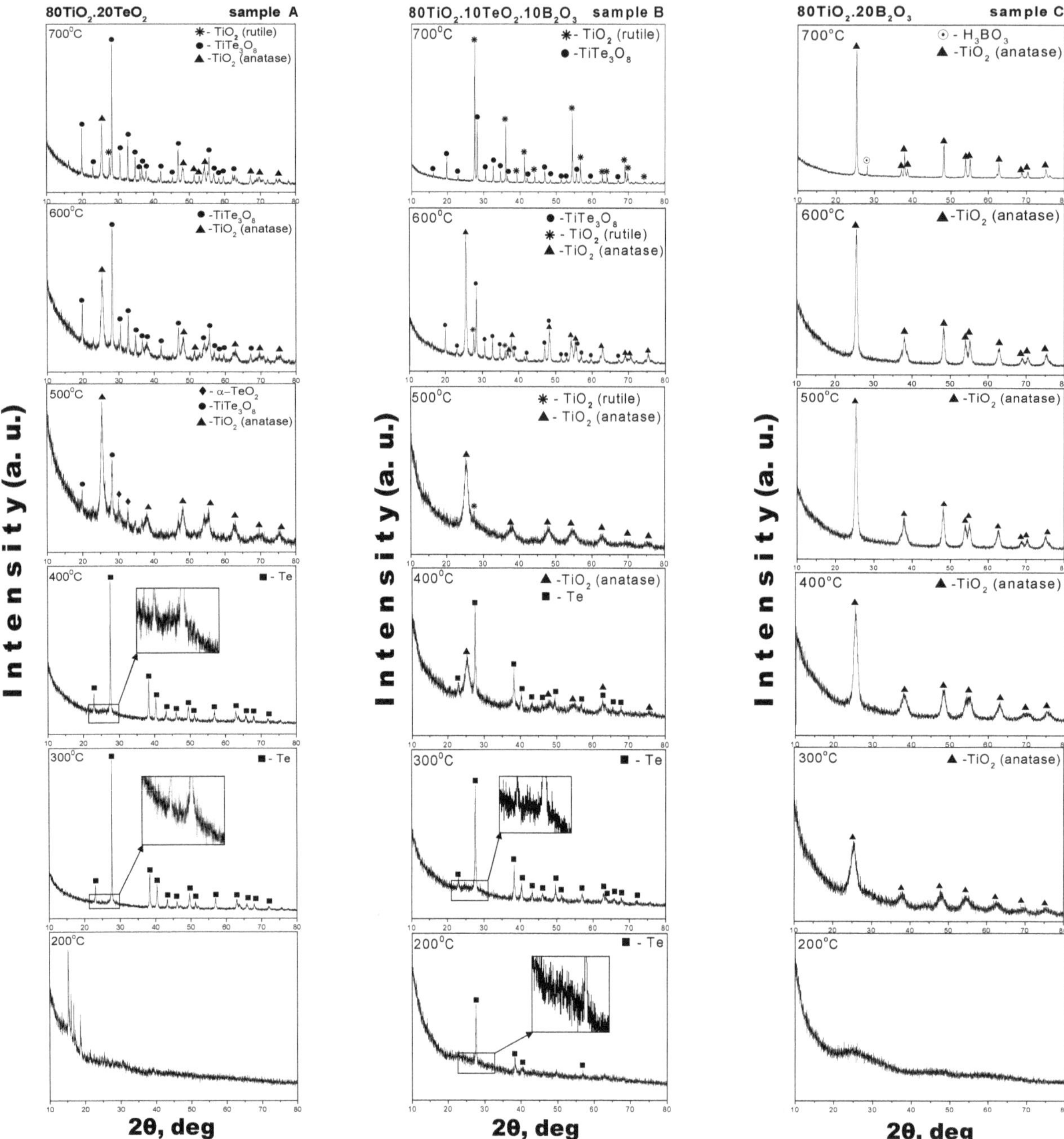

Figure 4. XRD patterns of samples containing 80 mol% TiO_2

than 40 mol% B_2O_3. For the binary B_2O_3–TeO_2 system, we did not obtain gels with the given experimental conditions using inorganic precursors (H_6TeO_6 and H_3BO_3).

According to the XRD patterns, gels heat treated up to 300°C exhibit a predominantly amorphous phase and only the formation of metallic tellurium is observed (JCPDS 78-2312) (Figures 4, 5, samples A, B, D, E, F and G). Tellurium is fully oxidized to TeO_2 at about 500°C. The amount of amorphous phase gradually decreases with increasing temperature, but it is still found up to 500°C. In the temperature range 200–500°C crystallisation also takes place. Since it is difficult to quantify the amount of the crystalline phases, it is not possible to give the real composition of the remaining amorphous phase. A small amount of H_3BO_3 (JCPDS 30-0199) is precipitated in the composition $40TiO_2.60B_2O_3$ (sample H, Figure 5) in the temperature range 200–700°C. In all ternary compositions H_3BO_3 is not detected. Anatase (JCPDS 78-2486) appears at lower temperature (300°C) in samples C ($80TiO_2.20B_2O_3$) and H ($40TiO_2.60B_2O_3$), and conversion to rutile (JCPDS 21-1276) is not observed up to 700°C. In the binary samples D ($40TiO_2.60TeO_2$) and A ($80TiO_2.20TeO_2$), anatase appears at 500°C and a small amount is converted to rutile at 700°C. At 400°C the average crystallite size (calculated using Sherrer's equation) of TiO_2 (anatase) in the powdered samples D and H is about 8 nm (Figure 5). In all ternary compositions, characteristic peaks of TiO_2 (anatase) are

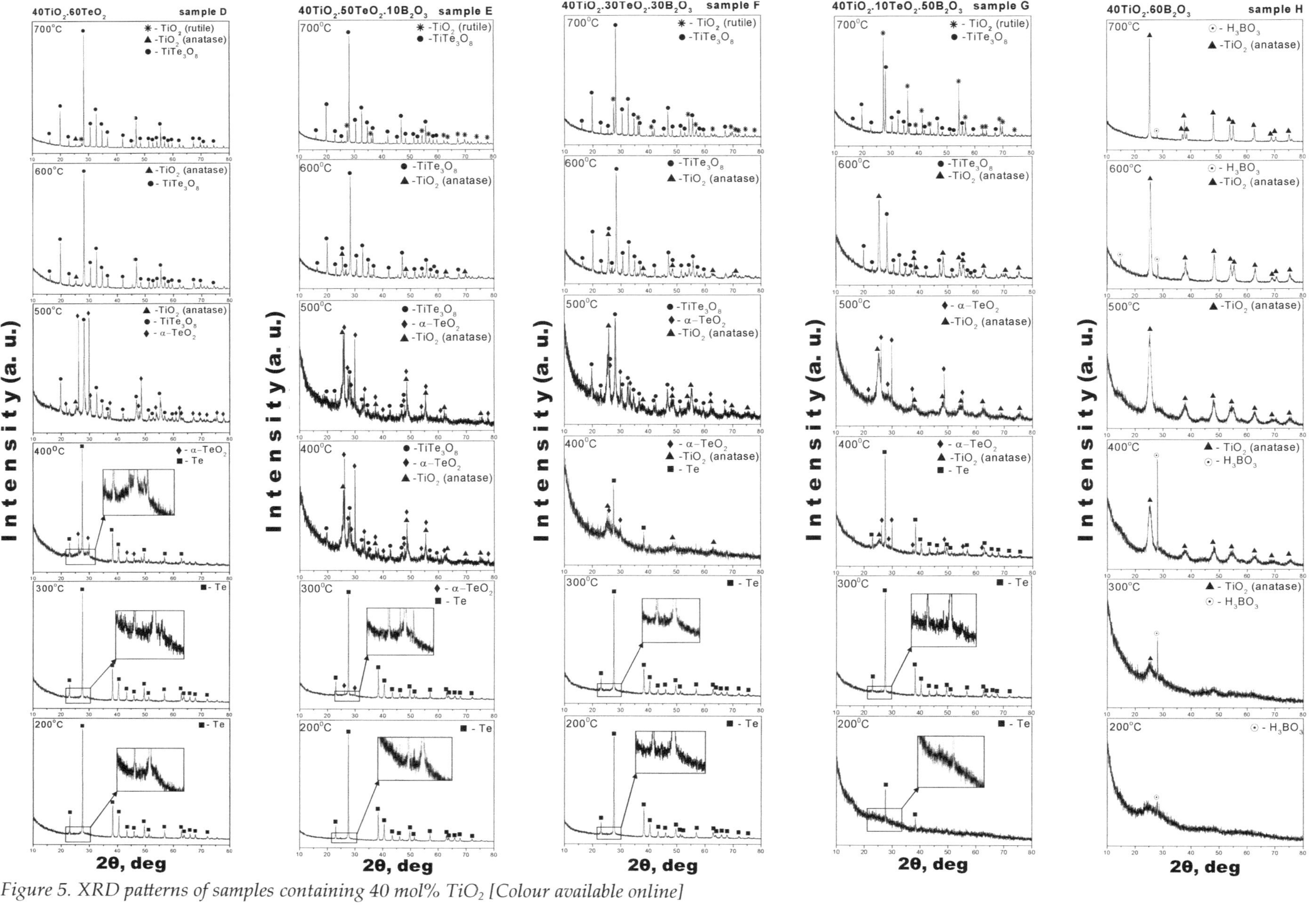

Figure 5. XRD patterns of samples containing 40 mol% TiO_2 [Colour available online]

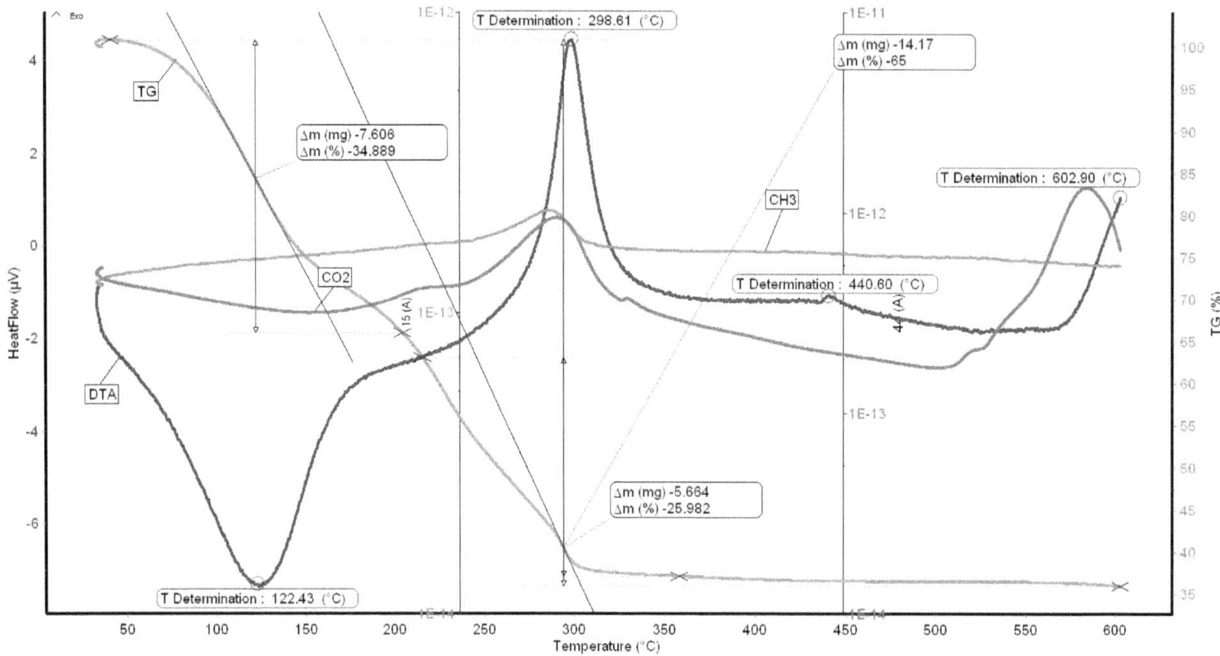

Figure 6. DTA/TG of the gel composition G ($40TiO_2.10TeO_2.50B_2O_3$) [Colour available online]

detected at 400°C. To complete the phase formation, additional heating was performed at higher temperatures (600–700°C). Anatase (TiO_2), α-TeO_2 and $TiTe_3O_8$ crystalline phases coexist in the prepared composite materials (Figures 4 and 5). These crystalline phases have been obtained at similar temperatures by other authors.[39,51]

In order to verify the thermal stability of the gels, simultaneous thermogravimetric (TG) and differential thermal analysis (DTA) were performed for a representative gel composition $40TiO_2.50B_2O_3.10TeO_2$ (sample G, Figure 6). As can be seen several stages are marked on the DTA/TG curves. The first stage is endothermic (~120°C), related to the evaporation of physically adsorbed water and/or organic solvent as ethylene glycol. A huge and sharp exothermic peak is observed at about 300°C, related to the intense combustion of the alkoxide groups bonded to Ti atoms. A very small exothermic peak is distinguished at about 440°C, probably attributed to the crystallisation of several phases, TeO_2 and TiO_2 (anatase) according to our XRD data. The last exothermic peak (at about 600°C) can probably be associated with the slow oxidation and release of CO_2 in the air atmosphere. According to the TG curve, strong weight loss (~65%) occurs and is finished at nearly 300°C when dehydration and combustion of the organics take place. Above this temperature no weight loss is observed in the

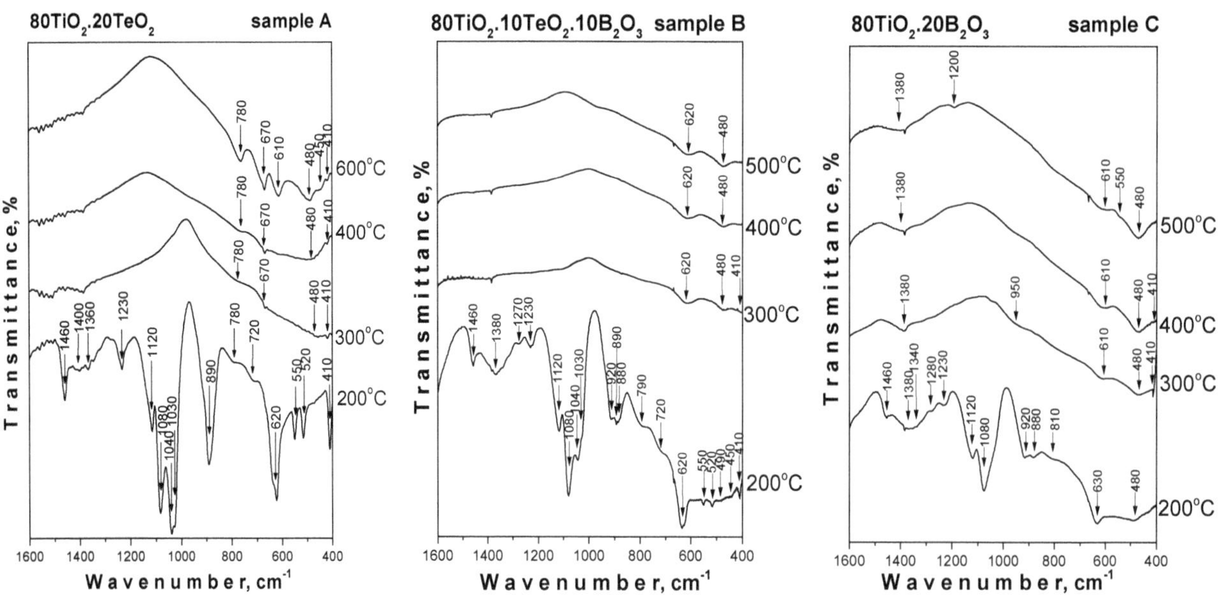

Figure 7. IR spectra of samples containing 80 mol% TiO_2

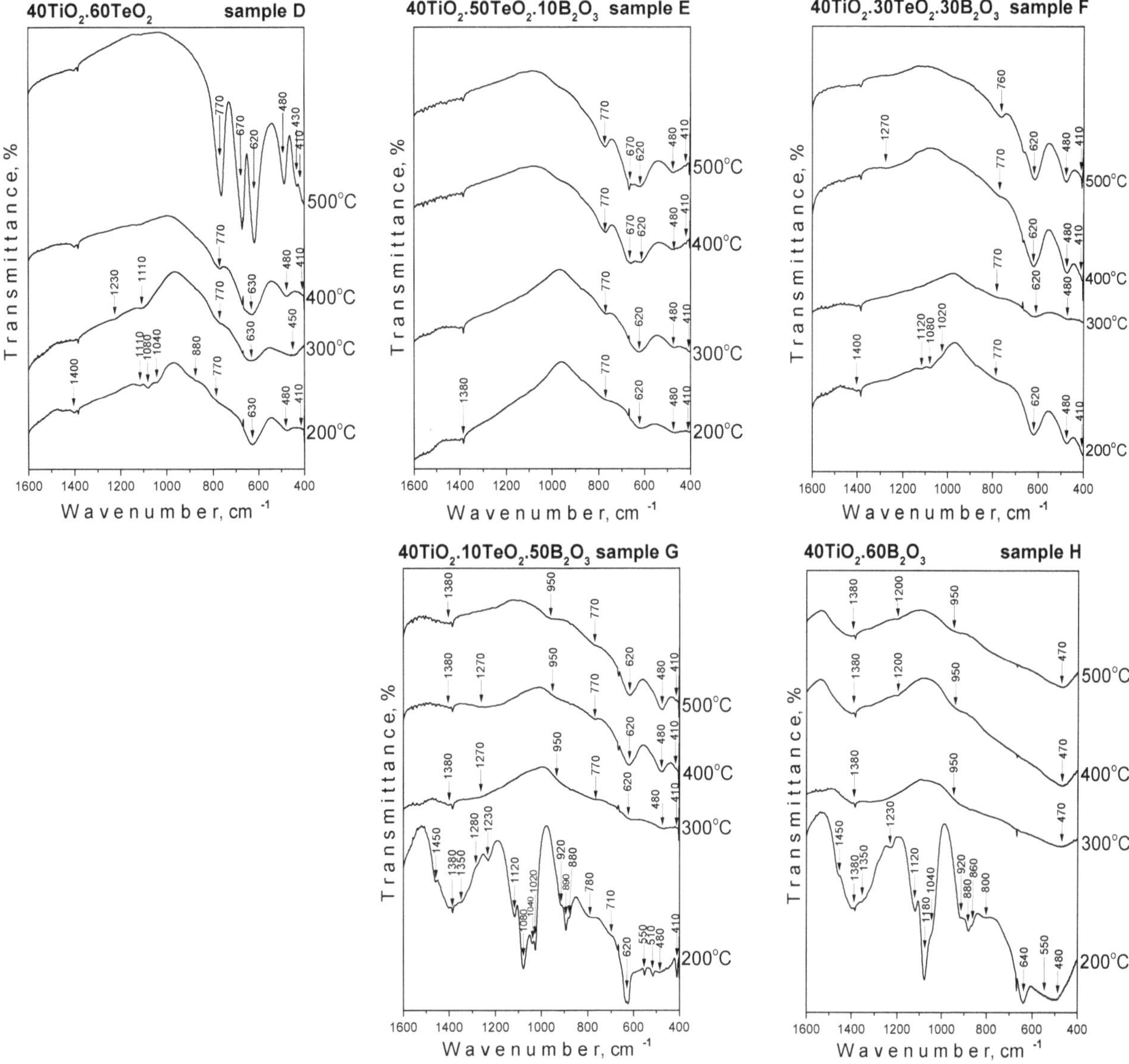

Figure 8. IR spectra of samples containing 40 mol% TiO_2

sample. For comparison, similar data are obtained by other authors.[4,52,53]

3.2. IR spectra

Figures 7 and 8 show the IR spectra of binary and ternary gels from the TiO_2–TeO_2–B_2O_3 system heated at different temperatures. The assignments of the vibration bands of the separate structural units is made on the basis of well known spectral data of the precursors and crystalline phases existing in the system. IR spectra of Ti(IV) n-butoxide, H_6TeO_6 and H_3BO_3 are presented in Figure 9. The IR spectrum of Ti butoxide is characterised by bands located between 1500–1300 cm^{-1} which are assigned to the bending vibrations of CH_3 and CH_2 groups (Figure 9(a)).[5,54] Bands in the range 1130–1040 cm^{-1} are related to the characteristic Ti–O–C stretching vibrations that can be used to determine the degree of hydrolysis.[4,54,55] Broad absorption bands below 1000 cm^{-1} correspond to C–H, C–O and Ti–O–C deformation vibrations.[54] The characteristic vibrations of TiO_6 structural units in TiO_2 (anatase) are at 640 and 450 cm^{-1}.[2,56–58] According to the local point symmetry approach, glasses containing deformed TeO_4 groups exhibit a strong absorption band at 635 cm^{-1} and a shoulder at 670 cm^{-1}. When the band at 700–670 cm^{-1} becomes stronger, symmetric TeO_3 units are formed.[40] More details on the IR spectral analysis of tellurite glasses are summarised in Mallawany's monograph.[35] The IR spectra of $TiTe_3O_8$ and other phases built up by symmetric TeO_4 building units possess strong absorption bands at 770, 670 and 620 cm^{-1}.[40,59,60] The vibrations of symmetric TeO_6 units from H_6TeO_6 acid are situated in the same absorption range (Figure 9(b), 700–600 cm^{-1}),[61,62] while bending vibrations of Te–OH are at 1220 and 1130 cm^{-1}.[63] The vibrational modes of the building units in binary $M_2O(MO)$–B_2O_3 glasses are active in the following infrared spectral regions: 1500–1200 cm^{-1} (B–O stretching for

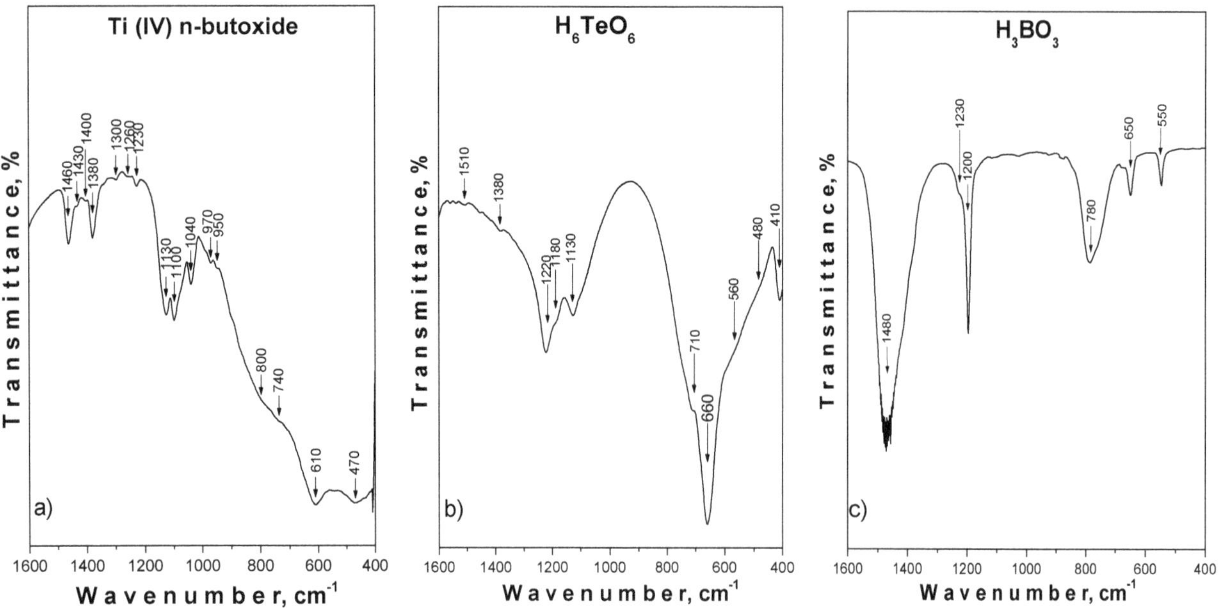

Figure 9. IR spectra of the used precursors: (a) Ti(IV) n-butoxide, (b) H_6TeO_6 and (c) H_3BO_3

trigonal BO_3 units), 1200–850 cm^{-1} (B–O stretching for tetrahedral BO_4 units) and 800–600 cm^{-1} (bending vibrations for various borate segments).[64–66] The strong absorption bands at 1480 and 1200 cm^{-1} can be assigned to the presence of B–OH bending vibrations (Figure 9(c)).[22,67]

In our IR spectra for samples heat treated up to 200°C, the absorption bands in the range 1600–400 cm^{-1} can be divided into three regions: 1500–1200, 1200–900 and 900–400 cm^{-1} (Figures 7 and 8). The intensity of absorption bands in the 1120–1040 cm^{-1} region (Ti–O–C stretching vibrations) decreases with increasing B_2O_3 content in samples containing a high amount of TiO_2 (80 mol%), (Figure 7, samples A, B, C), while in samples containing 40 mol% TiO_2 (Figure 8, samples D, E, F, G and H) the intensity of these bands decreases with increasing TeO_2 content. The bands above 1000 cm^{-1}, characteristic of the organic groups, are not visible above 300°C. The spectra in the temperature range 300–500°C are characterised mainly by bands below 900 cm^{-1}, typical for the inorganic units. These bands are broadened, with the low intensity that is a peculiarity of disordered systems. At lower temperatures (200, 300°C) the weak bands at 780–770 cm^{-1} and 630–620 cm^{-1} can be assigned to the vibrations of TeO_4 units (Figure 8, samples G, F, E, D). Well defined bands at 770, 670 and 620 cm^{-1} due to vibrations of TeO_4 are observed in the spectra of samples D and E heated at 400 and 500°C as a result of $TiTe_3O_8$ crystallisation. This is in accord with the XRD data. Over all temperature ranges (300–500°C), an increase in B_2O_3 content above 20 mol% (Figures 7 and 8, samples C, F, G, H) leads to the appearance of new absorption bands centred at 1380 and 1270 cm^{-1}, due to the vibrations of BO_3 units, as well as a very weak band about 950 cm^{-1}, caused by the vibrations of BO_4 groups incorporated in superstructural units. The low intensity band at 1200 cm^{-1} can be assigned to the precipitation of H_3BO_3.

3.3. UV–Vis absorption spectra

The diffuse reflectance absorption spectra of selected binary and ternary TiO_2–TeO_2–B_2O_3 gels (aged at room temperature), as well as of pure Ti butoxide gel, are illustrated in Figures 10(a)–(c). The observed absorption edges and calculated optical band gap values are given in Table 1. It is evident that both TeO_2 and B_2O_3 increase the UV and visible absorption in comparison to the pure Ti butoxide gel. B_2O_3 and TeO_2 cause a shift of the absorption edge of pure Ti butoxide gel towards a longer wavelength, the so called red shift (samples C, B, H, and F, Table 1). A red shift of the absorption edge is also observed in borate glasses as a result of the formation of BO_4 units.[7,68] It is noteworthy that the binary composition containing 60 mol% TeO_2 (sample D) exhibits a blue shift in the UV–Vis spectra (E_g of pure TeO_2 is 3·79 eV). The other peculiarity is the appearance of two absorption bands at 240–260 nm and 300–330 nm. It is well known that the transition metal oxides with d° electron configuration give absorption bands in the UV–Vis region due to oxygen–metal charge transfer.[69]

Table 1. Investigated binary and ternary gel compositions, observed cut-off and calculated optical band gap values (E_g)

No.	Compositions, (mol%)	Cut-off (nm)	E_g (eV)
1	Ti(IV) n-butoxide gel	389·71	3·18
	Section with 80% TiO_2		
2	$80TiO_2.20TeO_2$ (sample A)	402·86	3·08
3	$80TiO_2.10TeO_2.10B_2O_3$ (sample B)	431·09	2·88
4	$80TiO_2.20B_2O_3$ (sample C)	385·86	3·21
	Section with 40% TiO_2		
5	$40TiO_2.60TeO_2$ (sample D)	381·58	3·25
6	$40TiO_2.30TeO_2.30B_2O_3$ (sample F)	417·55	2·97
7	$40TiO_2.60B_2O_3$	408·8	3·08

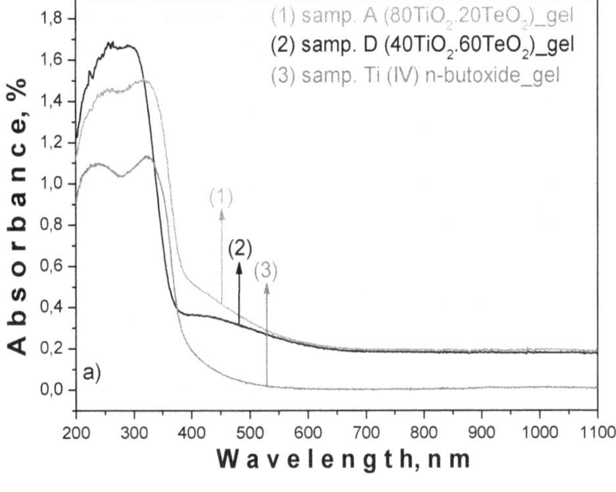

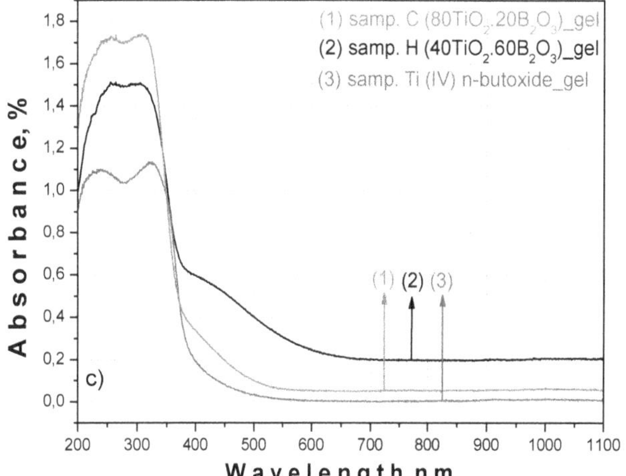

Figure 10. UV-Vis spectra of: (a) binary TiO_2–TeO_2 gels, (b) ternary TiO_2–TeO_2–B_2O_3 gels, (c) binary TiO_2–B_2O_3 gels compared to pure Ti(IV) n-butoxide gel [Colour available online]

The position of this electron transfer depends on the ligand field symmetry surrounding the metal (M) site. For oxygen ligands, a higher energy transition is expected for tetrahedral M (between 220 and 260 nm), compared to that of octahedral Me (between 250 and 360 nm).(69) For isolated TiO_4 units in TiO_2–SiO_2 gels, the ligand to metal charge transfer band is in the region 200–260 nm, while in a titania network (anatase) the charge transfer band of TiO_6 is at 300–600 nm.(70,71) Klein *et al*(72) observed a red shift of the absorption (above 270 nm) in the UV–Vis spectra of TiO_2–SiO_2 gels that is associated with an increase of the degree of polymerisation of Ti atoms.

4. Discussion

Considering the obtained results it may be concluded that the use of more stable and less expensive boric and telluric (VI) acids (compared to their alkoxides) is suitable for the sol-gel preparation procedure.

Useful information on the coordination number of Ti atoms in the as-prepared and dried at room temperature gels was obtained by UV–Vis spectroscopy. The pure Ti butoxide gel, binary and ternary gels contain TiO_4 (band at 240–260 nm) and TiO_6 units (band at 300–330 nm) (Figure 10(a)–(c)). TiO_4 groups are the main building units in the unhydrolysed Ti butoxide and the coordination geometry changes to TiO_6 as a result of polymerised Ti species (Ti–O–Ti links between TiO_6 units).(72) The highest absorption at 300–330 nm in the ternary TiO_2–TeO_2–B_2O_3 gels is an indication of the considerable $TiO_4 \rightarrow TiO_6$ transformation as compared to binary TiO_2–TeO_2 and TiO_2–B_2O_3 gels. This observation can be explained by the greater degree of hydrolysis and condensation processes in the ternary gels. The shifting of E_g values depending on the compositions is an interesting experimental fact. It is difficult to analyse more precisely without specialised experiments. It is well known that E_g depends on many factors: generally, disordered systems cause blue shift; increasing the crystallinity causes red shift; increasing the covalency of the bonds and the degree of polymerisation shift the absorption to the visible range.

Based on the XRD results, in the temperature range 200–500°C composite materials consisting of both an amorphous part and crystalline phases are obtained (Figures 4 and 5). Different phase transformations take place, depending on the heating temperature and the composition. This discussion is focused on the analysis of the structural units in the amorphous phases upon heating.

Firstly we consider the structural changes in gels heated at 200°C. The presence of unhydrolysed organic groups (bands at 1120, 1080 and 1040 cm^{-1}) and OH groups contribute to the formation of a mixed organic–inorganic amorphous structure for all compositions (Figures 7 and 8). The lower intensity of these bands in the spectrum of $80TiO_2.20B_2O_3$ (sample C), as compared to that of $40TiO_2.60B_2O_3$ (sample H), indicates that borate units in limited concentration can be incorporated in the titanate network, leading to a greater degree of hydrolysis at 200°C. The incorporation of high amounts of borate units ($40TiO_2.60B_2O_3$, sample H) is difficult; free B_2O_3 remains that reacts with air moisture, and as a result H_3BO_3 precipitates at room temperature (Figure 5). For the binary TiO_2–TeO_2 system, increasing tellurium content ($40TiO_2.60TeO_2$, sample D) drives the hydrolysis (weak bands at 1120–1040 cm^{-1}, Figure 8). This fact can be associated with the formation of more strong bridging Ti–O–Te links.

The transformation of the organic amorphous phase to an inorganic amorphous phase starts at about 300°C, and is discussed below. According to the DTA analysis, the combustion of the organic groups proceeds at 300°C (Figure 6), and this is accompanied by the strong diminishing of the IR bands at 1460 and 1360 cm^{-1}, which are characteristic for the organics (bending vibrations of CH_3 and CH_2 groups) (Figure 8).[5,54] It should be noted that in compositions with high B_2O_3 content (samples G and H) the absorption bands in the same spectral region are due to the vibrations of BO_3 units.

At 300°C the amorphous network for compositions containing high TiO_2 content (80 mol%) (samples A, B and C) is an inorganic titanate, built up mainly of TiO_6 structural units (Figure 7, bands at 620–610 and 480 cm^{-1}).[73,74] On the other hand, the structure of the amorphous phase in compositions with high TeO_2 content (Figure 8, $40TiO_2.60TeO_2$ - sample D and $40TiO_2.50TeO_2.10B_2O_3$ - sample E) may be considered as a tellurite, built up mainly of TeO_4 units (high intensity band at 630–620 cm^{-1}, along with the band at 780–770 cm^{-1}).[37,40,59,75] This result is in agreement with the Mössbauer and XANES data of Sabadel *et al.*[42] These authors did not observe significant structural modification in TiO_2–TeO_2 glasses (with up to 20 mol% TiO_2), indicating that TiO_2 acts as a network stabiliser. Hayakawa *et al* obtained TiO_2–TeO_2 films in which the presence of TeO_4 units was also proved.[39] This specific behaviour of TiO_2 at the formation of the short range order in a tellurite network is different from that of V_2O_5. Sinclair *et al*[76] and Arnaudov *et al*[75,77] established that the addition of a small amount of V_2O_5 (5 mol%) to TeO_2 leads to considerable $TeO_4 \rightarrow TeO_3$ transformations. Recently, on the basis of neutron diffraction and Raman spectroscopy, Barney *et al*[78] claimed that pure amorphous TeO_2 is built up of TeO_4 and TeO_3 units with 16% terminal oxygen. These data for the existence of terminal oxygen will initiate increased scientific interest in the development of new structural models.

In the temperature range 300–500°C, the amorphous network for compositions rich in B_2O_3 ($40TiO_2.60B_2O_3$ - sample H, $40TiO_2.10TeO_2.50B_2O_3$ - sample G) is built up of BO_3 triangles associated with BO_4 tetrahedra (broad bands centred at 1380 cm^{-1} and 1270 cm^{-1}, and a weak band at about 950 cm^{-1}, Figure 8).[64–66] Since only a small amount of anatase crystallises in these samples at 300°C (low peak intensity for anatase, Figure 5), we conclude that part of the TiO_2 is incorporated in the amorphous borate network. An indirect proof is the appearance of BO_4 units (band at 950 cm^{-1}) and the formation of weak Ti–O–B bridges. Regarding the influence of TeO_2 on boron coordination, several authors reported that TeO_2 contributes to the formation of four coordinated boron atoms in binary TeO_2–B_2O_3 glasses.[44–46,79] Based on the literature and our spectral data for the three-component amorphous network, we assumed that both tellurium and titanium lead to partial $BO_3 \rightarrow BO_4$ transformation, and amorphous phases are obtained from compositions situated outside the region for the formation of melt quenched glasses.

A hypothesis for the creation of amorphous networks which are part of the composite materials can be developed. The short range order of the synthesised amorphous phases is defined by TiO_6, BO_3, BO_4 and TeO_4 structural units. The presence of BO_4 units in B_2O_3-rich compositions is an indirect proof of the formation of Ti–O–B and Te–O–B bridges in the amorphous phase up to 500°C. These bonds are crucial for the formation of amorphous networks, but more experiments are necessary in order to elucidate their existence. As H_3BO_3 did not precipitate in the three-component compositions, obviously there is no free B_2O_3. This means that there is better connectivity between the building units in comparison with the binary TiO_2–B_2O_3 amorphous network. The experimental data that we have obtained prove that the sol-gel method is suitable for TiO_2, TeO_2 and B_2O_3 to be mixed in appropriate concentrations and to build up a disordered network.

5. Conclusions

Transparent monolithic gels were obtained over a wide composition range in the ternary TiO_2–TeO_2–B_2O_3 system. New gel-derived nanocomposite materials consisting of amorphous and various crystalline phases (anatase, rutile, paratellurite and $TiTe_3O_8$) were prepared. Nanosized anatase (8 nm) crystallised earlier (at 300°C) for B_2O_3-rich composition. Organic and OH groups participate in the amorphous organic–inorganic structure up to approximately 300°C. The short range order of the inorganic amorphous phases above 300°C is defined by TiO_6, BO_3, BO_4 and TeO_4 structural units in varying, ratio dependent on composition. The three component networks appear to have better connectivity compared to the binary TiO_2–B_2O_3 system. Titanium does not change the four coordination of tellurium (TeO_4), and in small concentration causes partial transformation $BO_3 \rightarrow BO_4$, which is preserved up to 500°C.

Acknowledgements

This study was performed with the financial support of The Ministry of Education and Science of Bulgaria's, Operational Program "Human Resources Development", co-financed by the European Social Fund of the European Union, contract: BGO51PO001-3.3.06-0050. The authors are also grateful to the financial support of The Ministry of Education and Science of Bulgaria, Contract No ДHTC/India 01/6, 2013: Bulgarian-Indian Inter-governmental Programme of Cooperation in Science and Technology.

References

1. Ganguli, D. *Bull. Mater. Sci.*, 1992, **15**, 421–430.
2. Sakka, S. *Handbook of sol-gel science and technology: processing, characterization and applications*, Volumes I, II and III, Kluwer Academic Publishers, The Netherlands, 2004.
3. Zarzycky, J. *J. Sol-gel Sci. Technol.*, 1997, **8**, 17–22.
4. Brinker, C. J. & Scherer G. W. *Sol-gel science: the physics and chemistry of sol-gel processing*, Academic Press Inc, San Diego, 1990.
5. Coradin, T. & Livage, J. In: *Encyclopedia of Inorganic Chemistry*, Ed. R. B. King, John Wiley & Sons, New York, 2005.
6. Mackenzie, J. D. & Ulrich, D. R. *Ultra-structure of Advanced Ceramics*, Wiley, New York, 1988, 589.
7. Vogel, W. *Chemistry of Glass*, American Ceramic Society, 1985.
8. Zarzycki, J. *Glasses and the vitreous state*, Cambridge University Press, 1991.
9. Dimitriev, Y., Ivanova Y. & Iordanova R. *J. Univ. Chem. Tech. Metall.*, 2008, **43** (2), 181–192.
10. Kakihana, M. *J. Sol-Gel Sci. Technol.*, 1996, **6**, 7–55.
11. Yi, G. & Sayer M. *J. Sol-Gel Sci. Technol.*, 1996, **6**, 65–74.
12. Coffman, P. R., Barlingay C. K., Gupta A. & Dey S. K. *J. Sol-Gel Sci. Technol.*, 1996, **6**, 83–106.
13. Yamashita, H., Yoko, T. & Sakka T. *J. Mater. Sci. Lett.*, 1990, **9**, 796–798.
14. Hirano, Sh.-i., Yogo, T., Kikuta, K. & Yamagiwa K. *J. Am. Ceram. Soc.*, 1992, **75** (9), 2590–2592.
15. Nogami, M. & Moriya, Y. *J. Non-Cryst. Solids*, 1982, **48**, 359–366.
16. Tohge, N., Matsuda, A. & Minami, T. *J. Am. Ceram. Soc.*, 1987, **70**, 13–15.
17. Villegas, M. A. & Fernandez Navarro, J. M. *J. Mater. Sci.*, 1988, **23**, 2464–2478.
18. Irwin, A. D., Holmgren, J. S. & Jonas, J. *J. Non-Cryst. Solids*, 1988, **101**, 249–254.
19. Abe, Y., Gunji, T., Kimata, K., Kuramata, M., Kasgoz, A. & Misono, T. *J. Non-Cryst. Solids*, 1990, **121**, 21–25.
20. Weinberg, M. C., Neilson, G. F., Smith, G. L., Dunn, B., Moore, G. S. & Mackenzie, J. D. *J. Mater. Sci.*, 1985, **20**, 1501–1508.
21. Tohge, N., Moore, G. S. & Mackenzie, J. D. *J. Non-Cryst. Solids*, 1984, **63**, 95–103.
22. Soraru, G. D., Babonneau, F., Gervais, C. & Dallabona, N. *J. Sol-Gel Sci. Technol.*, 2000, **18**, 11–19.
23. Ciceo, R. L., Trandafir, D. L., Radu, T., Pota, O. & Simon, V. *Ceram. Int.*, 2014, **40**, 9517–9524.
24. Gerganova, Ts., Ivanova, Y., Gerganov, T., Miranda Salvado, I. M. & Fernandes M. H. V. In: *The Sol-Gel Process: Uniformity, Polymers and Applications*, Ed. R. E. Morris, Nova Science Publishers, Hauppauge, 2010, Chapter 13.
25. Ota, R., Asagi, N., Fukunaga, J., Yoshida, N. & Fujil, T. *J. Mater. Sci.*, 1990, **25**, 4259–4265.
26. Xue, X., Wang, Y. & Yang, H. *Appl. Surf. Sci.*, 2013, **264**, 94–99.
27. Rehman, S., Ullah, R., Butt, A. M. & Gohar, N. D. *J. Hazard. Mater.*, 2009, **170**, 560–569.
28. Jung, K. Y., Park, S. B. & Ihm, S. K. *Appl. Catal. B: Environ.*, 2004, **51**, 239–245.
29. Moon, S. C., Mametsuka, H., Suzuki, E. & Nakahara, Y. *Catal. Today*, 1998, **45**, 79–84.
30. Arai Y., Itoh, K., Kohara, S. & Yu, J. *J. Appl. Phys.*, 2008, **103**, 094905–6.
31. Gupta, S. M. & Tripathi, M. *Cent. Eur. J. Chem.*, 2012, **10** (2), 279–294.
32. Arrachart, G., Cassidy, D. J., Karatchevtseva, I. & Triani, G. *J. Am. Ceram. Soc.*, 2009, **92**, 2109–2115.
33. Sanchez, C. & Livage, J. *New J. Chem.* 1990, **14**, 513–521.
34. Shubert, U. *J. Sol-Gel Sci. Technol.*, 2003, **26**, 47–55.
35. El-Malawany, R. *Tellurite glasses handbook: Physical properties and data*, CRC Press, Boca Raton, 2002.
36. Weng, L., Hodgson, S., Bao, X. & Sagoe-Crentsil, K. *Mater. Sci. Eng. B*, 2004, **107**, 89–93.
37. Hodgson, S. N. B. & Weng, L. *J. Mater. Sci.: Mater. Electron.*, 2006, **17**, 723–733.
38. Hubert-Pfalzgraf, L. G. *New J. Chem.*, 1987, **11**, 663–675.
39. Hayakawa, T., Koyama, H., Nogami, M. & Thomas, P. *J. Univ. Chem. Tech. Metall.*, 2012, **47**, 381–386.
40. Arnaudov, M., Dimitrov, V., Dimitriev, Y. & Markova, L. *Mater. Res. Bull.*, 1982, **17**, 1121–1129.
41. Kingery, W. D., Bowen, H. K. & Uhlmann, D. R. *Introduction to Ceramics*, Second ed., John Wiley, New York, 1976, p. 669.
42. Sabadel, J.-C., Armand P., Lippens, P. -E., Cachau-Herreillat D. & Philippot E. *J. Non-Cryst. Solids*, 1999, **244**, 143–150.
43. Kashchieva, E., Hinkov, P., Dimitriev, Y. & Miloshev, S. *J. Mater. Sci.*, 1994, **13**, 1760–1763.
44. Osaka, A., Jianrong, Q., Nanba, T., Takada, J., Miura, Y. & Yao, Y. *J. Non-Cryst. Solids*, 1992, **142**, 81–86.
45. Harris Jnr, I. A. & Bray, P. J. *Phys. Chem. Glasses*, 1984, **25**, 44–51.
46. Rada, S., Culea, M. & Culea, E. *J. Non-Cryst. Solids*, 2008, **354**, 5491–5495.
47. Kashchieva, E. P. Liquid – phase separation in tellurite systems, PhD Thesis, Higher Institute of Chemical Technology, Sofia, 1984 (in Bulgarian).
48. Lecomte, A., Bamiere, F., Coste, S., Thomas, P. & Champarnaud-Mesjard, J. C. *J. Eur. Ceram. Soc.*, 2007, **27**, 1151–1158.
49. Weng, L. & Hodgson, S. N. B. *Mater. Sci. Eng.*, 2001, **B 87**, 77–82.
50. Dharma, J. & Pisal, A. Perkin Elmer, Inc, Application note.
51. Weng, L. & Hodgson, S. N. B. *Opt. Mater.*, 2002, **19**, 313–317.
52. Saylikan, F., Asilturk M., Saylikan, H., Onal, Y., Akarsu, M. & Aprac, E. *Turk. J. Chem.*, 2005, **29**, 697–706.
53. Madarasz, J., Braileanu, A., Crisan, M. & Pokol, G. *J. Anal. Appl. Pyrolysis*, 2009, **85**, 549–556.
54. Velasco, M. J., Rubio, F., Rubio, J. & Oteo, J. L. *Spectr. Lett.*, 1999, **32**, 289–304.
55. Doeuff, S., Henry, M., Sanchez, C. & Livage, J. *J. Non-Cryst. Solids*, 1987, **89**, 206–216.
56. Uzunova-Bujnova, M., Dimitrov, D., Radev, D., Bojinova, A. & Todorovsky, D. *Mater. Chem. Phys.*, 2008, **110**, 291–298.
57. Beattie, I. R. & Gilson, T. R. *Proc. R. Soc. A*, 1968, **307**, 407–429.
58. Crisan, M., Zaharescu, M., Crisan, D., Ion, R. & Manolache, M. *J. Sol-Gel Sci. Technol.*, 1998, **13**, 775–778.
59. Yamaguchi, O., Tomihisa, D. & Shimizu, K. *J. Chem. Soc. Dalton Trans.*, 1988, **9**, 2083–2085.
60. Coste, S., Lecompte A., Thomas, P., Merle-Mejean, T. & Champarnaud-Mesjard, J. *J. Sol-Gel. Sci. Technol.*, 2007, **41**, 79–86.
61. Shashikala, M. N., Elizabeth, S., Cary, B. R. & Bhat, H. L. *Current Sci.*, 1987, **56**, 861–863.
62. Ktari, L., Dammak, M., Mhiri, T. & Kolsi, A. W. *Phys. Chem. News*, 2002, **8**, 01–08.
63. Nakamoto, K. *IR and Raman spectra of inorganic coordination compounds*, 1978, J. Wiley & Sons, Third edition, p.230.
64. Kamitsos, E. I., Patsis, A. P., Karakassides, M. A. & Chryssikos, G. D. *J. Non-Cryst. Solids*, 1990, **126**, 52–67.
65. Varsamis, C. P., Kamitsos, E. I. & Chryssikos, G. D. *Phys. Rev. B*, 1999, **60**, 3885–3898.
66. Kapoutsis, J. A., Kamitsos, E. I., Chryssikos G. D., Feller, H. A., Lower, N., Affatigato, M. & Feller, S. A. *Phys. Chem. Glasses*, 2000, **41** (5), 321–324.
67. Hsieh, C.-W., Chiang, A. S. T., Lee, C.-C. & Yang, S.-J. *J. Non-Cryst. Solids*, 1992, **144**, 53–62.
68. Rao, S., Ramadevudu, G., Shareefuddin, Md., Hameed, A., Chary, M. & Rao, M. L. *Int. J. Eng. Sci. Technol.*, 2012, **4** (4), 25–35.
69. Singh, R. S. & Singh, S. P. *J. Mater. Sci.*, 2001, **36**, 1555–1562.
70. Clarck, R. J. H. *Chemistry of titanium and vanadium*, Elsevier, 1968.
71. Gao, X. & Wachs, I. E. *Catal. Today*, 1999, **51**, 233–254.
72. Klein, S., Weckhuysent, B. M., Martens, J. A., Maier, W. F. & Jacobs, P. A. *J. Catal.*, 1996, **163**, 489–491.
73. Gegova, R., Dimitriev, Y., Bachvarova-Nedelcheva, A., Iordanova, R., Loukanov, A. & Iliev, T. *J. Chem. Technol. Metall.*, 2013, **48** (2), 147–153.
74. Iordanova, R., Bachvarova-Nedelcheva, A., Gegova, R. & Dimitriev, Y. *Bulg. Chem. Commun.*, 2013, **45** (4), 485–490.
75. Dimitriev, Y., Dimitrov, V., Gatev, E., Kashchieva, E. & Petkov, H. *J. Non-Cryst. Solids*, 1987, **95–96**, 937–944.
76. Sinclair, R. N., Wright, A. C., Bachra, B., Dimitriev, Y. B., Dimitrov, V. V. & Arnaudov, M. G. *J. Non-Cryst. Solids*, 1998, **232–234**, 38–43.
77. Arnaudov, M. & Dimitriev, Y. *Phys. Chem. Glasses*, 2001, **42** (2) 99–102.
78. Barney, E. R., Hannon, A. C., Holland, D., Umesaki, N., Tatsumisago, M., Orman, R. G. & Feller, S. *J. Phys. Chem. Lett.*, 2013, **4**, 2312–2316.
79. Rong, Q. J., Osaka, A., Nanba, T., Takada, J. & Miura, Y. *J. Mater. Sci.*, 1992, **27**, 3793–3798.

Structural investigation of aluminoborosilicate glasses containing Na_2MoO_4 crystallites by solid state NMR

Takahiro Ohkubo,[a,*] *Ryusuke Monden,*[a] *Yasuhiko Iwadate,*[a] *Shinji Kohara,*[c] *Kenzo Deguchi,*[b] *Shinobu Ohki*[b] *& Tadashi Shimizu*[b]

[a] *Faculty of Engineering, Department of Applied Chemistry and Biotechnology, Chiba University, 1-33 Yayoi-cho Inage-ku, Chiba 263-8522, Japan*
[b] *National Institute of Material Science, Sakura-site, 3-13 Sakura, Tsukuba, Ibaraki 305-0003, Japan*
[c] *Research and Utilization Division, Japan Synchrotron Radiation Research Institute/SPring8, 1-1-1 Kouto, Sayo-cho, Hyogo 679-5198, Japan*

Manuscript received 25 August 2014
Revised version received 30 November 2014
Accepted 5 February 2015

A structural investigation of aluminoborosilicate glasses containing MoO_3 (model nuclear waste glasses) was carried out using high energy x-ray diffraction (HEXRD), ^{29}Si, ^{27}Al, ^{11}B MAS and $^{11}B\{^{23}Na\}$ REDOR NMR experiments. HEXRD revealed that the opaque milk-white colour of the glass with 2 mol% MoO_3 results from Na_2MoO_4 crystallites in the glass. ^{11}B and ^{29}Si MAS NMR spectra showed that negatively charged BO_4^- and Si Q^3 units with one bridging oxygen were converted to BO_3 and Si Q^4 units without bridging oxygen by the introduction of Mo. From $^{11}B\{^{23}Na\}$ REDOR NMR experiments, the distance between Na and BO_4 in the partially crystallised glass (with Na_2MoO_4) is longer than that in the glass without MoO_3.

1. Introduction

Aluminoborosilicate glasses are widely applied in many technologies, such as optical components, sealing materials and nuclear waste immobilisation followed by geological disposal. For nuclear waste glass, one of the requirements is good chemical durability on exposure to groundwater, to prevent leaching out of radioactive elements. Molybdenum is one of the key elements that are added to nuclear waste glass, and it is important to consider its structural role and possible influence on the chemical durability.[1]

After reprocessing of spent nuclear fuel, a high level nuclear waste solution containing Mo is solidified within a highly durable aluminoborosilicate glass. Even though Mo is a non-radioactive element, it is known that the crystallisation of a poorly durable Mo-rich crystalline phase (called yellow phase, and containing water soluble alkali molybdate salts, such as Na_2MoO_4) with significant amounts of radioactive caesium can occur in the glass.[2,3] The formation of this yellow phase must be avoided during glass preparation, because it would lead to a waste form with lower long term performance than for homogeneous glasses. In order to develop a new material and understand the yellow phase, structural information on glasses containing Mo is required.[4–6]

For amorphous materials, solid state NMR is a powerful tool for obtaining element-specific structural information. Several structural studies of nuclear waste glasses have previously been published.[7–9] Magnin *et al* studied the environment of Mo in SiO_2–B_2O_3–Na_2O–CaO–MoO_3 by ^{95}Mo MAS (magic angle spinning) NMR,[9] and found that the Mo^{6+} cations in the glass are present as tetrahedral MoO_4^{2-}, as was also shown by the Mo K edge EXAFS spectra of French nuclear waste glass.[7] According to the NMR results, the MoO_4^{2-} environment becomes closer to the Ca^{2+} cations with increasing CaO in the glass.[9] As a result, the free Na^+ cations can compensate BO_4^- in the borosilicate network, but the remaining free Na^+ contributes to the crystallisation of molybdenum salts.[9] Double-resonance NMR was applied to SiO_2–B_2O_3–Na_2O–CaO–MoO_3 glasses to probe the degree of cation mixing.[8] ^{29}Si–^{11}B and ^{11}B–^{23}Na double-resonance experiments showed that the glass with lower Mo content had a homogeneous cation distribution and a higher B–Si mixing, but on the other hand, the crystallised glass with higher Mo content had a lower B–Si mixing, indicating phase separation. Raman spectroscopy also revealed that the Mo–O stretching vibration in SiO_2–B_2O_3–Na_2O–CaO–MoO_3 glass only slightly evolves with increasing B_2O_3 content, which could explain the nature of formation of molybdate salts.[5,10] However, current understanding of the structure of transparent glasses with and without Mo, and of glass ceramics in which

* Corresponding author. Email ohkubo.takahiro@faculty.chiba-u.jp
Original version presented at VIII Int. Conf. on Borate Glasses, Crystals and Melts, Pardubice, Czech Republic, 30 June–2 July 2014
DOI: 10.13036/17533562.56.4.139

Mo exceeds the solubility limit, is limited.

In this work, we focus on the influence of the glass composition on structure, and we consider glasses with MoO_3 content approaching its solubility limit in simplified nuclear waste glass. The composition of the simplified SiO_2–B_2O_3–Al_2O_3–Na_2O glass is a model for the nuclear waste glass used in Japan.(11) The structure of the glasses with and without Mo was probed by high energy x-ray diffraction (HEXRD), ^{29}Si, ^{27}Al, and ^{11}B MAS and $^{11}B\{^{23}Na\}$ rotational echo double resonance (REDOR) NMR experiments.(12)

2. Experimental

Three aluminoborosilicate glasses with and without MoO_3 were prepared. Table 1 summarizes the analytical compositions and density obtained by Archimedes' principle. The molar ratio of SiO_2, B_2O_3, Al_2O_3 and Na_2O is the same as the practical nuclear waste glass in Japan.(11) SiO_2, H_3BO_3, Al_2O_3, Na_2CO_3 and MoO_3 of reagent grade were used as raw materials for the sample preparation. In order to facilitate the recording of ^{29}Si MAS NMR spectra, a small amount of Fe_2O_3 (0·1 wt%) was added to the glass to reduce the longitudinal relaxation time. The starting materials were mixed and melted in a Pt crucible for 30 min at 1773 K. Then the Pt crucible was dipped into water to obtain a solidified glass. The melting and solidification procedures were repeated two times to obtain homogeneous samples. Finally, the prepared bulk sample was pulverised by ball milling for the solid state NMR experiments. Inductively coupled plasma atomic emission spectroscopy (ICP-AES) analysis was performed to confirm the difference between nominal and actual compositions, which was lower than 1% mass. Transparent glasses were obtained from compositions without Mo (Mo-0) and with 1 mol% MoO_3 content (Mo-1). On the other hand, the sample with 2 mol% MoO_3 (Mo-2) had an opaque milk-white colour.

Table 1. Notation and analytical composition for the samples in this study. The density, ρ, is determined by Archimedes' principle

Sample	SiO_2	B_2O_3	Al_2O_3	Na_2O	MoO_3	ρ (g/cm^3)
Mo-0	65·3	17·3	4·30	13·0	–	2·38
Mo-1	64·9	16·9	4·39	12·9	0·99	2·38
Mo-2	66·9	15·8	3·89	11·5	1·89	2·39

HEXRD experiments were carried out at SPring-8 (BL04B2 beam-line) using a two- axis diffractometer dedicated to glass materials.(13) The incident x-ray energy was 61·5 keV. The diffraction patterns of the samples were detected by a Ge detector. A vacuum chamber was used to suppress the air scattering around the sample. Datasets were corrected for absorption, background, and polarisation. Details of the data correction and normalisation procedures are given elsewhere.(13) After correction and normalisation, the total structure factor, $S(Q)$, was calculated from the coherently scattered intensity. We then obtained the pair correlation function, $G(r)$, by Fourier transforming the reciprocal space data.

^{29}Si, ^{27}Al, ^{11}B MAS and $^{11}B\{^{23}Na\}$ REDOR NMR experiments were carried out on a JEOL ECA-500 spectrometer at 11·75 T, equipped with a 4 mm H-XY Chemagnetics probe. ^{29}Si, ^{27}Al, and ^{11}B MAS NMR experiments were conducted with a pulse width of 4·5 μs (90° angle pulse), 0·30 μs (15° angle pulse), and 0·28 μs (15° angle pulse), respectively. Recycling delays for acquisitions were 30, 3, and 3 s for ^{29}Si, ^{27}Al, ^{11}B, which were determined to not affect the spectra. ^{29}Si, ^{27}Al, and ^{11}B chemical shifts were referenced to polydimethylsilane at −34 ppm,

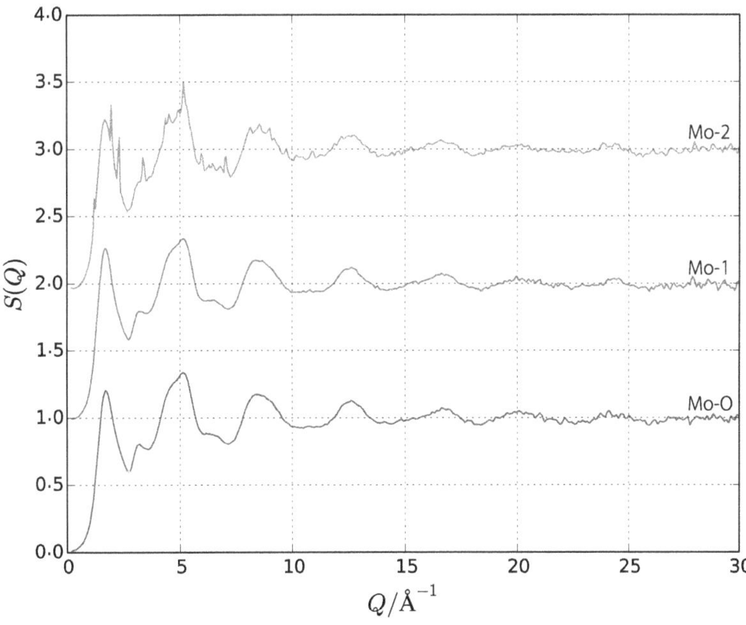

Figure 1. X-ray structure factors, S(Q), of the samples Mo-0, Mo-1, and Mo-2 [Colour available online]

1·0 M aqueous $AlCl_3$ solution at −0·10 ppm, and 1·0 M aqueous H_3BO_3 solution at 19·6 ppm. $^{11}B\{^{23}Na\}$ REDOR NMR experiments were carried out using standard Schaefer-Gullion sequence introducing the XY-4 phase cycling scheme(14) employing ^{11}B signal detection with a rotor-synchronised Hahn spin echo with 10 kHz spinning speed. Dephasing in the ^{23}Na dipolar field was accomplished by 180° pulses of 4 μs length applied to the ^{23}Na channel at the appropriate times required by the pulse sequence. Because the four- and three-coordinated B have significantly different 90° angle pulse, two sets of $^{11}B\{Na\}$ REDOR experiments were separately run using pulse lengths optimised for each of the resonances.

3. Results and discussion

The X-ray structure factors of samples Mo-O, Mo-1, and Mo-2 are shown in Figure 1. Typical amorphous oscillations were observed for Mo-0 and Mo-1, but additional Bragg peaks arising from crystallites were observed for Mo-2. These crystalline peaks are assigned to Na_2MoO_4. We note that laboratory XRD could not detect this small amount of crystalline Na_2MoO_4 peaks in Mo-2. It is clear that the opaque milk-white colour of Mo-2 results from Na_2MoO_4 crystallites in the glass. The x-ray pair correlation functions are shown in Figure 2; several peaks were found up to 6 Å. The first intense peak, ranging from 1·35 to 1·85 Å, arises from the overlap of contributions due to the first B–O, Si–O, and Al–O distances, and individual peaks for these three contributions are not observed. The three pair correlation functions have approximately the same form. The correlation functions for multicomponent glasses are relatively complex and may give less information than for a glass with a simpler composition.

Figure 3 shows ^{29}Si MAS NMR spectra of Mo-0, Mo-1, and Mo-2. The peak maximum of the ^{29}Si MAS NMR spectrum is slightly shifted toward high magnetic field by incorporating Mo. Calas *et al* reported that Mo is present in the glass as a negatively charged tetrahedral molybdate group, MoO_4^{2-}, according to EXAFS results for a silicate glass containing Mo.(7) It should be considered that the broadened line shape of the ^{29}Si MAS NMR spectra arises from a combination of the overlapping of various Si species, such as Q^3 and Q^4 units, centred at around −100 and −105 ppm,(15) respectively. (Here *n*, in Q^n, is the number of bridging oxygen atoms bonded to a given silicon atom.) Unfortunately, these multicomponent glasses are composed of various tetrahedral silicon species with different kinds of neighbours, which leads to complicated overlapping of the peaks in the ^{29}Si MAS NMR spectrum.(16) The observed trend, with a shift toward higher field on incorporation of Mo, is explained as essentially a decrease in the number of Q^3 units, rather than an increase in the number of

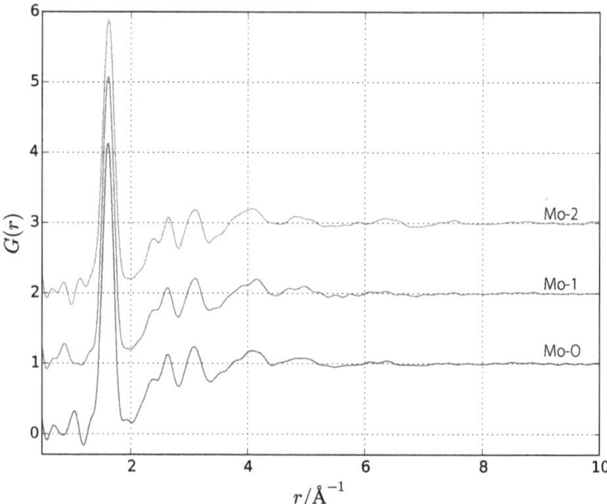

Figure 2. X-ray pair correlation functions, G(r), of the samples Mo-0, Mo-1, and Mo-2 [Colour available online]

Q^4 units. The Q^3 units have one nonbridging oxygen atom, which is charge compensated by the positive electrical charge of Na^+.(17) A simulation of the ^{29}Si MAS NMR spectra was carried out with two Gaussian lines centred at −99 and −107 ppm (associated with Q^3 and Q^4 units), shown in Figure 3. These chemical shift values were kept constant for the simulation of the spectra. This analysis gives a rough estimate of the relative Q^n populations, but it is sufficient to show the variation in Q^n population with Mo content. The obtained Q^3 populations (=$Q^3/(Q^3+Q^4)$) are 0·60, 0·40, and 0·36 for Mo-0, Mo-1, and Mo-2, respectively. The relative decrease in the Q^3 population was also

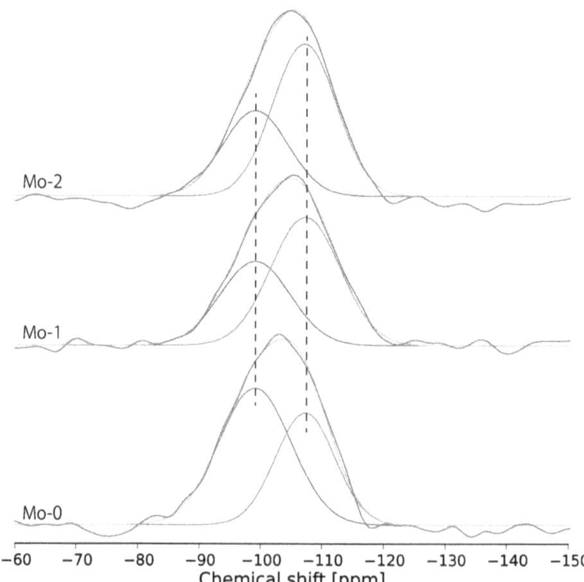

Figure 3. ^{29}Si MAS NMR spectra (magenta line) of the samples Mo-0, Mo-1, and Mo-2. Cyan lines are fitted results, composed of Q^3 (green line) and Q^4 (red line) components. Dashed lines indicate the chemical shift for Q^3 (−99 ppm) and Q^4 (−107 ppm), respectively [Colour available online]

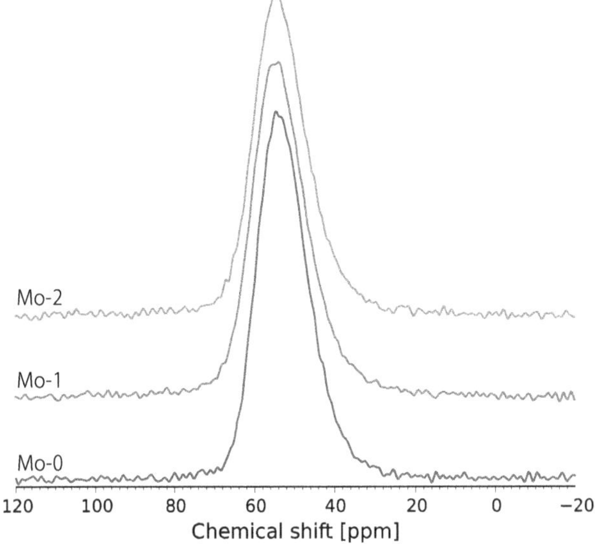

Figure 4. ^{27}Al MAS NMR spectra of the samples Mo-0, Mo-1, and Mo-2 [Colour available online]

established by deconvolution of the ^{29}Si MAS NMR spectra. The results show that in Mo-1 and Mo-2, Na^+ is predominantly compensated by negatively charged MoO_4^{2-}, rather than Q^3 units.

^{27}Al MAS NMR spectra for samples Mo-0, Mo-1, and Mo-2 are shown in Figure 4. All spectra have only a broadened asymmetric peak ranging from 70 to 30 ppm and corresponding to AlO_4 and AlO_5 units. The ^{27}Al MAS NMR spectra of these glasses show only a poorly defined AlO_5 shoulder on the main broad AlO_4 peak. These spectra exhibited the same line shape regardless of Mo content, indicating that the Al structure is not significantly modified by introducing Mo.

Figure 5 shows the ^{11}B MAS NMR spectra with deconvoluted lines, and Table 2 gives the NMR parameters obtained from fitting these spectra. Using the program dmfit2002,[18] all the ^{11}B MAS NMR spectra were deconvoluted into four lines, assigned to two three-coordinated B units (BO_3^{ring}, $BO_3^{non\text{-}ring}$) in rings and outside rings, and two four-coordinated B units (BO_4(3Si), BO_4(4Si)), corresponding to different numbers of neighbouring Si tetrahedron units.[19,20] The relative numbers of 4-coordinated B units (the summation of BO_4(3Si) and BO_4(4Si)), are 30·3, 26·7, and 23·9 for Mo-0, Mo-1, and Mo-2, respectively. Both silicon Q^3 units and negatively charged BO_4 units compensate the positive electrical charge of Na^+.[21] The crystallisation of sodium molybdate causes BO_4 units to be converted to BO_3 units due to the loss of Na^+ from the glass phase. However, Caurant *et al* reported that the BO_4 population in SiO_2–B_2O_3–Na_2O–CaO glasses was not significantly changed by increasing the Mo content.[5] This difference from the study reported here may be due to the effect of AlO_4^- in the glass network.

While the structural changes of Si and B units were

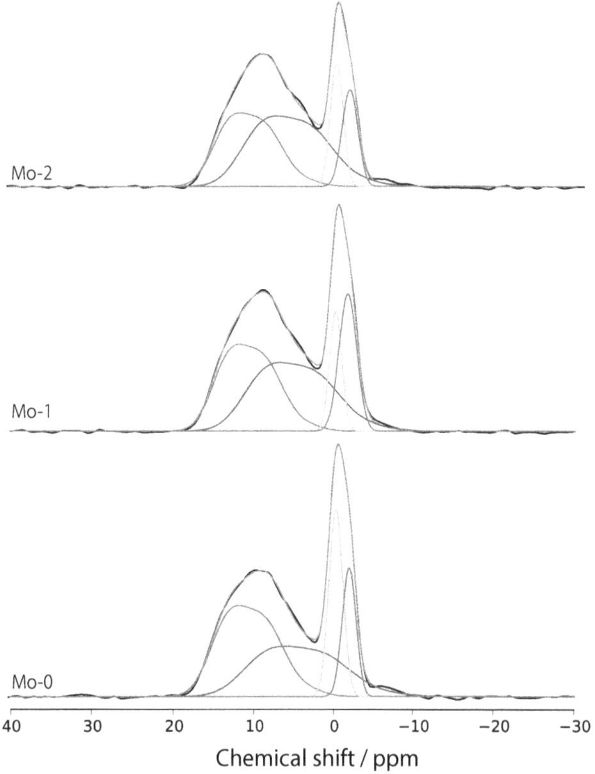

Figure 5. ^{11}B MAS NMR spectra (black line) of the samples Mo-0, Mo-1, and Mo-2. Red lines are the fitted spectrum. Green and blue lines are BO_3^{ring} and $BO_3^{non\text{-}ring}$ components, respectively. Cyan and magenta lines are BO_4(3Si) and BO_4(4Si) components respectively [Colour available online]

estimated from ^{29}Si and ^{11}B MAS NMR spectra, the spatial distribution of ^{23}Na around ^{11}B was investigated by $^{23}Na\{^{11}B\}$ REDOR experiments. The van Vleck second moments, $M_2(I–S)$, of heteronuclear dipolar couplings can be directly calculated from REDOR NMR experiments, which allow us to quantify the magnetic dipole–dipole interactions for each species

Table 2. ^{11}B NMR parameters and population of B species obtained from curve fitting of ^{11}B MAS spectra of the samples Mo-0, Mo-1, and Mo-2

	Mo-0			
	Pop.	δ_{iso} *(ppm)*	C_Q *(MHz)*	η
BO_3^{ring}	38·5	16·5	2·45	0·2
$BO_3^{non\text{-}ring}$	31·2	12·9	2·96	0·2
BO_4(3Si)	18·1	0·2	0·718	0
BO_4(4Si)	12·3	−1·46	0·712	0
	Mo-1			
	Pop.	δ_{iso} *(ppm)*	C_Q *(MHz)*	η
BO_3^{ring}	35·0	16·2	2·39	0·2
$BO_3^{non\text{-}ring}$	38·3	12·7	2·81	0·2
BO_4(3Si)	10·0	0·18	0·668	0
BO_4(4Si)	16·7	−1·13	0·803	0
	Mo-2			
	Pop.	δ_{iso} *(ppm)*	C_Q *(MHz)*	η
BO_3^{ring}	33·6	16·2	2·40	0·2
$BO_3^{non\text{-}ring}$	42·5	12·9	2·75	0·2
BO_4(3Si)	12·2	0·1	0·682	0
BO_4(4Si)	11·7	−1·44	0·757	0

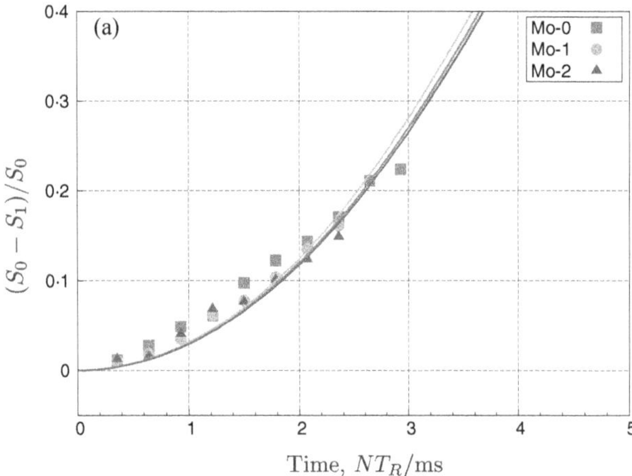

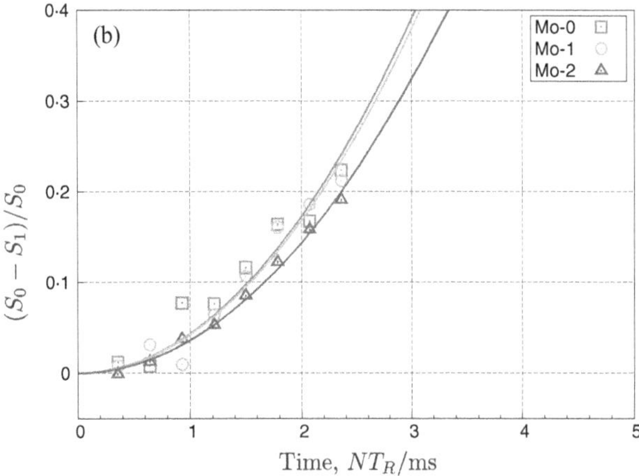

Figure 6. [11]*B{*[23]*Na} REDOR curves for the samples Mo-0, Mo-1, and Mo-2. (a)* BO_3 *and (b)* BO_4 *[Colour available online]*

of observed nuclide (^{11}B). The intensities of the reference signal without the π-pulse trains for the ^{23}Na channel (S_0), and the dipole recoupling signal with π-pulse (S_1), were observed with different dipolar evolution time, NT_r. S_0 and S_1 as a function of NT_r were calculated for BO_3 and BO_4 from the $^{11}B\{^{23}Na\}$ REDOR NMR spectra using the same deconvolution procedure as for the ^{11}B MAS NMR experiment. The values of $(S_0-S_1)/S_0$ for BO_3 and BO_4 units are shown in Figure 6 as a function of NT_r. For quadrupolar spin systems, such as ^{23}Na (I=3/2) and ^{11}B (S=3/2), $(S_0-S_1)/S_0$ values for short evolution time satisfying $(S_0-S_1)/S_0<0{\cdot}2$ are well described by the following expression,[22,23]

$$\frac{S_0 - S_1}{S_0} = \frac{f}{I(I+1)\pi^2}\left(NT_r\right)^2 M_2\left(I-S\right) \quad (1)$$

where f values are experimentally determined with f=0·263 for BO_4 and 0·216 for BO_3 units, respectively.[24] We modelled the REDOR data satisfying $(S_0-S_1)/S_0<0{\cdot}2$ by this simple parabolic function of NT_r, and derived $M_2(I-S)$ values for the BO_3 and BO_4 units. The distance between ^{11}B and ^{23}Na is then calculated from the $M_2(I-S)$ value using the following equation,[25]

$$M_2\left(I-S\right) = \frac{4}{15}\left(\frac{\mu_0}{4\pi}\right)^2 \hbar^2\gamma_I^2\gamma_S^2 S\left(S+1\right)\sum_j \frac{1}{r_j^6} \quad (2)$$

where γ_I and γ_S are the gyromagnetic ratios of the coupled nuclei, μ_0 is the vacuum permeability, and h is Planck's constant over 2π. $M_2(I-S)$ for a particular ^{11}B site is equal to the summation of the interactions with the ^{23}Na surrounding it at distance r. To calculate the ^{23}Na–^{11}B distance, we assumed an isolated ^{23}Na–^{11}B spin pair (j=1 in Equation (2)). Since the ^{11}B–^{23}Na distance calculated with j>1 resulted in a clearly unrealistic value, the assumption that j=1 is reliable in this study. Table 3 summarises the ^{23}Na–^{11}B distances for BO_3 and BO_4 in all samples. Although a wide distribution is expected for the ^{23}Na–^{11}B distances, the obtained BO_4–Na distance for Mo-2 is longer than that for the glass without Mo. On the other hand, the BO_3–Na distance is almost constant regardless of whether there is Na_2MoO_4 crystallisation in the glass. It is expected that the unscreened Coulomb interaction between BO_4^- and Na^+ is more effective than that for BO_4^- and Na_2MoO_4. As a result, BO_4^-–Na_2MoO_4. in Mo-0 is closer than in Mo-2. An interpretation of the longer BO_4–Na distance in Mo-2 is that Na around BO_4 in Mo-2 may be a neighbour Na in crystallised Na_2MoO_4, rather than an isolated Na^+.

Table 3. Calculated [11]*B–*[23]*Na distances, r, for* BO_3 *and* BO_4 *units in the samples Mo-0, Mo-1, and Mo-2, obtained from* [11]*B{*[23]*Na} REDOR NMR experiments*

Sample	*r, BO_3–Na (Å)*	*r, BO_4–Na (Å)*
Mo-0	3·04	2·96
Mo-1	3·02	2·97
Mo-2	3·05	3·05

4. Conclusions

Our structural study of the introduction of MoO_3 to SiO_2–B_2O_3–Al_2O_3–Na_2O glasses revealed an increase of the Q^4 and BO_3 populations, indicating that Na^+ preferentially compensates MoO_4^{2-}, rather than Q^3 and BO_4^-. $^{11}B\{^{23}Na\}$ REDOR NMR experiments revealed that the BO_3–Na distance is not significantly modified by introducing MoO_3, but the BO_4–Na distance in crystallised glass with Na_2MoO_4 is longer than that in the glass without Mo.

References

1. Short, R. J., Hand, R. J., Hyatt, N. C. & Möbus, G. Environment and oxidation state of molybdenum in simulated high level nuclear waste glass compositions. *J. Nucl. Mater.*, 2005, **340**, 179–186.
2. Caurant, D., Ma jerus, O., Fadel, E., Lenoir, M., Gervais, C. & Pinet, O. Effect of molybdenum on the structure and on the crystallisation of SiO_2–Na_2O–CaO–B_2O_3 glasses. *J. Am. Ceram. Soc.*, 2007, **90**, 774–783.
3. Short, R. J., Hand, R. J. & Hyatt, N. C. Molybdenum in nuclear waste glasses-incorporation and redox state. *Mater. Res. Soc. Symp. Proc.*, 2002, **757**, 141–146.
4. Farges, F., Siewert, R., Brown, G., Guesdon, A. & Morin, G. Structural environments around molybdenum in silicate glasses and melts. I. Influence of composition and oxygen fugacity on the local structure of molybdenum. *Can. Mineral.*, 2006, **44**, 731–753.

5. Caurant, D., Majérus, O., Fadel, E., Quintas, A., Gervais, C., Charpentier, T. & Neuville, D. Structural investigations of borosilicate glasses containing MoO_3 by MAS NMR and raman spectroscopies. *J. Nucl. Mater.*, 2010, **396**, 94–101.
6. Pinet, O., Dussossoy, J., David, C. & Fillet, C. Glass matrices for immobilizing nuclear waste containing molybdenum and phosphorus. *J. Nucl. Mater.*, 2008, **377**, 307–312.
7. Calas, G., Le Grand, M., Galoisy, L. & Ghaleb, D. Structural role of molybdenum in nuclear glasses: an EXAFS study. *J. Nucl. Mater.*, 2003, **322**, 15–20.
8. Martineau, C., Michaelis, V. K., Schuller, S. & Kroeker, S. Liquid-liquid phase separation in model nuclear waste glasses: A solid-state double-resonance NMR study. *Chem. Mater.*, 2010, **22**, 4896–4903.
9. Magnin, M., Schuller, S., Mercier, C., Trebosc, J., Caurant, D., Majérus, O., Angeli, F. & Charpentier, T. Modification of molybdenum structural environment in borosilicate glasses with increasing content of boron and calcium oxide by ^{95}Mo MAS NMR. *J. Am. Ceram. Soc.*, 2011, **94**, 4274–4282.
10. Caurant, D., Majérus, O., Fadel, E., Lenoir, M., Gervais, C., Charpentier, T. & Neuville, D. Structural characterization of SiO_2-Na_2O-CaO-B_2O_3-MoO_3 glasses. 2011, *arXiv preprint arXiv:1104.1862*.
11. Inagaki, Y., Mitsui, S., Makino, H., Ishiguro, K., Kamei, G., Kawamura, K., Maeda, T., Ueno, K., Banba, T. & Yui, M. Status of studies on HLW glass performance for confirming its validity in assessment. *J. Nucl. Fuel Cycle Environ.*, 2004, **10**, 69–84. (In Japanese.)
12. Pan, Y., Gullion, T. & Schaefer, J. Determination of C-N internuclear distances by rotational-echo double-resonance NMR of solids. *J. Magn. Reson.*, 1990, **90**, 330–340.
13. Kohara, S., Itou, M., Suzuya, K., Inamura, Y., Sakurai, Y., Ohishi, Y. & Takata, M. Structural studies of disordered materials using high-energy X-ray diffraction from ambient to extreme conditions. *J. Phys.: Condens. Matter*, 2007, **19**, 506101.
14. Gullion, T. & Schaefer, J. Elimination of resonance offset effects in rotational-echo double-resonance NMR. *J. Magn. Reson.*, 1991, **92**, 439–442.
15. MacKenzie, K. & Smith, M. *Multinuclear Solid-State Nuclear Magnetic Resonance of Inorganic Materials*, Elsevier Science Ltd. Oxford, 2002, pp. 206.
16. Mahler, J. & Sebald, A. Deconvolution of ^{29}Si magic-angle spinning nuclear magnetic resonance spectra of silicate glasses revisited-some critical comments. *Solid State Nucl. Magn. Reson.*, 1996, **5**, 63–78.
17. Maekawa, H., Maekawa, T., Kawamura, K. & Yokokawa, T. The structural groups of alkali silicate glasses determined from ^{29}Si MAS-NMR. *J. Non-Cryst. Solids*, 1991, **127**, 53–64.
18. Massiot, D., Fayon, F., Capron, M., King, I., Le Calve, S., Alonso, B., Durand, J.-O., Bujoli, B., Gan, Z. & Hoatson, G. Modelling one-and two-dimensional solid-state NMR spectra. *Magn. Reson. Chem.*, 2002, **40**, 70–76.
19. Du, L.-S. & Stebbins, J. F. Site preference and Si/B mixing in mixed-alkali borosilicate glasses: a high-resolution ^{11}B and ^{17}O NMR study. *Chem. Mater.*, 2003, **15**, 3913–3921.
20. Du, L.-S. & Stebbins, J. F. Nature of silicon-boron mixing in sodium borosilicate glasses: a high-resolution ^{11}B and ^{17}O NMR study. *J. Phys. Chem. B*, 2003, **107**, 10063–10076.
21. Dell, W., Bray, P. & Xiao, S. ^{11}B NMR studies and structural modeling of Na_2O–B_2O_3–SiO_2 glasses of high soda content. *J. Non-Cryst. Solids*, 1983, **58**, 1–16.
22. Bertmer, M., Züchner, L., Chan, J. & Eckert, H. Short and medium range order in sodium aluminoborate glasses. 2. Site connectivities and cation distributions studied by rotational echo double resonance NMR spectroscopy. *J. Phys. Chem. B*, 2000, **104**, 6541–6553.
23. Chan, J., Bertmer, M. & Eckert, H. Site connectivities in amorphous materials studied by double-resonance NMR of quadrupolar nuclei: High-resolution ^{11}B-^{27}Al spectroscopy of aluminoborate glasses. *J. Am. Chem. Soc.*, 1999, **121**, 5238–5248.
24. Epping, J. D., Strojek, W. & Eckert, H. Cation environments and spatial distribution in Na_2O–B_2O_3 glasses: New results from solid state NMR. *Phys. Chem. Chem. Phys.*, 2005, **7**, 2384–2389.
25. Van Vleck, J. The dipolar broadening of magnetic resonance lines in crystals. *Phys. Rev.*, 1948, **74**, 1168–1183.

Glass and glass-ceramics in the La_2O_3–Gd_2O_3–PbO–MnO–B_2O_3 system

R. Raykov,[1] A. Staneva,[1] Y. Dimitriev,[1] E. Kashchieva,[1] S. Slavov[1] & B. Blagoev[2]

[1] *University of Chemical Technology and Metallurgy, 8 Kl. Ohridski, 1756 Sofia, Bulgaria*
[2] *Institute of Electronics, Bulgarian Academy of Sciences, 72 Tzarigradsko Chaussee Blvd., 1784 Sofia, Bulgaria*

Manuscript received 12 August 2014
Revised version received 15 January 2015
Accepted 6 July 2015

This study is a continuation of our previous research on the system La_2O_3–PbO–MnO–B_2O_3. The purpose is the production of glass-ceramic materials based on glasses in the system La_2O_3–Gd_2O_3–PbO–MnO–B_2O_3 containing $(La_{1-x}Gd_x)_{1-y}Pb_yMnO_3$ crystalline particles. The studied compositions were located in a section of the phase diagram with constant B_2O_3 content (30 mol%). Glasses and glass-ceramic materials were obtained in a small part of the investigated 30 mol% B_2O_3 section. For a nominal composition $30B_2O_3.8{\cdot}75La_2O_3.8{\cdot}75Gd_2O_3.23{\cdot}33PbO.29{\cdot}17MnO$, which is outside the glass formation region, a glass-ceramic material was obtained directly from the supercooled melt. Its phase formation and microstructure were investigated by DTA, XRD and SEM. It was found that the main crystalline phase is a rare earth manganite.

1. Introduction

The crystallisation of glasses after heat treatment provides great potential to modify the properties and to create new materials for various applications.[1,2] An opportunity for the development of glass-ceramics was shown by the crystallisation of a rare earth manganese phase ($La(Sr)MnO_3$) in the glassy system MnO_2–SrO–La_2O_3–B_2O_3 (SiO_2) by a modified glass crystallisation method.[3,4] Our previous studies have focused on glass formation in manganese borate systems, and the structure and preparation of glass-ceramics containing $La_{0{\cdot}6}Sr_{0{\cdot}4}MnO_3$ and $La_{0{\cdot}6}Pb_{0{\cdot}4}MnO_3$.[5–8] Data on the rare earth manganite phases were first published in the original studies of Jonker & Van Santen.[9,10] Newer and more detailed information on the electrical and magnetic properties of the phases with composition $La_{1-x}M_xMnO_3$ (M=Pb, Sr, Ca, Ba) is given in original papers and comprehensive reviews.[11–19] The introduction of Pb^{2+} in $LaMnO_3$ causes an increase of the ratio Mn^{4+}/Mn^{3+}, a decrease in the synthesis temperature, and an improvement in the electroconductivity.[20–23] Furthermore, collosal magnetoresistance has been observed in the phase $La_{1-x}Pb_xMnO_3$.[24–27] Recently, the electrical and magnetic properties of this phase were verified.[28] Also multiferoic $GdMnO_3$ is of interest, because it shows room temperature ferroelectric behaviour associated with the frequency independent dielectric anomaly at 373 K (T_c), whereas antiferromagnetic behaviour is observed at around 40 K.[29–31] Gd doped $LaMnO_3$ has very complicated behaviour, depending on the doping concentration and temperature, and is either an antiferromagnetic insulator, or a feromagnetic metal, and also possesses a magnetocaloric effect.[32–35]

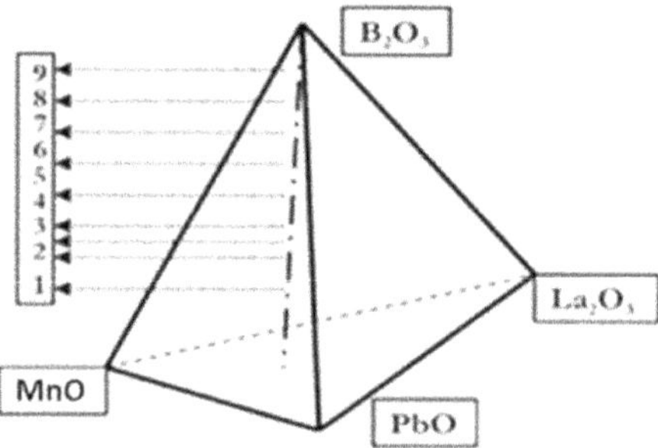

Figure 1. Selected compositions with constant ratio between MnO, PbO and La_2O_3, and increasing B_2O_3 content, and their graphical depiction[8]

No.	Composition	Appearance
1	$20B_2O_3.16La_2O_3.32PbO.32MnO$	Black glass-ceramic
2	$30B_2O_3.14La_2O_3.28PbO.28MnO$	Black glass
3	$35B_2O_3.14La_2O_3.28PbO.28MnO$	Black glass
4	$40B_2O_3.12La_2O_3.24PbO.24MnO$	Black glass
5	$50B_2O_3.10La_2O_3.20PbO.20MnO$	Black glass
6	$60B_2O_3.8La_2O_3.16PbO.16MnO$	Black glass
7	$70B_2O_3.6La_2O_3.12PbO.12MnO$	Violet glass
8	$80B_2O_3.4La_2O_3.8PbO.8MnO$	Two phase glass
9	$90B_2O_3.2La_2O_3.4PbO.4MnO$	Two phase glass

Our aim is the production of composite materials based on La_2O_3–Gd_2O_3–PbO–MnO–B_2O_3 glasses, containing $(La_{1-x}Gd_x)_{1-y}Pb_yMnO_3$ crystalline particles. The studied composition was located in a section of the phase diagram with constant B_2O_3 content (30 mol%).

2. Experimental

As can be seen from our previous research[5–8] in the system La_2O_3–PbO–MnO–B_2O_3 at constant ratio La:Pb:Mn=1:2:2, the addition of B_2O_3 increases the glass forming ability (Figure 1). On the basis of these

* Corresponding author. Email radi_pgtm@abv.bg
Original version presented at VIII Int. Conf. on Borate Glasses, Crystals and Melts, Pardubice, Czech Republic, 30 June–2 July 2014
DOI: 10.13036/17533562.56.4.145

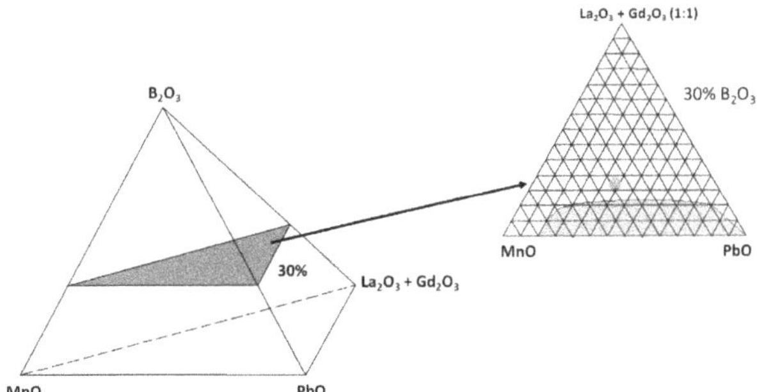

Figure 2. Glass formation region in a section of the phase diagram of La_2O_3–Gd_2O_3–PbO–MnO–B_2O_3 with a constant B_2O_3 content of 30 mol% [Colour available online]

results, we decided to prepare compositions with a constant B_2O_3 content of 30 mol%. The batches were prepared by mixing La_2O_3, Gd_2O_3, Pb_3O_4, MnO_2 and H_3BO_3 in an agate mortar. They were melted at 1473 K in a platinum crucible. The melts were cooled at a rate of 10^2 K/s by pressing between two aluminium plates. After we determined the glass formation region of this section of the phase diagram, the composition $30B_2O_3.8{\cdot}75La_2O_3.8{\cdot}75Gd_2O_3.23{\cdot}33PbO.29{\cdot}17MnO$ (mol%), which is located outside of the glass forming region, was melted with the same experimental conditions, and was subjected to more detailed studies.

Phase formation was studied by x-ray powder diffraction analysis using a Bruker D5005 diffractometer with Cu K_α radiation. The microstructure of the samples was studied by means of a Jeol JSM-840A scanning electron microscope (SEM). To examine the structure and chemical composition, polished cross-sections of the samples were made. The chemical composition of the samples was determined by x-ray microanalysis, using the energy dispersive spectroscopy (EDS) method on a LINK Analytical AN10000 system. DTA was performed using alumina crucibles in a Stanton Redcroft-STA-780 apparatus, with a heating rate of 10 K/min. The electrical conductivity of the materials was determined by V–A measurements at 70–270 K of sandwich structures with aluminium electrodes.

3. Results and discussion

The small glass formation region is shown in Figure 2. It is difficult to obtain glass at the cooling rate that was used, except for compositions situated near to the MnO–PbO side (Figure 2). A supercooled sample with initial nominal composition $30B_2O_3.8{\cdot}75La_2O_3.8{\cdot}75Gd_2O_3.23{\cdot}33PbO.29{\cdot}17MnO$ (mol%), was investigated by XRD analysis, as shown in Figure 3(a). The obtained material is glass-ceramic, because the XRD pattern exhibits both amorphous haloes and pronounced Bragg diffraction peaks. Comparison with the XRD patterns of $(La_{0{\cdot}9}Gd_{0{\cdot}1})_{0{\cdot}67}Sr_{0{\cdot}33}MnO_3$ and $(La_{0{\cdot}47}Gd_{0{\cdot}2})Sr_{0{\cdot}33}MnO_3$ shows that these Bragg peaks can reasonably be indexed to the pure perovskite-type crystal structure.[32,33] The pattern for a sample obtained by solid state reaction shows a similar result (Figure 3(b)).

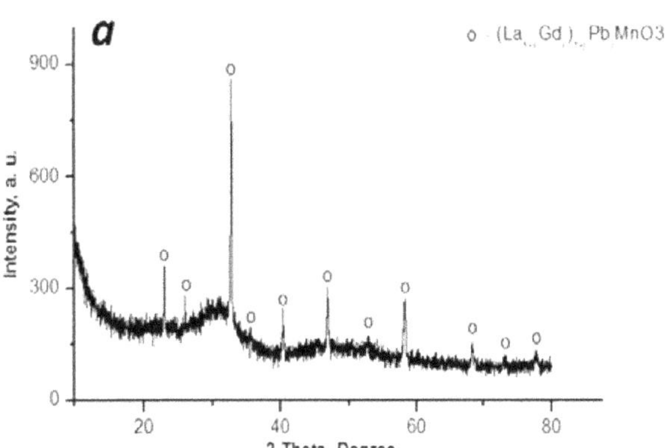

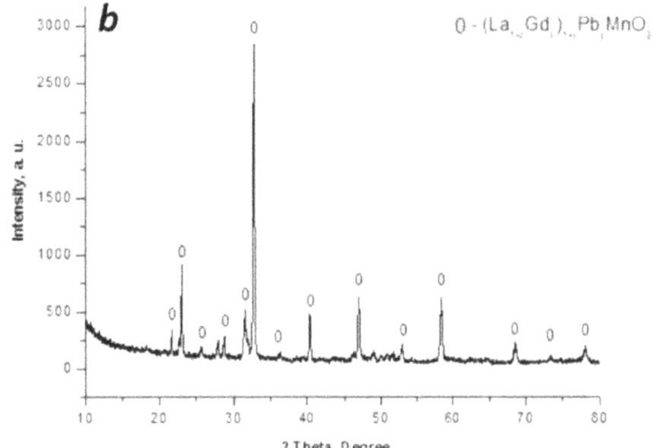

Figure 3. (a) XRD pattern of sample with nominal composition $30B_2O_3.8{\cdot}75La_2O_3.8{\cdot}75Gd_2O_3.23{\cdot}33PbO.29{\cdot}17MnO$, (b) XRD pattern of sample with nominal composition $12{\cdot}5La_2O_3.12{\cdot}5Gd_2O_3.33{\cdot}33PbO.41{\cdot}67MnO$ obtained by solid state reaction

The morphology and the particles size distribution were determined by SEM analysis, as shown in Figure 4. The average particle size is about 10 μm. In order to determine the chemical composition of the crystalline particles and of the amorphous matrix, micro-probe analysis was performed.

Figure 4 shows that the sample is microheterogeneous, and mainly two crystalline phases are observed. One of the crystallised phases is probably a rare earth borate containing La and Gd in the ratio 1:1, which is not detected by XRD analysis. The other crystalline phase is ~$LaGdMnPbO_3$ with a ratio La:Gd=3:1. The amorphous matrix includes mainly B_2O_3 and PbO, and a small amount of the other components.

Two not well resolved exo-effects are observed in the DTA curve near 600–700°C (873–973 K). The TG curve indicates an increase of mass, which is a result of the oxidation of the material. The overlapping of the two effects (crystallisation and oxidation) is the reason they are not clearly distinguished.

The electrical resistance changes considerably with increasing temperature, and the glass-ceramic mate-

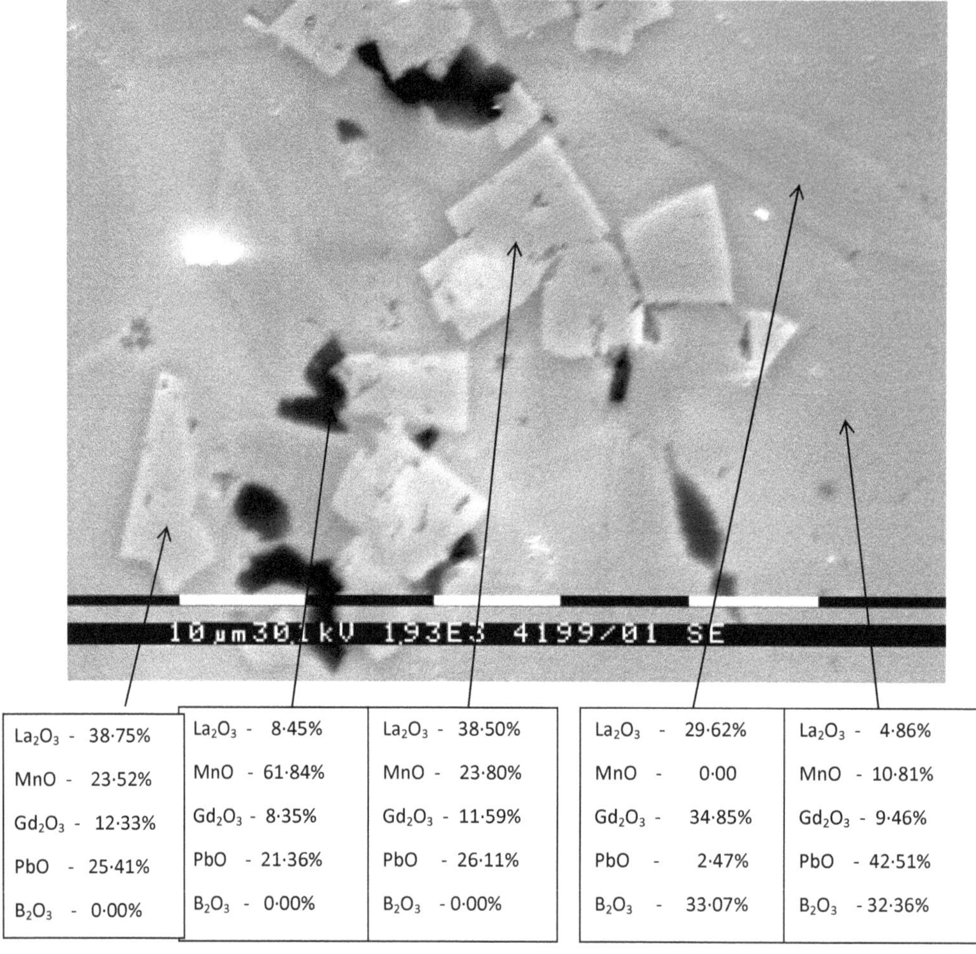

Figure 4. SEM micrograph of the glass-ceramic material with nominal composition $30B_2O_3.8{\cdot}75La_2O_3.8{\cdot}75Gd_2O_3.23{\cdot}33PbO.29{\cdot}17MnO$

rial behaves as a dielectric below room temperature in the range 70–270 K (Figure 6). The curve is compatible with the data for the resistance as a function of temperature for the phase $La_{0{\cdot}5-x}Gd_xSr_{0{\cdot}5}MnO_3$.[34] Visually, by contact of the material with a permanent magnet, some attraction was established, which indicates the presence of ferromagnetic properties.

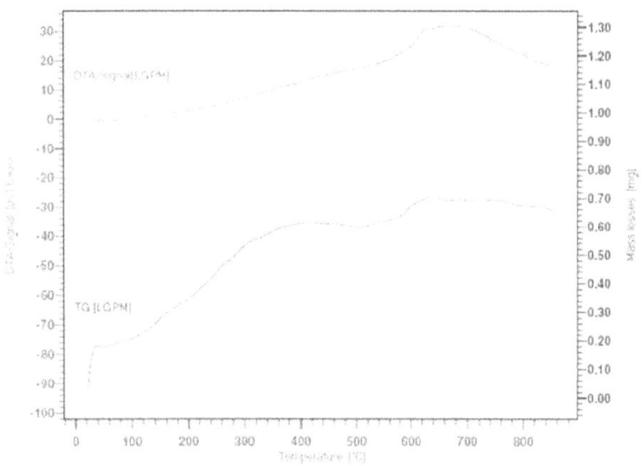

Figure 5. DTA on the material with nominal composition $30B_2O_3.8{\cdot}75La_2O_3.8{\cdot}75Gd_2O_3.23{\cdot}33PbO.29{\cdot}17MnO$ [Colour available online]

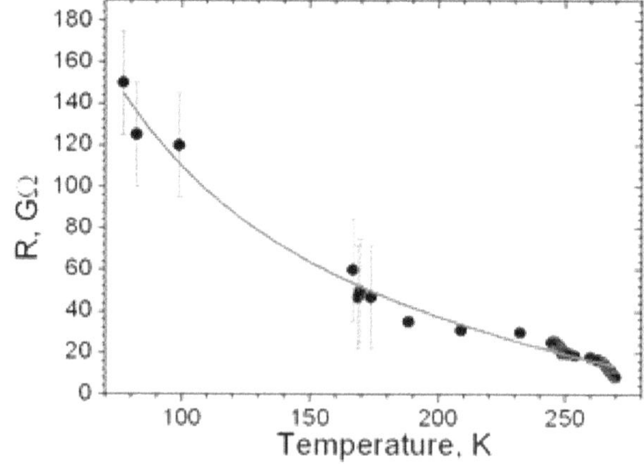

Figure 6. Electrical resistance in the temperature range 70–270 K of the glass-ceramic material with nominal composition $30B_2O_3.8{\cdot}75La_2O_3.8{\cdot}75Gd_2O_3.23{\cdot}33PbO.29{\cdot}17MnO$ [Colour available online]

4. Conclusions

Glasses and glass-ceramics have been obtained in the 30 mol% B_2O_3 section of the La_2O_3–Gd_2O_3–PbO–MnO–B_2O_3 phase diagram. It is shown that by melt quenching, glass-ceramic materials containing a

$(La_{1-x}Gd_x)_{1-y}Pb_yMnO_3$ phase as small crystals embedded in an amorphous matrix can be obtained without additional heat treatment.

Acknowledgements

The work was performed with the financial support of the Bulgarian National Scientific Foundation Contract BG051PO001-3.3.05-0001 Science & Business.

References

1. Strnad, Z. *Glass-ceramic materials*, Elsevier, Amsterdam, 1986.
2. Beregnoi, A. Sitali i fotositali, *Machinostroenie*, Moscow, 1981.
3. Muller, R. J., Schuppel, W., Eick, T. & Steinbeiss, E. LaSr-manganate powders by crystallization of a borate glass. *Magn. Magn. Mater.*, 2000, **217**, 155.
4. Muller, R., Eick, T., Steinmetz, H. & Steinbeiss, E. LaSr-manganate powders by crystallization of a glass *J. Eur. Ceram. Soc.*, 2001, **21**, 1941.
5. Staneva, A., Gattef, E. & Dimitriev, Y. Short range order of glasses in the systems $MnO–B_2O_3$ and $PbO–MnO–B_2O_3$. *Phys. Chem. Glasses*, 2000, **42**, 403.
6. Staneva, A., Gattef, E., Dimitriev, Y. & Mikhov, M. Magnetic materials containing LaSr manganite phase. *Solid State Sci.*, 2004, **6**, 47.
7. Staneva, A., Dulguerov, N., Kashchieva, E., Ivanova, Y., Mikli, V. M., Terzieva, S., Dimitriev, Y. & Andonov, A. Microstructure of Ferromagnetic Composites Containing Manganite Borate Glass Ceramics and $La_{0.6}Pb_{0.4}MnO_3$. In: *Nanoparticles, Nanoscience and Nanotechnology: Nanostructured Materials Application and Innovation Transfer*, Eds E. Balabanova & I. Dragieva, Heron Press, Sofia, 2007, p. 226.
8. Kashchieva, E., Dulguerov, N., Staneva, A. & Dimitriev, Y. Thermal Stability and Microstructure of Glasses and Glass-Ceramics in the System La_2O_3-PbO-MnO-B_2O_3. *Adv. Mater. Res.* 2008, **39–40**, 177.
9. Jonker, G. H. Magnetic compounds with perovskite structure IV Conducting and non-conducting compounds. *Physica*, 1956, **22**, 707.
10. Jonker, G. H. & Van Santen, J. Ferromagnetic compounds of manganese with perovskite structure. *Physica*, 1950, **16**, 337.
11. Rubinchik, I. In Soedinenia dvoinieh okislov, *redkozemlenhih elementov*, Minsk, 1974. (In Russian).
12. Rao, C. N. R. & Gopalakrishnan, G. *New Directions in Solid State Chemistry*, Cambridge University Press, 1997.
13. Rangavittal, N., Guru Row, T. N. & Rao, C. N. R. A study of cubic bismuth oxides of the type $Bi_{26-x}M_xO_{40-d}$ (M=Ti, Mn, Fe, Co, Ni or Pb) related to g-Bi_2O_3. *Eur. J. Solid. State. Inorg. Chem.*, 1994, **31**, 409.
14. Pena, O. Substitution effects in magnetic and superconducting materials. *Bol. Soc. Esp. Ceram. Vidrio*, 1999, **38**, 385.
15. Kuo, J. H., Anderson, H. U. & Sparlin, D. M. J. Oxidation-reduction behavior of undoped and Sr-doped $LaMnO_3$ nonstoichiometry and defect structure. *J. Solid State Chem.*, 1989, **83**, 52.
16. Nagaev, E. L. Lanthanum manganites and other giant-magnetoresistance magnetic conductors. *Usp. Fiz. Nauk*, 1996, **166**, 833.
17. Ibarra, M. R. & De Teresa, J. M. Colossal magne toresistance in manganites: Role of the electronic phase segregation. In: *Magnetism, Magnetic Materials and Their Applications*, Vol. 302–3, Ed. F. P. Missell, Transtec Publications Ltd, Zurich-Uetikon, 1999, p. 125.
18. Rao, C. N. R., Arulraj, A., Santosh, P. N. & Cheetham, A. K. Charge-ordering in manganates. *Chem. Mater.*, 1998, **10**, 2714.
19. Amaral, V. S., Lourenco, A. A., Araujo, J. P., Tavares, P. B., Alves, E., Sousa, J. B., Viera, J. M., da Silva, M. F. & Soares, J. C. Non-linear conduction in $LaCaMnO_3$ thin films: interface tunneling effects. *J. Magn. Magn. Mater.*, 2001, **226–230**, 942.
20. Morrish, A. H. & Evans, B. J. Studies of the ionic ferromagnet (LaPb) MnO_3 I. Growth and characteristics of single crystals. *Can. J. Phys.*, 1969, **47**, 2691.
21. Searle, C. W. & Wang, S. T. Studies of the ionic ferromagnet (LaPb) MnO_3 III. Ferromagnetic resonance studies. *Can. J. Phys.*, 1969, **47**, 2703.
22. Leung, L., Morrish, A. & Searle, C. Studies of the ionic ferromagnet $(LaPb)MnO_3$ II. Static magnetization properties from 0 to 800K. *Can. J. Phys.*, 1969, **47**, 2697.
23. Gaiduk, Y. S., Kharton, V. V., Naumovich, E. N., Nikolaev, A. V. & Samokhval, V. V. Physicochemical and electrochemical properties of $Li_{1-x}Pb_xMnO_3$ (x=0.1-0.6) solid-solutions. *Inorg. Mat.*, 1994, **30**, 1360.
24. Srinivasan, G., Brusca, T. E., Fisher, A. S., Babu, V. S. & Seehra, M. S. Magnetic and magnetoresistance studies on radio frequency sputtered La-Pb-Mn-O films. *J. Appl. Phys.*, 1996, **79**, 5185.
25. Anane, A., Dupas, C., Le Dang, K., Renard, J., Veillet, P., de Leon Guevarra, A., Millot, F., Pinsard, L. & Revcolevschi, A. Conductivity and magnetoresistance of $La_{1-x}Sr_xMnO_3$ and $La_{1-x}SrxMn_{1-y}Mg_yO_3$ single crystals. *J. Magn. Magn. Mater.*, 1997, **165**, 377.
26. Manoharan, S. S., Vasathachaya, N. Y., Hegde, M. S., Salyalakshmi, K. M., Prasad, Y. & Sabramanyam, S. V. Ferromagnetic $La_{0.6}Pb_{0.4}MnO_3$ thin films with giant magnetoresistance at 300K. *J. Appl. Phys.*, 1994, **76**, 3923.
27. Liu, J. Z., Chang, I. C., Irons, S., Klavins, P., Shelton, R. N., Song, K. & Wasserman, S. R. Giant magnetoresistance at 300K in single crystals of $La_{0.65}(PbCa)_{0.35}MnO_3$. *Appl. Phys. Lett.*, 1995, **66**, 3218.
28. Blagoev, B. S., Terzieva, S. D., Nurgaliev, T. K., Shivachev, B. L., Zaleski, A. J., Mikli, V., Staneva, A. D. & Stoyanova-Ivanova, A. K. Magnetic and transport characteristics of oxygenated polycrystalline $La_{0.6}Pb_{0.4}MnO_3$. *J. Magn. Magn. Mater.*, 2013, **329**, 34.
29. Samantaray, S., Mishra, D. K., Pradhan, S. K. & Mishra, P. Correlation between structural, electrical and magnetic properties of $GdMnO_3$ bulk ceramics. *J. Magn. Magn. Mater.*, 2012, **339**, 168.
30. Wang, X. L., Li, D., Cui, T. Y., Kharel, P., Liu, W. & Zhang, Z. D. Magnetic and optical properties of multiferroic $GdMnO_3$ nanoparticles. *J. Appl. Phys.*, 2010, **107**, 510.
31. Andreev, N., Abramov, N., Chichkov, V., Pestun, A., Sviridova, T. & Mukovskii, Ya. Fabrication and Study of $GdMnO_3$ Multiferroic Thin Films. *Acta Phys. Pol. A*, 2010, **117**, 218.
32. Debnath, J. C., Zeng, R., Strydom, A. M., Wang, J. Q. & Dou, S. X. Ideal Ericsson cycle magnetocaloric effect in $(La_{0.9}Gd_{0.1})_{0.67}$Sr0.33$MnO_3$ single crystalline nanoparticles. *J. Alloys Comp.*, 2013, **555**, 33.
33. Zhao Juan, Wang Gui Magnetocaloric effect in $(La_{0.47}Gd_{0.2})Sr_{0.33}MnO_3$ polycrystalline nanoparticles. *J. Magn. Magn. Mater.*, 2009, **321**, 43.
34. Seikh, M. M., Sudheendra, L. & Rao, C. N. R. Magnetic properties of $La_{0.5-x}Ln_xSr_{0.5}MnO_3$ (Ln=Pr, Nd, Gd and Y). *J. Solid State Chem.*, 2004, **177**, 3633.
35. Choithrani, R., Gaur, N. K. & Singh, R. K. Specific heat and transport properties of $La_{1-x}Gd_xMnO_3$ at $15K \leq T \leq 300K$. *Solid State Comm.*, 2008, **147**, 103.

Phys. Chem. Glasses: Eur. J. Glass Sci. Technol. B, October 2015, 56 (5), 177–182

A ^{10}B NMR study of trigonal and tetrahedral borons in ring structured borate glasses and crystals

Miles Faaborg,[1] Kyle Goranson,[1] Nathan Barnes,[1] Evan Troendle,[1] Rachel Rice,[2] Michael Chace,[1] Luke Montgomery,[1] Andrew Koehler,[1] Zachary Lindeberg,[2] Diane Holland,[3] Mark E. Smith,[3,4] Mario Affatigato,[1] Steve Singleton[1] & Steve Feller[1]

[1] *Physics Department, Coe College, Cedar Rapids, IA 52402, USA*
[2] *Mathematics and Physics Departments, Simpson College, Indianola, IA 50125, USA*
[3] *Department of Physics, University of Warwick, Coventry, CV4 7AL, UK*
[4] *Vice-Chancellor's Office, University House, Lancaster University, Lancaster, LA14YW, UK*

Manuscript received 1 November 2014
Revised version received 18 December 2014
Accepted 6 July 2015

We used ^{10}B NMR and our previously developed Spectrafit automatic fitting program to determine the best values of the quadrupole parameters of both trigonal and tetrahedral borons in intermediate range structures. Each boron site was characterised by the quadrupole coupling constant, C_Q, the asymmetry parameter, η, and their respective standard deviations, σ_{C_Q} and σ_η, assumed Gaussian. The results lead us to believe that η is especially sensitive to the length scale of the intermediate range order.

1. Introduction

Nuclear magnetic resonance (NMR) spectroscopy is a widely used technique for studying short range order in glass. In some cases, nuclei that have particularly strong interactions with bonding electrons may also be used to investigate intermediate range order (IRO) on a scale of 5 to 10 Å. Several NMR techniques may be used for this purpose, including the homo- and heteronuclear dipolar interactions[(1)] and the quadrupole interaction; this paper will focus on the latter, and in particular on the ^{10}B quadrupole interaction. ^{10}B is an appropriate isotope because of its strong quadrupole interaction and spin of three, which gives its powder patterns a high degree of detail, sensitive to both the quadrupole coupling constant, C_Q, and to the asymmetry parameter, η, and their distributions. Unfortunately, a full ^{10}B NMR spectrum from trigonal borons covers a wide frequency range, often exceeding 1·4 MHz, needing approximately one to two weeks of runtime to improve the signal-to-noise ratio to an acceptable level. These difficulties make ^{10}B NMR an unpopular choice in the field of glass science, particularly in industry; however, the detail available in a ^{10}B NMR spectrum provides an opportunity to reveal structural information at the IRO level.

Recently, we reported on an automated fitting procedure called *SpectraFit*[(2)] that we developed to determine both the average C_Q and η values present in borates as well as the standard deviations of those values, denoted by σ_{C_Q} and σ_η, respectively. In this work we have extended the fitting routine to allow rapid and precise fitting of two overlapping boron sites by adding a multimillion spectra library, in which C_Q and η were varied systematically in small steps. In addition, we developed algorithms to increase the efficiency of the searches, allowing rapid comparison of distributed powder patterns to experimental data.

We used the latest version of *SpectraFit* to examine ^{10}B NMR spectra from borates that exhibit two distinctive environments. We examined crystalline and vitreous cesium triborate ($Cs_2O.3B_2O_3$) composed of six-membered rings with two trigonal sites and one tetrahedral site, see Figure 1, and obtained two sets of quadrupole parameters from the trigonal and tetrahedral borons from each sample. This procedure was repeated for vitreous cesium diborate ($Cs_2O.2B_2O_3$) of a known IRO, different from that of cesium triborate,

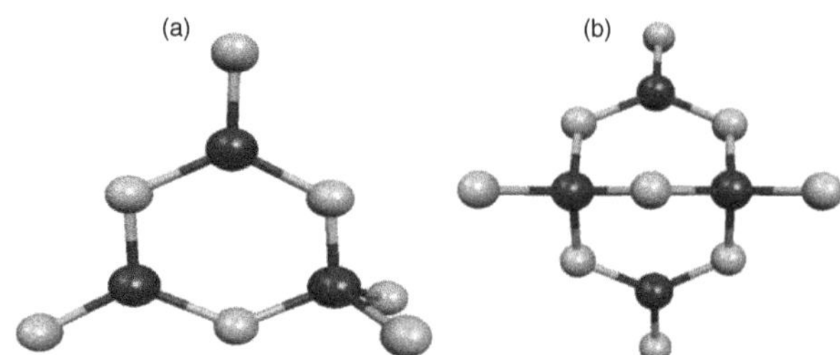

Figure 1. Intermediate borate ring structures: (a) triborate and (b) diborate (Colour available online)

* Corresponding author. Email SFELLER@coe.edu
Original version presented at VIII Int. Conf. on Borate Glasses, Crystals and Melts, Pardubice, Czech Republic, 30 June–2 July 2014
DOI: 10.13036/17533562.56.5.177

see Figure 1. In addition, we examined glassy cesium enneaborate ($Cs_2O.9B_2O_3$), a combination of boroxol and triborate rings. The quadrupole parameters from the trigonal borons were then compared to trigonal sites in glassy boron oxide and crystalline lithium orthoborate ($3Li_2O.B_2O_3$). The quadrupole parameters of the triborate and diborate tetrahedral borons were also determined.

2. Procedures

2.1 ^{10}B NMR

The pulsed NMR experiments were done statically at a central magnetic field of 7·0 T, corresponding to a Larmor frequency of 32·239 MHz, with a standard 90°–180° pulse echo, τ=3 ms, and 10–30 s pulse delay. To acquire such a wide spectrum, the magnetic field was varied by a field step unit[3] (a separate magnet was used to vary the main superconducting field). The final powder pattern is the sum of the spectra at each field. We found that about 25 steps were needed to avoid artefacts in the spectrum, and 1500 to 2500 acquisitions were needed at each magnetic field strength to yield an acceptable signal-to-noise ratio.

2.2 Sample preparation

Samples for NMR included fast, medium, and slow cooled vitreous boron oxide (B_2O_3), vitreous cesium enneaborate ($Cs_2O.9B_2O_3$), crystalline and vitreous cesium triborate ($Cs_2O.3B_2O_3$), and vitreous cesium diborate ($Cs_2O.2B_2O_3$), as well as crystalline lithium orthoborate ($3Li_2O.B_2O_3$). The samples were produced from boric acid enriched to ≥99% ^{10}B, provided by Eagle Picher, Sigma Aldrich, and Cambridge Isotopes. The samples were melted in platinum crucibles at 1000°C for 30 min. The fast cooled boron oxide was rapidly quenched using a roller quencher; the medium cooled boron oxide was prepared as droplets on a metal plate; and the slow cooled B_2O_3 glass was formed over the course of five days in a furnace contained in a nitrogen glove box to eliminate possible effects from the hygroscopicity of boron oxide. The samples of glassy cesium enneaborate, cesium triborate, and cesium diborate were prepared by mixing stoichiometric amounts of cesium carbonate and ^{10}B enriched boric acid and melting at 1000°C for 20–25 min. Weight loss was measured to confirm each of the binary glassy compositions. These samples were quenched using metal plates. Crystalline lithium orthoborate and cesium triborate were formed from ^{10}B-enriched boric acid and either lithium or cesium carbonate, melted at 1000°C, and slowly cooled in the crucible to room temperature. X-ray diffraction confirmed the sole presence of the orthoborate and triborate crystals.

Table 1. ^{10}B Quadrupole parameters for boron oxide and lithium orthoborate

Sample	C_Q *(MHz)* ±0·01 MHz	σ_{C_Q} *(MHz)* ±0·01 MHz	η ±0·01	σ_η ±0·01
Crystalline lithium orthoborate, $3Li_2O.B_2O_3$	Trigonal boron 5·55	0·11	0·06	0·00
Slow cooled (10^{-3} °C/s) glassy B_2O_3	Trigonal boron 5·12	0·26	0·10	0·02
Medium cooled (10^3 °C/s) glassy B_2O_3	Trigonal boron 5·42	0·16	0·13	0·06
Fast cooled (10^5 °C/s) glassy B_2O_3	Trigonal boron 5·40	0·26	0·12	0·06

2.3 Simulation of ^{10}B NMR spectra and boroxol ring structures

2.3.1 Simulation of ^{10}B NMR spectra

The powder pattern from quadrupole NMR does not follow a simple analytical mathematical form,[4] but its shape can be modelled theoretically using four variables to simulate the spectra: C_Q, η, σ_η, and σ_{CQ}. Finding simulated spectra that best match the experimentally produced NMR powder patterns is computationally expensive, but once found they provide insight into the longer range structures of glassy and crystalline borate materials.

Spectrafit[2] was based on the program *QuadFit*, developed by T. F. Kemp,[5] that produced distributed powder patterns that could then be visually compared to the experimental spectra. *Spectrafit*, a program initially developed by Khristenko *et al*,[2] implemented the ability to compute how closely theoretical and experimental spectra agree, based on a calculated normalised residuals per point (NRP) value. Further details of the operation of that program are given in the paper by Khristenko.[2] The differences between the initial *Spectrafit* and the version of *Spectrafit* modified for use for in this research are noteworthy. We created a library of spectra generated by varying the four primary variables (C_Q, σ_{CQ}, η, and σ_η) across a large range of practical values at relatively small step sizes. The result was a catalogue of possible spectra, which were quickly and automatically matched to the experimental data to find the best fit. For this research, the library of spectra was filled in at high resolution, changing each variable in steps of 0·01–0·025 MHz for C_Q and σ_{CQ} and 0·01–0·025 for η and σ_η. In addition, the current procedure was changed to only fit spectra based on the range of the NMR powder pattern that included features from complete $m \rightarrow m-1$ transitions; as noted earlier, ^{10}B NMR requires an extraordinarily wide range of frequencies, and as a result, the ends of the experimentally produced spectra do not always include all possible features in the theoretical powder patterns of ^{10}B. Also, a version of the program was modified

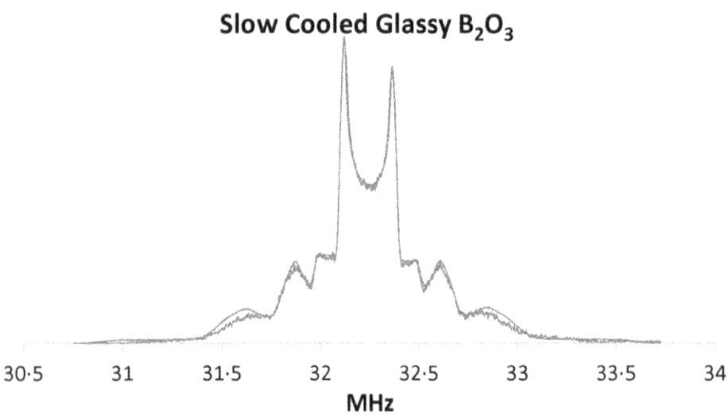

Figure 2. ^{10}B NMR spectrum from slowly cooled glassy boron oxide with the best fit spectrum superimposed in red (Colour available online)

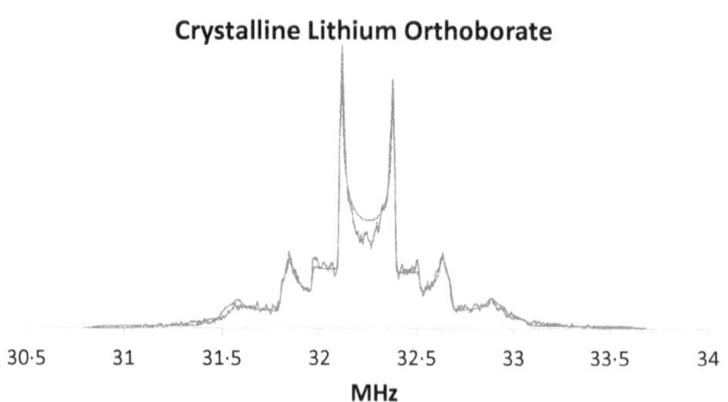

Figure 3. ^{10}B NMR spectrum from crystalline lithium orthoborate with the best fit spectrum superimposed in red (Colour available online)

to be multithreaded, so that multiple processors worked on multiple sections of the database at once. The improved resolution provided by the new library of spectra, as well as the improved fitting method, yielded simulated spectra with lower NRPs (average of 0·00393 with a standard deviation of 0·00102) and therefore stronger agreement with the experimentally generated spectra than in our earlier work.

2.3.2 Simulation of boroxol ring structures

Vitreous B_2O_3 fragments were modelled using *ab initio* methods employing the program Gaussian.[6] Fragments contained 40–80 atoms and the total number of borons in all seven fragments was 152. The fragments were constructed such that 75–80% of the boron nuclei were in rings, with an average of 77%, an arrangement that reflects the proportions of borons in rings observed in earlier reports.[7–15] Open valences on oxygen atoms were completed with hydrogen atoms, if necessary. Geometry optimisations and NMR electric field gradient tensor calculations were performed with the DFT-B3LYP/6-31G(d) model chemistry. The resulting tensors were used to calculate central values of η and C_Q, and their respective standard deviations σ_{C_Q} and σ_η, from the ensemble of borons used. All borons connected to hydroxyl groups were removed from the simulations.

3. Results

3.1 ^{10}B NMR spectra and simulations

Figures 2 to 7 show the experimental ^{10}B NMR spectra together with the simulated spectra acquired by minimising the NRPs obtained from *Spectrafit*, resulting in the quadrupole parameters given in Tables 1 and 2.

3.2 Computational models for glassy boron oxide

The calculated quadrupole parameters were obtained by minimising energy and hence optimising geometry for a variety of boron oxide fragments. The quadrupole parameters were determined by averaging over seven of these fragments (Table 3) containing 152 borons.

4. Discussion

We compare the present resultsfrom glassy boron oxide with literature values from ^{11}B NMR and NQR

Table 2. ^{10}B Quadrupole parameters for a variety of cesium borates

Sample	*C_Q (MHz) ±0·01 MHz (Trigonal boron) ±0·025 MHz (Tetrahedral boron)*	*σ_{C_Q} (MHz) ±0·01 MHz (Trigonal boron) ±0·025 MHz (Tetrahedral boron)*	*η ±0·01 (Trigonal boron) ±0·025 (Tetrahedral boron)*	*σ_η ±0·01 (Trigonal boron) ±0·025 (Tetrahedral boron)*
Cesium enneaborate glass, $Cs_2O.9B_2O_3$	Trigonal 5·38	0·31	0·16	0·09
	Tetrahedral 1·375	0·525	0·90	0·00
Cesium triborate glass, $Cs_2O.3B_2O_3$	Trigonal 5·54	0·44	0·24	0·13
	Tetrahedral 0·80	0·275	0·80	0·00
Cesium triborate crystal, $Cs_2O.3B_2O_3$	Trigonal 5·34	0·38	0·24	0·02
	Tetrahedral 0·38	0·08	0·58	0·25
Cesium diborate glass, $Cs_2O.2B_2O_3$	Trigonal 5·37	0·49	0·25	0·11
	Tetrahedral 1·25	0·525	0·125	0·075

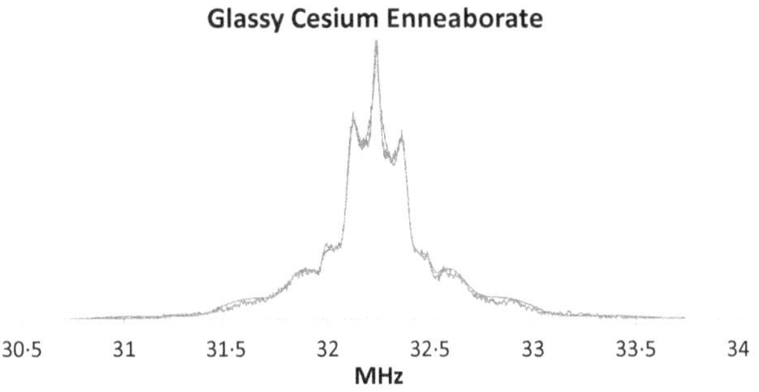

Figure 4. ^{10}B NMR spectrum from glassy cesium enneaborate with best fit spectrum superimposed in red (Colour available online)

by tabulating results for C_Q and η and for the quadrupole product, P_Q given by

$$P_Q=C_Q(1+\eta^2/3)^{0\cdot5} \tag{1}$$

Using the ratio of the quadrupole moments of ^{10}B and ^{11}B we can find $C_Q(^{11}B)$ from $C_Q(^{10}B)$:

$$C_Q\left(^{11}B\right)=\frac{Q^{^{11}B}}{Q^{^{10}B}}C_Q\left(^{10}B\right)=\frac{1}{2{\cdot}084}C_Q\left(^{10}B\right) \tag{2}$$

Table 4 gives the comparison of the present results to those from a variety of other NMR and NQR measurements.[9,10,13,15]

The three differently cooled boron oxide glasses each contain trigonal borons with bridging oxygens (BOs) whereas crystalline lithium orthoborate contains isolated boron–oxygen triangles with all nonbridging oxygens (NBOs). The asymmetry parameter for boron oxide is larger (0·12 on average) compared to that for lithium orthoborate (0·06), presumably due to the two kinds of oxygen (in-ring and non-ring) in B_2O_3. The in-ring oxygens are bonded to two in-ring borons and the non-ring oxygens are each bonded to one in-ring boron and one non-ring boron. This arrangement contrasts with the highly symmetric nonbridging oxygens in crystalline lithium orthoborate.

Overall, the values of η for trigonal borons show a systematic increase as the number of tetrahedral borons bonded through oxygens to trigonal borons increases (Table 5). The increasing number of trigonal–tetrahedral connections increasingly breaks the cylindrical symmetry of the trigonal borons. As a result it appears that η is especially sensitive to the particular ring structure the trigonal borons are in.

Table 3. Mean and standard deviation (σ) values for the asymmetry parameter and the quadrupole coupling constant of glassy B_2O_3 fragments compared with experimental results from boron oxide glass

η C_Q (MHz)	*Basis*	*Set#*	*Fragments**	*Mean σ*	*Mean σ*
6 31G(d)	7	0·148	0·062	5·498	0·384
Experimental averages from boron oxide glasses		0·12±0·01	0·05±0·01	5·41±0·01	0·22±0·01

\# with a DFT-B3LYP used for geometrical optimisation and the resulting NMR parameters

*Boron fragments range in size from 15 to 30 borons with an average of 77% of the borons in boroxol rings

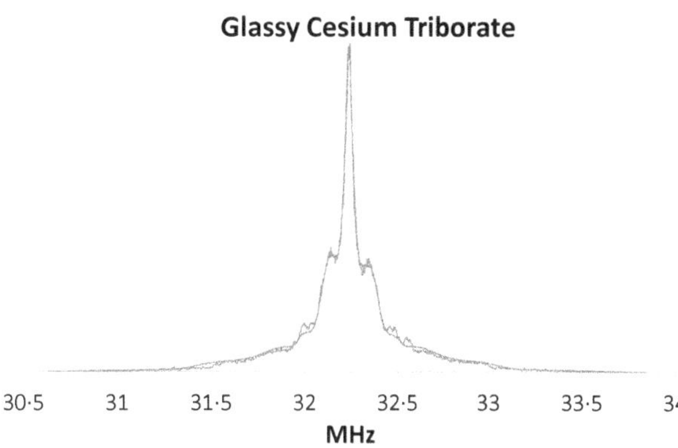

Figure 5. ^{10}B NMR spectrum from glassy cesium triborate with best fit spectrum superimposed in red (Colour available online)

In Table 2 we show a comparison of the quadrupole parameters from both vitreous and crystalline cesium triborate. The values for η from the trigonal borons from the two samples are both 0·24 indicating that the intermediate range order is similar between the crystal and the glass. As expected the distribution parameters (σ_{C_Q} and σ_η) are smaller for the crystal than for the glass.

The quadrupole coupling constant, C_Q, does not seem to correlate with intermediate range structure.

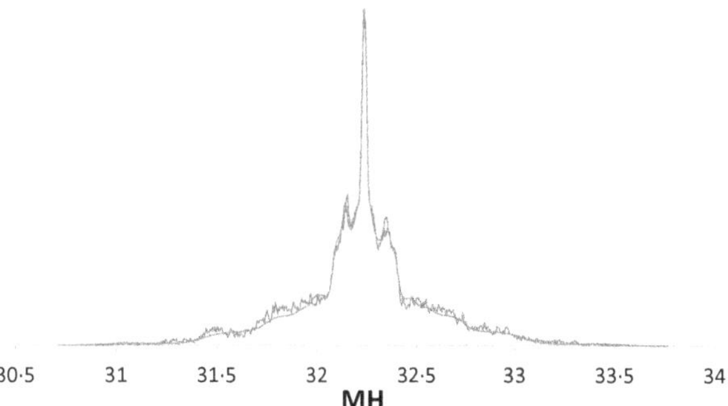

Figure 6. ^{10}B NMR spectrum from crystalline cesium triborate with best fit spectrum superimposed in red (Colour available online)

Table 4. Comparison of the present quadrupole parameter results from glassy boron oxide with those from ^{11}B NMR and NQR

Results from	*Method*	*C_Q (MHz)*	*η*	*P_Q (MHz)*
Present work	^{10}B pulsed NMR	2·60	0·13	2·61
Jellison, Panek, & Bray[9]	^{10}B wideline NMR	2·64	0·12	2·65
Hung *et al*[16]	^{11}B DOR NMR at 20 T			2·67 (Ring) 2·61 (Non-ring)
Kroeker, Neuhoff & Stebbins[17]	^{11}B MAS NMR	2·68	0·15	2·69
Gravina, Bray & Peterson[15]	^{11}B NQR			2·71 (Ring) 2·61 (Non-ring)
Hwang *et al*[13]	^{11}B MAS NMR and MQMAS NMR	2·68 2·60 (Non-ring)	0·16 0·19	2·69 (Ring) 2·62 (Non-ring)

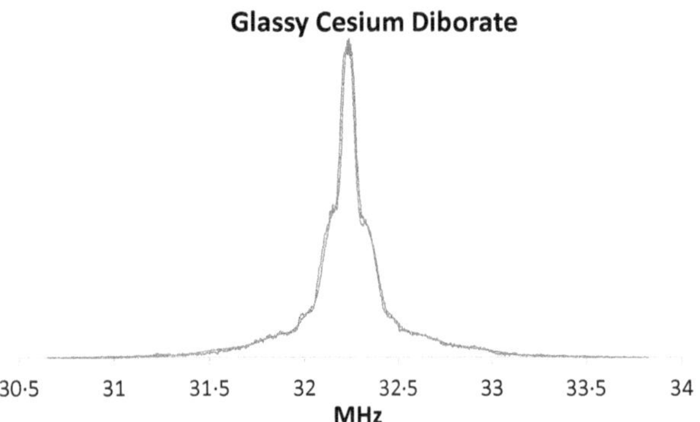

Figure 7. ^{10}B NMR spectrum from glassy cesium diborate with best fit spectrum superimposed in red (Colour available online)

We found precise values of C_Q for trigonal borons (Tables 1 and 2) spanning a small range from 5·34 MHz (2·56 MHz, ^{11}B) for crystalline $Cs_2O.3B_2O_3$, to 5·55 MHz (2·66 MHz for ^{11}B) for crystalline $3Li_2O.B_2O_3$; the latter value agrees reasonably well with an earlier result, 2·76 MHz, obtained using ^{11}B NMR.[11] The values of C_Q for the tetrahedral borons are considerably smaller than those for trigonal borons, ranging from 0·80 to 1·375 MHz with no clear pattern.

The standard deviations of η and C_Q show trends. The fits for glasses containing both trigonal and tetrahedral borons display a consistently larger σ_η for the trigonal borons (0·09 to 0·13, Table 2) than the fits for glasses containing only trigonal borons (0·02 to 0·06, Table 1), possibly due to increasing disorder caused by the added presence of tetrahedral borons. Similarly, σ_{C_Q} for trigonal borons is also larger for glasses with both types of sites (0·31 to 0·49 MHz, Table 2), compared with glasses with only trigonal borons (0·11 to 0·26 MHz).

The simulated boron fragments yielded quadrupole parameters comparable to the experimental parameters for boron oxide (Table 3).

5. Conclusions

We used an updated version of the program Spectrafit to determine accurate ^{10}B quadrupole parameters from a series of glasses and crystals, containing either trigonal and tetrahedral boron sites, or solely trigonal boron sites. The asymmetry parameter for the trigonal borons is especially sensitive to the intermediate range order of the boron rings. The quadrupole coupling constant was found to be close to 5·5 MHz for the trigonal borons without apparent pattern. It was found that the distributions of the quadrupole parameters of the trigonal borons were sensitive to the symmetry of these sites. The quadrupole coupling constants for the tetrahedral borons were about 4–5 times smaller than those for the trigonal borons. No apparent patterns were found within the parameter set for the tetrahedral borons.

Acknowledgements

We acknowledge the National Science Foundation of the United States under grants numbered NSF-DMR 1407404, NSF-DMR 1262325, NSF-REU 1358968, and NSF-REU 1004860. Mrs Janice Faaborg is thanked for help in editing the paper.

References

1. There are many references to studies that employ this technique. Three representative reference are: Ratai, E., Chan, J. C. C. & Eckert, H. *Phys. Chem. Chem. Phys.*, 2002, **4**, 3198. Epping, J. D., Stroijek, W. & Eckert, H. *Phys. Chem. Chem. Phys.*, 2005, **7**, 2384. Chen, B., Werner-Zwanziger, U., Nascimento, M. L. F., Ghussn, L., Zanotto, E. D. & Zwangziger, J. W. *J. Phys. Chem. C*, 2009, **113**, 20725.
2. Khristenko, V., Tholen, K., Barnes, N., Troendle, E., Crist, D., Affatigato, M., Feller, S., Holland, D., Kemp, T. & Smith, M. *Phys. Chem. Glasses: Eur. J. Glass Sci. Technol. B*, 2012, **53**, 121.
3. Poplett, I. J. F. & Smith, M. E. *Solid State Nucl. Magn.*, **1998, 11**, 211.
4. Jellison, Jr., G. E., Feller, S. & Bray, P. J. *J. Magn. Reson.*, 1977, **27** 121.
5. Kemp, T. F. & Smith, M. E. *Solid State Nucl. Magn.*, 2009, **35**, 243. http://www2.warwick.ac.uk/fac/sci/physics/research/condensedmatt/nmr/research/calculations/quadfit/
6. Gaussian 03, Frisch, M. J., Trucks, G. W., Schlegel, H. B., Scuseria, G. E., Robb, M. A., Cheeseman, J. R., Montgomery, Jr., J. A., Vreven, T., Kudin, K. N., Burant, J. C., Millam, J. M., Iyengar, S. S., Tomasi, J., Barone, V., Mennucci, B., Cossi, M., Scalmani, G., Rega, N., Petersson, G. A., Nakatsuji, H., Hada, M., Ehara, M., Toyota, K., Fukuda, R., Hasegawa, J., Ishida, M., Nakajima, T., Honda, Y., Kitao, O., Nakai, H., Klene, M., Li, X., Knox, J. E., Hratchian, H. P., Cross, J. B., Adamo, C., Jaramillo, J., Gomperts, R., Stratmann, R. E., Yazyev, O., Austin, A. J., Cammi, R., Pomelli, C., Ochterski, J. W., Ayala, P. Y., Morokuma, K., Voth, G. A., Salvador, P., Dannenberg, J. J., Zakrzewski, V. G., Dapprich, S., Daniels, A. D., Strain, M. C., Farkas, O., Malick, D. K.,

Table 5. Asymmetry parameter from various trigonal borons

Asymmetry parameter	*Sample*	*Environment*	*Tetrahedral borons bonded per trigonal boron*
0·06	Crystalline lithium orthoborate, $3Li_2O.B_2O_3$	BO_3 triangles with 3 NBOs	0
0·12*	Boron oxide glass, B_2O_3	BO_3 triangles, all BOs with boroxol ring and non-ring borons and oxygens	0
0·16	Cesium enneaborate glass, $Cs_2O.9B_2O_3$	BO_3 triangles, all BOs with boroxol ring and triborate ring borons and oxygens	0 for boroxol rings 1 for triborate rings, net average 0·33
0·24	Cesium triborate glass, $Cs_2O.3B_2O_3$	BO_3 triangles, all BOs with triborate ring borons and oxygens	1
0·24	Cesium triborate crystal, $Cs_2O.3B_2O_3$	BO_3 triangles, all BOs with triborate ring borons and oxygens	1
0·25	Cesium diborate glass, $Cs_2O.2B_2O_3$	BO_3 triangles, all BOs with diborate ring borons and oxygens	2

*Average of three samples produced at cooling rates that varied over 10^8 K/s

Rabuck, A. D., Raghavachari, K., Foresman, J. B., Ortiz, J. V., Cui, Q., Baboul, A. G., Clifford, S., Cioslowski, J., Stefanov, B. B., Liu, G., Liashenko, A., Piskorz, P., Komaromi, I., Martin, R. L., Fox, D. J., Keith, T., Al-Laham, M. A., Peng, C. Y., Nanayakkara, A., Challacombe, M., Gill, P. M. W., Johnson, B., Chen, W., Wong, M. W., Gonzalez, C. & Pople, J. A. Gaussian, Inc., Wallingford CT, 2004.

7. Sinclair, R. N., Wright, A. C., Wanless, A. J., Hannon, A. C., Feller, S., Mayhew, M. T., Meyer, B. M., Royle, M. L., Wilkerson, D. L., Williams, R. B. & Johanson, B. C. In: *Borate Glasses, Crystals and Melts,* Eds. A. C. Wright, S. A. Feller & A. C. Hannon, The Society of Glass Technology, Sheffield, 1997, pp. 140.
8. Hannon, A. C., Grimley, D. I., Hulme, R. A., Wright, A. C. & Sinclair, R. N. *J. Non-Cryst. Solids*, 1994, **177**, 299.
9. Jellison, J. E., Panek, L. W., Bray, P. J. & Rouse, G. B. *J. Chem. Phys.*, 1977, **66**, 802.
10. Joo, C., Werner-Zwanziger, U. & Zwanziger, J. W. *J. Non-Cryst. Solids*, 2000, **261**, 282.
11. Joo, C., Werner-Zwanziger, U. & Zwanziger, J. W. *J. Non-Cryst. Solids*, 2000, **271**, 265.
12. Lee, S. K., Mibe, K., Fei, Y., Cody, G. D. & Mysen, B. O. *Phys. Rev. Lett.*, 2005, **94**, 165507.
13. Hwang, S.-J., Fernandez, C., Amoureux, J. P., Cho, J., Martin, S. W. & Pruski, M. *Solid State Nucl. Magn. Reson.*, 1997, **8**, 109.
14. Youngman, R. E. & Zwanziger, J. W. *J. Non-Cryst. Solids*, 1994, **168**, 293.
15. Gravina, S. J., Bray, P. J. & Petersen, G. L. *J. Non-Cryst. Solids*, 1990, **123**, 165.
16. Hung, I., Howes, A. P., Parkinson, B. G., Anup, T., Samoson, A., Brown, S. P., Harrison, P. F., Holland, D. & Dupree, R. *J. Solid State Chem.*, 2009, **182**, 2402.
17. Kroeker, S., Neuhoff, P. S. & Stebbins, J. F. *J. Non-Cryst. Solids*, 2001, **293–295**, 440.
18. Feller, S. A. PhD thesis, Brown University, 1980, p. 83.

Phys. Chem. Glasses: Eur. J. Glass Sci. Technol. B, October 2015, 56 (5), 183–188

Density and refractive indices of alkali borate and borosilicate crystals and glasses: a comparative analysis

Maria G. Krzhizhanovskaya,[*,1] Peter Paufler,[2] Stanislav K. Filatov[1] & Rimma S. Bubnova[3,1]*

[1] *Saint Petersburg State University, Dept. of Crystallography, University Emb., 7/9, St. Petersburg, 199034, Russia*
[2] *Technische Universität Dresden, Institut für Strukturphysik, D-01062 Dresden, Germany*
[3] *Grebenshchikov Institute of Silicate Chemistry of the Russian Academy of Sciences, Adm. Makarov Emb., 2, St. Petersburg, 199034, Russia*

Manuscript received 26 September 2014
Revised version received 11 December 2014
Accepted 5 February 2015

This study is aimed at the comparative analysis of structure sensitive properties of alkali borate and borosilicate glasses and crystals. Among the properties under consideration are density and refractive index. The comparison of physical properties between crystals and glasses considers the dependence on alkali content and the size of the alkali cation. New experimental data on density and refractive index of potassium and rubidium borate and borosilicate glasses are presented. The mass density measurements were performed by employing the method of hydrostatic weighing. The refractive index measurements were carried out on a polarization microscope with a series of immersion liquids, using the Becke line method. The compared crystal density data were calculated from the crystal structures.

1. Introduction

The relation between the chemical composition, structure and properties of alkali borate, silicate and borosilicate glasses has been studied for many years. Excellent reviews describe the history of these studies.[1,2] The two classical glass structure theories,[3] Zachariasen's 'random network' theory and Lebedev's 'crystallite' theory, could not predict the physical properties of the glass. In 1958, Bray and coworkers first used NMR spectroscopy to investigate the structure of borate glasses.[1] Later Krogh-Moe[4] found structural groupings in the binary borate glass systems. It was assumed in Ref. 5 that the addition of alkali oxide above 33 mol% R_2O causes the formation of nonbridging oxygen ions in the borate matrix. Bray and co-workers used NMR techniques to measure the BO_4 content in borate glasses.[5,6] In 1972 Bottinga & Weill[7] suggested for the first time that the density of binary or ternary liquid silicate systems could be used to fit the coefficients of volume models for complex glass melts. Using the models and theories of Bray and co-authors and other theories, Feil & Feller[8] measured the density in the Na_2O–B_2O_3–SiO_2 glass system and performed modelling of the atomic arrangements. Their predictions were in good agreement with experimental results. After that the properties of borosilicate glasses were modelled by many authors.

* Corresponding author. Email krzhizhanovskaya@mail.ru
Original version presented at VIII Int. Conf. on Borate Glasses, Crystals and Melts, Pardubice, Czech Republic, 30 June–2 July 2014
DOI: 10.13036/17533562.56.5.183

On the basis of Krogh-Moe's structural model of glass, we agree with the idea of Bray and co-workers that the structures of some crystals are similar to those of glass; some characteristics of crystals may also conform to the corresponding glasses.[4] According to the principles of phase diagrams, a glass structure is a mixture of the structural units of the nearest congruently melting compounds, and glass structures can be calculated. In ternary alkali borosilicate (Na, K, Rb) systems, unusual extrema[9,10] in plots of glass transition temperature versus composition have previously been revealed in the neighbourhood of primary boroleucite crystallisation.[11,12] The purposes of this paper are to report an extensive comparative study of the density and refractive index of crystals and glasses in the potassium and rubidium borosilicate systems, and the consequent relationship of properties to the phase diagrams.

2. Experimental

The glasses were synthesised in Pt crucibles at temperatures of 900–1600°C, depending on composition. The chemical composition is indicated in the tables according to wet chemistry study of the resulting glasses.

Mass density measurements have been performed employing the method of hydrostatic weighing using dodecane $C_{12}H_{26}$ as the immersion liquid. For details of the method and a schematic presentation of the experimental set-up, see Refs 13 and 14. The procedure consisted of the following steps: (i) weighing the sample in air (buoyancy correction neglected) $\rightarrow m_a$,

(ii) weighing the sample in the liquid of density d_l=749·5 kg/m^3 (includes the mass of the liquid that has penetrated into the sample) $\rightarrow m_l^*$, (iii) removal of the immersion liquid adhering to the sample surface, (iv) weighing in air again (includes the mass of penetrated liquid) $\rightarrow m_a^*$, (v) extrapolation of the apparent mass of the sample in the liquid for the very first moment of immersion (correction for liquid penetration) $m_l = m_l^* - (m_a^* - m_a)$, (vi) volume of the sample $V = (m_l - m_a)/d_l$, density of the sample $d = m_a/V$. Weighing was performed using an ultramicrobalance of 2 g maximum load, 0·1 μg resolution and 0·5 μg standard deviation. The relative error of density measurement is 1%.

Refractive index measurements were carried out on a polarisation microscope with a series of immersion liquids, using the Becke line method. For the glasses an average refractive index, n, was defined, whereas for low symmetry crystals three principal indices of refraction were obtained. The uncertainty of the refractive index values is ±0·002. The refractive index values and also the optical sign were determined in specially oriented sections of crystals according to the standard rules of conoscopic measurements for biaxial crystals.

3. Density of K- and Rb-borosilicate glasses

The measured density values for potassium (55 compositions in the range 8–30 mol% K_2O and 0–80 mol% SiO_2) and rubidium (58 compositions in the range 4–33·3 mol% Rb_2O and 0–81 mol% SiO_2) borosilicate glasses are presented in Table 1 and in Figure 1 for K-borosilicate glasses.

3.1. K_2O–B_2O_3–SiO_2 densities

The density diagram of potassium borosilicate glasses differs from that of rubidium in the form of the contour lines as well as the absolute density values. The absolute values change in the narrow range from 1·94 (composition 8/0) up to 2·47 (20/60); the difference between maximum and minimum density values is about 0·5 g/cm^3 (Table 1). The data from Refs 15 and 16 were used together with our data for drawing the refractive index contour lines. It should be noted that the measurements from reference 16 were slightly higher (by about 0·02–0·03 g/cm^3), whereas those from Ref. 15 were a bit lower (by about 0·01–0·02 g/cm^3) than our density values. However, all measurements are in relative agreement with each other. The maximum of the values is located towards the right corner of the diagram (Figure 1), corresponding to the area of boroleucite ($KBSi_2O_6$) crystallisation.[11,12]

3.2. Rb_2O–B_2O_3–SiO_2 densities

The measured density of rubidium borosilicate glasses varies from 2·05 g/cm^3 (composition 8/0) up to 2·97 g/cm^3 (25/50) (Table 1). The density diagram was presented in our previous paper.[17] The density values increase with increasing rubidium and sili-

Table 1. Density of potassium and rubidium borosilicate glasses

Mol% K_2O/ mol% SiO_2	*D (g/cm^3)*	*Mol% K_2O/ mol% SiO_2*	*D (g/cm^3)*	*Mol% Rb_2O/ mol% SiO_2*	*D (g/cm^3)*	*Mol% Rb_2O/ mol% SiO_2*	*D (g/cm^3)*
8/0	1·94	20/60	2·47	4/60*	2·17	19/25*	2·65
8/10	2·025	22/25	2·33	5/15*	2·09	20/0*	2·50
10/80	2·40	22·5/0	2·17	5/25*	2·13	20/15*	2·61
12/20	2·11	24·3/5	2·23	6/30*	2·17	20/20*	2·66
12/30	2·13	24·3/53·9	2·45	8/0*	2·05	20/30*	2·715
12/60	2·33	24·6/2·5	2·215	6/50*	2·25	20/40*	2·74
12/70	2·42	25/0	2·20	6/70*	2·33	20/50*	2·82
12·5/75	2·43	25/20	2·33	8/10*	2·18	20/55*	2·83
15/70	2·45	25/40	2·43	8/30*	2·27	20/60*	2·88
16/20	2·18	25/50	2·45	8/50*	2·30	23·3/10*	2·63
16/60	2·43	25/70	2·43	10/10*	2·25	23·3/20*	2·79
16·6/66·7	2·44	26/0	2·22	10/30*	2·34	23·5/40*	2·94
17/10	2·14	26/44·1	2·45	10/80*	2·66	24/30*	2·87
17·6/50	2·40	26/16	2·27	12/30*	2·41	24·5/2·5	2·62
17·8/45	2·38	27/5	2·27	13/10*	2·34	25/0	2·68
18/40	2·35	27·2/0	2·25	13/15*	2·36	25/50	2·97
18·3/35	2·31	27·8/0	2·25	13/74*	2·76	26/10*	2·76
18·7/25	2·25	28/60	2·46	13/81	2·79	26/30*	2·94
18·7/75	2·42	28·6/0	2·25	14/20*	2·425	26·6/20*	2·88
19/15	2·20	29·5/15	2·23	14/30*	2·475	27/5*	2·87
19·4/45	2·40	30/0	2·25	14/40*	2·54	27/15*	2·84
19·5/78	2·405	30/2	2·25	14/50*	2·60	27·5/12·5	2·89
20/0	2·13	30/12·5	2·33	15/10*	2·40	27·5/15	2·92
20/10	2·185	30/20	2·37	16/0*	2·37	28/5	2·87
20/20	2·25	30/30	2·41	16·7/0*	2·41	28/15	2·92
20/30	2·34	30/40	2·45	16·7/66*	2·84	29·1/7·5	2·87
20/40	2·38	30/50	2·46	17/10*	2·46	29·5/10	2·88
20/50	2·44			17/15*	2·50	30/50	2·96
				18/35*	2·68	33·3/2·5	2·91

*density values from the density diagram presented previously.[17]

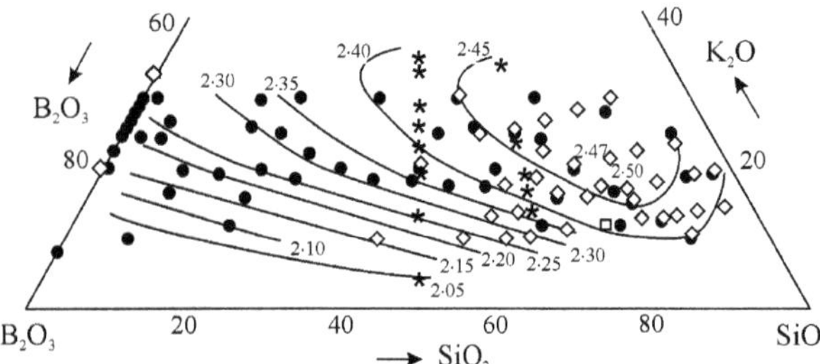

Figure 1. Density of potassium borosilicate glasses
● - present paper; ✱ - Takahashi et al;[15] ◇ - Akimov[16]

con content. It is notable that the density increases sharply with composition up to 30 mol% Rb_2O and much more weakly in the upper part of the diagram according to data from Feller *et al.*[18] Relatively high density values are observed in the right corner of the diagram,[17] corresponding to the field of crystallisation of boroleucite solid solutions.

4. Results of optical studies

4.1. Refractive index of K-borosilicate glasses

The refractive index of K-borosilicate glasses is presented in Table 2. Figure 2 shows the refractive index of 43 glasses of different compositions in the ternary diagram $K_2O–B_2O_3–SiO_2$. The values vary in a relatively narrow range, n=1·462–1·520, with the maximum value at approximately the composition $16·7K_2O.16·6B_2O_3.66·7SiO_2$ (mol%), corresponding to that of boroleucite $KBSi_2O_6$. This fact coincides with the observation of extrema in many structure-sensitive properties of crystals and glasses in the neighbourhood of the boroleucite composition.[9]

It is particularly remarkable for the $K_2O–B_2O_3–SiO_2$ system (Figure 2) that the refractive index contour lines at the bottom of the diagram run nearly horizontally. However, further up the diagram, their

Table 2. The refractive index of K-borosilicate glasses

Mol% K_2O/ mol% SiO_2	*n*	*Mol% K_2O/ mol% SiO_2*	*n*
2/60	1·462	18/40	1·503
3/70	1·466	18·7/75	1·502
4/40	1·467	19/15	1·490
4/60	1·466	19·5/78	1·499
5/50	1·467	20/40	1·511
6/60	1·473	20/70	1·510
7/40	1·473	24/60	1·518
7/50	1·475	25/50	1·519
7/70	1·480	25/55	1·518
8/36	1·475	25/70	1·507
8/60	1·484	28/60	1·515
9/50	1·483	29·5/15	1·509
10/80	1·501	30/30	1·511
10/60	1·501	30/40	1·518
11/40	1·487	30·2/40·5	1·518
11/50	1·499	32·7/3	1·506
12/70	1·507	35/20	1·511
12·5/75	1·507	35/30	1·490
15/70	1·515	35/40	1·510
16/60	1·515	40/20	1·510
16·6/66·7	1·520	40/40	1·508
17·6/50	1·510		

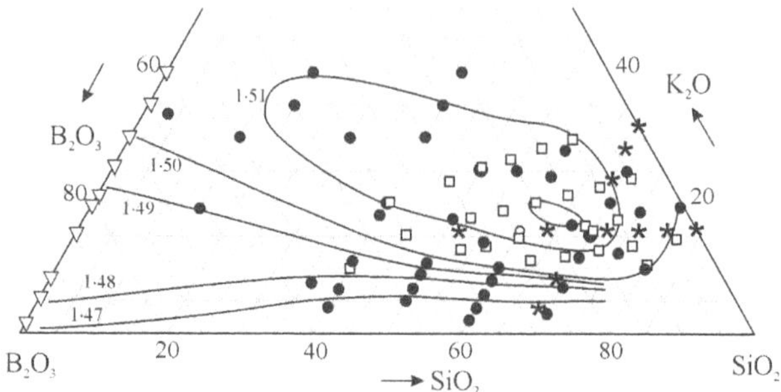

Figure 2. The refractive index of K-borosilicate glasses.
● - present paper; ✱ - Appen[19]; □ - Akimov[16]; ∇ - Mazurin[20] and SciGlass[21]

direction changes gradually to that of the bisector of the $B_2O_3–SiO_2–K_2O$ angle. This is evidence that the refractive index, n, is more sensitive to changes of K_2O content, less sensitive to the B_2O_3 content, and practically independent of the silica content in borosilicate glass.

4.2. Refractive index of Rb-borate crystals and glasses

4.2.1. $Rb_3B_7O_{12}$ (30 mol% Rb_2O, triclinic system)

In accord with its triclinic symmetry, $Rb_3B_7O_{12}$ has two optical axes. $Rb_3B_7O_{12}$ is biaxial, with negative orientation of the optical axes, and the principal indices of refraction are n_1=1·550(2), n_2=1·546(2), n_3=1·511(2), and the average value is n=1·537, with $2V$=45(7)°, and perfect cleavage takes place in the plane perpendicular to the acute bisector.

4.2.2. $Rb_2B_4O_7$ (33·3 mol% Rb_2O, triclinic)

According to conoscopic measurements, $Rb_2B_4O_7$ crystals are optically positive with principal indices of refraction n_1=1·562(2), n_2=1·550(2), n_3=1·542(2), and the average index is n=1·551. Both compounds have low birefringence.

4.2.3. Rb borate glasses

The results of crystal optical studies of Rb borate glasses are shown Figure 3; the measured values (mol% Rb_2O/n) are: 8/1·467; 20/1·486; 25/1·497; 30/1·508; 33·3/1·509. Figure 3 demonstrates the agreement of our present data on refractive index with that of other authors. Generally, the refractive indices of alkali borate glasses increase with increasing alkali metal content, and we observe the same behaviour for Rb glasses.

5. Correlation of properties

5.1. Refractive index of Rb borate crystals and glasses

Although the experimental data on Rb borates presented here are very limited, they are representative

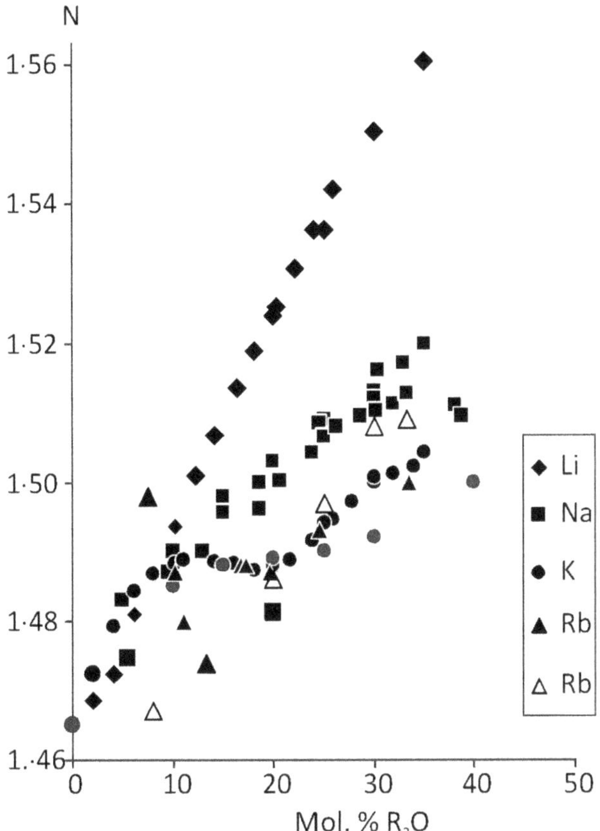

Figure 3. The refractive index of alkali borate glasses. Filled symbols are for data from Mazurin[20] and SciGlass,[21] open symbols are for the present work

for comparison. Thus, the refractive index, n, for the glass with composition $30Rb_2O.70B_2O_3$ is 1·508, while the average n of crystalline $Rb_3B_7O_{12}$ is 1·537; the difference is Δn=0·029, in favour of the crystal. The refractive index of another glass with composition 33·3 mol% Rb_2O is 1·509, while that of crystalline $Rb_2B_4O_7$ is 1·551; the difference is Δn=0·042. Averaging over these two compositions, we conclude that the refractive index of crystals exceeds that of glasses of equal composition by about Δn=0·035.

5.2. Density of alkali borate and borosilicate crystals and glasses

The data on alkali borate glass density reported by various authors, as collected by Mazurin[20] and SciGlass,[21] are shown in Figure 4. Despite the great quantity of literature data, they are in good agreement with each other. The figure demonstrates that the density increases both with increase of cation size and cation content as well.

The results in Figure 4 show that the density values of Li (M_{Li}=6·94; M atomic weight in atomic mass units), Na (M_{Na}=22·99) and K (M_K=39·1) glasses are very close to each other. Moreover, in the range 20–35% mol% R_2O, the densities of sodium glasses are higher than those of potassium glasses. Taking into account the fact that potassium is usually considered

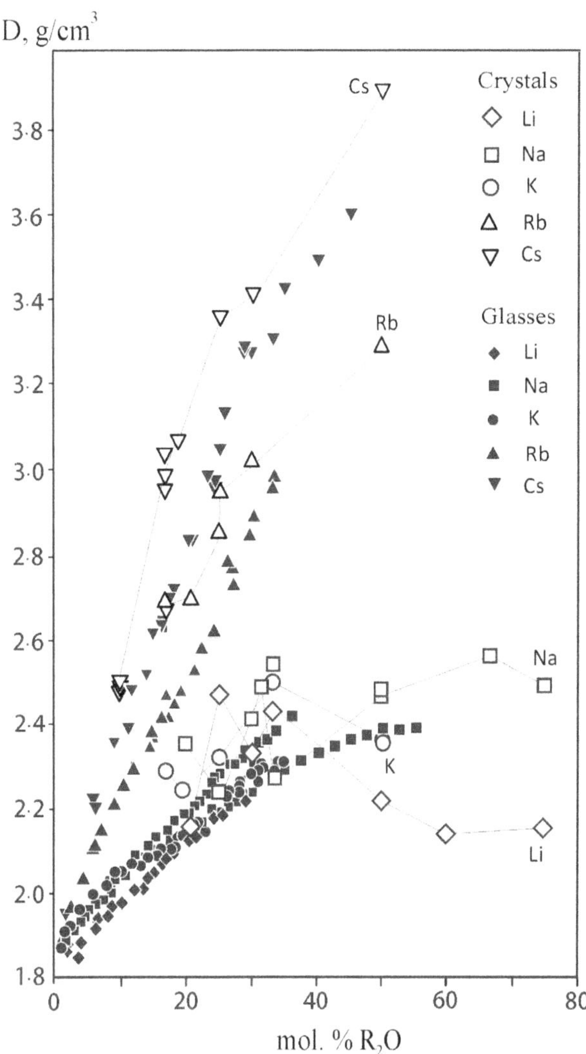

Figure 4. Density of alkali borate crystals and glasses. The open symbols denote crystals (ICSD 2012[22]), and the filled symbols denote glasses (Mazurin[20] and SciGlass[21])

as a "heavy" atom, and sodium in contrast belongs to "light" atoms, it would be normal to expect an increase of density on substitution of heavier atoms for light atoms. For rubidium (M_{Rb}=85·47) and caesium (M_{Cs}= 132.91), we observe a remarkable jump. For example, the density of Li–Na–K borate glasses with 30 mol% alkali oxide is within the range 2·25–2·35 g/cm³, while the densities of Rb and Cs borate glasses are 2·87 and 3·25 g/cm³, respectively. The dependence of density on composition is close to linear in the range of compositions 0–30 mol% alkali oxide. On further increase of alkali content, the slope of the curve changes. Unfortunately, the statistics in this part of the diagram are less good, and the data need to be more fully established.

Density data for crystals are most likely to be found in papers on the determination of crystal structures. These values are usually calculated from unit cell volume and the total mass of atoms in the unit cell, as derived from diffraction experiments. The calculated density data for alkali borate

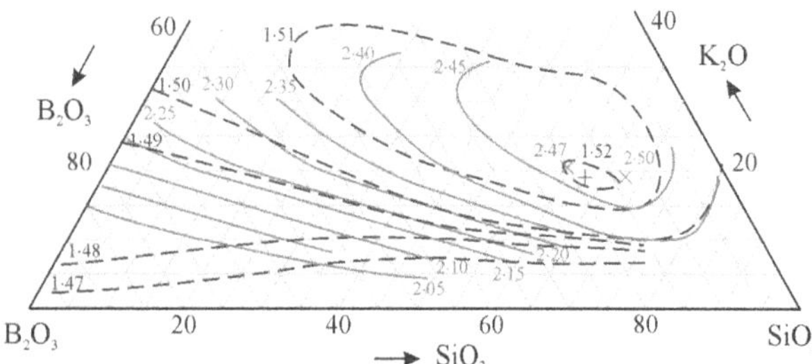

Figure 5. A comparison of density (solid line and "×") and refractive index (dashed line and "+") of potassium borosilicate glasses (Colour available online)

crystals (ICSD 2012[22]) are compared in Figure 4 to the densities of alkali borate glasses. Generally, the crystal density is higher than that of glass of the same composition. The data for the Rb borate crystals are taken from our studies. It is difficult to identify an obvious general dependency between the density of alkali borate crystals and their chemical composition, although in the Rb and Cs borates the density increases with alkali content. In particular, it should be noted that, for crystalline borates, Rb and Cs have greater similarity in density behaviour than for the other alkali elements. Furthermore, for Rb and for Cs, the dependences for crystals and glasses have a similar form. This may indicate some similarity of structural units in crystals and glasses for heavy alkali metals. On the other hand, the mass densities of crystal structures for the light elements, Li, Na, and K, are widely spread, and seem not to be useful to explain the property variations with different alkali oxide in borate glasses.

The density diagram of Rb-borosilicate glasses[17] differs from that of K-borosilicate glasses (Figure 1); in particular, it has no maxima at all. Looking again at the borate glass density data (Figure 4), a dramatic change in the form of the composition-dependence is observed between potassium and rubidium. An explanation can be found as follows: From the three types of cation, B, Si and K/Rb, the latter one, the big and heavy alkali metal, plays the leading role defining the density of the compound/glass. On the other hand, when a borate or borosilicate is formed by a light (Li, Na) or even intermediate (K) alkali metal, some details, like a maximum in the boroleucite region, appear in the density–composition diagram. However, when Rb or Cs contributes to the compound formation, the density values depend mainly on the content of this heavy atom.

5.3. A comparison of density and refractive index

Contours corresponding to density and refractive index values for K borosilicate glasses are shown together in the triangle diagram for the K_2O–B_2O_3–SiO_2 system in Figure 5. The region of maximal values is located near the SiO_2 corner; the form of the contours is very similar for both density and refractive index. This correlation is not particularly surprising because the refractive index of a substance is a function of its density, although the presence of physical property maxima in this part of the diagram cannot be explained by the chemical composition alone. As is known, this region is the field of boroleucite crystallisation, and it would be reasonable to assume a similarity of the boroleucite anomaly region both for crystals and glasses.

6. Conclusions

New data on refractive indices and mass density of potassium and rubidium borates and borosilicates have been presented. The refractive indices and density of alkali borate glasses were found to generally increase with increasing alkali metal content. In the glass property diagrams for light and intermediate alkali cations (Li, Na, K), there is a maximum in the structure-sensitive properties, while the properties of heavy Rb or Cs borate/borosilicate glasses are determined mostly by the heavy metal content. Comparing crystalline and glassy alkali borates and borosilicates of the same chemical composition shows higher values of both density and refractive index for the crystalline solid state. This may provide the basis for the construction of special optical elements.

Acknowledgments

We are sincerely grateful to I. G. Polyakova for providing the collection of glasses and to Yu. Schtuermer (St. Petersburg University, diploma work, 1997) for measuring potassium borosilicate glass refractive indices. The work is supported by the Russian Foundation for Basic Research (12-03-00981-a).

References

1. Wright, A. C. The great crystallite versus random network controversy: a personal perspective. *Int. J. Appl. Glass Sci.*, 2014, **5**, 31–56.
2. Wright, A. C., Dalba, G., Rocca, F. & Vedishcheva, N. M. Borate versus silicate glasses: why are they so different? *Phys. Chem. Glasses: Eur. J. Glass Sci. Technol. B*, 2010, **51**, 233–265 (and refs therein).
3. Shelby, J. E. *Introduction to glass science and technology*, Second edition, Royal Society of Chemistry, Cambridge, 2005.
4. Krogh-Moe, J. Structural interpretation of melting point depression in the sodium borate system. *Phys. Chem. Glasses*, 1962, **3**, 101–110.
5. Bray, P. J. & O'Keefe, J. G. Nuclear magnetic resonance investigations of the structure of alkali borate glasses. *Phys. Chem. Glasses*, 1963, **4**, 37–46.
6. Bray, P. J., Leventhal, M. & Hooper, H. O. Nuclear magnetic resonance investigations of the structure of lead borate glasses. *Phys. Chem. Glasses*, 1963, **4**, 47–66.
7. Bottinga, Y. & Weill, D. F. Densities of liquid silicate systems calculated from partial molar volumes of oxide components. *Am. J. Sci.*, 1970, **269**, 169–182.
8. Feil, D. & Feller, S. The density of sodium borosilicate glasses related to atomic arrangements. *J. Non-Cryst. Solids*, 1990, **119**, 103–111.
9. Polyakova, I. G. & Tokareva, E. V. Alkali borosilicate systems: phase diagrams and the structure of glass. In: *Borate Glasses, Crystals and Melts*, Eds A. C. Wright, S. A. Feller & A. C. Hannon, Society of Glass Technology, Sheffield, 1997, p.223–230.

10. Polyakova, I. G. Alkali borosilicate system: phase diagrams and properties of glasses. *Phys. Chem. Glasses*, **2000**, **41**, 247–258.
11. Voldan, J. Crystallisation of a three-component compound in the system $K_2O–SiO_2–B_2O_3$. *Silikaty*, 1979, **23**, 133–135.
12. Ihara, M. & Kamei, F. Crystal structure of potassium borosilicate, $K_2O.B_2O_3.4SiO_2$. *J. Ceram. Assoc. Jpn.*, 1980, **88**, 40–43.
13. Wolf, B., Paufler, P., Schubert, M., Rodig, C. & Fischer, K. Mass density evolution during manufacturing of Ag-sheathed BPSCCO tapes. *Supercond. Sci. Technol.*, 1996, **9**, 589–597.
14. Klöß, G. Lösung realstrukturbezogener Aufgabenstellungen mit Massendichtemessungen nach der Methode der hydrostatischen Wägung. *Thesis: Leipzig University*, 1987.
15. Takahashi, K., Osaka, A. & Furuno, R. Network of sodium and potassium borosilicate glass systems. *J. Non-Cryst. Solids*, 1983, 55, 15–26.
16. Akimov, V. V. Optical constants and density of potassium borosilicate glasses (in Russian). *Fiz. Khim. Stekla*, 1989, **15**, 758–760.
17. Bubnova, R. S., Polyakova, I. G., Krzhizhanovskaya, M. G., Filatov, S. K., Paufler, P. & Meyer, D. C. Structure-density relationship for crystals and glasses in the $Rb_2O–B_2O_3–SiO_2$. *Phys. Chem. Glasses*, 2000, **41**, 389–391.
18. Feller, S., Kottke, J., Welter, J., Nijhawan, S., Boekenhauer, R., Zhang, H., Feil, D., Paramswar, C., Budhwani, K., Affatigato, M., Bhatnagar, A., Bhasin, G., Bhowmik, S., MacKenzie, J., Royle, M., Kambeyanda, S., Pandikuthira P. & Sharma, M. Physical properties of alkali borosilicate glasses. In: *Borate Glasses, Crystals and Melts*, Eds A. C. Wright, S. A. Feller & A. C. Hannon, Society of Glass Technology, Sheffield, 1997, p.246–253.
19. Appen, A. A. & Gan Fuxi, Study of optical properties of sodium alumina-borosilicate glasses. *Zh. Prikl. Khim.*, 1959, **32**, 991–997. (In Russian.)
20. Mazurin, O. V. *Handbook of Glass Data*, Elsevier Science, 1985–1995.
21. SciGlass - Glass Property Information System. Distributed by ScienceServe.
22. Inorganic Crystal Structure Database, 2012, NIST/Karlsruhe.

Phys. Chem. Glasses: Eur. J. Glass Sci. Technol. B, October 2015, **56** (5), 189–196

High temperature behaviour of danburite-like borosilicates $MB_2Si_2O_8$ (M = Ca, Sr, Ba)

Liudmila A. Gorelova,[1,2] *Stanislav K. Filatov,*[*,1] *Maria G. Krzhizhanovskaya*[1] *& Rimma S. Bubnova*[1,2]

[1] *Department of Crystallography, St. Petersburg State University, University Emb 7/9, 199034 St. Petersburg, Russia*
[2] *Grebenshchikov Institute of Silicate Chemistry, Russian Academy of Science, Makarov Emb 2, 199034 St. Petersburg, Russia*

Manuscript received 26 September 2014
Revised version received 29 January 2015
Accepted 3 February 2015

The thermal behaviour of danburite-like phases $MB_2Si_2O_8$ (M = Ca, Sr, Ba) has been studied by in situ high temperature x-ray powder diffraction and heat treatments in a wide temperature range using a natural danburite (Ca) sample and synthetic samples of pekovite (Sr) and maleevite (Ba). Synthetic pekovite was obtained by solid state reactions at 900°C, whereas synthetic maleevite was synthesised by crystallisation from the melt. According to the high temperature x-ray powder diffraction study, the Ca, Sr, and Ba members of the danburite group demonstrate a weak anisotropy of thermal expansion with relatively low volume thermal expansion coefficients (19, 20, and 23×10^{-6} °C^{-1}, respectively). Both thermal expansion coefficients and the expansion anisotropy increase with increasing size of the M^{2+} cations.

1. Introduction

Among the anhydrous alkaline earth borosilicates with the general formula $MB_2Si_2O_8$ (M = Ca, Sr, Ba), there are three naturally occurring members of the danburite group: danburite itself (Ca),[1] pekovite (Sr) and maleevite (Ba).[2] Their orthorhombic structure (space group *Pnma*) is based upon a three-dimensional tetrahedral borosilicate framework, with an ordered distribution of the B and Si atoms. It is worthwhile to note that most of anhydrous borosilicates are structurally similar to their aluminosilicate analogues. In particular, the danburite structure type is identical to that of paracelsian $BaAl_2Si_2O_8$ (both orthorhombic and monoclinic modifications).[3] The crystal structure of danburite $CaB_2Si_2O_8$ has previously been studied by means of high temperature single crystal x-ray diffraction by Sugiyama & Takeuchi.[4]

Borosilicates presently attract considerable attention due to their various technological applications. Borosilicate glasses are used in many ways, from ordinary glassware to nuclear waste immobilisation, owing to their high chemical durability, thermal shock resistance, failure strength, etc. Recent interest in the alkaline earth borosilicate systems has been due to their modern application for producing low temperature co-fired ceramic (LTCC) materials,[5–7] widely used for microwave, wireless, sensor, and other devices.[8] Alkaline earth borosilicate materials with the $MB_2Si_2O_8$ stoichiometry are also of importance because of their potential as phosphors. For instance, Eu-doped $SrB_2Si_2O_8$,[9–10] $Sr_3B_2SiO_8$[11] and $BaB_2Si_2O_8$[12] demonstrate promising luminescent properties. Eu-doped $CaB_2Si_2O_8$ danburite-like phosphors have been obtained under hydrothermal conditions at 800°C and 140 MPa pressure.[13] Undoped $SrB_2Si_2O_8$ and $BaB_2Si_2O_8$ have been synthesised by solid state reactions.[12]

The focus of the present study is on the formation and thermal expansion of danburite-like $MB_2Si_2O_8$ phases (M = Ca, Sr, Ba). Polycrystalline samples of danburite from the Dalnegorsk borosilicate mineral deposit (Russia) and synthetic analogues of pekovite $SrB_2Si_2O_8$ and maleevite $BaB_2Si_2O_8$ have been studied by high temperature x-ray powder diffraction (HTXRD) and heat treatments in a wide temperature range. The specific focus is the principle of anisotropic thermal expansion of the danburite-like phases.

2. Experimental

2.1. Synthesis

Polycrystalline $BaB_2Si_2O_8$ borosilicate was obtained by solid state reactions from chemically pure $SrCO_3$, $BaCO_3$, H_3BO_3, and SiO_2, taken in the stoichiometric ratios. The carbonates were preliminarily calcined at 600°C for 2·5 h, and crushed together with the other ingredients in an agate mortar. The mixture was heated at 600°C for 3 h, and then mixed again in a mortar. Compressed tablets of the thoroughly mixed educts were put in platinum crucibles in an electric muffle furnace, and annealed in the temperature range 800–1100°C in air. Synthesis of $SrB_2Si_2O_8$ pekovite is described in detail elsewhere.[14]

* Corresponding author. Email filatov.stanislav@gmail.com
Original version presented at VIII Int. Conf. on Borate Glasses, Crystals and Melts, Pardubice, Czech Republic, 30 June–2 July 2014
DOI: 10.13036/17533562.56.5.189

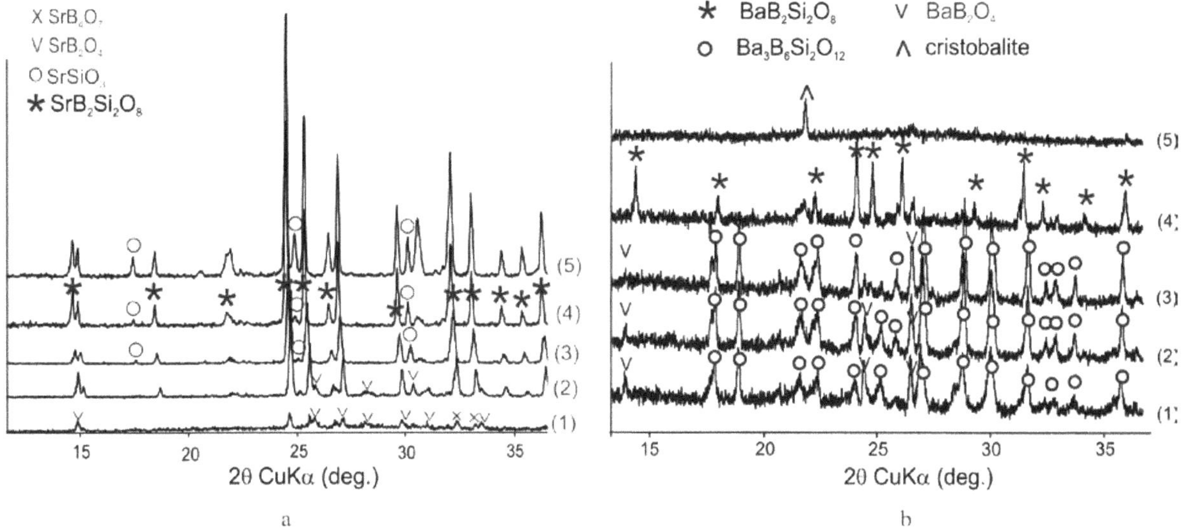

Figure. 1. XRD patterns of samples with nominal compositions $SrB_2Si_2O_8$ (a) and $BaB_2Si_2O_8$ (b) after different heat treatments, as follows. (a): (1) 900°C/1 h; (2) 900°C/5 h; (3) 900°C/72 h; (4) 900°C/127 h; (5) 900°C/127 h+1000°C/120 h. (b): (1) 800°C/258 h; (2) 800°C/317 h+900°C/28 h; (3) 800°C/317 h+900°C/28 h+950°C/3 h; (4) cooling from 1000 to 900°C for 2·5 h; (5) 1100°C/1 h

2.2. Room temperature x-ray diffraction study

Quenched polycrystalline samples of $BaB_2Si_2O_8$ after each heat treatment, and the samples of natural $CaB_2Si_2O_8$, were studied in air at room temperature in the 2θ range 5–60° using STOE STADI P (Cu $K\alpha_1$ radiation, 40 kV/30 mA, transmission geometry, position sensitive detector) and Bruker AXS D2 Phaser (Cu $K\alpha_{1+2}$ radiation, 30 kV/10 mA, Bragg–Brentano geometry, PSD Lynxeye) x-ray diffractometers. The XRD data for the unit cell refinement were corrected using the internal Si powder standard. The Rietveld method (Topas 4.2 software, Bruker) was applied for quantitative phase-composition analysis.

2.3. High temperature x-ray diffraction study

The thermal behaviour of the danburite-like phases upon heating in air was studied by high temperature x-ray powder diffraction by means of a Rigaku Ultima IV diffractometer (Cu $K\alpha_{1+2}$ radiation, 40 kV/40 mA, Bragg–Brentano geometry, PSD D-Tex Ultra), with a thermo-attachment, using a temperature range of 30–900°C, and a temperature step of 30–40°C. The α–β phase transition in quartz was used to check the calibration of the furnace chamber temperature before the high temperature measurement; the error did not exceed 5%. Thin powder samples were prepared on a Pt sample holder (20×12×1·5 mm) from an ethanol suspension.[15] The calculations of the unit cell parameters were performed using the program package Topas 4.2 (Bruker), whereas the calculation of the deformations and the tensor of thermal expansion were performed using an original program package TTT.[16] The temperature dependencies of the unit cell parameters were described by a quadratic polynomial function for the danburite and pekovite samples. A linear polynomial was used for maleevite, $BaB_2Si_2O_8$, since the sample contained an admixture of β-$BaSi_2O_5$ and its XRD pattern was of a lower quality.

3. Results

3.1. Formation of $SrB_2Si_2O_8$ and $BaB_2Si_2O_8$

Figures 1(a) and (b) show the room temperature XRD patterns for the quenched samples of $SrB_2Si_2O_8$ (pekovite) and $BaB_2Si_2O_8$ (maleevite) after annealing, respectively. As has been mentioned previously,[12,17] both $SrB_2Si_2O_8$ and $BaB_2Si_2O_8$ are metastable phases. During their synthesis, borate phases form first, whereas borosilicates start to crystallise at higher temperatures.

Synthetic pekovite, $SrB_2Si_2O_8$, mixed with trace amounts of $SrSiO_3$ and SiO_2 cristobalite (<5 wt%), was obtained by solid state reactions at 900°C for 72 h (Figure 1(a)). Its XRD pattern did not change for 127 h under further treatment at 900°C. After the initial treatment (900°C/1 h), the sample contained a significant quantity of intermediate borate phases, such as SrB_2O_4 and SrB_4O_7. In accordance with Ref. 16, $SrB_2Si_2O_8$ crystallised at 900°C/224 h and in 336 h started to decompose into SiO_2 and SrB_2O_4.

Polycrystalline maleevite, $BaB_2Si_2O_8$, admixed with β-$BaSi_2O_5$ (~15 wt%), was obtained by slow cooling of a stoichiometric melt from 1000 to 900°C for 2·5 h. After cooling the melt from 1100°C the compound does not exist any longer, and the XRD pattern contains only cristobalite peaks (Figure 1(b)). We were unable to obtain $BaB_2Si_2O_8$ at 800, 900 and 950°C by solid state reactions from a stoichiometric mixture. Instead of the formation of $BaB_2Si_2O_8$, we have obtained crystals of another borosilicate $Ba_3B_6Si_2O_{16}$,[18–19] in association with β-BaB_2O_4 (Figure

Table 1. Unit cell parameters for the danburite-like $MB_2Si_2O_8$ (M = Ca, Sr, Ba) borosilicates

	Ca		Sr		Ba		
	Present work	*Ref. 4*	*Present work*	*Ref. 2*	*Present work*	*Ref. 2*	*Ref. 12*
a (Å)	8·0419(5)	8·037(1)	8·1593(6)	8·155(2)	8·206(4)	8·141(2)	8·1726(2)
b (Å)	7·7310(9)	7·722(1)	7·9250(7)	7·919(1)	8·190(2)	8·176(2)	8·1837(2)
c (Å)	8·7610(6)	8·757(1)	8·9183(5)	8·921(1)	9·055(3)	9·038(2)	9·0428(2)
V (Å^3)	544·75(5)	543·46	576·68(5)	576·1(2)	608·6(3)	601·5(2)	604·8(1)

1(b)). $BaB_2Si_2O_8$[12] has previously been synthesised by solid state reactions with 50 mol% of excess boric acid with heating at 850°C/24 h.

A comparison of experimental unit cell parameters of the danburite-like borosilicates with literature data is given in Table 1. For the Ca and Sr borosilicates, the data on synthetic and natural materials agree well, whereas the *a* unit cell parameter for maleevite displays significant variation. It should be noted that the experimental data for maleevite were of lower quality compared to the other samples (see unit cell e.s.d.s in Table 1) because the synthetic sample contained a considerable admixture (~15 wt%) of β-$BaSi_2O_5$. On the other hand, natural maleevite[2] contained a considerable amount of Mg replacing Ba, which may explain the observed difference.[2]

3.2. Thermal expansion of $MB_2Si_2O_8$ borosilicates (M = Ca, Sr, Ba)

As expected, the diffraction maxima of the studied phases shift with increasing temperature, due to thermal expansion. No evidence of temperature induced phase transformations was observed for danburite and synthetic pekovite at temperatures up to 900°C. Synthetic maleevite starts to decompose above 800°C, with the formation of $Ba_3B_6Si_2O_{16}$, as described in section 3.1. This observation seems to be in good agreement with other data that describe this phase as the only stable one in the ternary BaO–B_2O_3–SiO_2 system.[18] The tetragonal↔cubic transition was observed for cristobalite present in the samples of the Sr and Ba compounds in the temperature range from 220 to 260°C, in good agreement with previously reported data.[20]

Table 2. The main values of thermal expansion coefficients ($\times10^6$ °C^{-1}) for $MB_2Si_2O_8$ (M = Ca, Sr, Ba)

M	*Ca*		*Sr*		*Ba*
T, °C	*30°C*	*900°C*	*30°C*	*900°C*	*30–900°C*
α_a	6·1(2)	10·1(2)	6·5(2)	10·6(2)	16·3(3)
α_b	4·2(3)	6·5(4)	5·9(2)	9·4(2)	5·0(3)
α_c	5·9(1)	9·5(2)	2·8(2)	5·4(2)	5·1(2)
α_V	16·3(4)	26·1(4)	15·1(3)	25·4(3)	26·3(5)

The dependence of the unit cell parameters of $MB_2Si_2O_8$ (*M* = Ca, Sr, Ba) borosilicates on temperature is shown in Figure 2; for $SrB_2Si_2O_8$ the data from Ref. 14 have been used. The main coefficients of the thermal expansion tensor are given in Table 2. In the orthorhombic system, the principal axes of the thermal expansion tensor are parallel to the crystallographic axes. The thermal dependences for danburite powder samples shown in Figure 2 demonstrate that $CaB_2Si_2O_8$ expands almost isotropically, as was previously reported from a single crystal x-ray diffraction study.[4] The expansion along the *a* and *b* axes of $SrB_2Si_2O_8$ is approximately the same, whereas the expansion along the *c* axis is about half as large. According to a polynomial approximation (see Appendix), the expansion of the Ca and Sr compounds increases slightly with increasing temperature, whilst $BaB_2Si_2O_8$ displays the most anisotropic character of thermal expansion, with the greatest expansion along the *a* axis, and the smallest expansion along the *c* axis.

4. Discussion

4.1. Compositional deformations of danburite-like structures

As can be seen from Figure 3, the dependence of the unit cell parameters of the danburite-like borosilicates

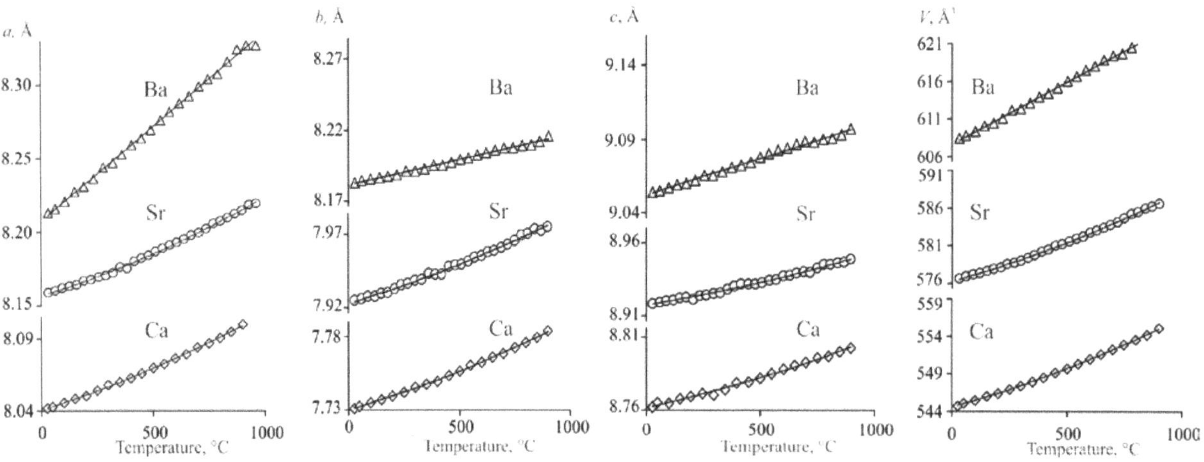

Figure 2. The unit cell parameters of $MB_2Si_2O_8$ (M = Ca, Sr, Ba) borosilicates at different temperatures

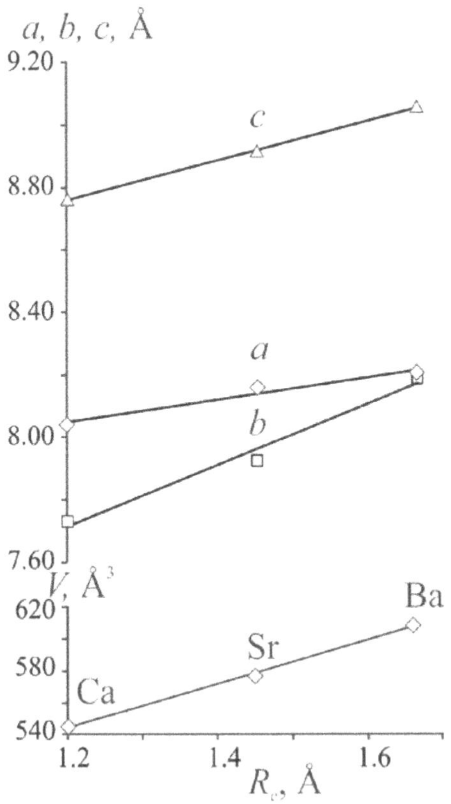

Figure 3. The dependence of the unit cell parameters on the cation radius of danburite-like phases $M^{2+}B_2Si_2O_8$ (M = Ca, Sr, Ba). Ca – danburite, Sr –pekovite, Ba –maleevite

on cationic substitution are similar to the temperature dependence (Figure 2), and both exhibit nearly isotropic behaviour.

In Figure 4 the changes of unit cell parameters in the danburite-paracelsian family are presented as a function of the sum of the radii of non-tetrahedral *M* cations and tetrahedral cations. It is of interest that the size of the unit cell is much more sensitive to the replacement of the tetrahedral atoms than to the replacement of non-tetrahedral cations, very similar to the tendency observed for leucite-like phases.[21] For example, the cell volume of $SrB_2Si_2O_8$ increases by approximately 5%, when Sr is replaced by Ba. In contrast, the unit cell volumes of $SrB_2Si_2O_8$ and $SrAl_2Si_2O_8$ differ by more than 15%. Such changes are obviously the result of the differences in atomic radii between Sr (1·31 Å) and Ba (1·52 Å), on one hand, and B (0·11 Å) and Al (0·39 Å), on the other.[22]

4.2. The hinge mechanism of thermal and compositional deformations of fourfold rings

Similar to aluminosilicates, borosilicates display a considerable flexibility of tetrahedral frameworks, due to the cooperative motions of tetrahedral groups, such as tilting and rotating, as first noted by Hazen & Finger.[23,24] It is assumed that the frameworks built from relatively rigid corner-sharing TO_4 tetrahedra (T = Si, Al) are flexible enough to be distorted under variable conditions. A rigid-unit modes model was applied by Dove and others[25–27] to elucidate the structural mechanisms of displacive phase transitions, and to understand such structure-dependent properties as anisotropic thermal expansion and, in particular, negative thermal expansion.

The idea of a hinge mechanism governing the tilting of rigid polyhedra has been proposed to explain negative thermal expansion by Sleight.[28] Usually, the high anisotropic deformation can be inferred from the behaviour of the unit cell parameters: two unit cell parameters change synchronously in opposite directions, whereas the third parameter and the unit cell volume remain almost unchanged. In general, the search for hinge-type points in the structure is not a trivial task.[29,30] As a rule, the hinge unit is a small

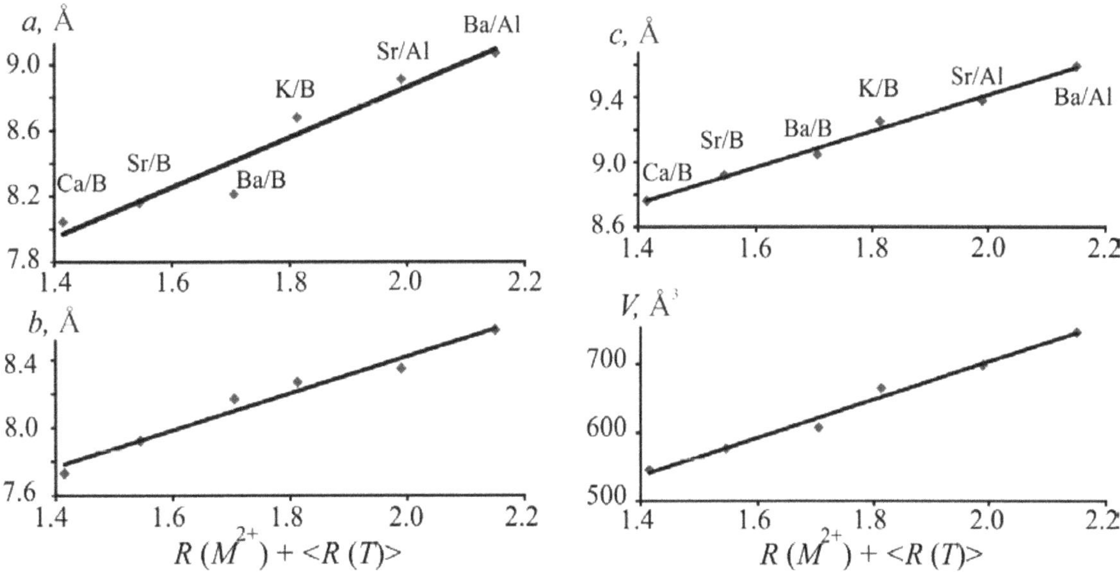

Figure 4. The unit cell parameters and the volume versus chemical composition (radius of non-tetrahedral cation M^+ average radius of tetrahedral atom T) of danburite-like borosilicates comparing to their aluminosilicate counterparts[3]

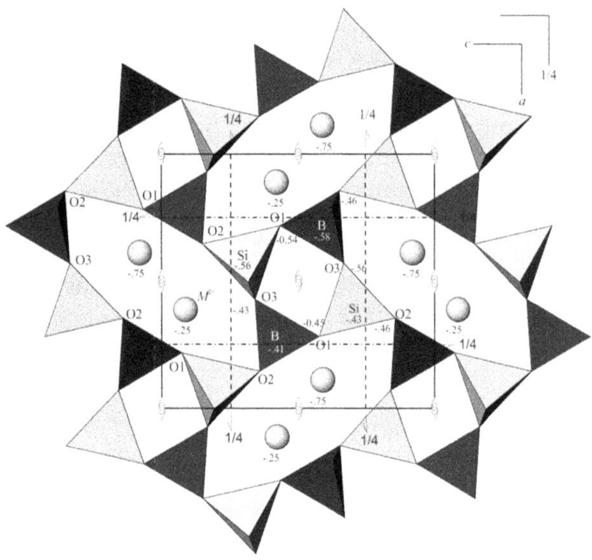

Figure 5. The crystal structure of danburite-like compounds

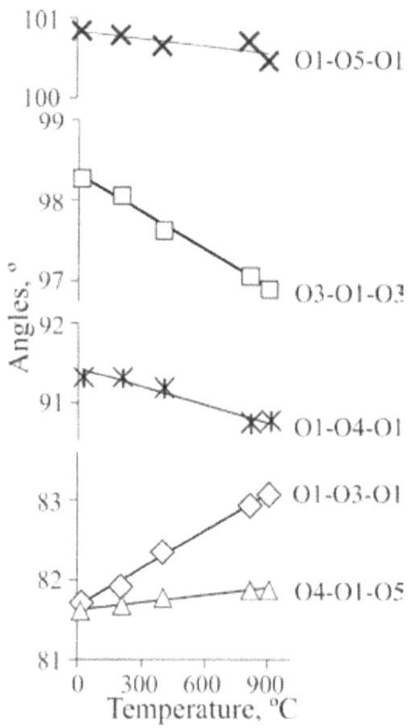

Figure 6. Temperature dependence of O–Ô–O angles in fourfold rings (hinge unit) of danburite $CaB_2Si_2O_8$. The calculations were performed using the atomic coordinates of $CaB_2Si_2O_8$[4]

fragment of the crystal structure, that may have the form of a distorted square or a parallelogram. In this parallelogram, the corners are atoms and the edges are interatomic distances (usually non-bonding). Several dramatic examples of hinge deformations have been described in the literature.[29,30]

As already mentioned, the danburite-like compounds do not possess a strong anisotropy of thermal expansion (Figure 2) and compositional deformations (Figure 3). However, in the danburite-like structures, the fourfold ring of TO_4 tetrahedra can be considered as a hinge unit (Figure 5) parallel to the *ac* plane. The O–O edges of the rigid BO_4 and SiO_4 tetrahedra are the sides of the hinge O1–O3–O1–O3 parallelogram, and the O atoms are the corners of the hinge (Table 3, Figures 6–8). The O1–Ô3–O1 angle inside the fourfold ring increases by 1·37° when the temperature increases from 20 to 900°C, whereas the O3–Ô1–O3 angle decreases by approximately the same value. Therefore, the thermal deformations of the ring (hinge unit) are sharply anisotropic: the O3–O3 diagonal distance decreases with increasing temperature, whilst the O1–O1 distance increases.

Table 3 demonstrates that the fourfold O3–O1–O3–O1 ring (the hinge unit) has the same character of

Table 3. Comparison of thermal and compositional changes of O–Ô–O angles in fourfold rings of two types. Calculations were executed using the atomic coordinates of danburite $CaB_2Si_2O_8$ (20, 208, 407, 817, 910°C),[4] *and pekovite $SrB_2Si_2O_8$ and maleevite $BaB_2Si_2O_8$ (20°C)*[2]

Thermal deformations of danburite structure $CaB_2Si_2O_8$

Angles (°)	*20°C*	*208°C*	*407°C*	*817°C*	*910°C*	*$\Delta°_{910-20}$*
			O3–O1–O3–O1 ring			
O1–Ô3–O1	81·72	81·93	82·36	82·94	83·09	1·37
O3–Ô1–O3	98·28	98·07	97·64	97·05	96·90	−1·38
			O4–O1–O5–O1 ring			
O4–Ô1–O5	81·65	81·68	81·76	81·87	81·90	0.25
O1–Ô5–O1	100·85	100·80	100·66	100·70	100·47	−0·38
O1–Ô4–O1	91·34	91·34	91·19	90·76	90·78	−0·56

Compositional (chemical) deformations of danburite-like phases at 20°C

Angles (°)	*Ca*	*Sr*	*Ba*	*$\Delta°_{Ba-Ca}$*
		O3–O1–O3–O1 ring		
O1–Ô3–O1	81·72	86·77	98·07	16·35
O3–Ô1–O3	98·28	93·23	81·93	−16·35
		O4–O1–O5–O1 ring		
O4–Ô1–O5	81·65	84·07	91·04	9·39
O1–Ô5–O1	100·85	97·04	85·18	−15·67
O1–Ô4–O1	91·34	87·50	77·07	−14·27

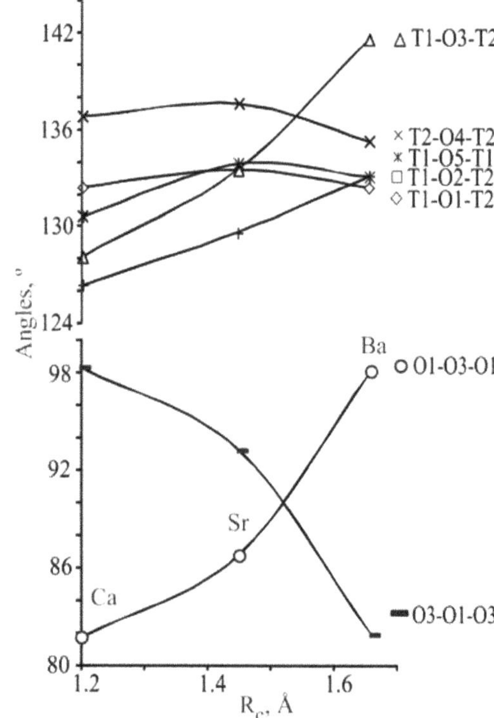

Figure 7. Dependence of the T–Ô–T and O–Ô–O angles on the cation radius in the series $M^{2+}B_2Si_2O_8$ (M = Ca, Sr, Ba)

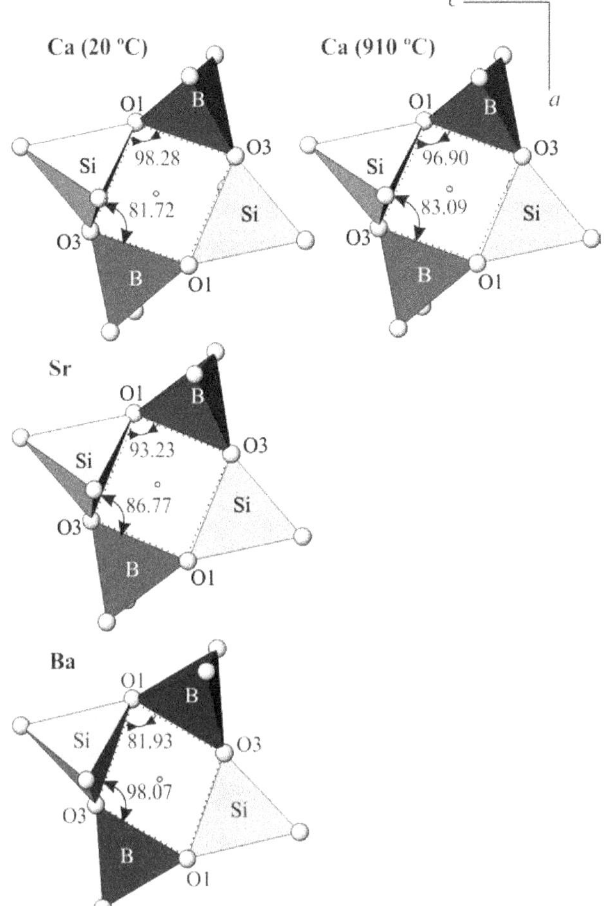

Figure 8. Changes of inter-tetrahedral angles in the fourfold B–O–Si–O–B–O–Si–O ring in the ac plane, with variation of the temperature (horizontal) and the chemical composition (vertical)

deformations induced by the Ca–Sr–Ba substitution as that induced by temperature. This is in agreement with the general statement that the structure deformations of different nature (and, in particular, thermal and compositional deformations) are usually similar:[23,24,31] the increasing temperature and the substitution of small ions by larger ones lead to very similar deformations of the crystal structure.

Considering the hinge O3–O1–O3–O1 unit as a square in the *ac*-plane, it is possible to calculate approximate coefficients for thermal (α) and compositional (γ) deformations of the ring using the formulae:

$$\alpha_1 = \frac{1}{l}\frac{l_2 - l_1}{t_2 - t_1}$$

and

$$\gamma_1 = \frac{1}{l}\frac{l_2 - l_1}{r_2 - r_1}$$

where t is temperature, r is the radius of the M^{2+} cation, l_1 is the initial length of a square diagonal at initial t_1 or r_1; l_2 is its final value at t_2 or r_2. The data employed in our calculations are shown in Table 4. The anisotropy diagrams for the thermal (solid line)

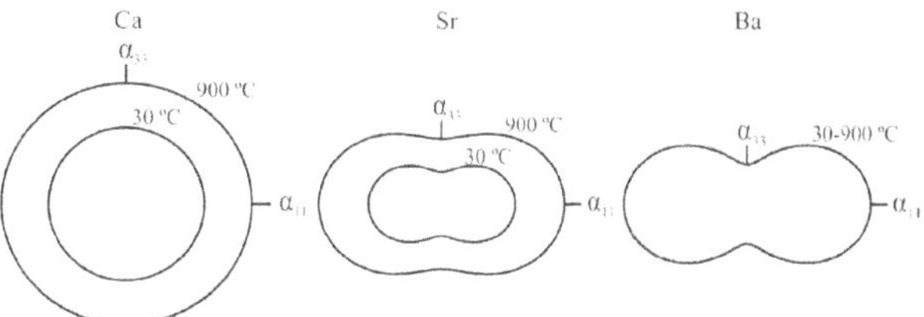

Figure 9. Anisotropy diagrams for the thermal expansion coefficients of danburite-like phases $M^{2+}B_2Si_2O_8$ (M = Ca (danburite), Sr (pekovite), Ba (maleevite))

and compositional (dotted line) deformation coefficients (Figure 10) illustrate the similarity of these types of structure deformation.

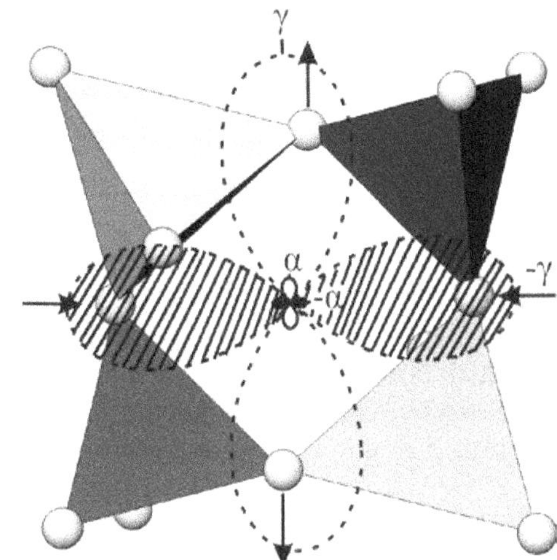

Figure 10. The hinge cell of danburite $CaB_2Si_2O_8$, together with the anisotropy diagrams for thermal (solid line at the centre, α) and compositional (dotted line, γ) deformations

4.3. Thermal expansion of the crystal lattice

The thermal expansion of the danburite-type crystal structure in the *ac* plane is dictated by the fourfold O1–O3–O1–O3 and the eightfold O2–O3–O2–O1–O2–O3–O2–O1 rings (Figure 5). The fourfold ring provides a particularly clear illustration of the similarity of thermal and compositional (Ca–Ba) deformations arising from the hinge mechanism (see section 4.2). In the *ac* plane, this ring has two different orientations; the anisotropy diagrams for the thermal expansion coefficients are shown in Figure 11. The

Table 4. Thermal and compositional (Ca–Ba) deformations of danburite crystal structure, $CaB_2Si_2O_8$

Bond	Diagonal distances O–O (Å) Ca/20°C	Ca/910°C	Ba/20°C	Diagonal distance changes (Å) Δ_{910-20}	Δ_{Ba-Ca}
Fourfold ring O1–O3–O1–O3					
O1–O1	3·32	3·37	3·84	0·05	0·52
O3–O3	3·84	3·80	3·84	−0·04	0
Eightfold ring					
O2–O2 (short diagonal)	3·18	3·23	3·59	0·05	0·41
O2–O2 (long diagonal)	8·80	8·82	8·63	0·02	−0·17

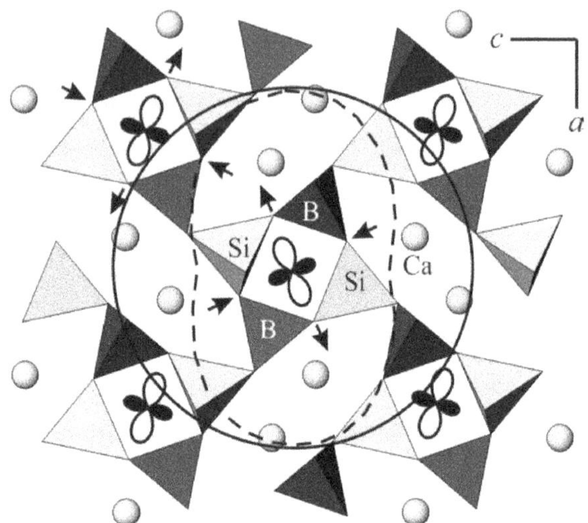

Figure 11. The structure of danburite $CaB_2Si_2O_8$, compared to the anisotropy diagrams for thermal (solid line) and compositional (Ca–Ba, dashed line) deformations

averaged deformations in the *ac* plane should be an expansion along the *a* axis and a contraction along the *c* axis (Figures 5, 11). It may seem that the distortion of the eightfold ring is completely determined by the four tetrahedral units. However, the O1–O2–O3 angles (Figure 5) do not belong to the fourfold rings, which makes it necessary to estimate the contribution of the eightfold ring to the thermal distortion in the *ac* plane as well. An approximate estimation of this contribution could be made using the values for the thermal deformation of the short (O2–O2) and the long (O2–O2) diagonals of the ring (Figures 5 and 11): In the temperature range 20–900°C, the short diagonal increases (nearly parallel to the direction of the fourfold ring contraction) by +0·05 Å, whereas the long diagonal increases by +0·02 Å. Taking into account that the short diagonal is about three times smaller than the long one, its relative elongation is 6 times greater than that of the long diagonal, which presumably compensate for the negative expansion caused by the fourfold rings.

Obviously, the model described above is strictly qualitative, and is intended to give an understanding of the nearly isotropic deformation of the danburite-type structure in the *ac* plane. In this case the high temperature structural data were used for $CaB_2Si_2O_8$.[4]

A similar situation is observed for the compositional deformations. Under Ca–Ba substitution, the short diagonal of the eightfold ring increases by 0·41 Å, whereas the long one decreases by 0·17 Å, which, in a similar way, compensates for a sharp anisotropy of the fourfold ring deformation.

Conclusions

In conclusion, in the structures of the danburite-like minerals and synthetic phases, the structural units such as fourfold rings demonstrate sharply anisotropic hinge deformations, which do not manifest themselves in the lattice deformations. These structural deformations constitute the basis of comparative (dynamic)[32] crystal chemistry of this group of materials. The effect of increasing temperature upon the crystal structure is generally similar to the effect of small-for-large chemical substitutions. The values of the thermal expansion coefficients increase with increasing cation size, and this rather unexpected increase of thermal expansion anisotropy with increasing cation size may be caused by cooperative motions that include the expansion of the eightfold rings.

Acknowledgements

This research was financially supported by the President grant for government support of the leading scientific schools of the Russian Federation no. 1583.2014.5. We thank Professor I. V. Pekov (Moscow State University) for providing us with the sample of danburite used in this study. We are grateful to Prof. S. V. Krivovichev for polishing the English. X-ray measurements were performed in XRD Research Centre of St. Petersburg State University.

References

1. Dunbar, C. & Machatschki, F. *Z. Kristallogr.*, 1931, **76**, 133.
2. Pautov, L. A., Agakhanov, A. A., Sokolova, E. V. & Hawthorne, F. C. *Can. Miner.*, 2004, **42**, 107.
3. Inorganic Crystal Structure Database (ICSD). Karlsruhe, Germany, 2012.
4. Sugiyama, K. & Takéuchi, Y. *Z. Kristallogr.*, 1985, **173**, 293.
5. Tummala, R. R. *J. Am. Ceram. Soc.*, 1991, **74**, 895.
6. Jean, J.-H., Fang, Yu.-C., Dai, S. X. & Wilcox, D. L. *J. Mater. Res.*, 2002, **17**, 1772.
7. Chiang, C.-C., Wang, S.-F., Wang, Y.-R. & Wie, W.-C. J. *Ceram. Int.*, 2008, **34**, 599.
8. Low Temperature Cofired Ceramic Publication List, http://www.ltcc.de/en/publications.php
9. Verstegen, J. M. P. J., Ter Vrugt, J. W. & Wanmaker, W. L. *J. Inorg. Nucl. Chem.*, 1972, **34**, 3588.
10. Wang, Y., Zhang, Z., Zhang, J. & Lu, Y. *J. Solid State Chem.*, 2009, **182**, 813.
11. Wang, L. & Wang, Y. *J. Lumin.*, 2011, **131**, 1479.
12. Saradhi, M. P., Boudin, S., Varadaraju, U. V. & Raveau, B. *J. Solid State Chem.*, 2010, **183**, 2496.
13. Juwhari, H. K. & White, W. B. *Mater. Lett.*, 2012, **88**, 16.
14. Gorelova, L. A., Krzhizhanovskaya, M. G. & Bubnova, R. S. *Fiz. Khim. Stekla*, 2012, **38**, 872. (In Russian.)
15. Filatov, S. K. *Kristall Technik*, 1971, **6**, 777.
16. Bubnova, R. S., Firsova, V. A. & Filatov, S. K. *Glass Phys. Chem.*, 2013, **39**, 347.
17. Baylor, R. & Brown, J. J. *J. Am. Ceram. Soc.*, 1976, **59**, 21.
18. Levin, E. M. & Ugrinic, G. M. *J. Res. Nat. Bur. Stand.*, 1953, **51**, 37.
19. Krzhizhanovskaya, M. G., Gorelova, L. A., Bubnova, R. S. & Filatov, S. K. *Z. Kristallogr.*, 2013, **228**, 544.
20. Hatch, D. M. & Ghose, S. *Phys. Chem. Miner.*, 1991, **17**, 554.
21. Bubnova, R. S. & Filatov, S. K. *Vysokotemperaturnaya Kristallokhimiya Boratov i Borosilikatov (High temperature crystal chemistry borates and borosilicates)*, Nauka, St. Petersburg, 2008, p 760. (In Russian.)
22. Shannon, R. D. *Acta Crystallogr.*, 1976, **32**, 751.
23. Hazen, R. M. & Finger, L. W. *Phase Transitions*, 1979, **1**, 1.
24. Hazen, R. M. & Finger, L. W. *Comparative crystal chemistry: temperature, pressure, composition and the variation of crystal structure*, London, 1982, p. 231.
25. Dove, M. T., Cool, Y., Palmer, D.C., Putnis, A., Salje, H. & Winkler, B.

Am. Miner., 1993, **78** (5/6), 486.
26. Dove, M. T., Pride, A. K. A. & Keen, D. A. *Miner. Mag.*, 2000, **64**, 267.
27. Palmer, D. C., Dove, M. T., Ibberson, R. M. & Powell, B. M. *Am. Miner.*, 1997, **82**, 16.
28. Sleight, A. W. *Endeavour*, 1995, **19(2)**, 64.
29. Filatov, S. K. & Bubnova, R. S. *Z. Kristallogr.*, 2007, **26**, 447.
30. Bubnova, R.S. & Filatov, S. K. *Z. Kristallogr.*, 2013, **228**, 395.
31. Filatov, S. K. Some structure-geometric regularities of crystal deformations under temperature, pressure and compositional changes. *Kristallographiya and kristallochimiya*, Leningrad State University, Leningrad, 1973, **2**, 5. (In Russian.)
32. Filatov, S. K. *Vysokotemperaturnaya Kristallokhimiya (High temperature crystal chemistry)*, Nedra, Leningrad, 1990, p. 288. (In Russian.)

Appendix

Approximate equations for the unit cell parameters of $MB_2Si_2O_8$ (M = Ca, Sr, Ba):
$CaB_2Si_2O_8$:
$a(T)$=8·041(1)+0·000048(1)T+0·000000018(2)T^2 Å
$b(T)$=7·730(1)+0·000045(1)T+0·000000016(1)T^2 Å
$c(T)$=8·761(1)+0·000036(3)T+0·000000012(3)T^2 Å
$V(T)$=544·56(3)+0·0086(2)T+0·0000032(2)T^2 Å^3

$SrB_2Si_2O_8$:
$a(T)$=8·158(1)+0·000052(1)T+0·000000020(2)T^2 Å
$b(T)$=7·923(1)+0·000046(2)T+0·000000016(2)T^2 Å
$c(T)$=8·918(1)+0·000024(2)T+0·000000014(2)T^2 Å
$V(T)$=576·39(4)+0·0085(2)T+0·0000036(2)T^2 Å^3

$BaB_2Si_2O_8$:
$a(T)$=8·205(1)+0·000134(3)T Å
$b(T)$=8·170(1)+0·000041(3)T Å
$c(T)$=9·044(1)+0·000046(2)T Å
$V(T)$=606·4(1)+0·0159(2)T Å^3

Phys. Chem. Glasses: Eur. J. Glass Sci. Technol. B, December 2015, **56** (6), 248–254

Structural units in alkali borate glasses and melts

Armenak A. Osipov, Leyla M. Osipova & Rimma T. Zainullina*

Laboratory of Experimental Mineralogy and Physics of Minerals, Institute of Mineralogy UB RAS, Miass, Russian Federation

Manuscript received 7 August 2014
Revised version received 27 December 2014
Accepted 3 March 2015

At present it is generally accepted that the regular network of borate crystals, the disordered network of borate glasses and the borate anions in borate melts consist of the same set of basic structural units. Recently, however, new boron–oxygen fragments in the form of monomeric $[BO_2]^-$ V-shaped anions were determined to be present in the structure of overheated borate melts with metaborate composition. The presence of such fragments in the structure of sodium, potassium and caesium borate melts with modifier oxide content less than 50 mol% is studied here by high temperature Raman spectroscopy. It is established that V-shaped groups form in the structure of the studied melts at modifier oxide contents of more than 25 mol%, for all alkali cations studied. The concentration of the V-shaped groups gradually increases with increase in M_2O (M=Na, K, Cs) content, and with increase in temperature. In addition, our studies have shown that these triatomic groups are absent in the structure of all glasses studied here.

Introduction

The absence of long range crystallographic order in glasses is their distinctive feature in comparison with crystalline solids. A similarity between glass and crystal structures, however, remains in the short range structure, i.e. the glass structure order is confined to coordination polyhedra corresponding to those in the crystals of analogous chemical composition. In other words, the random network of borate glasses and the regular lattice of borate crystals consist of the same building units (boron–oxygen polyhedra). In turn, it is suggested that the structure of rapidly quenched glasses reflects the structure of the melt with the same composition, "frozen" at a certain temperature (depending on the cooling rate). Extrapolation of the data on glass structure to the melt allows us to argue that the local structure of melts at high temperatures is qualitatively similar to the glass structure, and changes in concentrations of the basic structural units are explained by various disproportionation reactions. None of the currently known reactions propose the formation of other structural units in melts other than those that form the disordered network of glasses or the regular structure of crystals.

At present, it is generally accepted that five basic borate structural units can be distinguished: $B\varnothing_3$ symmetric triangle (Ø – bridging oxygen atom), $[B\varnothing_4]^-$ tetrahedron, $B\varnothing_2O^-$ metaborate triangle, $B\varnothing O_2^{2-}$ pyroborate unit, and BO_3^{3-} orthoborate anion (see Figure 1). Numerous studies of alkali borate glasses[1–8] have shown that these structural units are formed in the glasses in the following sequence

$$B\varnothing_3 \rightarrow [B\varnothing_4]^- \rightarrow B\varnothing_2O^- \rightarrow B\varnothing O_2^{2-} \rightarrow BO_3^{3-}$$

as modifier oxide content increases. In turn, this may be considered as a result of the interaction of modifier oxide with the boron–oxygen framework in the following sequence:[9]

$$1/2M_2O + B\varnothing_3 \rightarrow M^+ + [B\varnothing_4]^- \quad (1)$$

$$1/2M_2O + B\varnothing_3 \rightarrow M^+ + B\varnothing_2O^- \quad (2)$$

$$1/2M_2O + B\varnothing_2O^- \rightarrow M^+ + B\varnothing O_2^{2-} \quad (3)$$

$$1/2M_2O + [B\varnothing_4]^- \rightarrow M^+ + B\varnothing O_2^{2-} \quad (4)$$

$$1/2M_2O + B\varnothing O_2^{2-} \rightarrow M^+ + BO_3^{3-} \quad (5)$$

It is reasonable that the sequence of formation of these structural units in melts is the same as in glasses. At the same time, the concentrations of these units in the melt differ, more or less, from those in the glass with the same composition. This change in concentration of the structural units can be explained by the shift of isomerisation reaction.[10–12]

$$[B\varnothing_4]^- \Leftrightarrow BÆ_2O^- \quad (6)$$

or disproportionation reactions[8,10,13,14] (depending on melt composition)

$$2B\varnothing_2O^- \Leftrightarrow B\varnothing_3 + B\varnothing O_2^{2-} \quad (7)$$

$$2[B\varnothing_4]^- \Leftrightarrow B\varnothing_3 + B\varnothing O_2^{2-} \quad (8)$$

$$2B\varnothing O_2^{2-} \Leftrightarrow B\varnothing_2O^- + BO_3^{3-} \quad (9)$$

* Corresponding author. Email armik@mineralogy.ru
Original version presented at VIII Int. Conf. on Borate Glasses, Crystals and Melts, Pardubice, Czech Republic, 30 June–2 July 2014
DOI: 10.13036/17533562.56.5.248

toward the right with increasing temperature. As can be seen, none of these reactions propose the formation in the melt of other structural units than those presented in Figure 1. Hence, it follows that the set of structural units in melts should be the same as the set of structural units which form the random network of glasses. Recently, however, it was shown[15,16] that in addition to the structural units presented in Figure 1, other boron–oxygen groups, namely $[BO_2]^-$ triatomic V-shaped groups, are present in alkali borate melts with metaborate composition. The concentration of such groups increases with increasing temperature, and V-shaped groups are absent in the structure of alkali borate glasses and crystals. Thus, one more borate group appears in the list of structural units of alkali borate melts, at least, with metaborate composition. It is reasonable to expect that the $[BO_2]^-$ monomers are present not only in metaborate melts, but that they can also exist in the structure of other alkali borate melts with composition close to metaborate. Therefore, the aim of the present study is the investigation of the composition and temperature dependencies of V-shaped groups in the structure of sodium, potassium and caesium borate melts with modifier oxide content less than 50 mol%.

As was shown in Refs 15 & 16, $[BO_2]^-$ groups can be detected by high temperature Raman spectroscopy. For this reason, Raman spectroscopy was chosen as the main experimental method of the present study.

Experimental

Raman spectra of $Na_2O–B_2O_3$ and $Cs_2O–B_2O_3$ glasses and melts with alkali oxide content less than 50 mol% were measured previously, and the sample preparation procedure is given in Refs 17–19. Potassium borate samples were prepared by the traditional melt quenching technique. K_2CO_3 and H_3BO_3 were used as sources for the oxides in the samples. Appropriate amounts of boric acid and potassium carbonate were weighed and thoroughly mixed together in a mortar and pestle, transferred to a platinum crucible, and melted in an electric furnace in air at 950–1050°C (depending on the sample composition) for 1 or 2 h. The weight of the batch was 15 g in all experiments. Clear, bubble-free melt was then poured into a small Pt crucible and cooled in air at room temperature. The sample in the small Pt crucible was re-melted for ~20–30 min at the same temperature as the first melting and cooled in air at room temperature. All measurements were done immediately after the cooling of the melt. Specially designed high temperature equipment, based on a DFS-24 monochromator, was used for the Raman scattering measurements at various temperatures. A detailed description of the experimental setup and the recording conditions for the spectra is given in Ref. 20. To compare the spectra measured at various temperatures, all spectra

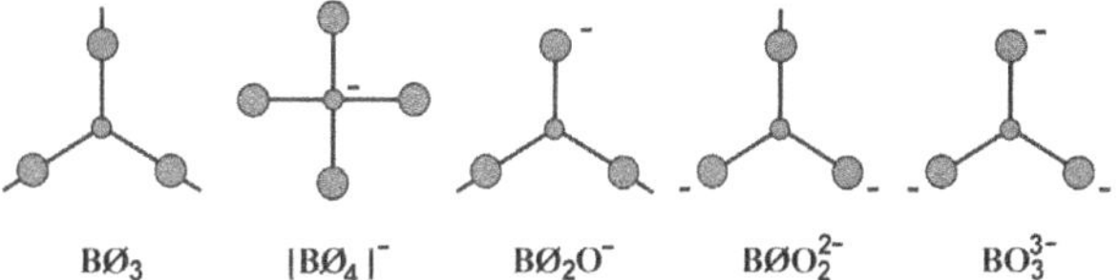

Figure 1. Basic structural units of vitreous and crystalline borates [Colour available online]

were corrected using the Bose–Einstein population factor:[21]

$$I_{red}(\nu) = I_{obs}(\nu)\left[1-\exp\left(-\frac{hc\nu}{kT}\right)\right]\frac{\nu\nu_0^3}{\left(\nu_0-\nu\right)^4} \tag{10}$$

where I_{obs} and I_{red} are the observed and reduced Raman intensities, respectively, n and n_0 are the Raman shift and the wavenumber of the excitation source, respectively, and h, k, c, and T represent Planck's constant, Boltzmann's constant, the speed of light, and the temperature, respectively.

It should be noted that Raman spectra of potassium and caesium borate melts with modifier oxide content more than 45 and 40 mol%, respectively, were not measured. A strong evaporation of these melts was observed during the high temperature measurements. This evaporation leads to the overgrowing of the top of electrical furnace with dendritic crystals. These crystals prevented penetration of the laser radiation into the melt.

The following designation of sample composition is used here: *XM*, where *X* is the modifier oxide content (mol%) and M=Na, K, or Cs.

Results and discussion

Several remarks should be emphasised before the discussion. As shown in Refs 15 & 16, two Raman lines near 1050–1065 cm^{-1} and 1210–1240 cm^{-1} are the specific indicators for the presence of $[BO_2]^-$ groups in the structure of metaborate melts. These two lines show synchronous increase in intensity with increasing temperature. The 1210–1240 cm^{-1} band, however, is also assigned to the symmetric stretching vibrations of the nonbridging oxygen atoms in $BØO_2^{2-}$ pyroborate units.[3,5,6,12,22–24] It may well be that $BØO_2^{2-}$ pyroborate units exist in alkali borate melts with metaborate composition, as well as in the structure of other alkali borate melts with composition close to metaborate. The presence of the pyroborate units can be explained by the shift of the equilibria (7) and (8) towards the right with increasing temperature. The results of molecular dynamics simulations[10] indicate that $BØO_2^{2-}$ units can exist in the structure of $Cs_2O–B_2O_3$ glasses and melts with Cs_2O contents larger than 20 mol%, and their concentration must increase with increasing temperature. On the other hand, it is well known that if $BØO_2^{2-}$ pyroborate units exist in alkali borate glasses or melts in the form of $B_2O_5^{4-}$ pyroborate groups,

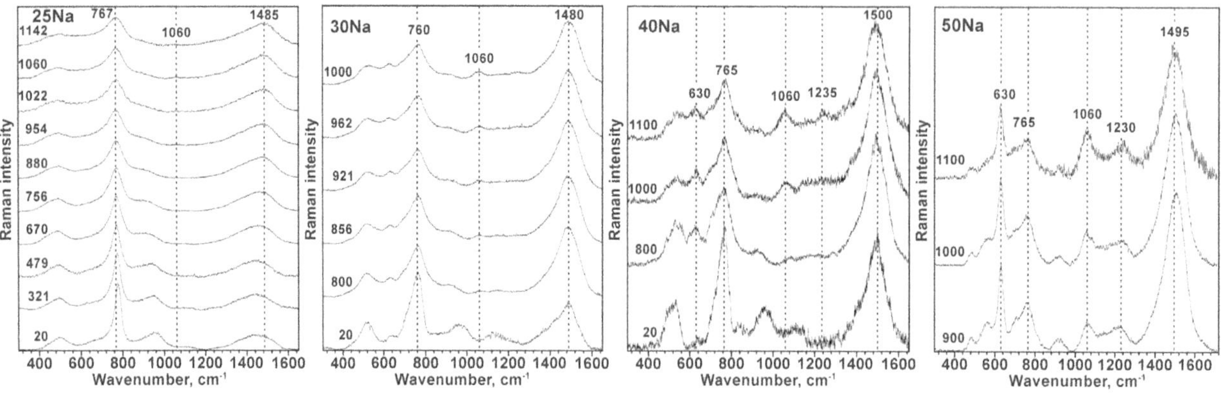

Figure 2. Raman spectra of $Na_2O–B_2O_3$ glasses and melts. (The temperature is given in °C)

then two Raman bands near 1210–1240 cm^{-1} and 820–860 cm^{-1} should be observed in the spectra (the 820–860 cm^{-1} band is due to the symmetric stretching vibrations of B–O–B bridges in the $B_2O_5^{4-}$ pyroborate dimer). However, $BØO_2^{2-}$ pyroborate units may exist not only in the form of $B_2O_5^{4-}$ pyroborate groups (see, for example, Ref. 10). They can also be present in the form of terminal elements of complex borate anions (not dimers). In this case the 820–840 cm^{-1} band may be absent in the Raman spectra, whereas the 1210–1240 cm^{-1} band will be observed. Thus, the 1210–1240 cm^{-1} band may have ambivalent origin ($BØO_2^{2-}$ pyroborate units as well as V-shaped fragments may contribute to the intensity of this band) and cannot be used as an unambiguous indicator to detect $[BO_2]^-$ V-shaped groups. In turn, the strong narrow line at 1050–1065 cm^{-1} is typical for the Raman spectra of alkali or alkaline earth carbonates.[25] This band is also observed in the Raman spectra of high alkali borate glasses and melts[6] if some amount of alkali carbonate has not been decomposed. This line, however, is not observed in the Raman spectra of alkali borate glasses with modifier oxide content less than 50% (the specific compositional range addressed in the present study).[6,23] As will be shown below, the 1050–1065 cm^{-1} line appears in the spectra of melts with modifier oxide content far less than 50 mol%. A study of the composition dependence of this band in the Raman spectra of $Na_2O–B_2O_3$ melts[17] has shown that for melts with Na_2O content less than 50 mol% this line has another structural origin than vibrations of CO_3^{2-} carbonate ions. Thus, taking into account the recently published results[15,16] and our results,[17] we conclude that the 1050–1065 cm^{-1} band can indeed be used to detect triatomic groups in the structure of borate melts. Therefore, the current work is focused on the study of this line in the Raman spectra of sodium, potassium and caesium borate glasses and melts.

The Raman spectra of xNa$_2$O.(100–x)B$_2$O$_3$ (x=25, 30, 40 and 50 mol% Na_2O) glasses and melts are shown in Figure 2. Clearly the 1050–1065 cm^{-1} band is absent in all spectra of the sample with the lowest concentration of sodium oxide. However, an increase in concentration of sodium oxide of only 5 mol% (30Na sample) leads to the appearance of this band in the high temperature Raman spectra. At the same time, the 1050–1065 cm^{-1} line is not observed in the 30Na glass spectrum, but its intensity increases with increasing temperature in the melt spectra. This band increases in intensity with further increase in Na_2O concentration. In addition, a new band at around 1210–1240 cm^{-1} appears in the Raman spectra of the high alkali samples (40Na and 50Na). Both the 1050–1065 and 1210–1240 cm^{-1} lines show an increase in intensity with increasing temperature that is especially evident in the spectra of the sodium borate melt with metaborate composition. Thus, it can be asserted

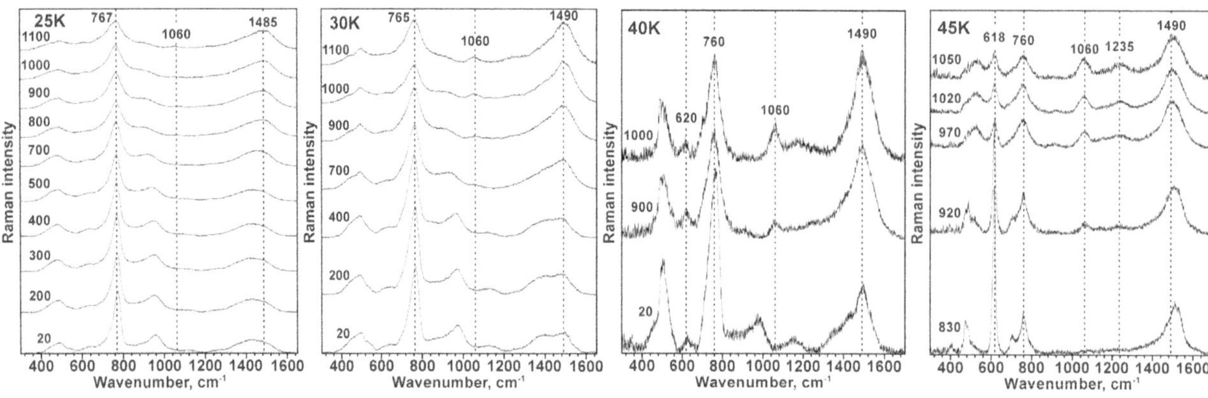

Figure 3. Raman spectra of $K_2O–B_2O_3$ glasses and melts. (The temperature is given in °C)

that the V-shaped groups start to form in sodium borate melts with Na_2O content more than 25 mol%, and that their concentration gradually increases with increasing sodium oxide and temperature.

The Raman spectra of potassium borate glasses and melts with K_2O contents from 25 to 45 mol% are shown in Figure 3. As before, the 1050–1065 cm^{-1} band is not observed in the 25K spectra measured at any temperature. This band first appears in the $30K_2O.70B_2O_3$ melt spectra, but it is absent in the low temperature, glass spectra. As the concentration of K_2O increases, the intensity of this line grows in the melt spectra. Moreover, the 1050–1065 cm^{-1} band increases in intensity with increasing temperature in all studied Raman spectra of potassium borate melts where this band is present. Nevertheless, this band is absent in all glass spectra, as seen in Figure 3. Hence it follows that the triatomic V-shaped groups form in the structure of potassium borate melts with K_2O content more than 25 mol% (as is also the case for sodium borate melts), and their amount increases with increasing melt temperature.

The same type of behaviour is also observed in the caesium borate spectra, which are shown in Figure 4. As before, the 1050–1065 cm^{-1} line is practically absent in the 25Cs Raman spectra, but its intensity quickly increases with increasing temperature and with increase in caesium oxide content. A similar behaviour is also observed for the 1210–1240 cm^{-1} line. It should be noted that the caesium borate spectra strongly differ from the Raman spectra of the sodium and potassium borate melts in the extremely high intensity of the 1050–1065 cm^{-1} and 1210–1240 cm^{-1} lines. The increase in intensity of the 1050–1065 cm^{-1} and 1210–1240 cm^{-1} lines with increasing temperature leads to the former becoming the most intense line, whilst the latter is comparable in intensity with the other bands in the Raman spectrum of molten $40Cs_2O.60B_2O_3$ measured at 1065°C.

Thus, visual analysis of the measured spectra shows that the 1050–1065 cm^{-1} band is absent in all studied spectra of glassy samples. This band is unambiguously detected in all melt spectra with modifier oxide content more than 25%, independent of the kind of alkali cation. This means that the V-shaped groups are untypical of the glass structure, but these groups can form during the melting of these glasses, as proposed in Refs 15 & 16. Our results show that the V-shaped groups form not only in overheated alkali borate melts with metaborate composition, but also in alkali borate melts with significantly lower modifier oxide content (≥25 mol%). In all cases, the concentration of these groups increases with increase in modifier oxide content and with increasing temperature.

The next important point concerns the structural reorganisation mechanisms that lead to the formation of the triatomic groups in the melt structure. According to Refs 15, 16, 26 & 27, the formation of the V-shaped groups is related to the destruction of the $B_3O_6^{3-}$ cyclic metaborate anions. It is easy to imagine the formation of the triatomic monomers at the expense of $B_3O_6^{3-}$ metaborate rings, for example, in the form of the following reaction:

$$3(B\text{Ø}_2O^-)_{ring} \Leftrightarrow B\text{Ø}O_2^{2-} + B\text{Ø}_3 + [BO_2]^-_{V\text{-}group} \quad (11)$$

One can readily see that equilibrium (11) can be considered as a result of simultaneous shift of the well-known reaction (7) and an isomerisation reaction of the form

$$B\text{Ø}_2O^- \Leftrightarrow [BO_2]^-_{V\text{-}group} \quad (12)$$

toward the right with increasing temperature.

A narrow line at about 610–630 cm^{-1} in the Raman spectra of the alkali borate glasses and melts corresponds to the breathing vibration of $B_3O_6^{3-}$ metaborate rings. Therefore, if the schemes given above are correct, then the increase in intensity of the 1050–1065 cm^{-1} line should be accompanied by a decrease in intensity of the 610–630 cm^{-1} band.

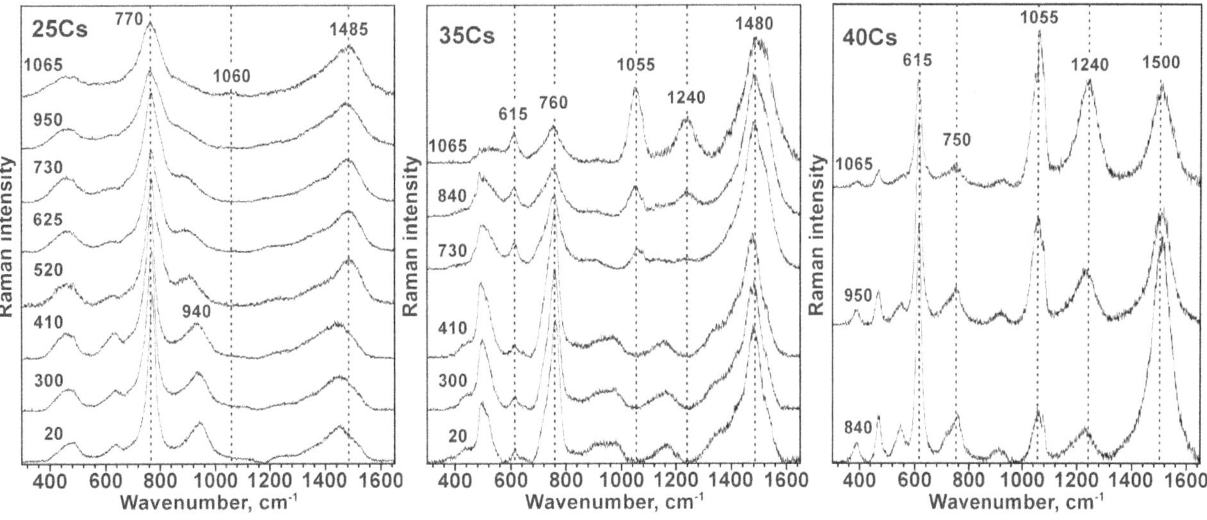

Figure 4. Raman spectra of Cs_2O–B_2O_3 glasses and melts. (The temperature is given in °C)

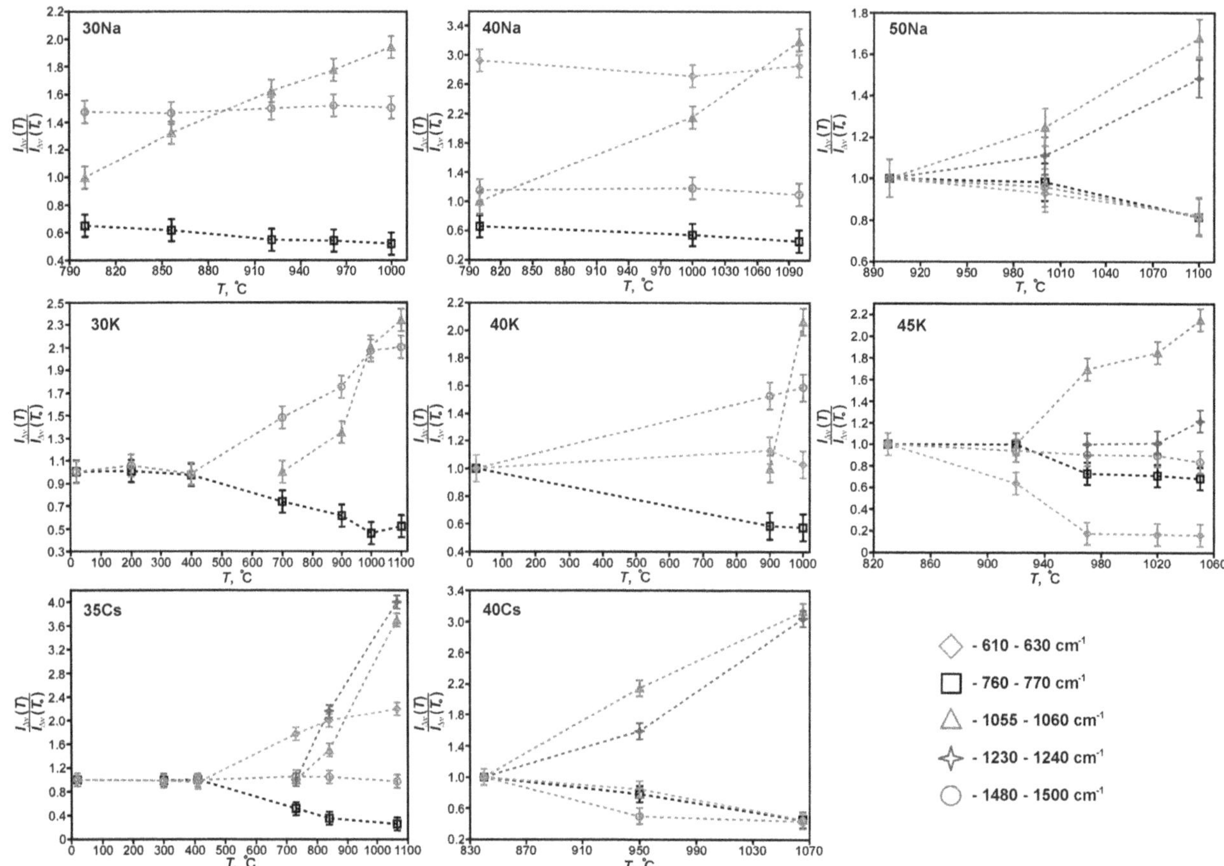

Figure 5. Relative intensity of the Raman bands in the studied spectra [Colour available online]

The relative intensity ($I_{\Delta n}^{T_x}/I_{\Delta n}^{T_0}$, where the subscript is the peak position, the superscript is the temperature, and T_0 is the lowest temperature for the given line) of some Raman bands in the high temperature spectra are shown in Figure 5. In other words, the band intensities in the high temperature spectra were measured relative to the intensities of the same lines in the Raman spectra obtained at lower temperatures. It is clear that an increase in intensity of the 1050–1065 cm^{-1} line with increasing temperature is typical of all measured spectra, whereas a decrease in intensity of the 610–630 cm^{-1} line is observed only in the Raman spectra of samples with relatively high concentration of modifier oxide (50Na, 45K and 40Cs). No significant changes in intensity of the 610–630 cm^{-1} line are observed in the Raman spectra of samples with lower M_2O contents. Moreover, the intensity of this line increases with increasing temperature in the 35Cs spectra. Thus, a destruction of the $B_3Ø_6^{3-}$ metaborate rings may indeed be responsible for the formation of the V-shaped groups in the melt structure, but only for composition in which the M_2O content exceeds 40 mol%.

The equilibrium (11) for the formation of the $[BO_2]^-$ anions proposes the production of $BØO_2^{2-}$ units in the form of terminal groups of the complex borate anions. Both the V-shaped groups and $BØO_2^{2-}$ units can contribute to the intensity of the 1210–1240 cm^{-1} line. Thus, a simultaneous growth of the 1050–1065 and 1210–1240 cm^{-1} bands is consistent with the suggested scheme, and the absence of the 820–860 cm^{-1} line (this line is due to the symmetric stretching mode of the B–O–B bridges in $B_2O_5^{4-}$ pyroborate anions[3,5]) unambiguously indicates that no $B_2O_5^{4-}$ anions are formed in the structure of the studied melts. It is important to note that the given mechanism does not exclude other mechanisms involving structural reorganisations of the high alkali borate melts, which are described by the well-known reactions (7) and (8). The new mechanism is supplementary to reactions (7) and (8), because none of them lead to the formation of the V-shaped groups in the melt structure. For example, a decrease in intensity of 750–765 cm^{-1} line indicates that the concentration of six-membered borate rings with two $[BØ_4]^-$ tetrahedra[5,23,28,29] decreases in the structure of the 50Na, 45K and 40Cs melts with increasing temperature. In turn, a decrease in the concentration of such rings may be understood on the basis of reaction (8), which describes a structural reorganisation in which $[BØ_4]^-$ tetrahedra are transformed into $BØO_2^{2-}$ pyroborate units and symmetric $BØ_3$ triangles.

As for the melts with lower modifier oxide content (30Na, 40Na, 30K, 40K and 35Cs), the 1050–1065 cm^{-1} line increases in intensity with increasing temperature, as for the high alkali melts (see Figure 5). At the same time, no significant changes are observed in the intensity of the 610–630 cm^{-1} line. Moreover, this band increases in intensity with increasing temperature in the 35Cs spectra. Consequently, there must be other mechanisms (in addition to those mentioned above) for the formation of the V-shaped groups in the structure of melts with relatively low M_2O concentration. An increase in intensity of the 1050–1065 cm^{-1} line is accompanied by a decrease in intensity of the 760–765 cm^{-1} in the 30Na, 40Na, 30K, 40K and 35Cs Raman spectra, and hence it can be assumed that $[B\varnothing_4]^-$ tetrahedra are also involved in the formation of the V-shaped groups in the structure of alkali borate melts with lower M_2O concentration. It is well known that symmetric $B\varnothing_3$ triangles, $[B\varnothing_4]^-$ tetrahedra and $B\varnothing_2O^-$ metaborate triangles are the typical structural units of alkali borate melts with modifier oxide contents from 25 to 45 mol%. A possible scenario for the formation of the V-shaped groups in the structure of overheated melts with these structural units is shown in Figure 6. In terms of the basic structural units, this scheme can be written as the following isomerisation reaction:

$$[B\varnothing_4]^- \Leftrightarrow [BO_2]^-_{\text{V-group}} \qquad (13)$$

The equilibrium (13) shifts to the right with increasing temperature. As before, the equilibrium (13) does not exclude other reactions (for example, reaction (6)) describing the structural reorganisations in alkali borate melts. Reaction (13) just explains the formation of the V-shaped groups in the structure of the overheated melts.

In summary it may be noted that it is impossible to predict the existence of the V-shaped groups in the structure of the overheated melts from the structural data for crystals and glasses. As shown above, none of the well-known reactions suggest the presence of any other structural groups in the melt structure, other than those found in crystals or glasses. Nevertheless, the existence of such "strange" groups is explainable. The V-shaped groups in the overheated melts may be considered as "predecessors" of the MBO_2 linear molecules in the gaseous metaborates.[30,31] The symmetric stretching vibration of the $[BO_2]^-$ radicals gives the Raman band located[31] at around 1030–1095 cm^{-1}, and this fact supports the stated assumption. It will be recalled that the 1050–1065 cm^{-1} line was used here to detect the V-shaped groups in the structure of the overheated melts. Moreover, a direct correlation between the intensity of the 1050–1065 cm^{-1} line and the evaporation of alkali borate melts was observed in our experiments. Obviously an increase in temperature leads to an increase in evaporation rate that correlates with the intensity of the 1050–1065 cm^{-1} line. Our data also showed that the evaporation rate depends on the sort of alkali cation, and increases in the Na→K→Cs direction. The 1050–1065 cm^{-1} peak intensity also increases in the same direction. Therefore, it is expected

• - Boron ● - Oxygen

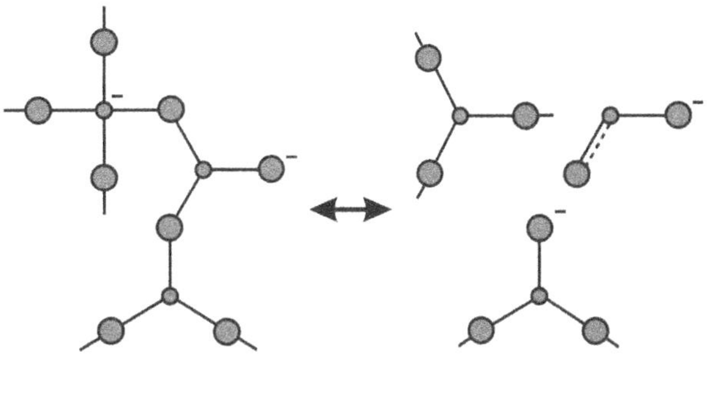

Figure 6. Schematic illustration of the formation of V-shaped groups in structure of alkali borate melts with M_2O ranged from 25 to 40 mol% [Colour available online]

that the concentration of the V-shaped groups in the melts correlates with the evaporation rate, if the triatomic groups are considered as "predecessors" of the MBO_2 molecules, which are formed in the vapour above the melts. This is precisely what we observed in our experiments.

Conclusions

The Raman spectra of sodium, potassium and caesium borate glasses and melts were measured for a wide range of temperatures. Based on analysis of these spectra, we studied the composition- and temperature-dependence of the V-shaped $[BO_2]^-$ groups in the structure of alkali borate glasses and melts. Our results show that the V-shaped groups start to form in the melt for M_2O content greater than 25%, independent of the sort of alkali cation. The concentration of the V-groups increases with increasing modifier oxide content and temperature in all studied samples. Within the limits of experimental error, the magnitude of the increase in the concentration of V-shaped groups depends on the sort of alkali cation, and it increases in the Na→K→Cs direction. The formation of the V-shaped groups can be related to the destruction of $B_3O_6^{3-}$ metaborate rings in the structure of alkali borate melts with relatively high concentration of modifier oxide. The V-shaped groups can form as a result of the transformation of $[B\varnothing_4]^-$ tetrahedra in the structure of alkali borate melts with M_2O content less than 40 mol%. In this case, a dynamic equilibrium between basic structural units can be described by the isomerisation reaction (13). Being one more possible mechanism for the structural transformations of alkali borate melts during heating, the formation of V-shaped groups does not exclude other well-known reactions (reactions (6)–(9)), which describe variations in the short range order structure. Finally, our data show that the V-shaped groups are present only in the structure of molten borates, and that they are absent in the glassy samples.

Acknowledgement

Partial support by the RFBR (Project No. 14-08-00323_a) is gratefully acknowledged.

References

1. Feller, S. A., Dell, W. J. & Bray, P. J. ^{10}B NMR studies of lithium borate glasses. *J. Non-Cryst. Solids*, 1983, **51**, 21–30.
2. Berryman, J. R., Feller, S. A., Affatigato, M., Kodama, M., Meyer, B. M., Martin, S. W., Borsa, F. & Kroeker, S. Thermal, acoustic, and nuclear magnetic resonance studies of cesium borate glasses. *J. Non-Cryst. Solids*, 2001, **293–295**, 483–489.
3. Dwivedi, B. P., Rahman, M. H., Kumar, Y. & Khanna, B. N. Raman scattering study of lithium borate glasses. *J. Phys. Chem. Solids*, 1993, **54** (5), 621–628.
4. Yun, Y. H. & Bray, P. J. B^{11} nuclear magnetic resonance studies of Li_2O-B_2O_3 glasses of high Li_2O content. *J. Non-Cryst. Solids*, 1981, **44**, 227–237.
5. Kamitsos, E. I. & Karakassides, M. A. Structural studies of binary and pseudo binary sodium borate glasses of high sodium content. *Phys. Chem. Glasses*, 1989, **30** (1), 19–26.
6. Kamitsos, E. I., Karakassides, M. A. & Chryssikos, G. D. Structure of borate glasses. Part 1. Raman study of caesium, rubidium, and potassium borate glasses. *Phys. Chem. Glasses*, 1989, **30** (6), 229–234.
7. Kamitsos, E. I., Patsis, A. P., Karakassides, M. A. & Chryssikos, G. D. Infrared reflectance spectra of lithium borate glasses. *J. Non-Cryst. Solids*, 1990, **126**, 52–67.
8. Wright, A. C., Dalba, G., Rocca, F. & Vedishcheva, N. M. Borate versus silicate glasses: why are they so different? *Phys. Chem. Glasses: Eur. J. Glass Sci. Technol. B*, 2010, **51** (5), 233–265.
9. Wright, A. C. Borate structures: crystalline and vitreous. *Phys. Chem. Glasses: Eur. J. Glass Sci. Technol. B*, 2010, **51** (1), 1–39.
10. Vegiri, A., Varsamis, C. P. E. & Kamitsos, E. I. Composition and temperature dependence of cesium-borate glasses by molecular dynamics. *J. Chem. Phys.*, 2005, **123**, 014508-1–014508-9.
11. Varsamis, C. P. E., Vegiri, A. & Kamitsos, E. I. Structure and dynamics of ionic borate glasses. *Phys. Chem. Glasses: Eur. J. Glass Sci. Technol. B*, 2006, **47** (4), 419–429.
12. Kamitsos, E. I. & Chryssikos, G. D. Borate glass structure by Raman and infrared spectroscopies. *J. Mol. Struct.*, 1991, **247**, 1–16.
13. Osipov, A. A. & Osipova, L. M. Structural specificity of high alkali lithium and sodium borate melts: a high temperature Raman study. *Phys. Chem. Glasses: Eur. J. Glass Sci. Technol. Part*, 2013, **54** (1), 1–14.
14. Osipova, L. M., Osipov, A. A. & Bykov, V. N. Structure of high-alkali melts in the lithium borate system from vibrational spectroscopic data. *Glass Phys. Chem.*, 2007, **33** (5), 486–491.
15. Voron'ko, Yu. K., Sobol', A. A. & Shukshin, V. E. Study of a boron-oxygen complexes in the molten and vapor states by Raman and luminescence spectroscopies. *J. Mol. Struct.*, 2012, **1008**, 69–76.
16. Voronko, Yu. K., Sobol, A. A. & Shukshin, V. E. Structure of boron-oxygen groups in crystalline, molten and glassy alkali-metal and alkaline-earth metaborates. *Inorg. Mater.*, 2012, **48** (7), 732–737.
17. Osipov, A.A. & Osipova, L. M. Structure of glasses and melts in the Na_2O-B_2O_3 system from high-temperature Raman spectroscopic data: I. Influence of temperature on the local structure of glasses and melts. *Glass Phys. Chem.*, 2009, **35** (2), 121–131.
18. Osipov, A. A. & Osipova, L. M. Structural studies of Na_2O-B_2O_3 glasses and melts using high-temperature Raman spectroscopy. *Phys. B*, 2010, **405**, 4718–4732.
19. Osipov, A. A. & Osipova, L. M. Structure of Cs_2O-B_2O_3 glasses and melts: the Raman spectroscopy data. *Glass Phys. Chem.*, 2014, **40** (4), 391–401.
20. Osipov, A. A. & Osipova, L. M. Raman scattering study of barium borate glasses and melts. *J. Phys. Chem. Solids*, 2013, **74**, 971–978.
21. Neuville, D. R. & Mysen, B. O. Role of aluminium in the silicate network: In situ, high-temperature study of glasses and melts on the join SiO_2-$NaAlO_2$. *Geochim. Cosmochim. Acta*, 1996, **60** (10), 1727–1737.
22. Dwivedi, B. P. & Khanna, B. N. Cation dependence of Raman scattering in alkali borate glasses. *J. Phys. Chem. Solids*, 1995, **56** (1), 39–49.
23. Meera, B. N. & Ramakrishna, J. Raman spectral studies of borate glasses. *J. Non-Cryst. Solids*, 1993, **159**, 1–21.
24. Santos, C. N., Meneses, D. S., Echegut, P., Neuville, D. R., Hernandes, A. C. & Ibanez, A. Structural, dielectric, and optical properties of yttrium calcium borate glasses. *Appl. Phys. Lett.*, 2009, **94**, 151901-1–151901-3.
25. Pasierb, P., Komornicki, S., Rokita, M. & Rekas, M. Structural proprieties of Li_2CO_3-$BaCO_3$ system derived from IR and Raman spectroscopy. *J. Mol. Struct.*, 2001, **596** (1–3), 151–156.
26. Tsvetkov, E. G. Model concept on the role of structure-forming cations in self-assembling of molten crystallization media with ionic-covalent interactions. *J. Cryst. Growth*, 2005, **275**, e53–e59.
27. Tsvetkov, E. G., Davydov, A. V., Ancharov, A. I. & Yudaev, I. V. From molten glass to crystallizable melt: The essence of structural evolution. *J. Cryst. Growth*, 2006, **294**, 22–28.
28. Kamitsos, E. I., Karakassides, M. A. & Chryssikos, G. D. Vibrational spectra of magnesium-sodium-borate glasses. 2. Raman and mid-infrared investigation of the network structure. *J. Phys. Chem.*, 1987, **91**, 1073–1079.
29. Yano, T., Kunimine, N., Shibata, S. & Yamane, M. Structural investigation of sodium borate glasses and melts by Raman spectroscopy. I. Quantitative evaluation of structural units. *J. Non-Cryst. Solids*, 2003, **321**, 137–146.
30. Lopatin, S. I. & Shugurov, S. M. Mass spectrometric study of stability, thermochemistry and structures of the gaseous oxyacid salts. *Open Thermodynam. J.*, 2013, **7**, 35–56.
31. Dement'ev, A. I., Rambidi, N. G., Simkin, V. Ya. & Topol', I. A. Molecular structure of the boron(III) oxide molecule in the SCF approximation. *J. Mol. Struct.*, 1980, **68**, 199–202.

Phys. Chem. Glasses: Eur. J. Glass Sci. Technol. B, April 2016, 57 (2), 59–67

Lasing transitions of Nd^{3+} ions in lead antimony borate glasses

B. Appa Rao[1] *& M. Chandra Shekhar Reddy*[2]

[1] *Department of Physics, Osmania University, Hyderabad, India-500 007**
[2] *Department of Physics, CMR College of Engineering & Technology, Kandlakoya, Hyderabad, India-501 401*

Manuscript received 17 August 2014
Revised version received 15 December 2015
Accepted 17 December 2015

Glasses of composition $30PbO.5Sb_2O_3.(45-x)B_2O_3.xNd_2O_3$, with x=0 to 1·0 in steps of 0·2, were prepared by the melt-quenching method. Various physical parameters, viz., density, molar volume and oxygen packing density were evaluated. Optical absorption and luminescence spectra of all the glasses were recorded at room temperature. From the observed absorption edges, the optical band gap and Urbach energies were calculated. Judd–Ofelt theory was applied to characterise the absorption and luminescence spectra of Nd^{3+} ions in these glasses. From the luminescence spectra, various radiative properties, such as transition probability, branching ratio and radiative lifetime were evaluated for different emission levels of Nd^{3+} ions. The radiative lifetime for the ${}^4F_{3/2}\rightarrow{}^4F_{11/2}$ multiplet was also evaluated from the recorded lifetime decay curves. Quantum efficiencies were estimated for all the glasses and found to increase with the concentration of Nd_2O_3. From the obtained results, conclusions were made about the possibility of using these glasses as laser material.

1. Introduction

Laser materials doped with Nd^{3+} ions[1] have received widespread interest because the Nd^{3+} ion has a large absorption coefficient, a wide absorption band, a long fluorescence lifetime, a very large fluorescence branching ratio, energy compactness and the possibility of lasing at different wavelengths at room temperature. It is the most investigated lanthanide ion, not only due to its NIR (near infrared) emission, but also because its sensitivity to a changing crystal field can be used to extrapolate the spectral properties of other Ln ions in similar matrices. Wilhelm *et al*[2] have reported the fluorescence lifetime enhancement of Nd^{3+} sol-gel glasses by Al co-doping and CO_2 laser processing. Karthikeyan & Mohan[3] have reported structural, optical and glass transition studies on Nd^{3+} doped lead borate glasses. Sen *et al*[4] have studied the spectroscopic properties of Nd^{3+} doped transparent oxyfluoride glass ceramics. Annapurna *et al*[5] have investigated the NIR emission and upconversion luminescence spectra of $ZnO-SiO_2-B_2O_3:Nd_2O_3$ glasses. Kumar *et al*[6] have explored the stimulated emission and radiative properties of Nd^{3+} ions in barium fluorophosphate glasses containing sulphate. Saisudha & Ramakrishna[7] have found large radiative transition probabilities in bismuth borate glasses doped with Nd^{3+} ions. Rosa-Cruz *et al*[8] have reported the results of a spectroscopic characterisation study of Nd^{3+} ions in barium phosphate glasses. Fernandez *et al*[9,10] have evaluated the upconversion losses in Nd^{3+} doped fluoroarsenate glasses.

The scientific interest in Nd^{3+}-containing glasses has increased gradually after the demonstration of lasing action in Nd-doped glasses by Snitzer.[11] Thereafter, considerable progress has been made, and now the less tunable, but high power terawatt lasers are all Nd-glass lasers.[12] The application of rare earth based glasses for these uses requires an extensive knowledge of the spectroscopic and radiative properties of Nd^{3+}-containing glasses.

Heavy metal oxide glasses, like $PbO-Sb_2O_3-B_2O_3$ glasses, have gained importance in recent years due to their nonlinear optical susceptibility,[13] χ^3, low phonon energy and high refractive index, which make possible many applications in various technological fields, such as ultrafast optical triggers, optically poled materials, power limiters and broad band optical amplifiers operating around 1·5 μm.[14,19] The efficiency of quantum emission from a given level depends strongly upon the phonon energy of the host medium. Because of the above mentioned characteristics, lead antimony borate glass offers a good environment for hosting rare earth ions, like Nd^{3+}, that give rich NIR emission. The objective of this investigation is to characterise the luminescence spectra of Nd^{3+} ions in the $PbO-Sb_2O_3-B_2O_3$ glass system, to characterise the spectra using Judd–Ofelt (J-O) theory and to comment on the advantages of this glass host over conventional glass systems.

2. Experimental methods

For the present study, glasses of composition $30PbO.25Sb_2O_3.(45-x)B_2O_3.xNd_2O_3$, where x=0, 0·2,

* Corresponding author. Email apparao.bojja@gmail.com
Original version presented at VIII Int. Conf. on Borate Glasses, Crystals and Melts, Pardubice, Czech Republic, 30 June–2 July 2014
DOI: 10.13036/17533562.57.2.035

Table 1. Physical parameters of $PbO–Sb_2O_3–B_2O_3$ glasses doped with Nd^{3+} ions

Physical parameter	N0	N2	N4	N6	N8	N10
Average MW (g/mol)	171·2	171·7	172·2	172·8	173·3	173·8
Density, ρ (g/cc) (±0·001)	4·972	5·101	5·212	5·316	5·481	5·567
Refractive index, n_d (±0·001)	1·503	1·512	1·517	1·521	1·524	1·529
Molar volume, Vm (MW/ϱ) (±0·01)	31·99	33·93	33·41	32·62	32·06	31·31
Molar Refraction, R_M (±0·001)	9·457	10·18	10·108	9·933	9·809	9·656
Polarisability, α_e ($\times10^{-24}$ cm^3) (±0·001)	3·75	4·034	4·005	3·936	3·886	3·826
Number of oxygen atoms per formula unit	2·4	2·4	2·4	2·4	2·4	2·4
Oxygen packing density, O_d (gm atom/L) (±0·01)	75·02	70·73	71·83	73·57	74·86	76·65
Nd^{3+} ion concentration, N_i ($\times10^{21}$/cc)(±0·01)	0	0·11	0·22	0·33	0·45	0·58
Inter-ionic distance, r_i (Å) (±0·001)	-	20·87	16·57	14·47	13·05	11·99
Polaron radius, r_p (Å) (±0·001)	-	0·146	0·158	0·165	0·171	0·176

0·4, 0·6 and 1·0 (all in mol%) were chosen. The samples were labelled as N0, N2, N4, N6, N8 and N10 respectively. Appropriate amounts of AR grade reagents of PbO, Sb_2O_3, B_2O_3 and Nd_2O_3 powders were thoroughly mixed in an agate mortar, and melted in a silica crucible in a programmable electrical furnace at a temperature in the range 900 to 950°C for 30 min, until bubble free liquid was formed. The resultant melt was poured into a brass mould, and subsequently annealed at 250°C for 2 h. The samples were then ground and optical polished to the dimensions of 1 cm×1 cm×0·2 cm. The refractive index, n_d, of the samples was measured (at λ=589·3 nm) using an Abbe refractometer with monobromo naphthalene as the contact layer between the glass and the refractometer prism. The optical absorption spectra of the samples were recorded at room temperature in the spectral wavelength range covering 300–2200 nm, with a spectral resolution of 0·1 nm, using a Jasco model V-670 UV-vis-NIR spectrophotometer. The luminescence spectra and lifetime measurements were carried out at room temperature, using a Jobin Yvon Fluorolog-3 spectrofluorometer with a xenon arc lamp as the radiation source, with a spectral resolution of 0·5 nm. Infrared transmission spectra were recorded on a Bruker-FTIR-Tensor27 spectrophotometer up to a resolution of 0·4 cm^{-1} in the spectral range 400–2000 cm^{-1}, using potassium bromide pellets (300 mg) containing pulverized sample (1·5 mg).

3. Results and discussion

Our visual examination, the absence of peaks in the x-ray diffraction pattern (Figure 1), and the existence of a glass transition temperature, T_g, (Figure 2) in the DSC traces, indicate that the samples prepared were glasses.

Various physical parameters (molar weight, MW,

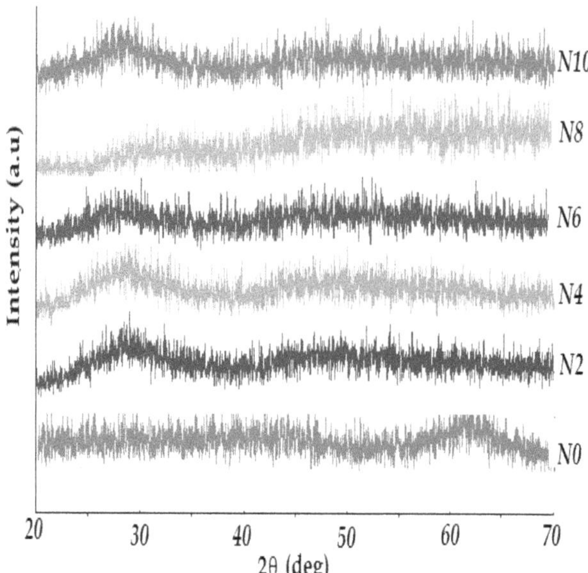

Figure 1. X-ray difractograms of $PbO–Sb_2O_3–B_2O_3$:Nd_2O_3 glasses [Colour available online]

density, ρ, molar volume, V_m, oxygen mol%, O, oxygen packing density, O_d, Nd^{3+} ion concentration, N_i, inter-ionic distance of Nd^{3+} ions, r_i, refractive index, n_d, polaron radius, r_p, molar refraction, R_M, and polarisability, α_e) were calculated using standard equations.[20]

As the concentration of Nd^{3+} ions increased, a considerable increase in the density or a considerable decrease in the molar volume of samples is observed.

Modification of the geometrical configurations of the glass network, change in coordination and the variation of dimensions of the interstitial holes can be considered as responsible for such a variation of density. Oxygen packing density is also found to increase with the increase in the concentration of Nd^{3+} ions. Such an increase indicates an increase in the structural compactness of the samples. Various physical parameters of $PbO–Sb_2O_3–B_2O_3$ glasses doped with Nd^{3+} ions are presented in Table 1.

The glass transition temperatures of all the glasses are presented in Figure 2. The DSC traces indicate

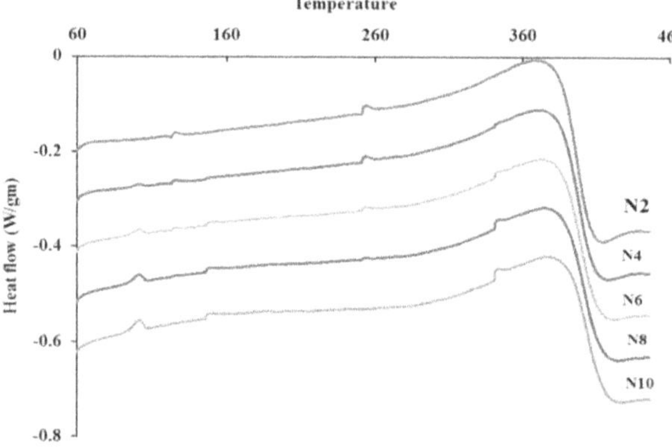

Figure 2. Variation of glass transition temperature of lead antimony borate glasses with increasing Nd^{3+} concentration [Colour available online]

Table 2. Glass transition temperatures of $PbO–Sb_2O_3–B_2O_3:Nd_2O_3$ glasses

Glass	T_g (°C)
N2	384
N4	386
N6	387
N8	391
N10	393

typical glass transitions, with inflection points between 384 and 393°C. Although the inflection points of all the samples appear to be nearly same, it is interesting that the glass transition temperature shows an increasing trend with increase in dopant concentration. The glass transition temperature for all the samples are given in Table 2.

The Fourier transform infrared spectra of Nd^{3+} doped $PbO–Sb_2O_3–B_2O_3$ glasses are shown in Figure 3. The spectra exhibit bands from borate groups: at 1200 to 1400 cm^{-1} due to asymmetric stretching of trigonal BO_3 units, at 1050 cm^{-1} due to stretching of tetrahedral BO_4 units, and another band at 688 cm^{-1} due to bending of B–O–B linkages in the borate network.[21] The ν_1 vibrational band of SbO_3 units is at 930 cm^{-1}. The ν_3 vibrational bands of SbO_3 units merge with the band due to bending vibrations of B–O–B linkages and form a common vibrational band due to B–O–Sb linkages.[22] In addition, a band due to PbO_4 structural groups[23] at about 462 cm^{-1} is observed in the spectra of all the samples. The bands at 1750 cm^{-1} are identified as being due to the stretching vibration of hydroxyl (OH) complexes (probably due to absorbed water molecules on the surface of the material).

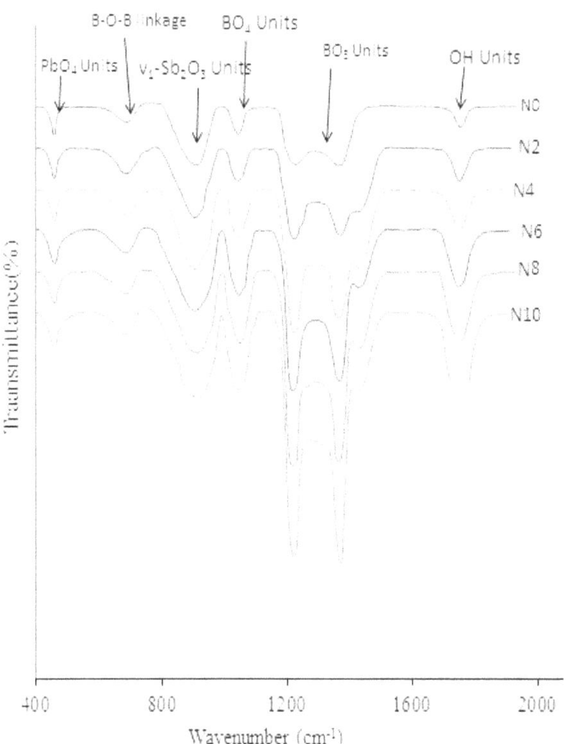

Figure 3. FTIR spectra of $PbO–Sb_2O_3–B_2O_3:Nd_2O_3$ glasses [Colour available online]

The optical absorption spectra (Figure 4) of Nd^{3+} doped $PbO–Sb_2O_3–B_2O_3$ glasses exhibit twelve well resolved bands viz., $^4I_{9/2}\rightarrow{}^4D_{5/2}$ (368 nm), $^2D_{5/2}+{}^2P_{1/2}$ (433 nm), $^4G_{11/2}$ (463 nm), $^2K_{15/2}+{}^2D_{3/2}+{}^2G_{9/2}$ (478 nm), $^4G_{9/2}$ (516 nm), $^4G_{7/2}$ (527 nm), $^4G_{5/2}+{}^2G_{7/2}$ (583 nm), $^2H_{11/2}$ (632 nm), $^4F_{9/2}$ (681 nm), $^4F_{7/2}+{}^4S_{3/2}$ (747 nm), $^4F_{5/2}+{}^2H_{9/2}$ (804 nm), $^4F_{3/2}$ (877 nm). The increase in the concentra-

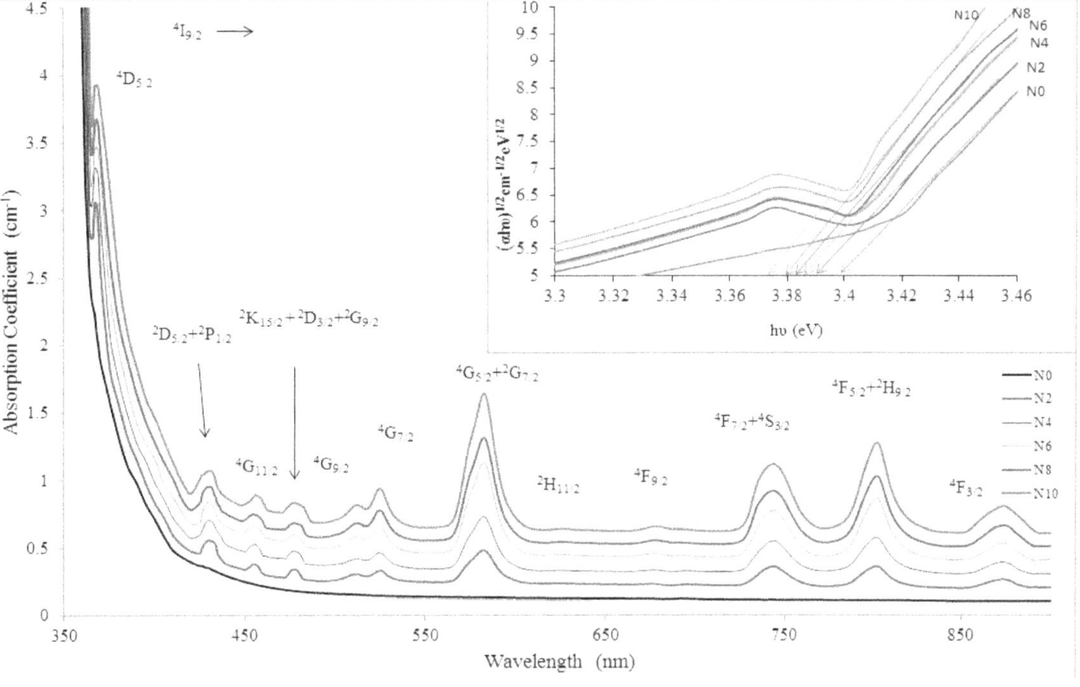

Figure 4. Optical absorption spectra of $PbO–Sb_2O_3–B_2O_3$ glasses doped with Nd^{3+} ions, recorded at room temperature. All transitions are from the ground state, $^4I_{9/2}$. The inset shows Tauc plots for evaluation of the optical band gap energy [Colour available online]

Table 3. Optical parameters of Nd^{3+} doped lead antimony borate glasses

Sample	N0	N2	N4	N6	N8	N10
Cutoff wavelength, λ (nm)(±1)	353	355	356	357	359	361
Optical bandgap energy, E_{opt} (ev) (±0·001)	3·399	3·391	3·386	3·383	3·38	3·374
Urbach energy, ΔE (eV) (±0·001)	0·511	0·523	0·529	0·538	0·551	0·563

tion of neodymium ions in the glass matrix does not alter the spectral positions of the absorption bands significantly, but it causes the absorption strength under each peak to increase.

The optical band gap energy is calculated from the Tauc plots of $(\alpha\hbar\omega)^{1/2}$ versus $\hbar\omega$ (inset of Figure 4). The optical band gap energy of the samples decreases with increasing concentration of Nd^{3+} ions. Urbach plots were obtained by plotting $\ln(\alpha)$ against $\hbar\omega$. Values for the Urbach energy, ΔE, were calculated by determining the reciprocal of the slope of the linear region of the curve. The Urbach energy indicates the degree of disorder, i.e. higher Urbach energy corresponds to higher disorder. The Urbach energy of the samples increases with increasing concentration of Nd^{3+} ions (Table 3).

Table 3 shows the optical parameters of PbO–Sb_2O_3–B_2O_3 glasses doped with Nd^{3+} ions. The experimental oscillator strengths (OSs) of the absorption transitions were estimated from the absorption spectra in terms of the area under an absorption peak. The numerical values of the OS were calculated using the following formula:

$$f_{exp} = 4{\cdot}32 \times 10^{-9} \int \varepsilon(\nu)\mathrm{d}\nu \quad (1)$$

$\varepsilon(\nu)$ is the molar absorptivity at frequency ν (expressed in cm^{-1}) and evaluated from Beer's law

$$\varepsilon(\upsilon) = \frac{1}{N_m t}\log\left(\frac{I_0}{I}\right) \quad (2)$$

where N_m is the concentration of Nd^{3+} ions in weight%, and t is the thickness of the sample in cm.

Using this oscillator strength, the original J-O parameters, T_λ, can be calculated by performing least square fit analysis using

$$f_{cal}(\psi J, \psi' J') = \frac{8\pi m c \upsilon}{3h(2J+1)}\left[\frac{(n^2+2)^2}{9n} S_{ed}(\psi J, \psi' J') + nS_{md}(\psi J, \psi' J')\right] \quad (3)$$

where h is Plank's constant, n is the refractive index of the medium, $(2J+1)$ is the multiplicity of the ground state and $(n^2+2)^2/9n$ is the local Lorentz field correction, S_{ed} is the electric dipole line strength, and S_{md} is the magnetic dipole line strength.

Though the majority of $4f^n$ intra-configurational transitions are induced electric dipole type, certain transitions are neither pure electric dipole (S_{ed}) type, nor pure magnetic dipole (S_{md}) type, but partial combinations of both. For these transitions, S_{ed} and S_{md} have to be calculated separately.

The electric dipole line strength (S_{ed}) was calculated using the expression:

$$S_{ed} = \sum_{\lambda=2,4,6} \left(\psi J \left\| U^\lambda \right\| \psi J'\right)^2 \Omega_\lambda \quad (4)$$

where Ω_λ (λ=2,4,6) are the host dependent J-O intensity parameters which represent the square of charge displacement due to induced electric–dipole transition. $\|U^\lambda\|^2$ are the doubly reduced matrix elements evaluated in the intermediate coupling approximation for the transition $\psi J \rightarrow \psi' J'$, and are considered to be host independent. Similarly, the magnetic dipole line strength[24] is given by

$$S_{md} = (\psi J, \psi' J') = \frac{e^2 h^2}{16\pi m^2 c^2}\left|\left\langle \psi J \left\| (L+2S) \right\| \psi' J' \right\rangle\right|^2 \quad (5)$$

The non-zero matrix elements are those of the diagonal in quantum numbers (α=L+2S), S and L. The selection rules on J is ΔJ=0, ±1. The experimental oscillator strengths, f_{exp}, and calculated oscillator strengths are computed by using Equations (1), (3), (4) and (5), respectively, and are presented in Table 4. Using the refractive index n_d, the parameters T_λ were transformed to Ω_λ using the equation[25]

$$\Omega_\lambda = \frac{3h(2J+1)}{8\pi^2 mc}\frac{9n_d}{(n_d^2+2)^2} T_\lambda \quad (6)$$

Table 4. The experimental ($f_{exp}\times10^{-6}$) and calculated ($f_{cal}\times10^{-6}$) spectral intensities of Nd^{3+} doped antimony lead borate glasses. All the transitions are from its ground state, $^4I_{9/2}$

Transitions from ground state $^4I_{9/2}$	N2 f_{exp}	N2 f_{cal}	N4 f_{exp}	N4 f_{cal}	N6 f_{exp}	N6 f_{cal}	N8 f_{exp}	N8 f_{cal}	N10 f_{exp}	N10 f_{cal}
$^4F_{3/2}$	2·3	2·12	2·34	2·4	2·44	2·4	2·67	2·61	2·73	2·72
$^4F_{5/2}+{}^2H_{9/2}$	7·01	6·66	7·28	7·46	7·7	7·46	8·44	8·07	8·82	8·31
$^4F7_{/2}+{}^4S_{3/2}$	9·1	9·3	9·6	10·35	10·22	10·35	11·21	11·17	11·14	11·45
$^4F_{9/2}$	0·38	0·69	0·58	0·77	0·63	0·77	0·69	0·83	0·73	0·85
$^2H_{11/2}$	0·13	0·19	0·14	0·21	0·15	0·21	0·17	0·23	0·19	0·24
$^4G_{5/2}+{}^2G_{7/2}$	21·29	21·3	22·77	24·14	24·14	24·14	26·33	26·33	26·71	26·71
$^4G_{7/2}$	3·8	3·49	3·88	3·95	4·0	3·95	4·43	4·3	4·45	4·42
$^4G_{9/2}$	0·67	0·7	0·68	0·78	0·77	0·78	0·86	0·84	1·13	0·86
$^2k_{15/2}+{}^2D_{3/2}+{}^2G_{9/2}$	0·76	0·76	1·17	1·17	1·27	1·38	1·38	1·38	1·46	1·46
$^4G_{11/2}$	0·84	0·98	0·86	1·1	0·98	1·1	1·09	1·2	1·07	1·24
$^2P_{1/2}+{}^2D_{5/2}$	0·47	0·47	0·72	0·54	0·94	0·54	0·97	0·59	1	0·61
$^4D_{5/2}$	3·92	5·1	4·67	5·79	5·21	5·79	5·41	6·31	5·56	6·56
RMS Deviation		0·4		0·6		0·24		0·32		0·39

Table 5. Judd–Ofelt parameters ($\Omega_\lambda \times 10^{20}$ cm^2) of Nd^{3+} doped lead antimony borate glasses, together with the reported trends for a number of other glass systems containing Nd^{3+} ions

Glass	Ω_2	Ω_4	Ω_6	Trend	Reference
N2	5·72	3·77	6·89	$\Omega_6>\Omega_2>\Omega_4$	Present work
N4	6·12	4·032	7·22	$\Omega_6>\Omega_2>\Omega_4$	Present work
N6	6·46	4·31	7·66	$\Omega_6>\Omega_2>\Omega_4$	Present work
N8	7·01	4·68	8·22	$\Omega_6>\Omega_2>\Omega_4$	Present work
N10	7·02	4·89	8·41	$\Omega_6>\Omega_2>\Omega_4$	Present work
$58{\cdot}5P_2O_5.17K_2O.14{\cdot}5MgO.9Al_2O_3.1Nd_2O_3$	6·22	5·95	6·83	$\Omega_6>\Omega_2>\Omega_4$	(28)
$20K_2O.30PbO.49{\cdot}5B_2O_3.0{\cdot}5Nd_2O_3$	5·90	5·71	7·10	$\Omega_6>\Omega_2>\Omega_4$	(29)
$30Li_2O.69B_2O_3.1Nd_2O_3$	4·20	3·89	4·74	$\Omega_6>\Omega_2>\Omega_4$	(30)
$3Nd_2O_3.97(30PbO.70B_2O_3)$	3·96	3·77	4·88	$\Omega_6>\Omega_2>\Omega_4$	(31)
$3Nd_2O_3.97(40PbO.60B_2O_3)$	3·59	3·50	5·26	$\Omega_6>\Omega_2>\Omega_4$	(31)
$3Nd_2O_3.97(50PbO.50B_2O_3)$	3·59	3·02	5·32	$\Omega_6>\Omega_2>\Omega_4$	(31)
$3Nd_2O_3.97(60PbO.40B_2O_3)$	3·61	3·03	5·33	$\Omega_6>\Omega_2>\Omega_4$	(31)
$3Nd_2O_3.97(70PbO.30B_2O_3)$	3·52	2·98	5·48	$\Omega_6>\Omega_2>\Omega_4$	(31)
$10Li_2SO_4.H_2O.39CdSO_4.50B_2O_3.1Nd_2(SO_4)_3.8H_2O$	34·51	10·17	36·89	$\Omega_6>\Omega_2>\Omega_4$	(32)
$10Na_2SO_4.H_2O.39CdSO_4.50B_2O_3.1Nd_2(SO_4)_3.8H_2O$	21·06	20·85	27·99	$\Omega_6>\Omega_2>\Omega_4$	(32)
B_2O_3	4·30	3·60	4·70	$\Omega_6>\Omega_2>\Omega_4$	(33)
$10Na_2O.90B_2O_3$	3·40	2·90	4·30	$\Omega_6>\Omega_2>\Omega_4$	(34)
$56{\cdot}15P_2O_5.17{\cdot}75K_2O.14{\cdot}28BaO.8{\cdot}37Al_2O_3.3{\cdot}45AlF_3.1Nd_2O_3$	6·60	6·36	7·30	$\Omega_6>\Omega_2>\Omega_4$	(35)
$30CaO.69B_2O_3.1Nd_2O_3$	4·4	3·7	4·6	$\Omega_6>\Omega_2>\Omega_4$	(36)
$35Bi_2O_3.30Na_2O.34B_2O_3.1Nd_2O_3$	4·72	2·12	3·93	$\Omega_2>\Omega_6>\Omega_4$	(37)
$35PbO.30Na_2O.34B_2O_3.1Nd_2O_3$	4·81	1·97	3·94	$\Omega_2>\Omega_6>\Omega_4$	(37)

where h is Planck's constant, m is the mass of an electron, c is the velocity of light, and J is the ground term J value.

The positions and the spectral intensities of certain transitions of rare earth ions are very sensitive to the environment of the rare earth ion.[26] These transitions follow selection rules, such as $\Delta J \leq 2$, $\Delta L \leq 2$ and $\Delta S=0$, and are called hypersensitive transitions. From Figure 3, it is noticeable that the transition ${}^4I_{9/2} \rightarrow {}^4G_{5/2} + {}^2G_{7/2}$ at about 583 nm is much brighter than the other transitions, and this is a hyper sensitive transition. The hyper sensitive transitions (bands) are normally associated with larger values of $||U^2||^2$, and hence intimately related to Ω_2. The intensity parameter Ω_2 indicates the covalency of the rare earth ligand bond, and increases with increase in the intensity of the hyper sensitive transition.[27] The same tendency was observed for the present glass system.

The J-O parameters of all the samples were calculated and are presented in Table 5. The intensity parameter Ω_2 is associated with the symmetry of ligand field around the rare earth ion, and hence the covalency of the rare earth ion. For the present study, Ω_2 follows the order N10>N8>N6>N4>N2. This trend indicates that the symmetry of the site associated with Nd^{3+} ion is highest for N2 glass and lowest for N10 glass.

3.1 Radiative properties

The J-O parameters, along with refractive index, n_d, were used to predict the radiative properties of the excited states of the rare earth ions. The radiative probability $A_R(\psi J, \psi' J')$ of a transition can be calculated using the equation

$$A_R(\psi J, \psi' J') = \frac{64\pi^4 \upsilon^3}{3h(2J+1)} \left[\frac{n_d(n_d^2+2)^2}{9} S_{ed}(\psi J, \psi' J') + n^3 S_{md}(\psi J, \psi' J') \right] \quad (7)$$

The total radiative transition probability, A_T, of an excited state is the sum of the $A_R(\psi J, \psi J')$ terms, calculated over all terminal states

$$A_T(\psi J) = \sum_{\psi' J'} A_R(\psi J, \psi' J') \quad (8)$$

$A_T(\psi J)$ is related to the radiative lifetime $\tau_R(\psi J)$ of an exited state by

$$\tau_R(\psi J) = \tau_{cal}(\psi J) = \frac{1}{A_T(\psi J)} \quad (9)$$

Strong emission probabilities, and more transitions from a level, lead to faster decay and shorter lifetimes. The theoretical lifetime, $\tau_R(\psi J)$, calculated from the J-O intensity parameter, Ω_λ, can be compared with the experimental lifetimes, $\tau_{exp}(\psi J)$.

The difference between the predicted and experimental lifetimes is due to non-radiative processes, W_{NR}, resulting from either the multi-phonon relaxation rate, W_{MPR}, or the energy transfer rate, W_{ET}. From the measured and calculated lifetimes, the quantum efficiency is estimated by the expression

$$\eta = \frac{\tau_{exp}}{\tau_R} = \frac{A_R}{A_R + W_{NR}} \quad (10)$$

The branching ratios were used to predict the relative intensities of all emission lines originated from a given exited state, ψJ. The experimental branching ratios were found from the relative areas of the emission bands. The branching ratios, β_R, corresponding to the emission from an excited level, ψJ, to its lower level, $\psi' J'$, is given by

$$\beta_R(\psi J, \psi J') = \frac{A(\psi J, \psi J')}{A_T(\psi J)} \quad (11)$$

Various radiative parameters, such as electric dipole line strength, S_{ed}, magnetic dipole line strength, S_{md}, radiative transition probability, A_R, total radiative

Table 6. The electric dipole line strengths, S_{ed}, magnetic dipole line strengths, S_{md}, radiative transition probability, A_R, branching ratios, β_R, total radiative transition probabilities, A, and radiative decay times, τ_R, of lead antimony borate glass doped with 1 mol% Nd_2O_3

Initial $^{l}(S,L)_J$	*Final* $^{l}(S',L')_J$	*Wave number* (cm^{-1})	$S_{ed}\times10^{22}$ (cm^2)	$S_{md}\times10^{22}$ (cm^2)	A_R (s^{-1})	β_R %
$^4G_{9/2}$	$^4G_{7/2}$	483	204·31	16·72	0·06	0
	$^2G_{7/2}$	2319	125·59	34·52	4·74	0·03
	$^4G_{5/2}$	2408	258·2	0	8·34	0·05
	$^2H_{11/2}$	3607	200·49	0·06	21·78	0·14
	$^4F_{9/2}$	4890	199·56	3·27	54·99	0·35
	$^4S_{3/2}$	6036	98·32	0	50·03	0·32
	$^4F_{7/2}$	6126	479·5	0·21	255·22	1·63
	$_2H_{9/2}$	7030	127·66	0·49	103·08	0·66
	$^4F_{5/2}$	7078	189·68	0	155·64	0·99
	$^4F_{3/2}$	8102	150·81	0	185·6	1·18
	$^4I_{15/2}$	13626	272·15	0	1593·29	10·16
	$^4I_{13/2}$	15690	894·2	0	7992·49	50·94
	$^4I_{11/2}$	17672	320·69	0·04	4096·14	26·11
	$^4I_{9/2}$	19531	67·7	0·01	1167·36	7·44
		$\sum A$ (s^{-1})=15688	τ_R (μs)=63			
$^4G_{7/2}$	$^2G_{7/2}$	1836	27·31	0·67	0·5	0
	$^4G_{5/2}$	1925	144·39	27·19	3·61	0·03
	$^2H_{11/2}$	3124	117·45	0	10·36	0·08
	$^4F_{9/2}$	4407	14·79	0·02	3·67	0·03
	$^4S_{3/2}$	5553	93·67	0	46·39	0·36
	$^4F_{7/2}$	5643	147·81	1·1	77·47	0·6
	$_2H_{9/2}$	6547	399·39	0	324·2	2·53
	$^4F_{5/2}$	6595	256·17	0·7	213·19	1·66
	$^4F_{3/2}$	7619	98·23	0	125·67	0·98
	$^4I_{15/2}$	13143	15·78	0	103·64	0·81
	$^4I_{13/2}$	15207	156·39	0	1590·89	12·41
	$^4I_{11/2}$	17189	487·87	0	7167·14	55·93
	$^4I_{9/2}$	19048	157·45	0	3147·53	24·56
		$\sum A$ (s^{-1})=12814	τ_R (μs)=78			
$^4G_{5/2}$	$^4S_{3/2}$	3628	88·1	0	16·23	0·08
	$^4F_{7/2}$	3718	217·72	0·13	43·19	0·21
	$_2H_{9/2}$	4622	10·91	0	4·16	0·02
	$^4F_{5/2}$	4670	250·43	0·02	98·38	0·47
	$^4F_{3/2}$	5694	364·55	0·03	259·58	1·25
	$^4I_{15/2}$	11218	3·7	0	20·16	0·1
	$^4I_{13/2}$	13282	57·02	0	515·28	2·49
	$^4I_{11/2}$	15264	221·81	0	3042·42	14·68
	$^4I_{9/2}$	17123	863·39	0	16717·54	80·67
		$\sum A$ (s^{-1})=20716	τ_R (μs)=48			
Initial $^{l}(S,L)_J$	Final $^{l}(S',L')_J$	wave number (cm^{-1})	$S_{ed}\times10^{22}$ (cm^2)	$S_{md}\times10^{22}$ (cm^2)	A_R (s^{-1})	β_R %
$^2H_{11/2}$	$^4F_{9/2}$	1283	92·78	4·1	0·4	0·09
	$^4S_{3/2}$	2429	28·1	0	0·78	0·17
	$^4F_{7/2}$	2519	129·37	0	3·99	0·88
	$_2H_{9/2}$	3423	250·83	15·23	20·72	4·57
	$^4F_{5/2}$	3471	21·03	0	1·7	0·37
	$^4F_{3/2}$	4495	7·2	0	1·26	0·28
	$^4I_{15/2}$	10019	118·32	0	229·47	50·61
	$^4I_{13/2}$	12083	13·01	0·93	47·79	10·54
	$^4I_{11/2}$	14065	12·49	0·26	68·59	15·13
	$^4I_{9/2}$	15924	10·11	0	78·73	17·36
		$\sum A$ (s^{-1})=453	τ_R (μs)=2205			
$^4F_{9/2}$	$^4S_{3/2}$	1146	2·05	0	0·01	0
	$^4F_{7/2}$	1236	140·32	28·68	0·75	0
	$_2H_{9/2}$	2140	38·43	19·22	1·36	0
	$^4F_{5/2}$	2188	124·08	0	3·01	0
	$^4F_{3/2}$	3212	100·8	0	7·73	0
	$^4I_{15/2}$	8736	657·82	0	1014·9	0·22
	$^4I_{13/2}$	10800	553·08	0	1612·27	0·35
	$^4I_{11/2}$	12782	331·65	0·16	1603·57	0·35
	$^4I_{9/2}$	14641	43·39	0·11	316·03	0·07
		$\sum A$ (s^{-1})=4559	τ_R (μs)=219			
$^4F_{5/2}$	$^4F_{3/2}$	1024	81·89	34·02	0·5	0·01
	$^4I_{15/2}$	6548	194·09	0	210·17	4·09
	$^4I_{13/2}$	8612	427·41	0	1052·89	20·49
	$^4I_{11/2}$	10594	112·26	0	514·81	10·02
	$^4I_{9/2}$	12453	451·22	0	3360·75	65·4
		$\sum A$ (s^{-1})=5139	τ_R (μs)=194			
$^4F_{3/2}$	$^4I_{15/2}$	5524	22·89	0	22·33	0·57
	$^4I_{13/2}$	7588	177·93	0	449·73	11·43
	$^4I_{11/2}$	9570	411·22	0	2085·1	53·01
	$^4I_{9/2}$	11429	159·39	0	1376·6	34·99
		$\sum A$ (s^{-1})=3934	τ_R (μs)=254			

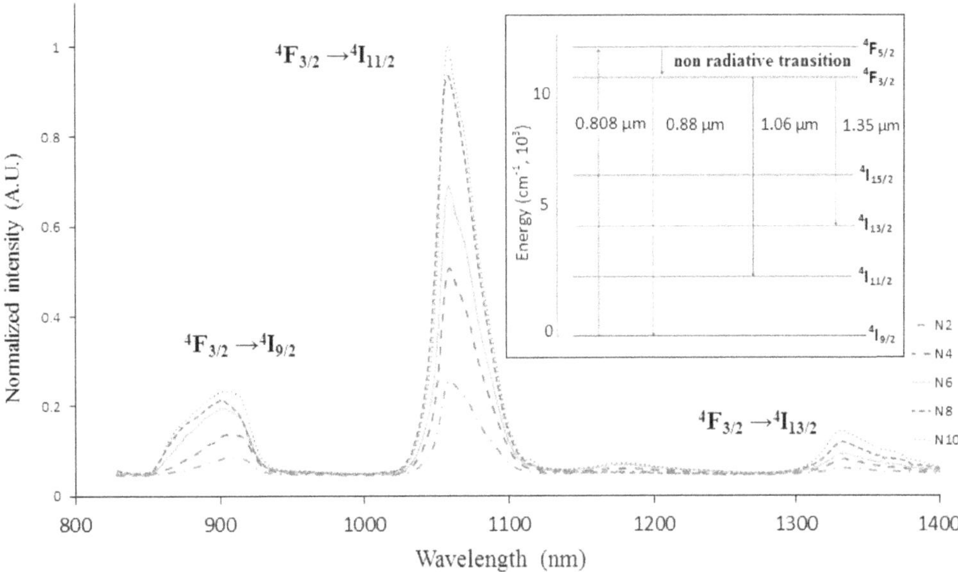

Figure 5. Normalised emission spectra of $PbO–Sb_2O_3–B_2O_3$ glasses doped with Nd^{3+} ions, recorded at room temperature. (The exciting wavelength is λ_{exc}=808 nm.) The inset shows the partial energy level diagram of Nd^{3+} doped lead antimony borate glasses [Colour available online]

transition probabilities, A_T, radiative decay times, τ_R, and branching ratios, β_R, of Nd^{3+} doped lead antimony borate glasses were calculated using Equations (7), (8), (9), (10), (11) and (13), respectively. The radiative parameters of lead antimony borate glass with 1 mol% Nd_2O_3 doping are presented in Table 6.

The values of the branching ratios, β_R, for the $^4F_{3/2}\rightarrow{}^4I_J$ (J=15/2, 13/2, 11/2 and 9/2) transitions depend upon Ω_4/Ω_6, since Ω_2 does not contribute in determining the intensity of these bands, as $||U^2||^2$ is zero for these transitions. In lead antimony borate glasses, approximately 35% of transitions terminate at the $^4I_{9/2}$ state, 53% at the $^4I_{11/2}$ state, 11% at the $^4I_{13/2}$ state, and 5% radiate to the $^4I_{15/2}$ state.

3.2 Photoluminescence

The fluorescence spectra of Nd^{3+} doped lead antimony borate glasses are shown in Figure 5. The spectra show a broad band at 898 nm ($^4F_{3/2}\rightarrow{}^4I_{9/2}$), a strong band at 1058 nm ($^4F_{3/2}\rightarrow{}^4I_{11/2}$), and another band at 1328 nm ($^4F_{3/2}\rightarrow{}^4I_{13/2}$). The intensity of the bands is observed to increase with concentration of Nd^{3+} ions, but no significant shift in position is observed.

3.3 Stimulated emission cross-section

An efficient laser transition is characterised by a large stimulated emission cross-section, while the induced emission cross-sections are characterised by J-O theory. The stimulated emission cross-section is given by the Füchtbauer–Ladenburg method,[38] using the equation

$$\sigma = \frac{\lambda_P{}^3}{8\pi c n^2 \Delta\lambda_{\text{eff}}} A_T(\psi J, \psi' J') \quad (12)$$

where λ_P is the peak wavelength, $\Delta\lambda_{\text{eff}}$ is the effective bandwidth of the emission band, n is the refractive index of the sample, and $A_T(\psi J,\ \psi' J')$ is the total spontaneous emission probability. The effective band width and emission cross-section values are presented in Table 7.

The inset of Figure 5 shows a partial energy level diagram of the excited manifolds of Nd^{3+} ions, and several relevant transitions corresponding to the excited $^4F_{3/2}$ state for absorption and emission. The three excited state absorption transitions correspond to the three laser levels at 1·35, 1·06 and 0·88 µm. The energy level separations between the $^4F_{3/2}$ and the upper $^4G_{7/2}$, $^2G_{9/2}$ and $^2P_{1/2}$ levels are in all cases close enough to the energies of possible laser transitions from the $^4F_{3/2}$ level to lower terminal levels, $^4I_{13/2}$, $^4I_{11/2}$ and $^4I_{9/2}$, respectively. But among the three transitions from $^4F_{3/2}$, the transition $^4F_{3/2}\rightarrow{}^4I_{11/2}$ is a potential lasing transition because of its higher stimulated emission cross-section (5·13×10^{-20} cm^2) and higher branching ratio (53%). An electric dipole transition between the $^4F_{3/2}$ and $^2G_{9/2}$ levels near 1·06 µm could reduce the net cross-section for stimulated emission between the $^4F_{3/2}$ and $^4I_{11/2}$ levels, as suggested by Vance.[39]

Table 7. Emission cross-sections of Nd^{3+} doped lead antimony borate glasses for the transitions $^4F_{3/2}\rightarrow{}^4I_J$ (J=13/2, 11/2,9/2)

Transition	λ_P (nm)	$\Delta\lambda_{eff}$ (nm)					σ_e ×10^{-20} (cm^2)				
$^4F_{3/2}\rightarrow$		N2	N4	N6	N8	N10	N2	N4	N6	N8	N10
$^4I_{13/2}$	1329	29	30	31	33	34	2·22	2·26	2·32	2·35	2·34
$^4I_{11/2}$	1059	25	26	27	28	29	4·77	4·84	4·96	5·07	5·13
$^4I_{9/2}$	896	48	49	50	53	54	0·81	0·85	0·89	0·91	0·93

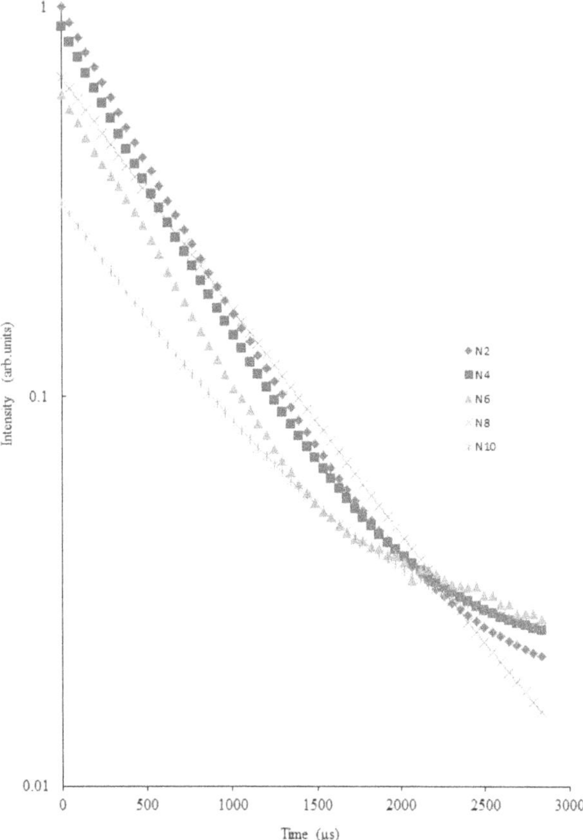

Figure 6. Luminescence decay profiles of the $^4F_{3/2}\rightarrow^4F_{11/2}$ transition of Nd^{3+} doped lead antimony borate glasses [Colour available online]

3.4 Decay curves

Figure 6 shows the decay profile of the $^4F_{3/2}\rightarrow^4I_{11/2}$ transition of Nd^{3+} ions in lead antimony borate glasses excited at a wavelength of 808 nm, recorded at room temperature. The experimental lifetimes, along with the branching ratios and emission cross-sections of all the glasses, are presented in Table 7.

When the Nd^{3+} ion is excited from its ground state to levels with energy higher than that of $^4F_{3/2}$, they decay non-radiatively to lower levels, down to the $^4F_{3/2}$ state, due to the very small energy gaps between the adjacent energy levels. As the energy gap between the metastable $^4F_{3/2}$ level and its lower level $^4I_{13/2}$ is sufficiently large (5600 cm^{-1} approximately), radiative transitions predominate here over non-radiative transitions. The intensities of these transitions depend on the values of the J-O parameters, especially on Ω_4 and Ω_6, as well as on the host refractive index. It represents a mean value over the different sites occupied by the Nd^{3+} ions in the glass matrix, and it is higher than that determined experimentally from the luminescence decay. This reduction in experimental lifetime can be explained by considering all the possible relaxation processes relevant to the excited Nd^{3+} ions.

Table 8. Experimental lifetimes, τ_{exp}, calculated lifetimes. τ_{cal}, quantum efficiency, η, branching ratios, β, and emission cross-sections, σ_e, of Nd^{3+} ions in lead antimony borate glasses for the transition $^4F_{3/2}\rightarrow^4I_{11/2}$

Sample	τ_{exp} (μs)	τ_{cal} (μs)	η (%)	β%	$\sigma_e \times 10^{-20}$ (cm^2)
N2	167	498·6	33·5	53·5	4·77
N4	189	469·6	39·4	53·4	4·84
N6	213	438·6	46·3	53·3	4·96
N8	235	405·7	57·9	53·2	5·07
N10	259	391·1	66·2	53·1	5·13

From Table 8, it appears that the branching ratios of the samples are almost constant, but the emission cross-sections and quantum efficiencies increase with increasing concentration of Nd^{3+} in the glasses. The results indicate that the glass N10 has a branching ratio of more than 50%, a quantum efficiency of 66%, and the highest emission cross-section, and it may therefore be suitable for efficient laser emission.

4. Conclusions

Lead antimony borate glasses with different concentrations of neodymium doping were prepared and their spectroscopic properties were investigated. The optical absorption spectra of the glasses exhibit twelve conventional absorption bands due to Nd^{3+} ion transitions. Judd–Ofelt theory was successfully applied to characterise the spectroscopic data. It was found that the Judd–Ofelt intensity parameters follow the trend $\Omega_6>\Omega_2>\Omega_4$. The fluorescence spectra of the glasses exhibit traditional near infrared bands due to $^4F_{3/2}\rightarrow^4I_{9/2}$, $^4I_{11/2}$, $^4I_{13/2}$ transitions. The intensity of these emission bands increases with the concentration of Nd^{3+} ions, indicating that there is no luminescence quenching in these glasses within the concentration range studied. It was found that the luminescence efficiency of all the transitions increases with increasing Nd_2O_3 concentration. In addition, it was observed that the emission cross-section for the $^4F_{3/2}\rightarrow^4I_{11/2}$ transition is more than that of the $^4F_{3/2}\rightarrow^4I_{9/2}$ and $^4F_{3/2}\rightarrow^4I_{13/2}$ transitions for all the glasses. The transition $^4F_{3/2}\rightarrow^4I_{11/2}$ (1·06 μm) has a branching ratio of 53%, a quantum efficiency of 66%, and an emission cross-section of $5{\cdot}13\times10^{-20}$ cm^2. Hence, the transition $^4F_{3/2}\rightarrow^4I_{11/2}$ is a potential lasing transition.

Acknowledgement

This work was financially supported by DST, New Delhi, under OU-DST-PURSE.

References

1. Kam, C. H. & Buddudu, S. *Opt. Laser Eng.*, 2001, **35**, 11.
2. Wilhelm, B., Romano, V. & Weber, H. P. *J. Non-Cryst. Solids*, 2003, **328**, 192.
3. Karthikeyan, B. & Mohan, S. *Physica B*, 2003, **334**, 298.
4. Cen, Y., Ma, E., Wang, Y. & Hu, Z. *Spectrochim. Acta A*, 2007, **67**, 709.
5. Annapurna, K., Dwivedi, R. N., Kunda, P. & Buddudu, S. *Mater. Lett.*, 2003, **57**, 2095.
6. Kumar, G. A., Martinez, A. & De La Rosa, E. *J. Lumin.*, 2002, **99**, 141.
7. Saisudha, M. B. & Ramakrishana, J. *Opt. Mater.*, 2002, **189**, 403.
8. De la Rosa-Cruz, E., Kumar, G. A., Diaz-Torres, L. A., Martínez, A. & Barbosa-García, O. *Opt. Mater.*, 2001, **18**, 321.

9. Fernández, J., Balda, R., Sanz, M., Lacha, L. M., Oleaga, A. & Adam, J.-L. *J. Lumin.*, 2001, **94**, 325.
10. Balda, R., Lacha, L. M., Mendioroz, A., Sanz, M., Fernández, J. Adam, J.-L. & Arriandiaga, M. A. *J. Alloys Comp.*, 2001, **323**, 255.
11. Sintzer, E. *Phys. Rev. Lett.*, 1961, **7**, 444.
12. Halzrichter, J. F. *Nature*, 1985, **309**, 361.
13. Terashima, K., Hashimoto, T., Uchino, T. & Yoko, T. *J. Ceram. Soc. Jpn.*, 1996, **104**, 1008.
14. Amano, M., Suzuki, K. & Sakatha, H. *J. Mater. Sci.*, 1997, **32**, 4325.
15. Kumar, D. & Chakravorty, D. *J. Phys. D*, 1982, **15**, 305.
16. Hufner, S. *Optical Spectra Transparent of Rare Earth Compounds*. Plenum Press, New York, 1968.
17. Dimitrov, V. & Dimitriev, Y. *J. Non-Cryst. Solids*, 1990, **122**, 133.
18. Ghosh, A. & Chaudhuri, B. K. *J. Non-Cryst. Solids*, 1986, **83**, 151.
19. Lines, E. *Principles and Applications of Ferroelectrics and Related Materials*. Clarendon Press, Oxford, 1979.
20. Weber, M. J. & Cropp, R. *J. Non-Cryst. Solids*, 1981, **4**, 137.
21. Rao, K. J. *Structural Chemistry of Glasses*. Elsevier, Amsterdam, 2002.
22. Srinivasarao, G. & Veeraiah, N. *J. Solid State Chem.*, 2002, **166**, 104.
23. Satyanarayana, T., Kityk, I. V., Piasecki, M., Bragiel, P., Brik, M. G., Gandhi, Y. & Veeraiah, N. *J. Phys.: Condens. Matter*, 2009, **21**, 245104.
24. Reisfeld, R. *Struct. Bond.*, 1975, **22**, 123.
25. Van der Ziel, A. *Solid State Physical Electronics*. Prentice-Hall of India, New Delhi, 1971.
26. Jorgensen, C. K. & Judd, B. R. *Molec. Phys.*, 1964, **8**, 281.
27. Peacock, R. D. *Struct. Bond.*, 1975, **22**, 83.
28. Surendrababu, S., Babu, P, Jayasankar, C. K., Joshi, A. S., Speghini, A. & Bettinelli, M. *J. Phys.: Condens. Matter*, 2006, **18**, 3975.
29. Mohan, S., Thind, K. S., Singh, D. & Gerward, L. *Glass. Phys. Chem.*, 2008, **34**, 265.
30. Takebe, H., Nageno, Y. & Morinaga, K. *J. Am. Ceram. Soc.*, 1995, **78**, 1161.
31. Saisudha, M. B. & Ramakrishana, *J. Opt. Mater.*, 2002, **18**, 403.
32. Jayashankar, C. K. & Ravi Kanth Kumar, V. V. *Physica B*, 1996, **226**, 313.
33. Weber, M. J., Ziegler, D. C. & Angell, C. A. *J. Appl. Phys.*, 1982, **53**, 4344.
34. Zahir, M., Olazcuaga, R., Parent, C., Le Flem, G. & Hagenmuller, P. *J. Non-Cryst. Solids*, 1985, **69**, 221.
35. Balakrishnaiah, R., Babu, P., Jayasankar, C. K., Joshi, A. S., Speghini, A. & Bettinelli, M. *J. Phys.: Condens. Matter*, 2006, **18**, 165.
36. Takebe, H., Nageno, Y. & Morinaga, K. *J. Am. Ceram. Soc.* 1994, **77**, 2132.
37. Karthieyan, B., Philip, R. & Mohan, S. *Opt. Commun.*, 2005, **246**, 153.
38. Digonnet, M. J. F. *Rare Earth Doped Fiber and Amplifiers*, Marcel Dekker, New York, 1993.
39. Vance, M. E. *IEEE J. Quantum Elect.*, 1970, **6**, 249.

Phys. Chem. Glasses: Eur. J. Glass Sci. Technol. B, December 2016, 57 (6), 233–244

Mechanical and tribological properties of boron oxide and zinc borate glasses

Fabiana Spadaro,[1] Antonella Rossi,[1,2] Emmanuel Lainé,[3] Joe Hartley[3] & Nicholas D. Spencer[1]

[1] *Laboratory for Surface Science and Technology, Department of Materials, ETH Zurich, Vladimir-Prelog-Weg 5, CH-8093 Zurich, Switzerland*
[2] *Dipartimento di Scienze Chimiche e Geologiche, Università degli Studi di Cagliari, Cittadella Universitaria di Monserrato, I – 09100 Cagliari, Italy*
[3] *Enabling Research, Infineum UK Ltd, Milton Hill, Steventon, Oxfordshire OX13 6BD, UK*

Manuscript received 7 December 2014
Revised version received 2 June 2016
Accepted 20 July 2016

Boron oxide plays a key role in many applications of glass that are of great technological significance, such as biomedical applications, bio-tribology and lubrication. In the present work, the mechanical and tribological properties of zinc borate glasses with different compositions are investigated, as well as the mechanical and tribological properties of the H_3BO_3 films that form spontaneously on boron oxide (B_2O_3) glasses when they are exposed to humid air. Atomic force microscopy (AFM), micro-indention measurements and pin-on-disk tribological tests were performed. The micro-indention measurements and AFM analysis show that the introduction of zinc oxide in the reagent mixture increases the hardness of the glasses and inhibits the formation of a H_3BO_3 film as the outermost layer. X-ray photoelectron spectroscopy XPS allowed the more covalent bonds in boric acid to be distinguished from the more ionic ones in borate glasses. In addition, it was found that the lubricating mechanism acting for H_3BO_3–B_2O_3 glass is completely different to that for zinc borate glasses. Increased temperature has a strong detrimental effect on the lubrication performance of H_3BO_3–B_2O_3 glass, while zinc borate glasses retain their good tribological performance at high temperature.

1. Introduction

Vitreous B_2O_3 is an example of a glass with intermediate range order (IRO).[1] It is widely acknowledged that boric oxide is an excellent glass former, due to the ability of boron to show both three- and four-oxygen coordination, as well as the high strength of B–O bonds. ZnO as a glass modifier enters into the glass network and breaks the B–O–B bridges, and hence its incorporation in borate glasses changes the properties of the glass system.[2,3] The glass forming ability of zinc borate is very low; in fact it is only possible to obtain homogeneous glass samples within a very narrow range of composition. It has been observed that the ZnO content has to be within the range 50–70 mol% in order to avoid phase separation.[4] The structure of boron oxide compounds consists mainly of boroxol rings, but with the introduction of the ZnO several changes occur in the network structure, resulting in the formation of polyborate groups. Boroxol rings are converted into groups such as tetraborate and diborate, containing four-coordinated boron atoms and only bridging oxygen (BO) atoms. Furthermore, the formation of groups containing both BO and nonbridging oxygen (NBO) is also detected, so that the following groups can be recognised: metaborate,

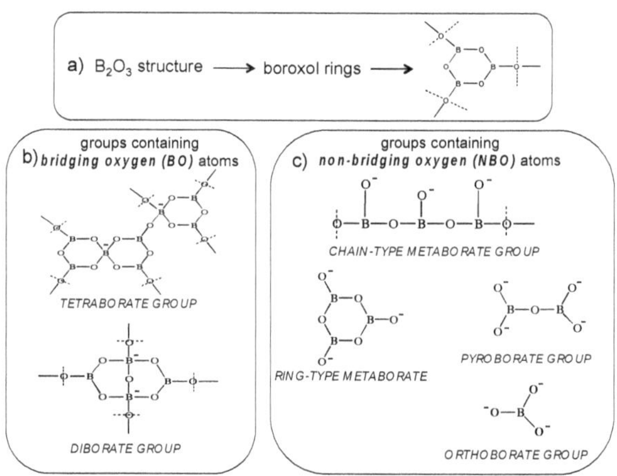

Figure 1. (a) Boroxol ring structure typical of pure boric oxide glass (B_2O_3), (b) and (c) various structural groups present in borate compounds as a consequence of the addition of a metal oxide additive, such as ZnO, which works as a network glass modifier

pyroborate and orthoborate. The structures of these groups are shown in Figure 1 and they are in agreement with the data in the literature.[5]

The properties of glasses are mainly determined by the degree of local order/disorder, so that investigations providing information about the local structure are important in order to understand the correlation between properties and structure. The present work is a contribution to the understanding of the correla-

* Corresponding author. Email antonella.rossi@mat.ethz.ch
Original version presented at VIII Int. Conf. on Borate Glasses, Crystals and Melts, Pardubice, Czech Republic, 30 June–2 July 2014
DOI: 10.13036/17533562.57.6.084

tion between the composition of the glasses and the mechanical properties, such as hardness (measured by micro-indentation), and their lubricating properties (determined by mechanical/ tribological tests).

The operating conditions, pressure and temperature, at which the tribological tests were carried out, were chosen to simulate those encountered in many mechanical systems.[6] Borate glass coatings can have an important role in the protection of all the parts in movements and the study of their performance can be of great technological interest. In the present work, the mechanical and tribological properties of boric oxide glass and of a series of different zinc borate glass compositions are presented and discussed.

2. Experimental

2.1 Synthesis

Three different compositions of zinc borate glass, $xZnO.(1-x)B_2O_3$, were prepared by the classical melt-quenching technique using Zn/B atomic ratios of ½, ¾, and 1, equivalent to 50, 60 and 67 mol% ZnO, respectively. In addition, pure boric oxide glass was also synthesised.

Zn/B=½ $1ZnO + 2B(OH)_3 \rightarrow (ZnO)*B_2O_3 + 3H_2O$ T=1000°C, t=2 h

Zn/B=¾ $3ZnO + 4B(OH)_3 \rightarrow (ZnO)_3*(B_2O_3)_2 + 6H_2O$ T=1000°C, t=2 h

Zn/B=1 $2ZnO+ 2B(OH)_3 \rightarrow (ZnO)_2*B_2O_3 + 3H_2O$ T=1100°C, t=2 h

B_2O_3 glass $2B(OH)_3 \rightarrow B_2O_3+ 3H_2O$ T=1100°C, t=2 h

The appropriate ratio of powdered, pure reagent-grade compounds $B(OH)_3$ (Fluka, Chemika, >99·8%) and ZnO (Alfa Aesar, 99·9%) were mixed in an agate mortar and melted in air in an alumina crucible (high density alumina produced by FRIALIT-DEGUSSIT, Switzerland) in a RHF 16/3 furnace (Carbolite®, Hope Valley, UK). The furnace temperature was increased at a rate of 2°C/min, to avoid the formation of small gas bubbles on the glass surface. To avoid alumina crucible contamination, different heating temperatures (T=1000 or 1100°C) were selected, taking account of different sample compositions. Once the desired temperature was reached, it was maintained for a heating time, t, of 2 h. Then the melt was quenched in a copper tray which had previously been cooled to −20°C. After quenching, the glass samples were annealed for at least 12 h (zinc borate glasses at 250°C and boric oxide glass at 200°C) to prevent splintering of the glass during the subsequent treatment steps (mechanical polishing and tribological tests). All the glass samples were discs, 2 cm in diameter and 5 mm thick. B_2O_3 glasses are colourless and transparent as synthesised, but they become dull and cloudy after a few minutes of exposure to humid air, because of the formation of a H_3BO_3 layer on the surface. Zinc borate glasses remain transparent and uncoloured, even after exposure to ambient conditions.

2.2 Differential thermal analysis

Differential thermal analysis (DTA) was carried on the glasses to determine their glass transition temperature, T_g, and melting point. The measurements were performed at a heating rate of 5°C/min under an inert argon atmosphere over a temperature range of 40–1100°C using a Netzsch STA 449 Jupiter calorimeter.

2.3 Mechanical polishing

The glass discs (both pure boric oxide and zinc borate glasses) were ground using grit 320, 600, 1200 and 2000 silicon carbide papers (Struers GmbH, Biermensdorf, Switzerland) on a rotating polishing wheel (Jeanwirtz Phoenix 4000, Wirtz-Buehler GmbH, Düsseldorf, Germany). Subsequent polishing was performed using diamond paste (Struers GmbH, Biermensdorf, Switzerland) of different grain sizes (3, 1 and ¼ μm) on polishing cloths (Struers GmbH, Birmensdorf, Switzerland). Reagent-grade ethanol was used for cooling, lubricating and cleaning the samples. The mechanical polishing was performed in order to obtain a smoother surface and a reproducible starting point as far as the roughness is concerned.

2.4 Atomic force microscopy

Atomic force microscopy (AFM) TappingMode™ images were acquired using an AFM microscope (Icon, Bruker, Santa Barbara, CA, USA) with a silicon cantilever (OMLCAC160- TS-R3, Olympus Micro Cantilevers, Japan) with a resonant frequency of 300 kHz and a spring constant of ~26 N/m (manufacturer value). These data were used to determine roughness. The roughness data were processed using NanoScope Analysis software, version 1.4.

2.5 Micro-indentation

The micro-indentation tests were carried out using a micro-Vickers hardness tester (MTX-α, Wolpert) at different loads (0·1, 0·25 1, 2, 3 and 5 N). Ten indentations per load were performed on each sample. The hardness was calculated measuring the length of the two diagonals of the indentation with the aid of a light microscope.

2.6 Tribological tests

A UMT-2 tribometer (Bruker (CETR), Campbell, CA, USA) was used in ball-on-disc configuration to investigate the tribological properties of the glasses.

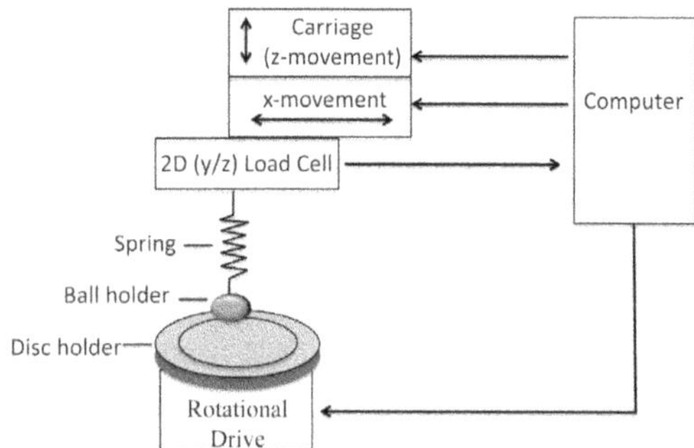

Figure 2. The main components of the tribometer

The tribological tests were performed in rotational motion (Figure 2).

The principal parts of the instrument, starting from the top to the bottom, are (see Figure 2): a carriage that allows both vertical (z-movement) and lateral movements (x-movement), a two-dimensional load cell that allows measurements of the normal load and the frictional forces, a spring that presses the ball onto the disc surface, the ball holder, the disc holder and the rotational driver to rotate the disc. The friction force, the load, the position of the carriage (x, z and rotation) and the time can be recorded during the experiment. Moreover, the instrument is supplied with a heating chamber, which allows the performance of tribological tests at high temperatures. Prior to each experiment, the procedure adopted is to check the planarity of the disc in order to ensure homogenous conditions over all the parts of the disc. A stylus is pressed (applied load=0·2 N) onto the disc, which is rotated at a low speed (1·4 rpm), while the carriage records the z-values as a function of the rotational position. If a z-variation higher than ±10 μm was measured, the planarity is adjusted by placing aluminium foil shims between the disc holder and the rotational drive. Prior to each tribological test, a running-in of the 100Cr6 hardened steel ball (radius=2 mm) was carried out, by applying a 5 N load at a speed of 141 mm/min for 10 min on 100Cr6 steel discs (hardened) at room temperature in Yubase 4, which is a common, commercial, hydrotreated, heavy paraffinic lubricating oil. The aim of the running-in is to obtain conformal surfaces on both sides of the contact: the disk and the steel ball. During this initial period, a rapid change of wear and friction coefficients occurs; this is followed by a period of stability, during which wear and friction coefficients remain unchanged. The radius of the worn area on the steel ball at the end of the running-in was measured by optical microscopy. From the image, the area of the flattened region of the ball, which corresponds to the initial contact area during the subsequent tribological test, could be calculated. On the basis of this value, a normal load was applied to reach the desired pressure. Afterwards, the steel ball was used for carrying out the tribotest on a polished glass disc of a specific composition in Yubase 4. All tribological tests were performed at an applied pressure of 1·04 GPa, at a sliding speed of 26 mm/min, for 2 h, at two different temperatures: ambient temperature (~25°C) and 100°C. The relative humidity was recorded during each test and always lay between 32 and 48% for the reported tests. Prior to XPS analysis, the sample was washed in n-hexane in an ultrasonic bath for 5 min and gently dried with argon.

Important parameters evaluated in order to estimate the tribological properties of these glass systems are: the coefficient of friction and the wear on the steel ball.

The coefficient of friction (*CoF*) is the ratio between the frictional force, F, and the normal load, L:[7,8]

$$CoF \quad -$$

The wear rate, W, is defined as the volume, V, of material removed per unit sliding distance, s:[9,10]

$$W=V/s$$

The wear on the disc would be another useful parameter to calculate, but under these operating conditions, the value was negligible and difficult to quantify, due to the very shallow scratches on the glass surface, which hardly constituted a wear track; therefore this value is not reported.

2.7 Optical microscopy

Optical microscopy images were taken using an AX10 Imager M1m (Carl Zeiss, Oberkochen, Germany) with objectives from 5× to 40× and equipped with a CCD camera.

2.8 Optical profilometry

A Sensofar PLu Neox (Sensofar-Tech, SL., Terrassa, Spain) 3D optical profiler was used for investigating the surface topography of tribological samples.

2.9 X-ray photoelectron spectroscopy

At the end of the tribological tests, the flattened area on the steel ball and points on the wear track on the glass disc were analysed (see Figure 3). X-ray photoelectron spectroscopy (XPS) analyses of the

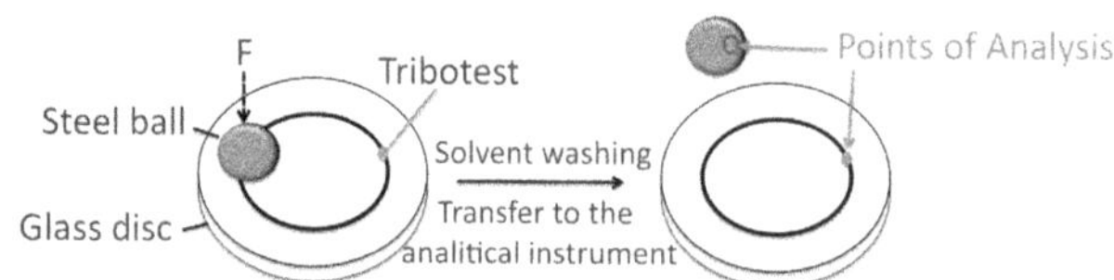

Figure 3. Typical ex-situ approach in tribological studies. At the end of the tribotest, the counter-faces are separated, washed, exposed to air during the transfer, and introduced into the analytical instrument (ultra-high vacuum) [Colour available online]

contact points were performed with a PHI Quantera SXM spectrometer (ULVAC-PHI, Chanhassen, MN, USA) equipped with a monochromatic Al K_{α} source whose beam size ranges from 9 to 200 μm. The emitted photoelectrons are collected by a gauze lens system at an emission angle of 45° and detected in a 32-channel detector. The spectrometer is also equipped with a high performance argon ion gun and an electron neutraliser for charge compensation. The calibration was performed using sputter-cleaned gold, silver, and copper as reference materials according to ISO15472:2001. All experiments were run at residual pressures below 5×10^{-7} Pa.

Survey spectra were acquired in fixed analyser transmission (FAT) mode, using a pass energy (PE) of 280 eV, while the high resolution spectra were collected with a PE of 69 eV; the full width at half maximum (FWHM) of the silver Ag $3d_{5/2}$ signal for high resolution spectra was 0·72 eV. To be sure to only collect the information within the worn area of interest, which was typically about 100 μm in diameter, the beam spot diameter was 20 μm. X-ray-excited, secondary-electron images (SXI) were used in order to determine the topography and, thus, to be able to collect small-area XPS spectra from the points of interest on the ball and on the disc. The electron neutraliser was used in order to compensate for sample charging. Sample charging was controlled with both the electron neutraliser and the ion gun control, through the ion neutraliser supplied by ULVAC-PHI. The extractor voltage was set to 30 V, and the emission current was set to 20 mA, while the bias was 1V. The FWHM of the carbon 1s signal of the ester group was 0·81 eV. The spectra were further corrected with reference to adventitious carbon at 284·8 eV.

3. Results

3.1 Differential thermal analysis

In Figure 4(a), the thermogram for H_3BO_3–B_2O_3 glass is reported. A series of endothermic peaks at temperatures above 100°C is observed, indicating the release of water and the formation of three different dehydration products: HBO_2(I), HBO_2(II) and HBO_2(III), as reported in the literature.[11] Kracek *et al* have investigated the equilibrium behaviour of the water–boron oxide system. Boric acid is stable to up about 158°C, between 158°C and 236°C three crystalline forms of metaboric acid (HBO_2(I), HBO_2(II), and HBO_2(III)) can form, and above 236°C only boron oxide is stable.[11] The measured B_2O_3 glass transition temperature is 259°C.

For zinc borate glasses, as shown in Figure 4(b), (c) and (d), it was observed that T_g decreases as the ZnO content is increased. Moreover, the thermal stability towards crystallisation of these glasses is strongly influenced by the presence of ZnO in the sample. It is not possible to clearly observe the devitrification peak for the Zn/B=1/2 glass, in which the proportion of ZnO is not high. The melting peaks of the other two glass compositions (Zn/B=3/4, 1) are very well defined (see Figures 4(c),(d)). This behaviour illustrates the tendency of these glasses to avoid crystallisation, but only when the ZnO is not present in large amounts.[12]

3.2 Surface roughness

The root-mean-square-roughness, R_q, of the glass surface, immediately after mechanical polishing, was measured by atomic force microscopy. The initial surface roughness of the boric oxide glass was 27·4 (1·9) nm, and an increase was clearly detectable over time. Immediately after polishing (T=25°C, relative humidity (RH)=43%), the surface was characterised by pits and protrusions, several nanometres in length (217(24) nm),[13] which became bigger until they coalesce (see Figure 5).

In the case of the zinc borate glasses, the surface roughness (R_q), immediately after polishing, was 3·02(0·03) nm, and no meaningful changes were

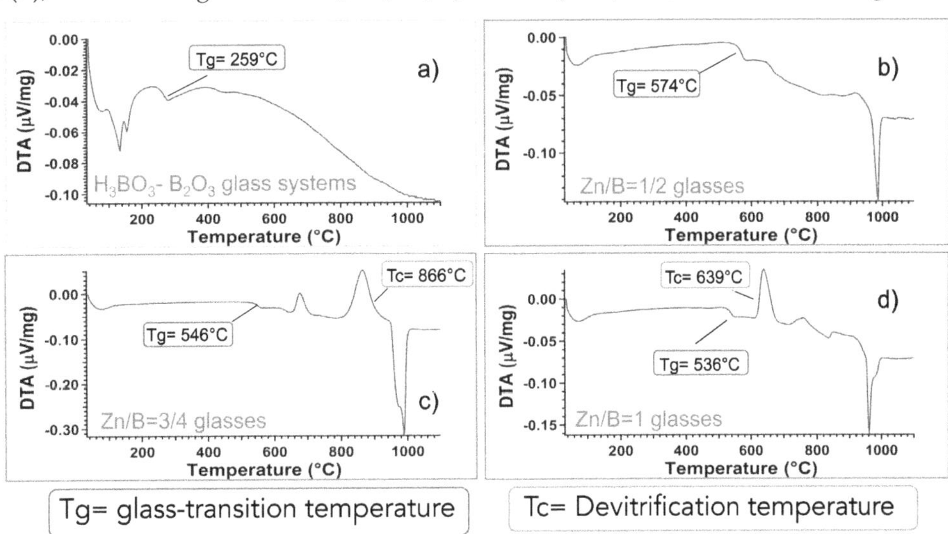

Figure 4. DTA analysis of (a) B_2O_3–H_3BO_3 glass, (b) Zn/B=½ glass, (c) Zn/B=¾ glass, (d) Zn/B=1 glass [Colour available online]

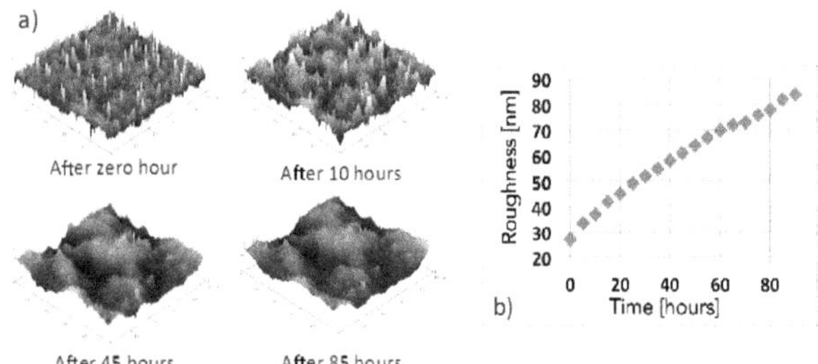

Figure 5. (a) AFM images: changes of B_2O_3 glass surface after polishing treatment with air exposure (T=25°C, RH=43%), (b) plot of B_2O_3 glass roughness versus time [Colour available online]

noted over time. Moreover, prismatic crystalline needles cannot be detected for the zinc borate glasses (Figure 6). They may not be able to form for these glass systems.

3.3 Vickers hardness

The Vickers hardness of the mechanically polished samples was measured as a function of load by means of micro-indentation. It was possible to use loads up to 10 N without causing brittle fracture of the borate surfaces.

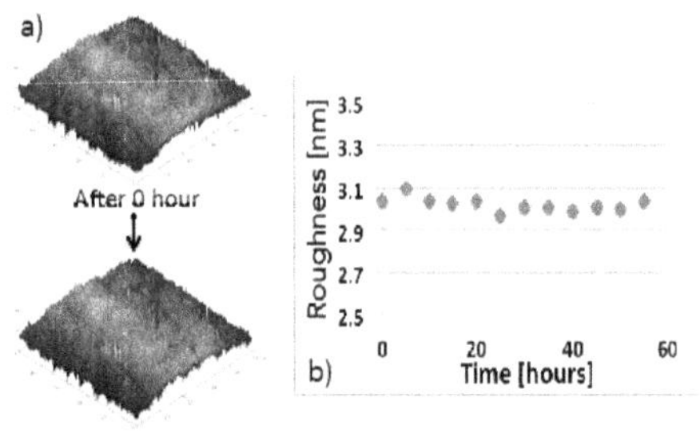

Figure 6. (a) AFM images: changes of ZnB=1/2 glass surface after polishing treatment with air exposure (T=25°C, RH=40%), (b) plot of ZnB=1/2 glass roughness versus time [Colour available online]

In Figure 7, the 3D topography of the H_3BO_3–B_2O_3 glass sample following indentation is shown as an example, while optical micrographs of the low load

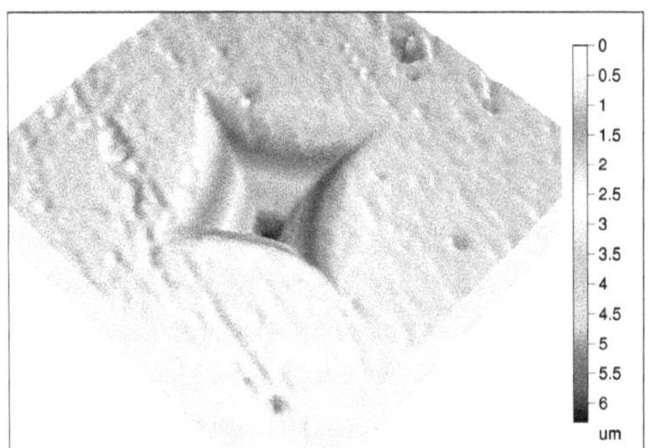

Figure 7. 3-D topography following microindentation of H_3BO_3–B_2O_3 glass, obtained by white light profilometry [Colour available online]

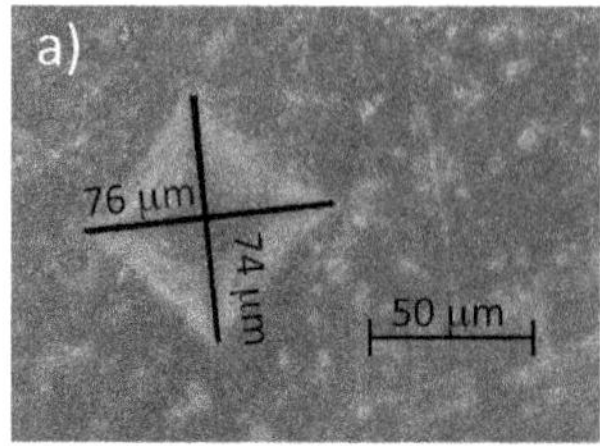

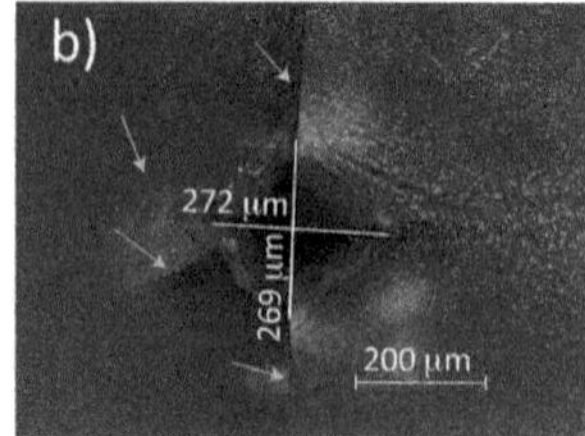

Figure 8. Impression shape on H_3BO_3–B_2O_3 glass, obtained by applying a load of (a) 10 N and (b) 100 N [Colour available online]

(10 N) indentation and high load (100 N) indentation of the H_3BO_3–B_2O_3 glass sample are shown in Figures 8(a) and (b), respectively.

In Figure 8(b), the blue arrows indicate the cracks radiating from the impression as a consequence of the high indentation load, as is typical for a brittle material with a high indentation load. The average values of hardness for the different borate glasses under investigation are summarised in Figure 9.

No significant differences in the hardness were observed between the different zinc borate glass compositions (Zn/B=1/2 glass samples 4·6(0·1) GPa, Zn/B=3/4 glass samples 4·7(0·1) GPa, Zn/B=1 glass samples 4·5(0·1)). On the other hand, H_3BO_3–B_2O_3 glass exhibits a lower hardness, 1·5(0·1) GPa, than zinc borate glasses. This value corresponds to the hardness of B_2O_3 glass and is in good agreement with the literature data.[14] Furthermore, since a H_3BO_3 layer grows spontaneously on B_2O_3 glass as soon it is exposed to the atmosphere, the Vickers hardness of a H_3BO_3 pellet was also measured. The hardness values of H_3BO_3 at two different loads, 0·1 and 0·25 N, were 0·06(0·01) GPa and 0·05(0·01) GPa, respectively. The Vickers hardness of H_3BO_3–B_2O_3 glass thus seems

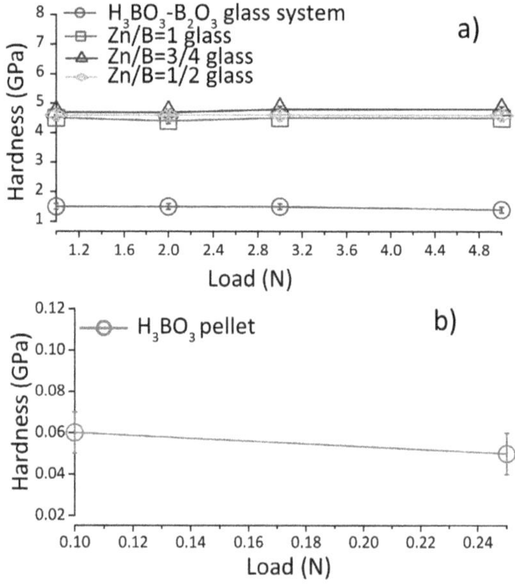

Figure 9. Vickers hardness as a function of the applied micro-indentation load, (a) for different compositions of borate glasses (load: 0·1, 0·25, 1, 2, 3, 5 N), and (b) for a H_3BO_3 pellet (load: 0·1 and 0·25 N) [Colour available online]

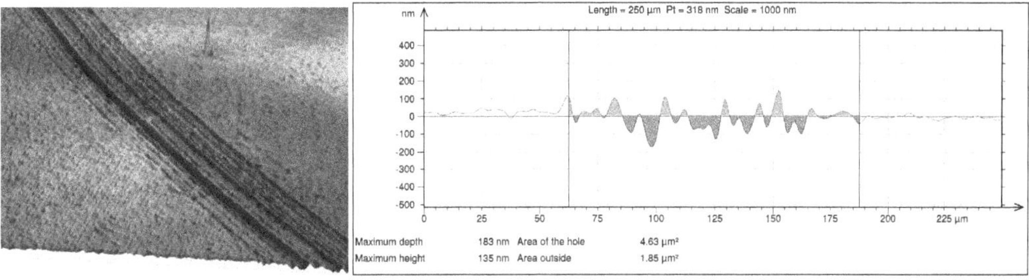

Figure 10. 3D-Topography and profile of a wear track after a tribological test on a zinc borate disc (Zn/B=1/2 at 25°C) obtained by white light profilometry [Colour available online]

not to be influenced unduly by the soft H_3BO_3 layer that is growing on the surface.

3.4 Tribological properties

As an example, the 3D-topography of a tribotrack, measured by means of a white light profilometer on a zinc borate disc is shown in Figure 10. In Figure 11, the results obtained from the tribological tests performed on boric oxide glasses at 25°C and at 100°C are shown, while in Figure 12 their corresponding steel ball wear rates are presented.

important to consider the relative humidity (RH) effects. The tribological tests at 100°C shown in Figure 10(b) were carried out at a RH of 30%, whereas those reported in Figure 12(c) were carried out at a RH of 45%. At lower RH, a sharp increase in the CoF was observed after ~15 min from the beginning of the test (Figure 11(b)), whereas the experiments, carried out at a higher RH are characterised by a slow increase in the CoF at the beginning, increasing again rapidly (Figure 11(c)) as soon as the lubricious film is broken at the interfaces. The wear rates measured on the steel balls for the tribological tests performed at 100°C

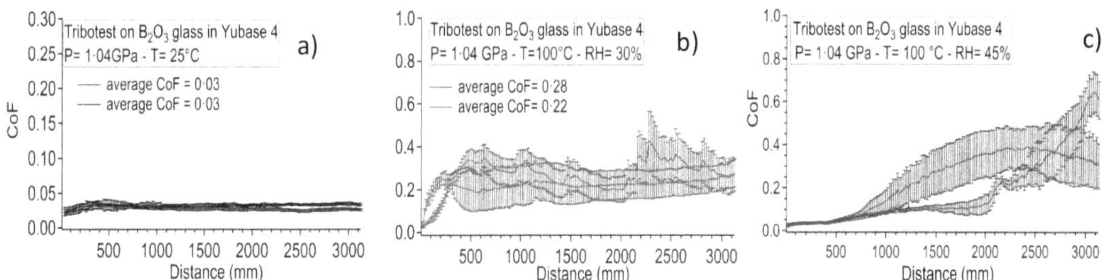

Figure 11. CoF of the tribological tests performed on a B_2O_3 glass disc: (a) at 25°C, (b) at 100°C and RH=30%, (c) at 100°C and RH=45% [Colour available online]

Boron oxide glasses show good lubricating properties at room temperature, with a CoF of 0·03, and a wear rate on the steel ball of $2{\cdot}1\times10^{-7}$ mm^3/m (see Figures 11(a) and 12(a)). At 100°C, they exhibit an increase in the CoF, which reaches an average value of 0·3 with sporadic spikes to 0·4, whereas the wear on the ball increases by three orders of magnitude to 1×10^{-4} mm^3/m (Figures 11(b) and 12(b)). It is also

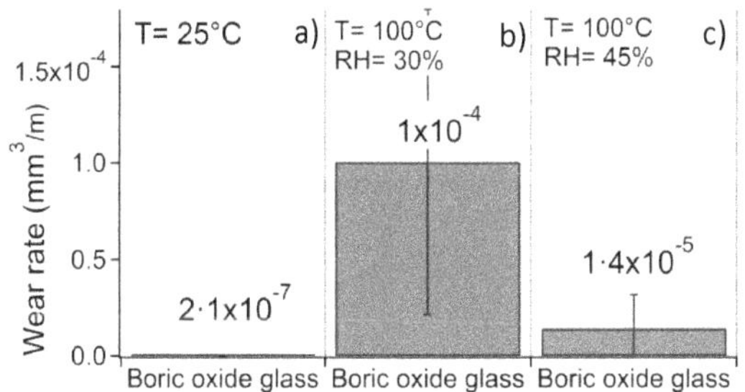

Figure 12. Wear rate of the steel ball for the tribotests performed on B_2O_3 discs: (a) at T=25°C, P=1·04GPa; (b) at T=100°C, P=1·04 GPa, and RH=30% and (c) at 100°C, P=1·04 GPa, and RH=45% [Colour available online]

and RH=45% (Figure 12(c)) are also quite low and closer to the values obtained for the tribological test performed at room temperature (Figure 12(a)). These outcomes, on boric oxide glasses, show that the RH may influence not only the CoF trends, but also the wear of the steel ball. In contrast, the results collected concerning the tribological behaviour of zinc borate glasses clearly demonstrate that the temperature and the relative humidity have negligible effects on the lubricant properties of these glass systems (see Figures 13 and 14).

The tribological experiments performed on the different zinc borate glass compositions at 25°C, did not show significant differences. Only the Zn/B=1/2 glass sample exhibits a slightly higher CoF (0·12) than the other two compositions. The Zn/B=3/4 glass sample and Zn/B=1 glass sample are characterised by CoF values of 0·1. Furthermore, higher temperatures induce a negligible increase of the CoFs in comparison to the values measured at low temperature (Figure 13). At 100°C, the CoFs for the different zinc borate

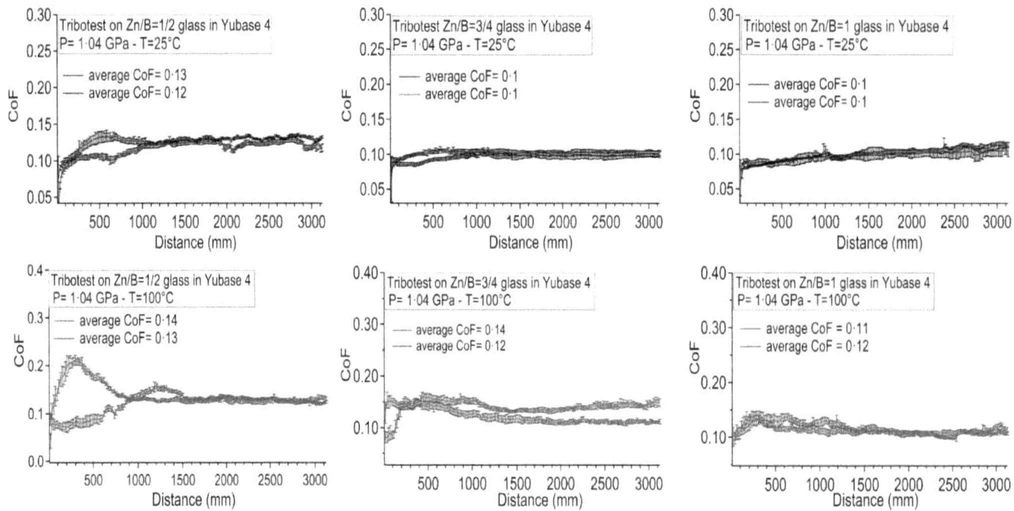

Figure 13. CoF for the tribological tests performed at 25°C and at 100°C on discs with Zn/B=1/2, Zn/B=3/4, and Zn/B=1 [Colour available online]

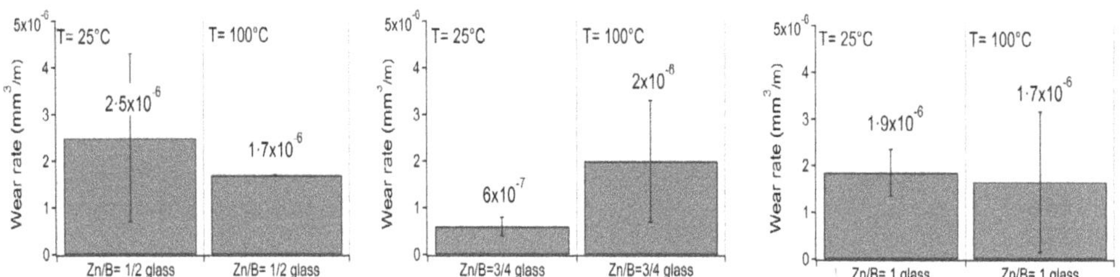

Figure 14. Wear rate on the steel ball at 25°C and at 100°C for the tribotests perfomed on discs with Zn/B=1/2, Zn/B=3/4, and Zn/B=1 [Colour available online]

glass samples are between 0·11 to 0·14. As for the wear of the steel ball at 100°C, surprisingly it was observed that temperature increase did not cause a big increase in the steel ball wear. In fact, at 100°C the wear on the balls was of similar magnitude to that for the tests performed at 25°C (Figure 14).

3.5 XPS results

3.5.1 Analysis of Zn/B=3/4 disc after tribotest

Small-area XPS spectra were recorded at the end of the tribological test in three different areas (Figure 15): in the contact area, at the border and in the non-contact area. Comparing the spectra in the three different areas, a slight shift of the boron B1s and of the zinc peak, $Zn2p_{3/2}$, towards lower binding energy (BE) was observed (Figure 15, Table 1).

Oxygen O1s signal: The most significant changes were observed in the O1s spectra (Figure 16). This signal was fitted with four characteristic peaks. The O1s(1) signal (BE 533·3 eV) is attributed to the boron–oxygen interaction in the boroxol rings (see Figure 16). The signal with the lowest BE, O1s(4) (BE 530·4 eV), is assigned to the oxygen–zinc interaction, involving small, highly charged borate groups such as orthoborates. The position of these peaks was constrained within ±0·2 eV of the values for the corresponding reference compounds, boron oxide glass and zinc oxide powder. The components O1s(2) and O1s(3) are related to oxygen in bridging (BO) and nonbridging (NBO) positions (Figure 16, Table 1), present in the polyborate groups of the glass system.

Table 1. XPS binding energies (in eV) for the Zn/B=3/4 disc after tribological testing with steel balls. The table provides the values obtained by analysing inside the wear track (in-contact area), at the border of the wear track (border) and outside the wear track (non-contact area). In addition values for the polished Zn/B=3/4 glass disc, for the as-received Zn/B=3/4 glass disc, for ZnO powder, and for the B_2O_3 glass disc are reported for reference (acquisition setup: spot size 100 μm, pass energy 26 eV, step size 0·05 eV)

Component	*Zn/B=3/4*			*Polished*	*Zn/B=3/4*	*ZnO*	*B_2O_3*
	Contact area	*Border*	*Non-contact area*	*Zn/B=3/4 glass disc*	*glass disc as received*	*powder*	*glass disc*
O1s (1)	533·4	533·3	533·4	533·3(0·02)	533·5(0·05)		533·3(0·2)
O1s (2)	532·3	532·4	532·5	532·3(0·01)	532·5(0·06)		
O1s (3)	531·4	531·4	531·6	531·4(0·01)	531·6(0·06)		
O1s (4)	530·4	530·4	530·1	530·4(0·01)	530·4(0·05)	530·2(0·04)	
B1s	192·2	192·3	192·8	192·3(0·05)	192·6(0·03)		193·7(0·2)
$Zn2p_{3/2}$	1022·3	1022·5	1022·8	1022·3(0·1)	1022·7(0·04)	1021·5(0·05)	

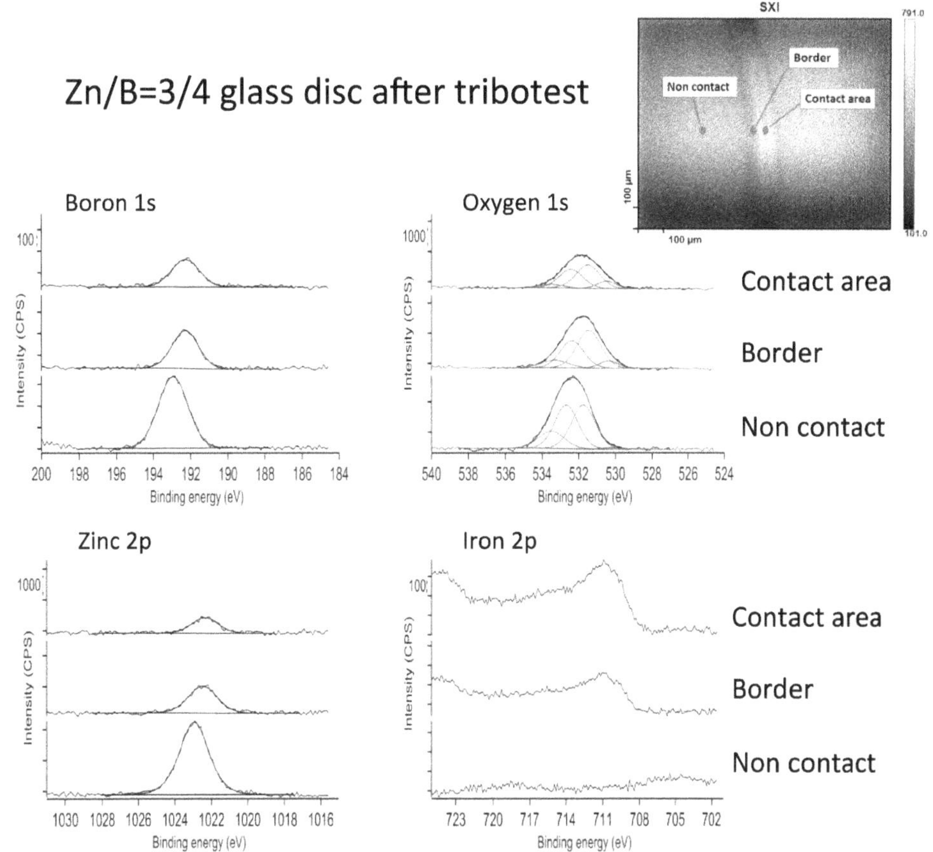

Figure 15. Small-area XPS spectra for the B1s, O1s, Zn2p, and Fe2p components, acquired inside, at the border and outside the wear track on the Zn/B=3/4 glass disc at the end of the tribological test (T=100°C, P=1·04 GPa). The x-ray-excited secondary-electron image (SXI) shows the points of analysis [Colour available online]

Covalent and ionic character: The O1s(1) and O1s(2) components are related respectively to oxygen atoms that link neighbouring B atoms (B–O–B) in boroxol rings, and to bridging oxygen, and they represent the covalent character of the compounds. Comparing the O1s spectra of the three areas (Figure 15), it was observed that the relative intensities of these two covalent-related components are significantly higher for the border and the non-contact area. The components O1s(3) and O1s(4), are related, respectively, to NBO atoms and to the $B–O^-$ interaction, and they represent the ionic character of the compounds. The relative intensities of these oxygen atoms with ionic character are highest in the wear track and lowest in the non-contact area (Figure 15). This change, from more covalent to more ionic character, from the non-contact to the contact area, can be quantified by calculating the percentage of oxygen signal intensity (Table 2); in the wear track the covalent character is the lowest and in the non-contact region it is the highest.

Influence of the tribostress: These results suggest that the glass disc undergoes depolymerisation in the wear track as a consequence of the tribological stress, and that smaller borate units are formed.

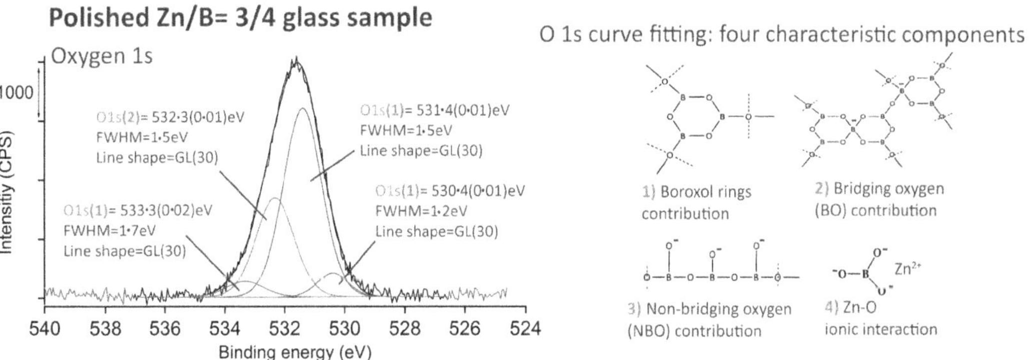

Figure 16. O1s high resolution spectrum acquired for a polished Zn/B=3/4 glass sample (left). The spectrum was collected with the following acquisition setup: spot size 100 µm, pass energy 26 eV, step size 0·05 eV. Also shown (right) are structural features to which the O1s components are ascribed, see text [Colour available online]

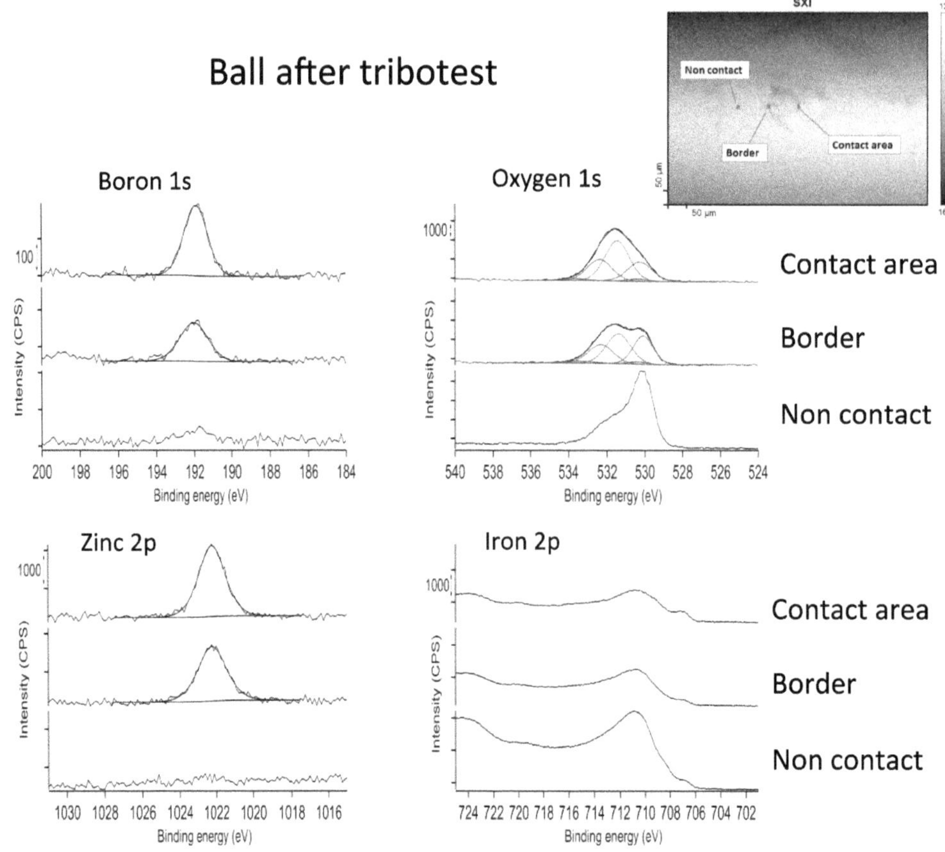

Figure 17. Small-area XPS spectra for the B1s, O1s, Zn2p, and Fe2p components, acquired in the contact area, at the border and in the non-contact area of the steel ball at the end of the tribological test on the Zn/B=3/4 glass disc (T=100°C, P=1·04 GPa). The x-ray-excited, secondary-electron image (SXI) shows the points of analysis [Colour available online]

Influence of exposure to high temperature: The binding energies of both B1s and Zn2p in the non-contact area of the Zn/B=3/4 zinc borate glass shift by 0·5 eV to more positive values after exposure to 100°C for 2 h (Table 1). Their binding energies are closer to the values measured for as-received Zn/B=3/4 glass disc (Table 1). These preliminary results indicate a possible structural rearrangement of the glass disc, but further investigations are required.

3.5.2 Analysis of the steel ball after tribotesting

The steel ball was characterised by small-area XPS at the end of the tribological test. In Figure 17, the points of analysis in the x-ray-excited secondary-electron image (SXI) are indicated by the red dots. The analysis was carried out inside, on the border and outside the flattened area of the steel ball. On the steel ball, there is only evidence of the glassy film on the ball/disc contact point. Boron, B1s, and zinc, $Zn2p_{3/2}$, were detected inside and at the border of the contact area on the steel ball, but no contribution of these species was observed in the non-contact region (Figure 17). The B1s peak is at slightly lower BE (191·9 eV, see Table 3) than measured inside the wear track on the glass disc (192·2 eV, see Table 1). This may indicate that a higher degree of depolymerisation occurred for the glass film transferred to the steel ball.

The O1s spectrum was fitted with five Gaussian/Lorentzian curves: the four characteristic components described above, and another one at lower BE due to oxygen bonded to iron.[15] The ratios between the

Table 2. Covalent/ionic character of the different regions of the Zn/B=3/4 glass disc after tribological testing. The ratio was calculated from the O1s small-area XPS spectra, using the sum of the areas of the covalent components, and the sum of the areas of the ionic components

	Covalent/ionic character		
	Inside	At the border	Outside
Tribotest on Zn/B= 3/4 glass disc at 100°C	0·75	0·8	1·4

Table 3. XPS binding energies (in eV) for the steel ball after the tribological test on Zn/B=3/4 disc. Values are given for the inside (in-contact) and the border (border) of the flattened area of the steel ball

Component	Zn/B=3/4 Contact area	Zn/B=3/4 Border
O1s (1)	533·7	533·51
O1s (2)	532·4	532·32
O1s (3)	531·5	531·4
O1s (4)	530·4	530·4
O1s (FeO_x)	530·3	530·1
B1s	191·9	192·0
$Zn2p_{3/2}$	1022·2	1022·2

Table 4. Covalent/ionic character for the glass transfer film on the contact point of the steel ball. The ratio was calculated from the O1s small-area XPS spectra for the steel ball, using the sum of the areas of the covalent components, and the sum of the areas of the ionic components

	Covalent/ionic character		
	Inside	*At the border*	*Outside*
Glass transfer film on the steel ball	0·6	0·7	Negligible

sum of the relative intensities of the two components 1 and 2, at higher BE, which are more representative of the covalent character of the glass, and the sum of the relative intensities of the other two components (3 and 4), which are more representative of the ionic character, were calculated (see Table 4). The calculated covalent/ionic ratios for the glass film (Table 4) are lower than those measured inside the wear track on the glass disc (Table 2). This may again indicate a higher depolymerisation for the glass transfer film on the steel ball.

The small shift of the B1s peak towards lower BE (Table 3), together with the measured ratios of the covalent/ionic character (Table 4), seem to suggest the formation of shorter chain lengths in the contact points. Furthermore, the O1s outside the contact area is mainly due to iron oxy-hydroxide contributions (Figure 17).

4. Discussion

4.1 Influence of relative humidity

Exposure of B_2O_3 glass to the environment induces significant changes. Following synthesis, it is colourless and fully transparent, but only a few minutes exposure to the laboratory atmosphere makes the glass dull (Figure 18). This behaviour is due to spontaneous reaction between B_2O_3 and moisture, leading to the formation of a boric acid layer:[13,16]

$$\tfrac{1}{2}B_2O_3 + 3/2H_2O \rightarrow H_3BO_3 \qquad \Delta H_{298}=-45{\cdot}1\ \mathrm{kJmol^{-1}}$$

This was experimentally confirmed by Raman spectroscopy (Figure 19): as well as the characteristic bands of the boroxol ring structure (1260 cm^{-1} and 808 cm^{-1}),[5] the presence of a less intense band at 875 cm^{-1} indicates the presence of boric acid.[17] As proposed in the literature,[18] the layer of boric acid formed on top of the B_2O_3 discs is characterised by a lamellar structure. Atoms lying within the same plane, oxygen, boron and hydrogen, are closely packed and strongly bonded to each other, whereas the layers themselves are widely spaced and held together by van der Waals' forces (Figure 20).

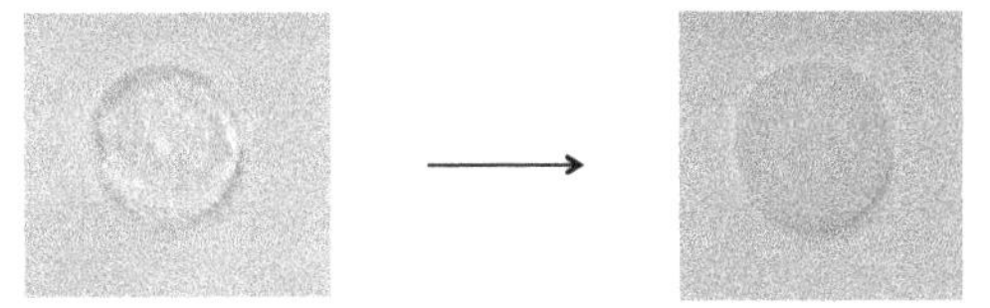

Figure 18. B_2O_3 glass immediately after preparation, and following exposure to ambient conditions

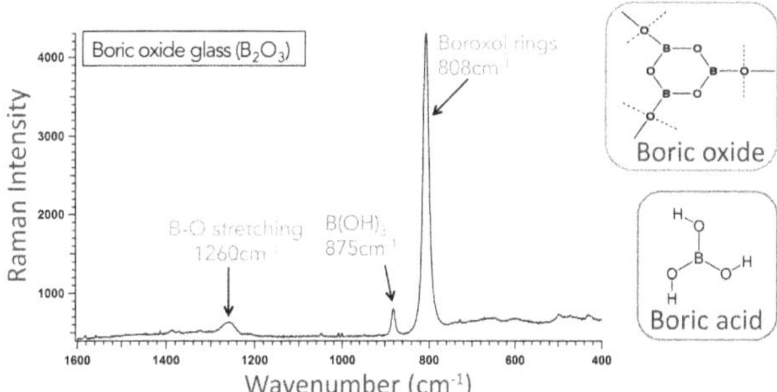

Figure 19. Raman spectrum of B_2O_3 glass with the presence of H_3BO_3 [Colour available online]

In addition, the thermogram collected for H_3BO_3–B_2O_3 glass (Figure 4(a)) indicates the release of water, and the formation of intermediate dehydration products as a consequence of the heating process, and thus the reversibility of this phenomenon. On the basis of these data, a preliminary interpretation of all the collected data may be possible. The development of pits and protrusions on top of the B_2O_3 glass may occur as a consequence of the formation of boric acid. Furthermore, H_3BO_3 layer affects the tribological properties of H_3BO_3–B_2O_3 glass because it is not only softer than the bulk material, B_2O_3, but also because it is characterised by a lamellar structure which helps the lubrication.

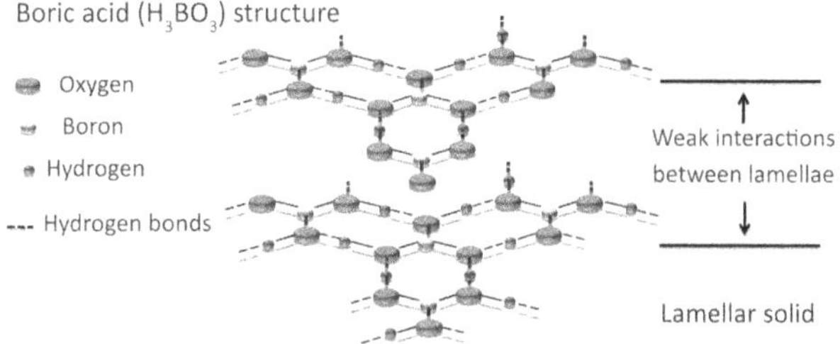

Figure 20. $B(OH)_3$ lamellar structure, adapted from Ref. 18 [Colour available online]

4.1.2 Zinc borate glasses

The AFM measurements carried out on zinc borate glasses show that their roughness remains unchanged after exposure to environmental air (see Figure 6). Moreover, their surfaces remain transparent and colourless in humid air (Figure 21). These findings confirm that the spontaneous reaction of these systems with moisture is inhibited.

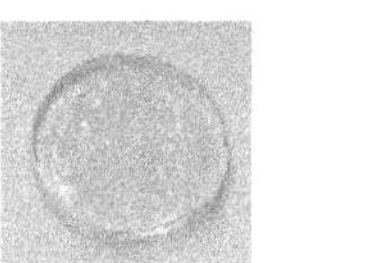
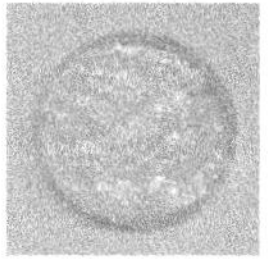
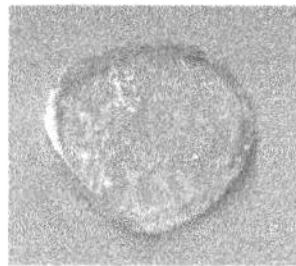

Figure 21. The different zinc borate glasses after exposure to ambient conditions (time=more than 3 months, T=25°C, RH=32–48%)

4.2 Influence of ZnO addition

All the investigated zinc borate glasses are characterised by a similar hardness (Figure 9). The improvement in the measured mechanical properties of the zinc borate glasses is correlated with the introduction of zinc oxide into the composition, and with the consequent structural rearrangements. In the glass matrix, the addition of zinc oxide leads to the destruction of boroxol rings, and their conversion into polyborate groups, as well as groups with nonbridging oxygen atoms.[5] Furthermore, as reported in the literature, for these compositions the local environment of the boron atom transforms from BO_4 tetrahedra to a BO_3 tetragonal planar structure with increasing zinc oxide content.[19] The strength of the glasses depends on the degree of connectivity. Moreover, the structural strength is not only given by the energy of the chemical bonds, but also by the ionic interactions between the chains, which are determined by the field strength of the modifier ions.[20] Thus, the higher structural strengths measured for the zinc borate glasses can be explained by the increased degree of ionic connectivity in the glass samples, as a consequence of the introduction of ZnO.

4.3 Lubrication mechanism

4.3.1 H_3BO_3–B_2O_3 glass

The tribological results indicate that temperature has a drastic influence on the lubrication properties of boric oxide glass. This behaviour is strictly related to the boric acid surface layer. Due to its lamellar structure, boric acid works as an excellent lubricant at room temperature, and makes this glass system one of the lowest friction solid lubricants known.[18,21] Moreover, even if the boric acid film is destroyed during the sliding, another one forms immediately, and is continuously replenished, so long as the experiment is performed in humid air at room temperature. Thus, the lubricating mechanism of B_2O_3 glass at 25°C is related to the formation of boric acid, which (due to its structure) is able to support the high applied pressure. Erdemir *et al* have found that, when the atomic layers of H_3BO_3 are subjected to a tangential force, they can align themselves parallel to the direction of relative motion, and then slide over one another, providing low friction.[16]

Tribological tests performed at 100°C, are characterised by higher values of friction and wear on the steel ball at the end of the tribotest. These results are understandable, considering the harsh conditions of these experiments; the high temperature and pressure cause the destruction, and hinder the renewal, of the lubricious boric acid film on the surface.

4.3.2 Zinc borate glasses

The results from the tribological tests on zinc borate glasses suggest a completely different lubricating mechanism than that proposed for H_3BO_3–B_2O_3 glass. The different zinc borate glass compositions (Zn/B=1/2, Zn/B=3/4 and Zn/B=1) are characterised by similar tribological properties.

The great advantage of the zinc borate glasses is that the temperature effects do not influence their lubricious performance significantly; they continue to show outstanding tribological properties at high temperatures. At 100°C, their CoFs show only a slight increase, and the wear rates on the steel balls are still of the same order of magnitude as those at room temperature (Figure 11). These results suggest another lubricating mechanism for these glass systems. The XPS analysis performed on the ball/disc contact points confirmed the formation of a transfer film; the small-area XPS spectra acquired on the flattened area of the steel ball exhibited the presence of zinc and boron, whilst the presence of iron oxide was detected inside the tribotrack of the glass disc. The B1s, O1s and Zn2p spectra acquired in the contact points are all slightly shifted towards lower BE (Tables 1 and 3). This trend corroborates the suggestion that glass depolymerisation occurred in these areas. From our previous investigations on zinc borate glasses,[22] it was observed that these peaks fall at higher BE in the presence of longer polyborate chains, because the electrons are delocalised along the structure, giving stability to the chains, but also increasing the effective positive charge on the B atom (higher BE). The depolymerisation reactions may be mainly caused by tribological stress, but, most likely, the presence of iron oxide (steel ball) inside the tribotrack of the glass disc can also act as an enhancer of the process. Furthermore, the negligible wear on the disc can be explained in terms of the formation of short-chain polyborates inside the tribotrack, which can be more wear-resistant. The presence of the transfer film on the steel ball is not only able to reduce the friction, but also to prevent further wear.[22] On the basis of the present results, the most likely hypothesis concerning the lubricating mechanism for zinc borate glasses is related to the formation of a transfer film made up of smaller, highly negatively charged borate groups, and of Zn^{2+} ions that are able to sustain the high loads, both at low and high temperatures.

In tribology, it is common to refer to the transfer film at the interface between two solids in moving contact as the "third body".[23,24] The third body is often composed of particles detached from the rubbing surfaces and trapped inside the contact. The third body is able to protect the materials rubbing against each other from further degradation. The third-body viewpoint, therefore, does not equate particle detachment with wear. The wear process, in fact, is constituted by three main steps:[25]

- particle detachment – a first-body particle is detached by any one of the well-established mechanisms (adhesion, abrasion, corrosion, fatigue, etc.),

- third body – once detached, the trapped particle is subject to the strenuous conditions at the interface along with other debris,
- particle ejection – finally, the particle can be ejected from both the contact and the wear track, and only then does it become a wear particle.

Elimination from the contact and from the wear track are both necessary conditions to produce a wear particle. The zinc borate case can be included in this kind of "third-body" lubricating process. The very low wear measured on the zinc borate glass discs, and on the steel ball, support the idea of the formation of a transfer film at the interface acting as a third-body layer. In tribology, the investigation of the properties of third bodies generated during sliding is an important goal in the design of long-lasting tribomaterials.[26]

5. Conclusions

The main goal of this study was to shed light on the mechanical and tribological properties of boric oxide glass and zinc borate glasses. On the basis of the results, the following conclusions can be drawn:

- Boric acid-covered boron oxide glass is characterised by low values of hardness, while higher values were measured for zinc borate glasses, as a consequence of the structural rearrangements induced by the addition of ZnO.
- Boron oxide glass shows an increasing surface roughness in ambient conditions, due to spontaneous reaction with moisture, and formation of a $B(OH)_3$ surface layer.
- The tribological properties of boron oxide glass at room temperature are strictly related to the formation of a lubricious film of boric acid.
- Temperature has a strong detrimental effect on the lubricating performance of boric oxide glass, because it inhibits the spontaneous and regeneration of the lubricious boric acid coating.
- The surface roughness of zinc borate glasses remains unchanged with humid air exposure.
- Zinc borate glasses retain their good tribological performance at high temperature, under the investigated experimental conditions.
- XPS analysis provides evidence of glass depolymerisation, and the formation of a zinc borate transfer film in the contact area.
- The third-body transfer film for zinc borate glasses is able to reduce the friction and wear of the steel ball and of the disc, both at room temperature and at high temperatures.
- More investigations are required on these glass systems to completely identify the lubricating mechanism. Nevertheless, these preliminary tribological results, collected under harsh conditions, appear to be promising, and encourage the potential use of these materials in future applications as protective coatings in many mechanical systems.

Supporting information

Further details of the study are available as Supporting Information in the electronic version of this paper.

Acknowledgements

The authors would like to thank Infineum UK Ltd. for funding this project.

References

1. Galeener, F. L. The structure and vibrational excitations of simple glasses. *J. Non-Cryst. Solids*, 1990, **123**(1–3), 182–196.
2. Mardhiah, K. A., Azhan, H. & Razali, W. A. W. Optical Characterisation of Erbium Doped Sodium Borate Glass. *Adv. Mater. Res.*, 2012, **622&623**, 191–194.
3. Abdel-Baki, M. & El-Diasty, F. Role of oxygen on the optical properties of borate glass doped with ZnO. *J. Solid State Chem.*, 2011, **184**(10), 2762–2769.
4. Koudelka, L. & Mošner, P. Borophosphate glasses of the $ZnO–B_2O_3–P_2O_5$ system. *Mater. Lett.*, 2000, **42**(3), 194–199.
5. Meera, B. N. & Ramakrishna, J. Raman Spectral Studies Of Borate Glasses. *J. Non-Cryst. Solids*, 1993, **159**(1–2), 1–21.
6. Patching, M. J., Evans, H. P. & Snidle, R. W. Conditions for Scuffing Failure of Ground and Superfinished Steel Disks at High Sliding Speeds Using a Gas Turbine Engine Oil. *J. Tribol.*, 1995, **117**(3), 482–489.
7. Barber, J. R. Multiscale Surfaces and Amontons' Law of Friction. *Tribol. Lett.*, 2013, **49**(3), 539–543.
8. Hähner, G. & Spencer, N. Rubbing and Scrubbing. *Phys. Today*, 1998, **51**(9), 22–27.
9. Telliskivi, T. Simulation of wear in a rolling–sliding contact by a semi-Winkler model and the Archard's wear law. *Wear*, 2004, **256**(7–8), 817–831.
10. Kato, K. & Adachi, K. *Wear Mechanisms, Modern Tribology Handbook*, Volume Two ed., CRC Press, 2000.
11. Kracek, F. C. & Merwin, H. E. The system, water–boron oxide. *Am. J. Sci.*, 1938, **235A**, 143–171.
12. Inoue, T., Honma, T., Dimitrov, V. & Komatsu, T. Approach to thermal properties and electronic polarizability from average single bond strength in $ZnO-Bi_2O_3-B_2O_3$ glasses. *J. Solid State Chem.*, 2010, **183**(12), 3078–3085.
13. Ma, X., Unertl, W. N. & Erdemir, A. The boron oxide–boric acid system: Nanoscale mechanical and wear properties. *J. Mater. Res.*, 1999, **14**(8), 3455–3466.
14. Mukhanov, V. A., Kurakevich, O. O. & Solozhenko, V. L. On the hardness of boron (III) oxide. J. Superhard Mater., 2008, **30**(1), 71–72.
15. Mangolini, F., Rossi, A. & Spencer, N. D. Chemical Reactivity of Triphenyl Phosphorothionate (TPPT) with Iron: An ATR/FT-IR and XPS Investigation. *J. Phys. Chem. C*, 2010, **115**(4), 1339–1354.
16. Erdemir, A., Bindal, C., Zuiker, C. & Savrun, E. Tribology of naturally occurring boric acid films on boron carbide. Surf. Coatings Technol., 1996, **86&87**(2), 507–510.
17. Ananthakrishnan, R. The Raman spectra of crystal powders. *Proc. Indian Acad. Sci. A*, 1937, **5**(3), 200–221 ().
18. Shah, F. U., Glavatskih, S. & Antzutkin, O. N. Boron in Tribology: From Borates to Ionic Liquids. *Tribol. Lett.*, 2013, **51**(3), 281–301.
19. Kajinami, A., Harada, Y., Inoue, S., Deki, S. & Umesaki, N. The Structural Analysis of Zinc Borate Glass by Laboratory EXAFS and X-Ray Diffraction Measurements. *Jpn. J. Appl. Phys.*, 1999, **38**(S1), 132–135.
20. Baikova, L. G., Pukh, V. P., Fedorov, Y. K., Sinani, A. B., Tikhonova, L. V. & Kireenko, M. F. Mechanical properties of phosphate glasses as a function of the total bonding energy per unit volume of glass. *Glass Phys. Chem.*, 2008, **34**(2), 126–131.
21. Sawyer, W. G., Ziegert, J. C., Schmitz, T. L. & Barton, T. In Situ Lubrication with Boric Acid: Powder Delivery of an Environmentally Benign Solid Lubricant. *Tribol. Trans.*, 2006, **49**(2), 284–290 ().
22. Crobu, M., Rossi, A., Mangolini, F. & Spencer, N. Tribochemistry of Bulk Zinc Metaphosphate Glasses. *Tribol. Lett.*, 2010, **39**(2), 121–134.
23. Godet, M. The third-body approach: A mechanical view of wear. *Wear*, 1984, **100**(1–3), 437–452.
24. Godet, M. Third-bodies in tribology. *Wear*, 1990, **136**(1), 29–45.
25. Berthier, Y. Experimental evidence for friction and wear modelling. *Wear*, 1990, **139**(1), 77–92.
26. Meierhofer, A., Hardwick, C., Lewis, R., Six, K. & Dietmaier, P. Third body layer—experimental results and a model describing its influence on the traction coefficient. *Wear*, 2014, **314**(1–2), 148–154.

REVIEW
Phosphate and fluoride phosphate optical glasses — properties, structure and applications

Doris Ehrt

Otto-Schott-Institut, Friedrich-Schiller-Universität Jena, Fraunhoferstr. 6 D-07743 Jena, Germany

Manuscript received 27 August 2014
Revised version received 29 January 2015
Accepted 10 March 2015

Our investigations of phosphate and fluoride phosphate optical glasses started in 1976. The aim was the development of optical glasses with high positive anomalous partial dispersions, making them desirable for lens designs that reduce the secondary spectrum in high performance optics to substitute for CaF_2 single crystals. A large variety of glasses have been prepared and investigated. The effect of cations and fluorine in phosphates on the refractive index, dispersion, thermal and chemical properties was studied. It was found that fluoride phosphate glasses based on AlF_3, MF_2, and P_2O_5, have the required optical properties. Fluor crown and phosphate crown optical glasses were developed. The structure and properties of these glasses depend mainly on the molar relation between fluorides and phosphates, which can be varied in a wide range between pure fluoroaluminate and phosphate glasses. The structure model can be described as chains of $Al(F,O)_6$ octahedra, bonding by mono- and diphosphate groups and cations. Their intrinsic transparency in the vacuum UV range is comparable with those of silica, and CaF_2. The absorption coefficients of possible trace impurities in different redox states were determined. The effect of UV lamp, UV laser, x-ray radiation, and laser writing of waveguides was studied. Together with colleagues from physics departments, we developed efficient laser and amplifier glasses with Nd^{3+}, Er^{3+}, and Yb^{3+}. The POLARIS (petawatt optical laser amplifier for radiation intensive experiments) system is based on a fluoride phosphate glass doped with Yb^{3+}, which is used as the active medium, pumped by light from laser diodes. Photoluminescence in glasses doped with ions in s^2 configuration (Sn^{2+}, Pb^{2+}, As^{3+}, Sb^{3+}), d^0 configuration (Ti^{4+}, Nb^{5+}, Ta^{5+}, Mo^{6+}, W^{6+}), or d^{10} configuration (Ag^+, Cu^+), which absorb strongly in the UV, has been demonstrated mainly in the UV and blue-green region. Efficient visible photoluminescence with different lifetimes was found in Mn^{2+} ($3d^5$), and in rare earth (f^n) doped glasses (Ce^{3+}, Pr^{3+}, Sm^{3+}, Eu^{2+}, Eu^{3+}, Tb^{3+}, Dy^{3+}, Ho^{3+}, Er^{3+}, Tm^{3+}), which can be used for various applications.

Introduction

In 1976 we started investigations of phosphate and fluoride phosphate optical glasses at the Otto Schott Insitute for Glass Chemistry at the University Jena, in close collaboration with scientists from JENAer GLASWERK and CARL ZEISS Jena. The aim was the development of optical glasses with high positive anomalous relative partial dispersions, making them suitable for lens designs that reduce the secondary spectrum (colour distortion) in high performance optics to substitute CaF_2 single crystals. The glasses should also have high refractive indices and high chemical resistance, and patents should be possible. It was clear from the literature that phosphate glasses with high fluorine content (so-called fluorophosphates), which have been known since the 1960s are promising candidates.[1–3] P–F bonding is required for positive anomalous dispersion.[4–7] However, the values for the anomalous partial dispersion of CaF_2 single crystals were not actually achieved until now, because the fluorine content in the glasses was too low. We started with investigations of glass formation, properties and structure of mainly alkaline earth glasses and the introduction of fluorine.

The crystallisation behaviour of glasses and solid state reactions during melting process were investigated. The structure of glasses was studied with various spectroscopic methods, e.g. UV-vis-IR, Raman, NMR and EPR, in collaboration with many colleagues of the Otto Schott Institute, the University in Jena and other external institutions. Many students have carried out their diploma and PhD works. The aim of this paper is a short review of the main results of the scientific work of all these people during a very long period in the field of fluoride phosphate glasses, which have a great variability, and are very interesting structurally and for different applications in the field of optics and photonics. After fluor crown glasses were developed and transferred into products, the materials were investigated in further directions, such as traces of impurities, radiation induced defects, and doping with rare earths and transition metals for active laser, amplifier, and photoluminescent glasses.

Corresponding author. Email doris.ehrt@uni-jena.de
Original version presented at Int. Conf. on Phosphate Glasses, Pardubice, Czech Republic, 2–4 July 2014. The conference was held in honour of Dr Doris Ehrt's contribution to the science of phosphate glasses.
DOI: 10.13036/17533562.56.6.217

Table 1. Measured properties of phosphate glasses

Glass	Melting temperature, T_m (°C)	T_g (°C) (5 K/min) ±3°C	T (°C) logη=4 (η in dPas) ±3°C	T (°C) logη=2 (η in dPas) ±3°C	TEC (ppm/K) (5 K/min) ±0·1	Density (g/cm³) ±0·001	[OH] (cm⁻¹) ±0·1	Refractive index n_e (546·06 nm) ±0·00002	ABBE number ν_e ±1	Vibration PO_2 (cm⁻¹)	Vibration M–O (cm⁻¹)
$NaPO_3$	800	265	380	480	24·0	2·51	~15	1·48585	65	1295	220
KPO_3	900	Part.cryst.									
$Mg(PO_3)_2$	1300	520	805	1065	7·5	2·43	5·2	1·4974	71	1300	390
$Ca(PO_3)_2$	1300	500	715	850	11·0	2·65	5·5	1·5469	66	1280	265
$Sr(PO_3)_2$	1300	485	670	845	13·0	3·15	5·7	1·55913	66	1245	180
$Ba(PO_3)_2$	1200	460	625	760	14·2	3·66	5·0	1·5899	64	1255	145
$Zn(PO_3)_2$	1150	400	710	945	7·3	2·88	7·5	1·5260	62	1275	
$Mn(PO_3)_2$	1300	470			11·5	2·92	6·0	1·57702	61	1245	240
$Pb(PO_3)_2$	800	350			14·6	5·00		1·7720	33	1215	
$La(PO_3)_3$	1500	570			7·8	3·26		1·6000	59	1275	
$Fe(PO_3)_3$	1350	590	925	1155	6·7	2·80	<2	n.m.*	n.m.*		
$Al(PO_3)_3$	1600	765	1230	1400	6·0	2·60	1·4	1·5310	70	1290	400
NSP ($10Na_2O.40SrO.50P_2O_5$)	1300	400	620	780	15·7	3·08	5·5	1·54686	66		
$Sr_{0·9}Mn_{0·1}(PO_3)_2$	1300	500			11·7	3·15	6·8	1·56352	66		
$Sr_{0·5}Mn_{0·5}(PO_3)_2$	1300	480			11·2	3·08	8·0	1·57501	62		
$Sr_{0·3}Mn_{0·7}(PO_3)_2$	1300	477			9·7	3·00	9·3	1·57394	61		

* not measured

Phosphate glasses

The formation of phosphate glasses has been well known for a very long time.[9–11] Many scientific papers on this subject have been published by many authors from all over the world. The main results on the properties of basic metaphosphate glasses that we have investigated from time to time during a long period, are given in Table 1. The cations have a large effect on the chemical and physical properties, due to different bonding and structure, e.g. the values for the glass transition temperature, T_g, vary between 265°C for $NaPO_3$ and 765°C for $Al(PO_3)_3$, and the values for the refractive index, n_e, vary between 1·486 for $NaPO_3$ and 1·772 for $Pb(PO_3)_2$. There are also large differences in the thermal expansion coefficients, TEC, with 24 ppm/K for $NaPO_3$ and 6 ppm/K for $Al(PO_3)_3$. The measured viscosity curves (Figure 1) show that the dependence on the modifier cation changes with temperature, e.g. for $Zn(PO_3)_2$ and $Ca(PO_3)_2$ at T>700°C. Unfortunately, the high crystallisation rates prevented measurement at high temperature in some cases, especially $Al(PO_3)_3$, $Fe(PO_3)_3$, $Ba(PO_3)_2$, and $Sr(PO_3)_2$. It was not possible to fit these curves with a simple VFT (Vogel–Fulcher–Tammann) equation. The temperatures for two viscosities (in dPa s) are given in Table 1 for comparison, logη=4 (working point) and logη=2 (melt). The large differences between the values for cations with nearly the same radius (and coordination number), Zn^{2+} and Mg^{2+}, Fe^{3+} and Al^{3+}, show that there is not a simple correlation with radius.

The UV absorption spectra of glasses are mainly affected by the content of trace impurities introduced by raw materials and melting technique. Figure 2(a) shows some examples from glasses prepared in high purity silica glass crucibles. $Zn(PO_3)_2$ (supplied by Chemische Fabrik Budenheim), $Sr(PO_3)_2$ and $Ba(PO_3)_2$ glass samples have the lowest UV absorption and an impurity content of Fe′<5 ppm (where the dash indicates all Fe ions, irrespective of redox state). $NaPO_3$

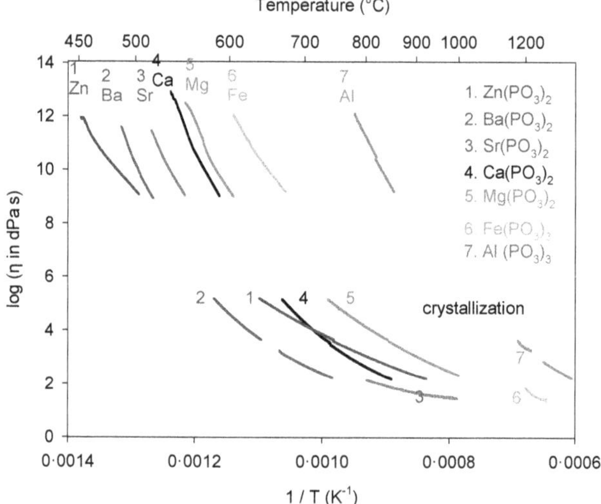

Figure 1. Measured viscosity curves of phosphate melts (using the beam bending method for η in the range 10^9 to 10^{11} dPas, and using the rotating cylinder method for η in the range $10^{1·5}$ to 10^5 dPas) [Colour available online]

has higher UV absorption with Fe′~30 ppm. $Zn(PO_3)_2$ (supplied by VEB Stickstoffwerke Piesteritz, from the time of the GDR, the German Democratic Republic) has the typical band (216 nm) for Pb^{2+} (~50 ppm) and Fe′ (~30 ppm). $Mg(PO_3)_2$ (~100 ppm Fe′), $Al(PO_3)_3$ and $Ca(PO_3)_2$ glass samples have much higher impurities. The silica content in the glasses introduced by the crucible was <1%. If the phosphate glasses were melted in a Pt crucible then the UV absorption of glasses was higher due to ionic Pt′ impurities (<10 ppm).[12] The OH content of glasses is important for luminescence and laser applications. It can be easily measured by IR absorption, and was done with all glasses prepared.[13,14] Figure 2(b) shows the spectra of the glasses from Figure 2(a) with the normalised OH content in cm⁻¹. The values from other phosphate glasses are also given in Table 1. The OH content is mainly dependent on the kind of raw material, e.g.

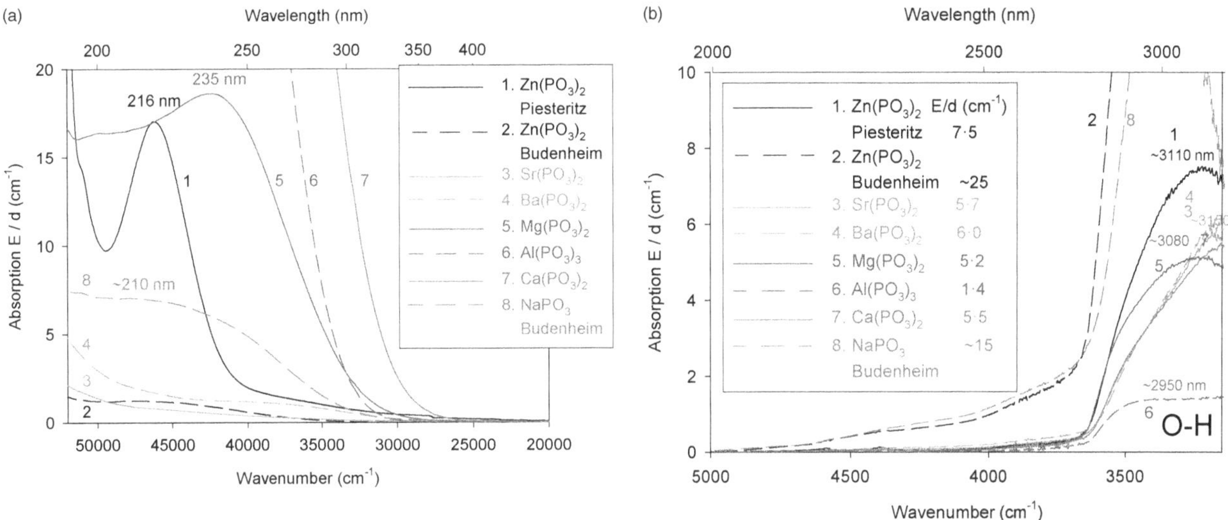

Figure 2. (a) Normalised UV absorption spectra of phosphate glasses; (b) Normalised IR absorption spectra. The inset table gives values for the OH content [Colour available online]

$Zn(PO_3)_2$ from Budenheim had water of crystallisation, and from the melting technology. An extreme effect of melting temperature on the thermal and chemical properties was found in $SnO–P_2O_5$ glasses, with different OH contents between ~0·1 and ~20 mol%, depending on melting temperature, 1200°C or 450°C.[15]

Glass formation and properties in fluoride phosphate systems

The conclusion in 1978, after studying the scientific literature[1–3] and patents, was that phosphate glasses with higher fluorine content should be promising candidates for optical glasses with positive anomalous relative partial dispersion, $\Delta P_{gF'}=(n_g-n_{F'})/(n_{F'}-n_{C'})$ with g=435·83 nm, F'=479·99 nm, C'=643·85 nm. The aim was the substitution of CaF_2 single crystals in high performance optics to reduce the chromatic aberration (secondary spectrum Δs, colour distortion) to obtain higher resolution with microscopes, e.g. for microlithography, telescopes or photo objectives. This is only possible with the unusual dispersion behaviour of optical materials. Silicate glasses, which are the usual optical glasses, were on the 'normal line' in the dispersion diagram, $P_{gF'}-\nu_e$ (see Figure 5, later). A deviation from this normal line occurs for optical materials with other UV and IR absorption bands, which change the dispersion curves in comparison with silicates; e.g. fluorides and phosphates have UV absorption bands in the UV at shorter wavelengths, and IR absorption bands at longer wavelengths, and borates have IR absorption bands at shorter wavelengths. We started with the experimental investigation[16,17] of the simple binary system $SrO–P_2O_5$ and substitution of SrF_2 for SrO. High purity raw materials were used, and samples without fluorine were melted in high purity silica glass crucibles, whilst Pt crucibles were used for fluoride phosphates. Raw materials and glasses were analysed by various methods. A special method was developed for high accuracy analysis of the fluorine content in raw materials and glasses.[8] The glasses were characterised using a variety of different methods: thermal analysis to determine the glass transformation, crystallisation, weight loss, thermal expansion and viscosity; optical measurements to determine refractive indices and dispersion. Glasses with high optical homogeneity, without striaes and bubbles ($\Delta n \leq 2\times10^{-5}$), are necessary to calculate the anomalous partial dispersion with high accuracy. Stable glass formation was obtained with a starting composition, 43 mol% SrF_2 and 57 mol% P_2O_5. However, the analysed fluorine in glass samples was at most 0·4 wt% fluorine. The glass samples did not have an anomalous dispersion. Glass formation regions in the systems $Al(PO_3)_3–SrF_2–BaF_2$ and $Al(PO_3)_3–MgF_2–BaF_2$ were already published (Figure 3(a)).[6] We found that the crystallisation rates of these melts are very high. Glasses could be obtained only by quenching the melts between brass plates. Optical properties could not be measured. The chemical analysis showed only 60–70% of the initial fluorine content remained in the glass samples. Glasses in the systems $Ba(PO_3)_2–AlF_3–MgF_2$ and $Ba(PO_3)_2–AlF_3–BaF_2$[7,16,17] have shown lower crystallisation tendency. Glass blocks, 300 g, could be obtained without crystals. MgF_2 could be substituted by ZnF_2. The regions of glass formation in Pb-systems are similar to Ba-systems (Figure 3(a)). However, the Pb-containing melts have lower viscosity and higher crystallisation rates. Glasses were opalescent due to precipitation of small crystals detected by electron microscopy.[17] Interestingly, two separated regions of glass formation with very different properties were detected. One region is near the metaphosphate composition, and the other region is near the fluorides with a molar ratio AlF_3/MF_2~2/3 (M=Ca, Sr, Ba). The phase diagram $CaF_2–AlF_3$[18] showed a low melting eu-

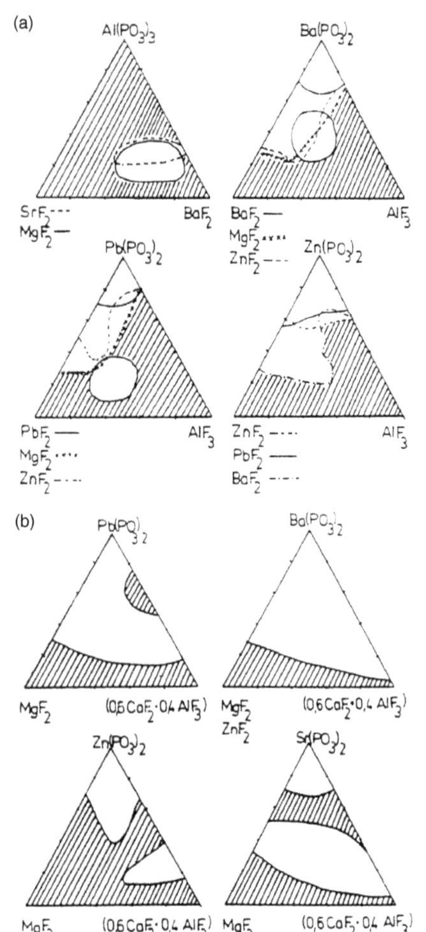

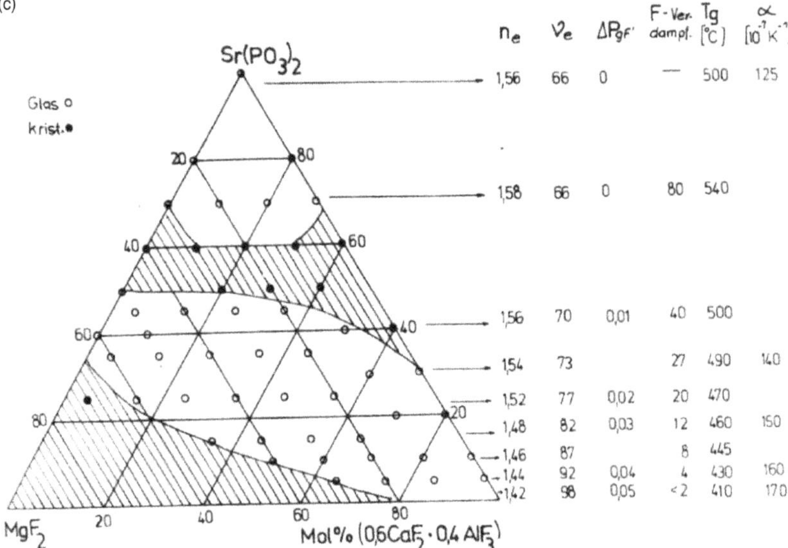

Figure 3. *(a) Glass formation in ternary phosphate-fluoride systems. (The shading indicates the region of no glass formation. The lines correspond to the glass formation boundary for each MF_2 fluoride, as indicated on the bottom left corner of the diagrams); (b) Glass formation in quaternary phosphate-fluoride systems; (c) Glass formation and measured properties in the system $Sr(PO_3)_2$–$(0{\cdot}6CaF_2.0{\cdot}4AlF_3)$–$MgF_2$ (n_e refractive index at 546nm, ν_e dispersion coefficient=Abbe number, $\Delta P_{gF'}$ anomalous partial dispersion, F-Verd. analysed fluorine loss during melting, T_g temperature of glass transformation, α thermal expansion coefficient in $10^{-7}K^{-1}$ T=100–300°C)*

tectic (828°C), $0{\cdot}6CaF_2.0{\cdot}4AlF_3$. We investigated glass formation in quaternary systems with this eutectic composition (Figure 3(b)) and found stable glasses with very interesting properties, shown in Figure 3(c) for the system $Sr(PO_3)_2$–$(0{\cdot}6CaF_2$–$0{\cdot}4AlF_3)$–MgF_2.(16,17) Phosphate rich glasses had very high fluorine loss, in contrast to fluoride rich glasses with low fluorine loss, which decreased with decreasing phosphate content. The refractive index and dispersion also decreased. The anomalous partial dispersion, $\Delta P_{gF'}$, increases with the fluorine content to ~0·05. Commercial CaF_2 single crystal has the values, ΔP_{gF}=0·047, n_e=1·435, ν_e=94·6. However, the problems were to minimise the crystallisation tendency of the interesting fluoride glasses with low phosphate content, and to develop a technology for stable preparation on a larger scale. We studied glass formation and crystallisation in various pure fluoride systems with AlF_3.(16,17,19–29) Figure 4 shows glass formation and crystal fields (I–VIII) of the ternary diagram CaF_2–SrF_2–AlF_3, T~750°C. For the binary system CaF_2–AlF_3, glass formation by quenching was found between the crystal compounds $CaAlF_5$ (T_m=830°C) and Ca_2AlF_7 (T_m=850°C). Glass formation could not be detected in the binary system SrF_2–AlF_3. A large immiscibility gap (ML), the crystals $SrAlF_5$ (T_m=870°C) and Sr_2AlF_7 (T_m=860°C), both forming substitution mixed crystals with Ca^{2+}, and the ternary compound $SrCaAlF_7$ (T_m=775°C), were detected. All crystalline fluoroaluminates have much lower melting temperatures than the simple fluorides, CaF_2 (T_m=1418°C), SrF_2 (T_m=1447°C), AlF_3 (sublimes at 1260°C). The best glass formation was in the eutectic of the ternary system. The crystalline compounds have similar low crystallisation temperatures. The hindrance for the crystallisation process

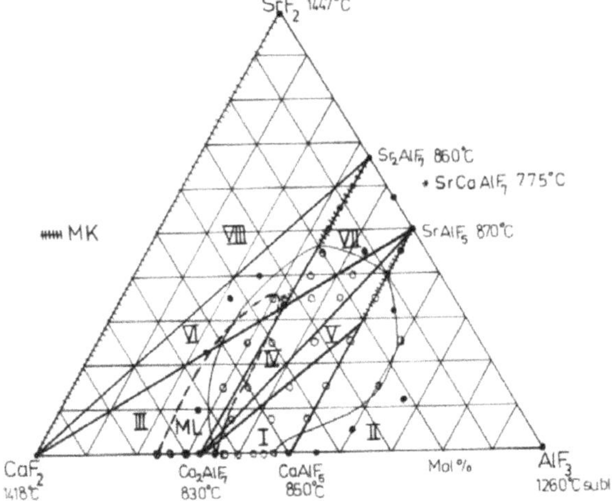

Figure 4. *Glass formation in the ternary fluoride system CaF_2–SrF_2–ALF_3 (open circles indicate glass formation, and closed circles indicate crystallisation), and crystal phases with melting temperatures (MK mixed crystals, ML immiscibility gap)*

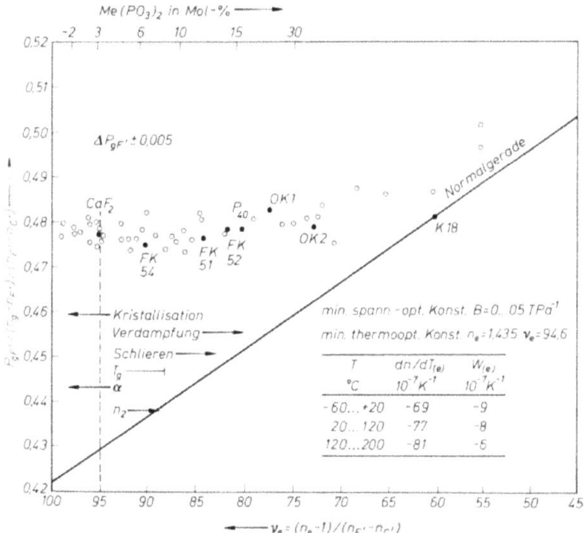

T °C	dn/dT(e) $10^{-7}K^{-1}$	W(e) $10^{-7}K^{-1}$
-60...+20	-69	-9
20...120	-77	-8
120...200	-81	-5

Figure 5. The state in 1982 of the development of optical fluoride phosphate glasses with anomalous partial dispersion (difference from the Normal Line) in the Otto-Schott-Institute of the University of Jena (O) in comparison with CaF_2 single crystals and developments from outside (•) (FK from SCHOTT Mainz and OK from the Soviet Union); insert (left) the changes of crystallisation tendency, volatilisation, striae formation, T_g, linear and nonlinear refractive indices, depending on fluoride/phosphate content; inset (right) the values measured for n_e, ν_e, stress- and thermo-optical constants of one glass similar to CaF_2

is very large. Glass formation during cooling of the melt has a great probability.

The introduction of further fluorides, e.g. MgF_2, BaF_2, LaF_3, YF_3, can enhance glass formation of fluoroaluminate melts. However, the addition of phosphate has the greatest effect. Pure fluoroaluminate glasses have a high crystallisation rate. Glasses are only obtained by quenching the melts between brass plates with a sample thickness around 5 mm. By adding 2 mol% phosphate, glass blocks can be obtained on ~100 g scale. The optical dispersion positions (n_e, ν_e, $\Delta P_{gF'}$) of these glass samples are more extreme than that of CaF_2 crystal (see Figure 5, diagram of the relative partial dispersion $P_{gF'}$ depending on dispersion coefficient ν_e with the 'normal line'). Large bulk glass samples (~5 kg) with 97 mol% fluorides and 3 mol% phosphate were prepared with high optical quality in a special Pt crucible with a down pipe. They have the same optical dispersion position as CaF_2 crystal (Figure 5). They also possess minimal stress constants, B~0–0·5 TPa^{-1}, and minimal thermo-optical constants, dn/dT and W. The values for one glass are given in the inset to Figure 5. The properties of these glasses are mainly dependent on the ratio between fluorides (F) and phosphates (P). Crystallisation, thermal expansion coefficients, and $\Delta P_{gF'}$ values decrease with increasing phosphate content, whereas vaporisation (POF_3, HF),[17,25–27,29] formation of striae, T_g, refractive indices, and dispersion all increase. The crystallisation rate decreases with increasing phosphate content to ~20 mol%. The phosphate groups hinder the crystallisation of fluoroaluminates and fluorides. Fluoride starting materials with low oxide and water content are important to avoid precipitation of $AlPO_4$ crystals. Glasses with >20 mol% phosphate show increasing crystallisation of mono- and diphosphates. This can also be recognised in the separation of the

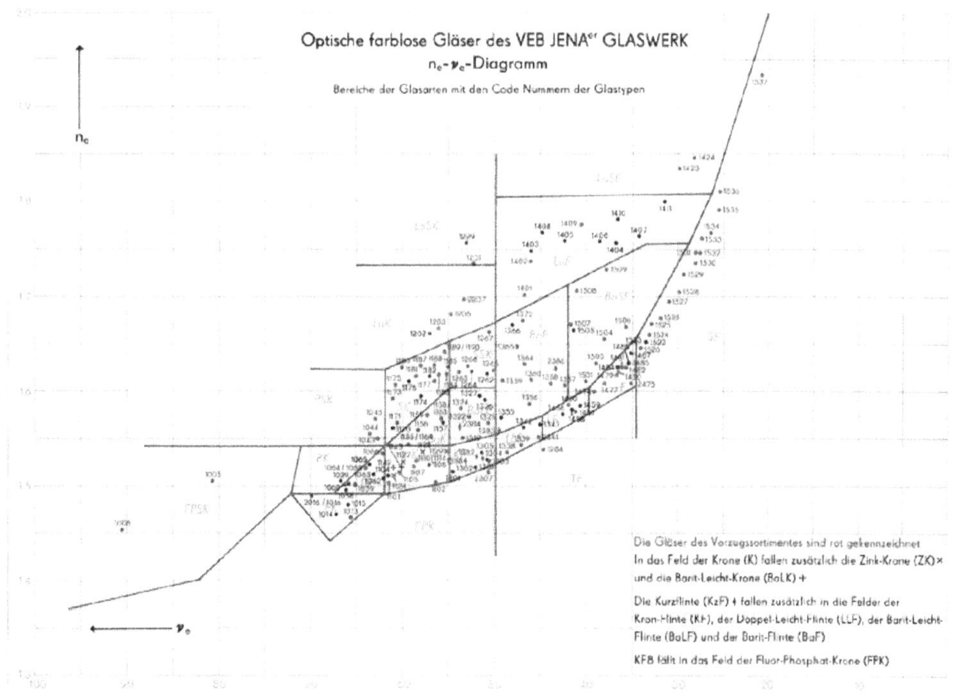

Figure 6. Refractive index (n_e)–dispersion (ν_e) diagram of commercial optical glasses produced by the JENAer GLASWERK 1984 with the new developed glasses FPSK1 (1005) and FPSK3 (1006) [Colour available online]

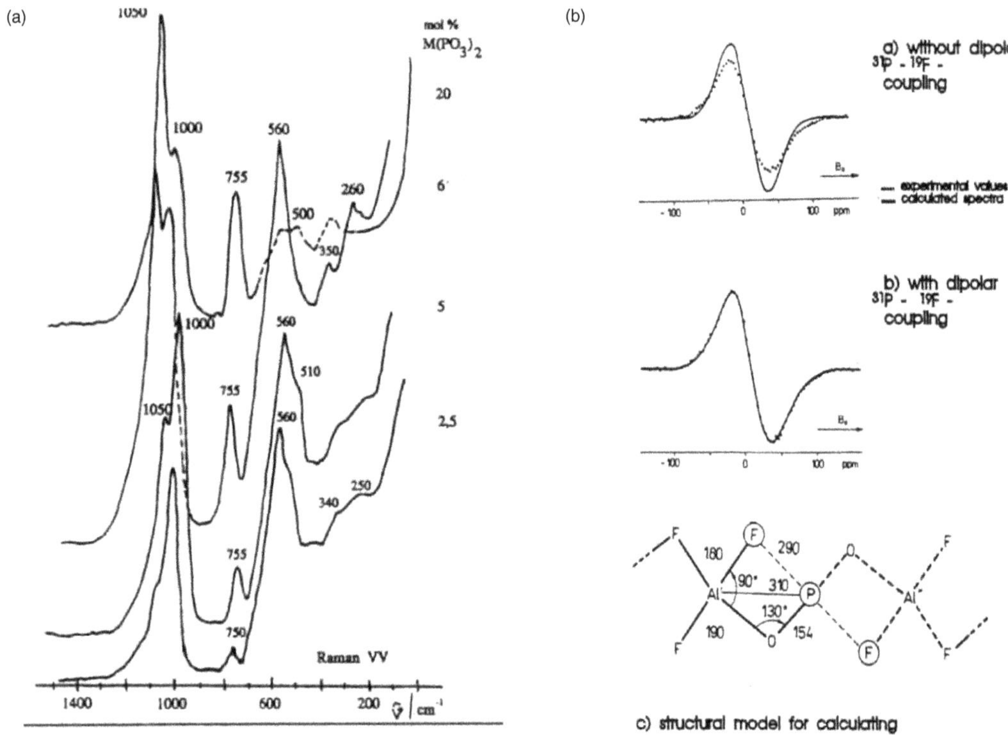

Figure 7. (a) Polarised Raman spectra of fluoride phosphate glasses with increasing phosphate content from 2·5 to 20 mol% phosphate (M=Sr);(b) Experimental and calculated broad line ^{31}P- NMR spectra of a fluoride phosphate glass with 5 mol% phosphate and 95 mol% (0·6CaF$_2$.0·4AlF$_3$), (a) without dipolar ^{31}P–^{19}F-coupling, (b) with ^{31}P–^{19}F-coupling and (c) structural model for calculations

glass forming regions (Figures 3(a), 3(b) and 4). FP glasses also have low nonlinear refractive indices (n_2), minimal stress and thermo-optical constants, which are important for high performance optics. Optimised glass samples were developed and as FPSK1 (n_e=1·50620, ν_e=80·7, ΔP_{gF}=0·027) and FPSK3 (n_e=1·45223, ν_e=90·3, ΔP_{gF}=0·038) were commercially produced by JENAer GLASWERK in 1984 (Figure 6).

Structure and dispersion of fluoride phosphate glasses

The first structural classification of crystalline fluoroaluminates was given by Pabst in 1950.[30] He assumed that these structures are based on AlF_6 octahedra, which may (depending on the F to Al ratio) either be discrete units, as in Na_3AlF_6, or joined in chains, sheets, frameworks (in which the diametrically opposite corners are shared) and other groupings, much as silicate structures are based on linked SiO_4 tetrahedra. He tried to compare and contrast the known results with the general features of silicate structures. MF_4 tetrahedra play only a minor role. The glass forming ability is very low because, in contrast to oxygen bonds, fluorine bonds are ionic, which causes very low viscosity at the melting point. Moreover, the variability of the bond angle between connected octahedra is lower than between connected tetrahedral groups, which promotes the crystallisation process of the melts.[17,31] We have assumed from different measurements of the structure that our fluoroaluminate glasses have a chain-like structure of AlF_6 octahedra, similar to the crystal phases.[16,17,28,29,31–38] However, it was surprising that the addition of very small amounts of phosphate affects the crystallisation behaviour so drastically. The polarised Raman spectra (Figure 7(a)) of fluoride phosphate (FP) glasses with 2·5 to 20 mol% $M(PO_3)_2$ have very strong bands at 1000 cm^{-1} associated with PO_4, at 1050 cm^{-1} and 755 cm^{-1} associated with P_2O_7, and at 560 cm^{-1} associated with $Al(F,Ø)_6$ groups, and the frequencies are shifted slightly compared to the corresponding bands for crystalline compounds. The intensity of the 560 cm^{-1} band decreases and the 1050 and 755 cm^{-1} bands increase with increasing phosphate content, and dominate at 20 mol% phosphate. Figure 7(b) shows a comparison of experimental and calculated broad line ^{31}P NMR spectra of a FP glass with 5 mol% $Sr(PO_3)_2$ and 95 mol% (0·6CaF$_2$.0·4AlF$_3$) without and with dipolar ^{31}P–^{19}F coupling. In the latter case, the experimental and calculated curves are identical. Structurally, this means that the phosphate and fluoroaluminate groups are joined directly.[17,28,29,34] This increases the glass forming ability, preventing the crystallisation of various fluorides. Figure 8 shows the typically extreme viscosity behaviour of a FP melt (FP10, which contains 90 mol% fluoroaluminates, and 10 mol% phosphate) in comparison with a phosphate melt P (NSP=10Na$_2$O.40SrO.50P$_2$O$_5$) and a sodium silicate melt NS (33Na$_2$O.67SiO$_2$). The T_g values for

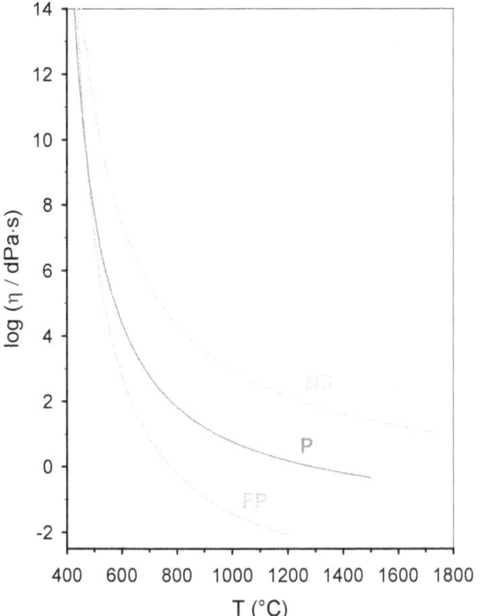
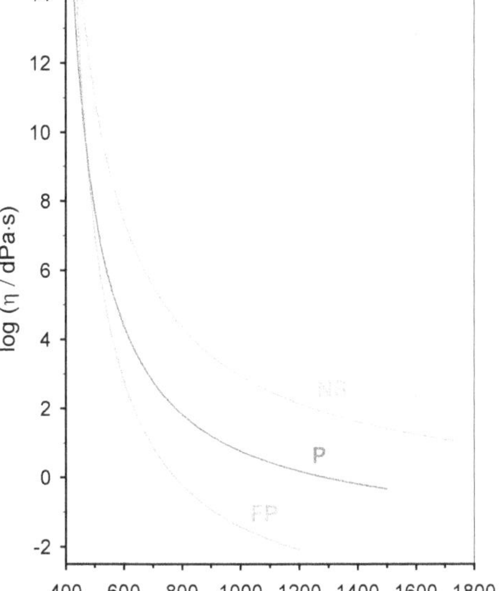

Figure 8. Viscosity curves for different melts: FP=FP10, P=NSP, NS=$Na_2O.2SiO_2$ [Colour available online]

the three glasses are very similar, but the viscosity values of their melts are very different. Fluoride melts are atypical glass forming systems; their strong ionic bonding character results in a very low viscosity of the melt at 1200°C, η~10^{-2} dPa s. This viscosity is comparable with that of water at room temperature. These melts have a very strong viscosity increase in a narrow temperature range at T<700°C, and a great tendency to devitrify during cooling. Crystallisation can be avoided by selected eutectic compositions. The viscosity values of the more covalent phosphate and silicate melts are much higher, η~10 dPa s, and η~10^2 dPa s, due to the different structures.

Table 2. Composition and measured properties of typical fluoride phosphate (FP) glasses

mol%	FP 0 (FA)	FP 3	FP 10	FP 20	P100
$Sr(PO_3)_2$	-	3	10	20	100
MgF_2	10	9·5	10	10	-
CaF_2	28	27·5	30	22	-
SrF_2	23	22·5	15	18	-
AlF_3	39	37·5	35	30	-
T_g (°C) ±5	400	420	440	480	490
n_e ±0·00002	1·4050	1·42721	1·46069	1·50455	1·56140
n_e ±1	105	97	89	80	66
r (g/cm^3) ± 0·001	3·42	3·50	3·46	3·52	3·20
VUV edge	150 nm	155 nm	160 nm	165 nm	180 nm
band gap	8·2 eV	8·0 eV	7·7eV	7·5 eV	6·8 eV

Typical FP glass compositions and properties are given in Table 2. The values for the VUV (vacuum ultraviolet) transmission edge and band gap were

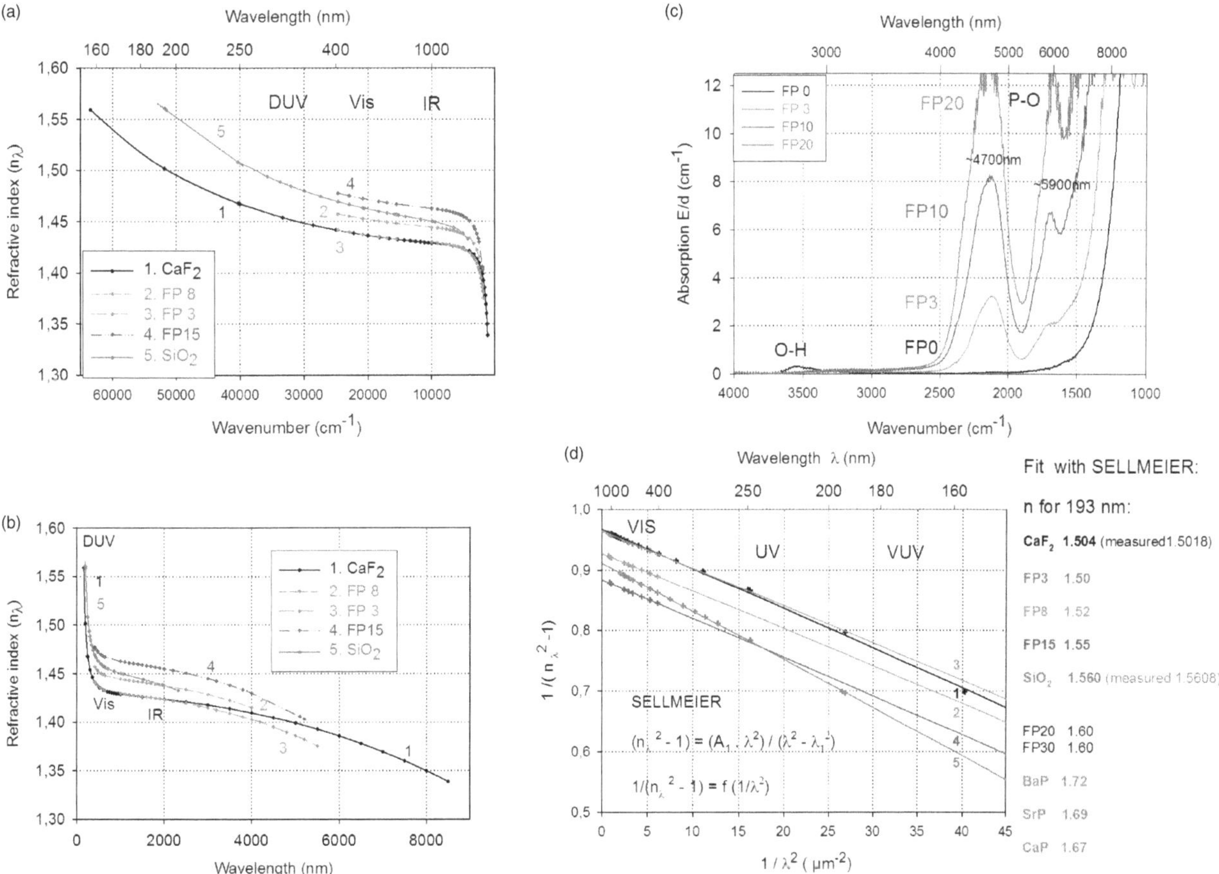

Figure 9. (a) Measured refractive indices for glass samples FP3, FP8 and FP15, compared with a commercial SiO_2 glass and a commercial CaF_2 single crystal as a function of wavenumber in the region from DUV to IR; (b) The same values from (a) as a function of wavelength from DUV to IR; (c) Normalised IR absorption of FP glass samples; (d) Sellmeier plot of measured n_λ values to determine the refractive index at the laser wavelength 193 nm for various fluoride phosphate (FP) and metaphosphate (P) glass samples [Colour available online]

determined by VUV measurements with very thin samples.[36–40] The VUV transmission edge of FP glasses shifts with increasing phosphate content from 150 nm (8·2 eV) in FP0 to 165 nm (7·5 eV) in FP20, and 180 nm (6·8 eV) in pure metaphosphate, $Sr(PO_3)_2$. The value for commercial single crystal CaF_2 (referred to below as CaF_{2c}) is 125 nm (10·0 eV), and the value for amorphous SiO_2 is 150 nm (8·2 eV). The measured dispersion curves from DUV (deep UV) to IR of CaF_{2c} and amorphous SiO_2 (SCHOTT Lithotec Jena) are compared with three FP glass samples as a function of energy in Figure 9(a), and as a function of wavelength in Figure 9(b). This comparison shows that the glass FP3 has identical dispersion to CaF_{2c} in the visible region, but a stronger decrease in the IR due to the P–O overtone and combination vibrations which increase with increasing phosphate content (Figure 9(c)). The OH content of FP glasses is very low, absorption coefficient (or extinction) $E_l \approx 3000$ nm<0·1 cm^{-1}, which has no significant effect on the dispersion curves. The measured dispersion curves can be fitted by a one-term Sellmeier dispersion formula,[37] with good agreement from NIR (~1 μm) to VUV (Figure 9(d)). It is possible to calculate the refractive index for shorter wavelength, e.g for the excimer laser wavelength 193 nm, which is important for microlithography equipment.

Deep UV transmission, trace impurities of polyvalent ions, and redox behaviour

The demand for deep UV (λ<300 nm) transmitting materials increased at end of the 1980s for lens systems in microlithography equipment, for windows, lens blanks and substrate materials for excimer laser system optics, for space applications, for fibre optics, and for solid state laser applications. Fluoride single crystals and vitreous silica are well known traditional materials for UV optics. Crystal sizes are limited and glass is better for fabricating optical surfaces with high accuracy. Vitreous silica can be produced with very high purity by the CVD process. However, temperatures greater than 2000°C are necessary. Moreover, for lens systems, a variety of glasses with different refractive indices and dispersions is required. Fluoride phosphate glasses are attractive candidates as UV transmitting materials. Their values for the UV resonance wavelength, λ_{UV}, using a two-term Sellmeier dispersion formula and the measured transmission edges of thin samples in the VUV, are comparable with those of silica and CaF_{2c}.[16,17,36–38] The intrinsic absorption in the VUV is due to electron transitions. However, the actual UV transmission of glass is frequently limited by trace impurities, mainly polyvalent ions in different redox states, introduced by raw materials and possibly contamination from the melting technique and method of processing. We have studied this effect very extensively, and determined the specific extinction (absorption) coefficients, ε_λ per ppm (weight), for many ions in different redox states in FP glass,[36,38,41,46] as given in Table 3. The UV absorptions are due to allowed charge transfer or s–p transitions of electrons with very high intensity. They are very sensitive to the surrounding matrix. In most cases, the ε values of FP glasses show a small increase with increasing phosphate content. For Pb^{2+} (s–p transition), a significant shift to longer wavelength was observed due to the higher covalence of the bonding character and higher optical basicity.[38] The d–d or f–f transitions of electrons in the visible and near IR range are forbidden and have intensities that are a few orders of magnitude lower than for the allowed electron transitions. The effect of the different redox states, Fe^{3+} and Fe^{2+} in the ppm range, on the UV absorption and transmission of FP glass is demonstrated in the Figures 10(a) and (b). For Fe^{3+}, ε has a peak value pf 0·18 $cm^{-1}ppm^{-1}$ at 260 nm, and this value is

Table 3. Absorption coefficients for the redox states of polyvalent ions in FP10 glass

Ion/electronic configuration	*Wavelength maximum λ (nm)*	*Wavenumber $1/\lambda$ (cm^{-1})*	*Energy (eV)*	*$\varepsilon_\lambda=E_\lambda/(cd)$* ($cm^{-1}ppm^{-1}\times10^4$)*
Ti^{4+} $3d^0$	225	44400	5·5	4500
Ti^{3+} $3d^1$	526	19000	2·4	
	670	15000	1·9	
V^{3+} $3d^2$	215	46500	5·8	1000–1500
	435	23000	2·9	~9
	678	14700	1·8	~5
V^{4+} $3d^1$	<200	>50000	>6·2	>1000
	670	15000	1·9	~3
	890	11200	1·4	~10
Cr^{3+} $3d^3$	175	57140	7·3	5000
	290	34500	4·3	weak
	440	22730	2·8	13
	640	15600	1·9	11
Fe^{2+} $3d^6$	170	58800	7·3	2000
	220	45450	5·6	55
	1050	9500	1·2	3
	1950	5100	0·6	2
Fe^{3+} $3d^5$	185	55500	6·9	2500
	260	38500	4·8	1800
Co^{2+} $3d^7$	<190	>52600	>6·5	
	525	19000	2·4	10
	570	17500	2·2	13
	1400	7100	0·9	weak
Ni^{2+} $3d^8$	170	58800	7·3	2000
	425	23500	2·9	8
	830	12000	1·5	2
	1500	6700	0·8	1
Cu^{+} $3d^{10}$	175	57140	7·1	4000
	230	43500	5·4	400
Cu^{2+} $3d^9$	180	55500	6·9	1500
	230	43500	5·4	1200
	830	12000	1·5	15
Ce^{3+} $4f^1$	208	48000	6·0	40
	220	45500	5·6	65
	245	40800	5·1	100
	260	38500	4·8	120
Ce^{4+} $4f^0$	270	37000	4·6	110–260?
Pb^{2+} $6s^2$	205	48800	6·0	900–1100

* ε_λ=specific absorption coefficient of the ion at the wavelength λ; E_λ=absorption or extinction at the wavelength λ; c=concentration in weight ppm; d=thickness of sample in cm

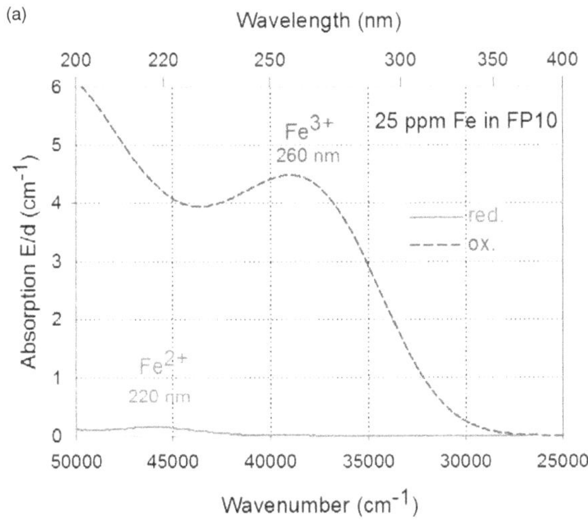

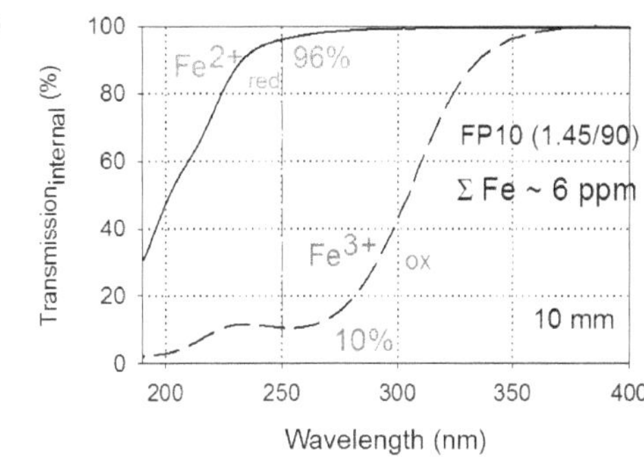

Figure 10. (a) The normalised UV absorption spectra of FP10 glass samples with 25 ppm Fe′ in different redox states as Fe^{3+}(ox.) and Fe^{2+}(red.) depending on melting conditions; (b) The effect of the redox state, Fe^{3+}/Fe^{2+}, on the UV transmission of a FP10 glass with a low total Fe′-content, 6 ppmw, and sample thickness 10 mm [Colour available online]

much higher than the peak value of 0·0055 $cm^{-1}ppm^{-1}$ at 220 nm for Fe^{2+}. This leads to large differences in the internal transmission, ranging from 10 to 96 % at around 250 nm, for 10 mm thick FP glasses with very low total Fe′ content, 6 ppm, depending on the oxidising or reducing melting conditions. Similar behaviour was found for Cu^{2+} and Cu^{+}. The redox behaviour of polyvalent ions in FP glasses is strongly dependent on the melting conditions and temperature.[38–49] If FP glasses are melted under reducing conditions in carbon crucibles under an argon atmosphere at 1000–1050°C, the redox state can be nearly completely shifted to Fe^{2+} and Cu^{+}, which were the main trace impurities of fluoride and phosphate raw materials, and a very high UV transmission can be reached. At higher melting temperatures, other reduced species can be formed, e.g. P^{3+} which also absorbs in the UV,[44] or Cu^{0} which causes different effects. The drastic effect of melting conditions and temperatures on FP10 glass samples doped with 0, 100, 500, 1000, 10000 and 20000 ppm Cu′ (as CuF_2) is shown in Figure 11.[45] The redox ratio of Cu^{2+}/Cu^{+} in glasses prepared under normal conditions (Pt crucible, air, 1000°C) is ~80/20 for all doping concentrations. Very strong changes were observed under reducing conditions (remelting, carbon crucible, argon atmosphere), depending on melting temperature, 900 to 1200°C. Cu^{0} and Cu_2O colloids, and metallic copper crystals were observed at temperatures >1100°C (red colour). Surprisingly, Cu_2O and Cu^{0} particles were aggregated on the surface of the glass samples at high Cu doping (10000 and 20000 ppm≡1 and 2 wt% Cu) at 1200°C; the glass inside was colourless and clear with similar low refractive index like the undoped glass.

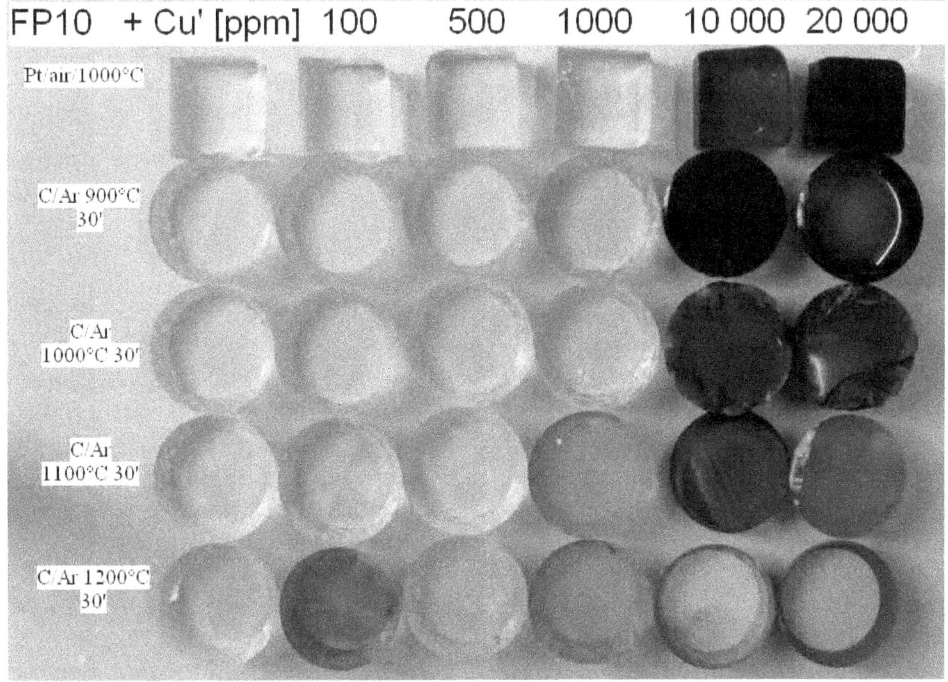

Figure 11. The effect of melting conditions on the Cu′ redox state of FP10 glass samples doped with 0, 100, 500, 1000, 10000 and 20000 ppmw Cu′ (sample thickness 10 mm) [Colour available online]

The redox behaviour of other polyvalent ions in glass forming melts was studied by various spectroscopic methods, e.g. optical absorption and photoluminescence, EPR, and Mössbauer. The thermodynamics of redox equilibria and diffusion of polyvalent ions was investigated in phosphate melts, NSP ($10Na_2O.40SrO.50P_2O_5$) doped with 0·5% ions.[49–51]

Radiation induced defects

When glasses are subjected to ionising radiation, the principal effects result from electronic processes. Electrons are excited to leave their normal position and move through the glass network. For each ionising electron, one electron deficient region or hole centre is formed. Absorption bands observed are due to centres formed by electron trapping (EC) or hole trapping (HC). If the absorption bands are in the visible region, the glass becomes coloured. Radiation induced defects are strongly dependent on glass composition, impurities, polyvalent ions, and radiation sources. This has been known for a long time and investigated by many authors from all over the world.[11,52]

We have studied these phenomena for different phosphate, fluoride phosphate, silicate, and borosilicate glasses, which have high intrinsic UV transmission, using different radiation sources, over a long period, of more than 15 years.[52–83] It is important for the applications of glasses in optics and photonics. Glass samples were prepared very carefully with very high purity raw materials, doped with different ions, and well characterised by various methods. X-ray radiation from a Cu-cathode with λ=0·154 nm was used to detect the principal possible radiation induced defects in the glass samples. High energetic x-rays and γ-rays usually cause partial rupture of chemical bonds, partial destruction of the network, reduction of ionisation of specific ions, and the introduction of all possible defects. Less energetic UV radiation can cause different effects by excitation of outer electrons, with a dependence on wavelength and intensity. UV-laser (XeCl 308 nm, KrF 248 nm, ArF 193 nm) can be used for selected excitation, e.g. in the absorption bands of polyvalent ions or the band gap of the glass network. The intensity of the laser radiation, pulse duration (ns, fs), and pulse number were varied over a wide range. The radiation induced optical absorption was measured, depending on time and intensities. The spectra were simulated by separation of optical absorption bands (Figure 12(a)). EPR measurements were carried out to detect paramagnetic centres and correlated with optical absorption bands.[44] Only phosphorous-related defects, POHC, EC, and oxygen defects, OHC, could be detected in phosphate and fluoride phosphate glasses. The kinetics of defect formation

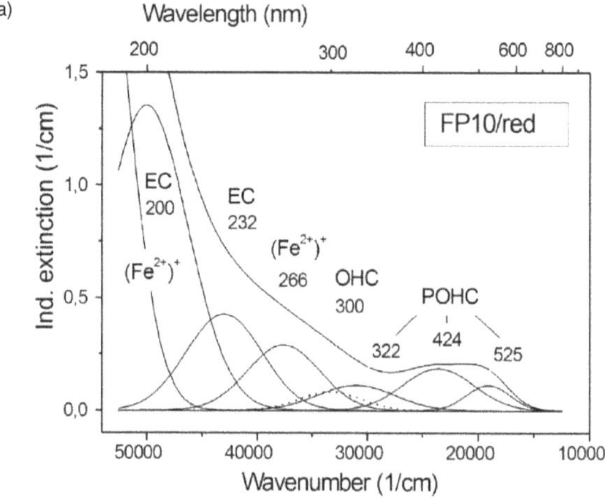

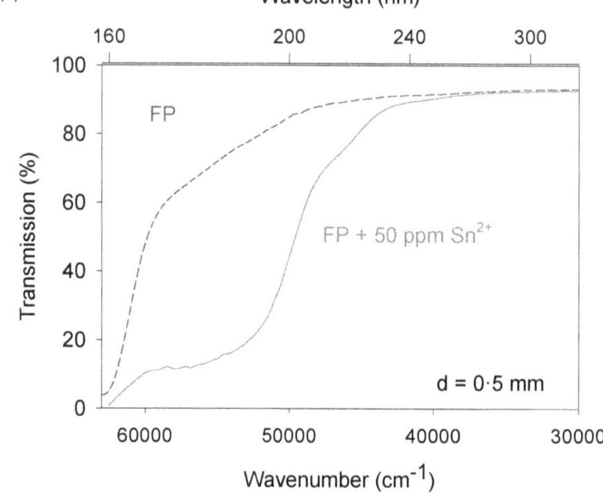

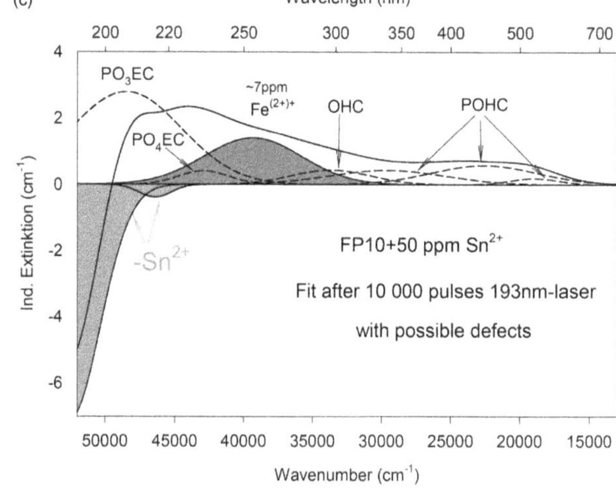

Figure 12. (a) The 193 nm laser induced absorption in a FP10 glass sample with simulation of various defect centres; (b) VUV transmission spectra of FP10 glass samples with and without 50 ppm Sn^{2+}; (c) 193 nm laser induced absorption in FP10 glass doped with 50 ppm Sn^{2+}, with defect centres [Colour available online]

and healing, by room temperature, thermal treatment or radiation, were investigated. The interaction between intrinsic and extrinsic defects was studied. Both the intrinsic UV transmission and the intrinsic UV radiation resistance of glasses can be correlated with the optical basicity (electron donor power) of the glass matrix. Pure fluoroaluminate glasses have the lowest optical basicity, the largest intrinsic UV transmission, and the largest UV radiation resistance. The radiation induced defects increase with increasing phosphate content of FP glasses. Usually, glasses prepared under reducing conditions showed stronger defects under UV radiation. Photo-oxidation of Fe^{2+} to $(Fe^{2+})^+$ hole centres was observed in all glasses increasing with increasing phosphate content in FP glasses, which causes an increase of absorption near 250 nm (Figure 12(a)). Sn^{2+} ions were photo-oxidised very fast, leading to a decrease of absorption near 200 nm, and to an increase near 250 nm (Figure 12(b) and (c)). Both Sn^{2+} and Sn^{4+} were involved in the radiation induced process. Extrinsic defects from impurities or doped ions affect the intrinsic defect formation by photo-ionisation. Intrinsic and extrinsic defects are competitors. The kinds of defect can change during formation and healing. The mechanisms are very complicated, with maxima and minima in the defect formation and healing curves. Detailed results are published in our papers.[52–80] Many of our results on photo-ionisation of polyvalent ions in phosphate, fluoride phosphate, and borosilicate glasses are summarised in Ref. 81. Laser and e-beam writing of integrated optical structures has been described in Refs 82 and 83 (Figure 13).

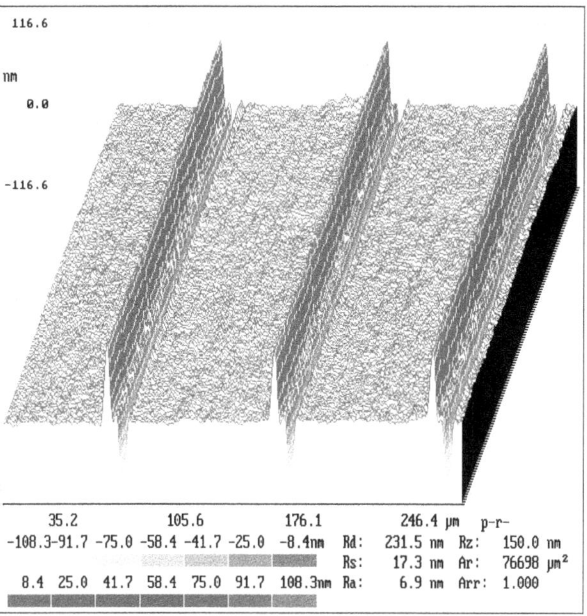

Figure 13. 3D shear image of refractive index modification of three fs-laser-induced tracks in the volume of a FP20 glass sample [Colour available online]

Laser and amplifier glasses

Phosphate and fluoride phosphate glasses doped with rare earth ions are interesting for applications as laser and amplifier glasses.[30,84–109] Lasers which emit in the 1·5–1·6 μm band can be used in optical communication, medicine (eye safety), meteorology, navigation, etc. One important possibility to generate such wavelengths is the use of the Er^{3+} $^4I_{13/2}$–$^4I_{15/2}$ electron transition in different crystal and glass hosts, a transition explored from the early days of solid state laser research.[30] Due to its three-level nature, lasing can be obtained only for rather low erbium concentration, <1%. Generally, f–f absorption bands are very weak because they are strongly forbidden by parity selection rules. Therefore, the weakly absorbing transitions of the low Er^{3+} content cannot ensure efficient absorption of the pump light, especially by flash lamp pumping. Co-doping with suitable sensitizers can increase the efficiency. CW-laser action of Er^{3+} at room temperature was obtained for the first time (to our knowledge) in our double sensitized fluoroaluminate glass by efficient two-step energy transfer pumping via Cr^{3+} (<0·2%)–Yb^{3+} (~10%)–Er^{3+} (<0·2%) in 1988.[84] Exploitation of the higher efficiency Nd^{3+}, $^4F_{3/2}$–$^4I_{11/2}$ transition,[88,94] lasing at 1·05 μm, was achieved by co-doping with Ce^{3+}–Cr^{3+} and energy transfer.[85,92,93] The disadvantages of flash lamp pumping, the low efficiency and thermal problems due to Cr^{3+} absorption, could be prevented by laser diode pumping, following progress in the development of these lasers. With laser diode pumping between 940 and 980 nm in the broad Yb^{3+} absorption (Figure 14(a)), and efficient energy transfer to Er^{3+}, efficient laser action at 1·5 μm with very large (~100 nm) and smooth tuning range was demonstrated in fluoride phosphate glasses. The lifetime of the upper level is 9·5 ms. A higher degree of line broadening and smoother line shape than with pure phosphate glass can be reached in fluoride phosphate glasses by mixed fluorine and oxygen surrounding of Er^{3+}. Fluoroaluminate glasses with erbium and ytterbium, and various phosphate content, offers good possibilities in broadband amplification with large tuning range, and are very attractive for use in wavelength division multiplexing systems. Erbium doping levels of up to 1×10^{20} cm^{-3} could be realised in FP20 glasses, before upconversion into higher levels quenches population inversion and efficiency for the 1·5 μm region considerably.[99,100] Upconversion increases with decreasing phosphate content in FP glasses and can be used for lasing in the visible range, ~540 nm (green).

Yb^{3+}-doped materials have attracted increasing interest for applications in diode pumped laser systems in the near infrared region, ~1 μm, similar to the Nd^{3+} ion. In comparison with other active rare earth ions, Yb^{3+} has the simplest energy diagram, avoiding such unwanted processes as excited state absorption, upconversion, and concentration quenching. Its long lifetime and very broad absorption and emission bands are advantages for the generation of short laser pulses. Efficient CW-laser operation of Yb-doped fluoride phosphate glasses at room temperature was demonstrated, in 1995.[101] Femto-

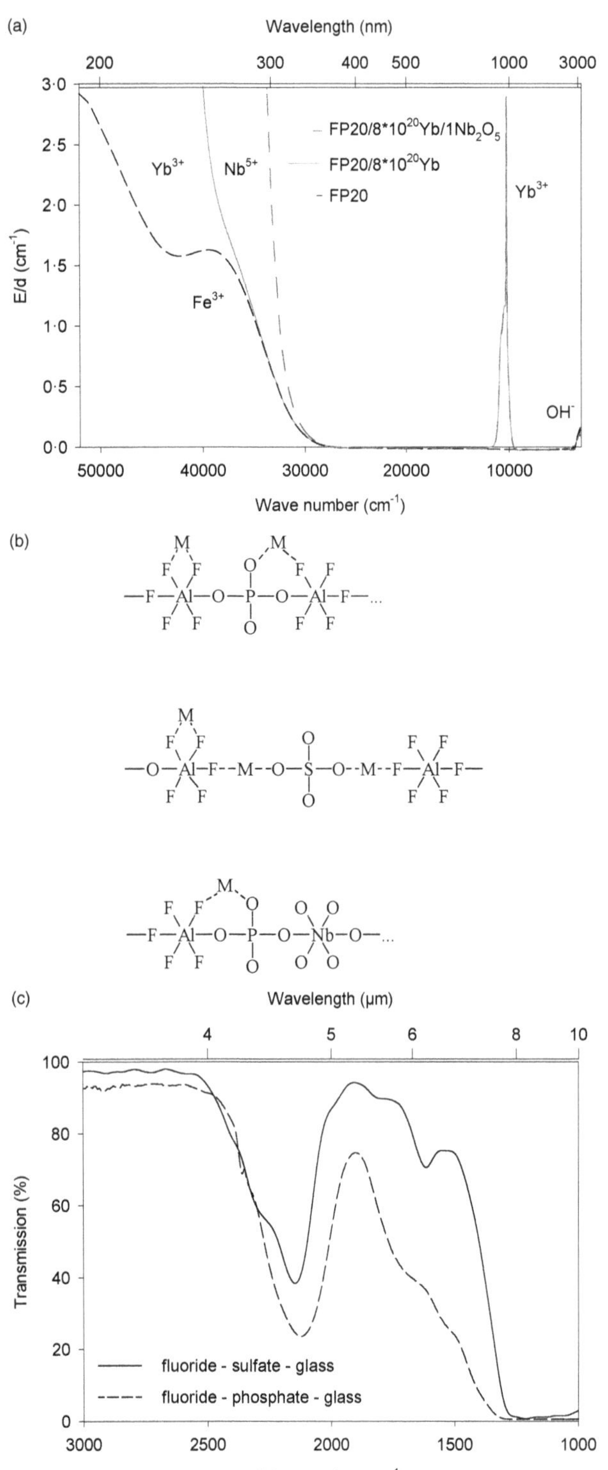

Figure 14. (a) Normalised absorption spectra of different FP20 glass samples; (b) Schematic structural models for several fluoride phosphate glasses: pure FP with incorporated mono-phosphate, sulphate, and Nb(V) group; (c) IR absorption spectra of a fluoride-sulphate (FS10, thickness 0·5 mm) and a fluoride phosphate glass sample (FP5, thickness 1 mm) [Colour available online]

second (fs) laser pulses with a bulk Yb^{3+}-FP glass were generated in 1997,[102] both to our knowledge for the first time. The preparation and properties of Yb-FP glasses of very high optical quality (optical homogeneity $\Delta n<10^{-6}$) on a medium scale were achieved by developing special technology.[31,103–106] The structure and properties could be changed by variation of the phosphate content, doping other components, like niobium oxide or sulphates which can substitute for $Al(F,\varnothing)_6$ or PO_4 groups and affect the surrounding of Yb^{3+} ions, and may increase the cross-section by enhancement of the asymmetry of the local environment of Yb^{3+} sites (Figure 14(a)–(c)). In contrast to many other solid state laser materials, high doping levels are possible without reduction of the upper state lifetime, 1–2 ms, depending on phosphate content (Figure 15(a)). Laser performance was investigated as well as laser tuning characteristics. Yb^{3+}-FP laser glasses have much larger tuning ranges, 44–48 nm, than commercial phosphate glasses QX (Kigre Inc.) with 38·5 nm (Figure 15(b)). The laser tuning range for Er^{3+}-FP glass is also much broader than commercial phosphate glasses. This is due to the mixed F/O ligands around Yb^{3+} ions in FP glasses. The appropriate Yb-FP laser glass for chirped pulse amplifier (CPA) experiments was selected. Nonlinear optical effects in laser glasses can easily be observed

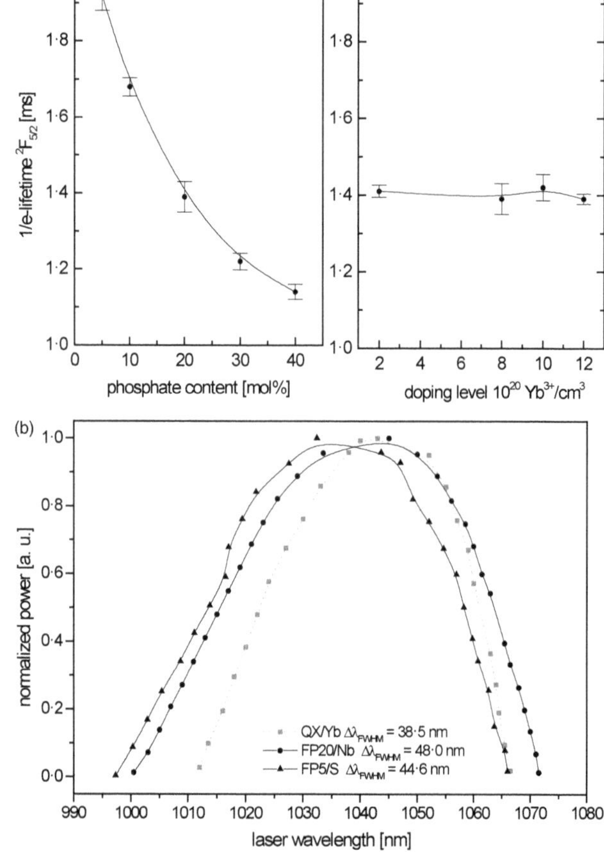

Figure 15. (a) Yb^{3+} fluorescence lifetime for FP glasses with varying phosphate content and constant doping level, 8×10^{20} Yb/cm^3, (left), and for FP20 glasses with several doping levels (right); (b) CW-tuning range (Yb^{3+}) for FP5/S, FP20/Nb, and a commercial phosphate glass, QX (Kigre Inc) [Colour available online]

when the laser intensity is increased. With shortening of the laser pulse duration from nanoseconds (ns) to picoseconds (ps) and femtoseconds (fs), extremely high laser energies can be reached, corresponding to peak power levels to 10^{15} W (petawatt). Laser induced instantaneous and permanent self focusing, multiphonon absorption, ionisation, and instantaneous and permanent point defect formation are possible. However, they must be avoided in materials for applications as high power laser systems with high efficiency, stability and reliability.[31] Yb^{3+}-FP glasses have many advantages, small linear and nonlinear refractive indices, large band gap, high radiation resistance, small phonon energy, long fluorescence lifetime, and low OH content, making them attractive for the generation of short pulses (fs) and high peak powers (petawatt), as required for numerous applications, e.g. in the POLARIS (petawatt optical laser amplifier for radiation intensive experiments) system.[105–109]

Photoluminescence in the UV–VIS region

Luminescence of solid state materials is an optical property that has been known for a long time, and has been used for many classical and new applications. In addition to the applied interest, this research field is also significant for its contribution to fundamental research on local structures. Luminescence can be excited by many types of energy. Photoluminescence is excited by electromagnetic, often ultraviolet, radiation. It is a very sensitive optical property. In most cases, the system consists of a host matrix and a luminescent centre, often called an activator, which absorbs the exciting radiation and is raised to an excited state, subsequently returning to the ground state by the emission of radiation. Phosphate and fluoride phosphate glasses are interesting hosts for luminescent species.[110–123] It can be demonstrated that ions, in the ppm concentration range, with s^2 electron configuration (As^{3+}, Sb^{3+}, Sn^{2+}, Pb^{2+}, Bi^{3+}), with d^0 configuration (Ti^{4+}, Nb^{5+}, Ta^{5+}, Mo^{6+}, W^{6+}), and with d^{10} configuration (Ag^+ and Cu^+) absorb strongly in the UV region and yield broad emission bands mainly in the ultraviolet and visible blue-green spectral region, depending on the electron configuration and optical basicity of the host glass composition (Figures 16(a)–(c)). The largest effect of the host glass matrix was found for the Mn^{2+} ($3d^5$) luminescence.[110–113,116,118] A strong green emission is provided in silicate, and borosilicate glasses (or crystals, e.g. willemite Zn_2SiO_4) with high optical basicity where Mn^{2+} ions are fourfold coordinated with oxygen. Phosphate and fluoride phosphate glasses with lower optical basicity favour a six-fold coordination of Mn^{2+} ions with fluorine and oxygen. The emission band is very broad and shifts from yellow-orange to red with increasing phosphate (Figure 17(a)) and Mn^{2+}

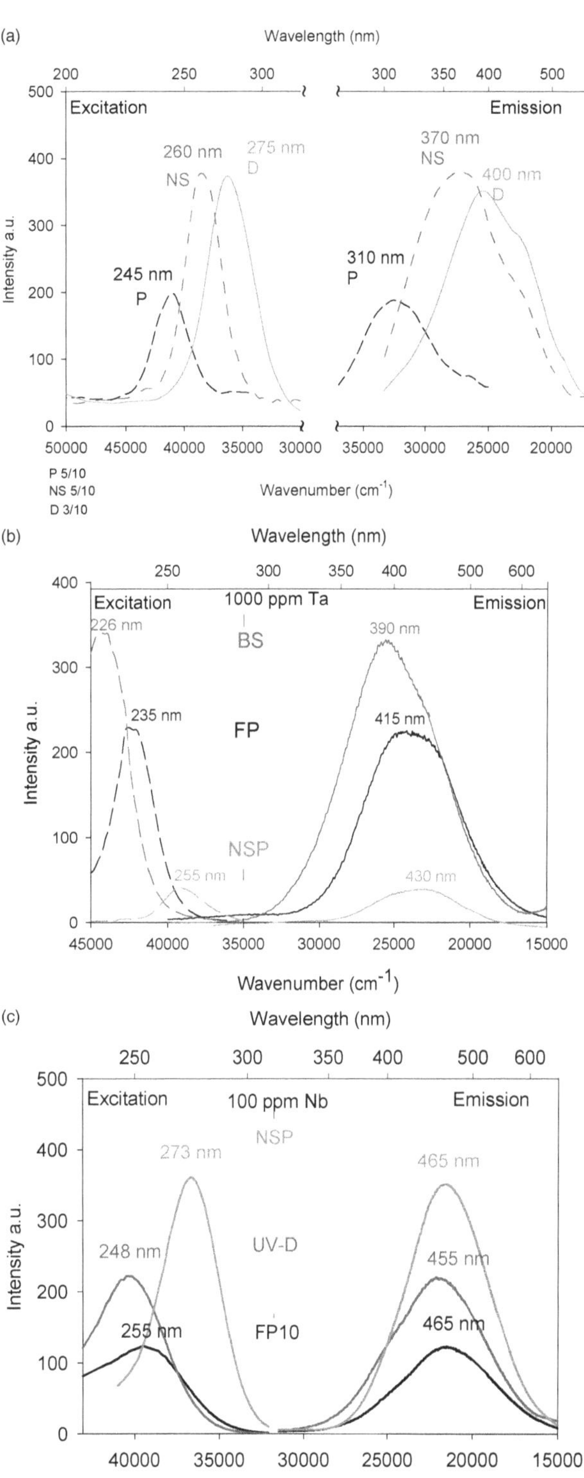

Figure 16. Luminescence excitation and emission spectra. (a) For $Pb^{2+}(6s^2)$ in various glass samples: P=$Sr(PO_3)_2$ doped with 50 ppm Pb^{2+}, NS=$15Na_2O$–$85SiO_2$ doped with 100 ppm Pb^{2+}, and D=borosilicate Duran® doped with 200 ppm Pb^{2+}; (b) For $Ta^{5+}(3d^0)$ 1000 ppm in NSP, FP10 and BS=Duran®; (c) For Nb^{5+} $(3d^0)$ 100 ppm in NSP, UV-Duran®, FP10 [Colour available online]

content. The emission decay curves measured do not have a simple exponential behaviour (Figure 17(b)),[113] due to a distribution of different local sites

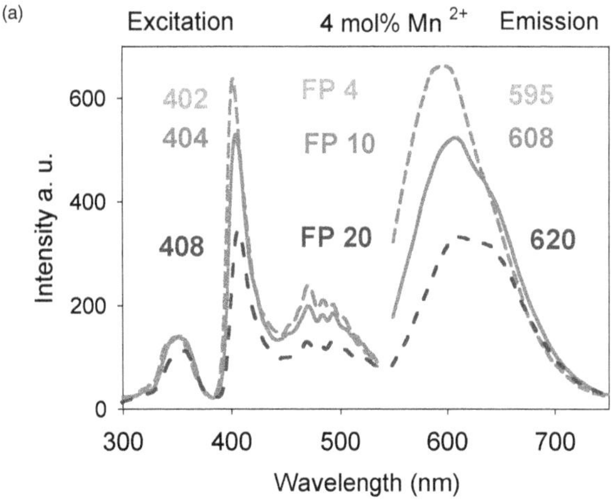

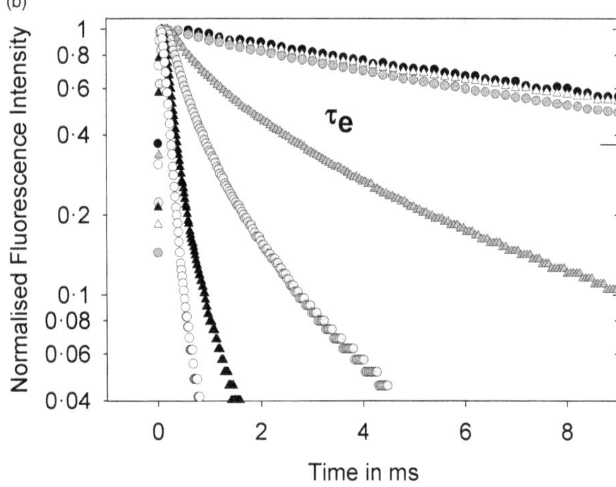

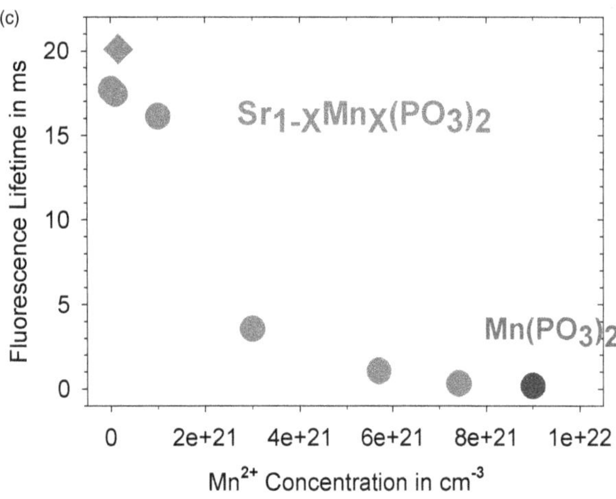

Figure 17. (a) Luminescence excitation and emission spectra of $Mn^{2+}(3d^5)$ in various FP glass samples with 4 mol% MnF_2; (b) Fluorescence decay curves for metaphosphate glasses $Sr_{1-x}Mn_x(PO_3)_2$ with increasing Mn^{2+} content measured at the emission maximum; (c) Dependence of the lifetime, τ_e, on Mn^{2+} content in $Sr_{1-x}Mn_x(PO_3)_2$ glass samples [Colour available online]

for the Mn^{2+} ions in the glass matrix. FP glasses have a very long Mn^{2+} lifetime, τ_e~22 ms, mainly due to spin- and parity-forbidden d–d transitions in the $3d^5$ configuration. The lifetime decreases with increasing phosphate and manganese content (Figure 17(c)). A very short lifetime, τ_e<0·5 ms, was measured at 680 nm in $Mn(PO_3)_2$ glass.

Phosphate and fluoride phosphate glasses have been doped with nearly all rare earth ions, and the static and time resolved photoluminescence have been investigated as a function of glass composition and doping concentration. Rare earth ions with a $4f^n$ electron configuration are characterised by incompletely filled 4f orbitals that lie inside the ions and are shielded from the surroundings by the filled $5s^2$ and $5p^6$ orbitals. Therefore the influence of the host matrix on the optical transitions within the $4f^n$ configuration is small but essential. The ligand-field splitting, only a few times 100 cm^{-1}, is very small in contrast to transition metal ions with a few times 10000 cm^{-1}. Optical absorption transitions are strongly forbidden. The f–f absorption and emission bands are very narrow, and have low intensities with low absorption coefficients. Furthermore, coupling with vibrations has only a

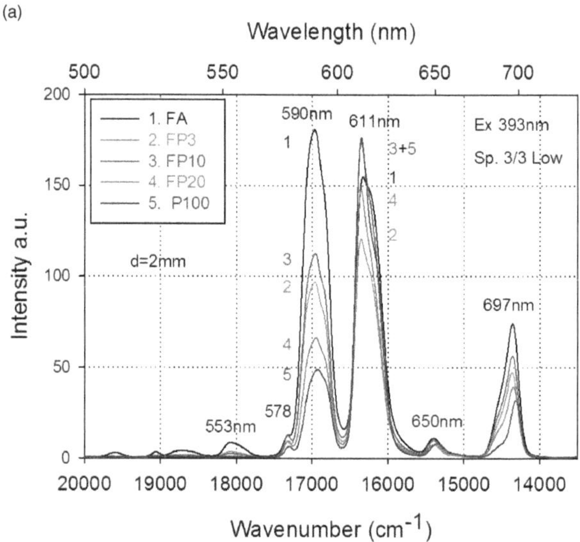

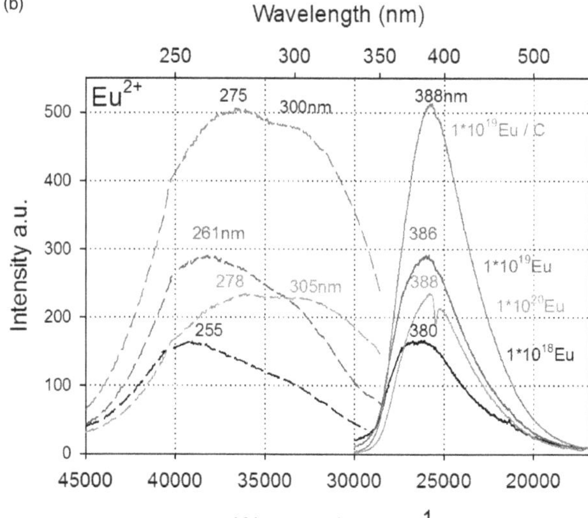

Figure 18. (a) Luminescence emission spectra of Eu^{3+} $(4f^6)$ in various glasses; (b) Luminescence excitation and emission spectra of Eu^{2+} $(4f^7)$ in FP10 glasses with different Eu' concentration [Colour available online]

very weak effect. The allowed f–d transitions, absorb in the UV, and are broader with much higher intensities, and also they provide broader and stronger emission in the UV–Vis region, e.g. Ce^{3+} ($4f^1$) or Eu^{2+} ($4f^7$) with strong blue emission and short lifetime (from nanoseconds to microseconds). The effect of varying the host glass composition (i.e. the fluoride phosphate content) on the photoluminescence of rare earth ions has been demonstrated in different cases, as described below.

It has long been known that Eu^{3+} is an efficient activator for red luminescence emission in crystals and glasses with very low concentration quenching. Figure 18(a) shows the luminescence emission spectra of Eu^{3+} ($4f^6$) in the red spectral region for glass samples with the same Eu^{3+} content. The ratio of the band intensities, 611 nm to 590 nm, changes strongly depending on the fluoride phosphate content. In the pure fluoroaluminate glass, FA, which has the highest intensity, the ratio is about one. With increasing phosphate content, FP3, FP10, FP20, to P200 (pure $Sr(PO_3)_2$), the intensity of the 590 nm band decreases, and the ratio increases to about four. Meanwhile, the lifetime decreases from 9 to 3 ms. Figure 18(b) shows the Eu^{2+} excitation and emission spectra depending on Eu′ concentration in FP10 glass samples. It was only with strong reducing conditions at low Eu′ doping, that it was possible to obtain only Eu^{2+}. In most cases, the glasses had Eu^{3+} ions, which is shown by the re-absorption at 396 nm (1×10^{20} Eu per cm^3). The measured lifetime for Eu^{2+} was 1·3 to 0·2 μs, decreasing with increasing europium and phosphate content.[115]

Tb^{3+} ($4f^8$) is known for green emission at ~540 nm. Figure 19 shows typical excitation and emission spectra of a Tb^{3+} doped FP10 glass sample. The excitation bands in the UV-A region at ~370 nm are very strong, which enables efficient excitation with a weak black lamp providing green emission at 542 nm. This band splits in phosphate glasses, P100. The lifetime for the green emission was 6–3 ms. The blue emission

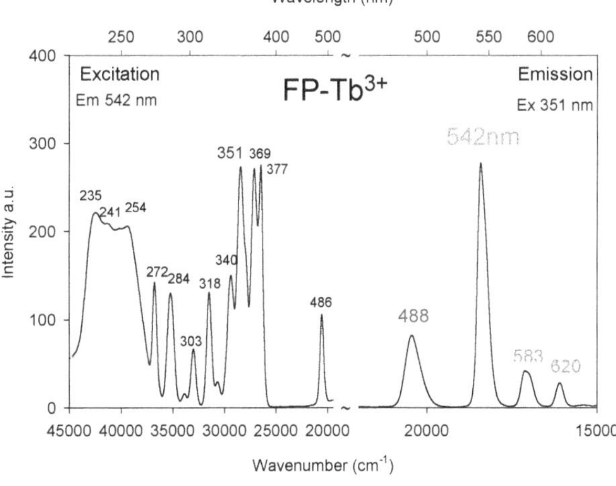

Figure 19. Luminescence excitation and emission spectra of Tb^{3+}($4f^8$) in FP10 glass [Colour available online]

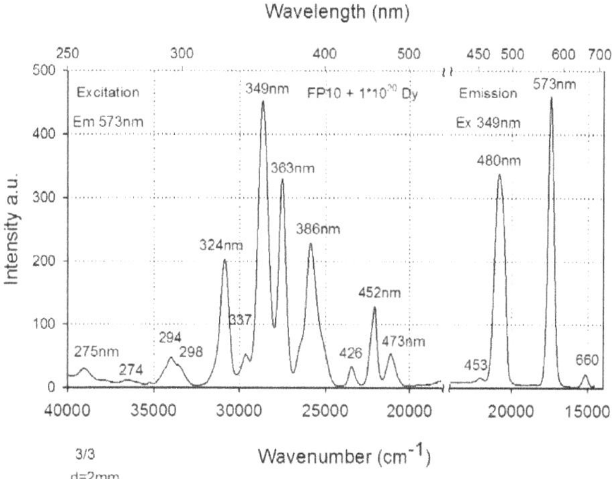

Figure 20. Luminescence excitation and emission spectra of Dy^{3+} ($4f^9$) in FP10 glass

at 488 nm, with much shorter lifetime (0·1–1·0 ms), decreases with increasing Tb^{3+} doping.[113]

Figure 20 shows the excitation and emission spectra of Dy^{3+} ($4f^9$) in FP10 glass. The intensity of the excitation bands at ~350 nm are strong, with typical strong emission in the blue (480 nm) and yellow (573 nm) spectral region. The intensity ratio of yellow to blue emission is strongly dependent on the host glass.[114] The ratio is about one in pure fluoroaluminate, glass and increases with increasing phosphate content to about two in pure phosphate (P100). The lifetime decreases with increasing phosphate and Dy^{3+} content in the range from 7 to 0·1 ms.

The green emission of Ho^{3+} ($4f^{10}$) and of Er^{3+} ($4f^{11}$) at 540 nm has been detected mainly in pure fluoride glasses.[110] Pure fluoroaluminate and fluorozirconate glasses doped with Er^{3+} provide very high green luminescence emission and long lifetime, ~600 μs, due to the surrounding fluorine ligands. The effect of phosphate addition decreases the luminescence intensity and lifetime drastically, due to more oxygen ligands and higher electron–phonon coupling. Strong luminescence quenching has been found with increasing Er^{3+} concentration.[117]

Doped FP glasses with different photoluminescence behaviour are attractive candidates as luminescence standards for analytical measurements using static and time resolved luminescence. They have high radiation resistance, high chemical and long time stability of the luminescence signals, and these properties are useful for many applications in photonics.

Conclusions

The structure and properties of these very interesting glasses depend mainly on the molar relation between fluorides and phosphates, which can be varied in a wide range between pure phosphate and fluoroaluminate glasses. They have a great potential for different applications in the field of optics and photonics. The structure model of fluoride phosphate glasses (FP) can be described as $Al(F,O)_6$ octahedral chains, bonding by mono- and diphosphate groups and cations. Fluoroaluminate glasses with low phosphate content have very interesting optical properties, but unusual chemical and physical properties, which were investigated in detail. In particular, their anomalous dispersion makes them suitable for lens designs in high performance optics with higher resolution. Special fluor-crown glasses were developed on a commercial scale. Their intrinsic VUV transparency is comparable with those of silica and CaF_2 single crystal. The absorption coefficients of possible trace impurities in different redox states were determined. The effect of UV lamp, UV laser, x-ray radiation, and laser writing of waveguides was studied.

Efficient active laser and amplifier glasses, doped with Nd^{3+}, Er^{3+}, and Yb^{3+}, were developed. Their properties were tailored over a wide composition range, with varying phosphate, fluoride and rare earth doping contents. They are attractive laser glasses for the generation of ultra-short pulses and high power amplifiers up to the petawatt-level with growing applications.

Photoluminescence in glasses doped with ions in the s^2 configuration (Sn^{2+}, Pb^{2+}, As^{3+}, Sb^{3+}), the d^0 configuration (Ti^{4+}, Nb^{5+}, Ta^{5+}, Mo^{6+}, W^{6+}), and the d^{10} configuration (Ag^+, Cu^+), which absorb strongly in the UV, has been demonstrated mainly in the UV and blue-green region. Efficient visible photoluminescence emission in the orange-red region with different lifetimes has been found in Mn^{2+} ($3d^5$), and in rare earth (f^n) doped glasses (Ce^{3+}, Pr^{3+}, Sm^{3+}, Eu^{2+}, Eu^{3+}, Tb^{3+}, Dy^{3+}, Ho^{3+}, Er^{3+}, Tm^{3+}) with blue, yellow, green, and red emission which can be used for various applications.

Detailed structural investigations of un-doped and doped phosphate and fluoride phosphate glasses were carried out with various methods and colleagues, especially with the group of Kamitsos and Eckert during the last decade to get a deeper knowledge on the relations between properties and structure.[124–129] My long experiences agree with the words: 'To every complex problem, there is a simple solution. And it's always wrong.' (H. L. Mencken, journalist and satirist, *Photonics Spectra*, Nov. 1998).

Acknowledgement

I wish to thank very much the co-workers, PhD and diploma students of my group: Rotraud Atzrodt, Thomas Kittel, Manfred Krauß, Werner Mikkeleit, Gerda Heidemann, Ulrike Penzlin, Christine Erdmann/Fuchs, Edith Carl, Wolfgang Seeber, Heike Ebendorff-Heidepriem, Annegret Brettschneider/Matthai, Michael Leister, Ute Natura, Polina Ebeling, Doris Möncke, Andreas Herrmann, Sandra Fibikar, Wolfram Wintzer, Matthias Carl, Matthias Müller, Michael Müller, Bernd Keinert, Gabriele Möller; all the other colleagues from the Otto Schott Institute, the Analytical Institute, the Institute for Optics and Quantum Electronics, etc, of the University of Jena, and the many external collaborators. They have performed excellent work over a very long time in the research team. I wish to give thanks for various financial support of different projects.

References

1. Jahn, W. Mehrstoffsysteme zum Aufbau optischer Gläser, Teil 3: Neue optische Gläser auf Fluoridbasis. *Glastech. Ber.*, 1961, **34**, 107–119.
2. Izumitani, T. & Nakakawa, K. Ursachen der anomalen Teildispersion von optischen Gläsern (engl.). *Proc. 7th Int. Congr. Glass*, Inst. Nat. Verre, Bruexelles, 1965, pp 1–12.
3. Käs, H. Änderung der Dispersion von Gläsern durch Einbau zusätzlicher Absorptionszentren. *Glastech. Ber.* 1972, **45**, 1–9.
4. Urosovskaja, L. N., Kostomarova, V. N. & Sinikas, R. I. Untersuchung der Glasbildung und Eigenschaften von Fluoroaluminiumphosphatgläsern, *Ž. prikl. chim.*, 1968, **41**, 500–504. (In Russian.)
5. Urosovskaja, L. N. Untersuchungen des Einflusses von Fluoriden auf die Glasbildung und die Eigenschaften von Aluminiumphosphatgläsern, *Ž. prikl. chim.*, 1968, **41**, 2361–2367. (In Russian.)
6. Golubcov, L. A., Chalilev, V. D., Evstropev, K. S. & Ivanova, E. N. Glasbildungsbereiche und Eigenschaften der Gläser in den Systemen $Al(PO_3)_3$-BaF_2-MeF_2(Me=Mg, Ca, Sr), *Ž. prikl. chim.*, 1971, **44**, 180–183. (In Russian.)
7. Chalilev, V. D., Petrovskaja, M. L. & Nikolina, G. F. Gesetzmäßigkeiten der Glasbildung in Phosphatsystemen mit Fluor, *Fiz. chim. stekl.*, 1975, **1**, 508–511. (In Russian.)
8. Mikkeleit, W. Fluoridanalysenmethoden, Otto Schott Institute, University of Jena, *unpublished*.
9. Van Wazer, J. R. *Phosphorus and its compounds*, Interscience, New York, 1958.
10. Rawson, H. *Inorganic glass-forming systems*, Academic Press, London, 1967.
11. Vogel, W. *Glass chemistry*, second edition, Springer-Verlag, Berlin, 1994.
12. Campbell, J. H., Wallerstein, E. P., Hayden, J. S., Sapak, D. L., Warrington, D. E. & Marker III, A. J. Effects of Melting Conditions on Platinum Inclusion Content in Phosphate Laser Glasses, *Glastech. Ber. Glass Sci. Technol.* 1995, **68**, 11–21.
13. Ebendorff-Heidepriem, H. & Ehrt, D. Dehydration of Phosphate Glasses, *J. Non-Cryst. Solids*, 1993, **163**, 74–80.
14. Ebendorff-Heidepriem, H. & Ehrt, D. Determination of the OH content of glasses, *Glastech. Ber. Glass Sci. Technol.*, 1995, **68**, 139–145.
15. Ehrt, D. Effect of OH-content on thermal and chemical properties of SnO-P_2O_5 glasses. *J. Non-Cryst. Solids*, 2008, **354**, 546–552.
16. Ehrt, D., Atzrodt, R. & Vogel, W. Struktureigenschaftsbeziehungen optischer Gläser mit anomaler Teildispersion. *Proc. Second Otto-Schott-Kolloquium*, Jena, July 1982, *Proc. Wiss. Z. FSU Jena, Math.-Naturwiss. R.*, 1983, **32**, 509–526.
17. Ehrt, D. Grundlagenuntersuchungen für die Entwicklung von Fluorophosphat- und Fluoroaluminatgläsern mit extremer positiver anomaler Teildispersion, Thesis B (Dr. sc. nat.), Friedrich Schiller University, Jena, 1984.
18. Holm, J. L. Phase Equilibria in the system CaF_2-AlF_3, *Acta Chim. Scand.*, 1965, **19**, 1512.
19. Ehrt, D., Krauß, M., Erdmann, Ch. & Vogel, W. Fluoroaluminatgläser (1)–(3). *Z. f. Chemie*, (1) Systeme CaF_2-AlF_3 und MgF_2-CaF_2-AlF_3, 1982, **22**, 315. (2) System CaF_2-SrF_2AlF_3, 1983, **23**, 37. (3) Einfluß von Phosphaten auf die Glasbildung im System MgF_2-CaF_2-SrF_2AlF_3, 1983, **23**, 111.
20. Krauß, M., Ehrt, D., Heide, K. & Vogel, W. Phasenanalytische Untersuchungen im System CaF_2-AlF_3. *Z. f. Chemie*, 1984, **24**, 247–250.
21. Ehrt, D., Fuchs, C. & Vogel, W. Kristallisation und optische Eigenschaften von Fluoroaluminatgläsern mit geringen Phosphatgehalten, *Silikattechnik* , 1984, **35**, I–IV.
22. Ehrt, D. & Vogel, W. Optische Gläser mit anomaler Teildispersion, *Feingerätetechnik*, 1982, **31**, 147–151.
23. Ehrt, D. & Vogel, W. Grundlagenuntersuchungen zu optischen Gläsern mit anomaler Teildispersion, *Silikattechnik*, 1883, **34**, 90–94.
24. Ehrt, D. & Vogel, W. Eigenschaften und Struktur von extremen optischen Gläsern. *9. Ibausil*, Weimar, 1985, publ. in *Wiss. Z. HAB Weimar*, 1985, **31**, 71–74.
25. Ehrt, D., Jäger, C. & Vogel, W. Reaktionsprozesse bei der Glasbildung

und Rekristallisation von Fluoroaluminatgläsern, *Third Otto-Schott-Kolloquium*, Juni 1986, Jena, *Proc. Wiss. Z. FSU Jena, Math.-Naturwiss. R.*,1987, **36**, 867–884.
26. Jäger, C. & Ehrt, D. ^{31}P NMR Investigations of Batch Reactions and Recrystallization Phenomena of Fluoroaluminate Glasses, *Exp. Technik d. Physik*, 1988, **63**, 349–384.
27. Jäger, C., Ehrt, D. & Haubenreißer, U. Investigation of solid state reactions of binary polyphosphate fluoride systems by means of thermal analysis, x-ray diffraction and NMR spectroscopy, part I–VI, *Z. Phys. Chemie, Neue Folge*, I. 1988, **159**, 75–87. II. 89–102, III. 103–112. IV. 1989, **162**, 97–107. V. 109–118. VI. 1989, **165**, 55–65.
28. Haubenreißer, U. Experimentelle und theoretische Untersuchungen zur magnetischen ^{31}P-Kernresonanz an polykristallinen Phosphaten und Beiträge der Aufklärung der Nahstruktur von Phosphatgläsern und Glaskeramiken, Thesis B (Dr. sc. nat.), Friedrich Schiller University, Jena, 1985.
29. Jäger, C. Experimentelle Untersuchungen von Reaktionen im Gemenge und von Kristallisationserscheinungen phosphatarmer Fluoroaluminatgläser sowie von Verbindungen des NASICO-Strukturtyps, Thesis B (Dr. sc.nat.), Friedrich Schiller University, Jena, 1987.
30. Pabst, A. A structural classification of fluoroaluminates, *Am. Mineral.*, 1950, **35**, 149–166.
31. Ehrt, D. Fluoroaluminate glasses for lasers and amplifiers, *Curr. Opin. Solid State Mater.*, 2003, **7**, 135–141.
32. Dubiel, M. & Ehrt, D. Nuclear Magnetic Resonance Investigations of the structure of Fluorophosphate Glass, *Phys. Stat. Sol. (a)*, 1987, **100**, 415.
33. Bärenwald, U., Dubiel, M., Matz, W. & Vogel, W. Structure Investigations on Ba(PO_3)$_2$ Glass with Neutron Diffraction and Wideline NMR Technique, *J. Non-Cryst. Solids*, 1988, **103**, 311–318.
34. Ehrt, D. Structure and properties of fluoride phosphate glasses, *Proc. SPIE*, 1992, **1761**, 213–222.
35. Bärenwald, U., Dubiel, M., Matz, W. & Ehrt, D. Structural models of fluoroaluminate glass system Ba(PO_3)$_2$-CaF_2-AlF_3. *J. Non-Cryst. Solids*, 1991, **130**, 171–181.
36. Ehrt, D. & Seeber, W. Glass for High Performance Optics and Laser Technology, *Fourth Int. Otto Schott Colloquium*, July 1990, Jena. publ. *J. Non-Cryst. Solids*, 1991, **129**, 19–30.
37. Ehrt, D., Mikkeleit, W., Burckhardt, W. & Mehner, H. UV transmission of fluoride phosphate glasses. *Proc. Ernst-Abbe-Conference (EAC), FSU Jena*, 1989, 146–160.
38. Ehrt, D., Carl, M., Kittel, T., Müller, M. & Seeber, W. High performance glass for the deep ultraviolet range. *PAC RIM* invited, 1993 Honolulu, USA, *J. Non-Cryst. Solids*, 1994, **177**, 405–419.
39. Carl, M., Heidenreich, E., Kittel, T., Müller, M. & Ehrt, D. Herstellung von Fluorophosphatgläsern für die Hochleistungsoptik unter Verwendung einer Glove-Box-Schmelzanlage, *Sprechsaal*, 1992, **125**, 571–582.
40. Müller, M., Carl, M., Kittel, T. & Ehrt, D. Carbon crucible technology melting optical glasses, *Glastech. Ber. Glass Sci. Technol.*, 1995, **68**, 312–317.
41. Ehrt, D. Redox behaviour of polyvalent ions in the ppm range, *J. Non-Cryst. Solids*, 1996, **196**, 304–308.
42. Ehrt, D., Ebendorff-Heidepriem, H., Albrecht, H.-St., Schröder, T. & Triebel, W. Charge transfer transitions and radiation effects in glasses in the deep-UV range, *Glastech. Ber. Glass Sci. Technol.*, 1995, **68** C1, 521–526.
43. Seeber, W. & Ehrt, D. Estimation of deep-uv and uv absorption coefficients of selected trace impurities in glasses. *Glastech. Ber. Glass Sci. Technol.*, 1997, **70**, 312–315.
44. Ehrt, D., Ebeling, P. & Natura, U. UV transmission and radiation-induced defects in phosphate and fluoride-phosphate glasses, *J. Non-Cryst. Solids*, 2000, **263&264**, 240–250.
45. Ehrt, D. & Brettschneider, A. Redox behaviour of copper ions in FP and P glass melts, *XVII Int. Congress on Glass*, Beijing, China, 1995. *Proc. Chinese Ceramic Soc. Beijing*, **3**, 157–162.
46. Ehrt, D., Leister, M. & Matthai, A. Redox Behaviour in Glass Forming Melts, *Molten Salt Forum*, 1998, **Vols. 5&6**, 347–354.
47. Ebendorff-Heidepriem, H. & Ehrt, D. Formation and UV absorption of cerium, europium and terbium ions in different valencies in glasses, *Opt. Mater.*, 2000, **15**, 7–25.
48. Ebendorff-Heidepriem, H. & Ehrt, D. Electron spin resonance spectra of $Eu^{2+}$and Tb^{4+} in glasses, *J. Phys.: Matter*, 1999, **11**, 7627–7634.
49. Ehrt, D., Leister, M. & Matthai, A. Polyvalent elements iron, tin and titanium in silicate, phosphate and fluoride glasses and melts, *Phys. Chem. Glasses*, 2001, **42**, 231–239.
50. Matthai, A., Claußen, O., Ehrt, D. & Rüssel, C. Thermodynamics of redox equilibria and diffusion of polyvalent ions in a phosphate glass melt, *Glastech. Ber. Glass Sci. Technol.*, 1998, **71**, 29–34.
51. Matthai, A., Ehrt, D. & Rüssel, C. Redox behaviour of polyvalent ions in phosphate glass melts and phosphate glasses, *Glastech. Ber. Glass Sci. Technol.*, 1998, **71**, 187–192.
52. Ehrt, D. & Vogel, W. Radiation Effects in Glasses, *Sixth Int. Conference: Radiation Effects in Insulators (REI'6) (invited)*, June 1991, Weimar. publ. in *Nucl. Instrum. Methods B*, 1992, **65**, 1–8.
53. Ebendorff-Heidepriem, H. & Ehrt, D. UV radiation effects in fluoride phosphate glasses, *J. Non-Cryst. Solids*, 1996, **196**, 113–117.
54. Ehrt, D., Natura, U., Ebeling, P. & Müller, M. Formation and healing of uv radiation defects in phosphate and fluoride phosphate glasses with high uv transmission, *XVIII. Int. Congress on Glass*, 1998, San Francisco, USA, *Proc. C9*, 1998, pp 15–20.
55. Ebeling, P., Ehrt, D. & Friedrich, M. Radiation-induced color centers in anion doped phosphate glasses, *Phosphorus Res. Bull.*, 1999, **10**, 484–489.
56. Ebendorff-Heidepriem, H. & Ehrt, D. Rare earth as indicators for radiation-induced defect center formation in phosphate containing glasses, *Phosphorus Res. Bull.*, 1999, **10**, 552–557.
57. Ebendorff-Heidepriem, H. & Ehrt, D. Rare earth as indicators for radiation-induced defect center formation in FP- and P-glasses, *Fifth ESG*, June 1999, Prague, Czech Republic, *Proc. C1*, pp 62–71.
58. Möncke, D., Natura, U. & Ehrt, D. Radiation defects in CoO and NiO doped glasses of different structure, *Fifth ESG*, June 1999, Prague, Czech Republic, *Proc. B4*, pp 49–56.
59. Möncke, D. & Ehrt, D. Radiation induced defects in CoO and NiO doped fluoride phosphate glasses, *Glastech. Ber. Glass Sci. Technol.*, 2001, **74**, 65–73.
60. Möncke, D. & Ehrt, D. Irradiation-induced defects in different glasses demonstrated on a metaphosphate glass, *Glastech. Ber. Glass Sci. Technol.*, 2001, **74**, 199–209.
61. Ehrt, D., Ebeling, P., Natura, U., Kolberg, U., Naumann, K. & Ritter, S. Redox Equilibria and UV Radiation Induced Defects in Glasses, *Proc. XIX Int. Congr. Glass* (invited), July 2001, Edinburgh, Scotland, **Vol. 1**, pp 84–93, and CD-ROM.
62. Ebendorff-Heidepriem, H. & Ehrt, D. Radiation-induced defects in europium and terbium doped glasses, *Proc. XIX Int. Congress on Glass* (invited), July 2001, Edinburgh, Scotland, **Vol. 2**, pp 891–892, and CD-ROM.
63. Ebendorff-Heidepriem, H. & Ehrt, D. Ultraviolet laser and x-ray induced valence changes and defect formation in europium and terbium doped glasses, *Phys. Chem. Glasses*, 2002, **43C**, 38–47.
64. Ebendorff-Heidepriem, H. & Ehrt, D. Effect of Tb^{3+} ions on x-ray-induced defect formation in phosphate containing glasses, *Opt. Mater.*, 2002, **18**, 419–430.
65. Ebendorff-Heidepriem, H. & Ehrt, D. Effect of europium ions on x-ray-induced defect formation in phosphate containing glasses, *Opt. Mater.*, 2002, **19**, 351–363.
66. Ebeling, P., Ehrt, D. & Friedrich, M. X-ray radiation induced effects in phosphate glasses, *Opt. Mater.*, 2002, **20**, 101–111.
67. Natura, U., Feurer, T. & Ehrt, D. Kinetic of uc laser radiation defects in high performance optics, *REI'10*, July 1999 in Jena, publ. in *Nucl. Instrum. Meth. B*, 2000, **166&167**, 470–475.
68. Natura, U. & Ehrt, D. Generation and healing behaviour of radiation induced absorption in fluoride phosphate glasses: dependence on UV radiation sources and temperature, *Nucl. Instrum. Meth. B*, 2001, **174**, 143–150,
69. Natura, U. & Ehrt, D. Modeling of excimer laser radiation induced defect generation in fluoride phosphate glasses, *Nucl. Instrum. Meth. B*, 2001, **174**, 151–158.
70. Möncke, D. & Ehrt, D. Radiation-induced defects in CoO- and NiO-doped fluoride, phosphate, silicate and borosilicate glasses, *Glass Sci. Technol.*, 2002, **75**, 243–253.
71. Ehrt, D. UV absorption and radiation effects in different glasses doped with iron and tin in the ppm range. *C. R. Chimie*, 2002, **5**, 679–692.
72. Ehrt, D. & Möncke, D. Charge transfer absorption of Fe^{2+} and Fe^{3+} complexes and UV radiation induced defects in different glasses, *Sixth ESG Conference*, June 2002, Montpellier, France, Proc.-CD.
73. Möncke, D., Marschall, R., Ehrt, D. EPR- and optical spectroscopy of irradiation induced defects in transition metal doped glasses, *Seventh Int. Otto Schott Colloquium*, July 2002, Jena, Germany, *Proc. Glass Sci. Technol.*, 2002, **75 C2**, 36–41.
74. Ebeling, P., Ehrt, D. & Friedrich, M. Influence of modifier cations on the radiation-induced effects of metaphosphate glasses, *Glass Sci. Technol.*, 2003, **76**, 56–61.
75. Möncke, D. & Ehrt, D. Irradiation induced defects in glasses resulting in the photoionization of polyvalent dopants, *Opt. Mater.*, 2004, **25**, 425–437.
76. Möncke, D., Ehrt, D. Photoionisation of As, Sb, Sn, and Pb in metatphosphate glasses, *J. Non-Cryst. Solids*, 2004, **345&346**, 319–322.
77. Möncke, D. & Ehrt, D. UV-light induced photoreduction in phosphate and fluoride-phosphate glasses doped with Ni^{2+}, Ta^{5+}, Pb^{2+}, and Ag^{+} - compounds, *Glass Sci. Technol.*, 2004, **77**, 239–248.
78. Möncke, D. & Ehrt, D. Photoinduced redox-reactions and transmission changes in glasses doped with 4d- and 5d-ions, *J. Non-Cryst. Solids*,

2006, **352**, 2631–2636.

79. Möncke, D. & Ehrt, D. Photoinduced redox-reactions of Zr, Nb, Ta, Mo, and W in glasses, *Phys. Chem. Glasses: Eur. J. Glass Sci. Technol. B*, 2007, **48**, 317–323.
80. Ehrt, D. Photoactive glasses (invited), *XXI Int. Congr. on Glass*, 2007, Strasbourg, France, Proc. CD.
81. Möncke, D. & Ehrt, D. Photoionization of Polyvalent Ions. In: *Materials Science Research Horizons*, Editor: P. Glick, Nova Science Publishers, Inc. New York, 2007, pp 1–56, and *Materials Science and Technologies Series*, 2009.
82. Ehrt, D., Kittel, T., Will, M., Nolte, S. & Tünnermann, A. Femtosecond-laser-writing in various glasses, *J. Non-Cryst. Solids*, 2004, **345&346**, 332–337.
83. Righini, G. C., Banyasz, I., Berneschi, S., Brenci, M., Chiasera, A., Cremona, M., Ehrt, D., Ferrari, M., Montereali, R. M., Nunci Conti, G., Pelli, S., Sebastiani, S. & Tosello, C. Laser Irradiation, ion implantation and e-beam writing of integrated optical structures, *Proc. SPIE*, 2005, **5840**, 649–657.
84. Heumann, E., Ledig, M., Ehrt, D., Seeber, W., Duczynski, E. W., v.d. Heide, H.-J. & Huber, G. CW laser action of Er^{3+} double sensitized fluoroaluminate glass at room temperature, *Int. Laser Sci. Conf. ILC4*, 1988, Atlanta, USA, publ. *Appl. Phys. Lett.*, 1988, **52**, 255.
85. Ehrt, D., Seeber, W., Ledig, M. & Heumann, E. Sensitized Nd^{3+} and Er^{3+} laser materials basing on fluoride phosphate glasses, *XV. Int. Congr. on Glass*, 1989 Leningrad, USSR, *Proc. Vol. 2b*, 190–194.
86. Ledig, M., Heumann, E., Seeber, W. & Ehrt, D. Spectroscopic and Laser Properties of Cr:Yb:Er-Fluoroaluminate Glass, *Opt. Quant. Electron.*, 1990, **22**, 107–122.
87. Seeber, W. & Ehrt, D. Lasermaterialien auf der Basis von Fluorid-Phosphate-Gläsern, *Silikattechnik*, 1990, **41**, 230–233.
88. Ebendorff-Heidepriem, H., Seeber, W. & Ehrt, D. Spectroscopic and Chemical Properties of Strontium Phosphate Glasses with different Nd^{3+} Concentrations, *phys. stat. sol. (a)*, 1992, **130**, 247–251.
89. Seeber, W., Arnold, P. & Ehrt, D. Luminescence Properties and Energy Transfer in Ce^{3+} and/or Tm^{3+} doped Fluoride Phosphate Glasses, *phys. stat. sol. (a)*, 1992, **130**, 243–246.
90. Ebendorff-Heidepriem, H., Seeber, W. & Ehrt, D. Spectroscopic and laser properties of Er^{3+} fluorescence at 540 nm and 1.5 μm in fluoride phosphate and phosphate glasses, *Glastech. Ber. Glass Sci. Technol.*, 1993, **66**, 235–244.
91. Ebendorff-Heidepriem, H., Ehrt, D. & Seeber, W. Er^{3+} fluorescence transitions in glasses, *Fifth Int. Otto Schott Colloquium*, 1994, Jena, Germany. *Proc. Glastech. Ber. Glass Sci. Technol.*, 1994, **67C**, 37–43.
92. Seeber, W., Ehrt, D. & Ebendorff-Heidepriem, H. Spectroscopic and laser properties of Ce^{3+} - Cr^{3+} - Nd^{3+} co-doped fluoride phosphate glasses, *J. Non-Cryst. Solids*, 1994, **171**, 94–104.
93. Seeber, W. Kombination experimenteller und theoretischer Methoden der Absorptions- und Luminenzspektroskopie zur Entwicklung amorpher Materialien für Optik und Quantenelektronik - Am Beispiel von Fluorid-Phosphatgläsern, *Habilitation Thesis*, 1995, Friedrich Schiller University, Jena.
94. Ebendorff-Heidepriem, H., Seeber, W. & Ehrt, D. Spectroscopic properties of Nd^{3+} in phosphate glasses, *J. Non-Cryst. Solids*, 1995, **183**, 191–200.
95. Ebendorff-Heidepriem, H., Ehrt, D., Bettinelli, M. & Speghini, A. Effect of glass composition on Judd-Ofelt parameters and radiative decay rates of Er^{3+} in fluoride phosphate and phosphate glasses, *J. Non-Cryst. Solids*, 1998, **240**, 66–78.
96. Ebendorff- Heidepriem, H., Ehrt, D., Bettinelli, M. & Speghini, A. Spectroscopic properties of rare earth ions in heavy metal oxide and phosphate containing glasses, *Proc. SPIE*, 1999, **3622**, 19–30.
97. Ebendorff-Heidepriem, H., Ehrt, D., Philipps, J., Töpfer, T., Speghini, A. & Bettinelli, M. Properties of Er^{3+} doped glasses for waveguide and fiber lasers, *Proc. SPIE*, 2000, **3942**, 29–39.
98. Philipps, J., Töpfer, T., Ebendorff-Heidepriem, H., Ehrt, D., Sauerbrey, R. & Borelli, N. F. Diode-pumped erbium-ytterbium-glass laser passively Q-switched with PbS semiconductor quantum-dot doped glass, *Appl. Phys. B*, 2001, **72**, 175–178.
99. Philipps, J., Töpfer, T., Ebendorff-Heidepriem, H., Ehrt, D. & Sauerbrey, R. Spectroscopic and lasing properties of Er^{3+}: Yb^{3+} -doped fluoride phosphate glasses, *Appl. Phys. B*, 2001, **72**, 399–405.
100. Philipps, J., Töpfer, T., Ebendorff-Heidepriem, H., Ehrt, D. & Sauerbrey, R. Energy transfer and upconversion in erbium-ytterbium-doped fluoride phosphate glasses, *Appl. Phys. B*, 2002, 233–236.
101. Mix, E., Heumann, E., Huber, G., Ehrt, D. & Seeber, W. Efficient CW-Laser Operation of Yb-doped Fluoride Phosphate Glass at Room Temperature, *Advanced Solid-State Lasers Topical Meeting*, 1995, Memphis, TN, USA, *Proc.* 1995, **WB5-1**, 230.
102. Petrov, V., Griebner, U., Ehrt, D. & Seeber, W. Femtosecond self mode looking in Yb:fluoride phosphate glass laser, *Opt. Lett.*, 1997, **22**, 408–410.
103. Töpfer, T., Hein, J., Philipps, J., Sauerbrey, R. & Ehrt, D. Tailoring the nonlinear refractive index coefficient of fluoride phosphate glasses for laser applications, *Appl. Phys. B*, 2000, **71**, 2003–2006.
104. Ehrt, D. & Töpfer, T. Preparation and structure of Yb^{3+} FP laser glasses, *Proc. SPIE*, 2000, **4102**, 13–22.
105. Töpfer, T., Hein, J., Wintzer, W., Ehrt, D. & Sauerbrey, R. Laser-glass, pump-laser-diodes, and amplifier for the POLARIS laser, *First International workshop on Glass and Photonics Revolution*, 2002, Bad-Soden, Germany, publ. *Glass Sci. Technol.*, 2002, **75 C1**, 223–234.
106. Ehrt, D., Wintzer, W., Töpfer, T. & Sauerbrey, R. Fluoroaluminate glasses for lasers and amplifiers, *XIII. Int. Symposium on Non-oxide Glasses and new Optical Glasses*, 2002, Pardubice, Czech Republic, *Proc.* p. 662–665.
107. Töpfer, T., Hein, J., Quednau, G., Hellwig, M., Philipps, J., Paoloni, S., Walther, H.-G., Theobald, W. Wintzer, W., Ehrt, D. & Sauerbrey, R. Scaling laser-diode pumped solid-state amplifiers to the petawatt level, *Quant. Electron.*, 2003, 37–40.
108. Paoloni, S., Hein, J., Töpfer, T., Walther, H.-G., Sauerbrey, R., Ehrt, D. & Wintzer, W. Laser beam induced optical aberrations in phosphate and fluoride phosphate glasses, *Appl. Phys. B*, 2004, **78**, 415–419.
109. Hein, J., Podleska, S., Siebold, M., Hellwig, M., Bödefeld, R., Sauerbrey, R., Ehrt, D. & Wintzer, W. Diode-pumped chirped pulse amplification to the joule level, *Appl. Phys. B*, 2004, **79**, 419–422.
110. Ehrt, D. Photoluminescence in the UV-VIS region of polyvalent ions in glasses, *J. Non-Cryst. Solids*, 2004, **345&346**, 319–322.
111. Ehrt, D. & Herrmann, A. Photoluminescence behaviour of active ions in different glasses, *XX. Int. Congress on Glass*, 2004, Kyoto, Japan, Proc. CD, and publ. in: *Verre*, 2005, **11**, 13–18.
112. Herrmann, A. & Ehrt, D. Hightech-Gläser für die Photonik, *Laser Photonik*, 2005, 12–14.
113. Herrmann, A. & Ehrt, D. Time-resolved fluorescence measurements of Tb^{3+} - and Mn^{2+} - doped glasses, *Glass Sci. Technol.*, 2005, **278**, 99–105.
114. Herrmann, A. & Ehrt, D. Time-resolved fluorescence measurements on Dy^{3+} and Sm^{3+} doped glasses, *J. Non-Cryst. Solids*, 2008, **354**, 916–926.
115. Herrmann, A., Fibikar, S. & Ehrt, D. Time-resolved fluorescence measurements on Eu^{2+} - and Eu^{3+} - doped glasses, *J. Non-Cryst. Solids*, 2009, **355**, 2093–2101.
116. Ehrt, D. Photoluminescence in glasses and glass ceramics, *IOP Conf. Ser.: Mater. Sci. Eng.*, 2009, **2**, 012001.
117. Herrmann, A. & Ehrt, D. Green and Red Er^{3+} Photoluminescence Behavior in Various Fluoride Glasses, *Int. J. Appl. Glass Sci.*, 2010, **1**, 341–349.
118. Ehrt, D. Photoactive glasses and glass ceramics, *IOP Conf. Ser.: Mater. Sci. Eng.*, 2011, **21**, 012001.
119. Ebendorff-Heidepriem, H. & Ehrt, D. Relationships between glass structure and spectroscopic properties of Eu^{3+} and Tb^{3+} doped glasses, *Ber. Bunsenges. Phys. Chem.*, 1996, **100**, 1621–1624.
120. Seeber, W. & Ehrt, D. Probe ions for local structure investigations of glass, *Ber. Bunsenges. Phys. Chem.*, 1996, **100**, 1593–1595.
121. Ebendorff-Heidepriem, H. & Ehrt, D. Spectroscopic properties of Eu^{3+} and Tb^{3+} ions for local structure investigations, *J. Non-Cryst. Solids*, 1996, **208**, 205–216.
122. Ebendorff-Heidepriem, H. & Ehrt, D. Tb^{3+} f-d absorption as indicator of the effect of covalency on the Judd-Ofelt parameter in glasses, *J. Non-Cryst. Solids*, 1999, **248**, 247–252.
123. Ebendorff-Heidepriem, H. Seltenerdionen als Indikatoren und aktive Ionen in Gläsern für die Photonik. *Habilitation Thesis*, 2000, Friedrich Schiller University, Jena.
124. Möncke, D., Ehrt, D., Velli, L. L., Versamis, C. P. E. & Kamitsos, E. I. Structure and properties of mixed phosphate and fluoride glasses, *Phys. Chem. Glasses*, 2005, **46**, 67–71.
125. Velli, L. L., Versamis, C. P. E., Kamitsos, E. I., Möncke, D. & Ehrt, D. Structural investigations of metaphosphate glasses, *Phys. Chem. Glasses*, 2005, **46**, 178–182.
126. Möncke, D., Ehrt, D., Velli, L. L., Versamis, C. P. E., Kamitsos, E. I., Elbers, S. & Eckert, H. Comparative spectroscopic investigation of different types of fluoride phosphate glasses, *Phys. Chem. Glasses: Eur. J. Glass Sci. Technol. B*, 2007, **48**, 399–402.
127. Velli, L. L., Versamis, C. P. E., Kamitsos, E. I., Möncke, D. & Ehrt, D. Optical basicity and refractivity in mixed oxyfluoride glasses, *Phys. Chem. Glasses: Eur. J. Glass Sci. Technol. B*, 2008, **49**, 182–187.
128. Konidakis, I., Versamis, C. P. E., Kamitsos, E. I., Möncke, D. & Ehrt, D. Structure and properties of mixed strontium-manganese metaphosphate glasses, *J. Phys. Chem. C*, 2010, **114**, 9125–9138.
129. Möncke, D., Kamitsos, E. I., Herrmann, A., Ehrt, D. & Friedrich, M. Bonding and ion-ion interaction of Mn^{2+} ions in fluoride-phosphate and boro-silicate glasses probed by EPR and fluorescence spectroscopy, *J. Non-Cryst. Solids*, 2011, **357**, 2542–2551.

Doris Ehrt – Glass chemical research in the spirit of Otto Schott

Doris Möncke

Otto-Schott-Institute of Materials Research, Friedrich-Schiller-University Jena, Fraunhoferstr. 6, D-07743 Jena, Germany

Manuscript received 3 November 2014
Accepted 1 April 2015

A brief review is given of the scientific career of HDoz. Dr. Doris Ehrt to whom the International Phosphate Conference 2014 is dedicated. The Phosphate Conference was held jointly with the Eighth International Conference on Borate Glasses, Crystals and Melts, in Pardubice. This concurrence is most fitting, since Doris Ehrt devoted a significant part of her research to borate and borosilicate glasses. Doris Ehrt (née Paul) was born shortly after the Second World War in Schkölen, Thuringia. She studied chemistry at the Friedrich Schiller University in Jena and joined the emerging glass group of Prof. Werner Vogel in 1969. Her initial research focused on phase separation in borosilicate glasses. In 1976, she began her research into phosphate and fluoride phosphate optical glasses with high positive anomalous partial dispersion and thus with a reduced secondary spectrum. These glasses were to be used in high performance optics in which glass lenses were to be substituted for CaF_2 single crystals. An abundance of multicomponent fluoroaluminate and fluoride phosphate glass systems were prepared and studied to investigate their structure and optical, thermal and chemical properties. Since Doris Ehrt summarises her work on fluoride phosphate glass systems in the previous pages, the current paper is focused more on her overall career, including all her non-phosphate related works and the transition from the socialist German Democratic Republic to the reunified Federal Republic of Germany.

Introduction

Doris Ehrt (Figure 1) is known for her research on optical properties of glasses. This research includes radiation induced defect formation and the study of UV transparent glasses. In this context, she has studied the absorption of impurities. She has also done research on all types of polyvalent dopants, including luminescent and laser ions. In her publications, she continually strives to correlate variations in the optical properties with variations in the structure of glasses and as a consequence, also of the ion site geometry.

In a summary paper in this volume, Doris Ehrt describes in detail the development of fluoride phosphate (FP) glasses.[1] She writes about the first fundamental studies of the glass forming regions, the subsequent characterisation of promising glasses, and, finally, the technological refinements in glass preparation. These refinements resulted in high quality rare earth doped fluoride phosphate glasses, which are used as fluorescence standards in microscopes[2] and which can, as laser glasses, even withstand the high energies realised in the POLARIS laser.[3]

In this paper a short biography of Doris Ehrt is presented, with a discussion of her considerable contributions to glass research. It is shown that Doris Ehrt's attention to optical glasses and her insistence on combining fundamental research with useful applications follows closely in the traditions of Werner Vogel[4] and Otto Schott[5] in particular, and Jena's vitreous heritage in general. Therefore, as Doris Ehrt covered many of the same subjects and contributed significant new insights to old questions, this paper will highlight the important questions which instigated the scientific study of glasses and which have driven the development of glass technology and glass science ever since.

Figure 1. Doris Ehrt in 2010

This brief appraisal of Doris Ehrt's life and achievements is illustrated by photographs and electron micrographs from the archives of the Otto Schott Institute and from her personal records. The history of the Otto Schott Institute is also based on previous papers, all of which have been co-authored by Doris Ehrt.[4,6–9]

Email doris.moencke@uni-jena.de
Original version presented at Int. Conf. on Phosphate Glasses, Pardubice, Czech Republic, 2–4 July 2014. The conference was dedicated to the life and work of HDoz. Dr. Doris Ehrt, and was held jointly with the Eighth International Conference on Borate Glasses, Crystals and Melts.
DOI: 10.13036/17533562.56.6.235

Early years

Doris Ehrt was born on 13th July, 1946, the third child of Frieda Paul and the baker Otto Paul. She spent her childhood in Schkölen, a small town in Thuringia, which, at this time, was part of Soviet occupied Germany. Doris was a pupil at the Maxim-Gorki-*Oberschule* (Primary School) of Schkölen during 1953–1961. She continued her schooling in 1961 at the Frederic-Joliot-Curie *Erweiterte Oberschule* (High School) in Eisenberg, until she graduated in 1965 with distinction. Figure 2 shows Doris in second grade (1954).

Figure 2. Doris Ehrt (née Paul) in second grade, 1954

The early death of her father that year did not quench her yearning for further education. Doris adhered to her plans to study chemistry at the Friedrich Schiller University (FSU) in Jena, where she moved in 1965.

As a student at the University

In September 1969, during her final year as an undergraduate student, Doris Ehrt started a research project on glasses in the Department of Glass Chemistry, under the supervision of Prof. Dr. Vogel. Werner Vogel was the first professor in Jena who was explicitly appointed to a chair of Glass Chemistry (1966). He was also the first director of the newly established Institute for Glass Chemistry (1976), which was renamed as the "Otto-Schott-Institute for Glass Chemistry" in 1969, shortly after Doris Ehrt joined the group.[4,6–9]

Phase separation in borosilicate glasses (PhD/Dr. rer. nat.)

In 1970 Doris Ehrt obtained her degree in chemistry with honours. Her diploma thesis, *Distribution of coloured ions on the micro-phases in the* Na_2O*–*B_2O_3*–*SiO_2 *glass system*,[10] focused on borosilicate glasses, a subject she continued in her PhD thesis, *Study of the distribution of CoO and* Fe_2O_3 *on the micro phases of the glass system* Na_2O*–*B_2O_3*–*SiO_2.[11] She continued studying these subjects throughout her career.[12–45] Doris Ehrt obtained her doctorate with distinction in 1973. Figure 3(a) is an electron micrograph of a borosilicate glass with a beautiful example of secondary phase separation. The process of secondary phase separation is explained in the corresponding schematic (Figure 3(b)). Both pictures are taken from Doris's PhD thesis.[11] Figure 4(a) and (b), also taken from her thesis, show two schematics of the temperature dependent phase separation in the borosilicate ternary.[11]

In 1969, in her final year as an undergraduate, she married Roland Ehrt. The birth of her daughter Anke in 1971 hardly slowed her dissertation, although she did apply for an extension of her PhD while caring for the newborn baby. The birth of their son Elko in 1975 completed the family.

Researcher at the university – establishing her own work group

FP-glasses with anomalous partial dispersion (Habilitation/Dr. sc. nat.)

After completing her PhD, Doris Ehrt remained as a researcher at the Otto Schott Institute. She wrote three papers about the phase separation of borosilicate glasses.[12–14] More importantly, in 1976 she began

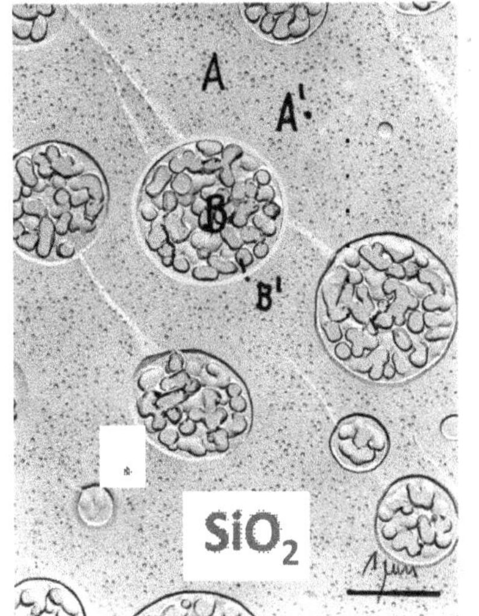

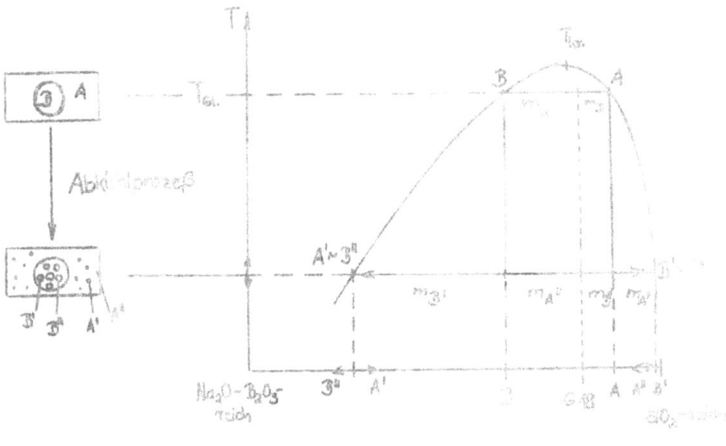

*Figure 3. (a) Electron microscopy picture of a phase separated borosilicate glass (4·5*Na_2O*–22·5*B_2O_3*–73*SiO_2*); clearly visibly are 1 µm large* Na_2O*–*B_2O_3 *rich droplets (B, as in the orange circle) in a* SiO_2 *matrix (A). Both show evidence of secondary phase separation in the form of* SiO_2 *rich regions (B') within the* Na_2O*–*B_2O_3 *rich droplets (B) and* Na_2O*–*B_2O_3 *rich droplets (A') in the* SiO_2 *matrix (A). (b) A schematic explaining multiple phase separation for the same glass as in (a), tempered at 720°C. Both from Doris Ehrt's PhD thesis*[10] *[Colour available online]*

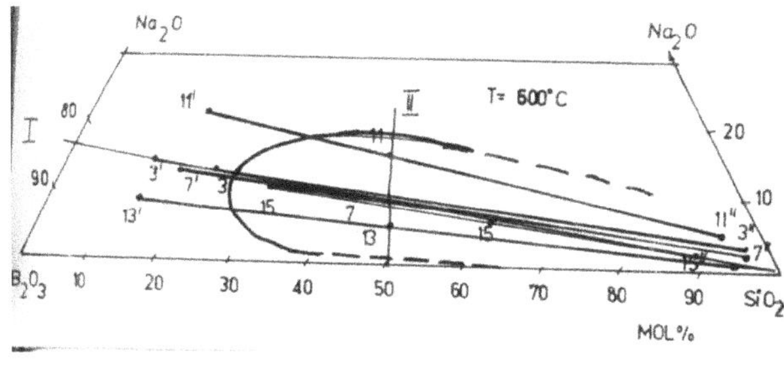

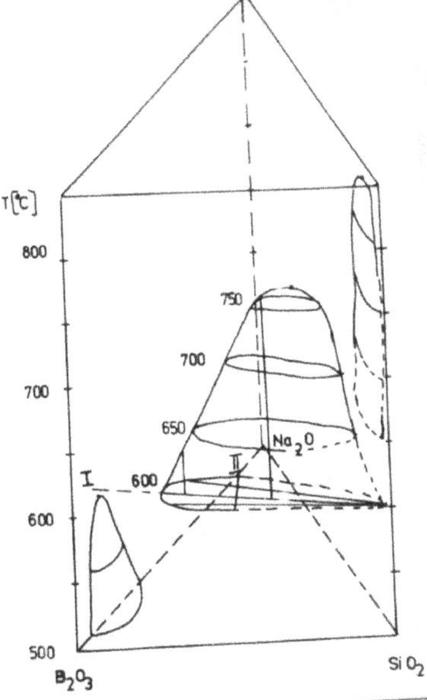

Figure 4. Schematic representation of phase separation in the Na_2O–B_2O_3–SiO_2 system. (a) shows the isotherms of the binodal phases at 600°C of the chemically analysed phases, and (b) shows the temperature dependence of the phases, as stated in Doris Ehrt's PhD thesis[10]

to work on a very different topic: the study of glass formation in fluoride glass systems. This work led to the development of fluoride phosphate-optical glasses by Doris Ehrt and her group. For more details on the scientific work, see the article by Doris Ehrt in this volume.[1]

The basis of this project was a so-called "state plan" of the government of the German Democratic Republic (the GDR, known informally as East Germany): Its goal was the improvement of existing and the development of new optical glasses. At the time, a close cooperation existed between the JENAer GLASWERK, CARL ZEISS Jena and the Otto Schott Institute of the University in Jena. Youth research collectives were obliged to take part in industrial practicals at the glass works. Prof. W. Vogel had worked at the JENAer GLASWERK while simultaneously pursuing his habilitation at Jena University. He was appointed as a lecturer at the university in 1963, while still working part time for the glassworks. When Doris Ehrt was appointed as one of the project leaders of this state plan, she continued this close collaboration with the JENAer GLASWERK and with the company of CARL ZEISS Jena. The goal of her work was to develop optical glasses with high positive anomalous partial dispersion. Such glass lenses could be substituted for CaF_2 single crystals in high performance optics, where the secondary spectra were to be avoided.[7] The project later included transferring these glasses to the production lines at the JENAer GLASWERK in the form of the fluorcrown glasses FPSK1 and FPSK3. Figure 5 shows a meeting between researchers of the Otto Schott Institute and the JENAer GLASWERK.

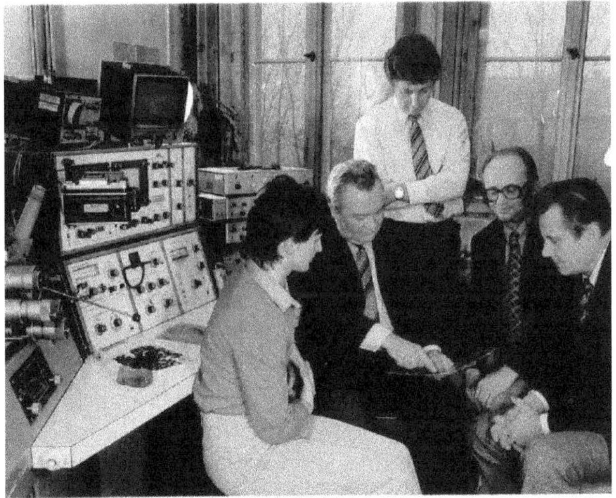

Figure 5. From left to right: Doris Ehrt, Prof. Werner Vogel (director of the Otto Schott Institute), Prof. Erich Heidenreich , Michael Gitter, (all researcher at the OSI) with Klaus Gerth from the JENAer GLASwerk in the Electron Microscopy Laboratory of the OSI in Jena, discussing the joint project on optical glasses in 1984 (Photograph from Jenaer Rundschau Vol.2, p.95) [Colour available online]

The research on FP glasses provided Doris Ehrt with enough material to complete her Dissertation B (Dr. sc. nat.), analogous to a habilitation, on the subject of *Fundamental research for the development of fluoride phosphate and fluoroaluminate glasses with extreme positive anomalous relative partial dispersion.*[46] She defended her thesis for the degree of Dr. sc. nat. successfully in 1984. In 1986, she also obtained the *Facultas docendi* in "Glass Chemistry". One year later, in 1987, she was appointed docent at the FSU, where she worked as a professor from 1994 until her retirement.

Parallels between Doris Ehrt's research and the history of glass science

While reviewing Doris Ehrt's publications and research, it soon became apparent that she worked on all the important questions which have driven the development of glass chemistry and glass technology since the first emergence of glass related research in

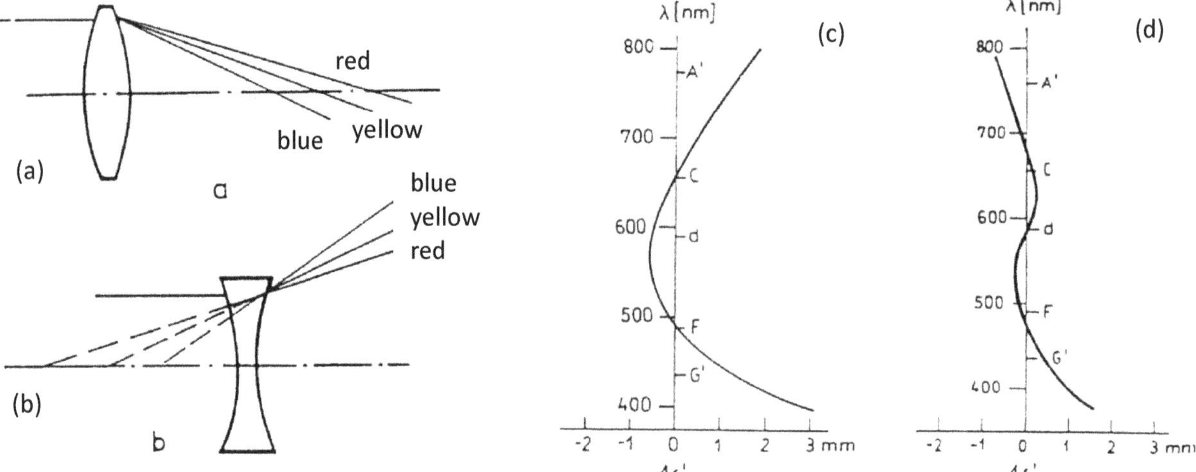

Figure 6. Chromatic aberration in (a) a converging lens, and (b) in a diverging lens. The wavelength dependence of (c) an achromat with colour correction at wavelengths C and F (the secondary spectrum, Δs', is caused by the remaining wavelengths), and (d) an apochromate for which achromasy is obtained for three wavelengths[47]

the 18th century.

Considering the long history of glass science in Jena, it is not surprising that Doris Ehrt herself wrote or co-authored several papers on the history of the glass industry. In particular, there were manuscripts about the Otto Schott Institute in Jena, and the life of her PhD supervisor and mentor Werner Vogel.[4,6–8]

Optical glasses and glass quality

The driving force in the early development of glass science was the need to improve diverse optical applications. In order to improve microscopes and telescopes, glasses with anomalous dispersion were needed for apochromatic lenses (see also Figure 6(a) and (b)).[5,7,47] It was therefore necessary to develop glasses with distinct optical properties. Even in the 18th and 19th centuries, after four to five millennia of empirical glass melting, only seven oxide components for glass preparation were known.[5] The Russian scientist Mikhail V. Lomonosov (1711–1765) began what were probably the first purely scientific studies of experimental glass melts in the middle of the 18th century. Sometime later, William V. Hartcourt (1789–1871), a British cleric and amateur chemist, introduced 20 more elements into glass making. However, the poor glass quality and high hygroscopic nature of many of his samples prevented further characterisation by his collaborator, the British mathematician and physicist George G. Stokes (1819–1903).[5] At the turn of the century, the Swiss glass technician Pierre Louis Guinand (1748–1824) had invented a special method for stirring glass melts. This new method considerably improved the homogeneity of glasses. He taught this technique to his collaborator, the optician Joseph Fraunhofer (1787–1826), who was the first person to actually measure the refractive index of glasses.[5,7,48] Their high quality glasses, technological development and scientific approach soon made their glass works and optical appliances very successful.

Similar to these early researchers, Doris Ehrt looked for glasses with an excellent sample quality, and she needed to develop glasses with special optical properties. Her glasses required a high positive anomalous partial dispersion in order to minimize the secondary aberration or secondary spectrum. The performance of systems based on optical lenses is often limited by an impeding secondary spectrum (see Figure 6(a) and (b)). After defining the fluoride phosphate glass systems which exhibit the required optical properties, Doris Ehrt's group developed a special inductive melting setup. The melt was stirred under a controlled atmosphere, and the temperature was carefully controlled during melting, fining and casting of the FP glass preforms (see also Figure 7).[49] As a result, Doris Ehrt and her group were able to prepare FP glasses with a refractive index homogeneity as high as the sixth decimal place.[49] These glasses proved their extraordinary quality in several high performance tests, starting in 1985 with Nd^{3+} and Er^{3+} FP laser glasses.[50,51] A highlight is the Yb^{3+} FP laser glasses, developed at the turn of the century,[52] which are still used in the further development of the POLARIS laser at the physics department of the FSU in Jena.[53,54]

Defect formation in glasses

In Fraunhofer's time, the British Crown asked the Royal Society and Board of Longitude to establish a *Committee for the Improvement of Glass for Optical*

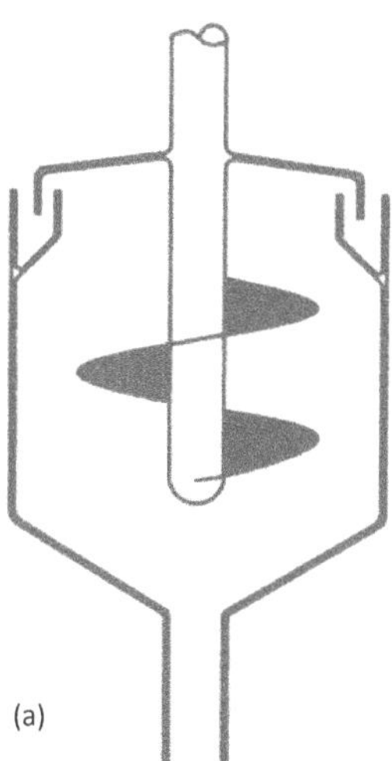

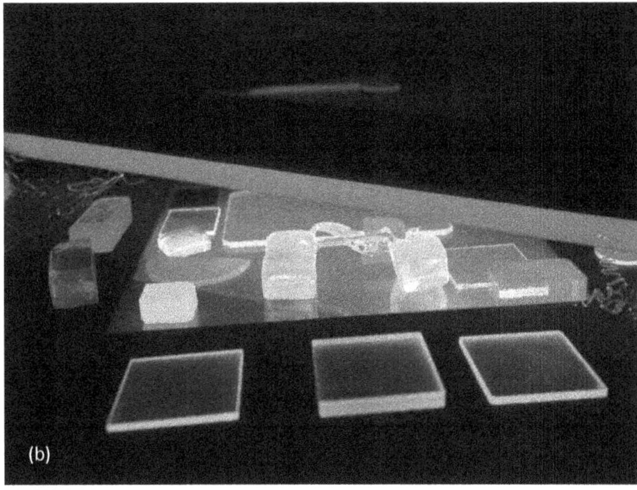

Figure 7. (a) The inductive heated melting setup with stirrer for the preparation of high performance FP laser glasses;[49] *(b) examples of fluorescent standards and laser glasses prepared by Doris Ehrt's group at the Otto Schott Institute [Colour available online]*

Purposes, headed by Michael Faraday (1791–1867).[55] The goal was to develop optical glasses with high optical quality and with new compositions, which could compete with the glasses made by Joseph Fraunhofer. Faraday participated in the first trials of characterisation and preparation of glasses for optical applications, and was the first to publish a paper on the solarisation of glass.[56] In this paper, he suggested that the darkening, which was observed in many glasses over time was due to the photo-oxidation of Mn^{2+} to $(Mn^{2+})^{+}$. This field of "radiation induced defect formation" became a central topic in the scientific career of Doris Ehrt.[7,16,20–23,25,28,34,49,57–75] Her research began with the general observation that irradiation resulted in a transmission loss in glasses.[7,9,57] This phenomenon was then studied systematically with regard to intrinsic defect formation in fluoride phosphate[7,59,61,70] and borosilicate (BS) glasses.[16,25,63] Several PhD theses were involved with these studies. Ute Natura studied the kinetics of defect formation and defect recovery.[62–65] Polina Ebeling identified and correlated the optical absorption bands with the electron spin resonance (ESR) signals of the paramagnetic defects in phosphate and FP glasses.[59,61,70,71] This work was soon extended to include extrinsic defects,[76] specifically, the photo-ionization of rare earth ions. This was the habilitation topic of Heike Ebendorff-Heidepriem.[57,68,69] The study was later extended to many other polyvalent elements, such as 3d, 4d, and 5d transition metal ions[20,21,23,28,34,66,67,73,75] and post transition metals, such as Sb, Pb, Sn or As.[73,74] More details on defect formation in FP glasses can be found in Doris Ehrt's own review of her work.[1] Doris Ehrt's research on defect formation included several industrial projects. She also instigated a close collaboration with the physics department of the University Jena by supplying them with the aforementioned FP laser glasses.[3,49,52–54,72,77–82]

Glass research in Jena – borosilicate glasses

Many publications exist on the development of glass research and the glass industry in Jena.[4–9] The chemist Johann W. Döbereiner (1780–1849) and the mechanic Friedrich J. C. Körner (1778–1847) were already trying to prepare new glass compositions with different properties in the 18th century. Their work was supported by none other than the poet, and at that time minister of state for the Duke of Weimar, Johann-Wolfgang von Goethe (1749–1832).[6,9] Apparently the experiments on glass melts in which CaO was to be replaced by BaO were not as effective as later attempts made in Jena by Otto Schott and his partners, Ernst Abbe and Carl Zeiss. The entrepreneurship, the research and development, and the social ideals and achievements of Abbe, Zeiss, and Schott, and their accomplishments for their workers and, in general, for the town of Jena, were quite remarkable.[5] Here, only the briefest summary can be given.[83] The mechanic and optician, Carl Zeiss (1816–1888), secured the scientific help of the professor of physics and mathematics, Ernst Abbe (1840–1905), in 1866. Abbe soon realised that a further improvement in the performance of the optical instruments assembled by Carl Zeiss could only be achieved by new glasses with certain optical properties not available at that time. Otto Schott (1851–1935) was the son of a glass workshop owner and himself a chemist and glass technologist. When he heard about the call for new glasses, he immediately sent samples of his private melting experiments to Ernst Abbe for characterisation.[5,83] After many letters in which samples and refractive index data were exchanged, Schott moved to Jena in 1882. Here, Abbe and Zeiss installed him in a small glass laboratory from which the *Jenaer Glaswerke Schott & Genossen* emerged in 1884. The initial collaboration between these three men originated in the need to improve optical glasses, but the first "third party funding", which actually provided the money to start the company, targeted mechanically improved glasses.[83] Specifically, the state gave a grant to improve and develop thermometer glass without detrimental zero-point-depression, which was a major problem for all types of scientific experiments at that time. Schott realised that one of his newly developed borosilicate glasses with a very low thermal expansion coefficient would solve the thermometer problem. These borosilicate glasses later became a mainstay of the success of the SCHOTT company.

Despite the commercial success of the borosilicate glass family, the chemistry and fundamental structure of these glasses was only poorly understood at that time. Almost a century later, Doris Ehrt studied phase separation and the temperature dependence of phase separation in borosilicate glasses in her diploma,[10] as well as in her dissertation thesis.[11–14] Even though she interrupted her studies of borosilicate glasses to study the fluoride phosphate glass system (see Doris Ehrt's article in this volume[1]), she resumed her research on borosilicate glasses for nearly the next 50 years. Interestingly, the number of her borate related publications increased again in the years prior to her retirement in 2011. In addition to the aforementioned study of radiation induced defects in borosilicate glasses,[16,21,25,28,34,63,75] Doris Ehrt looked at UV-transparent glasses:[84] the role of trace impurities especially in regard to absorption bands in the UV,[85] and, more generally, at the redox-equilibria of polyvalent ions in (boro-)silicate glasses.[17,24,29,32,37,86–90] The temperature dependence of the redox ratio of polyvalent ions[87–89] was studied, using a high frequency inductive heated melting setup. This procedure was also used in the study of alternative fining methods, including melting under vacuum.[91]

Other projects involved more fundamental questions regarding glass structure.[15] These questions included structural changes with thermal history[22,26,35] and the incorporation of Mn^{2+} ions in borosilicate glasses, for which phase separation once again becomes a significant issue.[43,45] As a member of the Technical Committee TC03 "Glass structure and properties" of the International Commission on Glass, Doris Ehrt provided four borosilicate glasses (NBS-A, -B, -C and -D) from homogenous industrial melts for a Round Robin NMR and modulation study.[92] Thus, she continued the study of some of the borosilicate glasses she knew so well from her PhD.[10–13] Previous results were augmented by viscosity and electrical conductivity measurements.[39] Nonlinear optical properties of thermally poled borosilicate glasses were also studied in collaboration with Efstratios I. Kamitsos in Athens.[41] These glasses are now being studied by the next generation of glass researchers. New studies are being conducted on the mechanical

Figure 8. Doris Ehrt and Oleg Mazurin in 2007 at the ICG in Strasbourg [Colour available online]

properties and the structural variations which are also probed by NMR spectroscopy.[93,94] Technical improvements promise radical new insights on the connectivities of the borate and silicate networks. At least three papers are currently under preparation. Speaking in the name of all their co-authors, we are very grateful for Doris Ehrt's active interest in our investigations, and for all the supportive discussions in which she has freely shared as yet unpublished information with us.

Binary phosphate and borate glasses

Otto Schott considered most elements of the periodic table when investigating all possible binary and mixed phosphate and borate glasses.[5,83] Continuing in his tradition, Doris Ehrt experimented with many different phosphate and borate compositions. The xMnO–(1−x)SrO–P_2O_5 system was thoroughly studied with regard to the optical, thermal and physical properties, and for correlations between the properties and the glass structure.[95] This glass series was selected by Doris Ehrt for photoluminescence studies.[30,31,96] Doris Ehrt was able to show that the OH-content had a significant influence on the glass structure and the glass transition temperature, T_g, in the SnO–P_2O_5 system.[97]

Tl_2O–B_2O_3, PbO–B_2O_3 and Bi_2O_3–B_2O_3 glasses were studied by Doris Ehrt for their nonlinear optical properties.[17] These initial studies were followed by more thorough studies of the structure and properties of ZnO–B_2O_3, La_2O_3–B_2O_3 and MnO–B_2O_3 glasses.[19,33,44,45] The most recent publication co-authored by Doris Ehrt is a paper on the luminescent properties of Nd^{3+} and Yb^{3+} doped $K_5Nd(MoO_4)_4$, $RbNd(WO_4)_2$ and $NdAl_3(BO_3)_4$ crystal matrices, published in 2015.[98]

Publications in the GDR

Many of Doris Ehrt's early publications are not accessible online, nor may they be found in most German Universities. During the existence of the GDR, it was difficult for East German scientists to publish in internationally recognized journals. First, they needed approval from the state in order to submit papers. Even with such approval, publication might still be denied by the "western" journal, not for scientific reasons, but because the researchers were considered to be too closely associated with the "eastern" competitor of a well-known German glass company. Doris Ehrt first learned about this problem years after reunification of Germany. Many of her older publications are written in German, but can now be downloaded from Doris Ehrt's page at the ResearchGate website. Over a dozen papers concerning glass formation in fluoroaluminate and fluoride phosphate glass systems are listed in the review paper by Doris Ehrt in this volume.(1) Two papers on phase separation and ion distribution in borosilicate glasses appeared in the GDR Jounal *Silikattechnik*.(12,13) However, some of the research which she did was printed in internationally renowned journals, such as the Russian journal *Fizika I Khimiya Stekla*.(14) The publication on phase separation in the $BaO–B_2O_3–SiO_2$ glass system was the result of a short research visit by Doris Ehrt at the Grebenshchikov Institute of Silicate Chemistry of the USSR Academy of Sciences, in Leningrad (St. Petersburg) with Evgeny A. Porai-Koshits and Oleg V. Mazurin in 1977. Figure 8 shows a picture of Doris Ehrt and Oleg Mazurin taken in 2007 on the occasion of the International Congress on Glass in Strasbourg, France.

International collaboration

During the existence of the GDR, scientific collaborations of the Otto Schott Institute were restricted to Russia and Eastern Europe. In addition to the collaboration with St. Petersburg, there were joint projects, including regular visits, with institutes in Bulgaria (e.g. to meet Ivan Gutzov, M. Marinov, and Bisserka Samuneva), Hungary and Poland. While Doris Ehrt never abandoned the old collaborations, she established new international collaborations after reunification of Germany. At the turn of the century, Jena and Aberdeen exchanged several PhD students after John A. Duffy extended the optical basicity concept to fluoride glasses.(99) In 2002, Doris began to collaborate with Efstratios I. Kamitsos in Greece.(35,41,43,100–103) Doris Ehrt is still a co-author of several publications which are currently under preparation, and structural investigations of her glasses by vibrational spectroscopy are being conducted in Athens.(45) Other collaborations extend to Poland,(98) and overseas to Nagaoka, Japan,(42) USA(19,104) and India.(38) In India in 2009, Doris Ehrt was on the advisory board and one of the Editors of the Conference Proceedings of the *International Seminar on Science and Technology of Glass Materials.*

Until recently, Doris Ehrt was one of the longest serving members of the International Organising Committee of the Phosphate Conference, as well as of the Borate Conference. She has supported the Balkan Conference on Glass Science and Technology from its very beginning, and hardly missed any of these conferences. Unquestionably, Doris Ehrt is marked by her experience as a participant and organiser of the International Otto Schott Colloqiua. These meetings have given glass scientists from the East and West the opportunity to meet in Jena and exchange ideas on glass research. The first meeting, known as "Jenaer Gespräche", was held in 1973 when Doris Ehrt was still a young researcher. She soon became one of the main organisers of this event, the last meeting of which (so far) was in 2006. The pictures in Figures 9 to 11 were all taken at earlier Otto Schott Colloquia, held between 1982 and 1990 in Jena. In Figure 9(a) and (b), Philip Bray and Doris Ehrt demonstrate the

Figure 9. Lectures at the International Otto Schott Colloquium in Jena. (a) Philip Bray in 1986 and (b) Doris Ehrt in 1990

Figure 10. Doris Ehrt discusses with Norbert Kreidl the structure of fluoride phosphate glasses, while on excursion to Schloß Großkochberg at the International Otto Schott Colloquium 1982

Figure 11. From left to right: Jaques Lucas, Ivan Gutzov, and Doris Ehrt, during the conference dinner at the Fuchsturm for the International Otto Schott Colloquium 1986

structure of borate and fluoride phosphate glasses, respectively. In Figure 10, Norbert Kreidl and Doris Ehrt discuss the structure of fluoride phosphate glasses. This structure was not as well understood in 1982 as it was when depicted on the blackboard in Figure 9(b); this picture was taken four years later. Coincidentally, Figure 10 reflects a pose often seen at conferences today: Doris Ehrt discussing the research of eager young students with the unquenchable curiosity and intensity so typical for her. Figure 11 gives an example of the cheerful character of the conference dinner of the Otto Schott Colloqium at the *Fuchsturm*. These dinners probably made the meetings in Jena as famous as the scientific discussions did.

Reunification of Germany

Doris Ehrt was appointed docent for glass chemistry at the FSU in 1987, not long before the fall of the Berlin wall in 1989. In 1990, Prof. W. Vogel retired. On 1st July 1990, the currency union between the FGR and GDR suspended all old regulations, ending all former GDR research contracts at midnight. Suddenly, new rules were set in place, though no-one was quite sure what these new rules were, and exactly how they would be implemented. After a short period of transition, in which the Otto Schott Institute was headed by several different directors, a steadier course was established in 1992 when Prof. Christian Rüssel assumed the post he still holds today. Doris Ehrt mastered the turbulence of the reunification well, and successfully procured funding for many interesting projects. She was appointed Hochschuldozent at the FSU with all the rights and duties of a professor in 1994. A direct quote from Doris Ehrt concerning the transition from GDR to FRG says: "*It was a very interesting time for me - no one knew how to proceed and there were so many possibilities ...*"

One reason for Doris's high success rate with grant proposals was her practical approach. She always combined fundamental research with practical technical applications. Furthermore, she has usually given a straightforward assessment of any experimental results and the potential of her research for industrial uses. As a consequence, she has shown herself to be a reliable partner. This quality has often resulted in long term collaborations on consecutive projects with the same industrial partner. The topics of her post-unification projects have covered a broad range. There have been seven projects on high temperature melts alone,[86–91,105] five focused on new glasses and materials,[33,106] and four on laser glasses.[3,52,54,72,78–81] The following series of ventures are also connected to laser glasses: seven projects related to UV-Vis induced defect formation in glasses[7,16,20–23,25,28,34,49,57–75] and two on the study of UV transmitting glasses.[85,90,107,108] A common element in all projects was the fundamental study of the structure and properties of these glasses,[15,17,19,26,35,39,41,45,95,109] for which potential applications, including the incorporation of polyvalent dopants,[24,29,32,37,86,110] the redox equilibrium,[18,28,34,73–75,87–89,105,111–113] bonding characteristics and site geometry,[21,22,31,43–45,95,96,114] were always considered.

One exceptional project stood out for its artistic character. The artist Susan Liebold contacted Doris Ehrt about a project involving fluorescing borosilicate glasses with low thermal expansion coefficients and a long working temperature, or long viscosity curve. The artist was looking for all types of interesting colours. Together they selected a range of low alkaline borosilicate glasses doped with Eu^{2+} (blue), Eu^{3+} (red), and Tb^{3+} (green-yellow), which fluoresce in the stated colours after excitation with black light (~370 nm). From these glasses, Susanne Liebold prepared the fluorescing tentacles of a model of the Siphnophora *Nanomia Cara*, depicted in Figure 12. The complete

Figure 12. Fluorescing tentacles of a model of the Siphnophora Nanomia Cara, which was prepared by the artist Susan Liebold from a low alkaline borosilicate glass (DURAN) melted by Doris Ehrt at the Otto Schott Institute. The red, blue, and green-yellow fluorescence after excitation by black light (~370 nm) is caused by the fluorescent dopants Eu^{2+}, Eu^{3+} and Tb^{3+} ions, respectively. (OZEANUM of Maritime Museum Stralsund, Germany ©2011)

model consists of 2600 pieces and, with a length of 2 m, is the size of a full grown organism. It can be seen at the OZEANEUM of the Maritime Museum Stralsund, Germany.

For references on fluoride phosphate glasses and rare earth doped glasses for luminescence studies see also the review paper by Doris Ehrt in this volume.[1]

The *Doktormutter*, her students and the work groups

Many of Doris Ehrt's ventures were also the topic of student projects. Between 1978 and 2011, Doris Ehrt supervised 20 Diploma theses, 14 PhD theses and two habilitation theses (Wolfgang Seeber in 1995 on *Amorphous materials for optics and optoelectronics on the example of Fluoride-Phosphate glasses*, and Heike Ebendorff-Heidempriem in 2000 on *Rare Earth ions as indicator and active ions in photonic glasses*). The last PhD thesis was successfully defended in 2014, after a growing family interrupted the student's research work. The author herself obtained her PhD on *Solarisation of Co- and Ni-doped glasses* in 2001 as a student of Doris Ehrt. She deeply treasures the continuing close collaboration with her former mentor and now friend.

Many projects were based on previous research, and students had not only to read and understand the preceding works, but their results and interpretations had to fit with those earlier findings. It was rumoured that this was considered a demanding task by other students, but all of Doris Ehrt's students enjoyed the interconnections between our projects, and the superb supervision of our *Doktormutter*. She would point us in the right direction, then leave us free to find our own solutions and pursue our own ideas. However, if she realised we were struggling, she was ready to help in whatever way she could.

Of course, such high quality research cannot be accomplished without the contribution of many treasured co-workers; all specialists in their fields. A few are shown in Figure 13 as representatives of the whole team. Each of the portrayed researchers and technicians has helped with very different aspects of the projects. Rotraud Atzrodt was Doris Ehrt's "right hand", and head of the laboratory. Hardly any words were required between them, as they seemed to have an unspoken understanding and agreement concerning experimental and analytical work. Even though Rotraud Atzrodt prepared most glasses, Doris Ehrt insisted on participating whenever new compositions were cast, as she wanted to get a feeling for the viscosity, or to watch phase separation and crystallization phenomena with her own eyes. Rotraudt Atzrodt, with her experience, her glass melting books and glass sample collection, was extremely helpful whenever any of us students needed information on glass preparation, or was looking for a certain dopant in a certain redox state in a certain glass matrix. Aptly, Rotraud Atzrodt and Doris Ehrt retired together in 2011. Bernd Keinert, a physicist, is shown characterising optical properties of the laser glass samples, i.e. the refractive indices and the partial dispersion. Gerda Heidemann is pictured while conducting fluorine analysis. She perfected this technique, and the wet chemical analysis of even ppm traces of iron impurities, over many years. As engineer, Wolfram Wintzer fine-tuned the inductive heated melting devices for laser glass production and helped to run the vacuum fining setup. Gabi Möller, a fine optician, has risen to every challenge, no matter how hygroscopic or strained the samples we give her might be, or whatever tricky sample dimensions we might think of.

Most of her former students, as well as her colleagues, still maintain close contact with Doris Ehrt. It makes no difference if they have retired, ventured into industry, or stayed in research. Even when they have drifted as far away from Jena as Heike Ebendorff-Heidepriem, who is now at the University of Adelaide, in Australia, close ties remain. Doris Ehrt's remarkable recall is treasured by all. She can remember incredible amounts of information about the structures and properties of the different glass systems that she has studied throughout the years. With her particular patience, she meticulously adds each new piece of structural information or newly obtained property to her knowledge of the

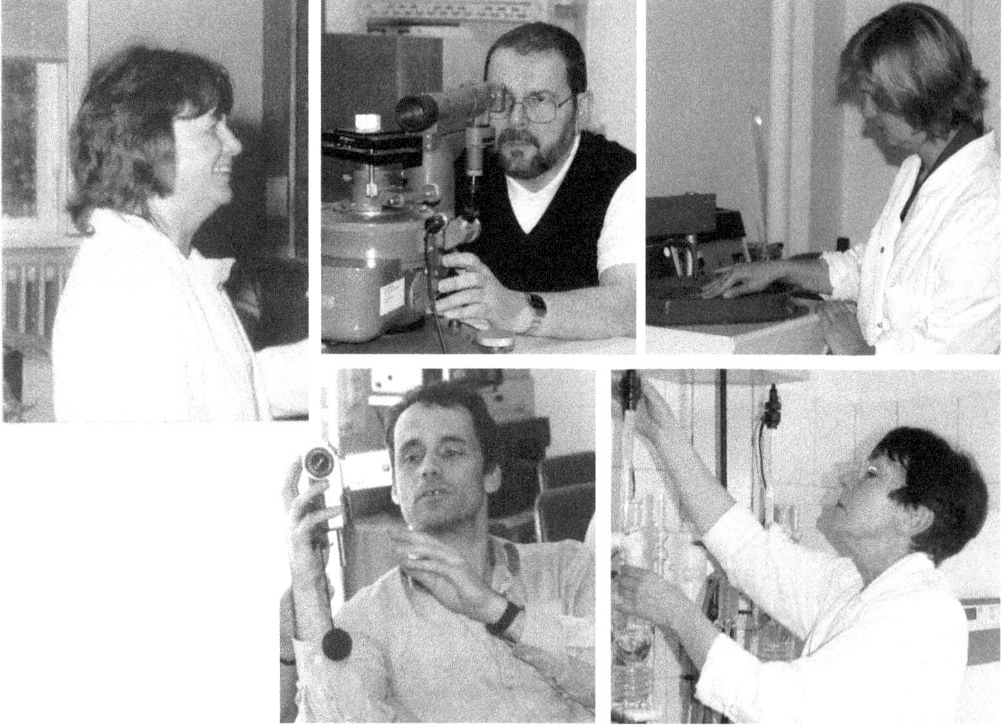

Figure 13. Standing in for the treasured staff of the Otto Schott Institute for Glass Chemistry: Rotraud Atzrodt (top left), the "right hand" of Doris Ehrt and head of the laboratory; Bernd Keinert (bottom left), during the optical characterization of laser glasses; Gabi Möller (top right) a fine optician, who rises to every challenge regarding sample preparation; Wolfram Wintzer (top centre), engineer who fine tuned the inductive heated melting setups for laser glass production; and Gerda Heidemann (bottom right), conducting fluorine analysis, which she perfected over the years

all-encompassing, never ending glass puzzle.

It is fitting that Doris Ehrt's profound research be acknowledged by her peers at the International Conference on Phosphate Glasses. Not only is she often invited for lectures, she has also been awarded distinguished prizes. In 1984, Doris Ehrt was given the ERNST-ABBE-PREIS, the most prestigious optical price of the CARL-ZEISS-STIFTUNG JENA. In 2000, she was awarded the Adolf Dietzel Industry Prize by the German Society of Glass Technology, one year before Heike Ebendorff-Heidepriem received the Woldemar Weyl Award 2001 for her habilitation research.

Life after retirement

In 2011, Doris Ehrt retired from the University and gave up her office at the Otto Schott Institute. Howev-

Figure 14. The audience of the Otto Schott Colloquium, 1982. Roland and Doris Ehrt can be identified easily, as can Matthias Müller, who operated the slide show, and was a great help in the preparation of this paper

er, she did not entirely retire from glass research. Her former students continue to ask for her advice, and many of her glass samples continue to be studied by the next generation of researchers at the Otto Schott Institute in Jena. Doris Ehrt also works as Associate Editor for the International Journal of Applied Glass Sciences, and thus keeps abreast of new scientific developments. Doris and Roland Ehrt continue to visit glass conferences together. Figure 14 shows them at the International Otto-Schott-Colloquium in 1982.

However, Doris Ehrt also enjoys her new found spare time. She nurses her herb garden, reads books for which she never before found the time, and takes dancing lessons with her husband. Furthermore, an increasing number of grandchildren, the fifth born in 2015, demand a fair share of their grandmother's time.

Acknowledgements

I am very grateful to Matthias Müller for helping with the structure of this paper. Together with Bernd Keinert, he also collected and provided many of the pictures from the OSIM archives. Doris and Roland Ehrt invited me to several memorable afternoons, during which they shared their recollections and anecdotes of the earlier years of the Otto Schott Institute with me. I was very grateful because I knew very little about this period in the history of the German Democratic Republic, as I was still growing up in West Germany at that time. I also want to thank Linda Bandilla and Alex Hannon for their time and effort in rephrasing the manuscript into a more readable English version.

References

1. Ehrt, D. Phosphate and Fluorophosphate Optical Glasses - Properties, Structure and Applications, *Phys. Chem. Glasses: Eur. J. Glass Sci. Technol. B*, 2015, 56 (6), 217–234.
2. Engel, A., Haspel, R., Resch-Renger, U., Hoffmann, K., Ehrt, D., Kolberg, U., Hayden, J. S., Summint, C. & Stelzel, M. Patent DE 10 2006 009 196 B4 2009.01.15: Verwendung eines Standards zur Referenzierung von Lumineszenzsignalen, Germany.
3. Töpfer, T., Hein, J., Wintzer, W., Ehrt, D. & Sauerbrey, R. Laser-Glass, Pump-Laser-Diodes, and Amplifier for the Polaris Laser, *Glass Sci. Technol.*, 2002, **75**, 223–234.
4. Ehrt, D. & Wright, A. C. Werner Vogel: Artistry with an Electron Beam, In: *Proceedings of the Fifth International Conference on Borate Glasses, Crystals and Melts*, Eds. G. Dalba & A. C. Wright, Society of Glass Technology, Sheffield, 2007, p. xix–xxxviii.
5. Zschimmer, E. *Chemical Technlogy of Glass*. Transl. M. Cable, Society of Glass Technology, Sheffield, 2013.
6. Vogel, W. & Ehrt, D. 25 Jahre Glaschemie in Jena, *Z. Chem.*, 1974, **14** (10), 396–404. (In German.). http://dx.doi.org/10.1002/zfch.19740141004
7. Ehrt, D. & Vogel, W. Radiation Effects in Glasses, *Nucl. Instrum. Meth. B*, 1992, **65** (1–4), 1–8. http://dx.doi.org/10.1016/0168-583X(92)95006-D
8. Rüssel, C. & Ehrt, D. Neue Entwicklungen in der Glaschemie, *Chem. unserer Zeit*, 1998, **32** (3), 126–135. (In German.) http://dx.doi.org/10.1002/ciuz.19980320303
9. Ehrt, D. Gläser Für die Hochleistungsoptik — ein Beispiel für industrienahe Forschung am Otto-Schott-Institut für Glaschemie der Universität Jena in den 1980er Jahren, In: *Monographie Band 43: Zeitzeugenberichte X - Chemische Industrie*. Eds. K. Krug & H. Bode, Gesellschaft Deutscher Chemiker, Frankfurt, 2011, p. 28-56 (in German).
10. Ehrt, D. Verteilung von Farbionen auf die Mikrophasen im Na_2O-B_2O_3-SiO_2 Glassystem, Diploma-Thesis, Friedrich-Schiller-University, Jena, 1970. (In German.)
11. Ehrt, D. Untersuchungen über die Verteilung von CoO und Fe_2O_3 auf die Mikrophasen des Glassystems Na_2O-B_2O_3-SiO_2, PhD-Thesis, Friedrich-Schiller-University, Jena, 1973. (In German.)
12. Ehrt, D., Reiß, H. & Vogel, W. Mikrostrukturuntersuchungen an CoO-haltigen Na_2O-B_2O_3-SiO_2-Gläsern, *Silikattechnik*, 1977, **28** (12), 359–364. (In German.)
13. Ehrt, D., Reiß, H. & Vogel, W. Einbau und Verteilung von Fe_2O_3 auf die Mikrophasen in Grundgläsern des Systems Na_2O-B_2O_3-SiO_2, *Silikattechnik*, 1976, **27** (9), 304–309. (In German.)
14. Averjanov, V. J., Mazurin, O. W., Porai-Koschiz, J. A., Reiß, H., Roskova, G. P., Vogel, W. & Ehrt, D. Influence of Temperature on the Conodes in Phase Separated Glass Melts of the System BaO-B_2O_3-SiO_2, *Fiz. Khim. Stekla*, 1979, **5** (6), 637–650. (In Russian.)
15. Kunath-Fandrei, G., Ehrt, D. & Jäger, C. Progress in Structural Elucidation of Glasses by ^{27}Al and ^{11}B Satellite Transition NMR-Spectroscopy *Z. Naturforsch. A*, 1995, **50** (4–5), 413–422.
16. Natura, U. & Ehrt, D. Formation of Radiation Defects in Silicate and Borosilicate Glasses Caused by UV Lamp and Excimer Laser Irradiation, *Glass Sci. Technol. - Glas Techn. Ber.*, 1999, **72** (9), 295–301.
17. Ehrt, D. Structure, Properties and Applications of Borate Glasses, *Glass Technol.*, 2000, **41** (6), 182–185.
18. Matthai, A., Ehrt, D. & Rüssel, C. Voltammetric Investigations of the Redox Behaviour of Fe, Ni, Co and Sn Doped Glass Melts of Ar® and Bk7® Type, *Glass Sci. Technol. - Glas Techn. Ber.*, 2000, **73** (2), 33–38.
19. Stone, C. E., Wright, A. C., Sinclair, R. N., Feller, S. A., Affatigato, M., Hogan, D. L., Nelson, N. D., Vira, C., Dimitriev, Y. B., Gattef, E. M. & Ehrt, D. Structure of Bismuth Borate Glasses, *Phys. Chem. Glasses*, 2000, **41** (6), 409–412.
20. Ehrt, D. UV-Absorption and Radiation Effects in Different Glasses Doped with Iron and Tin in the ppm Range, *Comptes Rendus Chimie*, 2002, **5** (11), 679–692. http://dx.doi.org/10.1016/s1631-0748(02)01432-7
21. Möncke, D. & Ehrt, D. Radiation-Induced Defects in CoO- and NiO-Doped Fluoride, Phosphate, Silicate and Borosilicate Glasses, *Glass Sci. Technol.*, 2002, **75** (5), 243–253.
22. Möncke, D. & Ehrt, D. Influence of Melting and Annealing Conditions on the Optical Spectra of a Borosilicate Glass Doped with CoO and NiO, *Glass Sci. Technol.*, 2002, **75** (4), 163–173.
23. Möncke, D., Marshall, R. & Ehrt, D. EPR- and Optical Spectroscopy of Irradiation Induced Defects in Transition Metal Doped Glasses, *Glass Sci. Technol.*, 2002, **75**, 36–41. http://dx.doi.org/10.1016/j.optmat.2003.11.001
24. Schütz, A., Ehrt, D., Dubiel, M., Yang, X., Mosel, B. & Eckert, H. Spectroscopic Characterisation of Fe-, Sb-, Sn-, Pb- and Ti-Complexes in Glasses, *Glass Sci. Technol.*, 2002, **75**, 442–445.
25. Ehrt, D. & Ebeling, P. Radiation Defects in Borosilicate Glasses, *Glass Technol.*, 2003, **44** (2), 46–49.
26. Möncke, D., Ehrt, D., Eckert, H. & Mertens, V. Influence of Melting and Annealing Conditions on the Structure of Borosilicate Glasses, *Phys. Chem. Glasses*, 2003, **44** (2), 113–116.
27. Ehrt, D. Photoluminescence in the UV-Vis Region of Polyvalent Ions in Glasses, *J. Non-Cryst. Solids*, 2004, **348** 22–29. http://dx.doi.org/10.1016/j.jnoncrysol.2004.08.121
28. Möncke, D. & Ehrt, D. Irradiation Induced Defects in Glasses Resulting in the Photoionization of Polyvalent Dopants, *Opt. Mater.*, 2004, **25** (4), 425–437. http://dx.doi.org/10.1016/j.optmat.2003.11.001
29. Schütz, A., Ehrt, D., Dubiel, M., Yang, X. C., Mosel, B. & Eckert, H. A Multi-Method Characterization of Borosilicate Glasses Doped with 1 up to 10 Mol% of Fe, Ti and Sb, *Glass Sci. Technol.*, 2004, **77** (6), 295–305.
30. Ehrt, D. & Herrmann, A. Photoluminescence Behaviour of Active Ions in Different Glasses, *Verre*, 2005, **11** 43–48.
31. Herrmann, A. & Ehrt, D. Time Resolved Fluorescence Measurements on Tb^{3+} and Mn^{2+} Doped Glasses, *Glass Sci. Technol.*, 2005, **78** (3), 99–105.
32. Yang, X. C., Dubiel, M., Ehrt, D. & Schütz, A. An Investigation of Valence State and Structural Geometry of Polyvalent Fe and Ti Ions in Sodium Borosilicate Glasses by X-Ray Absorption Spectroscopy, *Phys. Scr.*, 2005, **T115**, 445–447.
33. Ehrt, D. The Effect of ZnO, La_2O_3, PbO and Bi_2O_3 on the Properties of Binary Borate Glasses and Melts, *Phys. Chem. Glasses Eur. J. Glass Sci. Technol. B*, 2006, **47** (6), 669–674.
34. Möncke, D. & Ehrt, D. Photoinduced Redox-Reactions and Transmission Changes in Glasses Doped with 4d- and 5d-Ions, *J. Non-Cryst. Solids*, 2006, **352** (23–25), 2631–2636. http://dx.doi.org/10.1016/j.jnoncrysol.2006.03.034

35. Möncke, D., Ehrt, D., Varsamis, C.-P. E., Kamitsos, E. I. & Kalampounias, A. G. Thermal History of a Low Alkali Borosilicate Glass Probed by Infrared and Raman Spectroscopy, *Glas Techn. Ber. - Glass Sci. Technol.*, 2006, **47** (5), 133–137.
36. Herrmann, A., Ehrt, D., Time-Resolved Fluorescence Measurements on Dy^{3+} and Sm^{3+} Doped Glasses, *J. Non-Cryst. Solids*, 2008, **354** (10–11), 916–926. http://dx.doi.org/10.1016/j.jnoncrysol.2007.08.040
37. Yang, X. C., Dubiel, M., Ehrt, D. & Schütz, A. X-Ray Absorption Near Edge Structure Analysis of Valence State and Coordination Geometry of Ti Ions in Borosilicate Glasses, *J. Non-Cryst. Solids*, 2008, **354** (12–13), 1172–1174. http://dx.doi.org/10.1016/j.jnoncrysol.2006.11.034
38. Ehrt, D. Photoluminescence in Glasses and Glass Ceramics, *IOP Conf. Series: Mater. Sci. Eng.*, 2009, **2** (1), 012001.
39. Ehrt, D. & Keding, R. Electrical Conductivity and Viscosity of Borosilicate Glasses and Melts, *Phys. Chem. Glasses: Eur. J. Glass Sci. Technol. B*, 2009, **50** (3), 165–171.
40. Herrmann, A., Fibikar, S. & Ehrt, D. Time-Resolved Fluorescence Measurements on Eu^{3+}- and Eu^{2+}-Doped Glasses, *J. Non-Cryst. Solids*, 2009, **355** (43–44), 2093–2101. http://dx.doi.org/10.1016/j.jnoncrysol.2009.06.033
41. Möncke, D., Dussauze, M., Kamitsos, E. I., Varsamis, C. P. E. & Ehrt, D. Thermal Poling Induced Structural Changes in Sodium Borosilicate Glasses, *Phys. Chem. Glasses: Eur. J. Glass Sci. Technol. B*, 2009, **50** (3), 229–235.
42. Ehrt, D. Photoactive Glasses and Glass Ceramics, *IOP Conf. Series: Mater. Sci. Eng.*, 2011, **21** (1), 012001.
43. Möncke, D., Kamitsos, E. I., Herrmann, A., Ehrt, D. & Friedrich, M. Bonding and Ion-Ion Interactions of Mn^{2+} Ions in Fluoride-Phosphate and Boro-Silicate Glasses Probed by EPR and Fluorescence Spectroscopy, *J. Non-Cryst. Solids*, 2011, **357** (14), 2542–2551. http://dx.doi.org/10.1016/j.jnoncrysol.2011.02.017
44. Ehrt, D. Zinc and Manganese Borate Glasses – Phase Separation, Crystallisation, Photoluminescence and Structure, *Phys. Chem. Glasses Eur. J. Glass Sci. Technol. B*, 2013, **54** (2), 65–75.
45. Möncke, D., Ehrt, D. & Kamitsos, E. I. Spectroscopic Study of Manganese-Containing Borate and Borosilicate Glasses: Cluster Formation and Phase Separation, *Phys. Chem. Glasses: Eur. J. Glass Sci. Technol. B*, 2013, **54** (1), 42–51.
46. Ehrt, D. Grundlagenuntersuchungen für die Entwicklung von Fluorophosphat- und Fluoroaluminatgläsern mit extremer positiver anomaler Teildispersion, Dissertation B (Dr. sc. nat), Friedrich-Schiller-University, Jena, 1984. (In German.)
47. Ehrt, D., Atzrodt, R. & Vogel, W. Struktur-Eigenschaftsbeziehungen optischer Gläser mit anomaler Teildispersion, *2. Int. Otto-Schott-Colloquium*, Wiss. Z. d. Friedrich-Schiller-University, Jena, Germany, 1982, pp. 509–526. (In German.)
48. Vogel, W. *Glass Chemistry*, Second ed., Springer-Verlag, Berlin, 1992.
49. Ehrt, D. Fluoroaluminate Glasses for Lasers and Amplifiers, *Curr. Opin. Solid State Mater. Sci.*, 2003, **7** (2), 135–141. http://dx.doi.org/10.1016/s1359-0286(03)00049-4
50. Seeber, W., Ehrt, D. & Ebendorff-Heidepriem, H. Spectroscopic and Laser Properties of Ce^{3+}-Cr^{3+}-Nd^{3+} Co-Doped Fluoride Phosphate and Phosphate Glasses, *J. Non-Cryst. Solids*, 1994, **171** (1), 94–104. http://dx.doi.org/10.1016/0022-3093(94)90036-1
51. Seeber, W., Barth, S., Seifert, F., Ebendorff-Heidepriem, H. & Ehrt, D. New Yb-Doped Fluoride Phosphate Laser Glass-Structural Investigations Using Probe Ions, *J. Lumin.*, 1997, **72–74**, 449–450. http://dx.doi.org/10.1016/S0022-2313(96)00389-4
52. Philipps, J. F., Töpfer, T., Ebendorff-Heidepriem, H., Ehrt, D. & Sauerbrey, R. Spectroscopic and Lasing Properties of Er^{3+}:Yb^{3+}-Doped Fluoride Phosphate Glasses, *Appl. Phys. B*, 2001, **72** (4), 399–405.
53. Paoloni, S., Hein, J., Töpfer, T., Walther, H. G., Sauerbrey, R., Ehrt, D. & Wintzer, W. Laser Beam Induced Optical Aberrations in Phosphate and Fluoride Phosphate Glasses, *Appl. Phys. B*, 2004, **78** (3–4), 415–419. http://dx.doi.org/10.1007/s00340-004-1407-8
54. Rödel, C., Heyer, M., Behmke, M., Kübel, M., Jäckel, O., Ziegler, W., Ehrt, D., Kaluza, M. C. & Paulus, G. G. High Repetition Rate Plasma Mirror for Temporal Contrast Enhancement of Terawatt Femtosecond Laser Pulses by Three Orders of Magnitude, *Appl. Phys. B*, 2011, **103** (2), 295–302. http://dx.doi.org/10.1007/s00340-010-4329-7
55. Faraday, M. I. The Bakerian Lecture - on the Manufacture of Glass for Optical Purposes, *Phil. Trans. R. Soc. Lond.*, 1830, **120**, 1–57.
56. Faraday, M. Sur la coloration produit par la lumière, dans une espece particulière de carreaux de vitres, *Ann. Chim. Phys.*, 1825, **25**, 99–100. (In French.)
57. Ebendorff-Heidepriem, H. & Ehrt, D. UV Radiation Effects in Fluoride Phosphate Glasses, *J. Non-Cryst. Solids*, 1996, **196**, 113–117. http://dx.doi.org/10.1016/0022-3093(95)00561-7
58. Ebendorff-Heidepriem, H., Ehrt, D., Bettinelli, M. & Speghini, A. Effect of Glass Composition on Judd-Ofelt Parameters and Radiative Decay Rates of Er^{3+} Fluoride Phosphate and Phosphate Glasses, *J. Non-Cryst. Solids*, 1998, **240** (1–3), 66–78. http://dx.doi.org/10.1016/s0022-3093(98)00706-6
59. Ebeling, P., Ehrt, D. & Friedrich, M. Study of Radiation-Induced Defects in Fluoride-Phosphate Glasses by Means of Optical Absorption and EPR Spectroscopy, *Glass Sci. Technol. - Glas Techn. Ber.*, 2000, **73** (5), 156–162.
60. Ebendorff-Heidepriem, H. & Ehrt, D. Formation and UV Absorption of Cerium, Europium and Terbium Ions in Different Valencies in Glasses, *Opt. Mater.*, 2000, **15** (1), 7–25. http://dx.doi.org/10.1016/s0925-3467(00)00018-5
61. Ehrt, D., Ebeling, P. & Natura, U. UV Transmission and Radiation-Induced Defects in Phosphate and Fluoride-Phosphate Glasses, *J. Non-Cryst. Solids*, 2000, **263** (1–4), 240–250. http://dx.doi.org/10.1016/s0022-3093(99)00681-x
62. Natura, U., Feurer, T. & Ehrt, D. Kinetics of UV Laser Radiation Defects in High Performance Glasses, *Nucl. Instrum. Methods B*, 2000, **166**, 470–475. http://dx.doi.org/10.1016/s0168-583x(99)00698-9
63. Natura, U., Ehrt, D. & Naumann, K. Formation of Radiation Defects in High-Purity Silicate Glasses in Dependence on Dopants and UV Radiation Sources, *Glass Sci. Technol. - Glas Techn. Ber.*, 2001, **74** (2), 23–31.
64. Natura, U. & Ehrt, D. Modeling of Excimer Laser Radiation Induced Defect Generation in Fluoride Phosphate Glasses, *Nucl. Instrum. Meth. B*, 2001, **174** (1–2), 151–158. http://dx.doi.org/10.1016/s0168-583x(00)00450-x
65. Natura, U. & Ehrt, D. Generation and Healing Behavior of Radiation-Induced Optical Absorption in Fluoride Phosphate Glasses: The Dependence on UV Radiation Sources and Temperature, *Nucl. Instrum. Meth. B*, 2001, **174** (1–2), 143–150. http://dx.doi.org/10.1016/s0168-583x(00)00449-3
66. Möncke, D. & Ehrt, D. Radiation-Induced Defects in CoO- and NiO-Doped Fluoride-Phosphate Glasses, *Glass Sci. Technol. - Glas Techn. Ber.*, 2001, **74** (3), 65–73.
67. Möncke, D. & Ehrt, D. Irradiation-Induced Defects in Different Glasses Demonstrated on a Metaphosphate Glass, *Glass Sci. Technol. - Glas Techn. Ber.*, 2001, **74** (7), 199–209.
68. Ebendorff-Heidepriem, H. & Ehrt, D. Effect of Europium Ions on X-Ray-Induced Defect Formation in Phosphate Containing Glasses, *Opt. Mater.*, 2002, **19** (3), 351–363. http://dx.doi.org/10.1016/S0925-3467(01)00237-3
69. Ebendorff-Heidepriem, H. & Ehrt, D. Effect of Tb^{3+} Ions on X-Ray-Induced Defect Formation in Phosphate Containing Glasses, *Opt. Mater.*, 2002, **18** (4), 419–430. http://dx.doi.org/10.1016/s0925-3467(01)00182-3
70. Ebeling, P., Ehrt, D. & Friedrich, M. X-Ray Induced Effects in Phosphate Glasses, *Opt. Mater.*, 2002, **20** (2), 101–111. http://dx.doi.org/10.1016/s0925-3467(02)00052-6
71. Ebeling, P., Ehrt, D. & Friedrich, M. Influence of Modifier Cations on the Radiation-Induced Effects of Metaphosphate Glasses, *Glass Sci. Technol.*, 2003, **76** (2), 56–61.
72. Ehrt, D., Kittel, T., Will, M., Nolte, S. & Tünnermann, A. Femtosecond-Laser-Writing in Various Glasses, *J. Non-Cryst. Solids*, 2004, **345–346**, 332–337. http://dx.doi.org/10.1016/j.jnoncrysol.2004.08.039
73. Möncke, D. & Ehrt, D. UV Light Induced Photoreduction in Phosphate and Fluoride-Phosphate Glasses Doped with Ni^{2+}, Ta^{5+}, Pb^{2+}, and Ag^{+} Compounds, *Glass Sci. Technol.*, 2004, **77** (5), 239–248.
74. Möncke, D. & Ehrt, D. Photoionization of As, Sb, Sn, and Pb in Metaphosphate Glasses, *J. Non-Cryst. Solids*, 2004, **345**, 319–322. http://dx.doi.org/10.1016/j.jnoncrysol.2004.08.036
75. Möncke, D. & Ehrt, D. Photoinduced Redox Reactions in Zr, Nb, Ta, Mo, and W Doped Glasses, *Phys. Chem. Glasses: Eur. J. Glass Sci. Technol. B*, 2007, **48** (5), 317–323.
76. Möncke, D. & Ehrt, D. Photoionization of Polyvalent Ions, In: *Materials Science Research Horizons*, Ed. H.P.Glick, Nova Science Publishers Inc., New York, 2007, pp. 1–56.
77. Petrov, V., Griebner, U., Ehrt, D. & Seeber, W. Femtosecond Self Mode Locking of Yb:Fluoride Phosphate Glass Laser, *Opt. Lett.*, 1997, **22** (6), 408–410. http://dx.doi.org/10.1364/ol.22.000408
78. Philipps, J. F., Töpfer, T., Ebendorff-Heidepriem, H., Ehrt, D., Sauerbrey, R. & Borrelli, N. F. Diode-Pumped Erbium-Ytterbium-Glass Laser Passively Q-Switched with a PbS Semiconductor Quantum-Dot Doped Glass, *Appl. Phys. B*, 2001, **72** (2), 175–178. http://dx.doi.org/10.1007/s003400000415
79. Hein, J., Podleska, S., Siebold, M., Hellwing, M., Bodefeld, R.,

Sauerbrey, R., Ehrt, D. & Wintzer, W. Diode-Pumped Chirped Pulse Amplification to the Joule Level, *Appl. Phys. B*, 2004, **79** (4), 419–422. http://dx.doi.org/10.1007/s00340-004-1586-3
80. Philipps, J. F., Töpfer, T., Ebendorff-Heidepriem, H., Ehrt, D., Sauerbrey, R. & Borrelli, N. F. Diode-Pumped Erbium-Ytterbium-Glass Laser Passively Q-Switched with a PbS Semiconductor Quantum-Dot Doped Glass (Vol 72, Pg 175, 2001), *Appl. Phys. B*, 2002, **74** (3), 285–285. http://dx.doi.org/10.1007/s003400100687
81. Philipps, J. F., Töpfer, T., Ebendorff-Heidepriem, H., Ehrt, D. & Sauerbrey, R. Energy Transfer and Upconversion in Erbium–Ytterbium-Doped Fluoride Phosphate Glasses, *Appl. Phys. B*, 2002, **74** (3), 233–236. http://dx.doi.org/10.1007/s003400200804
82. Possner, T., Ehrt, D., Sargsjan, G. & Unger, C. Stripe Waveguides by Cs^+- and K^+-Exchange in Neodymium-Doped Soda Silicate Glasses for Laser Application, *Proc. SPIE*, Glasses for Optoelectronics II, 1991, **1513** 378–385.
83. Kühnert, H. *Forschungen zur Geschichte des Jenaer Glaswerks Scott & Genossen*, Böhlau Verlag, Köln, 2012. (In German.)
84. Seeber, W. & Ehrt, D. Estimation of Deep-UVand UV Absorption Coefficients of Selected Trace Impurities in Glasses, *Glastech. Ber.-Glass Sci. Technol.*, 1997, **70** (10), 312–315.
85. Ehrt, D. Redox Behaviour of Polyvalent Ions in the ppm Range, *J. Non-Cryst. Solids*, 1996, **196** 304–308. http://dx.doi.org/10.1016/0022-3093(95)00604-4
86. Ehrt, D., Leister, M. & Matthai, A. Polyvalent Elements Iron, Tin and Titanium in Silicate, Phosphate and Fluoride Glasses and Melts, *Phys. Chem. Glasses*, 2001, **42** (3), 231–239.
87. Leister, M. & Ehrt, D. The Influence of High Melting Temperatures on the Behaviour of Polyvalent Ions in Silicate Glasses, *Glass Sci. Technol. - Glas Techn. Ber.*, 2000, **73**, 194–203.
88. Leister, M., Ehrt, D., von der Gönna, G., Rüssel, C. & Breitbarth, F. W. Redox States and Coordination of Vanadium in Sodium Silicates Melted at High Temperatures, *Phys. Chem. Glasses*, 1999, **40** (6), 319–325.
89. Leister, M. & Ehrt, D. Redox Behavior of Iron and Vanadium Ions in Silicate Melts at Temperatures up to 2000 Degrees C, *Glass Sci. Technol. - Glas Techn. Ber.*, 1999, **72** (5), 153–160.
90. Ehrt, D., Leister, M. & Matthai, A. Redox Behaviour in Glass Forming Melts, *Molten Salt Forum*, 1998, **5–6**, 547–554.
91. Wintzer, W., Romhild, S. & Ehrt, D. Effect of Vacuum on Fining Behaviour of Glass Melts, *Glass Sci. Technol.*, 2002, **75** 485–488.
92. Report of Technical Committee 3 of the International Commission on Glass http://www.icglass.org/technical_committees/?id=1&committe e=TC03:_Glass_Structure
93. Winterstein-Beckmann, A., Möncke, D., Palles, D., Kamitsos, E. I. & Wondraczek, L. Raman Spectroscopic Study of Structural Changes Induced by Micro-Indentation in Low Alkali Borosilicate Glasses, *J. Non-Cryst. Solids*, 2014, **401**, 110–114. http://dx.doi.org/10.1016/j.jnoncrysol.2013.12.038
94. Möncke, D., Tricot, G., Winterstein-Beckmann, A., Wondraczek, L. & Kamitsos, E. I. On the Connectivity of Borate Tetrahedra in Borate and Borosilicate Glasses, *Phys. Chem. Glasses Eur. J. Glass Sci. Technol. B*, 2015, **56** (5), 203–211.
95. Konidakis, I., Varsamis, C.-P. E., Kamitsos, E. I., Möncke, D. & Ehrt, D. Structure and Properties of Mixed Strontium-Manganese Metaphosphate Glasses, *J. Phys. Chem. C*, 2010, **114** (19), 9125–9138. http://dx.doi.org/10.1021/jp101750t
96. Herrmann, A. & Ehrt, D. Hightech-Gläser für die Photonik, *Laser+Photonik*, 2005, 12–14. (In German.)
97. Ehrt, D. Effect of OH-Content on Thermal and Chemical Properties of $SnO-P_2O_5$ Glasses, *J. Non-Cryst. Solids*, 2008, **354** (2–9), 546–552. http://dx.doi.org/10.1016/j.jnoncrysol.2007.07.092
98. Piasecki, M., Mandowska, E., Herrmann, A., Ehrt, D., Majchrowski, A., Jaroszewicz, L. R., Brik, M. G. & Kityk, I. V. Tailoring Nd^{3+} Luminescence Characteristics by Yb^{3+} Doping in $K_5Nd(MoO_4)_4$, $RbNd(WO_4)_2$ and $NdAl_3(BO_3)_4$ Crystal Matrices, *J. Alloys Compd.*, 2015, **639**, 577–582. http://dx.doi.org/10.1016/j.jallcom.2015.03.018
99. Duffy, J. A. A Common Optical Basicity Scale for Oxide and Fluoride Glasses, *J. Non-Cryst. Solids*, 1989, **109**, 35–39.
100. Möncke, D., Ehrt, D., Velli, L. L., Varsamis, C. P. E. & Kamitsos, E. I. Structure and Properties of Mixed Phosphate and Fluoride Glasses, *Phys. Chem. Glasses*, 2005, **46** (2), 67–71.
101. Velli, L. L., Varsamis, C. P. E., Kamitsos, E. I., Möncke, D. & Ehrt, D. Structural Investigation of Metaphosphate Glasses, *Phys. Chem. Glasses*, 2005, **46** (2), 178–181.
102. Möncke, D., Ehrt, D., Velli, L. L., Varsamis, C. P. E., Kamitsos, E. I., Elbers, S. & Eckert, H. Comparative Spectroscopic Investigation of Different Types of Fluoride Phosphate Glasses, *Phys. Chem. Glasses: Eur. J. Glass Sci. Technol. B*, 2007, **48** (6), 399–402.
103. Velli, L. L., Varsamis, C. P. E., Kamitsos, E. I., Möncke, D. & Ehrt, D. Optical Basicity and Refractivity in Mixed Oxyfluoride Glasses, *Phys. Chem. Glasses: Eur. J. Glass Sci. Technol. B*, 2008, **49** (4), 182–187.
104. Glebova, L., Ehrt, D. & Glebov, L. Luminescence of Dopants in PTR Glass, *Phys. Chem. Glasses: Eur. J. Glass Sci. Technol. B*, 2007, **48** (5), 328–331.
105. Ehrt, D. Redox Behaviour of Sn^{4+}/Sn^{2+} in Akali Free Aluminosilicate Glasses and Melts, *Phys. Chem. Glasses: Eur. J. Glass Sci. Technol. B*, 2008, **49** (2), 68–72.
106. Kolberg, U., Curdt, A., Gierke, M., Winkler-Trudewig, M., Kron, G. & Ehrt, D. Patent DE 10 2007 063 463 B4 2010.06.10: Kernglas im Alkali-Zink-Silikat-Glassystem für einen faseroptischen Lichtleiter und faseroptische Lichtleiter mit diesem Kernglas, Germany.
107. Ehrt, D. & Seeber, W. Glass for High Performance Optics and Laser Technology, *J. Non-Cryst. Solids*, 1991, **129** (1–3), 19–30. http://dx.doi.org/10.1016/0022-3093(91)90076-I
108. Ehrt, D., Carl, M., Kittel, T., Müller, M. & Seeber, W. High-Performance Glass for the Deep Ultraviolet Range, *J. Non-Cryst. Solids*, 1994, **177**, 405–419. http://dx.doi.org/10.1016/0022-3093(94)90555-X
109. Ebendorff-Heidepriem, H. & Ehrt, D. Determination of the OH Content of Glasses *Glastech. Ber. -Glass Sci. Technol.*, 1995, **68** (5), 139–146.
110. Seeber, W. & Ehrt, D. Spectroscopic Properties of Cu^+ and Tm^{3+} in Glasses, *phys. status solidi (a)*, 1992, **130** (2), K215–K217. http://dx.doi.org/10.1002/pssa.2211300248
111. Müller, M., Carl, M., Kittel, T. & Ehrt, D. Carbon Crucible Technology for Optical-Glass Melting *Glastech. Ber. - Glass Sci. Technol.*, 1995, **68** (10), 312–317.
112. Matthai, A., Claussen, O., Ehrt, D. & Rüssel, C. Thermodynamics of Redox Equilibria and Diffusion of Polyvalent Ions in a Phosphate Glass Melt, *Glastech. Ber. - Glass Sci. Technol.*, 1998, **71** (2), 29–34.
113. Matthai, A., Ehrt, D. & Rüssel, C. Redox Behaviour of Polyvalent Ions in Phosphate Glass Melts and Phosphate Glasses, *Glastech. Ber. - Glass Sci. Technol.*, 1998, **71** (7), 187–192.
114. Seeber, W. & Ehrt, D. Probe Ions for Local Structure Investigations of Glasses, *Ber. Bunsen Phys. Chem.*, 1996, **100** (9), 1593–1595.

Phys. Chem. Glasses: Eur. J. Glass Sci. Technol. B, December 2014, **55** (6), 280–287

The structure of a borosilicate and phosphosilicate glasses and its evolution at temperatures above the glass transition temperature: lessons from *in situ* MAS NMR

S. Venkatachalam,[1] *C. Schröder,*[2] *S. Wegner*[3] *& L. van Wüllen*[1,*]

[1] *Institute of Physics, Augsburg University, Universitätsstr. 1, D86159 Augsburg, Germany*
[2] *Institut für Physikalische Chemie, Westfälische Wilhelms-Universität Münster, Correnssstr. 30 D48149 Münster Germany*
[3] *Bruker Biospin GmbH, Silberstreifen, D76287 Rheinstetten Germany*

Manuscript received 2 April 2014
Accepted 8 September 2014

In this report, we present a study of the network organization of phosphosilicate glasses and a borosilicate glass at ambient temperature and its evolution upon heating and annealing. A range of advanced solid state NMR approaches including REDOR (rotational echo double resonance) NMR spectroscopy, 2D heteronuclear correlation spectroscopy (HETCOR), radio frequency driven recoupling (RFDR) as well as in situ MAS and REDOR NMR spectroscopy up to temperatures of 665°C was employed to evaluate the network organization at ambient temperatures and its evolution upon heating. For the borosilicate glass, the results reveal a cleavage of SiO_4–BO_4 linkages at temperatures above the glass transition with subsequent crystallisation of binary alkali silicate phases. For the phosphosilicate glasses, the six coordinated silicon present in the glasses is being converted to five coordinate silicon upon heating.

Introduction

Owing to the prospect to tune the physical or chemical key properties via variations of e.g. composition or processing conditions such as melting and annealing temperature, amorphous solids represent a rather important branch of materials science. Among the oxidic glasses, phosphate and borate based glasses find a widespread range of applications as biomaterials,[1–4] sealing glasses,[5] separators, laser hosts[6] or nuclear waste storage.[7–10] The macroscopic properties at ambient temperatures are governed by the microscopic structure of the material which in turn significantly depends on the processing conditions (e.g. phase separation, network organization). This is mainly due to the slow kinetics of the equilibria at temperatures around and above the glass transition temperature, which govern the final glass structure at ambient temperatures. The faster a glass melt is cooled, the less time is available for the system to structurally relax and adjust to the changing equilibria. At the glass transition, then, a state of high entropy is frozen, corresponding to a high fictive temperature. To analyse the structural changes in a glass network which occur at high temperatures, two routes may principally be followed. In the *ex situ* approach, a set of glasses with identical composition is prepared employing a range of quench rates (or annealing temperatures (T_a)), then, the structural evolution is being monitored in subsequent experiments, performed at room temperature. However, due to the limited range of quench rates available for routine glass making, this approach covers a limited temperature window. The alternative route, the *in situ* approach, offers to monitor the changes in the network organization at temperatures around and above the glass transition temperature as they happen. This opens the way to e.g. elucidate reversible changes in the network organization (structural relaxation) or to characterise the intrinsic details of crystallisation processes.

Due to the lack of translatorial periodicity diffraction techniques are of only limited values for the structural characterization of amorphous solids, thus entailing the necessity to resort to alternative techniques such as XANES, EXAFS, XPS, Raman/IR spectroscopy or NMR spectroscopy. Among these, especially solid state NMR spectroscopy has emerged as a rather powerful technique for the investigation of structural motifs on short and intermediate length scales. The *short range order* (1–2 Å) can be successfully analysed employing magic angle spinning NMR spectroscopy, aiding in the identification and assignment of specific silicate, borate or phosphate polyhedra with different degrees of connectivity (local building units). Information about the *intermediate*

Corresponding author. Email leo.van.wuellen@physik.uni-augsburg.de
Original version presented at Int. Conf. on Phosphate Glasses, Pardubice, Czech Republic, 2–4 July 2014

range order (2–8 Å) in the glasses may be obtained via an exploration of the homo- and heteronuclear dipolar couplings between two nuclei which scale with the inverse cube of the internuclear distance. Here, one can tap from a large inventory of solid state NMR strategies to quantitatively determine homo- and heteronuclear dipolar couplings and to analyse these with respect to the spatial distribution of a given chemical species and connectivity patterns. Approaches to evaluate homonuclear dipolar couplings such as 2D-RFDR[11,12] and 2D-double quantum-NMR experiments as well as heteronuclear dipolar couplings (REDOR and related approaches[13–19]) have contributed enormously to the identification of connectivity patterns in amorphous solids at ambient temperatures, that is the spatial arrangement of the local building polyhedra towards an extended three-dimensional network.[20–23]

Whereas a multitude of studies has been devoted to the structural characterization of oxide glasses and the evolution of the structural details with temperature employing the *ex situ* approach, only a rather limited amount of work following the *in situ* approach has appeared over the last two decades. Apart from *in situ* XANES studies and *in situ* Raman scattering again *in situ* solid state NMR techniques have played a pivotal role to follow the evolution of the structure of amorphous solids with temperature. The introduction of *in situ* MAS NMR spectroscopy techniques for temperatures up to 700°C is inevitably linked to J. F. Stebbins and co-workers who provided important contributions about structural relaxation processes and dynamic species exchange in silicate glasses,[24–26] borate and borosilicate glasses.[27,28] Contributions from our laboratory include *in situ* studies on (alumino)phosphate[29,30] and borosilicate glasses.[31]

In this report, we present a study of the network organization of phosphosilicate glasses and a borosilicate glass at ambient temperature and its evolution upon heating and annealing. For the *borosilicate glass*, the network structure and its evolution with temperature was analysed employing ^{11}B- and ^{29}Si-MAS-NMR as well as ^{11}B{^{29}Si}REDOR-NMR spectroscopy[20,32] to evaluate the amount of mixing of borate and silicate units as a function of temperature. Among the family of silicate glasses *phosphosilicate glasses* present a special case, since in these glasses silicon may adopt an octahedral coordination by oxygen.[33–37] In addition, some authors report the existence of five-coordinated silicon in these systems. Of special interest is the fate of the six-coordinated silicon at temperatures above T_g. Dupree *et al* find a significant reduction of this species with increasing fictive temperature in earth alkali phosphosilicate and binary phosphosilicate glasses.[36] We have employed a range of advanced solid state NMR techniques including ^{29}Si{^{31}P}- and ^{31}P{^{29}Si}-REDOR-NMR spectroscopy, 2D-cross-polarization heteronuclear correlation spectroscopy (2D CP-HETCOR) and ^{29}Si- and ^{31}P-RFDR-NMR experiments to establish the network organization in phosphosilicate glasses at ambient temperatures. The structural changes at temperatures above the glass transition temperature have been analysed employing *in situ* ^{29}Si- and ^{31}P- MAS-NMR and ^{29}Si{^{31}P}-REDOR-NMR spectroscopy up to temperatures of 650°C.

Table 1. Nominal compositions of the studied glasses

Glass	Na_2O/%	Cs_2O/%	P_2O_5/%	SiO_2/%	B_2O_3/%
PS-1	16·7		50	33·3	
PS-2	35		55	10	
PS-3		35	55	10	
BS-1	45			45	10

Experimental

Sample preparation

Transparent glasses were prepared using Na_2CO_3, Cs_2CO_3, P_2O_5, $^{29}SiO_2$ and B_2O_3 as starting material. B_2O_3 was melted and cooled back to ambient temperatures prior to being used for the synthesis. The glasses were prepared employing the standard melt quenching method. Materials were mixed in appropriate amounts and then heated in a Pt crucible to 600–800°C to remove CO_2, then, melting was performed at 1300–1450°C for 30–60 min with subsequent quenching on a copper plate. The nominal compositions of the glasses studied in this work are given in Table 1. All glasses were doped with 0·001–0·005 mol% $MnCO_3$ to reduce the long spin lattice relaxation times for ^{29}Si and ^{31}P. All glasses were prepared using $^{29}SiO_2$.

NMR experiments

The *in situ* NMR experiments were performed on a Bruker Avance III NMR spectrometer operating at 7 T with resonance frequencies of 59·6 MHz, 96·2 MHz and 121·5 MHz for ^{29}Si, ^{11}B and ^{31}P, respectively, employing a 7 mm XY-Bruker LASER-MAS probe. The bottom-less MAS rotor is equipped with an inner container made from aluminium nitride (AlN) which carries the sample. Heating of the sample is achieved using a 200 W diode laser operating at 976 nm. The laser beam is fed through an optical fibre into the probe, the fibre ending ca. 1 cm underneath the stator, and then directed to the AlN container. Temperature calibration was performed using the ^{79}Br-MAS-NMR signal of KBr (cf. Figure 1).The chemical shift exhibits a temperature dependence given by[38]

$$T\,/\,°C = \frac{\delta_{iso}}{0{\cdot}025} + 295$$

with δ_{iso} (295 K)=0. Thurber *et al*[38] used this approach in the limited temperature range 200<*T*<350 K, however, the linearity between chemical shift and laser current/temperature extends to *T*=950 K. In addition,

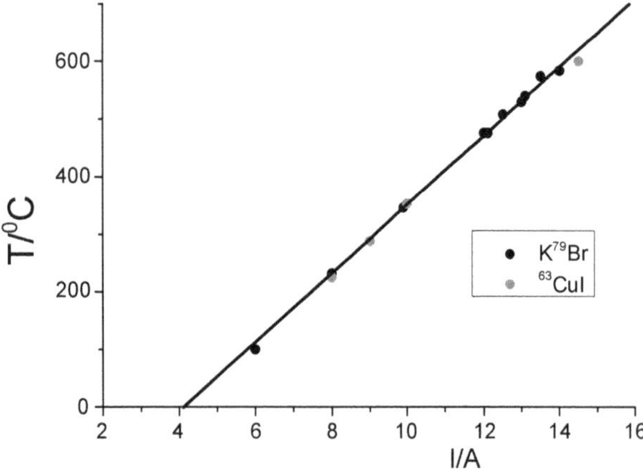

Figure 1. Temperature calibration of the LASER-MAS probe using the 79Br resonance of KBr and the 63Cu resonance of CuI [Colour available online]

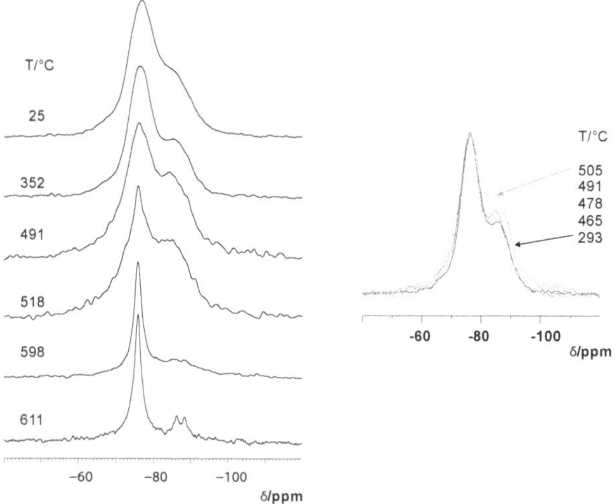

Figure 2. In situ ^{29}Si-MAS-NMR spectra of the Na_2O:B_2O_3:SiO_2 glass BS-1 [Colour available online]

the calibration was double checked employing the temperature dependence of the ^{63}Cu-MAS-NMR signal of ^{63}Cu in α-CuI[39,40] and the melting point of CuI. MAS was performed at 4000 Hz for the *in situ* MAS NMR experiments. For the ^{29}Si- and ^{31}P-RFDR-MAS-NMR experiments, a 4 mm Bruker MAS probe and MAS frequencies of 10 kHz was used. The REDOR and CPHETCOR experiments at ambient temperatures were performed on a Varian DNMR spectrometer operating at 11·7 T with resonance frequencies of 99·5, 132·3 and 202·3 MHz for ^{29}Si, ^{11}B and ^{31}P, respectively, employing a 1·6 mm Varian T3-MAS probe. The REDOR experiments were performed at a rotation speed of 6000 Hz, the CPHETCOR at 10000 Hz. Further details of the experiments are given in the respective figure captions.

Results and discussion

Borosilicate glass

The *in situ* ^{29}Si-MAS-NMR spectra as a function of temperature for glass BS-1 are collected in Figure 2. Three different local environments with isotropic chemical shifts of −69·3, −76·5 and −85·2 ppm can be identified in the spectra, which can be assigned to $Q^{(1)}$-, $Q^{(2)}$- and $Q^{(3)}$-units, respectively. The intensity of the signal at −85·2 ppm increases considerably with temperature. At temperatures of 518°C crystallization of first Na_2SiO_3 (signal at −76·2 ppm[41]), then ß-$Na_2Si_2O_5$ (signals at −86·7 and −88·6 ppm[42]) occurs. In the *in situ* ^{11}B-MAS-NMR spectra, collected in Figure 3, the signals for tetrahedrally coordinated BO_4-units (narrow signal around −0·5 ppm) and trigonally planar coordinated BO_3-units (broad signal extending from 10 to −10 ppm, broadened by the strong second order quadrupolar interaction) can be identified. However, the spectra suffer from a dominant background signal, which originates from the stator made from boron nitride and from the MACOR rings (Macor contains trigonally planar borate units) which are needed for heat isolation. The contribution from the BN can be effectively suppressed performing a rotor synchronized spin echo experiment, the contribution from the second source however has to be eliminated by a blank experiment. Due to the low boron content of the glass studied here (10%), the subtraction process entails a severe deterioration of the signal/noise ratio and was therefore abandoned. Thus, the spectra plotted in Figure 3 contain the contribution from the Macor-rings. Since the signal for the BO_4 units however is not affected by the background signal, we concentrate on the evolution of this signal with temperature to analyze the data. Although this entails a slight reduction in the information content of the spectra, it is clear that the relative fraction of the BO_4 signal considerably decreases beginning at a temperature of 480°C. This reduction proves to be irreversible as exemplified by a comparison of spectra taken at 353°C during heating and during

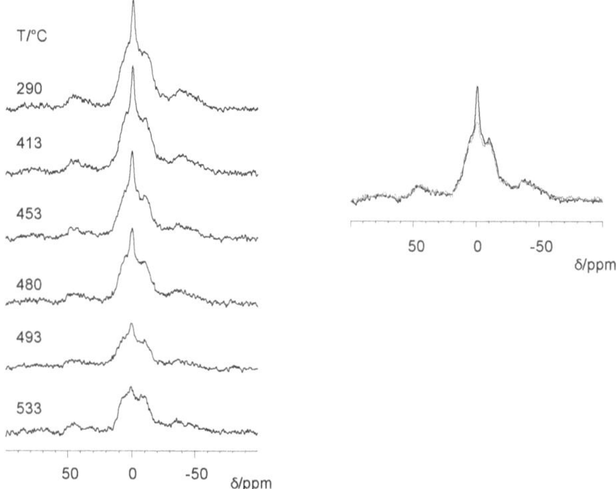

Figure 3. Left: In situ 11B-MAS NMR spectra of the sodium borosilicate glass for the indicated temperatues; right: spectra at 353°C acquired during the heating (black) and cooling (red) cycle [Colour available online]

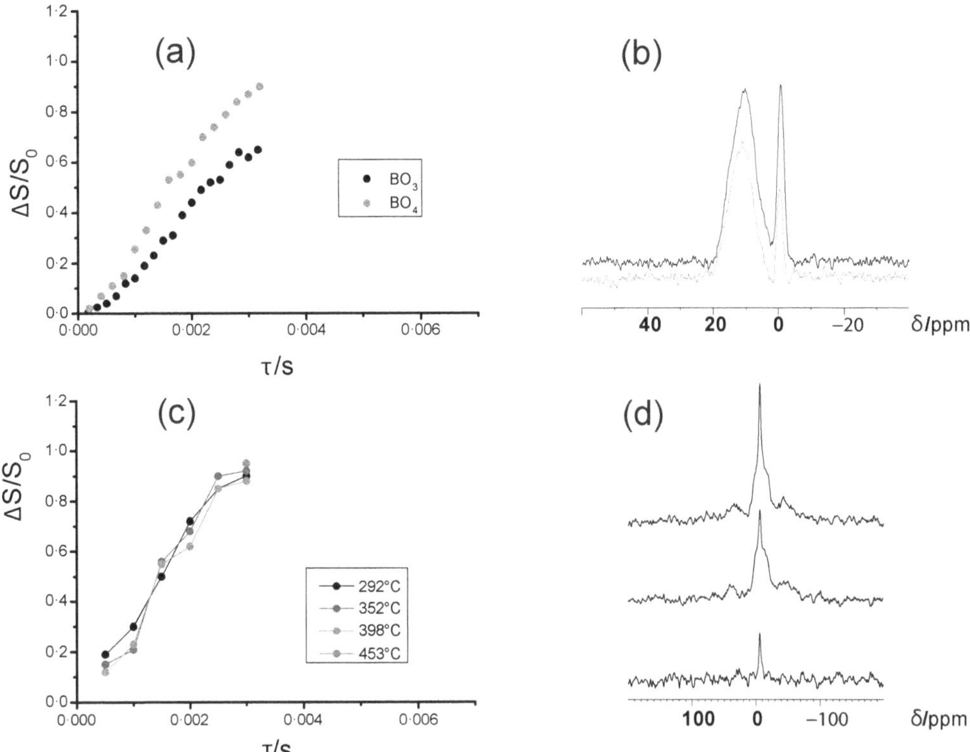

Figure 4. Results of a ^{11}B{^{29}Si}-REDOR-NMR experiment at room temperature employing a 11·7 T magnet (a) and under in situ conditions employing a 7 T magnet (c). In (b), a spin-echo spectrum (black) and the corresponding REDOR-spectrum (red) for an evolution time of 1·33 ms is shown. The spectra in (d) (from top to bottom: spin-echo sepctrum, REDOR spectrum and difference of the two) were acquired in situ at T=400°C at an evolution time of 1·5 ms at 7 T [Colour available online]

cooling. Upon annealing the glass at temperatures of 450–550°C, a slight increase (several percent) of the relative fraction of the BO_4 units is observed (data not shown). This is compatible with an equilibrium $BO_4 \rightarrow BO_3$, which is shifted to the right with increasing temperatures. At ambient temperature the glass is exhibiting a BO_3/BO_4 ratio corresponding to the above equilibrium at the fictive temperature T_f. Upon annealing at $T_a > T_g$, the time scale of the glass dynamics allows for an adjustment of the BO_3/BO_4 ratio according to the equilibrium at the annealing temperature T_a.

The ^{11}B{^{29}Si}-REDOR-NMR experiments were performed to obtain information about the mixing of borate and silicate units. As pointed out by Stebbins,[43] the chemical shift of the identified ^{29}Si-NMR resonances *per se* indicate only marginal mixing between B and Si. The experiments were performed at RT employing a field of 11·7 T, since at this field, the signals from trigonal BO_3 and tetrahedral BO_4 units are well separated. The results of the ^{11}B{^{29}Si}-REDOR-NMR experiment are collected in Figure 4. The REDOR evolution curves for the BO_3 and BO_4 units clearly indicate that the ^{11}B nuclei in both borate polyhedra present in the glass experience a significant dipolar coupling to SiO_4 tetrahedra with the effect for the tetrahedrally coordinated borate species being considerably stronger. Thus, the degree of mixing between borate and silicate units within the network seems to be larger than to be expected from the chemical shift values. For the *in situ* ^{11}B{^{29}Si}-REDOR-NMR experiments employing the LASER-MAS NMR probe only the evolution of the BO_4 signal was followed. In Figure 4(c), the REDOR evolution curves for the temperatures indicated are collected, in Figure 4(d), the rotor-synchronized ^{11}B-MAS-NMR spectrum (top), the ^{11}B{^{29}Si}-REDOR-NMR spectrum (middle) and the difference of the two (bottom) are shown. The results allow for the conclusion that the average heteronuclear ^{11}B–^{29}Si dipolar coupling does not change significantly up to a temperature of 505°C, meaning that the number of neighbouring silicate units for the BO_4 polyhedra is not altered. In addition, this shows that no dynamic process on the time scale of the REDOR experiment (ca. 3 ms) is present up to these temperatures since this would entail an averaging of the dipole coupling and hence lead to a reduction of the REDOR effect. Our findings allow for an extended view on the glass dynamics in the borosilicate glass. According to our results, the two network formers experience considerable mixing (predominantly via SiO_4–BO_4 bonding) even at temperatures far above T_g (377°C). The *in situ* ^{29}Si-MAS-NMR data may be interpreted in terms of an exchange between the various silicate units present, but the resulting exchange rates obtained from our data are considerably smaller

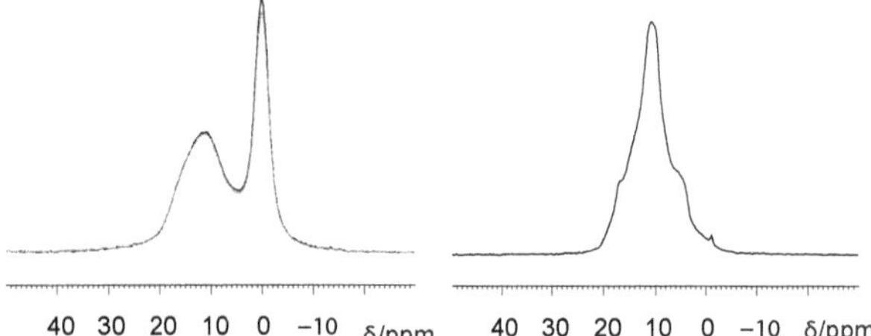

Figure 5. Left: 11B-MAS-NMR spectra for the melt quenched borosilicate glass (black) and the glass prepared via double roller quenching (red). Right: 11B-MAS-NMR spectrum of the sample annealed at 600°C [Colour available online]

than those published in literature and are close to the values determined for the BO_4–BO_3 exchange.[43] Further, our *in situ* ^{29}Si-MAS-NMR spectra do not exhibit a broadening and smearing of the individual lines as usually observed for dynamic exchange. An alternative explanation for the observed evolution of the *in situ* ^{29}Si-MAS-NMR spectra involves a reorganization according to $Q^2 \rightarrow Q^1 + Q^3$ or $Q^3(1B) \rightarrow Q^3(0B)$. Since according to the REDOR results, most of the direct Si–O–B linkages are SiO_4–BO_4 linkages, this reaction is linked to a coordination change $BO_4 \rightarrow BO_3$, which is corroborated by the evolution of the ^{11}B-MAS-NMR spectra. This coordination change was also observed in binary borate glasses[27] and borosilicate glasses.[44] This view is supported by a comparison of ^{11}B-MAS-NMR spectra prepared employing different quench rates (normal melt quench techniques and double roller quenching). The fast cooled glass, with a higher fictive temperature, exhibits a slightly lower BO_4 content (cf. Figure 5(left)), as also observed for aluminoborosilicate glasses.[45,46] The separation into borate and silicate rich aggregates is also supported by the observation that, as soon as sufficient B–O–Si linkages are broken, the precipitation of binary crystalline sodium silicates, Na_2SiO_3 and β-$Na_2Si_2O_5$ is observed. Annealing the glass at 600°C then leads to a residual glass matrix containing almost exclusively trigonally planar coordinated boron (cf. Figure 5(right)). Thus, the presented results are consistently pointing toward a mixed borosilicate network which is phase separating and then crystallizing upon heating above the glass transition.

Phosphosilicate glasses

The ^{29}Si- and ^{31}P-MAS-NMR spectra of glass PS-1 are collected in Figure 6. The two signals in the ^{31}P-MAS-NMR spectrum can be assigned to Q^3- and Q^2-environments.[47] In the ^{29}Si-MAS-NMR spectrum three different signals can be identified which represent SiO_4- (−116·6 ppm), SiO_5- (−162·8 ppm) and SiO_6- (−211·9 ppm) units. To obtain information about the interconnection of the identified basic building units we performed a series of dipolar based experiments.

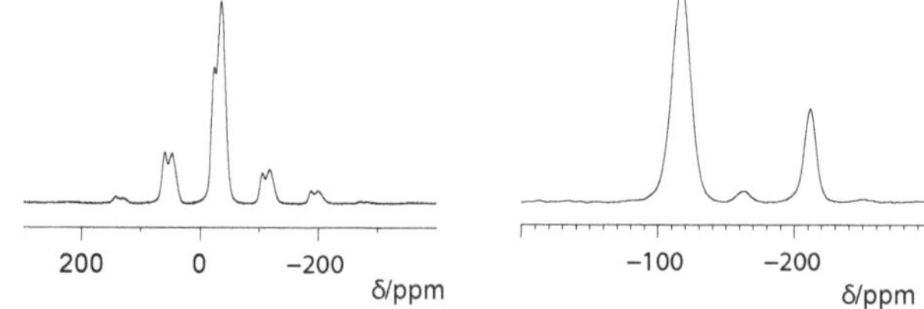

Figure 6. ^{31}P- (left) and ^{29}Si-MAS-NMR spectra (right) of glass PS-1

The 2D ^{29}Si{^{31}P}CPMAS-HETCOR-NMR experiment[48,49] provides qualitative information about the connectivity pattern between phosphate and silicate units. Since the intensity of the SiO_5 signal proves too low we have to restrict our analysis to the signals for the SiO_4- and SiO_6-units. The spectrum (cf. Figure 7) indicates that the six-coordinated silicon species obtain their magnetization preferably from Q^3-phosphate units, whereas the tetrahedral SiO_4-units are polarized from Q^3- and Q^2-phosphate groups.

A quantitative evaluation of the heteronuclear dipole coupling between ^{29}Si and ^{31}P is possible employing ^{29}Si{^{31}P}-REDOR-NMR and ^{31}P{^{29}Si}-REDOR-NMR. The results of the corresponding experiments are collected in Figure 8 together with simulations using the SIMPSON software package. The REDOR-evolution curve for the SiO_6 signal can be simulated assuming a SiP_6-7-spin interaction with Si–P distances

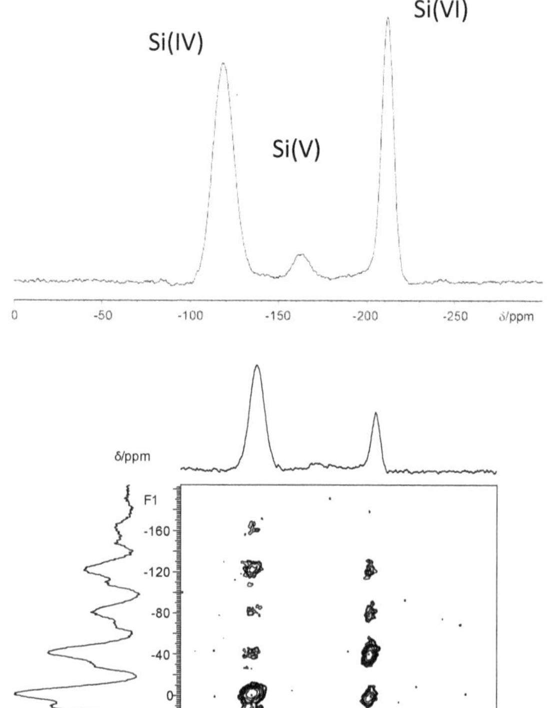

Figure 7. ^{29}Si{^{31}P}-CPMAS-NMR (top) and ^{29}Si{^{31}P}-CPMAS-HETCOR spectrum (bottom) for glass PS-1

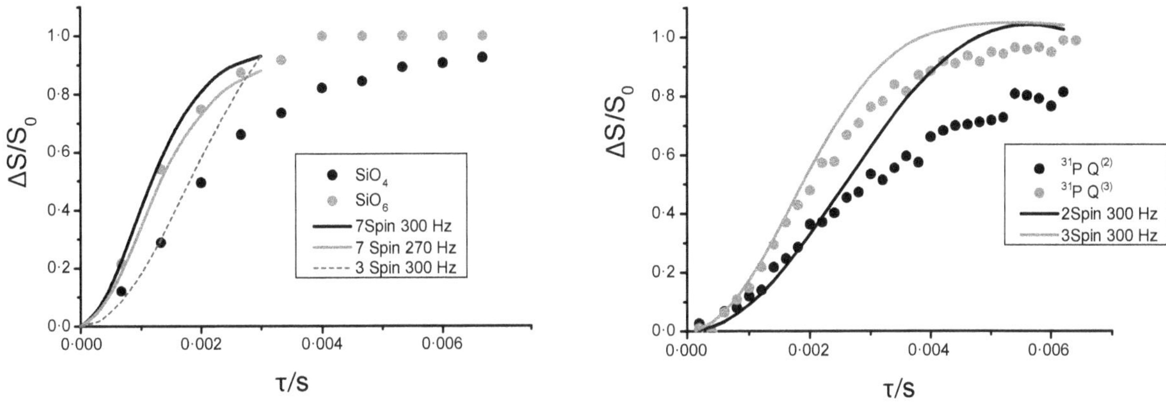

Figure 8. ^{29}Si{^{31}P}-REDOR NMR (left) and ^{31}P{^{29}Si}-REDOR NMR data (right) for glass PS-1. The solid curves in the left figure are the results of a SIMPSON-simulation assuming a SiP_6-7-Spin system with Si–P distances of 3·18 Å (black curve) and 3·3 Å (red curve); the dashed curves represents the results of a simulation assuming a SiP_2-three spin system with Si–P distances of 3·18 Å. The curves in the right figure were generated assuming a P-Si_2-three spin- (red curve) and a P–Si-two spin-system (black curve) with Si–O distances of 3·18 Å [Colour available online]

of 3·2 Å; for the SiO_4 data the assumption of a SiP_2-3-spin system produces the best simulations. For the reverse experiment, ^{31}P{^{29}Si}REDOR-NMR, the data for the Q^3-units can be simulated assuming ca. 2 Si-neighbours per Q^3-unit (P–Si_2-3-spin-system), whereas a simulation of the data for the Q^2-signal produces an average number of one Si-neighbour (P-Si-2-spin-system). These findings are corroborated by the results of ^{31}P- and ^{29}Si-RFDR-NMR experiments, (cf. Figure 9) evaluating the homonuclear dipolar couplings. Only marginal build-up of cross intensity is observed in the ^{29}Si-RFDR-NMR experiment even at mixing times of 28 ms, whereas for the ^{31}P-RFDR NMR experiment, cross peak build-up is observed for mixing times as low as 12 ms. From the ^{29}Si-RFDR-NMR experiment we thus conclude that no direct SiO_4–SiO_6 connections exist in the glass, in accordance with the results from the ^{29}Si{^{31}P}-REDOR-NMR experiment. The results of the ^{31}P-RFDR-NMR experiment on the other hand suggest that connections between Q^3- and Q^2-phosphate units constitute an important structural motif in the glass.

Increasing the P/Si ratio entails an increase in the relative fraction of SiO_6. In glass PS-2 with the composition $0{\cdot}35Na_2O{:}0{\cdot}55P_2O_5{:}0{\cdot}1SiO_2$ we find Si exclusively in six coordinate environments. Owing to the increased Na/P ratio the ^{31}P-MAS-NMR spectrum features an increased fraction of Q^2-^{31}P-environments (cf. Figure 11, top spectra).

Combining the results of the NMR experiments a network model may be suggested. According to our findings, an SiO_6 octahedron connected to Q^3 phosphate units constitutes the central building block. These are interconnected via Q^2-chains. With decreasing P/Si ratio at some point (P/Si=2) not enough phosphorous is present to coordinate the SiO_6 units, leading to the formation of tetrahedral SiO_4 (and a minor fraction of SiO_5).

The evolution of the glass structure with temperature was studied employing *in situ* ^{29}Si- and ^{31}P-MAS-NMR and ^{31}P{^{29}Si}-REDOR-NMR. In the *in situ* ^{29}Si-MAS-NMR spectra for glass PS-1, collected in Figure 10, we find a clear reduction in the relative fraction of SiO_6-units, which completely disappear at temperatures of 630°C. In the region around −160

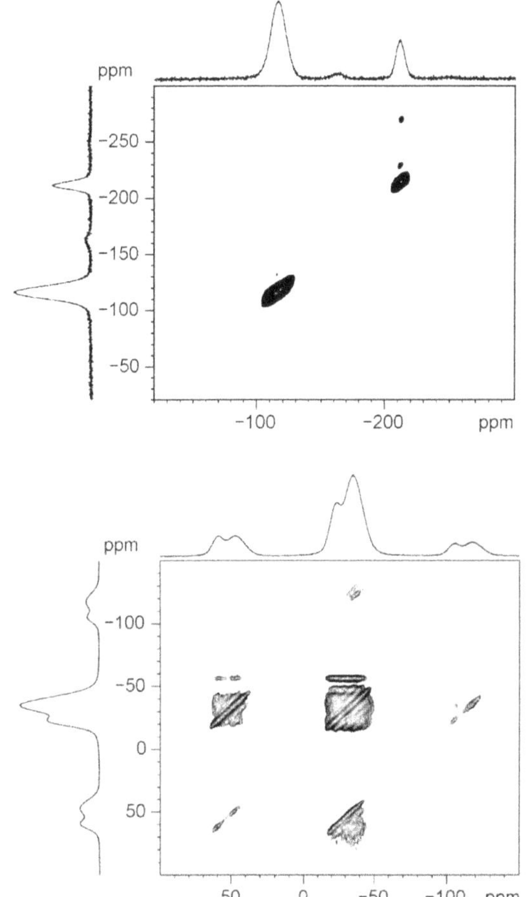

Figure 9. ^{29}Si- (top) and ^{31}P- (bottom) RFDR-NMR spectra acquired with mixing times of 12 ms (^{29}Si) and 28 ms (^{31}P), respectively [Colour available online]

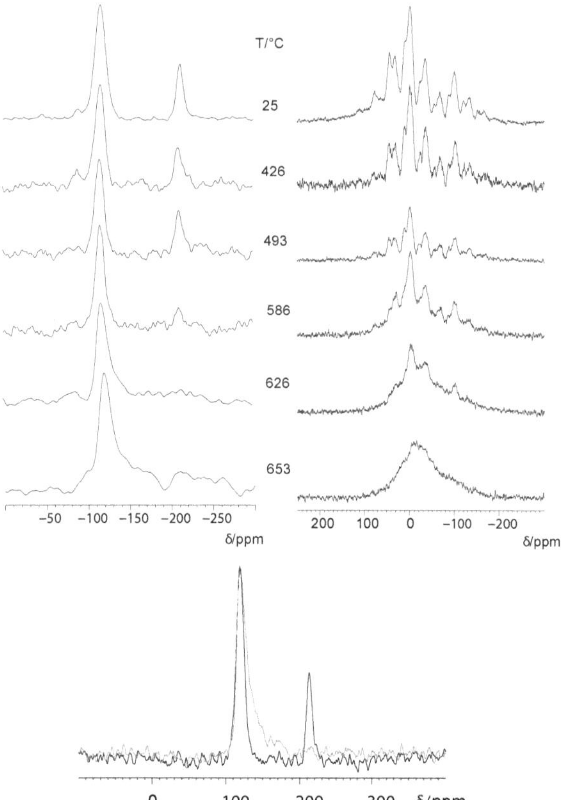

Figure 10. Top: ^{29}Si- (left) and ^{31}P- (right) in situ MAS NMR spectra for glass NPS-1, obtained on a 7 T-magnet system; bottom: ^{29}Si- in situ MAS-NMR spectra obtained at 450°C (black) and 670°C (red), obtained on a 9 T magnet at Bruker Biospin, Karlsruhe [Colour available online]

ppm some intensity emerges, which can however not be further analyzed due to its extreme width and limited S/N ratio. The spectra are thus compatible with either the assumption of conversion of SiO_6 into SiO_4 units as well as into SiO_5 units. Additional experiments employing a 9·4 T magnet however clearly reveal that the disappearance of the SiO_6 signal is accompanied by an emerging signal around −160 ppm, thus supporting the latter conversion reaction. In the *in situ* ^{31}P-MAS-NMR spectra a broadening of the signal at −22 ppm (Q^2 units) can be identified at temperatures of 580°C, which may be explained by the beginning of a coalescence ($\tau^{-1}=\omega_{MAS}$). For the signal at −35 ppm the spinning sidebands are visible up to temperatures of 626°C, indicating a somewhat lower mobility. A dynamic exchange between the two phosphate species is occurring only at the highest measured temperature (653°C). The *in situ* MAS-NMR spectra for glass PS-2, here T_g is considerably lower due to the increased Na_2O and the decreased SiO_2 content, are collected in Figure 11. Again, the ^{31}P-MAS-NMR spectra show a broadening of the Q^2 signal, indicating a coalescence due to the onset of the dynamics of the Q^2 chains. The Q^3 signal (together with the corresponding spinning sidebands) exhibit this coalescence behaviour only at a higher temperature. Beginning at T=532°C exchange between the Q^2- and Q^3-units is setting in.

In the corresponding ^{29}Si-MAS-NMR spectra we observe a shift of the signal to lower field to −160 ppm beginning at 532°C. This indicates a conversion of six-coordinated silicon to SiO_5 units. The line widths in the temperature range 532<T<665°C however are rather large (ca. 40 ppm), indicating an considerable distribution of bond lengths and/or angles within the SiO_5-polyhedra. The observation that the ^{31}P-Q^2-signals broaden first, indicates that the dynamics in the glass commences with the chains (dominated by Q^2 groups) connecting the $Si(OP)_6$ centres. Upon further temperature increase, the $Si(OP)_6$-centres are beginning to participate in the glass dynamics, as concluded from the observed $Si^{(VI)}$–O–P(Q^3)-bond break-

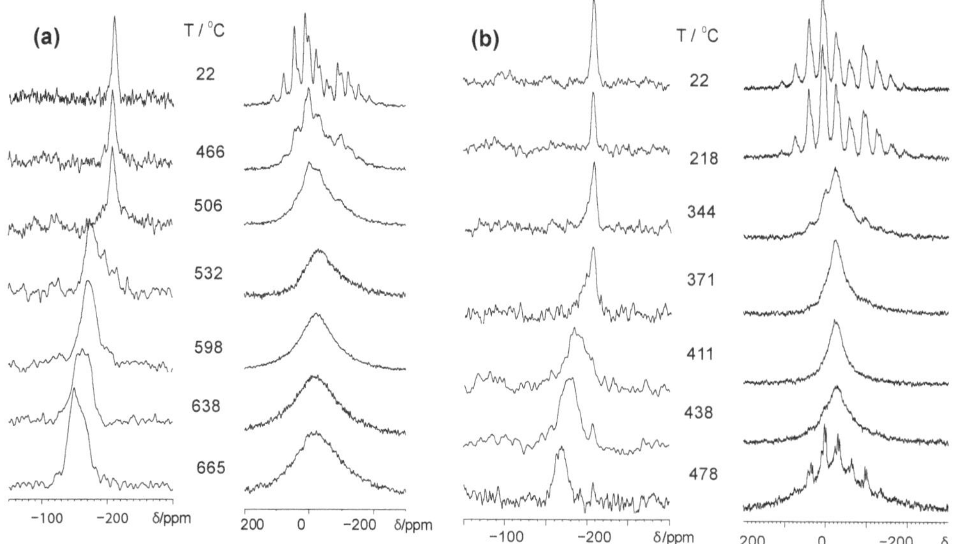

Figure 11. (a) ^{29}Si- (left) and ^{31}P- (right) in situ MAS-NMR spectra for glass NPS-2; (b) ^{29}Si- (left) and ^{31}P- (right) in situ MAS-NMR spectra for glass NPS-3

ing, in accordance with the concomitant decrease in the fraction of SiO_6 polyhedra and the increasing mobility of the phosphate Q^3 units. To confirm that the observed downfield shift in the ^{29}Si-MAS-NMR spectra is not due to some artefacts (e.g. temperature dependence of shim-unit), we repeated the experiments on a third glass, in which Na_2O was replaced by Cs_2O. In this glass, PS-3, the onset of the downfield shift of the ^{29}Si-MAS-NMR signal and the Q^2–Q^3 phosphate exchange set in at temperatures of 371°C and are more or less finished at 411°C (compared to an onset temperature of 532°C in the Na_2O based glass PS-2). In addition, crystallization is observed in the Cs_2O based glass beginning at 438°C. Taken together, the *ex situ* and *in situ* NMR experiments can be consistently discussed as follows. The glasses are dominated by $Si^{(VI)-}$ units which are predominantely coordinated via six ^{31}P-Q^3-polyhedra. If not enough phosphorous is present to serve the full coordination of the SiO_6-octahedra, excess silicate units are incorporated as SiO_4 polyhedra within the phosphate chains. This view is compatible with the observation that SiO_4- and SiO_6-environments do not exchange; instead a coordination change $Si^{VI} \rightarrow Si^{V}$ is observed. With increasing temperatures, glass dynamics starts with the Q^2 dominated chains connecting the Si(OP)a centres. Upon further temperature increase, a breaking of the SiO_6–P-Q^3-linkages is observed. In this picture, the SiO_4 units stay more or less unchanged during heating, whereas the $Si^{(VI)}$ environments are converted into $Si^{(V)}$-species.

An alternative explanation of the evolution of the ^{29}Si-MAS-NMR spectra with temperature, an exchange Si^{IV}–Si^{VI}, seems less likely, since the broadened line should extend over the chemical shift range for the Q^4-species.

Conclusions

Employing a combination of *ex situ* and *in situ* NMR techniques, the network organization in alkali phosphosilicate glasses and a borosilicate glass together with its evolution at temperatures above T_g were analyzed. The phosphosilicate glasses are dominated by SiO_6 units which are coordinated via six phosphate polyhedra, predominantely ^{31}P-Q^3-groups. Excess silicate units are incorporated as SiO_4 polyhedra within the phosphate chains. With increasing temperatures, the network dynamics commences with the Q^2 dominated chains connecting the $Si(OP)_6$ centres. Upon further temperature increase, a breaking of the SiO_6–P-Q^3-linkages is observed, converting the SiO_6 environments into $Si^{(V)}$-species. For the borosilicate glass, a separation of a mixed borosilicate matrix into borate rich and silica rich fractions is observed, with binary sodium silicates crystallizing from the silicate parts.

Acknowledgements

Financial support by the Deutsche Forschungsgemeinschaft is gratefully acknowledged.

References

1. Kokubo, T. *Biomaterials*, 1991, **12**, 155–163.
2. Hench, L. L. *J. Am. Ceram. Soc.*, 1991, **74**, 1487–1510.
3. Vogel, W., Holand, W., Naumann, K. & Gummel, J. *J. Non-Cryst. Solids*, 1986, **80**, 34–51.
4. Vogel, J., Wange, P. & Hartmann, P. *Glastechn. Ber.: Glass Sci. Technol.*, 1997, **70**, 220–223.
5. Brow, R. K. & Tallant, D. R. *J. Non-Cryst. Solids*, 1997, **222**, 396–406.
6. Weber, M. J. *J. Non-Cryst. Solids*, 1990, **123**, 208–222.
7. Weber, W. J., Ewing, R. C., Angell, C. A., Arnold, G. W., Cormack, A. N., Delaye, J. M., Griscom, D. L., Hobbs, L. W., Navrotsky, A., Price, D. L., Stoneham, A. M. & Weinberg, W. C. *J. Mater. Res.*, 1997, **12**, 1946–1978.
8. Sales, B. C. & Boatner, L. A. *Science*, 1984, **226**, 45–48.
9. Lee, W. E., Ojovan, M. I., Stennett, M. C. & Hyatt, N. C. *Adv. Appl. Ceram.*, 2006, **105**, 3–12.
10. Donald, I. W., Metcalfe, B. L. & Taylor, R. N. J. *J. Mater. Sci.*, 1997, **32**, 5851–5887.
11. Bennett, A. E., Ok, J. H., Griffin, R. G. & Vega, S. *J. Chem. Phys.*, 1992, **96**, 8624.
12. Griffiths, J. M. & Griffin, R. G. *Anal. Chim. Acta*, 1993, **283**, 1081–1101.
13. Gullion, T. Schaefer, J. *J. Magn. Resonance*, 1989, 81.
14. Gullion, T. *Concepts Magn. Resonance*, 1998, **10**, 277–289. 15. Gullion, T. *Ann. Rep. NMR Spectr.*, 2009, **65**, 111–137.
16. Gullion, T. & Vega, A. J. *Progr. Nucl. Magn. Resonance Spectr.*, 2005, **47**, 123–136.
17. Gullion, T. *J. Magn. Resonance*, 1999, **139**, 402–407.
18. Gullion, T. *Chem. Phys. Lett.*, 1995, **246**, 325–330.
19. Chopin, L., Vega, S. & Gullion, T. *J. Am. Chem. Soc.*, 1998, **120**, 4406–4409.
20. van Wüllen, L., Müller, U. & Jansen, M. *Angew. Chem.-Int. Ed.*, 2000, **39**, 2519.
21. Eckert, H., Elbers, S., Epping, J. D., Janssen, M., Kalwei, M., Strojek, W. & Voigt, U. In: *New Techniques in Solid-State NMR, Vol. 246*, 2005, pp. 195–233.
22. van Wüllen, L. & Jansen, M. *J. Mater. Chem.*, 2001, **11**, 223–229.
23. Wegner, S., van Wüllen, L. & Tricot, G. *J. Non-Cryst. Solids*, 2008, **354**, 1703–1714.
24. Stebbins, J. F., Sen, S. & Farnan, I. *Am. Mineral.*, 1995, **80**, 861–864.
25. Farnan, I. & Stebbins, J. F. *Science*, 1994, **265**, 1206–1209.
26. Farnan, I. & Stebbins, J. F. *J. Am. Chem. Soc.*, 1990, **112**, 32–39.
27. Sen, S., Xu, Z. & Stebbins, J. F. *J. Non-Cryst. Solids*, 1998, **226**, 29–40.
28. Stebbins, J. F. & Ellsworth, S. E. *J. Am. Ceram. Soc.*, 1996, **79**, 2247–2256.
29. van Wüllen, L., Wegner, S. & Tricot, G. *J. Phys. Chem. B*, 2007, **111**, 7529–7534.
30. Wegner, S., van Wüllen, L. & Tricot, G. *J. Phys. Chem. B*, 2009, **113**, 416–425.
31. Wegner, S., van Wüllen, L. & Tricot, G. *Solid State Sci.*, 2010, **12**, 428–439.
32. van Wüllen, L. & Jansen, M. *J. Mater. Chem.*, 2001, **11**, 223–229.
33. Grimmer, A. R., Vonlampe, F. & Magi, M. *Chem. Phys. Lett.*, 1986, **132**, 549–553.
34. Prabakar, S., Rao, K. J. & Rao, C. N. R. *J. Mater. Res.*, 1991, **6**, 592–601.
35. Prabakar, S., Rao, K. J. & Rao, C. N. R. *Mater. Res. Bull.*, 1991, **26**, 285–294.
36. Dupree, R., Holland, D., Mortuza, M. G., Collins, J. A. & Lockyer, M. W. G. *J. Non-Cryst. Solids*, **1989**, **112**, 111–119.
37. Dupree, R., Holland, D., Mortuza, M. G., Collins, J. A. & Lockyer, M. W. G. *J. Non-Cryst. Solids*, 1988, **106**, 403–407.
38. Thurber, K. R. & Tycko, R. *J. Magn. Resonance*, 2009, **196**, 84–87.
39. Becker, K. D. *J. Chem. Phys.*, 1978, **68**, 3785–3793.
40. Wu, J. S., Kim, N. & Stebbins, J. F. *Solid State Nucl. Magn. Resonance*, 2011, **40**, 45–50.
41. Clark, T. M., Grandinetti, P. J., Florian, P. & Stebbins, J. F. *J. Phys. Chem. B*, 2001, **105**, 12257–12265.
42. Heidemann, D., Hubert, C., Schwieger, W., Grabner, P., Bergk, K. H. & Sarv, P. *Z. Anorg. Allgemeine Chem.*, 1992, **617**, 169–177.
43. Stebbins, J. F. & Sen, S. *J. Non-Cryst. Solids*, 1998, 224, 80–85.
44. Osipov, A. A., Osipova, L. M. & Eremyashev, V. E. *Glass Phys. Chem.*, 2013, **39**, 105–112.
45. Wu, J. S. & Stebbins, J. F. *J. Non-Cryst. Solids*, 2010, **356**, 2097–2108.
46. Wu, J. S., Potuzak, M. & Stebbins, J. F. *J. Non-Cryst. Solids*, 2011, **357**, 3944–3951.
47. Miyabe, D., Takahashi, M., Tokuda, Y., Yoko, T. & Uchino, T. *Phys. Rev. B*, 2005, **71**.
48. Coelho, C., Azais, T., Bonhomme-Coury, L., Maquet, J. & Bonhomme, C. *Comptes Rendus Chim.*, 2006, **9**, 472–477.
49. Lejeune, C., Coelho, C., Bonhomme-Coury, L., Azais, T., Maquet, J. & Bonhomme, C. *Solid State Nucl. Magn. Resonance*, 2005, **27**, 242–246.

Author Index

DOI: 10.13036/17533562.56.2.063 Phys. Chem. Glasses: Eur. J. Glass Sci. Technol. B, April 2015, 56 (2), 63–66

Thermodynamic model and structure of $ZnO–MoO_3–P_2O_5$ glasses

Marek Liška,[1] *Jan Macháček,*[2] *Mária Chromčíková*[1] *& Ondrej Gedeon*[2]

[1] *Vitrum Laugaricio – Joint Glass Center of IIC SAS, TnUAD, and FChPT STU, Študentská 2, Trenčín, SK-91150, Slovakia*
[2] *Institute of Chemical Technology, Technická 5, Prague, CZ-166 28, Czech Republic*

Manuscript received 6 September 2014
Accepted 15 December 2014

In the present work the Shakhmatkin & Vedishcheva thermodynamic model was constructed for the $ZnO–MoO_3–P_2O_5$ glasses. On the basis of equilibrium phase diagrams and the crystal structural data the glass was considered as an ideal solution of three oxides and nine compounds representing different Q^n units. For the components considered in the model no thermodynamic data were available in contemporary thermodynamic databases. Therefore new method of parameterization of the thermodynamic model was proposed based on the known structural data. Four compositional series were considered. The first series (A) has equimolar content of $ZnO/P_2O_5–(0{\cdot}5–x/2)[ZnO.P_2O_5].xMoO_3$. The other three series have a constant content of one of the components, (B) $0{\cdot}1ZnO.yMoO_3.(0{\cdot}9–y)P_2O_5$, (C) $zZnO.0{\cdot}2MoO_3.(0{\cdot}8–z)P_2O_5$ and (D) $(0{\cdot}5–t)ZnO.tMoO_3.0{\cdot}5P_2O_5$. For these compositional series the Q^n distribution was obtained from the ^{31}P MAS NMR spectra by Šubčík et al.[1] *Using these experimental data the temperature independent reaction Gibbs energies of formation of compounds considered in the thermodynamic model were estimated by minimization of the sum of squared deviations between experimental and calculated relative abundances of Q^n units.*

1. Introduction

Phosphate glasses with additions of tungsten oxide or molybdenum oxide offer prospective applications for electro-optical applications due to their electrochromic properties and high ionic conductivity.[1] They can be also used for nuclear waste vitrification. The composition–structure–property relationships needed for the tailoring of the glass composition for particular practical application can be rationalized within the proper thermodynamic model. Mainly in the field of silicate glasses the thermodynamic model of Shakhmatkin & Vedishcheva was successfully applied in previous years.[2–10] This model considers glasses and melts as a solution formed from salt-like products of equilibrium chemical reactions between the simple chemical entities (oxides, halogenides, chalcogenides, ...) and from the original (un-reacted) entities. These salt-like products (also called associates, groupings or species) have the same stoichiometry as the crystalline compounds, which exist in the equilibrium phase diagram of the system considered. The model does not use adjustable parameters only the standard Gibbs energies of formation of crystalline compounds and the analytical composition of the system considered are used as input parameters. The contemporary databases of thermodynamic properties (like the FACT computer database[11,12]) enable the routine construction of the Shakhmatkin & Vedishcheva model for most of important multicomponent silicate systems. However in most cases of non-silicate multicomponent glass systems the needed thermodynamic data are not found in any contemporary thermodynamic database. Therefore the aim of the present work is to propose and validate a method for obtaining estimates of missing thermodynamic parameters by reproducing the structural information acquired by solid state ^{31}P MAS NMR.

Corresponding author. Email liska@tnuni.sk
Original version presented at Int. Conf. on Phosphate Glasses, Pardubice, Czech Republic, 2–4 July 2014

2. Method

The system's Gibbs energy is expressed supposing the state of the ideal solution:

$$G(n_1, n_2, \dots n_N) = \sum_{i=1}^{N} n_i G_{m,i} + RT \sum_{i=1}^{N} n_i \ln \frac{n_i}{\sum_{j=1}^{N} n_j} \qquad (1)$$

where N is the number of components, n_i is the molar amount of i-th component, T is the system temperature (i.e. the glass transition temperature, T_g, for particular glass) and $G_{m,i}$ is the molar Gibbs energy of pure i-th component at the pressure of the system and temperature T. The system components are ordered such way that X_i (i=1, 2,...M<N) are pure oxides (halogenides, chalcogenides,...) and X_i (i=M+1, M+2,...N) are compounds formed from basic entities by reversible reactions

$$X_i \leftrightarrow \sum_{j=1}^{M} \nu_{i,j} X_j,\ i = M+1, M+2, \dots N \qquad (2)$$

The glass composition is given by the molar fractions of individual oxides $x_{g,i}$ (i=1, 2,...M). The starting system composition ($n_{0,i}$, i=1, 2...N) is represented by the pure unreacted oxides with the molar amounts numerically equal to theirs molar fractions in glass:

$$n_{0,i}/mol = x_{g,i}, \quad i=1, 2...M \tag{3}$$

$$n_{0,i}=0, \quad i=M+1, M+2...N \tag{4}$$

Equation (1) can be rewritten in terms of reaction Gibbs energies, $\Delta_r G_{m,i}$:

$$\begin{aligned} G(n_1, n_2, ...n_N) &= \sum_{i=1}^{M} n_{0,i} G_{m,i} + \sum_{i=M+1}^{N} n_i \Delta_r G_{m,i} \\ &+ RT \sum_{i=1}^{N} n_i \ln \frac{n_i}{\sum_{j=1}^{N} n_j} \\ &= \sum_{i=M+1}^{N} n_i \Delta_r G_{m,i} \\ &+ RT \sum_{i=1}^{N} n_i \ln \frac{n_i}{\sum_{j=1}^{N} n_j} + const. \end{aligned} \tag{5}$$

where reaction Gibbs energies are defined by:

$$\Delta_r G_{m,i} = G_{m,i} - \sum_{j=1}^{M} \nu_{i,j} G_{m,i} \tag{6}$$

The minimization of the system's Gibbs energy constrained by the overall system composition has to be performed with respect to the molar amount of each system component to reach the equilibrium system composition.[13]

The material balance constrains are represented by:

$$n_{0,i} = n_i + \sum_{j=M+1}^{N} \nu_{j,i} n_j, \quad i = 1,2...M \tag{7}$$

Generally the structural information, $P_{j,k}$ (e.g. the relative abundance of Q^j structural unit), for the k-th glass composition can be expressed by equilibrium molar amounts of system components:

$$_{j\,k} \quad \frac{\sum \; _{j,i} \;\; _{i,k}}{\sum \; _{i} \;\; _{i\,k}} \tag{8}$$

where $a_{j,i}$ and b_i are constants, and $n_{i,k}$ is the equilibrium molar amount of i-th component in the k-th glass composition. Then the estimate of unknown reaction Gibbs energies can be found by minimization of the sum of squared deviations between calculated experimental structural data:

$$\begin{aligned} U(\Delta_r G_{m,M+1}, \Delta_r G_{m,M+2}, ...\Delta_r G_{m,N}) \\ = \sum_{k=1}^{N_g} \sum_{j=0}^{N_P} \left(P_{j,k}^{calc} - P_{j,k}^{exp} \right)^2 = \min \end{aligned} \tag{9}$$

Such obtained estimates of reaction Gibbs energies are temperature independent. However on the level of equilibrium constants of formation of system components, K_i, the temperature dependence is introduced via:

$$\ln K_i = \frac{-\Delta_r G_{m,i}}{RT}, \; i = M+1, M+2...N \tag{10}$$

On the other hand this procedure leads to the effective (i.e. not true) values of reaction Gibbs energies and can therefore count (compensate) the non-ideal state of the system.

Table 1. Glass compositions and glass transition temperatures[1]

No.	Series	x_g(ZnO)	$x_g(MoO_3)$	$x_g(P_2O_5)$	T_g / K
1	D	0·00	0·50	0·50	803
2	D	0·10	0·40	0·50	809
3	B,C	0·10	0·20	0·70	762
4	B	0·10	0·40	0·50	811
5	B	0·10	0·50	0·40	760
6	B	0·10	0·60	0·30	688
7	B	0·10	0·70	0·20	664
8	A	0·15	0·70	0·15	641
9	A	0·20	0·60	0·20	661
10	A	0·25	0·50	0·25	686
11	C,D	0·30	0·20	0·50	765
12	A	0·30	0·40	0·30	703
13	A	0·35	0·30	0·35	716
14	A,C	0·40	0·20	0·40	721
15	A	0·45	0·10	0·45	725
16	A,D	0·50	0·00	0·50	733
17	C	0·50	0·20	0·30	722

3. Results and discussion

Four compositional series were considered. The first series (A) has equimolar content of ZnO/P_2O_5–$(0·5-x/2)[ZnO.P_2O_5].xMoO_3$. The other three series have a constant content of one of the components, (B) $0·1ZnO.yMoO_3.(0·9-y)P_2O_5$, (C) $zZnO.0·2MoO_3.(0·8-z)P_2O_5$ and (D) $(0·5-t)ZnO.tMoO_3.0.5P_2O_5$. For these compositional series the Q^n distribution was obtained from the ^{31}P MAS NMR spectra by Šubčík *et al.*[1] The considered glass compositions are summarized in Table 1 together with the glass transitions temperatures.

On the basis of equilibrium phase diagrams and the crystal structural data[14] nine binary compounds representing different Q^n units were found. One system component, namely M3P, was added to

Table 2. Considered system components and number of Q^n units

No	Component	Abbreviation	Q^n	$-\Delta_r G$ / kJmol^{-1}
1	ZnO	-	-	-
2	MoO_3	-	-	-
3	P_2O_5		$2Q^3$	-
4	$3ZnO.P_2O_5$	Z3P	$2Q^0$	90·1
5	$2ZnO.P_2O_5$	Z2P	$2Q^1$	76·9
6	$ZnO.P_2O_5$	ZP	$2Q^2$	51·2
7	$ZnO.2P_2O_5$	ZP2	$2Q^2+2Q^3$	43·5
8	$3ZnO.2MoO_3$	Z3M2	-	34·7
9	$ZnO.MoO_3$	ZM	-	25·8
10	$MoO_3.2P_2O_5$	MP2	$2Q^1$	78·6
11	$MoO_3.P_2O_5$	MP	$2Q^2$	47·1
12	$3MoO_3.P_2O_5$	M3P	$2Q^0$	97·9

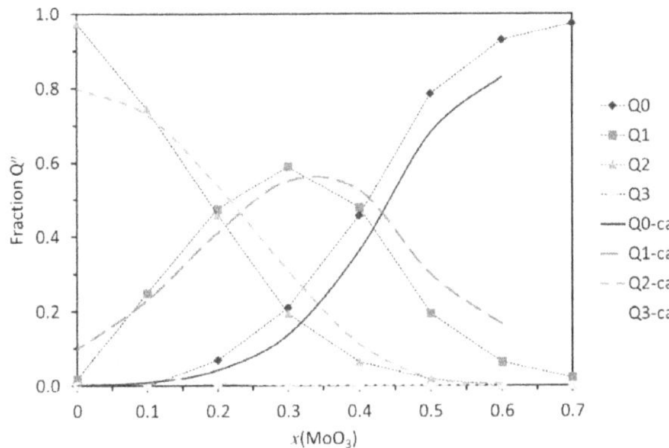

Figure 1. Comparison of experimental (points) and calculated (line) Q-distribution for glasses of the A-series ZnO/P_2O_5–(0·5–x/2)[$ZnO.P_2O_5$].$xMoO_3$ [Colour available online]

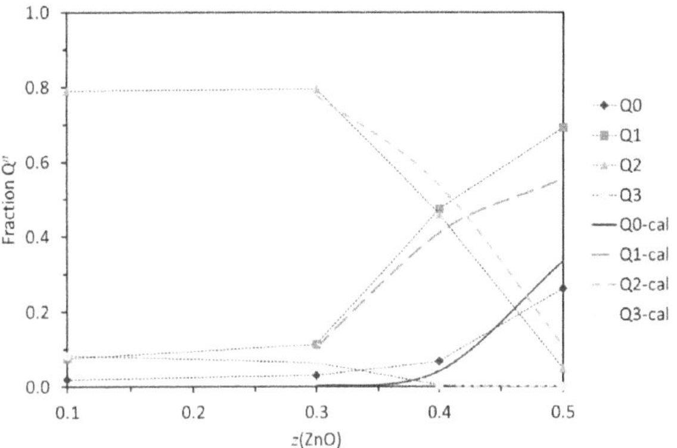

Figure 3. Comparison of experimental (points) and calculated (line) Q-distribution for glasses of the C-series: $zZnO.0{\cdot}2MoO_3.(0{\cdot}8-z)P_2O_5$ [Colour available online]

the set obtained from the structural database. The reason was the absence of representative of the Q^0 unit in the binary subsystem MoO_3–P_2O_5. The same situation can be found in case of glasses of binary system CaO–SiO_2 where the stable crystalline phase representing the Q^3 units is missing.[6] Also in this case the $CaO.2SiO_2$ was added to reach the ability of reproducing the experimental structural data by the thermodynamic model.

Thus the glass was considered as an equilibrium solution of three unreacted oxides and nine binary compounds. On the basis of structural data the Q^n units were attributed to system components (see Table 2).

$$P_{0,k} = x(Q^0) = \frac{2n_k(\mathrm{Z3P}) + 2n_k(\mathrm{M3P})}{2x_{g,k}(\mathrm{P_2O_5})}$$

$$P_{1,k} = x(Q^1) = \frac{2n_k(\mathrm{Z2P}) + 2n_k(\mathrm{M2P})}{2x_{g,k}(\mathrm{P_2O_5})}$$

$$P_{2,k} = x(Q^2) = \frac{2n_k(\mathrm{ZP}) + 2n_k(\mathrm{ZP2}) + 2n_k(\mathrm{MP})}{2x_{g,k}(\mathrm{P_2O_5})}$$

$$P_{3,k} = x(Q^3) = \frac{2n_k(\mathrm{P_2O_5}) + 2n_k(\mathrm{ZP2})}{2x_{g,k}(\mathrm{P_2O_5})}$$

where $n_k(X)$ denotes the equilibrium molar amount of the component X in k-th glass composition.

The minimization of the target function was performed by the evolutionary algorithm[15] in EXCEL version 2010. The huge set of optimized parameters was used consisting of nine reaction Gibbs energies and 204 (=17×12) equilibrium molar amounts of system components. The weighted composed target function was used in the form of the sum of U function, the squared differences between the equilibrium constants calculated from molar amounts and from Gibbs energies, plus 10^3 times multiplied sum of squared violations of material balance expressed as:

$$T(n_1, n_2, \ldots n_N) = \sum_{k=1}^{N_g} \sum_{i=1}^{M} \left[n_{0,i,k} - n_{i,k} + \sum_{j=M+1}^{N} \nu_{j,i} n_{j,k} \right]^2 \quad (11)$$

Due to the great number of optimized parameters the typical computation time on the standard PC reached two weeks. The obtained estimates of reaction Gibbs energies are summarized in Table 2. In Figures1–4 the experimental and calculated Q-distributions are compared for glass compositional series A through D. Taking into account the experimental error of MAS NMR determination of Q^n relative abundance we can consider the results plotted in Figures1–4 as acceptable. However some other tools of validation

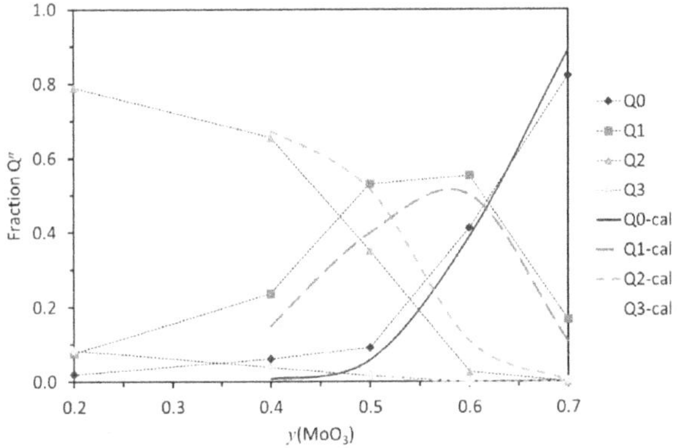

Figure 2. Comparison of experimental (points) and calculated (line) Q-distribution for glasses of the B-series: $0{\cdot}1ZnO.yMoO_3.(0{\cdot}9-y)P_2O_5$ [Colour available online]

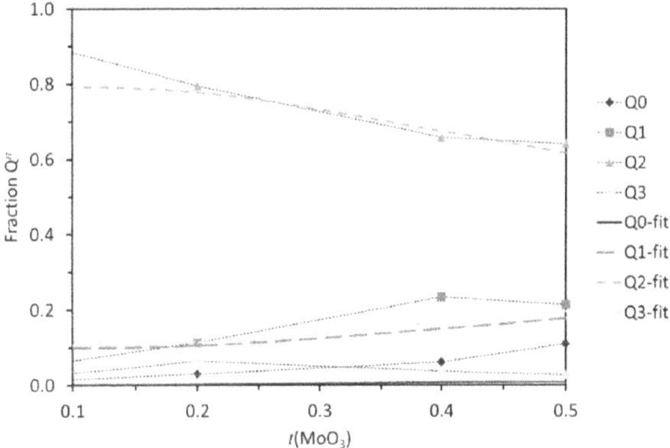

Figure 4. Comparison of experimental (points) and calculated (line) Q-distribution for glasses of the D-series: $(0{\cdot}5-t)ZnO.tMoO_3.0{\cdot}5P_2O_5$ [Colour available online]

of obtained thermodynamic model parameterization are still needed. Therefore the results obtained were used for the interpretation of Raman spectra of binary $ZnO–P_2O_5$ glasses. It was found that the spectral decomposition obtained by using the results of thermodynamic model fit perfectly the results obtained by the multivariate curve resolution (MCR) method.[16]

4. Conclusions

The proposed method of finding the estimates of reaction Gibbs energies led to the construction of the thermodynamic model of glasses of the studied ternary system that reproduces the glass structure (i.e. Q-distribution) with acceptable accuracy.

Acknowledgement

This work was supported by the Slovak Grant Agency for Science under the grant VEGA 1/0006/12, and by the Slovak Research and Development Agency project APVV-0487-11. This work was also supported by the Grant Agency of the Czech Republic through the grant No. P108/10/1631.

References

1. Šubčík, J., Koudelka, L., Mošner, P., Montagne, L., Tricot, G., Delevoye, L. & Gregora, I. Glass-forming ability and structure of $ZnO–MoO_3–P_2O_5$ glasses. *J. Non-Cryst. Solids*, 2010, **356**, 2509–2516.
2. Shakhmatkin, B. A., Vedishcheva, N. M., Shultz, M. M. & Wright, A. C. The thermodynamic properties of oxide glasses and glass-forming liquids and their chemical structure. *J. Non-Cryst. Solids*, 1994, **177**, 249–256.
3. Vedishcheva, N. M., Shakhmatkin, B. A., Shultz, M. M. & Wright, A.C. The thermodynamic modelling of glass properties: a practical proposition? *J. Non-Cryst. Solids*, 1996, **196**, 239–243.
4. Shakhmatkin, B. A., Vedishcheva, N. M., Wright, A. C. In: A. C. Wright, S. A. Feller, A. C. Hannon (Editors), *Borate Glasses Crystals and Melts*. Soc. Glass. Technol., Sheffield 1997, p.189.
5. Shakhmatkin, B. A., Vedishcheva, N. M. & Wright, A. C. Can Thermodynamics Relate the Propertiers of Melts and Glasses to their Structure? *J. Non-Cryst. Solids*, 2001, **293–295**, 220–236.
6. Vedishcheva, N. M., Shakhmatkin, B. A. & Wright, A. C. Thermodynamic modelling of the structure of glasses and melts: single-component, binary and ternary systems. *J. Non-Cryst. Solids*, 2001, **293–295**, 312–317.
7. Vedishcheva, N. M., Shakhmatkin, B. A. & Wright, A. C. Thermodynamic modelling of the structure of sodium borosilicate glasses. *Phys. Chem. Glasses*, 2003, **44**, 191–196.
8. Vedishcheva, N. M., Shakhmatkin, B. A. & Wright, A. C. The structure of sodium borosilicate glasses: thermodynamic modelling vs. experiment. *J. Non-Cryst. Solids*, 2004, **345&346**, 39–44.
9. Shakhmatkin, B. A., Vedishcheva, N. M. & Wright, A.C. Thermodynamic modelling of the structure of oxyhalide glasses. *J. Non-Cryst. Solids*, 2004, **345&346**, 461–468.
10. Liška, M. Studying Structure and Thermal Properties of Oxide Glasses, In: *Some Thermodynamic, Structural and Behavioral Aspects of Materials Accentuating Non-crystalline States*. University of West Bohemia in Pilsen, Nymburg 2009, pp. 344–362.
11. High Temperature Glass Melt Property Database for Process Modeling. T. P. Seward III & T. Vascott (Editors), Amer. Ceram. Soc., Westerville, Ohio 2005.
12. http://www.crct.polymtl.ca/fact/
13. Vonka, P. & Leitner, J. Calculation of Chemical Equilibria in Heterogeneous Multicomponent Systems. *Calphad*, 1995, **19**, 25–36.
14. Inorganic Crystal Structure Database, Release 2003. Fachinformationszentrum Karlsruhe, Germany, and the US Department of Commerce.
15. Frontline Systems, Inc.: www.solver.com
16. Liška, M., Zemanová, V., Lissová, M., Plško, A., Chromčíková, M., Gavenda, T. & Macháček, J. Thermodynamic model and Raman spectra of $ZnO-P_2O_5$ glasses. *Book of Abstracts of 11th European Symposium on Thermal Analysis and Calorimetry* - www.estac11.fi, p. 130, Espoo, Finland, August 2014.

DOI: 10.13036/17533562.56.2.071 Phys. Chem. Glasses: Eur. J. Glass Sci. Technol. B, April 2015, 56 (2), 71–75

Bismuth silver phosphate glasses as alternative matrices for the conditioning of radioactive iodine

Thomas Lemesle, Lionel Montagne, François O. Méar, Bertrand Revel*

Université Lille Nord de France – UCCS - UMR CNRS 8181, USTL, F-59655, Villeneuve d'Ascq, France

Lionel Campayo & Olivier Pinet

CEA, DEN, DTCD/SECM/LDMC - Marcoule, F-30207 Bagnols-sur-Cèze, France

Manuscript received 1 October 2014
Accepted 15 December 2014

In the context of the management of radioactive wastes, iodine constitutes a special case owing to its volatility that does not enable its vitrification into conventional borosilicate glasses. Silver phosphate glasses are good candidates, since they can be melted at low temperature and they can accommodate large AgI quantity within their glass network. To improve their thermal characteristics, Al_2O_3 is added in the glass formulation in order to increase the glass network reticulation. But only a limited amount can be used, since phase separation is observed through the formation of $Al(PO_3)_3$ crystals. We show that Bi_2O_3 can be used as an efficient alternative to Al_2O_3. DSC curves indicate that the crystallization peak is almost supressed, and ^{31}P NMR shows the formation of bismuth phosphate groups but no formation of any crystalline phase. These effects are discussed in terms of cationic field strength and polarisability.

1. Introduction

The management of radioactive iodine is an important technological question in the frame of the global radioactive waste management policy. ^{129}I is a fission product with a very long radioactive half-life of 15·7 million years, and it is classified as a long-life radioactive waste with intermediate activity in the French classification. In most of radioactive waste treatment processes, iodine is trapped from exhaust gases with silver nitrate, $AgNO_3$, and hence recovered as silver iodide, AgI. Different matrices are currently under development to find a definitive conditioning solution for AgI, like cements,[1] ceramics[2] or glasses.[3] Vitrification in a classical borosilicate glass is not possible owing to the quite low volatilization temperature of AgI, close to 600°C at room pressure. This vitrification process indeed involves temperatures in the range of 1000 to 1300°C, even with the latest cold crucible technology.[4] On the contrary, phosphate glasses are known to be able to be made at low temperature,[5] which enabled applications as low temperature sealing for the packaging of microelectronic components,[6] or as glass–polymer composites.[7] The key point of this advantageous and specific property comes from the pentavalency of phosphorus, which does not enable a large polymerization degree of the glass network like in silicate or borosilicate glasses.[8] The obvious counterpart is a weak resistance to hydrolysis by water, leading to severe limitations of applicability of these phosphate glasses. Fortunately, several strategies could be developed to enhance the phosphate glass durability, while preserving their lower melting temperature ("melting" is used here as the elaboration temperature). They are based on the modification of the phosphate glass network polymerization degree, for instance by the addition of another network forming oxide like B_2O_3[9] or by the substitution of oxygen by nitrogen.[10] Another strategy, often used in combination with the previous ones, is to add multivalent oxides which are able to reticulate the glass network. Spectacular results could be obtained with Al_2O_3[11] or Fe_2O_3.[12] In the case of Fe_2O_3, the improvement of durability enabled to formulate glass matrices for radioactive wastes, with the further advantage that the phosphate glasses are able to incorporate a high loading of wastes.[13]

Our strategy for the development of specific matrices for iodine conditioning is based on silver phosphate glasses, in which Al_2O_3 is added in order to increase their durability. Silver phosphate glasses were expected to be able to dissolve a high amount of AgI. Indeed, previous works[14] aiming at developing ionic conductor reported that these glasses are able to incorporate up to 70 mol% of AgI. In a previous paper,[15] we have shown that a good compromise could be found, since we proposed compositions with 1–3 mol% of Al_2O_3, that could be vitrified at a temperature of 650°C, that contained up to 30 mol% of AgI (the loading in borosilicate glasses is less than

* Corresponding author. Email lionel.montagne@univ-lille1.fr
Original version presented at Int. Conf. on Phosphate Glasses, Pardubice, Czech Republic, 2–4 July 2014

1%), and without volatilization during the process.

Nevertheless, we observed that these compositions are limited in their Al_2O_3 content by its limited solubility in silver phosphate glasses. In the present paper, we propose to substitute Al_2O_3 by Bi_2O_3. Bi_2O_3 is indeed reported to enable the preparation of glasses with low to moderate T_g,[16] which suggests an acceptable melting temperature to limit iodine volatilization. Moreover, the trivalent character of Bi_2O_3 suggests that it will confer to silver phosphate glasses a good chemical durability, in the same way as Al_2O_3.

2. Experimental

The glasses were prepared from analytical grade AgI (Alfa Aesar, 99·9%), $AgNO_3$ (Roth, 99·9%), $Al(OH)_3$ (Riedel-de-Haën, 99%), Bi_2O_3 (Riedel-de-Haën, 99%), and $(NH_4)_2HPO_4$ (Acros, 99%). First, reactants (without AgI) were heated in a Pt/Au crucible to 600°C with a heating rate of 1°C/min, and calcined for 3 h, to remove nitrate, hydroxyl and ammonium groups. Al_2O_3 or Bi_2O_3 were then added to the resulting batch, which was subsequently melted at 900°C under air for 1 h. The melt was finally cast onto a brass plate cooled by a water stream. To prepare iodine-bearing glasses, the previous glass was ground, mixed with silver iodide, then introduced in a furnace and heated at 650°C for 1h. The glass was also cast onto a cooled brass plate. The volatilization losses checked by weighting were not significant, hence iodine volatilization is considered to be negligible (owing to the low melting temperature), and the batch compositions (reported in Table1) can be considered as reflecting actual compositions. The thermal behaviour of the glasses was studied with DTA at a heating rate of 5°C/min. The measurements were carried out with ca. 100 mg powder samples. ^{31}P MAS NMR spectra were measured at 9·4T on a 400 MHz spectrometer with a 4 mm probe. The spinning speed was 12·5 kHz and relaxation (recycling) delay was 180 s. The chemical shifts of ^{31}P nuclei are given relative to H_3PO_4 at 0 ppm. We notice that ^{27}Al NMR spectra are not shown in this paper since they did not bring more information in complement to ^{31}P NMR.

3. Results and discussion

In this study, we prepared four glass samples (Table 1). These glasses are based on the silver metaphosphate composition ($AgPO_3$), in which we added 2 mol% of Al_2O_3 or Bi_2O_3 ($98AgPO_3–2Al_2O_3$, or $98AgPO_3–2Bi_2O_3$). This amount of Al_2O_3 or Bi_2O_3 is limited to a low value since we have shown that the solubility of alumina is low in silver phosphate glasses.[15] Then, AgI has been added in both compositions, with a concentration of 19 mol%, leading to the following molar compositions: $79·4AgPO_3–1·6Al_2O_3–19AgI$, or $79·4AgPO_3–1·6Bi_2O_3–19AgI$. Weight losses were measured after the melting process of the glasses, in order to check for volatilizations of glass components. The base glasses ($98AgPO_3–2Al_2O_3$ and $98AgPO_3–2Bi_2O_3$) showed very limited weight losses, lower than 1%. This weight loss is commonly observed during the melting of phosphate glasses at a temperature of 900°C. Hence, we can consider

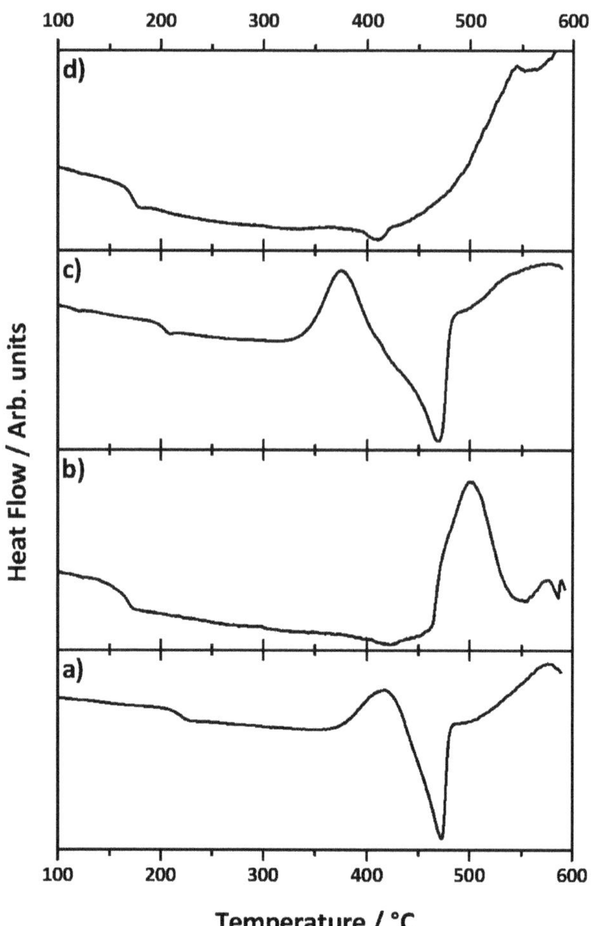

Figure 1. DSC curves of the glasses (a) $98AgPO_3–2Al_2O_3$, (b) $79·4AgPO_3–1·6Al_2O_3–19AgI$, (c) $98AgPO_3–2Bi_2O_3$, (d) $79·4AgPO_3–1·6Bi_2O_3–19AgI$

Table 1. Batch glass compositions (in mol%), weight loss (Δm) measured during glass preparation, glass transition temperature (T_g) and crystallization temperature (T_x)

Sample	Ag_2O	P_2O_5	AgI	Al_2O_3	Bi_2O_3	Δm (%) ±0·5	T_g (°C) ±5°C	T_x (°C) ±5°C
$98AgPO_3–2Al_2O_3$	49	49	-	2	-	0·5	210	375
$79·4AgPO_3–1·6Al_2O_3–19AgI$	38·7	38·7	19	1·6	-	1·6	163	470
$98AgPO_3–2Bi_2O_3$	49	49	-	-	2	0·4	200	325
$79·4AgPO_3–1·6Bi_2O_3–19AgI$	38·7	38·7	19	-	1·6	1·5	160	-

that the batch composition reflects the actual composition of the glasses. When loaded with AgI, the measured weight losses after the melting processes are quite larger than for the base glasses (Table1), but they remain very limited and at an acceptable level. Here again, in the following discussion we will consider that the batch composition reflects the actual composition of the glasses. We notice that the AgI concentration in our glasses is much larger than the amount that can be loaded in borosilicate glasses (typically around 1 w% or less).[17] On the other hand, it is known that much larger AgI content can be incorporated in silver phosphate glasses, for instance for the elaboration of ionic conductors, in which a 70 mol% AgI content can be reached.[14] Nevertheless, we selected an intermediate amount, which is already very large compared to borosilicate glasses, because beyond this limit we could not prepare glasses with a small weight loss during the melting process.

The DSC curves of the glasses are presented on Figure 1. For the $98AgPO_3$–$2Al_2O_3$ glass (Figure 1(a)), the glass transition temperature (T_g) is observed at 210°C and the crystallization temperature T_x at 375°C (considering the onset of the crystallization process). It is followed by a large endothermic peak with a minimum at 470°C, due to the melting of the crystalline phase. When AgI is added to this base glass (Figure 1(b)), we observe a large decrease of the T_g value, now located at 163°C. Thus, a spectacular decrease in T_g (of 47°C) is obtained by the addition of AgI in the base glass. We also observe a shift to higher temperature of the crystallization peak, now located at 470°C. A very small endothermic peak can be seen at 420°C, whose assignment remains unclear. Its low intensity may indicate that some crystallization occurred, either during glass preparation, or even during the DSC experiment. Anyway, the addition of AgI in the base glass results in two favourable effects, i.e. a decrease of the T_g temperature (which indicates that the glass can be melted at a lower temperature), and an increase of the T_x temperature (which indicates an improved resistance to devitrification). Finally, the addition of AgI leads to an increase of the glass stability expressed as (T_x–T_g), from 165 to 307°C.

The structure of the alumina-bearing glasses has been examined using ^{31}P solid-state NMR, as shown on Figure 2(a),(b). The spectrum of the base glass (Figure 2(a)) contains broad resonances, as expected for glasses. ^{31}P NMR chemical shift is indeed sensitive to variations in bond angles and lengths, whose distribution in amorphous compounds leads to Gaussian distributions. A main resonance is observed at −16 ppm, assigned to Q^2 groups (i.e. middle chain groups). This large intensity is expected since only a small amount of Al_2O_3 has been added in the silver metaphosphate glass, which is constituted of phosphate chains of theoretically infinite length. In a previous work related to silver aluminophosphate

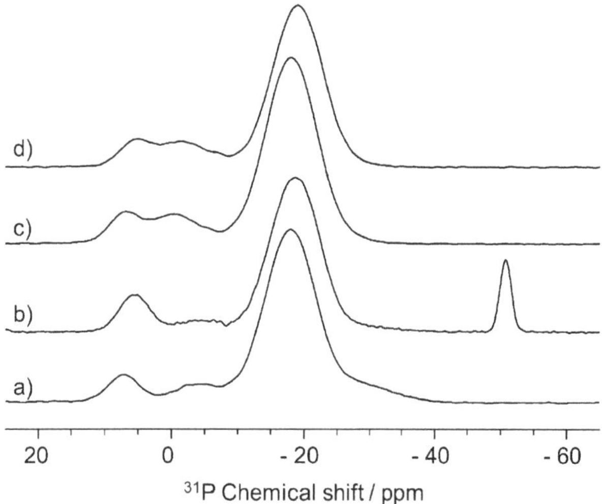

Figure 2. ^{31}P MAS-NMR spectra of the glasses (a) $98AgPO_3$–$2Al_2O_3$, (b) $79{\cdot}4AgPO_3$–$1{\cdot}6Al_2O_3$–$19AgI$, (c) $98AgPO_3$–$2Bi_2O_3$, (d) $79{\cdot}4AgPO_3$–$1{\cdot}6Bi_2O_3$–$19AgI$

glasses,[15] we have shown with a ^{31}P–^{27}Al cross-polarization NMR experiment that the aluminophosphate groups are located at −8 and −35 ppm. They are assigned to aluminate groups bonded to Q^1 and Q^2 phosphate groups, respectively. Their intensity are quite weak on the spectrum in Figure 2(a), which is in accordance with the low alumina amount in the glass. Finally, a resonance is observed at 15 ppm, which is not connected to alumina (as shown by a ^{31}P–^{27}Al cross-polarization NMR experiment), and which is thus assigned to Q^1 groups charged compensated only by silver ions.

When AgI is added in the $98AgPO_3$–$2Al_2O_3$ glass, we observe on the ^{31}P NMR spectrum (Figure 2(b)) a shift of the Q^1 and Q^2 resonances, from 15 to 10·5 ppm and from −16 to −19·5 ppm, respectively. This is assigned to the modification of bonds lengths and angles of the phosphate network, due to the presence of the large iodide anions (I^-). More importantly, we observe that the resonances assigned to the aluminophosphate groups decrease in intensity (resonance at −9 ppm), or even disappear (resonance at −35 ppm). This clearly indicates that the aluminophosphate connectivity decreases in the glass network after AgI addition. Moreover, a new resonance appears at −50·8 ppm on Figure 2(b), assigned to the aluminium metaphosphate phase, $Al(PO_3)_3$ [A]. The narrow shape of this resonance indicates that this phase crystallized from the glass, despite it could not be detected on the x-ray diffractogram (not shown), probably owing to the low Al_2O_3 content in the glass.

The DSC curves of the Bi_2O_3-containing glasses are shown on Figure 1(c),(d). The thermogram of the $98AgPO_3$–$2Bi_2O_3$ glass (Figure 1(c)) shows a T_g at 200°C, a T_x at 325°C, which is followed by a large endothermic melting peak at 415°C. These characteristic temperatures are similar to that of the $98AgPO_3$–$2Al_2O_3$ glass (Figure 1(a)), but they are observed at lower values. It thus confirms that Bi_2O_3 enables to decrease the thermal characteristics of the glass, as compared to Al_2O_3. This is a favourable property in order to limit iodine volatilization during glass preparation. After addition of AgI, the thermogram of the resulting glass is strongly modified, as shown in Figure 1(d). A decrease in T_g is observed, from 200 to 160°C, and the crystallization peak is almost suppressed. A large deviation of the baseline upwards may however indicate that some crystallization occurs during the thermal treatment, but it remains an ill-defined process compared to the glass without AgI (Figure 1(c)) or to the glass with both alumina and AgI (Figure 1(b)). Similar to the thermogram of the latter glass, a very weak endothermic and unassigned peak is detected at 395°C.

The ^{31}P NMR spectra of the glasses containing Bi_2O_3 are shown in Figure 2(c),(d). As expected, on the spectrum of the $98AgPO_3$–$2Bi_2O_3$ glass (Figure 2(c)), Q^1 and Q^2 resonances are observed at 14 and −18 ppm, respectively, the latter having again the largest intensity. Phosphate groups connected to bismuth, which would be equivalent to the aluminophosphate groups, are observed only as a single resonance at −3 ppm, no resonance is observed in the −35 ppm region as for the Al_2O_3-containing glass (Figure 2(a)). This means that either there is no bismuth connected to Q^2 groups in the glass network, or the resonance is located below the main Q^2 resonance at −18 ppm. Unfortunately the low Larmor frequency and the very large quadrupolar constant of the Bi nucleus does not enable to record cross-polarization NMR spectra as for aluminophosphate glasses. Hence, we cannot conclude on the presence of bismuth-Q^2 groups near −18 ppm. Concerning the Q^1 site connected to Bi^{3+}, its resonance is located at a larger chemical shift value (−3 ppm) compared to the one in the aluminophosphate glass (−9 ppm). This is due to the much lower electrostatic field value (defined as z/a^2, z being the cationic charge and a the ionic radius) of Bi^{3+} compared to Al^{3+} (0·53 and 0·99 10^{-20} m^{-2}, respectively).[18] It is indeed known that the ^{31}P chemical shifts are dependant to the cationic field strength.[19]

After addition of AgI in the $98AgPO_3$–$2Bi_2O_3$ glass, there is almost no modification of the ^{31}P NMR spectrum (Figure 2(d)). A shift is observed for the resonances of the Q^1 and Q^2 groups (from 14 to 10 ppm, and −18 to −20·5 ppm, respectively), due to the expansion of the glass network induced by the presence of the large I^- anions. The most interesting feature is that there is no modification of the intensity of the Q^1 sites bonded to Bi^{3+} (near −4 ppm), and no narrow peak appears as in the case of the aluminophosphate glass, for which the resonance of $Al(PO_3)_3$ crystal is observed at −50·8 ppm (Figure 2(b)). This suggests that the bismuth silver phosphate glass enables a better incorporation of a large amount of AgI within the glass network, as compared to the equivalent aluminium silver phosphate glass. Indeed, no crystallization is detected and the relative intensity of the different resonances present in the glass without AgI are preserved after AgI incorporation. We notice that all the glasses have been analysed with x-ray diffraction, but no peak could be observed on the diffractograms (not shown), and anyway the low amount of Al_2O_3 or Bi_2O_3 would not enable to detect them.

Conclusion

Silver phosphate glasses formulated with Al_2O_3 or Bi_2O_3 were prepared, in which large amount of AgI could be incorporated without significant volatilization. DSC curves indicate that AgI addition induces a large decrease in T_g and a decrease of the crystallization tendency, and even almost a suppression of the crystallization in the case of the bismuth containing glass. This is confirmed with solid state ^{31}P NMR, which shows that after AgI addition the phosphate groups bonded to bismuth are preserved and no crystal is detected, contrary to the aluminophosphate glass. This difference in behaviour is probably due to the larger polarisability of Bi^{3+} ions, which are more able than Al^{3+} ions to accommodate the network distortion induced by the incorporation of AgI in large amount. This result will enable to formulate glasses with a good thermal stability during the long term disposal of radioactive wastes. Further work is underway to optimize the glass compositions, based on the concept develop in this paper.

Acknowledgements

T.L. acknowledges CEA for its PhD grant. The FEDER, Region Nord Pas-de-Calais, Ministère de l'Education Nationale de l'Enseignement Superieur et de la Recherche, CNRS, and USTL are acknowledged for funding of NMR spectrometers.

References

1. Clark, W. E. & Thompson, C. T. Immobilization of iodine in concrete, *US Pat.*, 1977, 4017417.
2. Audubert, F., Carpena, J., Lacout, J. L. & Tetard, F. Elaboration of an iodine-bearing apatite Iodine diffusion into a $Pb_3(VO_4)_2$ matrix, *Solid State Ionics*, 1997, **95**, 113–119.
3. Darab, J. G., Meiers, E. M. & Smith, P. A. Behavior of simulated hanford slurries during conversion to glass, *MRS Proc.*, 1999, **556**, 215.
4. Do Quang, R., Petitjean, V., Hollebecque, F., Pinet, O., Flament, T. & Prod'homme, A. Vitrification of HLW Produced by Uranium/Molybdenum Fuel Reprocessing in COGEMA's Cold Crucible Melter, *Proc. 9th Int. Conf. on Radioactive Waste Management and Environmental Remediation*, 2003, pp. 1585–1591.
5. Ray, N. H. Composition–property relationships in inorganic oxide glasses, *J. Non-Cryst. Solids*, 1974, **15** (3), 423–434.
6. He, Y. & Day, D. E. Development of a low temperature phosphate sealing glass, *Glass Technol.*, 1992, **33**, 214–219.
7. Guschl P. C. & Otaigbe, J. U. Crystallization kinetics of low-density polyethylene and polypropylene melt-blended with a low-Tg tin-based phosphate glass. *J. Appl. Polym. Sci.*, **90**, 3445–3456.
8. Van Wazer, J. R. & Holst, K. A. Structure and Properties of the Con-

densed Phosphates. I. Some General Considerations about Phosphoric Acids, *J. Am. Chem. Soc.*, 1950, **72** (2), 639–644.
9. Ducel, J. F., Videau, J. J. & Suh, S. K. ^{31}P MAS and ^{11}B NMR study of sodium rich borophosphate glasses, *Phys. Chem. Glasses*, 1994, **35** (1), 10–16.
10. Brow, R. K., Reidmeyer, M. R. & Day, D. E. Oxygen bonding in nitrided sodium- and lithium-metaphosphate glasses, *J. Non-Cryst. Solids*, 1988, **99** (1), 178–189 .
11. Brow, R. K. Nature of Alumina in Phosphate Glass: I, Properties of Sodium Aluminophosphate Glass. *J. Am. Ceram. Soc.*, 1993, **76**, 913–918.
12. Day, D. E., Wu, Z., Ray, C. S. & Hrma, P. Chemically durable iron phosphate glass wasteforms, *J. Non-Cryst. Solids*, 1998, **241** (1), 1–12.
13. Kim, C. & Day, D. E Immobilization of Hanford LAW in iron phosphate glasses, *J. Non-Cryst. Solids*, 2003, **331** (1–3), 20–31.
14. Martin, S.W. Ionic Conduction in Phosphate Glasses. *J. Am. Ceram. Soc.*, 1991, **74**, 1767–1784.
15. Lemesle, T., Méar, F. O., Campayo, L., Pinet, O., Revel, B. & Montagne, L. Immobilization of radioactive iodine in silver aluminophosphate glasses, *J. Hazard. Mater.*, 2014, **264** (15), 117–126.
16. Montagne, L., Palavit, G. & Mairesse, G. ^{31}P MAS NMR and FTIR analysis of $(50-x/2)Na_2O.xBi_2O_3.(50-x/2)P_2O_5$ glasses, *Phys. Chem. Glasses*, 1996, **37** (5), 206–211.
17. Ojovan, M. I. & Lee, W. E. *An Introduction to Nuclear Waste Immobilisation*, Elsevier Science, Amsterdam, The Netherlands, 2005, pp. 77–78.
18. Volf, M. B. *Chemical apporach to glass*, Elsevier Science: Amsterdam, The Netherlands, 1984, p. 118.
19. Brow, R. K., Phifer, C. C., Turner, G. L. & Kirkpatrick, R. J. Cation Effects on ^{31}P MAS NMR Chemical Shifts of Metaphosphate Glasses, *J. Am. Ceram. Soc.*, 1991, **74**, 1287–1290.

DOI: 10.13036/17533562.56.2.067 Phys. Chem. Glasses: Eur. J. Glass Sci. Technol. B, April 2015, 56 (2), 67–70

Melting conditions impact on the properties of copper containing phosphate glasses for high power and high energy amplifiers

V. I. Arbuzov,[1,2] Yu. K. Fyodorov,[1] S. I. Nikitina,[1] R. V. Smirnov,[1,2] V. M. Volynkin[1] & M. V. Voroshilova[1]*

[1] *Laboratory of Laser Glasses, Research and Technological Institute of Optical Material Science, 192171 Saint-Petersburg, Russia*
[2] *Department of Opto-Information Technologies and Materials, ITMO University, 197101 Saint-Petersburg, Russia*

Manuscript received 2 October 2014
Accepted 15 December 2014

Russia has started construction of the 192 channel pulse amplifier facility UFL-2M with an output energy of 2·8 MJ. Large disc active elements (DAE) with a light aperture of 400×400 mm for the UFL-2M are made from a neodymium phosphate glass KNFS3 of the system Na_2O–K_2O–MgO–CaO–SrO–BaO–B_2O_3–SiO_2–P_2O_5 with a neodymium concentration of $3{\cdot}5\times10^{20}$ cm^{-3}. One of the rather effective routes that output energy decreases is the so-called parasitic generation in a DAE caused by luminescence quanta propagating along big polished planes of a DAE. The phenomenon of parasitic generation is known to lead to a significant (by tens percents) decrease in the signal gain. To suppress this phenomenon, cladding plates should be pasted to the DAE side edges that are made usually of copper-containing glasses as they have an intense absorption band with a maximum at approximately 880 nm caused by Cu^{2+} ions. The cladding glass, the photo-resistant glue and the DAE glass must be assembled in such a way that the refractive index, n, meets the following condition $n_{clad}>n_{glue}>n_{DAE}$. To eliminate thermal strains at the junction of the cladding plates and the DAE, the difference in values of coefficients of thermal expansion $|\alpha_{clad}-\alpha_{DAE}|$ should be close to 0. The decimal coefficient of radiation absorption at the lasing wavelength 1053 nm, a_{1053}, must be in the limits from 1·2 to 2·0 cm^{-1}. The paper shows how such copper-containing glasses were developed that satisfy the requirements on refraction index, coefficient of thermal expansion and absorption coefficient at the lasing wavelength. Their parameters are n_e=1·544±0·002, n_{1053}=1·534±0·002, α_{20-300}=(117±2)×10^{-7} K^{-1} as compared with that of the glass KNFS3 n_e=1·540±0·002, n_{1053}=1·530±0·002, α=(116±2)×10^{-7} K^{-1}. The developed copper containing glasses can be used in the production of large disk active elements for the amplifier facility UFL2M.

Introduction

Neodymium phosphate glasses are widely used all over the world in the production of disc active elements (DAE) of high energy and high pick power laser facilities such as the NIF (National Ignition Facility in the USA) and the LMJ (Laser Mega-Joule in France)[1,2] intended for research in the fields of laser controlled thermonuclear fusion, interaction of laser radiation with matter, physics of high energy density, and plasma physics. Russia recently started construction of the 192 channel pulse amplifier facility UFL-2M with an output energy of 2·8 MJ.[3] Large DAEs with a light aperture of 400×400 mm (DAE400) for the UFL-2M are made of sodium potassium aluminum strontium barium metaphosphate glass KNFS3 (one of the glasses of the system Na_2O–K_2O–Al_2O_3–SrO–BaO–P_2O_5 with small additions of CaO, B_2O_3, and SiO_2[4]) with a neodymium concentration of $3{\cdot}5\times10^{20}$ cm^{-3}. Physical and technical principles of the UFL-2M construction were developed previously by the example of the four channel amplifier facility "Luch" (Russia) on the basis of a DAE with light aperture of 200×200 mm (DAE200, Figure 1) made of neodymium metaphosphate glass KGSS 0180/35 with the same neodymium concentration.[5,6] To initiate the fusion reaction in a nuclear fuel (for example, in a mixture of deuterium and tritium) ultra

Figure 1. Disc active element (light aperture 200×200 mm) with cladding [Colour available online]

Corresponding author. Email arbuzov@goi.ru, viarb@yandex.ru
Original version presented at Int. Conf. on Phosphate Glasses, Pardubice, Czech Republic, 2–4 July 2014

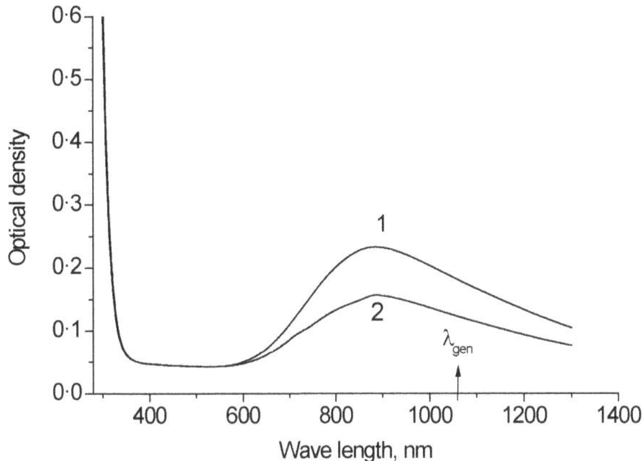

Figure 2. Absorption spectra of glasses with 0·25 wt% CuO melted in oxidizing (1) and reducing (2) conditions. Sample thickness is 1 mm

high energies are required, therefore laser glasses for such facilities should have a high quantum yield of luminescence (at least 75%). Achieving such values of the quantum yield is complicated by the fact that even small impurities of hydroxyl groups of the structural water in the glass can reduce the quantum yield of its luminescence by several times. Therefore, in the melting process, the glass melt is dehydrated by blowing dry oxygen through it.[5,6] However, obtaining high values of quantum yield in the dehydration process of the melt is complicated by the fact that, for a given concentration of neodymium, migration of excitation energy on the activator ions greatly increases the range of each hydroxyl group. This means that both laser glass manufacturers and amplifier facilities designers should eliminate all possible ways of reducing energy output.

One of these rather effective ways is the so called parasitic generation in a DAE that is known to lead to a significant (by tens of percents) decrease in the signal gain.[7,8] To suppress this phenomenon, cladding plates should be pasted to the DAE side edges (see dark-blue cladding plates in Figure 1) that are made usually of copper-containing glasses as they have an intense absorption band with a maximum at approximately 880 nm (Figure 2) caused by Cu^{2+} ions. The aim of the paper is to show how such copper-containing glass can be developed for the DAE made of the glass KNFS3.

Requirements to the cladding glass

To understand the phenomenon of parasitic generation in a DAE, let us consider an energy scheme that consists of only two laser levels with the energy difference ΔE between them (Figure 2). Under the action of pumping lamp radiation, activator ions in a DAE find themselves at the upper laser level. There are three ways for electrons (black circles at the upper level) to drop back to a lower laser level: (1) spontaneous transitions that are responsible for luminescence

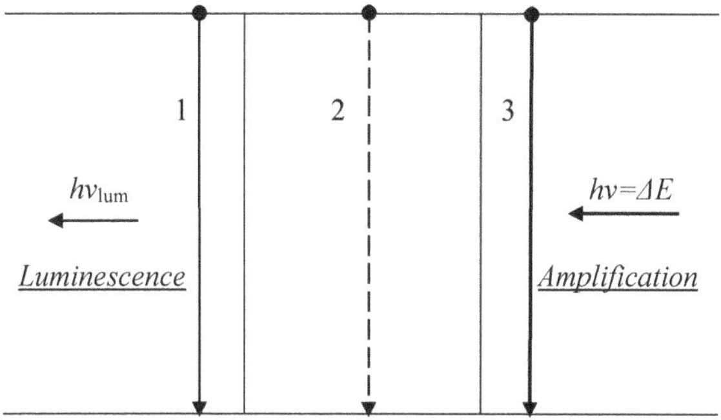

Figure 3. Electron transitions between laser levels in the ensemble of excited activator ions

emitted in every directions; (2) non-radiation transitions as a consequence of excited electron interaction with phonons and hydroxyl groups of structural water that lead to luminescence quenching; and (3). induced transitions that take place under the action of radiation with the quanta energy $h\nu=\Delta E$ and that lead to amplification of incident radiation in its direction.

In the case of action of intense radiation with quanta energy $h\nu=\Delta E$ on the system of excited activator ions, all the electron transitions in the system of excited activator ions should be, seemingly, induced ones. Nevertheless, spontaneous transitions go on simultaneously with induced transitions. Because luminescence quanta have the same energy $h\nu=\Delta E$ they can also be responsible for the induced electron transitions in the ensemble of excited activator ions. The longer the optical way of such quanta in the DAE, the bigger is the number of electron transitions induced by luminescence quanta. Thus, propagation of these quanta along big polished planes of a DAE leads to a maximal release of inverse population density in the ensemble of excited activator ions and, as a result, to a decrease in the radiation energy that is amplifying by the laser in the required direction. Luminescence quanta that are reflected or scattered from side edges of the DAE continue to reduce the inverse population density. Therefore, to exclude luminescence quanta reflections and scattering from lateral sides of the DAE, plates of special glass (cladding) should be pasted to them by special photoresistant glue. As already noted, these plates are made usually of copper-containing glass with a decimal coefficient of radiation absorption at the lasing wavelength 1053 nm, a_{1053}, which, for various DAE types, must be in the limits from 1·2 to 2·0 cm^{-1}. Luminescence quanta enter into the cladding plates that, at a given thickness (10–12 mm), entirely absorb luminescence quanta and, therefore, reduce the efficiency of their reflection and scattering at the junction of the cladding plates and the DAE and suppress further development of parasitic generation in a DAE.[8]

In the case of a DAE with cladding, the cladding glass, the photoresistant glue and the DAE glass must be fitted together on the refraction index, n, in such a way that the following condition $n_{clad}>n_{glue}>n_{DAE}$

should be met. Furthermore, because of luminescence light absorption, cladding plates are heated up somewhat more intensely than the DAE. It means that the thermal expansion of cladding plates is somewhat larger than that of the DAE. To eliminate thermal strains at the interface of the cladding plates and the DAE, the difference in values of coefficients of thermal expansion $|\alpha_{clad}-\alpha_{DAE}|$ should be close to 0 K^{-1}.

Development of copper containing glass

Both refractive index and coefficient of thermal expansion strongly depend on the glass composition. The glass KNFS3 itself is characterized by the values of n_e=1·540±0·002, n_{1053}=1·530±0·002, α_{20-300}=(116±2)×10^{-7} K^{-1}. To fulfill the above requirement on refractive index, the KNFS3 glass composition should be changed by increasing the contribution of components with high refraction (SrO, BaO, Nd_2O_3).[9] At the same time, these changes in the glass composition should be as low as possible to fulfill the above requirement on the coefficient of thermal expansion.

It is clear that technology of glass melting should be characterized by the reproducibility of glass parameters. At the same time, as copper is a variable valence element, redox conditions of glass melting must have a strong impact on the quantitative ratio of Cu^+/Cu^{2+}.[10–12] Therefore, it is desirable, when passing from melting to melting, either to ensure the same quantitative ratio of Cu^+/Cu^{2+} at neutral conditions of melting or to shift the chemical equilibrium $Cu^+ \leftrightarrow Cu^{2+}$ as far as possible to the right due to glass melt oxidation. To establish limits of the ratio changes, cladding glasses with addition of 0·15 to 0·28 wt% CuO were melted in reducing (additions of carbon or aluminium powder to the glass batch), neutral and oxidizing (blowing oxygen through the melt) conditions. A pipe for oxygen blowing was made of a ceramic based on SiO_2 (Stecryt). Waste of especially pure silica glass ground to a fine powder is used in making Stecryt by slip casting. A semi-liquid mixture of the powder and distilled water is put in into a plaster mould of the future article that is an assembly of a few parts. After plaster has absorbed water from the slip, the mould is disassembled, the article is dried in air and then fired in a furnace. The solubility of Stecryt constituents in phosphate glass melts proved to be at the level of 0–0·5 wt% that allows one to get highly homogeneous glasses.

A mixture of dry metaphosphates of the main glass components (K, Na, Al, Sr, Ba), silicon pyrophosphate and oxides of neodymium and copper was used for batch preparation. Before melting, reagent samples were calcined in a special furnace for determining their water content. Results of calcining were taken into account when preparing batch for 800 ml of glass. The glass batch was melted in a laboratory electrical furnace in Stecryt crucibles at the temperature of 1280°C. Homogenization of the glass melt was carried out by a Stecryt stirrer. Melting proceeded for about 4 h, by the end, the temperature in the furnace went down to 800–750°C and the glass melt was poured into a warmed graphite mould. For annealing, the mould with glass melt was then placed in a muffle furnace heated up to 500°C. Over an hour the temperature was reduced to 400°C. Then the power was switched off, and, within 20 h, the furnace slowly cooled to room temperature. Formation of crystals neither on the surface of castings, nor in their volume was observed when decreasing the temperature of the glass melt or solid casting from 1200 to 600°C. Samples 1, 2, 3 and 5 mm thick were made of the obtained glass to measure their absorption spectra and coefficients of radiation absorption at the lasing wavelength. A spectrophotometer, Shimadzu UV3600, was used to carry out these measurements. Samples with the dimensions 4×4×50 mm were used to measure the coefficient of thermal expansion in the temperature range from 20 to 300°C by a vertical quartz dilatometer DKV-1. Refractive index of glasses under study was determined by the Obreimov method.

In the case of glasses with 0·2 wt% CuO melted in reducing conditions (addition of 0·5 g carbon per 100 g of glass), the a_{1053} value was equal to 1·30 cm^{-1}, whereas for glasses melted in neutral and oxidizing conditions they were equal to 1·47 and 1·54 cm^{-1}, respectively. The ratio 1·54/1·30 of a_{1053} values for glasses melted in oxidizing and reducing conditions is equal to 1·185. This proved to be rather low, which apparently indicates insufficient severity of oxidizing and reducing conditions when melting glass with 0·2 wt% CuO. Indeed, when reducing the CuO concentration to 0·1 wt%, this ratio can reach 1·63 at the same redox conditions of melting. Virtually the same result (1·92/1·15=1·67) is observed when melting glasses with 0·25 wt% CuO in oxidizing (curve 1 in Figure 2) and in more severe reducing conditions (adding 0·3 g of aluminium per 100 g of glass) (curve 2 in Figure 2). At the same time it should be noted that sometimes glasses with aluminium as a reducer become black in the course of annealing but they turn dark-blue after remelting.

Absorption bands of Cu^+ ions in phosphate glasses of various compositions are situated in the range 200 to 290 nm.[10] This absorption seems to be rather high as indicated in the Figure 2. Optical density of a 1 mm thick glass sample with 0·25 wt% CuO in the range between 250 and 260 nm is of the order of 4 that does not allow one to measure it properly. The ratio Cu^+/Cu^{2+} in aluminium potassium barium metaphosphate glasses melted in neutral and oxidizing conditions proved to be equal to 60/40.[12] As glasses under our study are close in their compositions (taking into account the main components K_2O, Al_2O_3, BaO, P_2O_5) to the glass KNFS3 it is not excluded that they can be characterized approximately by the same ratio Cu^+/Cu^{2+}.

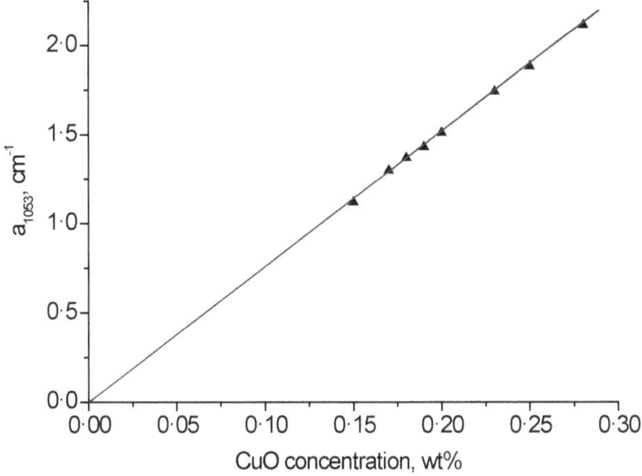

Figure 4. Dependence of absorption coefficient at the lasing wave length on CuO concentration in glasses melted in oxidizing conditions

Probably, this was the reason that all our attempts to measure absorption spectra of Cu^+ and Cu^{2+} ions simultaneously using the same glass samples failed because specific absorption of Cu^+ ions seems to be multiply higher than the specific absorption of Cu^{2+} ions. But such a task was not proposed in this work. However, it is important to know how the decrease in absorption of Cu^+ ions is quantitatively related with the increase in absorption of Cu^{2+} ions in the glass with a particular concentration of CuO when passing from reducing conditions of melting to oxidizing ones as this allows one to determine the quantitative ratio of Cu^+/Cu^{2+} and the values of specific absorption coefficients of Cu^+ and Cu^{2+} ions in the developed glasses. An aim of future work will be to investigate glasses with much smaller CuO concentrations.

A transition from oxidizing conditions to neutral ones in the course of the same melting experiment did not reduce the a_{1053} value obtained for the oxidized glass. This fact was taken into account when elaborating industrial technology for melting cladding glass.

Hereafter, all the practical glasses were melted in oxidizing conditions. The dependence of the a_{1053} on the CuO concentration in laboratory glasses is shown in Figure 4. One can see from this figure that this dependence is described by a line that is, firstly, linear and, secondly, passes through the origin of the coordinates. That is to say that the redox conditions of glass melting remained the same and the developed technology is characterized by the required reproducibility of glass parameters. At the same time it should be noted that, if reagents contained some reducers even at the impurity level, a_{1053} can be lowered by several tens of per cents as compared with that in the glass melted from the batch that does not contain any reducing impurities. This means that, before using reagents in glass melting, it is necessary to carry out close examination of them due to the presence of such reducing agents.

Conclusions

The copper-containing cladding glasses for disk active elements made of the neodymium phosphate glass KNFS3 were developed. The glasses satisfy all the requirements of refractive index, coefficient of thermal expansion and absorption coefficient at the lasing wavelength. Their parameters are: n_e=1·544±0·002, n_{1053}=1·534±0·002, α_{20-300}=(117±2)×10^{-7} K^{-1}, compared with that of the glass KNFS3: n_e=1·540±0·002, n_{1053}=1·530±0·002, α=(116±2)×10^{-7} K^{-1}. The developed glasses can be used in the production of large disk active elements for the amplifier facility UFL2M.

Acknowledgement

This work was financially supported by the Russian Scientific Foundation (Agreement # 14-23-00136).

References

1. Campbell, J. H., McLean, M. J., Hawley-Fedder, R., Saratwala, T., Ficini-Dorn, G. & Trombert, J. H. Development of Continuous Glass Melting for Production of Nd-doped Phosphate Glasses for the NIF and LMJ Laser Systems. *SPIE*, 1999, **3492**, 778–786.
2. Campbell, J. H. & Suratwala, T. I. Nd-doped Phosphate Glasses for High-energy/High-peak-power Lasers. *J. Non-Cryst. Solids*, 2000, **263–264**, 318–431.
3. Avakyants L. I., Arbuzov V. I., Babina T. O., Fyodorov Yu. K., Ignatov A. N., Krekhova E. Yu., Pozdnyakov A. E. et al. Phosphate glasses for making rod and disc active elements of lasers and high-pick-power/high-energy radiation amplifiers. *The 23 International Congress on Glass. Prague. 2013. Book of Abstracts.* P. 75-76.
4. Patrikeev, A. P., Belousov, S. P., Gerasimov, V. M., Ignatov, A. N., Pozdnyzkov, A. E., Surkova, V. F. & Avakyants, L. I. Laser phosphate glass. *Patent of Russian Federation. Int. class. C03C, H01S. Published 27.11.2013.*
5. Arbuzov, V. I., Charukhchev, A. V., Fyodorov, Yu. K., Lunter, S. G., Nikitina, S. I., Shashkin, V. S., Shashkin, A. V. & Sirazetdinov, V. S. Neodymium phosphate glasses for high-energy and high-pick-power lasers. *Glasstechn. Ber. Glass Sci. Technol.*, 2002, **75C2**, 209–214.
6. Arbuzov V. I., Lunter S. G., Nikitina S. I., Petrovskii G. T., Semenov A. D., Fedorov Yu. K., Shashkin V. S., Shashkin A. V., Volynkin V. M., Ponomarev V. Ya., Sirazetdinov V. S., Charukhchev A. V. Large disc-shaped active elements made from neodymium phosphate glass elements made from neodymium phosphate glass. *Opticheskii Zhurnal (Journal of Optical Technology)*. 2003. **70**. 361-369 (in Russian).
7. Remontet, C., Bouchut, J. M., Licchesi, V. & Duchesne, J. LMJ Cladding Industrialization. *SPIE*, 1989, **3492**, 851–858.
8. Arbuzov V. I., Vakhmyanin K. P., Volynkin V. M., Nikitina S. I., Potapova N. I., Fyodorov Yu. K., Tsvetkov A. D., Charukhchev A. V., Shashkin A. V. Absorbing coatings for large disk-shaped active elements made from KGSS-0180/35 neodymium phosphate glass for laser amplifiers. *Opticheskii Zhurnal (Journal of Optical Technology)*. 2002. **69**. 13 – 17 (in Russian).
9. Shchavelev O. S., Babkina V. A. System of calculating optical and thermooptical properties of phosphate glasses according to their chemical composition. *Fizika i Khimiya Stekla (Glass Physics and Chemistry)*. 1977. **3**. 519 – 523 (in Russian).
10. Debnath, R., Chaudhury, J. & Chandra Bera, S. Optical properties and nature of coordination of Cu^+ ions in calcium metaphosphate glass. *Phys. Stat. Solids*, 1990, **157**, 723–733.
11. Arbuzov, V. I., Gusev, G. E. & Semyonov, A. D. Influence of melting conditions on the balance of valent forms of copper ions in aluminum-potassium-barium-phosphate glass. *Opt. Zh. (J. Opt. Technol.)*, 2007, **74**, 55–59. (In Russian.)
12. Arbuzov, V. I. & Gusev, P. E. Influence of the melting temperature and rate of cooling the melt to the glass making temperature on the redox equilibria $Fe^{2+}\leftrightarrow Fe^{3+}$ and $Cu^+\leftrightarrow Cu^{2+}$ in aluminum potassium barium phosphate glasses. *Glass Phys. Chem.*, 2009, **35** (5), 463–467.

Phys. Chem. Glasses: Eur. J. Glass Sci. Technol. B, August 2015, 56 (4), 121–127

Glass-forming ability, structure and physical properties of $ZnO–In_2O_3–P_2O_5$ glasses

Petr Mošner, Ladislav Koudelka, Antonín Račický*

Department of General and Inorganic Chemistry, Faculty of Chemical Technology, University of Pardubice, 532 10 Pardubice, Czech Republic

Lionel Montagne & Bertrand Revel

Université Lille Nord de France – UCCS - UMR CNRS 8181, USTL, 59655, Villeneuve d'Ascq, France

Manuscript received 3 September 2014
Accepted 15 December 2014

Glasses in the ternary system $ZnO–In_2O_3–P_2O_5$ were studied, and the glass-forming region has been determined. Glasses containing up to 20 mol% In_2O_3 were prepared by cooling of the melt in air. The structure and properties of the glasses were primarily studied for the compositional series $xZnO.10In_2O_3.(90–x)P_2O_5$. The basic characteristic parameters of the glasses (density, molar volume, index of refraction) were determined. Structural studies were carried out using Raman and ^{31}P NMR spectroscopies. The methods of differential thermal analysis, hot stage microscopy and dilatometry were also applied for thermal studies. The glass transition temperature of the studied glasses increases with increasing indium oxide content, and decreases with increasing zinc oxide content. ^{31}P NMR spectra of the glasses reveal that the replacement of P_2O_5 by ZnO results in the transformation of Q^2 units into Q^1 units, which is also shown by the Raman spectra.

1. Introduction

The addition of trivalent oxides to phosphate glasses usually leads to an increase in their chemical durability and thermal stability.[1] A number of studies have been devoted to aluminophosphate glasses and borophosphate glasses over recent years.[2–7] Phosphate glasses containing gallium oxide are also able to form over relatively broad concentration regions.[8–10] Studies of phosphate glasses containing indium oxide, however, are relatively limited[11–14] and the content of In_2O_3 in these glasses was mostly less than 5 mol%.

Sales & Boatner have prepared and studied Pb–In–P–O glasses and Pb–Sc–P–O glasses containing 3–6 wt% In_2O_3 and 2–6 wt% Sc_2O_3.[11] Suzuya *et al* studied the structure of the glass $56{\cdot}3PbO.39{\cdot}5P_2O_5.4.2In_2O_3$ with neutron diffraction,[12] and determined coordination numbers for indium and lead of 5·5 and 5·3, respectively. They also came to the conclusion that the improved chemical durability of In_2O_3-containing glasses could be ascribed to the presence of octahedrally coordinated InO_6 species in these glasses.

In our recent study we investigated the zinc phosphate glass series $(50–x)ZnO.xIn_2O_3.50P_2O_5$, with In_2O_3 replacing ZnO.[15] We were able to prepare glasses containing 0–20 mol% In_2O_3. We studied the physical properties of these glasses and used Raman and NMR spectroscopy to study their structure. The replacement of ZnO by In_2O_3 in phosphate glasses results in a depolymerisation of the metaphosphate chains, due to an increase in the O/P ratio, since In_2O_3 supplies three oxygen atoms to the glass network, in comparison to only one oxygen atom for ZnO. From the compositional dependence of the molar volume of zinc borophosphate glasses, $30ZnO.(20–x)B_2O_3.xIn_2O_3.50P_2O_5$, with In_2O_3 replacing B_2O_3, we came to the conclusion that indium oxide in these glasses behaves as a glass modifier and provides oxygen atoms to the structural network, so that compensating In^{3+} cations are present in the glass structure.

In this paper we have determined the glass-forming region in the $ZnO–In_2O_3–P_2O_5$ ternary system, and studied the properties and structure of glasses containing a maximum of 65 mol% P_2O_5, which we consider to be the limit for hygroscopic stability.

2. Experimental

$ZnO–P_2O_5–In_2O_3$ glasses were prepared by melting analytical grade ZnO, In_2O_3 and H_3PO_4, using a total batch weight of 10 g. The homogenised starting mixtures were slowly calcined up to 600°C to remove the water. The reaction mixtures were then heated to 1200–1350°C under ambient air, in a platinum crucible covered with a lid. The melt was held at maximum temperature for 20 min, and subsequently poured into a preheated graphite mould ($T<T_g$). The obtained glasses were then cooled to room temperature. The

*Corresponding author. Email Petr.Mosner@upce.cz
Original version presented at Int. Conf. on Phosphate Glasses, Pardubice, Czech Republic, 2–4 July 2014
DOI: 10.13036/17533562.56.4.121

weight loss measurements indicated that the volatilisation losses were not significant, even at the highest temperature, and hence the batch compositions can be considered as reflecting actual compositions. The amorphous character of the prepared glasses was checked with x-ray diffraction.

The glass density, ρ, was determined on bulk samples by Archimedes' method, using toluene as the immersion liquid. The molar volume, V_M, was calculated as $V_M=M/r$, where M is the average molar weight of the glass composition aZnO.$b$$P_2O_5$.$c$$In_2O_3$ calculated for $a+b+c=1$.

The linear refractive indices were measured with the prism coupling method, using a Metricon Model 2010/M at 453, 532 and 637 nm. Linear refractive index values n_d (587·6 nm) were obtained from a dispersion curve calculated with Metricon software.

The thermal behaviour of the glasses was studied with a Netzsch DTA 404 PC operating in DSC mode, at a heating rate of 10°C/min over the temperature interval 30–900°C. The measurements were carried out with 100 mg powder samples (the average particle size was 10 μm) in a platinum crucible under a N_2 atmosphere. The glass transition temperature, T_g, (the midpoint of the change in the heat capacity, c_p, in the glass transition region) and the values of the crystallisation temperature, T_c, (the onset of the first crystallisation peak) were determined from the DSC curves.

The glass transition temperature was also determined from dilatometric curves obtained from bulk samples with dimensions of 10×5×5 mm using a dilatometer DIL 402 PC (Netzsch) and a heating rate of 5°C/min. The evaluation of dilatometric curves was carried out with Proteus software. The glass transition temperature, T_g, was determined from the change in the slope of the elongation versus temperature, the dilatometric softening temperature, T_d, from the maximum of the expansion trace corresponding to the onset of viscous deformation, and the coefficient of thermal expansion, α, as a mean value over the temperature range 150–250°C.

Thermal properties were also studied with hot stage microscopy (HSM, from HESSE Gmbh), which was carried out with powder samples (~10 μm) pressed into cylinders (3 mm in diameter and height) with a hand press. The specimen was placed on a corundum sample holder and the measurement was carried out with a heating rate of 5°C/min from room temperature to the flow temperature, T_f, (the first temperature at which the sample is melted to a third of its original height) under a static air atmosphere. The projected area and the height of the pressed powder sample were monitored with a CCD camera during heating.

The Raman spectra in the range 1760–60 cm^{-1} were measured on bulk samples at room temperature using a Horiba-Jobin Yvon LaBRam HR spectrometer. The spectra were recorded in back-scattering geometry under excitation with Nd:YAG laser radiation (532 nm) on the sample. The exposure time was 3 s, the accumulation was 10×, and the grid had 600 streaks/mm.

^{31}P MAS NMR spectra were measured at 9·4 T on a 400 MHz BRUKER Avance spectrometer with a 4 mm probe. The spinning speed was 12·5 kHz and the relaxation (recycling) delay was 180 s. The chemical shifts of ^{31}P nuclei are given relative to H_3PO_4 at 0 ppm.

3. Results and discussion

We have prepared and studied 12 homogeneous glass samples in the ternary system ZnO–In_2O_3–P_2O_5, the composition of which is shown in Figure 1. We were able to prepare glasses containing 0–20 mol% In_2O_3 and the glass forming region is shown for glasses prepared by free cooling to room temperature of the melt in air. All the glasses were transparent and colourless. Table 1 contains the basic characterisation data of the prepared glasses: density, molar volume, glass transition temperature, dilatometric softening temperature, crystallisation temperature, thermal expansion coefficient and the index of refraction. We have also shown the approximate glass-forming

Table 1. Composition, density, ρ, molar volume, V_M, index of refraction, n_d, glass transition temperature, T_g^a (onset on DSC curve), T_g^b (measured by dilatometry), crystallisation temperature, T_c, dilatometric softening temperature, T_d, and the coefficient of thermal expansion, α, of ZnO–P_2O_5–In_2O_3 glasses

Sample no.	ZnO [mol %]	P_2O_5	In_2O_3	O/P	$\rho\pm0{\cdot}02$ g/cm	$V_M\pm0{\cdot}5$ cm^3/mol	$n_d\pm0{\cdot}005$	$T_g^a\pm2$ [°C]	$T_c\pm2$ [°C]	$T_g^b\pm2$ [°C]	$T_d\pm2$ [°C]	$\alpha\pm0{\cdot}3$ [ppm/°C]
1	50	50	0	3·00	2·84	39·3	1·523	452	523	436	462	9·6
2	45	50	5	3·10	3·09	39·3	1·553	479	-	479	506	8·8
3	47·5	47·5	5	3·15	3·19	37·6	1·565	473	508	480	496	8·4
4	25	65	10	2·92	3·03	46·3	1·548	593	632	609	631	7·9
5	30	60	10	3·00	3·08	44·6	1·551	576	634	584	609	8·1
6	35	55	10	3·09	3·18	42·2	1·559	556	633	554	587	7·9
7	40	50	10	3·20	3·29	39·9	1·583	509	735	520	547	7·9
8	45	45	10	3·33	3·52	36·4	1·600	493	672	505	524	8·2
9	50	40	10	3·50	3·75	33·4	1·625	508	628	505	522	8·1
10	35	50	15	3·30	3·53	39·9	1·592	549	756	541	585	7·8
11	20	60	20	3·16	3·52	44·6	1·575	652	756	610	629	8·3
12	30	50	20	3·40	3·74	40·4	1·617	572	699	572	602	7·8

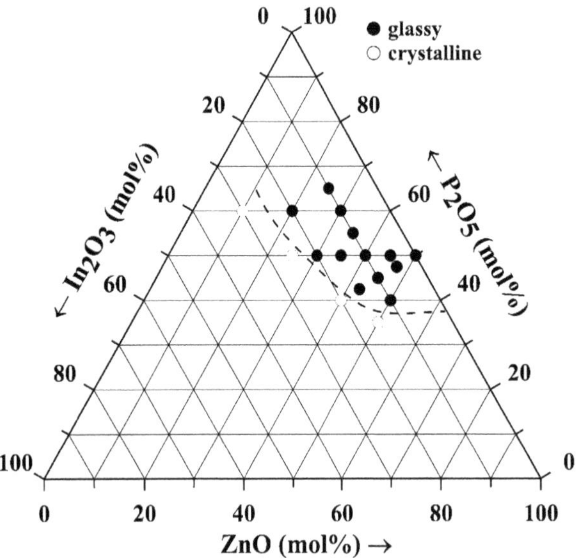

Figure 1. Glass forming region and composition of the glasses studied in the $ZnO–In_2O_3–P_2O_5$ system

region in Figure 1. We have only studied glasses with a maximum content of 65 mol% P_2O_5 and therefore the glass forming region is incomplete and is only limited to glasses with ≤65 mol% P_2O_5. We were able to prepare glasses containing 10 mol% In_2O_3 in a wide compositional range 25–50 mol% ZnO, and therefore the studied changes in properties and structure are primarily for this compositional series. For the sake of completeness we have also covered, however, certain other samples from the entire glass-forming region. Glasses containing up to 20 mol% In_2O_3 can be prepared in this ternary system, particularly in the region with higher P_2O_5 content.

As shown in Table 1, the glass density increases for either increasing In_2O_3 or increasing ZnO content. The molar volume decreases when ZnO replaces P_2O_5 in the series with 10 mol% In_2O_3, because the total number of atoms in the molar formula decreases when seven atoms in P_2O_5 are replaced by only two atoms in ZnO. On the other hand, when ZnO is replaced by In_2O_3 at constant P_2O_5 content, the density increases, but the molar volume does not change significantly. In this case two atoms of ZnO are replaced by five atoms of In_2O_3. Therefore, we assume that it is necessary to consider also the bonding relations in these glasses and the effect of covalent or ionic radii of zinc and indium. The character of In–O chemical bonds is more ionic than that of Zn–O bonds, and the effect of these differences also plays a significant role in determining the molar volume.

The DSC curves of the glass series xZnO.10In_2O_3.(90–x)P_2O_5 are shown in Figure 2. Exothermic crystallisation peaks can be seen for all the glass samples and in some curves endothermic melting peaks (of crystalline phases) can be seen as well. Diffraction patterns of crystalline phases, formed in the glasses after 2 h of annealing of powdered samples

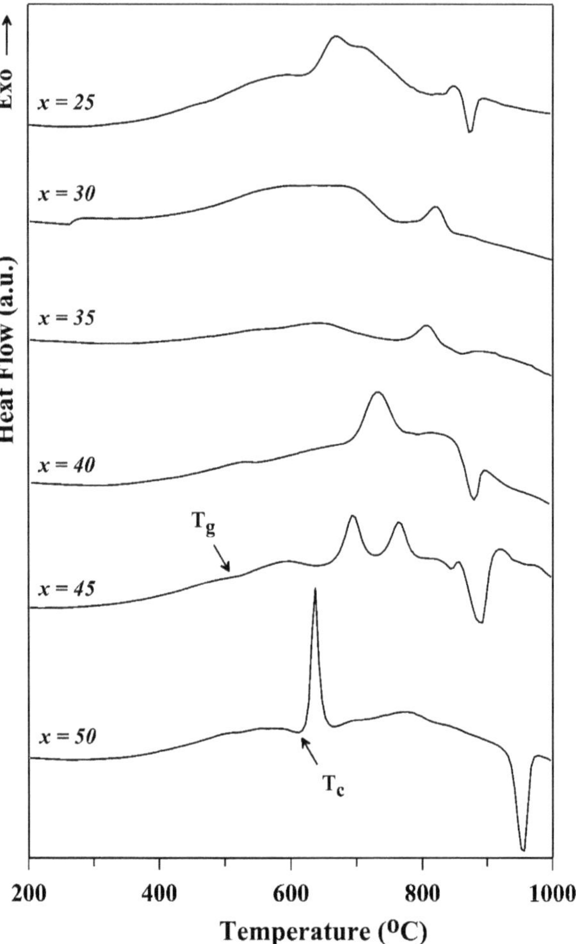

Figure 2. DSC curves of xZnO.10In_2O_3.(90–x)P_2O_5 glasses

at 700–850°C, show that the observed crystallisation peaks in the DSC curves (Figure 2) correspond to the formation of monoclinic and hexagonal modifications of zinc metaphosphate $Zn(PO_3)_2$, monoclinic and orthorhombic modifications of zinc diphosphate $Zn_2P_2O_7$, InP_3O_9, and $Zn_3(PO_4)_2$. Crystalline samples formed for compositions outside the glass forming region (see Figure 1) contain $Zn_3(PO_4)_2$, $Zn_2P_2O_7$ and $InPO_4$ compounds. The determination of the glass transition temperature from these curves is tedious, however, as the change in the heat capacity, c_p, in the glass transition region of these samples is relatively small. The compositional trend in T_g can be seen more clearly from the HSM measurements (Figure 3). There is a correlation between the onset of the first deformation of the pressed powder sample and the glass transition temperature; T_g is lower by about 30°C than the onset of the first change in the area and the height of the sample. We also used dilatometry to measure the glass transition temperature, and the T_g values obtained from these measurements are provided in Table 1. Both methods revealed that the glass transition temperature in the series xZnO.10In_2O_3.(90–x)P_2O_5 (which has a constant In_2O_3 content) decreases with increasing ZnO content as

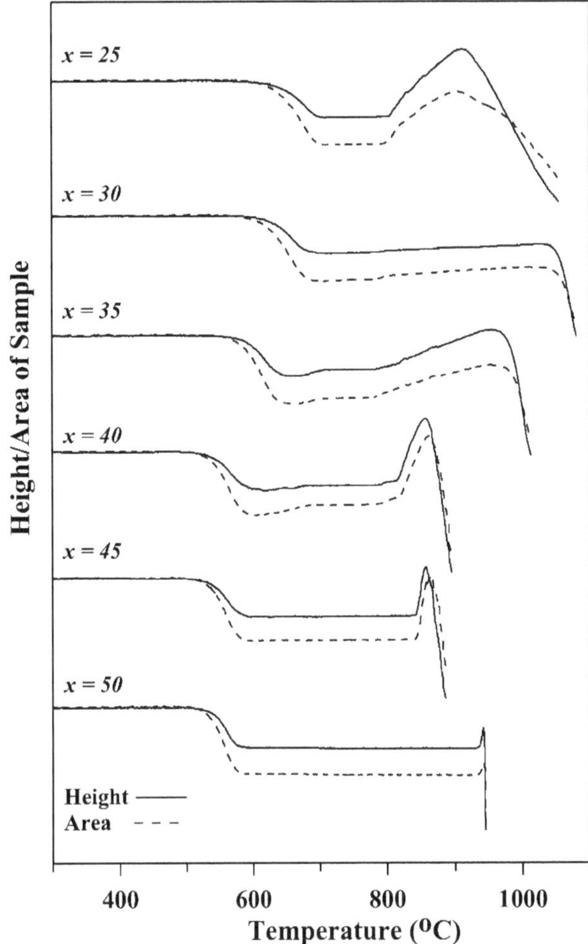

Figure 3. Hot-stage microscopy curves of $xZnO.10In_2O_3.(90-x)P_2O_5$ glasses

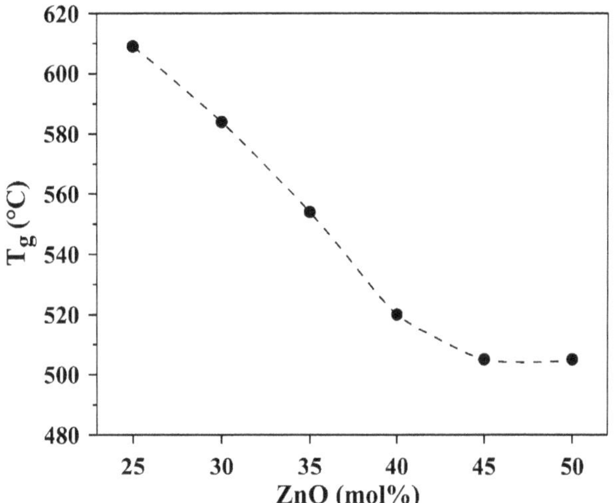

Figure 4. Compositional dependence of the glass transition temperature, T_g, (from dilatometric measurements) in the series $xZnO.10In_2O_3.(90-x)P_2O_5$. The line is only a guide for the eye. The error in the values is smaller than the symbol size

shown in Figure 4. On the other hand, when we compare glasses with constant P_2O_5 content, we find that their T_g values increase with increasing In_2O_3 content, in agreement with Koudelka *et al.*[15] From the dilatometric curves we have also determined values for the dilatometric softening temperature, T_d, and mean values for the thermal expansion coefficient, α, within the range 150–250°C. As shown in Table 1, T_d values are 16–44°C above the T_g values obtained from dilatometric curves. Compositional trends for T_d are similar to the trends for T_g. The values for the thermal expansion coefficient, α, vary in the range 7·8–9·6 ppm/°C, and do not change significantly with composition. The highest value of α (9·6 ppm/°C) was found for zinc metaphosphate glass ($50ZnO.50P_2O_5$), and the incorporation of In^{3+} ions into the phosphate network results in a small decrease of the thermal expansion. We can also see from the HSM curves (Figure 3) that the crystallisation of a glass leads to a decrease in the glass viscosity, and thus the HSM curves exhibit a plateau between glass softening and the melting temperature of the crystallised glass. This can be seen for the glass samples with x=45 and 50 mol% ZnO, where pronounced crystallisation peaks can be observed in the DSC curves (Figure 2).

We have also measured the index of refraction of the studied glasses at three wavelengths, and the extrapolated n_d values for the wavelength 587·6 nm are shown in Table 1. Glasses in the series $xZnO.10In_2O_3.(90-x)P_2O_5$ show an increase in n_d with increasing ZnO content, with values ranging from 1·548 to 1·625 for ZnO contents in the range 25–50 mol%. The index of refraction also increases with increasing In_2O_3 content for glasses with the same P_2O_5 content.

The ^{31}P MAS NMR spectra of the glass series $xZnO.10In_2O_3.(90-x)P_2O_5$ are shown in Figure 5. The NMR spectrum of the glass with 25 mol% ZnO is dominated by the signal of Q^2 units with chemical shift δ=−33 ppm. This is close to the value of −32 ppm obtained for the chemical shift of Q^2 phosphate units in zinc metaphosphate glass by Brow *et al.*[16] With increasing ZnO content in this glass series, the intensity of this resonance decreases, and another resonance appears in the chemical shift region from −14 ppm to −17 ppm. These chemical shift values are characteristic of Q^1 structural units, and as can be seen in Figure 5, the intensity of this resonance increases with increasing ZnO content. Moreover, in the spectrum for the sample with 50 mol% ZnO, another resonance appears at about −3 ppm, which can be assigned to Q^0 phosphate units. According to a previous study of $(50-x)ZnO.xIn_2O_3.50P_2O_5$ glasses, indium oxide behaves as a modifying oxide, giving its oxygen atoms to the structural network, and behaving as In^{3+} ions. Increasing the O/P ratio in ternary $ZnO-In_2O_3-P_2O_5$ glasses leads to a gradual transformation of Q^2 units into Q^1 and Q^0 units.

It was possible to decompose the ^{31}P MAS NMR spectra into individual components belonging to different phosphate structural units, and in this way to obtain quantitative data on the proportions of Q^n

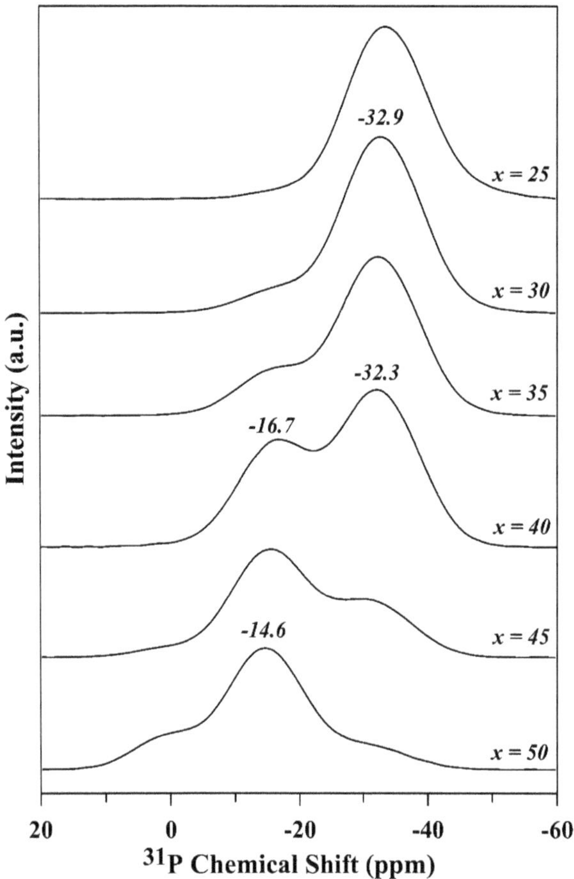

Figure 5. ^{31}P MAS NMR spectra of $xZnO.10In_2O_3.(90-x)P_2O_5$ *glasses*

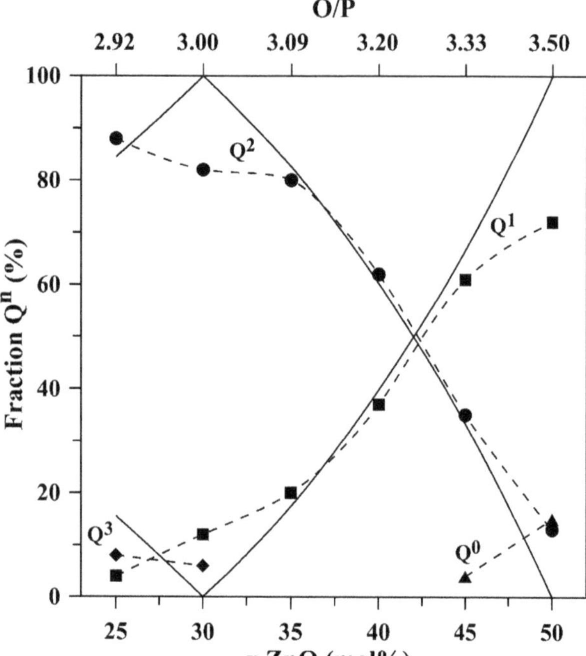

Figure 6. Compositional dependence of the relative number of Q^n *units in* $xZnO.10In_2O_3.(90-x)P_2O_5$ *glasses. The error in the values is smaller than the symbols. The dashed lines are only a guide to the eye. The solid lines in the distribution figure represent theoretical amounts of* Q^n *units calculated according to Brow*[1]

units in nine $ZnO-In_2O_3-P_2O_5$ glass samples. These data are provided in Table 2. In Figure 6 we have plotted the dependence on the ZnO content of the relative amounts of individual phosphate structural units for the compositional series $xZnO.10In_2O_3.(90-x)P_2O_5$. This dependence demonstrates that in the glasses with 25–40 mol% ZnO, the metaphosphate units, Q^2, are the dominant structural units, whereas for the glasses with 45–50 mol% ZnO, Q^1 units play the dominant role. We have also calculated the theroretical proportions of Q^n units using the relations given by Brow,[1] based on the assumption that In_2O_3 behaves as a modifier and gives its oxygen atoms to the phosphate network. The calculated dependences are plotted in Figure 6 as solid lines.

As can be seen from Figure 6, there is a good agreement of experimental and theoretical values for the proportions of Q^n units within the composition range 35–45 mol% ZnO. Differences between these values can be seen for glasses with 25–30 mol% ZnO (60–65 mol% P_2O_5), i.e. in the ultraphosphate region, where according to Brow *et al*[18] phosphate glasses prepared in open crucibles contain some water, which depolymerises the phosphate network by forming P–OH bonds and reducing the proportion of Q^3 units. Another deviation from the theoretical values occurs for the glass containing 50 mol% ZnO, where only Q^1 units should be present. It seems that in this ternary glass the disproportionation of Q^1 units according to the equation

$$Q^1 \rightarrow Q^2 + Q^0$$

takes place, as in zinc polyphosphate glasses.[16]

Table 2. Chemical shifts and relative amounts of phosphate structural units, Q^n*, measured with* ^{31}P *MAS NMR*

Sample no.	ZnO (mol%)	P_2O_5	In_2O_3	Q^3 ppm (±0·2)	Q^3 % (±2·0)	Q^2 ppm (±0·2)	Q^2 % (±2·0)	Q^1 ppm (±0·2)	Q^1 % (±2·0)	Q^0 ppm (±0·2)	Q^0 % (±2·0)
1	50	50	0	-	-	−32·0	97·2	−14·1	2·8	-	-
4	25	65	10	−43·6	8·2	−33·4	87·7	−18·4	4·1	-	-
5	30	60	10	−40·8	6·5	−32·6	82·0	−17·2	11·5	-	-
6	35	55	10	-	-	−32·5	80·0	−16·0	20·0	-	-
7	40	50	10	-	-	−32·4	62·0	−16·0	37·0	0·9	1·0
8	45	45	10	-	-	−31·1	34·9	−15·2	61·1	0·7	4·1
9	50	40	10	-	-	−31·0	13·1	−14·6	71·8	0·9	15·0
11	20	60	20	-	-	−33·5	56·6	−18·0	40·5	−3·6	2·9
12	30	50	20	-	-	−32·8	21·4	−17·4	65·4	−3·0	13·2

The Raman spectra of the glass series xZnO.10In_2O_3.(90–x)P_2O_5 are shown in Figure 7. The Raman spectrum of the glass with 25 mol% ZnO is not very different from the Raman spectrum of zinc metaphosphate glass,[15,16] and its dominant band at 1205 cm^{-1} can be ascribed to the symmetric stretching motion of nonbridging oxygen atoms bonded to phosphorus atoms in Q^2 units, whereas the band at 696 cm^{-1} can be ascribed to the symmetric stretching motion of bridging oxygen atoms between two phosphorus atoms in Q^2 phosphate tetrahedra.[17] The Raman spectrum of the glass with 25 mol% ZnO is consistent with the ^{31}P MAS NMR spectrum, revealing the dominant role of Q^2 units in this glass. The shape of these two dominant Raman bands changes substantially with increasing ZnO content in this glass series. The main vibrational band (due to stretching vibration of nonbridging oxygen atoms) shifts to lower wavenumber, and new bands appear in the region 900–1100 cm^{-1}. These bands can be ascribed to symmetric stretching vibrations of nonbridging oxygen atoms in Q^1 units (1048 cm^{-1})[16] and Q^0 units (970 cm^{-1}).[16] The second dominant band (due to P–O–P vibrations) also changes its shape, as a result of the appearance of the band at 754 cm^{-1}, arising from the symmetric stretching vibration of P–O–P bonds between Q^1 units.[16] These changes in the Raman spectra are also consistent with the ^{31}P MAS NMR spectra (Figures 5 and 6), which show that the dominant structural role in xZnO.10In_2O_3.(90–x)P_2O_5 glasses is transferred from Q^2 to Q^1 units with increasing ZnO content. All the observed changes in the Raman spectra are similar to the evolution of the Raman spectra of zinc polyphosphate glasses studied by Brow *et al*,[16] as indium oxide contributes to the changing O/P ratio and its In…O interactions have a prevailing ionic character, as concluded previously.[15]

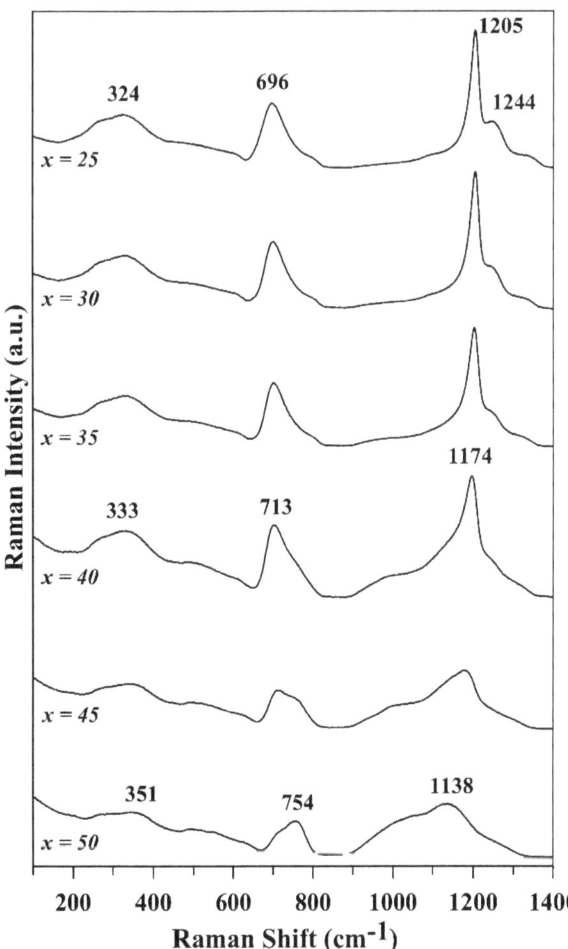

Figure 7. Raman spectra of xZnO.10In$_2$O$_3$.(90–x)P$_2$O$_5$ glasses

Conclusion

We have determined the glass-forming region in the ternary system ZnO–In_2O_3–P_2O_5, and demonstrated that for free cooling of the melt in air, glasses containing up to 20 mol% In_2O_3 can be prepared. The glass-forming region has not been determined for P_2O_5 content higher than 65 mol%, because such glasses are usually hygroscopic and we were interested in more stable glasses. We have determined the basic properties of glasses in the glass-forming region and found that the ratio of O/P is important for their properties and structure. In agreement with our previous study, the obtained results also indicate that indium oxide behaves in this ternary system as a glass modifier. It gives oxygen atoms to the structural network, and is present in the glass structure as a charge compensating cation, In^{3+}. The presence of In^{3+} ions in the glasses is confirmed also by the formation of InP_3O_9 in the products of glass crystallisation.

Acknowledgements

The Czech authors are grateful for the financial support from the Grant Agency of the Czech Republic (Grant No. 13-00355S). The FEDER, Region Nord Pas-de-Calais, Ministère de l'Education Nationale de l'Enseignement Superieur et de la Recherche, CNRS, and USTL are acknowledged for funding of NMR spectrometers.

References

1. Brow, R. K. Review: the structure of simple phosphate glasses. *J. Non-Cryst. Solids*, 2000, **263-264**, 1–28.
2. Brow, R. K. & Tallant, D. Structural design of sealing glasses. *J. Non-Cryst. Solids*, 1997, **222**, 396–406.
3. Koudelka, L. & Mošner, P. Borophosphate glasses of the ZnO-B_2O_3-P_2O_5 system. *Mater. Lett.* 2000, **42**, 194–199.
4. Brow, R. K. An XPS study of oxygen bonding in zinc phosphate and zinc borophosphate glasses. *J. Non-Cryst. Solids*, 1996, **194**, 267–273.
5. Brow, R. K. Nature of alumina in phosphate glass: I, Properties of sodium aluminophosphate glass. *J. Am. Ceram. Soc.*, 1993, **76**, 913–918.
6. Smith, C. E., Brow, R. K., Montagne, L. & Revel, B. The structure and properties of zinc aluminophosphate glasses. *J. Non-Cryst. Solids*, 2014, **386**, 105–114.
7. Tsuchida, J., Schneider, J., Rinke, M.T. & Eckert, H. Structure of ternary aluminum metaphosphate glasses. *J. Phys. Chem.*, 2011, **115**, 21927–21941.
8. Berthet, P., Bretey, E., Berthon, J., d'Yvoire, F., Belkebir, A., Rulmont,

A. & Gilbert, B. Structure and ion transport properties of Na_2O-Ga_2O_3-P_2O_5 glasses. *Solid State Ionics*, 1994, **70–71**, 476–481.

9. Ilieva, D., Jivov, B., Petkov, C., Penkov, I. & Dimitriev, Y, J. Infrared and Raman spectra of Ga_2O_3-P_2O_5 glasses. *Non-Cryst. Solids*, 2001, **283**, 195–202.
10. Subbalakshmi, P. & Veeraiah, N. Dielectric dispersion and certain other physical properties of PbO-Ga_2O_3-P_2O_5 glass system. *Mater. Lett.*, 2002, **56**, 880–888.
11. Sales, B. C. & Boatner, L. A. Optical, structural and chemical characteristics of lead-indium phosphate and lead-scandium phosphate glasses. *J. Am. Ceram. Soc.*, 1987, **70**, 615–621.
12. Suzuya, K., Loong, C. K., Price, D. L., Sales, B. C. & Boatner, L. A. The structure of lead-indium phosphate and lead-scandium phosphate glasses. *J. Non-Cryst. Solids*, 1999, **258**, 48–56.
13. Baskaran, G. S., Flower, G. L., Rao, D. K. & Veeraiah, N. Structural role of In_2O_3 in PbO-P_2O_5-As_2O_3 glass system by means of spectroscopic and dielectric studies. *J. Alloys Compd.*, 2007, **431**, 303–312.
14. Sharma, M. V. N. V. D., Sarma, A. V. & Rao, R. B. Electrical characterization and relaxation behavior of lithium-indium-phosphate glasses via impedance spectroscopy. *Turk. J. Phys.*, 2009, **33**, 87–100.
15. Koudelka, L., Račický, A., Mošner, P., Rösslerová I., Montagne, L. & Revel, B. Behavior of indium oxide in zinc phosphate and borophosphate glasses. *J. Mater. Sci.*, 2014, **49**, 6967–6974.
16. Brow, R. K., Tallant, D. R., Myers, S. T. & Phifer, C. C. The short-range structure of zinc polyphosphate glass. *J. Non-Cryst. Solids*, 1995, **191**, 45–55
17. Bobovich, Y. S. Issledovanije struktury stekloobraznych fosfatov s pomoshtchju spektrov kombinacionnovo rassejanija sveta. *Opt. Spektrosk.*, 1962, **13**, 492–497.
18. Brow, R. K., Kirkpatrick, R. J. & Turner G. L. The short range structure of sodium phosphate glasses I. MAS NMR studies. *J. Non-Cryst. Solids*, 1990, **116**, 39–45.

Phys. Chem. Glasses: Eur. J. Glass Sci. Technol. B, December 2015, **56** (6), 278–284

Phosphorus incorporation into silica during modified chemical vapour deposition combined with solution doping

F. Lindner,[*1] S. Unger,[1] A. Kriltz,[2] A. Scheffel,[1] A. Dellith,[1] J. Dellith[1] & H. Bartelt[1]*

[1] *Leibniz Institute of Photonic Technology, Albert-Einstein-Str. 9, 07745 Jena, Germany*
[2] *Institute of Physical Chemistry, Friedrich-Schiller-University Jena, Lessingstr. 10, 07743 Jena, Germany*

Manuscript received 29 August 2014
Revised version received 20 February 2015
Accepted 10 March 2015

It is well known that the efficiency of silica-based erbium/ytterbium-doped fibre lasers and amplifiers can be greatly enhanced by a high level of phosphorus codoping due to an improved Yb to Er transfer efficiency. The manufacture of these types of rare earth (RE)-doped silica fibres with a high P concentration is principally carried out via modified chemical vapour deposition (MCVD) technology in combination with solution doping. The supply of RE ions for the P-doped active core of the fibre preform occurs via the liquid phase during the multi-stage preparation process. The incorporation of phosphorus into the silica matrix is determined and strongly influenced by the chemical equilibrium, evaporation by formation of gaseous phosphorus oxides, and diffusion during the process steps. The knowledge and understanding of these interaction processes are very important for the optimisation of the fabrication process for high power laser and amplifier fibres. Here, we report on a systematic investigation of phosphorus incorporation into the silica matrix during the MCVD process in combination with a solution doping technique. The P_2O_5-doped silica soot material of the individual steps was prepared with a wide range of different process parameters (gas concentration of the starting compounds $POCl_3$ and $SiCl_4$, pre-sintering temperatures, porosity). The samples were investigated concerning their porosity, morphological characterisation, and composition via scanning electron microscopy, Fourier transform infrared spectroscopy and energy dispersive x-ray spectroscopy.

1. Introduction

MCVD (modified chemical vapour deposition) technology in combination with solution doping[1–3] is the leading technology for the manufacture of rare earth-doped high power silica laser fibre preforms. Here, a high phosphorus content promotes the stable performance of Yb-doped fibres with low pump-induced losses (photodarkening effect)[4] and an efficient energy transfer from Yb ions to Er ions in Er/Yb-doped fibres. High output powers in the multi-kW range are realised in RE-doped fibres via large mode area (LMA) structure. These fibres have a large core with a low numerical aperture (NA). One alternative to achieve this low core NA is the incorporation of a rare earth (RE) together with phosphorus and aluminum. The photodarkening and the refractive index are strongly reduced (compared to pure silica) for compositions with a P to Al ratio of one.[5,6] MCVD/solution doping technology is a multi-stage process, starting from reverse deposition of a P_2O_5-doped porous silica layer, followed by the pre-sintering of this layer to a defined relative density, d_r, RE/Al solution penetration of the material, drying and cleaning steps, and finally sintering to a glassy layer.[7] The amount of Al and RE incorporated is determined by the relative density. In order to get a sufficient stability for the following preparation steps and an efficient incorporation of the dopants, the relative density should be in the order of 0·2, compared to the density of the vitrified layer. This is realised by a pre-sintering step. The pre-sintering temperature depends on the composition of the porous layer and the concentration of the dopants.[3] The phosphorus incorporation in the silica matrix is determined and strongly influenced by chemical equilibrium, evaporation by formation of gaseous phosphorus oxides and diffusion during the different steps.[7,8] Knowledge and understanding of these interaction processes are very important for the optimisation of the preparation process to realise defined refractive index profiles.

In this work, the use of FTIR (Fourier transform infrared spectroscopy) and EDX (energy dispersive x-ray spectroscopy) measurements of porous, phosphorus-doped MCVD layer materials to characterise their structure and assess their phosphorus content is presented. In phosphorus silicate glass the silicon–oxygen (SiO_4) and phosphorus–oxygen ($O{=}PO_3$) tetrahedra are randomly arranged in a three-dimensional network. Silicon atoms are linked to four other cations by oxygen bridges, while phosphorus has just three

Corresponding author. Email florian.lindner@ipht-jena.de
Original version presented at Int. Conf. on Phosphate Glasses, Pardubice, Czech Republic, 2–4 July 2014
DOI: 10.13036/17533562.56.5.278

oxygen bridges and one double bond to oxygen.[9,10] The Raman spectra of phosphorus silicate glasses prepared using MCVD shows an intense band with a maximum at 1320 cm^{-1} caused by stretching vibrations of O=P bonds. The intensity of the O=P band grows with increasing P_2O_5 content in $(P_2O_5)_x(SiO_2)_{1-x}$ glasses.

In principle there are difficulties in the interpretation of the infrared spectra of glasses. The complex IR profiles of glassy substances are connected with the specific structure of the glassy state. The infrared spectra of silicate glasses show many defined characteristics belonging only to the glassy phase, but the spectra of the corresponding crystalline structure are close to them. The changes in shape, absorption maximum, and peak position of the bands are the result of the distribution of bond lengths and angles, as well as of a disorder in the relative orientation of the structural units. Thus, it is necessary to separate the complex spectra into their single components. An overview of the specific bands of silicate glasses is presented in Ref. 11.

2. Samples and methods

2.1. Preparation of the specimens

The systematic investigation of the incorporation of phosphorus into the silica matrix during the MCVD/solution doping process was performed by studying the dependence on the process parameters (composition of the starting gas phase ($POCl_3$ and $SiCl_4$), pre-sintering temperature, porosity) in four series of samples. The P_2O_5-doped silica soot layers were deposited in quartz glass carrier tubes with inner and outer diameters of 11 mm and 14 mm, respectively. In contrast to the "normal" MCVD process, the deposition was carried out by an opposite movement of the gas flow and the burner (reverse deposition) at a temperature of 1600°C. The total gas flow of the gaseous halides and the carrier oxygen gas was set to 800 sccm (standard cubic centimetres per minute). The initial gas phase conditions are specified by the ratio of the molar fractions of $POCl_3$ (x_{POCl3}) and $SiCl_4$ (x_{SiCl4}). The process parameters used are listed in Table 1.

Samples in series A are a reverse deposited P-doped silica layer with starting gas phase composition x_{POCl3}/x_{SiCl4}=0·1–0·88, without any further treatment (i.e. "untreated" soot material). The relative density, d_r, of these soot layers is low, about 0·03 compared to the density of the vitrified layer. Samples in series B were prepared under similar conditions, with starting gas phase composition x_{POCl3}/x_{SiCl4}=0·1–0·88. Then the layer was pre-sintered under an O_2 atmosphere, to a relative density of about 0·2. To achieve this relative density for each sample, it was necessary to decrease the pre-sintering temperature with increasing phosphorus content (or starting gas phase composition, x_{POCl3}/x_{SiCl4}) as shown in Figure 1.

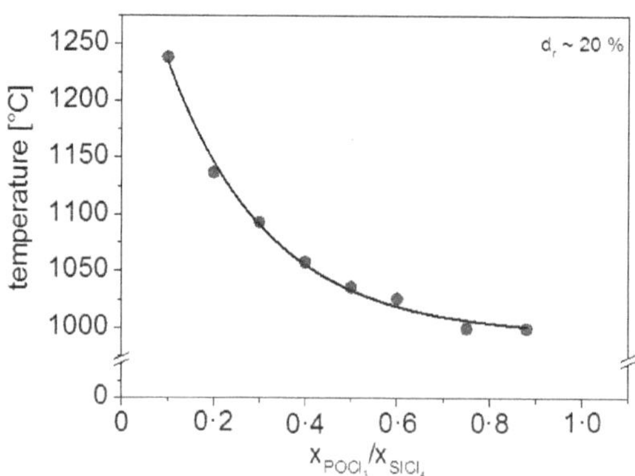

Figure 1. Dependence of the pre-sintering temperature on the composition of the starting gas phase composition to achieve a relative density of 0·2 [Colour available online]

Samples in series C were prepared with a constant starting gas phase composition x_{POCl3}/x_{SiCl4}=0·3, and a pre-sintering under oxygen atmosphere with different temperatures to achieve different relative densities, as listed in Table 1. The sample of series D was prepared with starting gas phase composition x_{POCl3}/x_{SiCl4}=0·3, pre-sintering under oxygen atmosphere and relative density of d_r=0·2, and was penetrated with pure water for one hour, followed by a drying and chlorination step at 1000°C to reduce the OH content and eliminate other impurities. This simulates the complete solution doping procedure. The soot samples were characterised for their P concentration, porosity, and morphology using the following methods.

2.2. Microscopy

Morphological characterisation of the surface of the different samples was made using a JEOL JSM 6300F scanning electron microscope (SEM). The silica tube with internal deposition of P-doped silicate soot was cut into small slices to investigate the material

Table 1. Composition of the starting gas phase of $SiCl_4$ and $POCl_3$ (x_{POCl_3}/x_{SiCl_4}), the measured phosphorus concentration, $c_{P_2O_5}$, pre-sintering temperature, T, and relative density, d_r, of the different series

Samples	x_{POCl_3}/x_{SiCl_4}	$c_{P_2O_5}$ [mol%]
series A	0·10	4·3
untreated	0·20	8·4
	0·30	9·7
d_r~0·03	0·40	10·3
	0·50	11·0
	0·60	13·6
	0·75	13·6
	0·88	14·3

Samples	x_{POCl_3}/x_{SiCl_4}	T [°C]	$c_{P_2O_5}$ [mol%]
series B	0·10	1238	3·7
pre-sintered	0·20	1137	6·2
	0·30	1093	7·3
d_r~0·2	0·40	1058	7·6
	0·50	1036	9·4
	0·60	1026	9·7
	0·75	1000	9·9
	0·88	1000	10·2

Samples	T [°C]	d_r	$c_{P_2O_5}$ [mol%]
series C	-	0·03	9·7
x_{POCl3}/x_{SiCl4}	950	0·06	8·1
=0·3	1000	0·07	7·4
	1050	0·11	7·8
	1100	0·23	7·2
	1135	0·37	7·1
	1175	0·75	6·0

at several tube positions. A sputtered carbon layer protected the samples against environmental influences, and avoided charging effects of the samples during measurement.

2.3. Energy dispersive x-ray spectroscopy

The chemical composition of the samples was measured using electron microprobe analysis (a Jeol JXA-8800L spectrometer with a Bruker silicon drift detector). The soot was extracted from the inner side of the tube following the different preparation steps. Samples were prepared for the measurement by manually pressing the material into thin slices of about 2 mm in diameter directly on the sampler holder and sputtering of a carbon protection layer. Three to five sample preparations and measurements of every soot material sample were performed to achieve a significant result.

2.4. Vibrational spectroscopy

The infrared (IR) measurements were performed on the soot material using a Thermo Scientific Nicolet iS10 FTIR spectrometer and a single reflection diamond cuvette. The attenuated total reflection (ATR) method was used to characterise the porous layer material in the range from 6000 cm^{-1} to 600 cm^{-1} with a 2 cm^{-1} resolution. Each spectrum represents an average of 100 scans. The significant bands of the P_2O_5–SiO_2 soot were registered between 1400 cm^{-1} and 700 cm^{-1}.

A small amount of the layer material was placed directly on the ATR diamond and a stamp was used to press the soot into a thin layer. The different morphological consistency of the materials induced different sample contacts and changes in the level of absorbance. Multiple measurements on the same samples were made to reduce measurement errors and eliminate the influence of the compacting pressure on the spectra. The absorbance level of the soot material is at least five times lower than that of a polished glass sample because of the different sample contact. The measured spectra are pure ATR measurements without advanced ATR correction for the depth of indentation, d_P, of the evanescent field, which depends on the wavelength, λ, the angle of incidence, θ, and the refractive indices of the ATR crystal, n_2, and the material, n_1:

$$d_p = \frac{\lambda}{2\pi n_1 \sqrt{\sin^2\theta - \left(n_2 / n_1\right)^2}} \qquad (1)$$

Even with knowledge of the P_2O_5 concentration, estimating the refractive index of a pressed soot layer is difficult. Depending on the different phosphorus content of the samples, the shift in the corresponding bands after such a correction could be unbalanced, which would lead to an additional error in the band analysis.

3. Results and discussion

3.1. Microscopic investigation of the layer

A morphological characterisation of the fractured surface of the different samples clarifies the changes

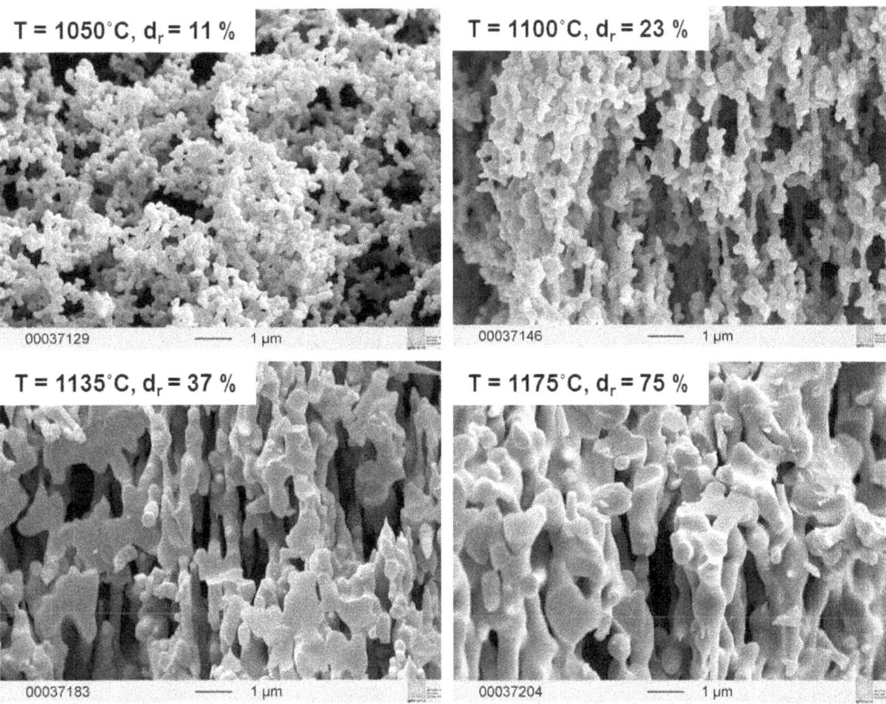

Figure 2. SEM micrograph of the phosphorus-doped silica layer at different pre-sintering temperatures with the associated relative density

that occur after heat treatment. The untreated material has small and highly porous structure (d_r~0·03), while the structure of the pre-sintered material becomes larger, straightened, and more compact (d_r~0·2). Figure 2 illustrates the changes in the structure with increasing pre-sintering temperature and associated relative density. The alignment of the structure in one direction occurs because of the rotation of the process tube during the pre-sintering step. The penetration of a layer with a low relative density (d_r<0·2) will remove the porous soot completely, while a very high relative density (d_r>0·2) avoids an efficient incorporation of RE and Al.

3.2. EDX analysis

Previous investigations[7,8] of direct vitrified phosphorus-doped silica layers show that the incorporation of phosphorus depends on the composition of the starting gas phase (x_{POCl_3}/x_{SiCl_4}) and the deposition temperature. The experimental results show that at higher concentrations the incorporation of phosphorus has a tendency for saturation. Instead of complete incorporation of P_2O_5, a thermodynamic equilibrium of the condensed phase (P_2O_5) and gaseous phosphorus oxides (P_4O_{10}, P_4O_6, PO_2, PO) occurs. Thermodynamic calculations show that at temperatures higher than 1800°C, gaseous PO_2 dominates the equilibrium, while only gaseous P_4O_{10} is present below 1500°C:

T>1800°C: PO_2 (g) dominates

$$PO_{2\cdot5}(cond) \Leftrightarrow PO_2(g) + 0\cdot25O_2(g) \quad (2)$$

T<1500°C: P_4O_{10} (g) dominates

$$PO_{2\cdot5}(cond) \Leftrightarrow 0\cdot25P_4O_{10}(g) \quad (3)$$

Using EDX analysis, a phosphorus concentration in the range from 4·3 to 14·3 mol% P_2O_5 was measured for the "untreated" layer material (series A). This shows that the dependence of the P_2O_5 content on the composition of the starting gas phase (x_{POCl_3}/x_{SiCl_4}) is similar to that of direct vitrified layers (Figure 3). However, the measurement error for the "untreated" phosphorus-doped soot material could be more than 20% because of its strongly hygroscopic behaviour, as illustrated at x_{POCl_3}/x_{SiCl_4}=0·4 in Figure 3.

Both depositions take place at 1600°C where the first equilibrium (Equation (2)) with gaseous PO_2 dominates. Phosphorus is not completely incorporated in the case of the reverse deposition because of the equilibrium between the condensed phosphorus species dissolved in silica and the gaseous phosphorus oxides. These experimental results show that the mechanism for incorporation of phosphorus into vitreous and reversely deposited ("untreated") porous silica material is subject to similar conditions.

The next process step, the pre-sintering of the phosphorus-doped silica layer material, is a heat treatment that leads to the evaporation of phosphorus. Phosphorus concentrations in the range from 3·7 to 10·3 mol% P_2O_5 were measured by EDX analysis of the pre-sintered material (series B). The dependence of the phosphorus content on the composition of the starting gas phase (x_{POCl_3}/x_{SiCl_4}) is shown in Figure 4. In comparison to the untreated soot material, the trend of the curve is similar but at a lower phosphorus concentration. Because of the low pre-sintering temperatures required to achieve a relative density of 0·2, the second equilibrium (Equation (3)) with gaseous P_4O_{10} is dominant. The pre-sintered phosphorus-doped material shows only very weak hygroscopic behaviour, which leads to a small measurement error, illustrated at x_{POCl_3}/x_{SiCl_4}=0·4 in Figure 4.

3.3. Infrared spectroscopy

The infrared spectra of the phosphorus-doped silica layer material of series A ("untreated" soot) are

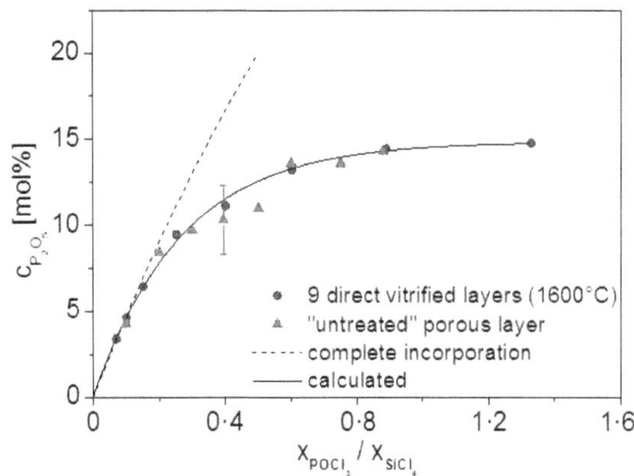

Figure 3. Incorporation of phosphorus during the "normal" MCVD process of direct vitrified layers, compared with porous "untreated" layer material (series A), showing the dependence on the starting gas phase composition. A typical measurement error is shown at x_{POCl_3}/x_{SiCl_4}=0·4 [Colour available online]

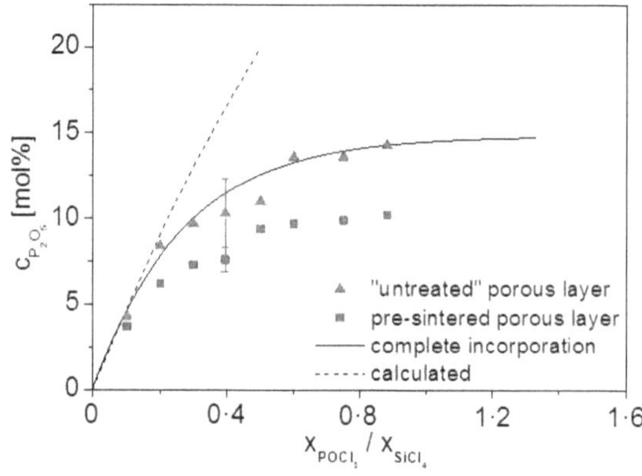

Figure 4. Phosphorus incorporation of porous "untreated" layer material (series A), compared with porous pre-sintered layer material (series B), showing the dependence on the composition of the starting gas phase. A typical measurement error is shown at x_{POCl_3}/x_{SiCl_4}=0·4[Colour available online]

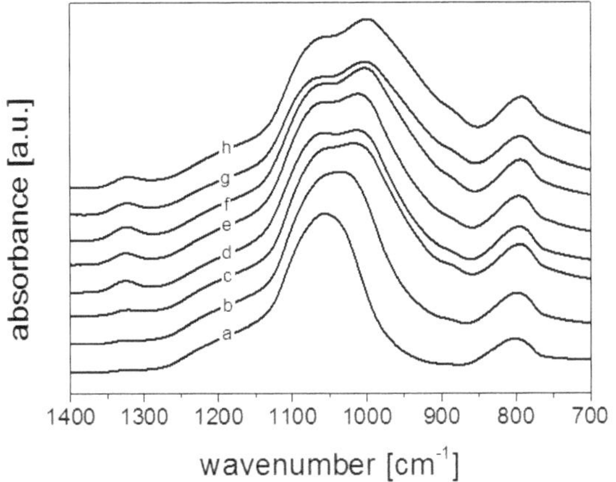

Figure 5. Infrared spectra of series A (untreated layer material) with increasing P_2O_5 concentration from 4·3 to 14·3 mol% P_2O_5 (a) to (h)

shown in Figure 5. With increasing P_2O_5 concentration from 4·3 mol% (a) to 14·3 mol% (h), several bands in the spectra related to phosphorus increase, while the bands associated with silica seem to remain unchanged. In particular, the band at 1325 cm^{-1} relates to the P=O bond.(9,10,12,13) A previous NMR measurement of silicophosphate glasses with P_2O_5 concentrations of 8 mol% and higher(14) shows that phosphorus enters the silica network only as Q^3 groups with one P=O bond, and hence the band at 1325 cm^{-1} is assigned to this group.

The bands at around 1180 cm^{-1} and 1060 cm^{-1} are due to asymmetric Si–O–Si stretching vibrations,(11,15) while the band at around 800 cm^{-1} is due to symmetric Si–O–Si stretching vibrations. The absorption bands at around 1000 cm^{-1} and 900 cm^{-1} also increase with higher phosphorus concentration, but the assignment of these bands to a specific phosphorus bond is undetermined. Due to the overlap of the different Si–O–Si, Si–O–P and P–O–P bands between 1250 cm^{-1} and 900 cm^{-1}, a detailed analysis of every single band is difficult. On the one hand, the low level of absorbance influences the shape of the spectra; on the other hand, ATR measurement without advanced ATR correction shows small shifts in several bands compared to peak positions known from literature.

The infrared spectra of series B, the pre-sintered phosphorus-doped layer material, show a similar behaviour. With increasing P_2O_5 content from 3·7 (a) to 10·2 mol% P_2O_5 (h), the bands associated with phosphorus grow while the Si–O–Si bands seem to remain unchanged (Figure 6).

The main differences between the ATR spectra of the "untreated" and pre-sintered soot material are the different absorbance levels, the changes observable in the range between 1100 cm^{-1} and 800 cm^{-1}, and the significantly larger band at 1325 cm^{-1} in the case of the pre-sintered material (Figure 7). The evaporation of phosphorus during heat treatment is one reason for the absence of the band at around 900 cm^{-1} and for a weaker band at around 1000 cm^{-1} in the case of pre-sintered soot compared to the "untreated" material. Another reason could be the strong increase of the P=O band at 1325 cm^{-1} (Q^3 group). The fact that the intensity of this band increases, while the P_2O_5 content strongly decreases by the pre-sintering step, leads to the assumption that the generation of phosphorus double bonds reduces the intensity in the range of 1250–850 cm^{-1} because of the transformation of non-bridging oxygen and bridging oxygen bonds to Si or P into terminal P=O bonds . Certainly the absence of the band at 900 cm^{-1} and the considerably larger P=O band in the infrared spectra of the pre-sintered soot material shows that there is a significant change in the structure of the material from the "untreated" to the pre-sintered state.

The morphological characterisation described in Section 3.1 explains the different sample contact and the variation in the level of absorbance of the infrared ATR measurements for series A and B. While a small and strongly porous material ("untreated") can easily be pressed to a thin layer, the larger, straightened, and more compact pre-sintered structure cannot create such a compressed layer, leading to a different

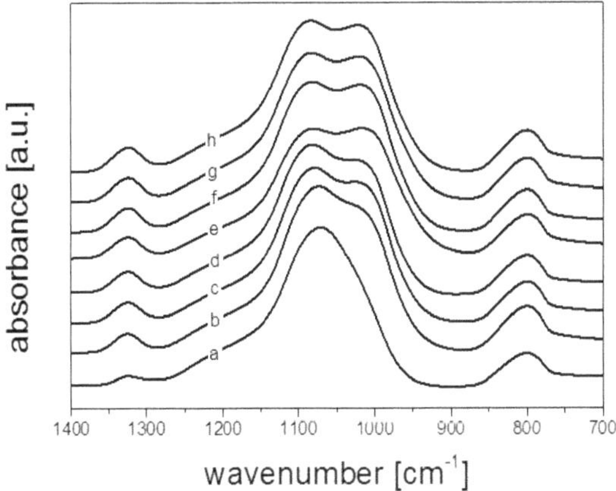

Figure 6. Infrared spectra of series B (pre-sintered layer material) with increasing P_2O_5 concentration from 3·7 to 10·2 mol% P_2O_5 (a) to (h) [Colour available online]

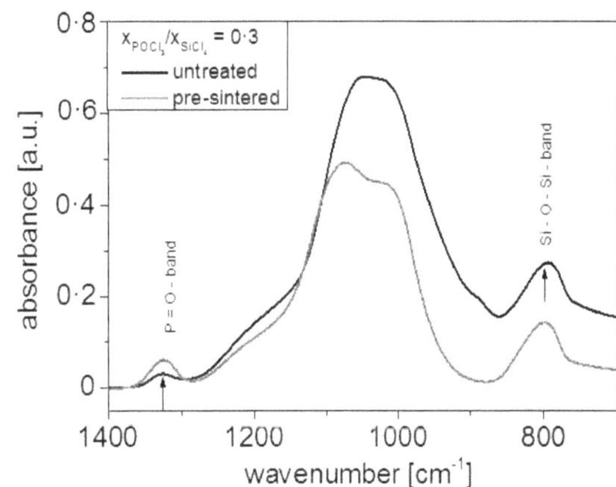

Figure 7. Infrared spectra of "untreated" soot material compared to pre-sintered material with a starting gas phase composition x_{POCl_3}/x_{SiCl_4}=0·3 [Colour available online]

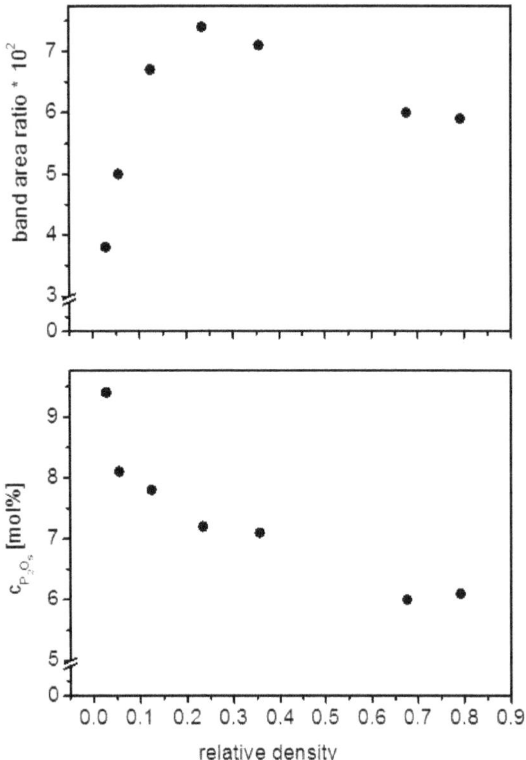

Figure 8. Band area ratio (upper part of figure) for the P=O band (1325 cm^{-1}) and the Si–O–Si band (800 cm^{-1}), and concentration of P_2O_5 (lower part of figure), as a function of the relative density of the phosphorus-doped silica soot material of series C

sample contact. The absorbance of the pre-sintered material reaches at most 0·5, while the maxima of the "untreated" soot vary between 0·5 and 0·8. These differences complicate a direct comparison of the infrared spectra of the porous MCVD layer material. To avoid this measurement variation, the integrated absorbance of the Si–O–Si band between 856 cm^{-1} and 767 cm^{-1} and of the P=O band between 1338 cm^{-1} and 1309 cm^{-1} were used. The P=O band rises with increasing level of $POCl_3$ in the starting gas phase composition (x_{POCl_3}/x_{SiCl_4}), while the Si–O–Si band seems to remain unchanged. Both bands show only a small overlap with other bands in this range and can be analysed without band decomposition. The band area ratio of the P=O band (1325 cm^{-1}) to the Si–O–Si band (800 cm^{-1}) can be used for all of the samples and enables a comparison of the development of the P=O bond for the different process states. Figure 8 shows the dependence of the band area ratio and the related phosphorus concentrations on the relative density

Table 2. Phosphorus concentration ($c_{P_2O_5}$) of series D during the different process steps of MCVD in combination with solution doping

Sample	*Soot state*	$c_{P_2O_5}$ *[mol%]*
series D	"untreated"	9·2
x_{POCl_3}/x_{SiCl_4}=0·3	pre-sintered	7·2
d_r~0·2	after drying	7·0
	after chlorination	6·6

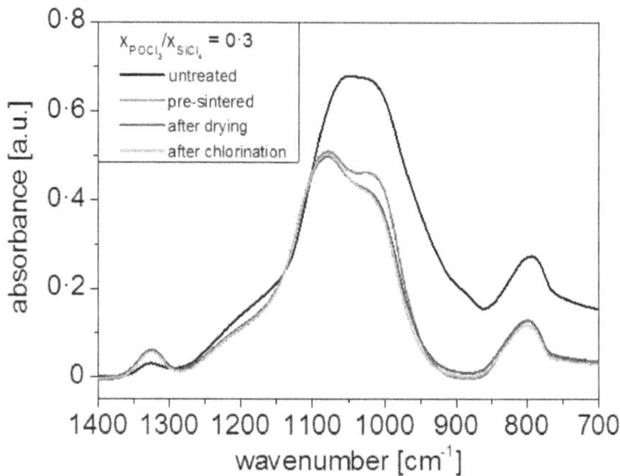

Figure 9. Infrared spectra of soot material with a starting gas phase composition x_{POCl_3}/x_{SiCl_4}=0·3 after the reverse deposition ("untreated"), pre-sintering, drying, and chlorination steps [Colour available online]

of the soot material of series C. For a low relative density or with low pre-sintering temperatures, a strong increase in the P=O band intensity takes place, with a maximum P=O bond formation at a relative density of around 0·2. Afterwards the band area ratio decreases and exhibits a tendency toward saturation. The measured P_2O_5 concentration decreases in the whole range because of increasing phosphorus evaporation at higher pre-sintering temperatures. Previous experimental results[2] have shown that the relative density of the porous layer is a key parameter for the solution doping process. The incorporated concentration of Al and RE depends on the solution concentration and the relative density. The lower the density the higher is the incorporated amount. But the density should be not lower than 0·2 in order to achieve a stable state of the layer for penetration without peeling from the inner surface of the tube. In this way, an efficient incorporation can be realised with a relative density of 0·2, and this may be related to the distinct maximum of the P=O groups of the soot material for this density. This topic is not fully clarified yet and needs further investigation.

The pre-sintering of the reverse deposited ("untreated") phosphorus-doped silica layer causes strong changes in the material structure. The further steps – penetration with a solution, drying and chlorination – were performed with the sample of series D. Table 2 gives the measured phosphorus concentration, $c_{P_2O_5}$, and Figure 9 shows the infrared spectra of the sample at the different process stages. Drying and chlorination do not strongly affect the characteristics of the material because of the low process temperature (1000°C). The ATR spectra after these stages show only a small decrease at around 1000 cm^{-1} compared to the pre-sintered state, while the P=O band (1325 cm^{-1}) is similar. The differences in phosphorus concentration are also small. Therefore, the pre-sintering of the phosphorus-doped silica material represents the most important step with the strongest structural changes for the MCVD process in combination with the solution doping technique.

Conclusions

In this paper, the morphologic and spectroscopic properties of phosphorus-doped silica soot material with a P_2O_5 concentration up to 14 mol%, manufactured by MCVD in combination with solution doping, were investigated. Fourier transform infrared spectroscopy, scanning electron microscopy, and energy dispersive x-ray spectroscopy have been used for the characterisation of the samples. The different process steps (reverse deposition, pre-sintering, penetration, drying, and chlorination) affect the phosphorus-doped silica layer in a different way. The phosphorus content of the reverse deposited layers ("untreated" material) depends on the ratio of $SiCl_4$ and $POCl_3$ in the starting gas phase, similar to that for directly vitrified layers. This means that the incorporation of phosphorus in both cases is subject to similar conditions. The heat treatment of this layer (pre-sintering) causes strong phosphorus evaporation, while further steps (penetration, drying, and chlorination) have only a small effect on the phosphorus content of the layer.

The original raw infrared (ATR) spectra of the material show a rise in the phosphorus-associated bands with increasing phosphorus content. The "untreated" soot is characterised by an additional phosphorus-related band at around 900 cm^{-1}, and a higher level of absorbance because of better sample contact. Subsequent pre-sintering leads to an enhancement of P=O bonds (1325 cm^{-1}), dependent on the sintering temperature and on the relative density. At a relative density of 0·2, a maximum intensity of the P=O peak is reached. Here, incorporation of Al/RE proves most effective. Further process steps (penetration, drying, and chlorination) do not strongly affect the infrared spectra of the soot material compared to the pre-sintered material. Thus, the strongest change in the phosphorus-doped silica layer during the MCVD/solution doping process is caused by the pre-sintering step. Here the temperature is the most important parameter to achieve a specific relative density, obtain a straightened and more compact structure, and affect the evaporation of phosphorus. The infrared measurement of soot material in combination with the evaluation of the band area ratio of the P=O band (1325 cm^{-1}) and the Si–O–Si band (800 cm^{-1}) was used to optimise and control the phosphorus-doped silica fibre preform preparation during the different process steps. Successful decomposition of the spectra would explain the phosphorus structures at the different process stages, and enable assessment of the phosphorus content of the material without EDX analysis.

Acknowledgements

The authors are grateful for financial support from the Leibniz Institute of Photonic Technology, Jena, and for participation in the International Conference on Phosphate Glasses.

References

1. Townsend, J. E., Poole, S. B. & Payne, D. N. Solution-doping technique for fabrication of rare-earth-doped optical fibres. *Electron. Lett.*, 1987, **23**, 329–331.
2. Kirchhof, J., Unger, S. & Schwuchow, A. Fiber Lasers: Materials, Structures and Technologies. *Proc SPIE*, 2003, **4957**, 1–12.
3. Kirchhof, J., Unger, S., Schwuchow, A., Grimm, S. & Reichel, V. Materials for high-power fiber lasers. *J. Non-Cryst. Solids*, 2006, **352**, 2399–2403.
4. Unger, S., Schwuchow, A., Jetschke, S, Reichel, V., Scheffel, A. & Kirchhof, J. Optical properties of Yb-doped laser fibers in dependence on codopants and preparation conditions. *Proc SPIE* 2008, **6890**, 6890016.
5. Unger, S., Schwuchow, A., Jetschke, S., Reichel ,V., Leich, M., Scheffel, A. & Kirchhof, J. Influence of aluminium-phosphorus codoping on optical properties of ytterbium-doped laser fiber. *Proc SPIE* 2009,**7212**, 72121B-1.
6. Jetschke. S., Unger, S., Schwuchow, A., Leich, M. & Kirchhof, J. Efficient Yb laser fibers with low photodarkening by optimisation of the core composition. *Opt. Express*, 2008, **16**, 15540–5
7. Kirchhof, J.,Unger, S., Schwuchow, A., Jetschke, S. & Knappe, B. Spatial distribution effects and laser efficiency in Er/Yb doped fibers. *Proc SPIE*, 2004, **5350**, 222–223.
8. DiGiovanni, D. J., Morse, T. F. & Cipolla Jr, J. W. Theoretical Model of Phosphorus Incorporation in Silica in Modified Chemical Vapor Deposition. *J. Am. Ceram. Soc.*, 1988, **71**, 914–923.
9. Wong, J. Vibrational spectra of vapor-deposited binary phosphosilicate glasses. *J. Non-Cryst. Solids*, 1976, **20**, 83–100.
10. Plotnichenko, V. G., Sokolov, V. O., Klotashev, V. V. & Dianov, E. M. On the structure of phosphosilicate glasses. *J. Non-Cryst. Solids*, 2002, **306**, 209–226.
11. Handke, M., Mozgawa, W. & Nocun, M. Specific features of the IR spectra of silicate glasses. *J. Non-Cryst. Solids*, 1994, **325**, 129–36.
12. Dayanand, C., Bhikshamaiah, G., Jaya Tyagaraju, V., Salagram, M. & Krishna Murthy, A. S. R. Structural investigations of phosphate glasses: a detailed infrared study of the x(PbO)-(1-x)P_2O_5 vitreous system. *J. Mater. Sci.*, 1996, **31**, 1945–1967.
13. Corbridge, D. E. C. & Lowe, E. J. Infrared spectra of some inorganic phosphorus compounds. *J. Chem. Soc.*, 1954, 493–502.
14. Youngman, R., Hogue, C. & Aitken, B. Crystallization of Silicon Pyrophosphate from Silicophosphate Glasses as Monitored by Multi-Nuclear NMR. *MRS Proc.*, 2006, **984**, MM12-03.
15. Shibata, N., Horigudhi, M. & Edahiro, T. Raman spectra of binary high-silica glasses and fibers containing GeO_2, P_2O_5 and B_2O_3. *J. Non-Cryst. Solids*, 1981, **45**, 115–126.

Structural characterisation of tin fluorophosphate glasses doped with Er_2O_3

*J. Trimble,[1] R. Golovchak,[1] J. Oelgoetz,[1] C. Brennan[2] & A. Kovalskiy[1]**

[1] *Department of Physics and Astronomy, Austin Peay State University, Clarksville, TN 37044, USA*
[2] *Department of Chemistry, Austin Peay State University, Clarksville, TN 37044, USA*

Manuscript received 30 September 2014
Revision received 23 January 2015
Manuscript accepted 30 April 2015

EXAFS and confocal Raman microscopy have been used to study $50SnF_2.(20\text{-}x)SnO.30P_2O_5.xEr_2O_3$ (x=0, 0·1, 0·25) glasses. EXAFS data reveal an average coordination of Sn to O of 1·5 in both undoped glass and Er-doped glass samples. The first coordination sphere of Er in glasses doped with Er_2O_3 was found to have 9 F atom neighbours at an average bond length of 2·292±0·005 Å, indicating that Er preferentially bonds with F. Raman spectra clearly show the emergence of orthophosphate Q^0 units on addition of Er_2O_3. The saturation of Er solubility is found to be between 0·25 and 0·5 mol% Er_2O_3. An increase of the glass transition temperature from ~80°C in undoped glass to ~87°C in samples doped with 0·25 mol% Er_2O_3 was also observed. This process is accompanied by an increase in the difference between the crystallisation and glass transition temperatures, which is usually associated with improved thermal stability of the glass.

1. Introduction

Fluorophosphate glasses have been studied as promising media for many applications, such as optical amplifiers, upconverting optical devices[1,2] and low temperature sealing agents.[3] Compositions of these glasses have also been developed as a host matrix for organic dopants for use in optical signal processing devices.[4] However, these materials have exhibited poor chemical durability, which has limited their technological application. The addition of rare earth elements or other dopants can improve the chemical stability of the glass,[1,5] as well as introducing additional functionality for active optical applications.[1] A variety of network modifying cations may be added to fluorophosphate glass compositions to engineer their properties.[1–3,5,6] One such element is Sn, which lowers the melting temperature of fluorophosphate glasses. The most durable compositions among tin fluorophosphate glasses have been found to have ~30 mol% P_2O_5; this is because less than 5% of O in these glasses is involved in P–O–P bridging oxygen bonds.[5] These bonds are probably more susceptible to reacting with water molecules from the air.[5]

The atomic structure of these glasses has been studied previously using Raman, XPS and infrared spectroscopy.[5,7–9] Phosphorus cations are believed to be four-fold coordinated and primarily exist in tetrahedral geometry as FPO_3 monomers or P_2O_7 dimers in glasses with 30 mol% P_2O_5.[5,9] Sn is expected to be three-fold[7] or four-fold[10] coordinated, in trigonal pyramidal or tetrahedral geometry.

It was previously shown that the addition of Er_2O_3 to the matrix of phosphate and fluorophosphate glasses strengthens network connectivity and improves their chemical durability.[1,11,12] Also, the addition of Er_2O_3 or ErF_3 significantly contributes to the efficiency of energy conversion processes in fluorophosphate glasses.[1,13] However, the mechanism of Er incorporation, the specifics of structural organisation, and the dependence on Er concentration, are not fully clear, especially in the tin fluorophosphate matrix.

The atomic or molecular structure of Er-doped fluorophosphate glass is still unclear, despite previous studies.[6,14] In the present study, advanced experimental methods, such as EXAFS (extended x-ray absorption fine structure) and Raman microscopy, were used to investigate the structure of undoped and Er-doped tin fluorophosphate glasses. The main goal of this work is to clarify the coordination of Sn and Er in the glass matrix, and to verify the molecular structure of the glasses, especially in Er-doped compositions.

2. Experimental procedure

Samples of $50SnF_2.(20-x)SnO.30P_2O_5.xEr_2O_3$ (x=0, 0·1, 0·25) glass have been investigated. Twelve gram melts of reagent grade SnF_2, SnO, P_2O_5 and Er_2O_3 were mixed in an argon atmosphere and melted in alumina crucibles at 330°C for 2 h, and 400°C for an additional hour in air. The samples were stirred after 2 h and again 15 min before being cast. Glass disks of ~2 cm diameter and 0·5 cm thickness were quenched onto copper plates and stored in desiccators.

The Sn:P and O:F ratios were obtained from energy dispersive x-ray spectroscopy using a Hitachi TM-

* Corresponding author. Email kovalskyya@apsu.edu
Original version presented at Int. Conf. on Phosphate Glasses, Pardubice, Czech Republic, 2–4 July 2014
DOI: 10.13036/17533562.57.1.069

1000 scanning electron microscope. The experimentally determined Sn:P ratios in the final composition were ~10% higher in the undoped glass and ~1–5% lower in doped glass than the nominal ratio (~1·16). The O:F ratio in the final glass compositions was found to be ~10% higher (undoped), or ~15% higher (doped samples) than the nominal composition ratio (~1·7), indicating the loss of F during synthesis probably through the evolution of HF gas. Aluminium impurities from the crucible were detected at ~1 wt%.

Undoped samples and samples doped with 0·1 mol% Er_2O_3 were transparent and colourless, while the samples doped with 0·25 mol% Er_2O_3 have a faint pink tint. Data from samples doped with 0·5 mol% Er_2O_3 and higher, which are a translucent pink with visible phase separation, were excluded from consideration. All measurements were made before any corrosion became noticeable. Within one week, a strong corrosion was observed in the undoped samples exposed to a normal atmosphere, with a white surface film covering the glass. The effect of this corrosion is significantly diminished with increasing Er_2O_3 concentration.

The EXAFS measurements were performed at the X18B x-ray beamline at the National Synchrotron Light Source at Brookhaven National Laboratory. The samples were powdered and glued onto Kapton tape, and data were collected at the Sn K-edge (29·2 KeV) and the Er L_3-edge (8·358 KeV) in fluorescent mode, using a Passivated Implanted Planar Silicon (PIPS) detector.

In general, the EXAFS signal represents modulation in the absorption coefficient, μ_i, as a function of x-ray energy, $E=\hbar\omega$. Using the one-electron approximation of Fermi's Golden Rule for $\mu_i(\omega)$, under the plane wave approximation, the EXAFS equation for isotropic materials can be expressed as[15]

$$\mu_i(k) = \frac{\mu_1(k)}{\mu_0(k)} = \sum_j \frac{n_j S_{0j}^2(k) F_j(k) \sin(2kR_j + \varphi_j(k))}{kR_j^2 \exp\left(\frac{2R_j}{\Lambda} + 2k^2\sigma_{0j}^2\right)} \quad (1)$$

where $\mu_1(k)$ and $\mu_0(k)$ are the terms contributed by single scattering and background, respectively (multiple scattering terms are negligible for the EXAFS region). $k \approx 0{\cdot}512(E-E_0)^{1/2}$ Å^{-1} is the photoelectron wavevector, E_0 is the threshold energy, S_{0j} is the passive electron reduction factor, n_j is the degeneracy of the path, $N_j = nS_{0j}^2(k)$ is the number of neighbours in the jth shell at an average distance R_j, $F_j(k)$ is the effective amplitude of the backscattered electron wave, $\phi_j(k)$ is the effective phase shift between backscattered and outgoing electron wave, Λ is the mean free path of photoelectrons, and σ_{0j}^2 is the the Debye–Waller factor related with disorder.

The standard Athena-Artemis software[16] package was used to process the experimental EXAFS spectra. A Kaiser-Bessel type of window function[17] was applied for restricting the EXAFS data in k-space. The Artemis software was used to generate an input file for FEFF calculations,[16] using crystallographic data for $Sn_3(PO_4)_2$[19] and ErF_3.[20] The calculated scattering paths for the nearest neighbours were used to fit the first shell in k-space of Sn and Er, exploiting the Levenberg–Marquardt method of nonlinear least-squares minimisation[18] implemented in Artemis (the energy shift, ΔE_0, was fixed during the fitting).

Raman spectroscopy measurements were taken with an Xplora (Horiba Jobin-Yvon) Raman confocal microscope using 785 nm 52 mW excitation and an 1800 groove grating. 150 acquisitions of 12 s exposures using a 300 μm hole and a 200 μm slit were averaged. Scans with spectral resolution of ~3·5 cm^{-1} were performed at room temperature.

Differential scanning calorimetry (DSC) measurements were made with a TA Instruments 2920 differential scanning calorimeter. A heating rate of 5°C/min was used to scan samples of ~22 mg in a nitrogen atmosphere. The glass transition temperature T_g was taken as the onset value of the endothermic shift of the baseline, whereas the crystallisation temperature T_x was determined from the peak value of the corresponding exotherm. Based on previous studies using this instrument, we estimate the experimental uncertainty of the reported T_g and T_x values as ±2°C.

3. Results and discussion

Raman spectra of the $50SnF_2.(20-x)SnO.30P_2O_5.xEr_2O_3$ (x=0, 0·1, 0·25) glasses are shown in Figure 1. Peaks in the high frequency region of the spectrum have previously[5,9,21] been identified as follows: the peak at 1035 cm^{-1} is assigned to the P–O symmetric stretching vibration of the pyrophosphate Q^1 unit[5,21] (where Q^n denotes a PO_4 tetrahedron with n bridging oxygen atoms) or the fluorophosphate monomer, FPO_3.[5,9] The band near 860 cm^{-1} is associated with the P–F symmetric stretch,[5] and the band near 740 cm^{-1} is related to a P–O–P symmetric stretching mode.[5,21]

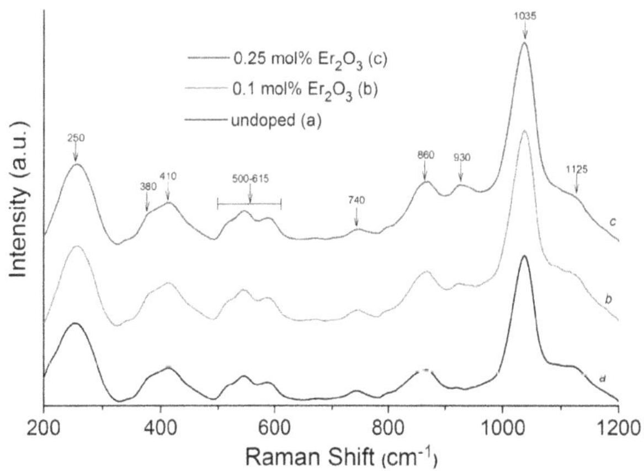

Figure 1. Raman spectra of $50SnF_2.(20-x)SnO.30P_2O_5.xEr_2O_3$ (x=0, 0·1, 0·25) glasses [Colour available online]

The peak near 1125 cm^{-1} is associated with the P–O symmetric stretch of the metaphosphate Q^2 unit,[21] which exists residually in our undoped glass. In the low frequency region of the spectrum, the band near 250 cm^{-1} has been associated with $Sn–O_2$ bending modes.[22] The bands between 500–615 cm^{-1} have not been described in detail but this region has previously been associated with Sn and P vibrations without specific assignment.[5,23,24] Based on our measurements of the spectrum of crystalline SnF_2 (not shown), we assign the peaks near 380 and 410 cm^{-1} to Sn–F fragments. This assignment is in good agreement with theoretical calculations of Sn–F vibrations.[25]

The effect of the Er dopant on the Raman spectra of the tin fluorophosphate glass is shown in Figure 1(b) and (c). The band near 930 cm^{-1} that emerges with Er_2O_3 addition is assigned to the P–O symmetric stretch of isolated orthophosphate Q^0 tetrahedra.[5] There is also a slight decrease in intensity of the band near 1125 cm^{-1} that is associated with Q^2 units. From the experimental spectra we cannot make trustworthy conclusions about the changes of the peak near 1035 cm^{-1} that is associated with FPO_3 or Q^1 units, or the band near 740 cm^{-1}, but nevertheless we can note the tendency of the former peak to decrease and the latter to increase. These changes in the Raman spectra are consistent with the depolymerisation of the phosphate chain as the [P]/[O] ratio decreases when Er_2O_3 replaces SnO. The P–F band near 860 cm^{-1} narrows with Er addition, but does not change significantly in intensity. No bands are observed in any of the measured Raman spectra in the region from 1125 to 1500 cm^{-1}.

Our EXAFS studies allow the determination of average coordination numbers, bond lengths and Debye-Waller factors for both Sn and Er atoms (Table 2, Figure 2). The first coordination shell of Sn was fitted with the Sn→O path only, since including the Sn→F path, derived from the known crystalline structures, did not lead to convergence of the fit. The possible reason could be that Sn–F interatomic distances in known crystal structures differ too much from those in the glass structure, or that the EXAFS oscillations corresponding to the Sn→F path in the investigated glasses significantly overlap with the background and, thus, cannot be extracted with confidence. Another explanation could be that the Sn–F bond lengths in the glass are very irregular (covering a broad range), which prevents a steady pattern in the modulation of the absorption coefficient (and, thus, the related EXAFS oscillations to occur) as is the case for the second and further coordination spheres in disordered solids (no peaks corresponding to these coordination spheres could be observed in the partial radial distribution function). So, it is possible that the observed spectrum is dominated by the EXAFS oscillations of Sn→O path, where the Sn–O bond length is relatively well defined throughout the structure. The obtained Sn–O bond length is 2·114±0·005 Å in the undoped sample, increasing to 2·119 and 2·121±0·005 Å in the samples doped with 0·1 and 0·25 mol% Er_2O_3, respectively. The Sn–O coordination is found to be 1·5 in the undoped sample, and remains constant (within the experimental uncertainty) in the two doped samples. If we assume the total coordination for Sn to be three, we can speculate that Sn is bonded to 1–2 oxygen atoms and the remaining bonds are with F. This is in good agreement with the structural model proposed by Brow *et al.*[7] Missing paths could be a reason for the slight mismatch of the experimental data and the fitting at higher *k* values (Figure 2(b)) in the Sn K-edge EXAFS spectra. The Er L_3-edge EXAFS spectra are noisier because of the low concentration of Er in the investigated samples. Nevertheless, it was possible to obtain an acceptable goodness of fit with an Er coordination of approximately nine in the sample doped with 0·25 mol% Er_2O_3 and an Er–F bond length of 2·292±0·005 Å. This Er–F bond length is slightly greater than the value 2·280 Å, as measured by EXAFS for Er–F bonds in crystalline

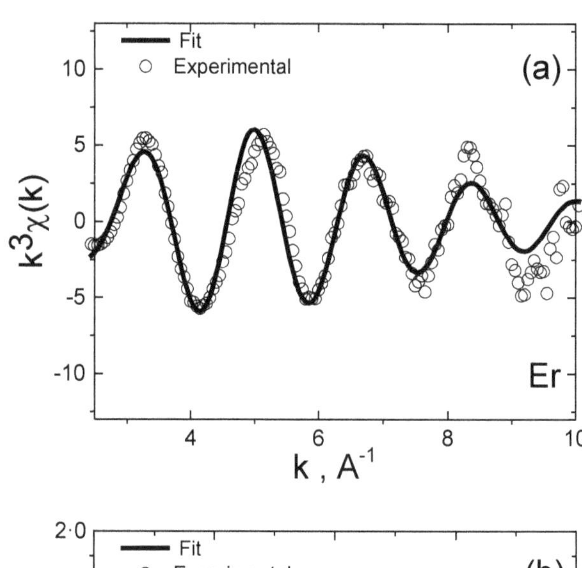

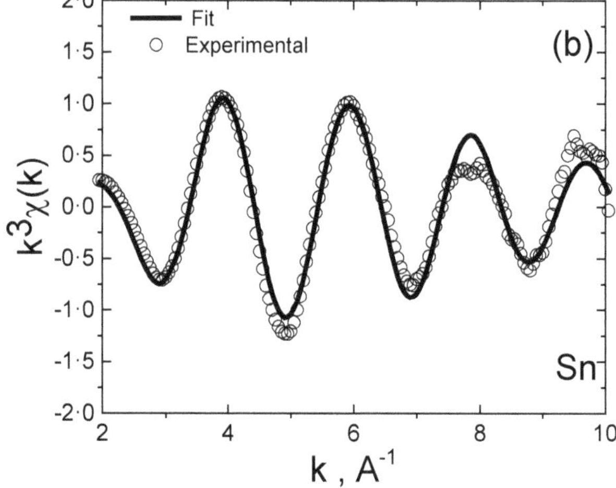

Figure 2. Examples of k^3-weighted EXAFS oscillations (symbols) and fits (solid lines) for (a) the Er L_3-edge of the $50SnF_2.19{\cdot}75SnO.30P_2O_5.0{\cdot}25Er_2O_3$ glass, and (b) the Sn K-edge of the undoped $50SnF_2.(20–x)SnO.30P_2O_5$ glass

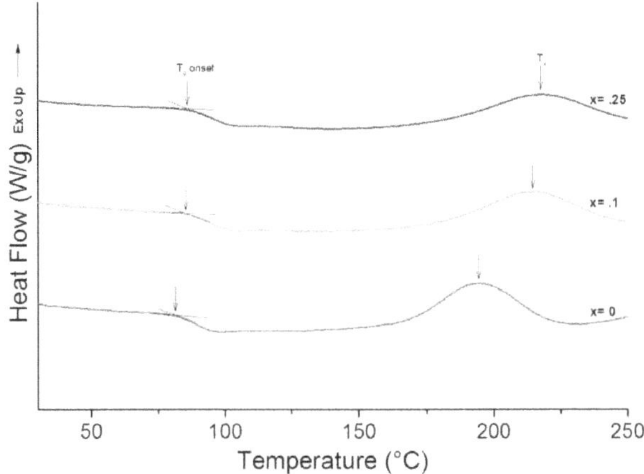

Figure 3. DSC thermograms of $50SnF_2.(20-x)SnO.30P_2O_5.xEr_2O_3$ (x=0, 0·1, 0·25) glasses, heated at 5°C/min [Colour available online]

ErF_3.[26] Note that each Er in crystalline ErF_3 has nine first neighbour F atoms at six different lengths in the range of 2·248–2·589 Å.[20]

DSC thermograms of the glass samples are shown in Figure 3, and the measured values for T_g, T_x, and T_x-T_g, are given in Table 1. The glass transition temperature for the undoped sample of 80°C increases by 5–7°C with Er addition. The Dietzel thermal stability criterion[27] (calculated as T_x-T_g) improves with Er doping. It should be noted that the value T_g=95°C previously reported for the undoped glass[8] (measured at 4°C/min) was obtained for samples with a different thermal history (quenching temperature of 500°C versus 400°C). Differences in T_g values such as this have been reported for fluorophosphate glasses before, and have been attributed to variations in F content in the final glass composition.[28] The increase of T_g and T_x in the doped samples indicates that the addition of Er_2O_3 strengthens the glass network.[1,12] Stronger ionic bonds between Er^{3+} and neighbouring anions replace weaker bonds to Sn^{2+}, which improves the thermal properties. The increase in characteristic temperatures may also be partially attributed to the greater density of network linkages that occurs when a nine coordinated Er–F environment replaces mixed three coordinated Sn–[F,O] bonding.

The structural model of Brow *et al*[7] is based on XPS analysis of tin fluorophosphate glass with high SnF_2 content and a ratio $[SnO+SnF_2]:P_2O_5>1$. This model is valid for the glass composition studied in the present paper. In this model, phosphorus cations are four-fold coordinated and exist in tetrahedral geometry as FPO_3. Sn is three-fold coordinated and is at the apex of a trigonal pyramid with a lone pair of electrons. Sn is linked to phosphorus via bridging oxygen, and is linked to another Sn via a bridging F. The third neighbour of Sn in this model alternates between O and F, which on average gives a Sn–O coordination of 1·5, in good agreement with our data. This finding is supported by previous research that found both P–F and Sn–F bonds are features of these glasses.[8,29] The presence of FPO_3 units for phosphate glass compositions with ~30 mol% P_2O_5 has been well studied.[5,9] Aside from this work, we know of no EXAFS data for tin fluorophosphate glasses that give experimental confirmation of the Sn coordination sphere. Thus the results presented in this paper confirm the Sn–O coordination in the structural model given by Brow *et al.*[7]

The coordination number of ~9 for Er atoms in $50SnF_2.19{\cdot}75SnO.30P_2O_5.0{\cdot}25Er_2O_3$ glass is consistent with the ErF_3 crystal structure (orthorhombic cell, space group *pnma*[20]). In this crystal, Er has a total of 13 neighbours, with nine closest neighbours followed by two more at a slightly longer bond length, and another two more at an even longer bond length. This result suggests that Er_2O_3 acts as a network modifier in our glasses, and that F migrates to Er sites to preserve charge neutrality, which would be consistent with the formation of Q^0 fragments observed in the Raman spectra. The preference for F coordination to the rare earth in tin fluorophosphate glass may result from the formation of P–F bonds reducing the polarisability of the corresponding P–O bonds, which typically fill the coordination of rare earths with O.[6,30] Preferential covalency of Er bonding with O in fluorophosphate glass has been reported for Al-containing compositions[6,30] where only a few P–F bonds form. In tin fluorophosphate glass, Sn–F bonds are not preferred (only forming in a limited number to prevent Sn–O–Sn linkages[5]), whereas, for example, in Al–F–P–O glass strong Al–F bonds form.[31] The affinity of the cation modifier to bond with F in fluorophosphate glass determines the extent of P–F bond formation, which in turn determines the preference of coordination of rare earths to O or F. It should be noted that Er may still be involved in interchain bonding in structures such as Sn–F–Er–F–Sn, while Er–F terminal fragments are also likely to be present.

4. Conclusions

It has been found that Er_2O_3 can replace SnO in $50SnF_2.30P_2O_5.20SnO$ glasses up to 0·25 mol% without the formation of a crystalline phase. EXAFS and Raman data support the hypothesis that Er bonds

Table 1. T_g, T_x and T_x-T_g values for the doped and undoped tin fluorophosphate glasses

Batch Composition (mol%)	*Appearance*	T_g (°C)	T_x (°C)	T_x-T_g (°C)
$50SnF_2.30P_2O_5.20SnO$	Transparent, colourless	80	194	114
$50SnF_2.30P_2O_5.19{\cdot}9SnO.0{\cdot}1Er_2O_3$	Transparent, colourless	85	216	131
$50SnF_2.30P_2O_5.19{\cdot}75SnO.0{\cdot}25Er_2O_3$	Transparent, light pink	87	218	131

Table 2. Bond lengths, Debye–Waller factors, and coordination numbers of Sn and Er in undoped and doped $50SnF_2.30P_2O_5.20SnO$ glasses

Composition	Edge					
	Er L_3			Sn K		
	r (Å) (±0·005)	σ_0 (Å^2) (±0·0014)	N_{Er} (±1·3)	r (Å) (±0·005)	σ_0 (Å^2) (±0·0012)	N_{Sn} (±0·10)
Undoped				2·114	0·0074	1·50
Doped with 0·1 mol% Er_2O_3				2·119	0·0082	1·55
Doped with 0·25 mol% Er_2O_3	2·292	0·0103	9·0	2·121	0·0083	1·59

principally with F in the glass, with an Er coordination number of ~9 in the doped glass samples, as determined by EXAFS, consistent with the ErF_3 structure. Preferential Er-F bonding in tin fluorophosphate glass may be attributed to the presence of P–F bonds, lowering the polarisability of the nonbridging P–O bonds, which would otherwise fill the coordination of Er with O. The EXAFS data also suggest that the coordination number of Sn to O is close to 1·5, and remains constant in doped samples within the uncertainty of the method. The Raman spectra confirm the formation of isolated Q^0 orthophosphate units on addition of Er_2O_3. It was confirmed that T_g increases with the addition of Er in doped glass samples, and that their thermal stability improves due to an increase in the value of T_x–T_g (according to the Dietzel criterion).

Acknowledgements

We would like to thank the NSF International Materials Institute for New Functionality in Glass (NSF Grant No. DMR 0844014) and the NASA Tennessee Space Grant Consortium for funding. The Horiba Fluorolog 3 and Horiba Confocal Raman Microscope were purchased under NASA grant/cooperative agreement NNX10AJ04G. EXAFS data were taken at the X18B NSLS beamline at Brookhaven National Laboratory. The scientific and technical support of Dr. Syed Khalid (NSLS), and Chatree Saiyasombat (Lehigh University) are highly appreciated. The authors would like also to acknowledge the intellectual contribution of former undergraduate students at Austin Peay State University: Nichole Boyer, Tristan Harper and James York-Winegar.

References

1. Yung, S. W., Lin, H. J., Lin, Y. Y., Brow, R. K., Lai, Y. S., Horng, J. S. & Zhang, T. Concentration effect of Yb^{3+} on the thermal and optical properties of Er^{3+}/Yb^{3+}-codoped ZnF_2–Al_2O_3–P_2O_5 glasses. *Mater. Chem. Phys.*, 2009, **117**, 29–34.
2. Zhang, L., Sun, H., Xu S., Li, K. & Hu, L. Broadband amplification and upconversion luminescence properties of Er^{3+}/Yb^{3+} co-doped fluorophosphate glass. *Solid State Commun.*, 2005, **135**, 449–54.
3. Aitken, B. G., Koval, S. E. & Qiuesada, M. A. Durable tungsten-doped tin-fluorophosphate glasses. Corning Incorporated, assignee. Patent EP 2061727 A2, 27 May 2009.
4. Tick, P. A. & Hall, D. W. Nonlinear optical effects in organically doped low melting glasses. *Diff. Defect Data*, 1987, **53–54**, 179–88.
5. Xu, X. J., Day, D. E., Brow, R. K. & Callahan, P. M. Structure of tin fluorophosphate glasses containing PbO or B_2O_3. *Phys. Chem. Glasses*, 1995, **36** (6), 264–271.
6. Tanabe, S., Yoshii, S., Hirao, K. & Soga, N. Upconversion properties, multiphonon relaxation, and the local environment of rare-earth ions in fluorophosphate glasses. *Phys. Rev. B*, 1992, **45**, 4620–5.
7. Brow, R. K., Phifer, C. C., Xu, X. J. & Day, D. E. An x-ray photoelectron spectroscopy study of anion bonding in tin(II) fluorophosphate glass. *Phys. Chem. Glasses*, 1992, **33**, 33–9.
8. Xu, X. J. & Day, D. E. Properties and structure of Sn–P–O–F glasses. *Phys. Chem. Glasses*, 1990, **31** (5), 183–7.
9. Videau, J. J. & Portier, J. Raman spectroscopic studies of fluorophosphate glasses. *J. Non-Cryst. Solids*, 1982, **48**, 385–92.
10. Tick, P. A. Water durable glasses with ultra low melting temperatures. *Phys. Chem. Glasses*, 1984, **25** (6), 149–54.
11. Shih, P. Y. Thermal, chemical and structural characteristics of erbium-doped sodium phosphate glasses. *Mater. Chem. Phys.*, 2004, **84**, 151–6.
12. Shyu, J. J. & Chiang, C. C. Effects of Er_2O_3 doping on the structure, thermal properties, and crystallisation behavior of SnO–P_2O_5 glass. *J. Am. Ceram. Soc.*, 2010, **93**, 2720–5.
13. Liao, M., Hu, L., Duan, Z., Zhang, L. & Wen, L. Spectroscopic properties of fluorophosphate glass with high Er^{3+} concentration. *Appl. Phys. B*, 2007, **86**, 83–9.
14. Tian, Y., Xu, R., Hu, L., & Zhang, J. Effect of chloride ion introduction on structural and 1.5 μm emission properties in Er^{3+} -doped fluorophosphate glass. *J. Opt. Soc. Am. B*, 2011, **28**, 1638–44.
15. Stern, E. A. *X-ray Absorption: Basic Principles of EXAFS, SEXAFS and XANES*. Edited by D. C. Koningsberger & R. Prins, John Wiley & Sons, New York, 1988.
16. Ravel, B. & Newville, M. Athena, Artemis, Hephaestus: data analysis for x-ray absorption spectroscopy using IFEFFIT. *J. Synchrotron Radiat.*, 2005, **12**, 537–41.
17. Kaiser, J. F. Digital Filters. In *System Analysis by Digital Computer*. F. F. Kuo & J. F. Kaiser, Wiley, New York, 1966, ch. 7.
18. Vetterling, W. T., Teukolsky, S. A. & Press, W. H. *Numerical Recipes: Example Book* (C). Second Edition, Cambridge University Press, 1992.
19. Mathew, M., Schroeder, L. W. & Jordan, T. H. The crystal structure of anhydrous stannous phosphate, $Sn_3(PO_4)_2$. *Acta. Cryst. B*, 1977, **33**, 1812–16.
20. Krämer, K., Romstedt, H., Guedel, H. U., Fischer, P., Murasik, A. & Fernandez-Diaz, M. T. Three dimensional magnetic structure of ErF_3. *Eur. J. Sol. State Inorg. Chem.*, 1996, **33**, 273–83.
21. Lim, J. W., Yung, S. W. & Brow, R. K. Properties and structure of binary tin phosphate glasses. *J. Non-Cryst. Solids*, 2011, **357**, 2690–4.
22. Mcguire, K., Pan, Z. W., Wang, Z. L., Milkie, D., Menéndez, J. & Rao, A. M. Raman studies of semiconducting oxide nanobelts. *J. Nanosci. Nanotechno.*, 2002, **2**, 1–4.
23. Rouse, G .B., Miller, P. J. & Risen, W. M. Mixed alkali glass spectra and structure. *J. Non-Cryst. Solids*, 1978, **28**, 193–207.
24. Galeener, F. L. & Mikkelsen, J. C. The Raman spectra of pure vitreous P_2O_5. *Solid State Comm.* 1979, **30**, 505–10.
25. York-Winegar, J., Harper, T., Brennan, C., Oelgoetz, J. & Kovalskiy, A. Structure of SnF_2–SnO–P_2O_5 Glasses. *Phys. Proc.*, 2013, **44**, 159–65.
26. Terrasi, A., Priolo, F., Franzo, G., Coffa, S, D'Acapito, F. & Mobilio, S. EXAFS analysis of Er sites in Er–O and Er–F co-doped crystalline Si. *J. Lumin.*, 1999, **80**, 363–7.
27. Dietzel. A. Glass structure and glass properties. *Glasstech. Ber.*, 1948, **22**, 41–50.
28. Ehrt, D. Effect of OH-content on thermal and chemical properties of SnO–P_2O_5 glasses. *J. Non-Cryst. Solids*, 2007, **354**, 546–52.
29. Osaka, A. & Miura, Y. Bonding state of fluorine in lead-tin oxyfluorophosphate glasses. *J Non-Cryst. Solids*, 1990, **125**, 87–92.
30. Ebendor-Heidepriem, H., Ehrt, D., Bettinelli & Speghini, M. A. Effect of glass composition on Judd-Ofelt parameters and radiative decay rates of Er^{3+} in fluoride phosphate and phosphate glasses. *J Non-Cryst. Solids*, 1998, **240**, 66–78.
31. Möncke, D., Ehrt, D., Velli, L. L., Varsamis, C. P. E., Kamitsos, E. I., Elbers, S. & Eckert, H. Comparative spectroscopic investigation of different types of fluoride phosphate glasses. *Phys. Chem. Glasses: Eur. J. Glass Sci. Technol. B*, 2007, **48** (6), 399–402.

Phys. Chem. Glasses: Eur. J. Glass Sci. Technol. B, February 2016, **57** (1), 32–36

The Raman spectra and structure of $PbO–WO_3–P_2O_5$ glasses

Mária Chromčíková,[1] Branislav Hruška,[1] Jana Holubová,[2] Magdaléna Lissová[1] & Marek Liška[1]*

[1] *Vitrum Laugaricio – Joint Glass Center of IIC SAS, TnUAD, and FChPT STU, Študentská 2, Trenčín, SK-91150, Slovakia*

[2] *Department of General and Inorganic Chemistry, Faculty of Chemical Technology, University of Pardubice, Studentská 573, Pardubice, CZ-532 10, Czech Republic*

Manuscript received 14 September 2014
Revision received 21 March 2015
Manuscript accepted 17 April 2015

The Raman spectra of glasses from the compositional series $(0{\cdot}5-x/2)PbO.xWO_3.(0{\cdot}5-x/2)P_2O_5$ (x=0·1, 0·2, 0·3, 0·4, 0·5 and 0·6) were studied. Three components were identified in the set of baseline-subtracted and temperature-corrected Raman spectra by principal component analysis. The thermodynamic model of Shakhmatkin & Vedishcheva, with four components (PbP_2O_6, $W_2P_2O_{11}$, $PbWO_4$, and WP_2O_8) corresponding to the stable crystalline phases, was constructed. Correlation analysis between the scores obtained by multivariate curve analysis of the Raman spectra and the equilibrium molar amounts obtained from the thermodynamic model was performed. On this basis, the value of the equilibrium constant for the $2W_2P_2O_{11}+PbP_2O_6=3WP_2O_8+PbWO_4$ reaction was estimated as logK=−1·2.

1. Introduction

Phosphate glasses containing tungsten oxide are a special group of glasses, known for their electrochromic and photochromic properties, which lead to a wide range of applications. Their specific properties include larger thermal expansion coefficients, lower viscosities and softening temperatures than silicate glasses.[1,2] The aim of the present study is to determine the effect of WO_3 incorporation on the spectral properties and the structure of $PbO–WO_3–P_2O_5$ glasses. The main message of the present paper is that useful thermodynamic data (e.g. reaction Gibbs energies or equilibrium constants) can be – even for relatively complex glass forming systems – obtained by combining the 'blind' statistical treatment of spectral data (Principal Component Analysis and Multivariate Curve Resolution) with the thermodynamic hypothesis of chemical reactions/equilibria taking place in the system studied. The coincidence between the experimental and calculated spectra can be considered as the confirmation of this hypothesis.

2. Method

The Raman spectra were analysed by means of principal component analysis (PCA).[3,4] The number of spectrally active, non-correlated components was determined on the basis of the so-called indicator function, and the Malinowski significance level. Furthermore, the Multivariate Curve Resolution (MCR) method[5,6] was used, resulting in the spectra of quasi pure components (so called loadings), and the relative abundances of these components (so called scores). The MCR result was then combined with the result of thermodynamic modeling to determine the values of unknown parameters of the thermodynamic model (e.g. reaction Gibbs energies or equilibrium constants).

The PCA analysis was performed using the factor analysis toolbox in Matlab software,[3] the MCR calculations were done using the Solo-Mia software of Eigenvector Inc.,[6] and the thermodynamic model was constructed using our own Fortran software.

For the studied glasses, the thermodynamic model of Shakhmatkin & Vedishcheva[7–10] was used, considering the glass as an ideal solution of four components: PbP_2O_6 (PbP), WP_2O_8 (WP), $W_2P_2O_{11}$ (W2P), and $PbWO_4$ (PbW). It should be stated here that the model constructed is in fact incomplete, because not all the stable crystalline phases were considered. Of the known lead tungstate compounds, only $PbWO_4$ was considered, whilst Pb_2WO_5[11] was neglected because its Raman intensity is significantly lower than for the phosphate compounds, and hence is not sufficiently visible in the Raman spectra of the glasses. Moreover, the ternary compound $Pb(WO_2)_2(WO_4)_2$[12] was not taken into account, because the stoichiometry of the studied glass compositional series is well covered by the considered components. Furthermore, due to the negative values of the reaction Gibbs energies of the considered system compounds, and due to the stoichiometry of the studied glasses, only negligible amounts of unreacted oxides are present at equilibrium in the studied glasses.

Corresponding author. Email maria.chromcikova@tnuni.sk
Original version presented at Int. Conf. on Phosphate Glasses, Pardubice, Czech Republic, 2–4 July 2014
DOI: 10.13036/1753-3562.57.1.056

In the first step, the three following chemical reactions between pure oxides were considered:

$$PbO+P_2O_5=PbP_2O_6 \text{ (PbP)} \tag{1}$$

$$2WO_3+P_2O_5=W_2P_2O_{11} \text{ (W2P)} \tag{2}$$

$$PbO+WO_3=PbWO_4 \text{ (PbW)} \tag{3}$$

and after that WP_2O_8 (WP) was formed by the equilibrium reaction

$$2W_2P+PbP=3WP+PbW \quad (\xi) \tag{4}$$

where ξ is the reaction extent and the equilibrium constant is given by

$$K=\frac{x^3(\mathrm{WP})x(\mathrm{PbW})}{x^2(\mathrm{W2P})x(\mathrm{PbP})}=\frac{n^3(\mathrm{WP})n(\mathrm{PbW})}{n^2(\mathrm{W2P})n(\mathrm{PbP})}\frac{1}{\sum_i n_i} \tag{5}$$

where $x(i)/n(i)$ is equilibrium molar fraction/molar amount of component i.

Furthermore, it was assumed that there is no significant abundance of unreacted oxides in the studied glasses. Then, after applying the mass conservation law

$$n(\mathrm{PbP})+n(\mathrm{PbW})=n_0(\mathrm{PbO}) \tag{6}$$

$$2n(\mathrm{W2P})+n(\mathrm{PbW})=n_0(\mathrm{WO_3}) \tag{7}$$

$$n(\mathrm{PbP})+n(\mathrm{W2P})=n_0(\mathrm{P_2O_5}) \tag{8}$$

Only one parameter, i.e. the equilibrium constant K, is needed for obtaining the equilibrium molar amounts of the system components. For a given value of K, the corresponding extent of the reaction, ξ, can be found by solving the equation

$$\begin{aligned}&\xi^4(-27-4K)+\xi^3(-27C-4KD+4KB+4KA)\\&+\xi^2(4KBD+4KAD-4KAB-KA^2)\\&+\xi(-4KABD-KA^2D+KA^2B)+KA^2BD=0\end{aligned} \tag{9}$$

where

$$A=[n_0(\mathrm{WO_3})+n_0(\mathrm{P_2O_5})-n_0(\mathrm{PbO})]/3 \tag{10}$$

$$B=[2n_0(\mathrm{P_2O_5})+n_0(\mathrm{PbO})-n_0(\mathrm{WO_3})]/3 \tag{11}$$

$$C=[n0(\mathrm{WO_3})-2n0(\mathrm{P_2O_5})+2n0(\mathrm{PbO})]/3 \tag{12}$$

$$D=A+B+C \tag{13}$$

The fourth power polynomial of the reaction extent, ξ (Equation (9)), was obtained by inserting the equilibrium molar amounts given by the Equations (14)–(17) into the definition of equilibrium constant K given in Equation (5). The maximum value of the extent of reaction, ξ_{max}, is given by the glass composition as

$$\xi_{max}=\min\{A/2,B\}$$

The system equilibrium composition is then given by:

$$n(\mathrm{W2P})=A-2\xi \tag{14}$$

$$n(\mathrm{PbP})=B-\xi \tag{15}$$

$$n(\mathrm{WP})=3\xi \tag{16}$$

$$n(\mathrm{PbW})=C+\xi \tag{17}$$

The estimate of the value of the equilibrium constant, K, can be found by maximising the correlation between the scores and the corresponding equilibrium molar amounts of spectrally active components.

Table 1. Composition of studied glasses in mol fractions of pure oxides, $x_g(i)$, and the maximum extent of reaction, ξ_{max} (Equation (4))

Glass	$x_g(PbO)$	$x_g(WO_3)$	$x_g(P_2O_5)$	ξ_{max}
A10	0·45	0·10	0·45	0·0167
A20	0·40	0·20	0·40	0·0333
A30	0·35	0·30	0·35	0·0500
A40	0·30	0·40	0·30	0·0667
A50	0·25	0·50	0·25	0·0833
A60	0·20	0·60	0·20	0·0000

3. Experimental

The composition of the studied glasses is summarised in Table 1, together with the maximum ξ values. The studied glasses were prepared from analytical grade purity PbO, WO_3, and H_3PO_4 water solution (85 wt%). In the first stage, the reaction mixture was heated slowly in a platinum crucible up to 600°C, with the final calcination at the maximum temperature for 2 h to remove water. After the calcination, the temperature of the reaction mixture was increased by a heating rate of 12·5°C/min up to a temperature in the range 900–1450°C, depending on the composition. The melt was held at the maximum temperature for 20 min, and then poured into a preheated graphite mould. The glasses obtained were then transferred to

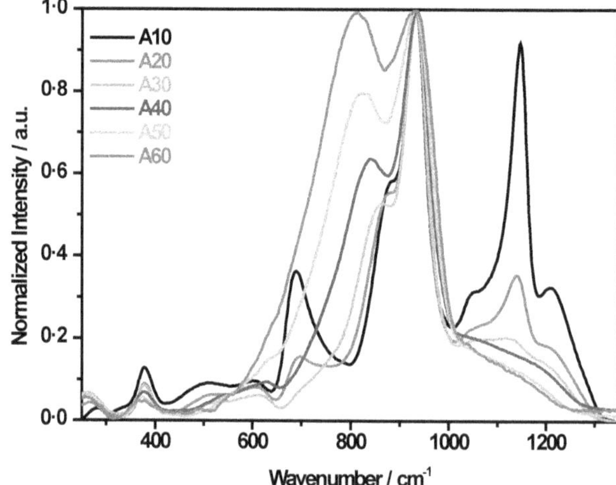

Figure 1. Baseline-subtracted and normalised Raman spectra [Colour available online]

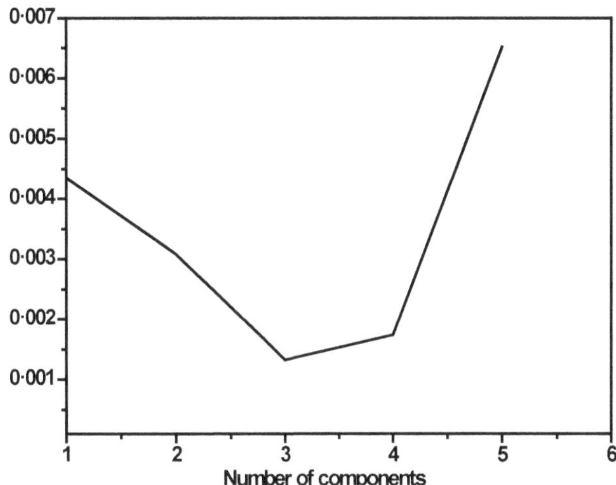

Figure 2. Principal component analysis indicator function

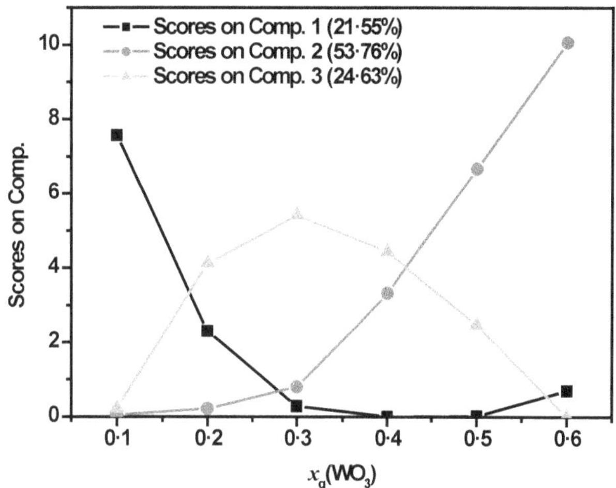

Figure 4. Multivariate curve resolution scores [Colour available online]

an annealing furnace, and kept for 30 min at a temperature ~5°C below the glass transition temperature, T_g. The annealing furnace was then switched off, and the samples were cooled down to room temperature. The amorphous character of the obtained glasses was checked by x-ray powder diffraction.

The Raman spectra were recorded in the wavenumber range 200–1500 cm^{-1}, using a Renishaw inVia Reflex Raman spectrometer with a Leica DM2500 microscope. A 514 nm Ar$^+$ ion laser with 10 mW power was used as the excitation source for a spot of about 1 mm diameter. The spectra were baseline-subtracted using the software WiRE ver. 3.3 provided by Renishaw with the Raman spectrometer. After the baseline subtraction, thermal correction was performed.[13] However, it is worth noting that the thermal correction does not affect the results of the principal component analysis.

4. Results and discussion

After baseline-subtraction and thermal-correction, the Raman spectra (Figure 1) were analysed by principal component analysis. Three spectrally active components were found on the basis of minimal value of the indicator function (Figure 2), and by the Malinowski significance level (Figure 3) falling below 5%. These three components were attributed to phosphate containing compounds PbP (PbP_2O_6), WP (WP_2O_8), and W2P ($W_2P_2O_{11}$). It is assumed that, since the PbW ($PbWO_4$) component does not contain phosphate Q-units, it is not observed in the studied spectral range.

The results of MCR analysis are summarised in Figures 4 and 5, whilst the experimental and MCR calculated spectra are compared in Figure 6. It can be seen that for all studied glass composition the spectra virtually coincide. The very small differences can be attributed to neglected lead tungstate components. The scores for component 1 decrease with increasing WO_3 content across the studied composition range (Figure 4), whilst the scores of component 2 increase monotonically, and the scores of component 3 reach a local maximum. From the glass composition and the ξ_{max} values (Table 1), it can be seen that the glass

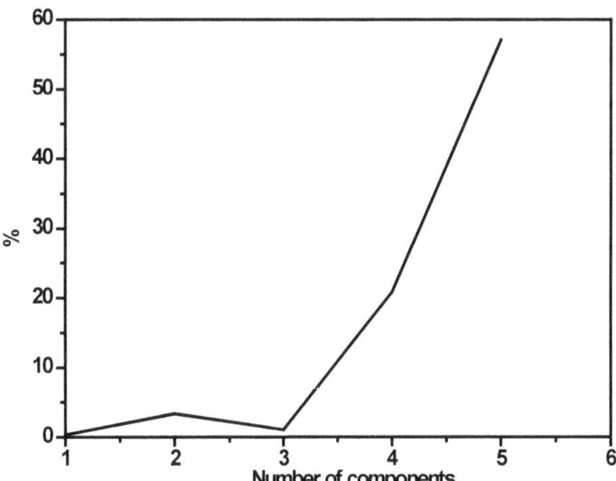

Figure 3. Principal component analysis Malinowski significance level (%)

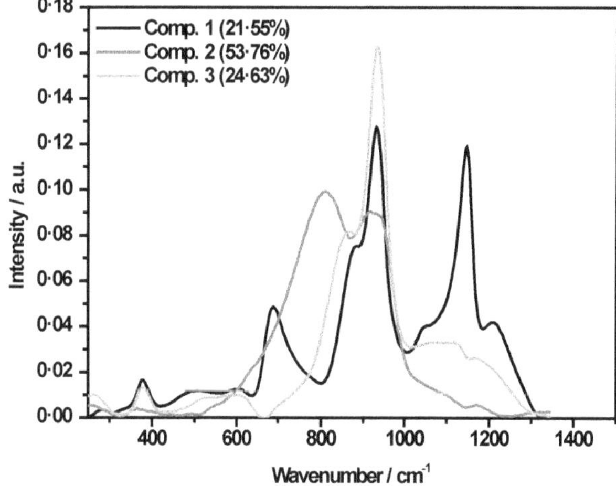

Figure 5. Multivariate curve resolution loadings [Colour available online]

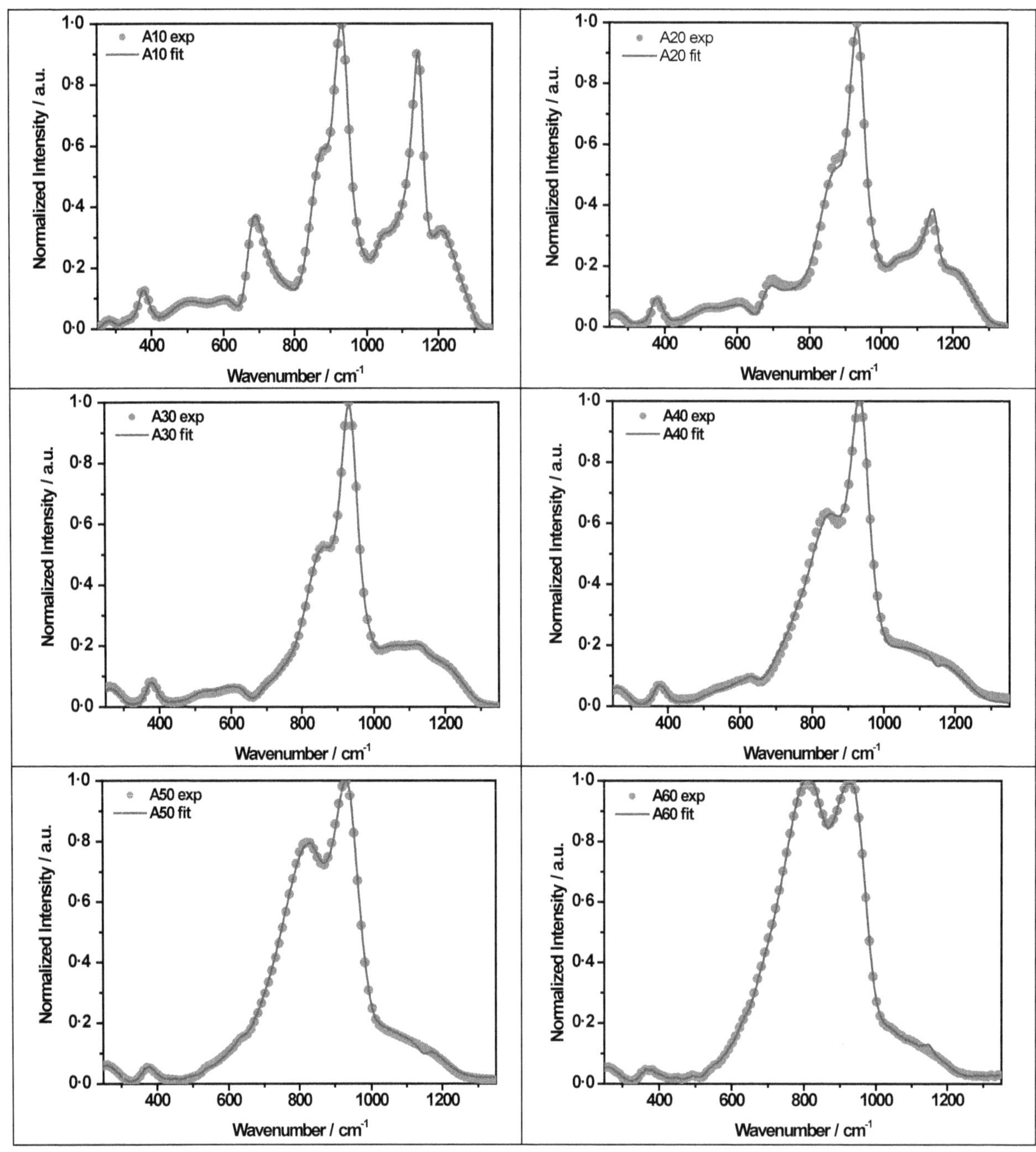

Figure 6. Comparison of experimental (points) and MCR calculated (lines) normalised Raman spectra [Colour available online]

A60 is composed of W2P and PbW only, while the glass A10 consists mainly of PbP. It can therefore be deduced that the MCR component 1 corresponds to PbP, component 2 to W2P, and component 3 to WP.

Table 2. Equilibrium molar amounts of glass components (mol) obtained for logK=−1·2

Glass	n(PbP)	n(WP)	n(W2P)	n(PbW)
A10	0·4056	0·0331	0·0113	0·0444
A20	0·3166	0·0503	0·0331	0·0834
A30	0·2302	0·0593	0·0605	0·1198
A40	0·1462	0·0615	0·0924	0·1538
A50	0·0651	0·0547	0·1302	0·1849
A60	0·0000	0·0000	0·2000	0·2000

The spectra of the MCR components are plotted in Figure 5. The system composition was calculated for log$K$ values ranging from −10·0 to 5·0 with a step of 0·1. For each K value, the coefficients of correlation between equilibrium molar amounts and corresponding scores were evaluated (r_1 for Score 1 versus n(PbP), r_2 for Score 2 versus n(W2P), and r_3 for Score 3 versus n(WP)). The overall correlation r was described by the product

$$r=r_1r_2r_3 \quad (18)$$

The dependence of r on log$K$, plotted in Figure 7, reaches a local maximum for log$K$=−1·2. The

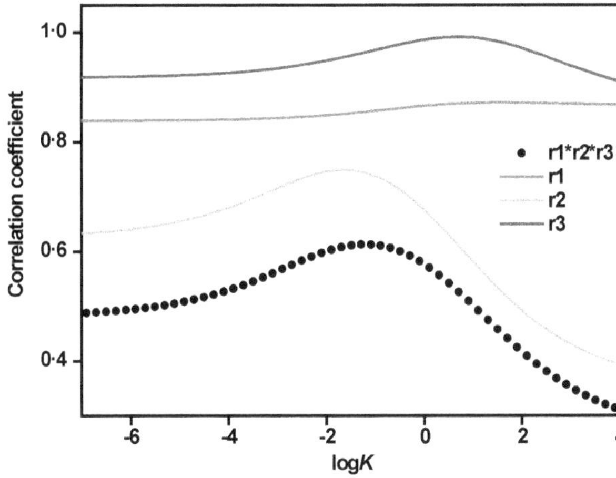

Figure 7. Coefficients of correlation between scores and equilibrium molar amounts of the components [Colour available online]

equilibrium molar amounts of glass components obtained for this log$K$ value for the studied glasses are summarised in Table 2. However, the dependences r_1(log$K$), r_2(log$K$), and r_3(log$K$) behave in different ways. Specifically, the r_1 correlation does not have any local maximum, and the log$K$ coordinate of the local maximum for r_3 is significantly higher compared with the r and r_2 maxima. Thus the log$K$=–1·2 value can be considered only as a rough estimate.

5. Conclusions

The thermodynamic model of Shakhmatkin and Vedishcheva with only four components corresponding to the stable crystalline phases was constructed for $PbO–WO_3–P_2O_5$ glasses. A method of evaluation of the thermodynamic model with only one adjustable parameter was proposed. On the basis of correlation analysis between the scores obtained by multivariate curve analysis of the measured Raman spectra and the equilibrium molar amounts obtained from the thermodynamic model, the value of the equilibrium constant for the $2W_2P_2O_{11}+PbP_2O_6=3WP_2O_8+PbWO_4$ reaction was estimated. However, the obtained value, log$K$=–1·2, can be considered only as a rough estimate.

Acknowledgement

This work was supported by the Slovak Grant Agency for Science under the grant VEGA 1/0006/12, by the Slovak Research and Development Agency, Project ID: APVV-0487-11.

References

1. Šubčík, J., Koudelka, L., Mošner, P., Montagne, L., Tricot, G., Delevoye, L. & Gregora, I. Glass-forming ability and structure of $ZnO–MoO_3–P_2O_5$ glasses. *J. Non-Cryst. Solids*, 2010, **356**, 2509–16.
2. Lissová, M. PhD Thesis, A. Dubček University of Trenčín, 2013.
3. Factor analysis Toolbox for MATLAB®. Applied Chemometrics, www.chemometrics.com.
4. Malinowski, E. R. *Factor Analysis in Chemistry*. Third Edition, J. Wiley & Sons, New York, 2002.
5. Ruckebusch, C. & Blanchet, L. Multivariate curve resolution: A review of advanced and tailored applications and challenges. *Anal. Chim. Acta*, 2013, **765**, 28–36.
6. http://www.eigenvector.com/courses/EigenU_MCR.html, 2014.
7. Vedishcheva, N. M., Shakhmatkin, B. A. & Wright, A. C. The structure of sodium borosilicate glasses: thermodynamic modelling vs. experiment. *J. Non-Cryst. Solids*, 2004, **345&346**, 39–44.
8. Vedishcheva, N. M., Shakhmatkin, B. A., Shultz, M. M. & Wright, A. C. The thermodynamic modelling of glass properties: a practical proposition? *J. Non-Cryst. Solids*, 1996, **196**, 239–43.
9. Shakhmatkin, B. A., Vedishcheva, N. M. & Wright, A. C. Can Thermodynamics relate the properties of melts and glasses to their structure? *J. Non-Cryst Solids*, 2001, **293–295**, 220–6.
10. Vedishcheva, N. M., Shakhmatkin, B. A. & Wright, A. C. Thermodynamic modelling of the structure of glasses and melts: single-component, binary and ternary systems. *J. Non-Cryst. Solids*, 2001, **293–295**, 312–17.
11. Fujita, T. & Muramatsu, K. High temperature form of Pb_2WO_5 and transformation phenomena to its low form. *Mater. Res. Bull.*, 1979, **14**, 5–12.
12. Rösslerová, I., Koudelka, L., Černošek, Z., Mošner, P. & Beneš, L. Study of crystallisation of $PbO–WO_3–P_2O_5$ glasses by thermoanalytical and spectroscopic methods. *J. Non-Cryst. Solids*, 2014, **384**, 41–6.
13. McMillan, P. F. & Wolf, G. H. Vibrational spectroscopy of silicate liquids. In *Structure, Dynamics and Properties of Silicate Melts*. Eds J. F. Stebbins, P. F. McMillan & D. B. Dingwell, Mineralogical Society of America, Washington DC 1995, 247–315.

Nanocrystallisation in vanadate phosphate and lithium iron vanadate phosphate glasses

Tomasz K. Pietrzak,[1] *Jerzy E. Garbarczyk, Marek Wasiucionek & Jan L. Nowiński*

Faculty of Physics, Warsaw University of Technology, Koszykowa 75, 00-662 Warsaw, Poland

Manuscript received 22 August 2014
Revision received 23 June 2015
Manuscript accepted 23 June 2015

In this paper we have summarised our recent research on nanocrystallisation in vanadate-phosphate (VP) and lithium-iron-vanadate-phosphate (LFVP) glasses. These materials are amorphous analogues of crystalline V_2O_5 or $LiFePO_4$ cathode materials for Li-ion batteries. The influence of synthesis and further thermal treatment conditions on electrical, thermal and electrochemical properties has been studied. In particular, a significant increase in electronic conductivity (up to $0{\cdot}7\times10^{-1}$ and 7×10^{-3} S/cm for VP and LFVP nanomaterials, respectively) was observed as a result of thermal nanocrystallisation. The microstructure has been observed and studied by a combination of XRD, SEM/TEM and EDX methods. Densely packed, small (15–30 nm and 5–15 nm in size) grains were observed in VP and LFVP materials, respectively. The significant increase in the conductivity is discussed in terms of Mott's model of electron hopping and a core-shell model.

1. Introduction

Nowadays, there is a high demand for a wide variety of advanced materials whose properties should fulfil given requirements, depending on potential applications. It is commonly known that nanomaterials usually exhibit different properties than similar materials in bulk, and this holds for cathode materials for Li-ion batteries. As has been shown a number of times, a nanostructured morphology of electrode (anode/cathode) materials leads to enhanced ionic/electronic transport,[1] and to improved electrochemical performance.[2] Such nanostructured materials can be prepared by a variety of methods, including very sophisticated and expensive ones, such as MBE (molecular beam epitaxy), MOCVD (metalorganic chemical vapour deposition) or nanolithography. We have proposed and explored another, much cheaper way to produce nanostructured electronic or mixed electronic–ionic conductors to be used as cathode materials in Li-ion cells. This involves carrying out a nanocrystallisation process with a glass of an appropriate composition. Therefore materials with improved electrical conductivity and good electrochemical performance are expected as a result of these studies.

Vanadium oxides have for a long time been interesting as alternative cathode materials for lithium and Li-ion batteries.[3] Recently Whittingham *et al* used some vanadium oxide additives to improve the electrical and electrochemical properties of $LiFePO_4$ olivines.[4,5] Crystalline forms of vanadates have been investigated for decades and are well characterised. Much less is known about the physical properties of their amorphous or nanostructured analogues.[6,7] Most of our studies have been focused on materials with nominal composition $90V_2O_5.10P_2O_5$ (VP).[7,8] The addition of P_2O_5, a supporting glass former, was necessary to prepare fully amorphous samples via a typical glass-making process like melt quenching.

Another compound very important as a cathode material for Li-ion batteries is $LiFePO_4$ olivine, known for its excellent electrochemical performance (Goodenough[9]) and intensively studied all over the world. Its main disadvantage, low electrical conductivity, has usually been circumvented by using thin carbon coating of individual grains of the materials. We have attempted to increase the conductivity of that material by thermal nanocrystallisation of glassy analogues of $LiFePO_4$ olivines. Due to the poor glass-forming properties of pure lithium iron phosphates, we had to add a fraction of a modest network former (V_2O_5), which could also contribute to the electronic conduction in the final nanocrystallised composites.

In this paper, we summarise our results on both VP and LFVP (Li_2O–FeO–V_2O_5–P_2O_5) glasses and nanomaterials. Even though the motivation is for applications, our main goal was to study the phenomenon of nanocrystallisation in glasses, and to gain an in-depth understanding of this process.

2. Nanocrystallisation in glassy systems

First reports on enhanced conductivity of thermally nanocrystallised vanadates were published in the late 1970s, when Limb and Davis[10] reported the electrical conductivity for V_2O_5–P_2O_5–B_2O_3 glasses after crystallisation at several different temperatures. The authors, however, did not measure the phenom-

Corresponding author. Email topie@if.pw.edu.pl
Original version presented at Int. Conf. on Phosphate Glasses, Pardubice, Czech Republic, 2–4 July 2014
DOI: 10.13036/17533562.57.3.038

enon *in-situ*, but only gave the correlation between conductivity and heat treatment.

Probably the first observation of the influence of nanocrystallisation in glass on its electrical properties was reported by Adams[11] in 1994. A suitable heat treatment of $AgI–Ag_2O–V_2O_5$ glass led to a significant increase in ionic conductivity on crystallisation. Annealing at temperature slightly greater than the glass transition temperature, T_g, led to the formation of $Ag_8I_4V_2O_7$ nanocrystallites, whilst annealing at a higher temperature led to a conductivity drop. It was shown later that the best performance was reached when the estimated surface-to-volume ratio was highest.[12]

Our studies on nanocrystallisation in $Li_2O–V_2O_5–P_2O_5$ glasses, which exhibit mixed ionic-electronic conductivity, started in 2004.[13] It was shown that heating of $15Li_2O.70V_2O_5.15P_2O_5$ glassy samples up to 385°C led to an increase in the conductivity by two orders of magnitude and a decrease in the activation energy. Similar preliminary studies on $90V_2O_5.10P_2O_5$ glasses[14,15] resulted in one order of magnitude increase in the conductivity. Further development of this technique led to nanocrystalline materials with electrical conductivity equal to ca. 10^{-3} S/cm.[7,8] Later, our studies showed that it is possible to obtain highly conductive V_2O_5 nanomaterials by a twin roller technique.[16] Furthermore, a correlation between the temperature of the melt and the electronic conductivity in nanocrystalline samples was observed. Higher synthesis temperatures resulted in materials that are better conducting. This is in good agreement with Szörenyi *et al*,[17] who reported greater oxygen loss in V_2O_5 melted at higher temperatures, causing greater concentration of V^{4+} ions, which plays an important role for hopping conductivity. This observation was applied to further improve the electrical properties of VP samples. These results are presented in this paper.

First attempts to obtain glassy and nanocrystalline analogues of $LiFePO_4$ olivines were made by Garbarczyk, Julien, and co-workers.[18–20] However, samples of NASICON (sodium super ionic conductor) structure were obtained, and the increase in the conductivity due to crystallisation was only one order of magnitude.[21] Nanocrystallisation in a simpler binary $Fe_2O_3–P_2O_5$ system (i.e. $FePO_4$) was also observed, but with moderate effect.[22] Further studies have shown that the addition of V_2O_5 to the ternary $Li_2O–FeO–P_2O_5$ system may have positive impact on electrical conductivity enhancement.[23] Preliminary experiments resulted in samples with conductivity approximately equal to 10^{-6} S/cm at room temperature (RT), three orders of magnitude better than the initial quaternary glass. Our recently published studies have shown that similar olivine-like materials with added V_2O_5 can exhibit electrical conductivity better than 10^{-3} S/cm at RT.[24,25]

Similar observations of the nanocrystallisation phenomenon were also made in a wide variety of glasses by others. El-Desoky and co-workers reported that heat-treated glassy samples of composition $75V_2O_5.10BaO.15Fe_2O_3$ with different sulphur additions exhibit conductivity ca. two orders of magnitude greater than initial samples.[26] The core-shell concept described in Ref. 7 was applied to explain a grain-size effect in $10BaTiO_3.70V_2O_5.20Bi_2O_3$ glasses.[27] Furthermore, DTA and impedance spectroscopy measurements were shown for glasses and nanomaterials of this system.[28] The effect of nanocrystallisation in $10ZnO.30Fe_2O_3.60P_2O_5$ glasses was studied by Moguš-Milanković *et al.*[29] Variation in the heat treatment affected the concentration of Fe^{2+}/Fe^{3+} hopping centres, and therefore the electronic conductivity of the samples, which varied from ca. 9×10^{-13} to 3×10^{-10} S/cm. Other interesting studies of electrical (ionic) conductivity at temperatures close to T_g were conducted by Rathore & Dalvi. A comparative investigation between glasses and nanomaterials in the system $Li_2SO_4–LiPO_3$, was reported in reference 30. In contrast to our research on electronic conductivity, Rathore and Dalvi reported that crystallisation in the $Li_2SO_4–Li_2O–P_2O_5$ glassy ionic system led to a decrease in the total conductivity of samples.[31] This was, however, a valuable study for deeper understanding of the nanocrystallisation process itself.

3. Experimental

3.1. Synthesis of vanadate phosphate glass

Amorphous samples of nominal composition $90V_2O_5.10P_2O_5$ were prepared from commercial predried chemicals, V_2O_5 (ABCR, 99·5%) and $(NH_4)H_2PO_4$ (POCh – Polish Chemicals, 99·5%), which were ground and mixed in an agate mortar, and divided into six batches, labelled as P_1,... P_6. Alumina crucibles filled with the powders were placed in an electric furnace, heated in air to a temperature in the range 800–1300°C, and held at that temperature for 15–60 min. The details of this procedure are presented in Table 1(a). The molten mixtures were rapidly poured onto a stainless steel plate held at RT, and immediately pressed with another identical plate. The average thickness of the resulting samples was ca 0·7 mm.

3.2. Synthesis of lithium iron vanadate phosphate glass

The synthesis of glassy samples from the $Li_2O–$

Table 1(a). Synthesis details of the VP samples

Batch ID	Melt temperature/°C	Melting time/min
P1	800	15
P2	950	15
P3	1100	15
P4	1100	60
P5	1200	60
P6	1300	60

$FeO–V_2O_5–P_2O_5$ system, with nominal compositions $LiFe_{1-2\cdot5x}V_xPO_4$, (x=0·08, 0·10 and 0·20; see Table 1(b) for details) was carried out in two stages. In the first stage, the starting reagents, Li_2CO_3 (Aldrich, 99·9%), $FeC_2O_4.2H_2O$ (Aldrich, 99·9%), V_2O_5 (ABCR, 99·5%) and $(NH_4)H_2PO_4$ (POCh – Polish Chemicals, 99·5%), were homogenised and calcined under a flow of nitrogen (purity 99·999%) in two steps: (i) at 300°C for 2 h, and (ii) at 570°C for the next 2 h. In the second stage, the ground powders were melted at 1300°C for 15 min, and quenched using stainless steel plates held at RT. In order to prevent batches from oxidising, a double-crucible technique was used. Further details of this method can be found elsewhere.[32]

3.2. Sample characterisation

X-ray diffraction (XRD) patterns were collected using a Philips X'Pert Pro diffractometer, equipped with a copper anode and a Ni filter. Cu K_α radiation of wavelength λ=1·542 Å was used in the experiments. Diffraction patterns confirmed the amorphous nature of the as-prepared material, and provided information on the average crystallite size in the samples after thermal treatment (by applying the Scherrer formula to the width of the reflections).

Thermal events occurring in the VP samples were observed using a TA Instruments Q200 differential scanning calorimeter, and DSC scans were carried out in a non-isothermal mode from ca. 40 to 550°C. For thermal analysis of LFVP samples, a TA Instruments SDT Q600 calorimeter was used in temperature range up to 800°C. In both cases, measurements were carried out at a series of heating rates (in the range 1–40°C/min).

The electrical conductivity measurements were carried out by impedance spectroscopy. The setup was based on a Solartron 1260 Gain Phase/Impedance Analyser, integrated with an oven and modules for temperature programming and stabilisation (Eurotherm 2404). The samples were heated from RT up to a wide variety of temperatures, and subsequently cooled down to RT. The electronic conductivity was measured *in-situ* during this heat treatment, in order to observe the influence of the nanocrystallisation on the electronic conductivity and to correlate the electrical results with the thermal analyses. For the LVFP samples, each value of the conductivity at a particular temperature was determined from an impedance spectrum (e.g. Figure 16). To collect a spectrum of good quality, the measurement was taken at strictly isothermal conditions, with the temperature stabilised for at least 40 min prior to the electrical measurement. The acquisition of each spectrum took another ca. 30 minutes. At the very end of spectrum acquisition, a brief measurement (1 Hz–1 MHz, 2 points per decade) was taken to ensure that there had been no changes in the impedance response during the whole measurement. Then the temperature of the sample was changed with a 25°C step. For the VP purely electronically conducting samples, each value of the conductivity at a particular temperature was measured at one fixed frequency (100 Hz). This made it possible to keep a nearly constant heating rate of 1°C/min with a much shorter measurement than in the case of the LVFP samples. The samples were kept at maximum temperature no longer than a few minutes, before the cooling stage started. In our earlier paper we showed that this measurement procedure gives exactly the same results as the measurement of complete impedance spectra at step-like stabilised temperatures.[7] In addition to this, we have shown that the time for which the samples are kept at maximum temperature does not matter for temperatures above 255°C.[33] The conductivity of some of the best-conducting samples was also measured in the low temperature range (down to −165°C). For this experiment, a nitrogen cryostat was used. The temperature was stabilised by a Lakeshore 331 controller.

Table 1(b). Synthesis details of the LFVP samples

Batch ID	*Nominal composition*	*Normalised composition*
F08	$LiFe_{0\cdot80}V_{0\cdot08}PO_4$	$27\cdot1Li_2O.43\cdot6FeO.2\cdot2V_2O_5.27\cdot1P_2O_5$
F10	$LiFe_{0\cdot75}V_{0\cdot10}PO_4$	$27\cdot8Li_2O.41\cdot6FeO.2\cdot8V_2O_5.27\cdot8P_2O_5$
F20	$LiFe_{0\cdot50}V_{0\cdot20}PO_4$	$31\cdot25Li_2O.31\cdot25FeO.6\cdot25V_2O_5.31\cdot25P_2O_5$

Preliminary galvanostatic charge–discharge cycling tests were performed for the VP sample with the highest conductivity (i.e. P_6). About 5 g of the powdered glassy sample was heated to 450°C in a tube furnace, with a thermal program similar to the one used in the electrical measurements. Then the sample was ball-milled with a Retch PM100 planetary mill. The cathodes for these tests were prepared using 70 wt% active material, 20 wt% carbon black (CB) and 10 wt% binder (PVDF – polyvinylidene difluoride). Metallic lithium plates were used as anodes. Details of the procedure can be found elsewhere.[16] The current is given in C units, where C corresponds to the current that would fully discharge a battery of a maximum theoretical capacity in one hour. The maximum theoretical capacity of VP samples (Q_{max}=398 mAh/g) was calculated assuming that lithium ions can be intercalated only into V_2O_5, but not P_2O_5. Therefore $C/10$ rate corresponds to 39·8 mA current per 1 g of active material in the cathode. Several initial charge/discharge cycles were carried out at a C/20 current rate in the range 1·5–4·0 V using an Autolab PGSTAT30. In order to test the cyclability of the cell, the rate was then increased to C/10 and tens of cycles were performed. Additionally, another cell was discharged with $C/20$ current without a voltage cut-off, until maximum theoretical capacity was reached. This was to determine the maximum amount of lithium that can be intercalated into the structure.

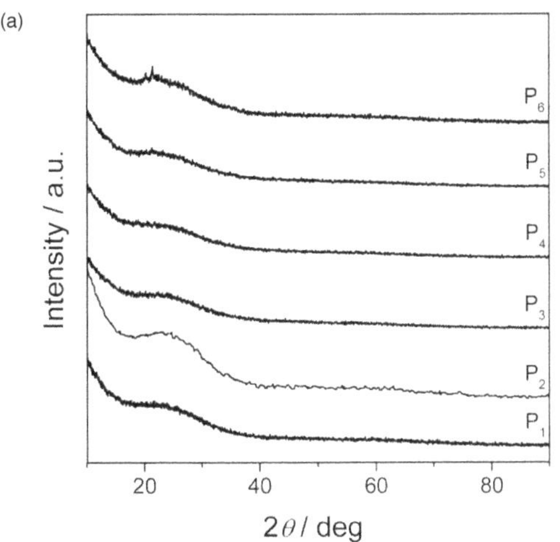

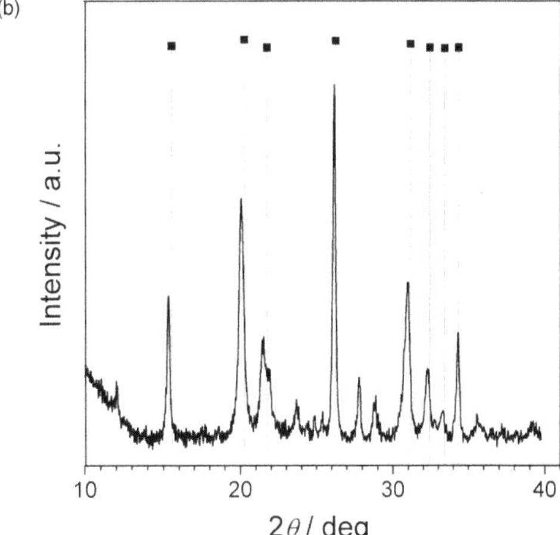

Figure 1. XRD patterns of (a) the as-prepared VP samples, and (b) sample P6 after nanocrystallisation. Vertical lines indicate peak positions for the orthorhombic V_2O_5 structure

4. Results

4.1. Vanadate-phosphate samples (VP)

4.1.1 XRD

Powder diffraction patterns of as-prepared samples (Figure 1(a)) proved that the samples were fully amorphous. One can observe a wide halo at low angles related to short-range order in glassy materials. No Bragg reflections were observed, except for a tiny peak in the case of batch P_6.

The diffraction pattern of the active material is typical for a nanocrystalline material (Figure 1(b)). The peaks are broad and have relatively high intensity. The main peaks were identified with the orthorhombic V_2O_5 structure (JCPDS-ICDD card no. 00-041-1426). Additional minor peaks may be ascribed to the monoclinic VO_2 phase. The grain size, as determined from the broadening of several lines,[34] varies from 15 to 30 nm. The position, *hkl* indices, and respective grain size are summarised in Table 2. The average grain size is 21 nm.

Table 2. Peak position, hkl indices and respective grain size estimated using the Scherrer formula, for the active material of the VP cathode

Peak position (degrees)	hkl	Estimated grain size (nm)
15·412	200	28
20·290	010	15
21·539	101	13
26·480	110	31
30·935	400	17
47·113	600	18

4.1.2. DSC

In the DSC curves for all the as-prepared samples (Figure 2), a discernible baseline shift (characteristic of a glass transition) was followed by an exothermic peak (crystallisation). The temperatures of these transitions, T_g and T_c, respectively, increased nearly linearly with increasing synthesis temperature (Figure 3).

It is commonly known that the temperatures of the thermal events in DSC measurements depend on the heating rate. This behaviour has also been shown in our recent papers.[7,8,22] There is a very simple, but efficient formula, proposed by Lasocka *et al*,[35] describing the behaviour of T_g versus heating rate:

$$T_g=A+B\ln\theta \quad (1)$$

where A and B are empirical constants, and θ is the heating rate in °C/min. A practical advantage of this expression, in addition to its simplicity, is that the fitting parameter A equals the glass transition temperature for a heating rate θ=1°C/min. Experimental data presented in appropriate coordinates (Figure 4) show that Equation (1) can be successfully applied to describe the glass transition temperature versus the heating rate in this case. Furthermore, the slope of the fitted lines is comparable for nearly all the as-prepared samples.

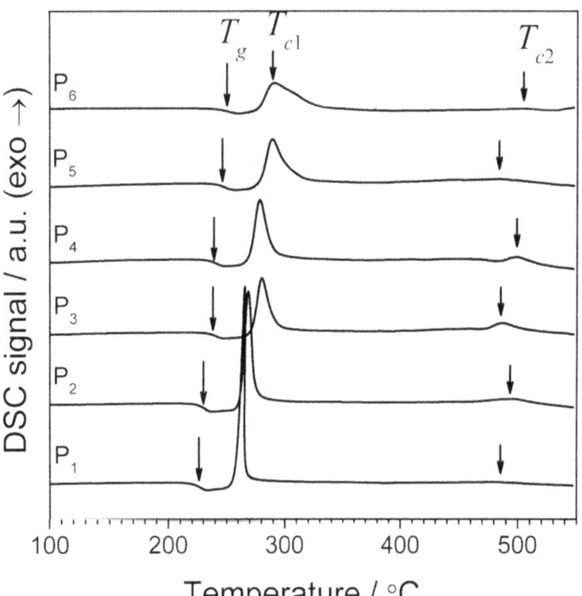

Figure 2. DSC curves of the VP samples (heating rate 10°C/min)

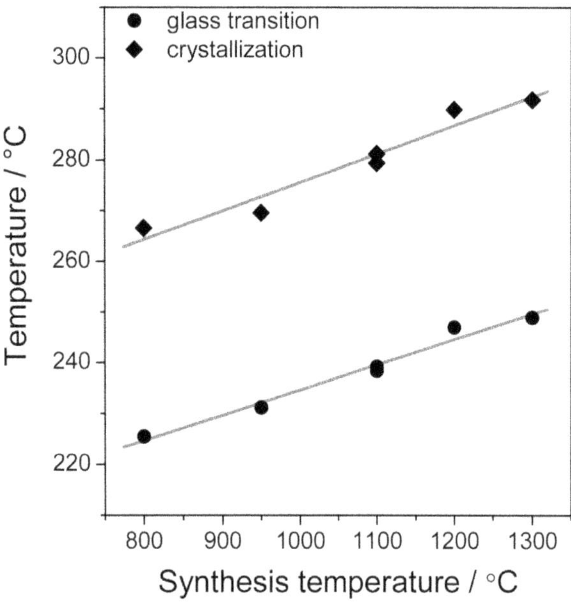

Figure 3. Experimental dependencies of the glass transition and crystallisation temperatures on the glass synthesis temperature for the VP samples

Crystallisation phenomena occurring during non-isothermal heating of glasses have been successfully represented by the Kissinger formula:[36]

$$\ln\left(\frac{\theta}{T_{\mathrm{m}}^{2}}\right) = -\frac{Q}{k_{\mathrm{B}}T} + C \tag{2}$$

where T_m is the temperature of the maximum of the crystallisation peak, k_B is Boltzmann's constant, Q is the activation energy of the crystallisation process, and C is a constant. In some cases, the Kissinger

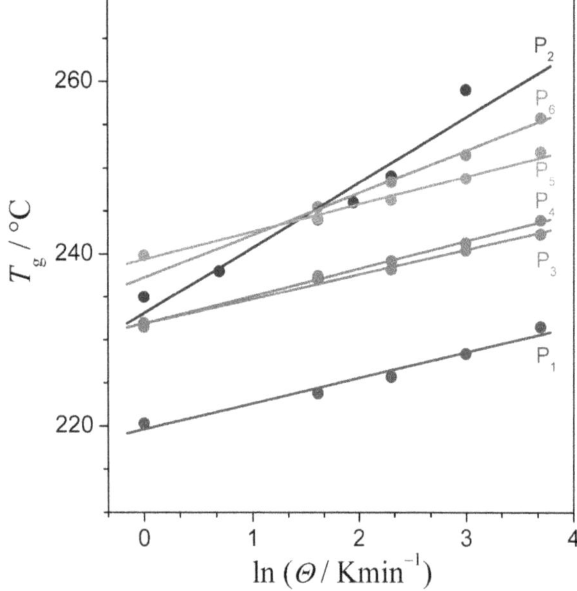

Figure 4. The glass transition temperatures of the VP samples determined from DSC measurements versus the heating rate θ. Linearity of the data is evidence that Lasocka's equation can describe the variation of T_g in DSC experiments [Colour available online]

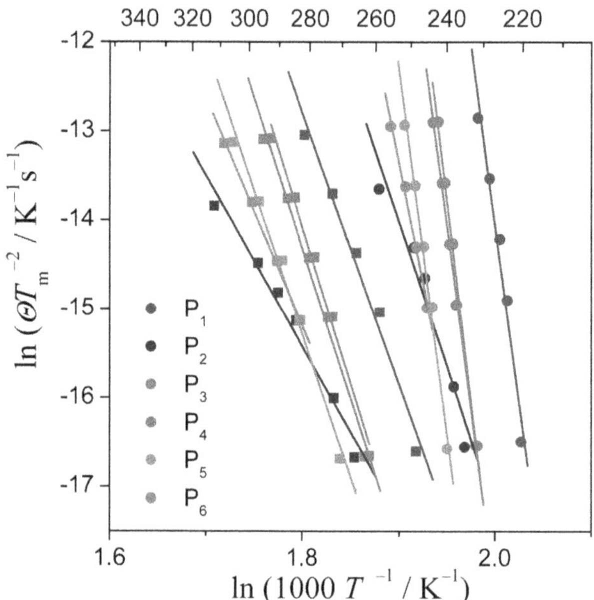

Figure 5. Kissinger plot for VP samples of the glass transition (circles) and crystallisation (squares) temperatures observed in DSC measurements. Slopes of the fitted lines yield the activation energy of the corresponding thermal processes [Colour available online]

Table 3. Activation energy for glass transition and crystallisation calculated according to Kissinger's formula

Batch ID the glass	*Activation energy of transition/eV*	*Activation energy of crystallisation/eV*
P_1	6·9(7)	2·60(20)
P_2	2·8(4)	1·66(13)
P_3	7·6(6)	3·02(18)
P_4	6·94(13)	2·91(18)
P_5	7·1(4)	2·77(18)
P_6	4·7(3)	2·21(8)

formula may also be used to describe the dependence of the glass transition temperature on the heating rate.[37] For the VP glasses studied in this work, the Kissinger equation applies satisfactorily to both the glass transition and the crystallisation phenomena (Figure 5). The values of the activation energies for the glass transition and for crystallisation are given in Table 3.

4.1.3. Electronic conductivity

The conductivity of as-prepared glassy samples was measured at RT for several samples of each batch. The average values of the conductivity (20–80 μS/cm) are presented in Figure 6 and in Table 4. The electronic conductivities of the glasses increase linearly with synthesis temperature (Figure 6).

As mentioned in the Experimental section, each as-prepared sample was heated to a certain maximum temperature and then cooled down to RT. The maximum temperature varied from 300 to 550°C. The final values of the conductivity at RT of nanocrystalline samples, which underwent different thermal treatment, are shown in Figure 7. The values were

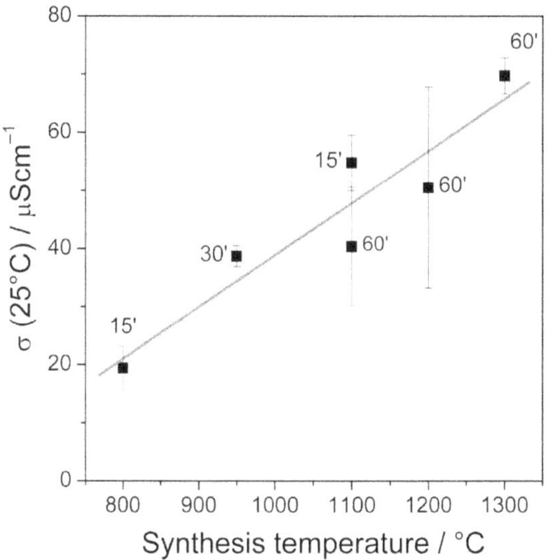

Figure 6. Dependence of the electronic conductivity at RT of the as-prepared VP samples (glasses) on synthesis temperature

Table 4. Electronic conductivity for glassy and the best nanocrystallised samples from all the batches. The maximal temperature of optimal nanocrystallisation is given

Batch ID	Electronic conductivity of glass at 25°C (mS/cm)	Electronic conductivity of nanomaterial at 25°C (mS/cm)	Maximal temperature of optimal nanocrystallisation (°C)
P1	0·019(4)	0·86	450
P2	0·039(2)	2·0	450
P3	0·055(5)	6·9	425
P4	0·04(1)	3·2	425
P5	0·05(1)	3·7	475
P6	0·070(3)	73	550

calculated (by extrapolation) from the temperature dependencies of electronic conductivity measured *in-situ* upon cooling. In general, the greatest increase was observed for sample P_6, synthesised at 1300°C.

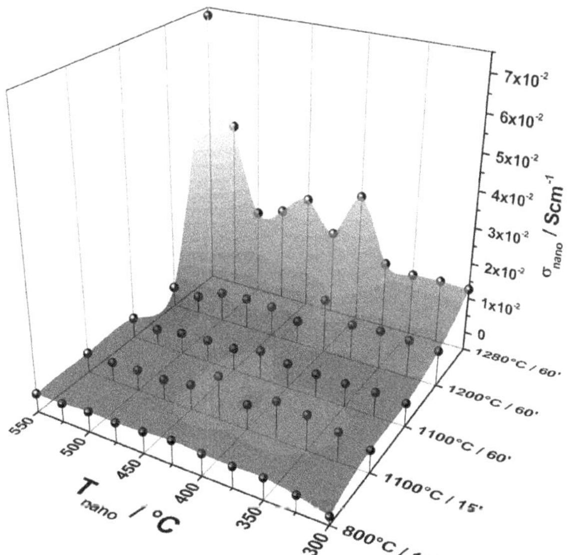

Figure 7. The electronic conductivity (at 25°C) of VP samples after thermal nanocrystallisation [Colour available online]

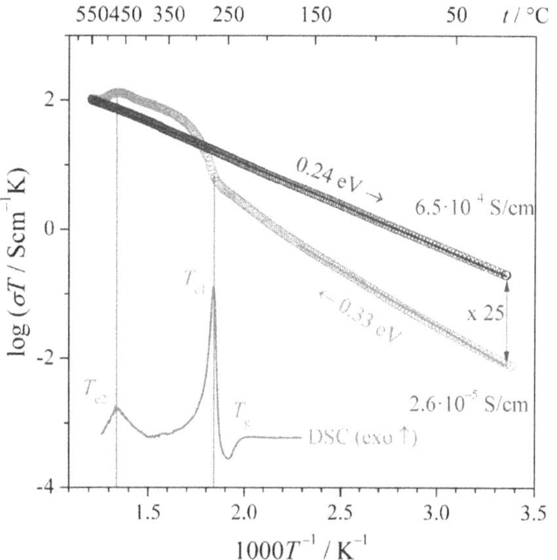

Figure 8. Temperature dependence of the electronic conductivity of sample P_5 upon heating to 550°C and cooling. The conductivity drop is clearly correlated with the second crystallisation peak [Colour available online]

Moreover, the ratio exhibited a sort of a plateau for moderate thermal treatment temperatures (ca. 400°C).

The temperature dependence of the electronic conductivity follows the Arrhenius formula:

$$\sigma(T) = \frac{\sigma_0}{T}\exp\left(-\frac{E_a}{k_B T}\right) \quad (3)$$

where E_a is an activation energy and σ_0 is a constant. The sample P_6, heated up to 550°C and then cooled, exhibited the highest conductivity of all the samples, 0·07 S/cm at 25°C. The significant increase in the conductivity of all the samples is correlated with the

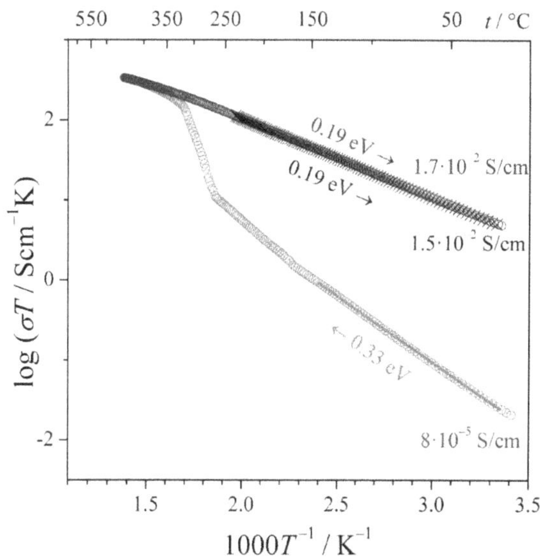

Figure 9. Temperature dependence of the electronic conductivity for (i) sample P_6, upon heating to 460°C (1°C/min) and cooling down to RT during the electric measurements (circles), and (ii) active material for electrochemical measurements (crosses) [Colour available online]

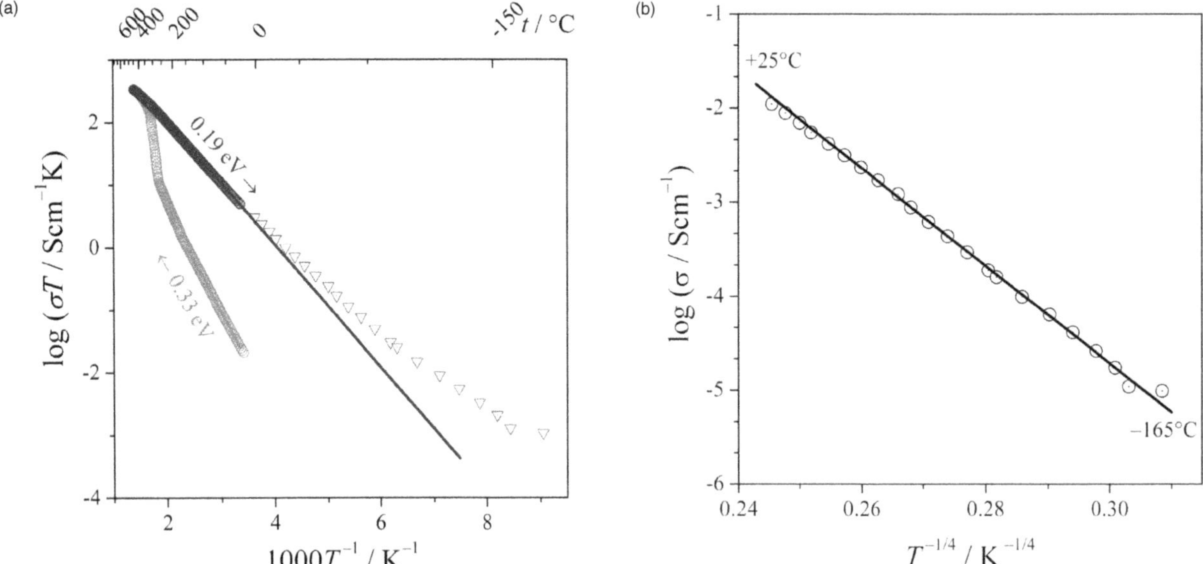

Figure 10. Electronic conductivity for sample P6 heated up to 460°C and cooled down to −165°C: (a) data presented in an Arrhenius plot – systematic deviation from linear behaviour is observed; (b) data presented in a Mott plot [Colour available online]

crystallisation process observed in the DSC traces. The thermal treatment was repeated for several samples, with a very good reproducibility of the results. However, for sample P_5, which was synthesised at 1200°C, one can observe a distinct conductivity drop at high temperatures (Figure 8). This overheating phenomenon was also observed in other batches (P_1–P_4) synthesised at lower temperatures.

The reproducibility of the nanocrystallisation described previously for small samples was checked for larger amount of the sample P_6 (ca. 5 g) in another tube furnace. The conductivity (at RT) and the activation energy of the sample heated up to 460°C in this tube furnace were $1{\cdot}5\times10^{-2}$ S/cm and 0·19 eV, respectively – very similar to the sample heated to the same temperature during impedance measurement (Figure 9).

Below +25°C, the dependence of the conductivity deviates from the Arrhenius formula (Figure 10(a)). These data, when plotted as logσ versus $T^{-1/4}$, lie on a straight line (Figure 10(b)). This is strong evidence of good agreement with Mott's formula for conductivity based on variable range hopping:[38]

$$\sigma(T)=\sigma_{VRH}\exp(-B/T^{-1/4}) \qquad (4)$$

4.1.4. SEM images

In SEM images of sample P_6 after nanocrystallisation at 400°C (Figure 11(a)), one can see nanocrystallites with a relatively uniform size below 100 nm. The shells seem to be strongly distorted, and contacts between the grains are good. In addition to this, some interesting nanorod-like features were observed in several samples (Figure 11(b)).

4.1.5. Electrochemical properties

The gravimetric capacity determined in the first discharge cycle at a C/20 rate was 225 mAh/g, and it

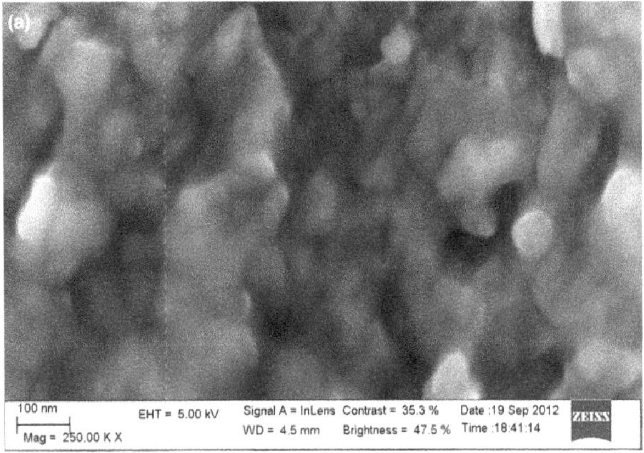

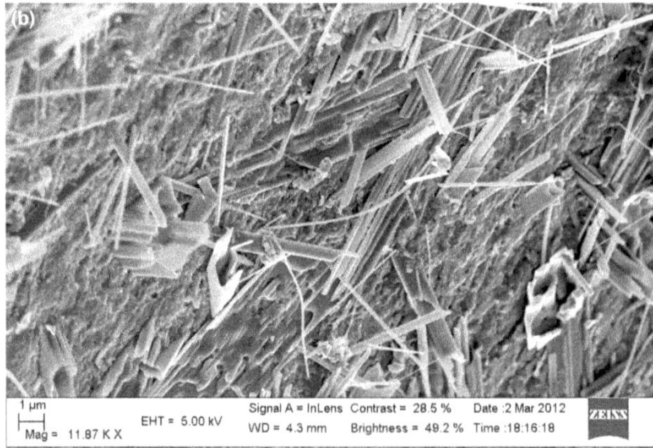

Figure 11. (a) SEM image of sample P_6 after nanocrystallisation at 400°C during impedance measurements; (b) V_2O_5 nanorods observed in a SEM image of sample P_6 after nanocrystallisation at 550°C during impedance measurements

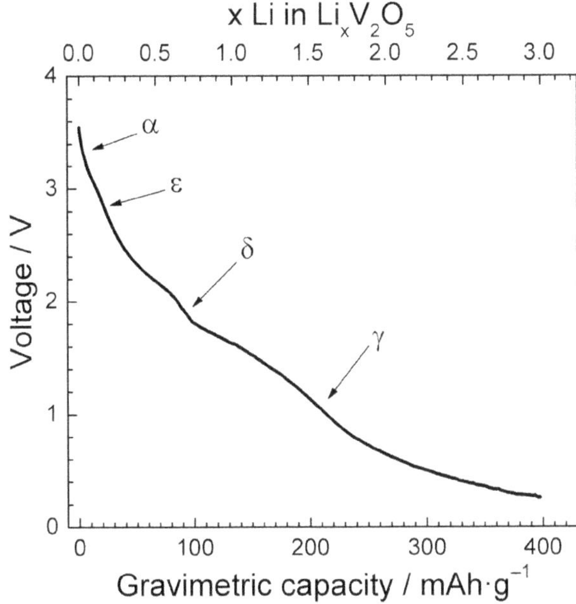

Figure 12. Discharge curve with C/20 current for a laboratory cell with nanocrystallised P_6 sample as a cathode. V_2O_5 phase transitions are observable. Phases α, ε, δ, γ were ascribed on the basis of Chernova et al[3] and Delmas et al[39]

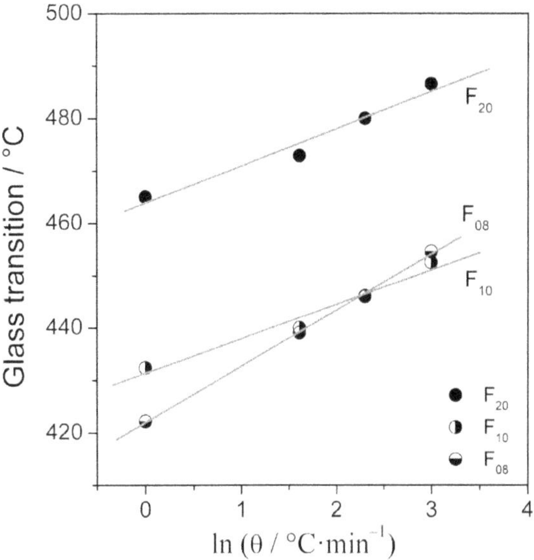

Figure 14. The glass transition temperatures in LFVP glasses determined from DSC measurements versus the heating rate, θ

dropped to 140 mAh/g after three cycles. The gravimetric capacity for C/10 current was ca. 90 mAh/g. However, the reversibility of the process for C/20 and C/10 was very good.

A full discharge cycle (Figure 12) showed that the maximum theoretical gravimetric capacity (398 mAh/g) can be reached for C/20 current. However, this caused a substantial voltage drop down to 0·4 V. Several smooth steps are observed in the discharge curve. This is characteristic for V_2O_5 crystallographic phase transitions reported for V_2O_5 upon intercalation of lithium.[3,39]

4.2. Lithium iron vanadate-phosphate samples

4.2.1. XRD and DTA

XRD measurements confirmed that all batches were entirely amorphous. For nanocrystalline samples, broad reflections were observed, and two phases were identified: $LiFePO_4$ triphylite and $Li_3V_2(PO_4)_3$ NASICON phase.

DTA curves for LFVP samples F_{08}, F_{10} and F_{20}, recorded at various heating rates, are shown in Figs. 13(a)–(c). The T_g values are in good agreement with Lasocka's formula (Figure 14) and both T_g and T_c can be represented in a Kissinger plot (Figures 15(a)–(c)). The activation energy of the glass transition is significantly greater than the activation energy for crystallisation (e.g. 6·36 and 3·98 eV, respectively, for F_{10}).

4.2.2. Electrical conductivity

Impedance spectra of the samples upon heating and cooling are presented in Figure 16, and show that the samples nanocrystallised after heating exhibit higher conductivity than the starting amorphous samples. Additionally, impedance spectra of the nanocrystallised samples contain a low frequency spur, absent in

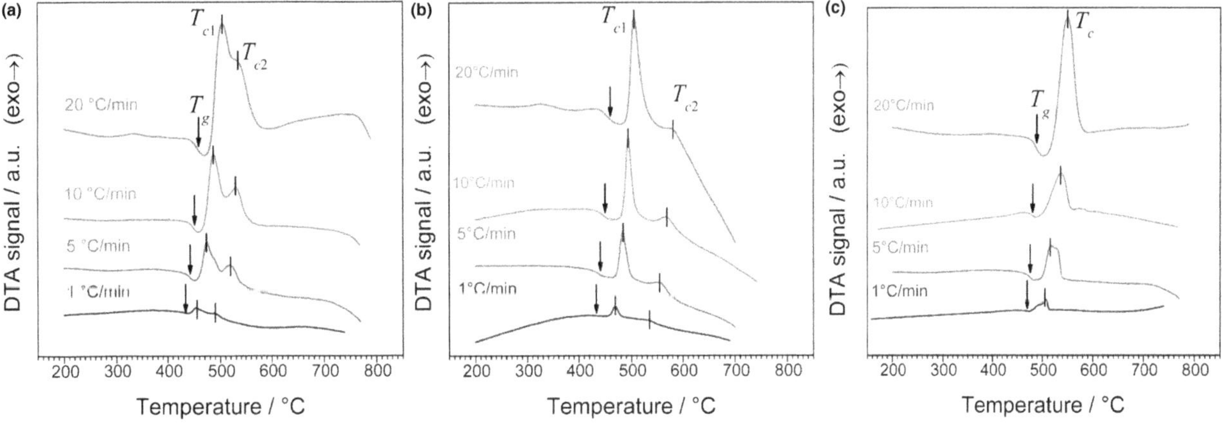

Figure 13. DTA scans for LFVP glasses F08 (a), F10 (b), and F20 (c) [Colour available online]

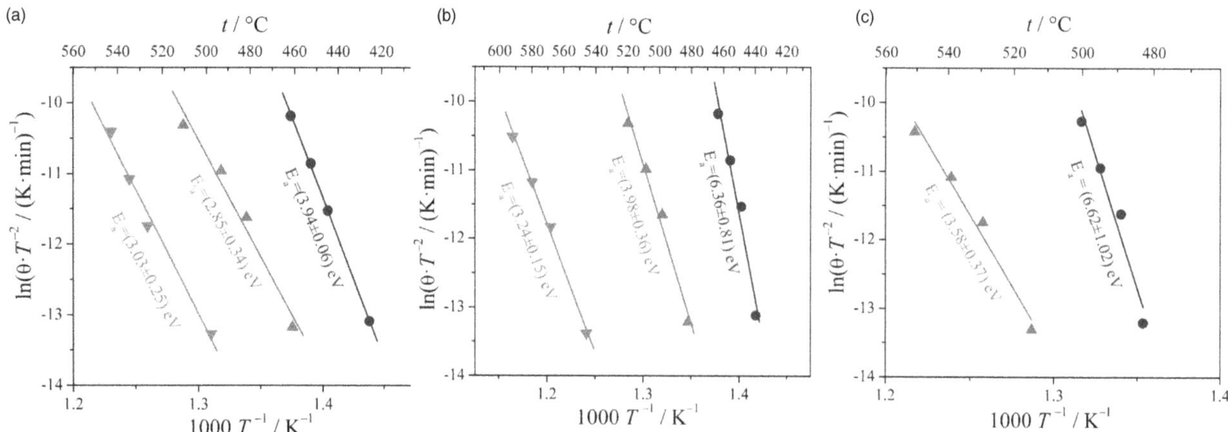

Figure 15. Kissinger plot for the glass transition (circles) and crystallisation (triangles) temperatures observed in DTA measurements on LFVP samples F08 (a), F10 (b), and F20 (c) [Colour available online]

the original spectra. This is due to blocking of lithium ions at the sample/electrode interface. We conclude that nanocrystallisation in the sample not only improved electronic conductivity, but also increased lithium ion mobility.

There is a big difference in the effect of nanocrystallisation for samples F_{08} and F_{20}. The detailed results for sample F_{10} have already been published,[24] but they are much more similar to those for sample F_{08}. The conductivity of the best-conducting samples, F_{08} and F_{10} after nanocrystallisation, was as high as $7{\cdot}3\times10^{-3}$ S/cm (E_a=0·11 eV) and 3×10^{-3} S/cm (E_a=0·17 eV[24]), respectively, whereas the best conducting F_{20} sample exhibited a modest conductivity of $1{\cdot}3\times10^{-6}$ S/cm (E_a=0·53 eV).

4.2.3. SEM/TEM images

The microsctructure of well conducting LFVP samples (F_{08} and F_{10}) consists of $LiFePO_4$ nanograins of size 5–15 nm, embedded in a highly disordered glassy matrix. Images for sample F_{10} have already been published.[24] Here we present the STEM (scanning transmission electron microscope) image of sample F_{08} after nanocrystallisation (Figure 17). One can see randomly but densely distributed 5–10 nm grains, which are in good contact. Such a structure is strongly required for good electronic conductivity. For sample F_{20}, larger crystalline grains were observed.

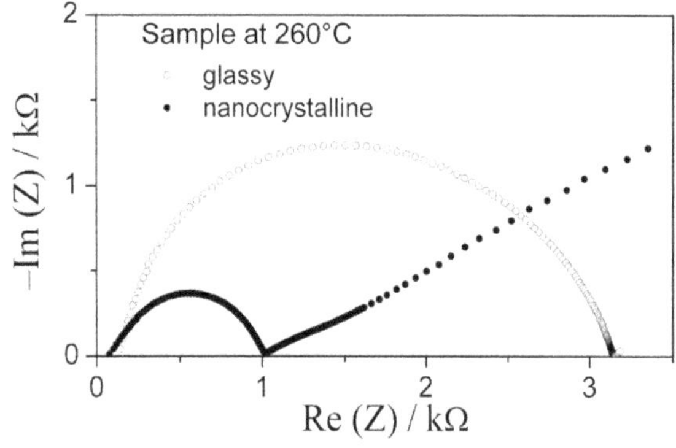

Figure 16. Impedance spectra taken at 260°C upon heating a LFVP glassy sample, and cooling the nanocrystalline sample F_{08}

5. Discussion

It is known that keeping batches of molten V_2O_5 at very high temperatures leads to oxygen loss according to:[17]

$$V_2O_5 \leftrightarrow V_2O_{5-x} + \frac{x}{2}O_2 \uparrow \qquad (5)$$

This is accompanied by reduction of a fraction of V^{5+} to V^{4+} ions, and therefore has a positive impact on the concentration of V^{5+}/V^{4+} hopping centres. One may therefore expect that higher melting temperature will result in a larger concentration of hopping centres.

The expression for electronic conductivity in glasses containing transition metal oxides (in particular, vanadium) was given by Mott[38,40]

$$\sigma = \nu_{el} c(1-c)\frac{e^2}{Rk_BT}\exp(-2\alpha R)\exp\left(-\frac{E_a}{k_BT}\right) \qquad (6)$$

where R is the average distance between hopping centres, α, is the inverse localisation length of the

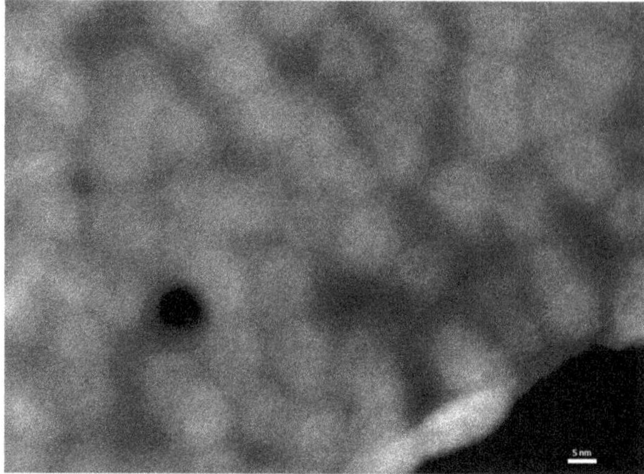

Figure 17. STEM image of the LFVP sample F08 after nanocrystallisation. The scale bar corresponds to 5 nm

electron wave function, $c=[V^{4+}]/([V^{4+}]+[V^{5+}])$ is the fraction of occupied hopping sites for electrons, and E_a is the activation energy for electronic conduction. Assuming that all the pre-exponential factors are constant, one obtains the Arrhenius formula Equation (3).

The values of the electronic conductivity, glass transition temperature, and crystallisation temperature for VP batches synthesised at various temperatures are in a good agreement with our predictions and previous studies. An increase in the synthesis temperature led to an increase in the concentration of V^{4+} ions. This causes an increase in the $c(1-c)$ term in Equation (6), and consequently results in higher electronic conductivity. The presence of V^{4+} ions was also confirmed by the XRD patterns of the nanocrystallised samples (e.g. Figure 1(b)), in which minor peaks corresponding to the monoclinic VO_2 phase were observed.

Thermal events for glassy samples upon heating were shifted towards higher temperatures for larger heating rates. This behaviour is satisfactorily described by Kissinger's formula. For the glass transition temperature, fitting with Lasocka's formula also gave good results. The reason why glasses melted at higher temperatures exhibit higher T_g and T_c is not completely clear. Doupovec *et al* have shown that the presence of Fe^{2+}/Fe^{3+} aliovalent ions has a significant influence on T_c.[41] This behaviour may also be adopted in the case of V^{4+}/V^{5+} ions. Additionally, melting at high temperatures may have resulted in sample contamination with Al_2O_3. However, its presence was not measured in a quantitative way. Al_2O_3 may result in higher T_g and T_c, but in this case it is rather unlikely to be the only reason for such behaviour in the VP samples.

The XRD and SEM measurements confirm that the highly conducting VP samples contain V_2O_5 nanocrystallites. The larger mean grain size observed by SEM than by XRD can be explained in terms of a core-shell model.[7,8] In this concept, grains consist of a crystalline core (observed by XRD) covered by a disordered shell (observed in SEM images). This issue has been discussed elsewhere in more detail, e.g. in Pietrzak *et al*.[7] However, the shape of the crystallites is not spherical, but rather cuboidal. Glushenkov *et al*[42] have shown that V_2O_5 nanostructures prefer to grow along the [010] axis. The variation of grain sizes in different planes presented in Table 2 is in reasonable agreement with this idea, as the estimated grain size in the (200) plane is twice as big as in the (010) plane. The presence of structures that resemble nanorods in some of the samples is not surprising, as V_2O_5 exhibit a readiness to crystallise in a wide variety of nanostructures, e.g. VONTs (vanadium oxide nanotubes),[3] nanobelts,[43] nanorods,[42] or even nanoflowers.[44] It would be interesting to determine the conditions under which the probability of producing such nanorods is high, but this is beyond the scope of this research.

The influence of nanocrystallisation on the electronic conductivity in vanadate-phosphate glasses has been discussed by us previously.[7,8,33] We have shown that Equations (4) and (6), proposed by Mott for glassy materials, can also be applied to nanomaterials obtained by nanocrystallisation of these glasses. In our model, the concentration of hopping pairs (i.e. V^{5+}/V^{4+}) is greater in the highly disordered surfaces of nanograins. Therefore tightly packed small grains with a high volume fraction of surface areas that are in good contact (i.e. giving conditions for fast electronic transport) result in high electronic (i.e. hopping) conductivity. Previously we studied nanocrystallisation in $90V_2O_5.10P_2O_5$ glasses synthesised only at one fixed temperature (950°C). This approach resulted in a sample with a conductivity of 2×10^{-3} S/cm. In this paper, further optimisation of the nanocrystallisation process included variation of the synthesis temperatures. As we had expected, higher synthesis temperature (which resulted in a higher concentration of V^{4+} hopping centres) led to higher conductivity enhancement after nanocrystallisation. As a result, our systematic research led to a further increase in conductivity for nanocrystalline VP material up to 0·07 S/cm at RT. Further systematic studies resulted in the synthesis of highly-conducting LFVP nanomaterials (F_{08} and F_{20}).

The conductivity enhancement in LVFP samples induced by nanocrystallisation is attributed to the appearance of an extended network of interfacial regions around the 5 nm grains following the nanocrystallisation process. While $LiFePO_4$ grains should be a good place for lithium storage and exchange, the surrounding matrix with a high concentration of Fe^{2+}/Fe^{3+} and V^{4+}/V^{5+} hopping centres is essential for good electronic conductivity. The presence of another phase, $Li_3V_2(PO_4)_3$, cannot account for the increase, because this phase is relatively poorly conducting (2×10^{-7} S/cm[45]). Therefore, it is evident that the microstructure plays a decisive role in the effect. For nanocrystallised glasses, a morphology involving a dense packing of small nanograins in a disordered matrix is required. These two requirements (small, densely distributed nanograins) are successfully met in the nanocrystallised samples F_{08} and F_{10}, resulting in high electrical conductivity. The nanograins are more densely packed in F_{08} than in F_{10}, which results in an even better electrical conductivity for the F_{08} sample. This hypothesis is also a good explanation for the modest conductivity of sample F_{20}, whose microstructure does not consist of small nanograins.

The temperature dependence of the electrical conductivities of the best conducting nanocrystalline samples, P_6, F_{08}, F_{10} (from Pietrzak *et al*[24]) and F_{20} are compared in Figure 18. An immense increase in the conductivity in various samples (e.g. $LiFePO_4$) is often ascribed to the presence of metallic or carbon impuri-

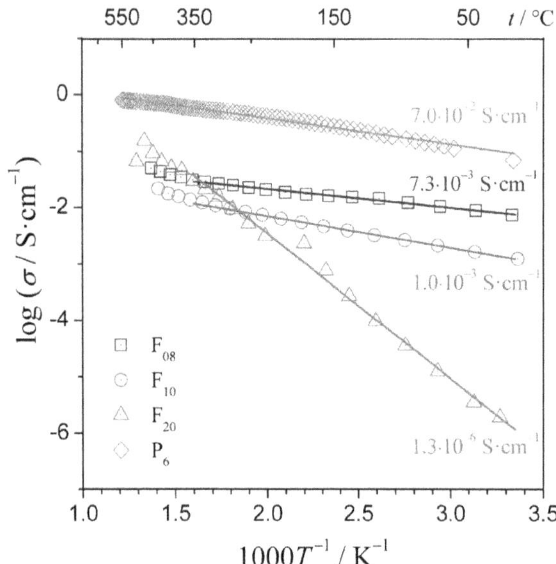

Figure 18. Comparison of electrical conductivity of the best-conducting VP and LFVP samples [Colour available online]

ties, originating from preparation routines.[46,47,48] To reject this hypothesis for the samples under study, their conductivity was measured down to −165°C. The results showed a systematic deviation from the Arrhenius formula. All data points measured below RT are well described by Mott's formula for variable range hopping. Such behaviour is characteristic of samples with hopping, and not metallic conductivity.[16]

While disordered materials are generally disregarded as cathodes, due to structural limitations of lithium diffusion, some authors (e.g. Ceder[49,50]) report that disordered materials can actually exhibit better electrochemical performance than ordered polycrystalline ones.

The main goal of the research reported here was to better understand the nanocrystallisation process in vanadate-phosphate and lithium iron vanadate-phosphate glasses, and to obtain high conductivity material, rather than to focus on optimising the electrochemical performance of laboratory cells. Nonetheless, preliminary electrochemical measurements have shown that the VP sample exhibits readiness to work as a cathode for lithium ion batteries, and three lithium ions per V_2O_5 can be intercalated into the microstructure. The advantage of the obtained nanomaterial is that voltage drops corresponding to phase transitions of vanadium pentoxide, which are distinct for crystalline V_2O_5,[3,39] in this case are smooth (Figure 12). The electrochemical characterisation of LFVP nanomaterials is expected in the near future.

6. Conclusions

The most important outcomes of this research can be summarised as follows:

- The temperature and the melting time of the samples have significant influence on the electronic conductivity and the glass transition temperature, T_g, and crystallisation temperature, T_c, of vanadate-phosphate (VP) glass samples. Higher synthesis temperature leads to a better electronic conductivity of the glass, and higher T_g and T_c.
- A high electronic conductivity, σ(25°)=0·07 S/cm, and low activation energy, E_a=0·13 eV, were observed for a VP sample synthesised at 1300°C and nanocrystallised at 550°C. The best conducting nanocrystalline lithium iron vanadate-phosphate (LFVP) sample, F_{08}, had $\sigma(25°)\approx7\times10^{-3}$ S/cm. The conductivity dependence at low temperature can be described by Mott's theory for variable range hopping. This proves that the high conductivity cannot be ascribed to metallic impurities or internal short circuit.
- The microstructure of the samples consists of nanocrystallites with size less than 100 nm. The average size of nanograins in the sample P_6 selected for electrochemical measurements was 20(7) nm. In some VP samples, structures similar to nanorods were observed. Randomly distributed, close-packed grains of size 5–15 nm were observed in the best-conducting LFVP sample. These self-consistent observations confirm that the presence of nanograins is a key factor for enhanced electronic conductivity.
- Preliminary electrochemical measurements showed good capability of a high conducting sample to intercalate three Li ions per V_2O_5 into the microstructure. The gravimetric capacity for 1·5–4·0 V range was moderate and equal to ca. 225 mAh/g for C/20 current. This is probably due to high loading of the cathode layer, and in all likelihood can be optimised.
- While the results presented here are for nanocrystallisation of selected glassy systems, we believe that this technique can be extended to a wide variety of systems in order to obtain nanomaterials with improved properties.

Acknowledgements

This work has been financed by the National Science Centre: project OPUS-4 no. DEC–2012/07/B/ST5/03184 (2013–2016). We are thankful to Dr Aldona Zalewska (Faculty of Chemistry, Warsaw University of Technology) for providing us with $LiPF_6$ electrolyte for electrochemical measurements. We wish to acknowledge Łukasz Pawliszak, Agata Dorau, Anna Kaleta and Przemysław P. Michalski for their participation in this research.

References

1. Malik, R., Burch, D., Bazant, M. & Ceder, G. Particle size dependence of the ionic diffusivity. *Nano Lett.*, 2010, **10**, 4123–7.
2. Chou, S.-L., Wang, J.-Z. Wang, Sun, J.-Z., Wexler, D., Forsyth, M., Liu,

H.-K., MacFarlane, D. R. & Dou, S.-X. High capacity, safety, and enhanced cyclability of Lithium metal battery using a V_2O_5 nanomaterial cathode and room temperature ionic liquid electrolyte. *Chem. Mater.*, 2008, **20**, 7044–51.
3. Chernova, N. A., Roppolo, M., Dillon, A. C. & Whittingham, M. S. Layered vanadium and molybdenum oxides: batteries and electrochromics. *J. Mater. Chem.*, 2009, **19**, 2526–52.
4. Hong, J., Wang, C. S., Chen, X., S. Upreti, M.S. Whittingham, Vanadium modified $LiFePO_4$ cathode for Li-ion batteries. *Electrochem. Solid St.*, 2009, **12**, A33–8.
5. Omenya, F., Chernova, N. A., Upreti, S., Zavalij, P. Y., Nam, K.-W., Yang, X.-Q. & Whittingham, M. S. Can vanadium be substituted into $LiFePO_4$? *Chem. Mater.*, 2012, **23**, 4733–40.
6. Murawski, L., Sanchez, C., Livage, J. & Audiere, J. P. Small polaron transport in amorphous V_2O_5 films. *J. Non-Cryst. Solids*, 1990, **124**, 71–75.
7. Pietrzak, T. K., Garbarczyk, J. E., Gorzkowska, I., Wasiucionek, M, Nowiński, J. L., Gierlotka, S. & Jóźwiak, P. Correlation between electrical properties and microstructure of nanocrystallised V_2O_5–P_2O_5 glasses. *J. Power Sources*, 2009, **194**, 73–80.
8. Pietrzak, T. K., Garbarczyk, J. E., Wasiucionek, M., Gorzkowska, I., Nowiński, J. L. & Gierlotka, S. Electrical properties vs. microstructure of nanocrystallised V_2O_5–P_2O_5 glasses — an extended temperature range study. *Solid State Ionics*, 2011, **192**, 210–214.
9. Padhi, A. K., Nanjundaswamy, K. S. & Goodenough, J. B. Phospho-olivines as positive-electrode materials for rechargeable lithium batteries. *J. Electrochem. Soc.*, 1997, **144**, 1188–94.
10. Limb, Y. & Davis, R. F. Electronic conductivity and related properties of amorphous and crystallised V_2O_5-based glasses. *J. Am. Ceram. Soc.*, 1979, **62**, 403–10.
11. Adams, S., Hariharan, K. & Maier, J. Crystallisation in fast ionics glassy silver oxysalt systems. *Solid State Phenomena*, 1994, **39–40**, 285–288.
12. Adams, S., Hariharan, K. & Maier, J. Interface effect on the silver ion conductivity during the crystallisation of AgI–Ag_2O–V_2O_5 glasses. *Solid State Ionics*, 1995, **75**, 193–201.
13. Garbarczyk, J. E., Jóźwiak, P., Wasiucionek, M. & Nowiński, J. L. Enhancement of electrical conductivity in lithium vanadate glasses by nanocrystallisation. *Solid State Ionics*, 2004, **175**, 691–4.
14. Garbarczyk, J. E., Jóźwiak, P., Wasiucionek, M. & Nowiński, J. L. Effect of nanocrystallisation on the electronic conductivity of vanadate–phosphate glasses. *Solid State Ionics*, 2006, **177**, 2585–8.
15. Garbarczyk, J. E., Jóźwiak, P., Wasiucionek, M. & Nowiński, J. L. Nanocrystallisation as a method of improvement of electrical properties and thermal stability of V_2O_5-rich glasses. *J. Power Sources*, 2007, **173**, 743–7.
16. Pietrzak, T. K., Maciaszek, M., Nowiński, J. L., Ślubowska, W., Ferrari, S., Mustarelli, P. Wasiucionek, M., Wzorek, M. & Garbarczyk, J. E. Electrical properties of V_2O_5 nanomaterials prepared by twin rollers technique. *Solid State Ionics*, 2012, **225**, 658–62.
17. Szörenyi, T., Bali, K. & Hevesi, I. Structural characterisation of vanadium phosphate glasses: density and molar volume data. *Phys. Chem. Glasses*, 1982, **23**, 42–4.
18. Ait Salah, A., Jóźwiak, P., Garbarczyk, J., Benkhouja, K., Zaghib, K., Gendron, F. & Julien, C. M. Local structure and redox energies of lithium phosphates with olivine- and Nasicon-like structures. *J. Power Sources*, 2005, **140**, 370–5.
19. Jóźwiak, P., Garbarczyk, J. E., Wasiucionek, M., Gorzkowska, I., Gendron, F., Mauger, A. & Julien, C. M. DTA, FTIR and impedance spectroscopy studies on lithium–iron–phosphate glasses with olivine-like local structure. *Solid State Ionics*, 2008, **179**, 46–50.
20. Jóźwiak, P., Garbarczyk, J., Gendron, F., Mauger, A., Julien, C. M. Disorder in Li_xFePO_4: From glasses to nanocrystallites. *J. Non-Cryst. Solids*, 2008, **354**, 1915–25.
21. Jóźwiak, P., Garbarczyk, J. E., Wasiucionek, M., Gorzkowska, I., Gendron, F., Mauger, A. & Julien, C. The thermal stability, local structure and electrical properties of lithium-iron phosphate glasses. *Mater. Sci.–Poland*, 2009, **27**, 307–18.
22. Pietrzak, T. K., Wewiór, Ł., Garbarczyk, J. E., Wasiucionek, M., Gorzkowska, I., Nowiński, J. L. & Gierlotka, S. Electrical properties and thermal stability of $FePO_4$ glasses and nanomaterials. *Solid State Ionics*, 2011, **188**, 99–103.
23. Pietrzak, T. K., Gorzkowska, I., Nowiński, J. L., Garbarczyk, J. E. & Wasiucionek, M. Preparation of triphylite-like glasses and nanomaterials in the $LiFePO_4$–V_2O_5 system and study on their electrical conductivity. *Funct. Mater. Lett.*, 2011, **4**, 143–5.
24. Pietrzak, T. K., Wasiucionek, M., Gorzkowska, I., Nowiński, J. L. & Garbarczyk, J. E., Novel vanadium-doped olivine-like nanomaterials with high electronic conductivity. *Solid State Ionics*, 2013, **251**, 40–6.
25. Garbarczyk, J. E., Pietrzak, T. K., Wasiucionek, M., Kaleta, A., Dorau, A. & Nowiński, J. L. High electronic conductivity in nanostructured materials based on lithium-iron-vanadate-phosphate glasses. *Solid State Ionics*, 2015, **272**, 53–9.
26. Hassaan, M. Y., Ebrahim, F. M., Mostafa, A. G. & El-Desoky, M. M. Effect of sulfur addition and heat treatment on electrical conductivity of bariumvanadate glasses containing iron. *Mater. Chem. Phys.*, 2011, **129**, 380–84.
27. Al-Assiri, M. S. & El-Desoky, M. M. Grain-size effects on the structural, electrical propertiesand ferroelectric behaviour of barium titanate-basedglass–ceramic nano-composite. *J. Mater. Sci.: Mater. Electron.*, 2013, **24**, 784–92.
28. Al-syadi, A. M., El Sayed, Y., El-Desoky, M. M. & Al-Assiri, M. S. Impedance spectroscopy of V_2O_5–Bi_2O_3–$BaTiO_3$ glasse–ceramics. *Solid State Sci.*, 2013, **26**, 72–82.
29. Moguš-Milanković, A., Sklepić, K., Skoko, Ž., Mikac, L., Musić, S. & Day, D. E. Influence of nanocrystallisation on the electronic conductivity of zinc iron phosphate glass. *J. Am. Ceram. Soc.*, 2012, **95**, 303–11.
30. Rathore, M. & Dalvi, A. Electrical transport in Li_2SO_4–Li_2O–P_2O_5 ionic glasses and glass–ceramic composites: A comparative study. *Solid State Ionics*, 2013, **239**, 50–55.
31. Rathmore, M. & Dalvi, A. Crystallisation in Li_2SO_4–Li_2O–P_2O_5 glassy ionic system: An assessment through electrical transport. *J. Non-Cryst. Solids*, 2014, **402**, 79–83.
32. Hirose, K., Honma, T., Benino, Y. & Komatsu, T. Glass–ceramics with $LiFePO_4$ crystals and crystal line patterning in glass by YAG laser irradiation. *Solid State Ionics*, 2007, **178**, 801–7.
33. Pietrzak, T. K., Wasiucionek, M., Nowiński, J. L. & Garbarczyk, J. E. Isothermal nanocrystallisation of vanadate-phosphate glasses. *Solid State Ionics*, 2013, **251**, 78–82.
34. Patterson, A. L. The Scherrer formula for X-ray particle size determination. *Phys. Rev.*, 1939, **56**, 978–82.
35. Lasocka, M. The effect of scanning rate on glass transition temperature of splat-cooled $Te_{85}Ge_{15}$. *Mater. Sci. Eng.*, 1976, **23**, 173–7.
36. Kissinger, H. E. Reaction kinetics in differential thermal analysis. *Anal. Chem.*, 1957, **29**, 1702–6.
37. Vázquez, J., Wagner, C., Villares, P., Jimenez-Garay, R. Glass transition and crystallisation kinetics in $Sb_{0.18}As_{0.34}Se_{0.48}$ glassy alloy by using non-isothermal techniques. *J. Non-Cryst. Solids*, 1998, **235–237**, 548–53.
38. Austin, I. G. & Mott, N. F. Polarons in crystalline and non-crystalline materials. *Adv. Phys.*, 1969, **18**, 41–102.
39. Delmas, C., Cognac-Auradou, H., Cocciantelli, J. M., Ménétrier, M. & Doumerc, J. P. The $Li_xV_2O_5$ system: An overview of the structure modifications induced by the lithium intercalation. *Solid State Ionics*, 1994, **69**, 257–64.
40. Mott, N. F. Electrons in disordered structures. *Adv. Phys.*, 1967, **16**, 49–144.
41. Doupovec, J., Sitek, J. & Kákoš, J. Crystallisation of iron phosphate glasses. *J. Thermal Anal.*, 1981, **22**, 213–19.
42. Glushenkov, A. M., Stukachev, V. I., Hassan, M. F., Kuvshinov, G. G., Liu, H. K. & Chen, Y. A novel approach for real mass transformation from V_2O_5 particles to nanorods. *Cryst. Growth Des.*, 2008, **8**, 3661–5.
43. Wang,Y. & Cao, G. Developments in nanostructured cathode materials for high-performance lithium-ion batteries. *Adv. Mater.*, 2008, **20**, 2251–69.
44. Parida, M. R. Vijayan, C., Rout, C. S., Sandeep, C. S. S., Philip, R. & Deshmukh, P. C. Room temperature ferromagnetism and optical limiting in V_2O_5 nanoflowers synthesised by a novel method. *J. Phys. Chem. C*, 2011, **115**, 112–17.
45. Liu, H., Gao, P., Fang, J. & Yang, G. $Li_3V_2(PO_4)_3$/graphene nanocomposites as cathode material for lithium ion batteries. *Chem. Commun.*, 2011, **47**, 9110–12.
46. Ravet, N., Abouimrane, A. & Armand, M. Correspondence from our readers. *Nat. Mater.*, 2003, **2**, 702–3.
47. Subramanya Herle, P., Ellis, B., Coombs, N. & Nazar, L. F. Nano-network electronic conduction in iron and nickel olivine phosphates. *Nat. Mater.*, 2004, **3**, 147–52.
48. Ait Salah, A., Mauger, A., Julien, C. M. & Gendron, F. Nanosized impurity phases in relation to the mode of preparation of $LiFePO_4$. *Mater. Sci. Eng. B*, 2006, **129**, 232–44.
49. Kang, B. & Ceder, G., Battery materials for ultrafast charging and discharging. *Nature*, 2009, **458**, 190–3.
50. Lee, J., Urban, A., Li, X., Su, D., Hautier, G. & Ceder, G. Unlocking the potential of cation-disordered oxides for rechargeable lithium batteries. *Science*, 2014, **343**, 519–22.

Phys. Chem. Glasses: Eur. J. Glass Sci. Technol. B, August 2016, 57 (4), 161–165

Radiation induced luminescence properties of pure and Sn-doped 60ZnO.40P_2O_5 glass

T. Yanagida*

Nara Institute of Science and Technology, 8916-5 Takayama-cho, Ikoma, Nara, 630-0192 Japan

Y. Fujimoto

Tohoku University, 6-6-07 Aoba, Aramaki, Aoba-ku, Sendai-shi 980-8579, Japan

H. Masai

Institute for Chemical Research, Kyoto University, Gokasho, Uji, Kyoto 611-0011, Japan

Manuscript received 31 August 2014
Revised version received 24 May 2015
Accepted 1 June 2015

We have investigated the optical properties and the ionising radiation induced luminescence of xSnO.(60–x)ZnO.40P_2O_5 (xSZP, x=0, 2·5 and 5) glasses prepared by a conventional melt-quenching method. In optical transmittance measurements, they exhibit 80–90% transparency from UV to visible wavelengths. The photoluminescence quantum yield of the 2.5SZP glass is 95%, with microsecond order decay constants, while that of the Sn-free glass is 3%, with a fast decay time of a few nanoseconds. The peak emission wavelength in scintillation spectra under x-ray irradiation is similar to that in photoluminescence, and appears at around 400 nm. The scintillation decay times of Sn-doped and Sn-free samples under x-ray exposure are a few μs and a few ns, respectively. The thermally stimulated luminescence glow curve was evaluated. For the Sn-doped sample, glow peaks were observed at 210 and 300°C, while the Sn-free sample showed a peak at 150°C. By conducting an initial rise method for trap analysis, the trap depth was found to be 5–10 meV. Optically stimulated luminescence was also examined, and a luminescence peak appeared at around 500 nm with 630 nm stimulation. Based on these results, we propose an energy level diagram of radiation induced luminescence in these glasses.

Introduction

Ionising radiation induced luminescence has been used to convert invisible ionising radiation to visible photons. From the viewpoint of the energy transfer process, there are several technical terms for ionising radiation induced luminescence, such as scintillation,[1] thermally stimulated luminescence (TSL),[2] optically stimulated luminescence (OSL),[3] and radiophotoluminescence (RPL).[4] Because the luminescence can be used in various kinds of fields, such as medical imaging,[5] security systems in airports,[6] oil-logging,[7] astrophysics,[8] and particle physics,[9] industrially applicable materials have been enthusiastically examined. In most radiation detectors, bulk and transparent materials are required in order to absorb high energy radiation efficiently, and to lead photons to photodetectors with as little loss as possible. Until now, most commercially available materials for ionising radiation detectors are in single crystal form. Therefore, single crystals have been well studied. On the other hand, investigation of glasses for radiation detectors has not been adequate, even though they glassy materials exhibit a wide chemical composition range and good formability. There are actually only two commercially available glass-based products, Li-glass scintillator (Saint Gobain),[10] and Ag-doped Na_2O–$AlPO_4$ for RPL dosimeters (Chiyoda Technol. Corp.).[11] Thus glasses can also be used for ionising radiation detectors, and there remains large scope for the study of optical glasses for ionising radiation detector applications.

Recently, we have demonstrated high photoluminescence (PL) efficiency of RE (rare earth)-free phosphate[12,13] and borate glasses.[14] The emission is due to the Sn^{2+} centre, which is the most conventional and harmless among ns^2-type centres possessing the allowed $(5s)^2$–5s5p transition. In addition, UV-excited white light emission was also demonstrated in SnO- and MnO-codoped zinc phosphate glasses, using energy transfer from Sn^{2+} to Mn^{2+}.[13] The Sn-doped oxide glass phosphor is a fascinating material, not only because it contains no rare earth cation, but also because it exhibits transparency and industrially-adoptable good formability. More recently, we have reported that Sn^{2+}-containing zinc phosphate glasses also exhibit scintillation behaviour.[15] However, the

* Corresponding author. Email t-yanagida@ms.naist.jp
Original version presented at Int. Conf. on Phosphate Glasses, Pardubice, Czech Republic, 2–4 July 2014
DOI: 10.13036/17533562.57.4.048

correlation between the amount of Sn^{2+} and radioluminescence (RL) in the glass system has not yet been clarified.

The aim of this study is to the investigate ionising radiation induced luminescence properties of xSnO.(60–x)ZnO.40P_2O_5 (xSZP, x=0, 2·5 and 5) glasses together with the optical properties. Since the energy of ionising radiation (typically in the keV–MeV range) is higher than the band gap of insulator materials, intrinsic emission of the host can be observed and it is important to the examine nondoped (host) material.

Experimental

The xSZP (x=0, 2·5 and 5) glasses, whose starting materials were SnO, ZnO, and $(NH_4)_2HPO_4$, were prepared by a conventional melt-quenching method, using a 50 cm^3 platinum crucible.[16] The weight of batches, melting temperature, and duration of melting were about 10 g, 1100°C, and 30 min, respectively. The glass melt was quenched on a stainless steel plate kept at 200°C, and then annealed at the glass transition temperature, T_g, for 1 h. The samples were mechanically polished to obtain a mirror surface. In order to check whether crystalline phases appeared or not, x-ray diffraction (XRD) analyses were performed at 40 kV and 40 mA using a RINT2000 (Rigaku) diffractometer.

In-line transmittance was evaluated by using a JASCO V670 spectrometer. Photoluminescence quantum yield (QY) was also evaluated using a Hamamatsu Quantaurus-QY quantum yield spectrometer. The absolute QY was calculated by the following equation, $QY=N_{emit}/N_{absorb}$, where N_{emit} and N_{absorb} are the numbers of emitted and absorbed photons, respectively. Under 250 nm excitation, emissions were accumulated from 300 to 700 nm. PL decay time profiles were recorded by using a Hamamatsu Quantaurus-t fluorescence lifetime spectrometer. The excitation and monitoring wavelengths were 280 and 415 nm, respectively.

Scintillation spectra were collected by using a CCD-based spectrometer under x-ray exposure.[17] Scintillation decay time profiles were observed by using a pulse x-ray equipped afterglow characterisation system with a fast mode.[18] As dosimeter properties, the OSL was evaluated using the spectrum mode of the Quantaurus-t after x-ray exposure, since the light source of this instrument is an LED, which is relatively stable compared to the conventional Xe lamp found in normal PL systems. The TSL was investigated by using a Nanogray TL-2000 reader. Detailed procedures for OSL and TSL measurements have been described previously.[19,20] The x-ray dose in these evaluations was 10 Gy. Compared with general dosimeter evaluations for personal dosimetry, this exposed dose is larger since these samples were not as bright as a dosimeter.

(a)

(b)

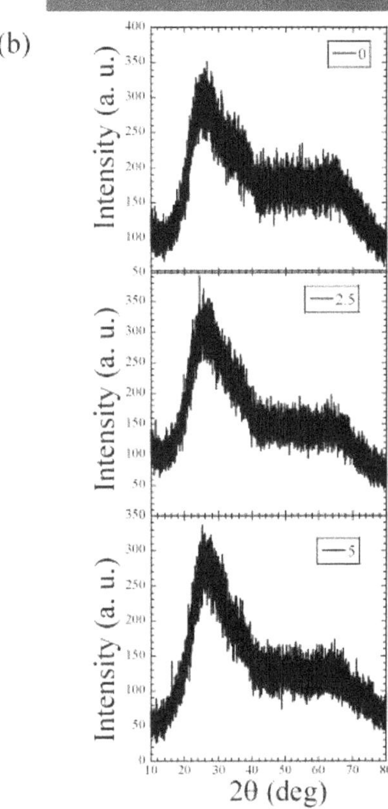

Figure 1. (a) From left to right, photograph of xSZP (x=0, 2·5 and 5) glasses. (b) From top to bottom, XRD patterns of xSZP (x=0, 2·5 and 5) glasses

Results and discussion

Figure 1(a) shows a photograph of the xSZP (x=0, 2·5 and 5) glasses. The non-doped and 2·5SZP glass samples had dimensions of 10×10×1 mm^3, while the 5SZP sample was 5×5×1 mm^3. Except for the radioluminescence (scintillation) spectra, the sample size did not affect the experimental results in this work. For radioluminescence, we generally use only the wavelength-dependence information. The sample size affects the radioluminescence intensity, and so the size difference is not a problem in this work. All glasses were visibly transparent, as shown in Figure 1(a). The XRD patterns are shown in Figure 1(b). The diffraction patterns exhibit a halo pattern, indicating that these glasses are amorphous without the precipitation of any crystallites.

Optical in-line transmittance results are shown in Figure 2. For the Sn-doped samples, absorption due to Sn^{2+} appears around 250 nm, while the absorption edge of nondoped glass is at around 200 nm. For all samples, 80–90% transmittance was achieved for

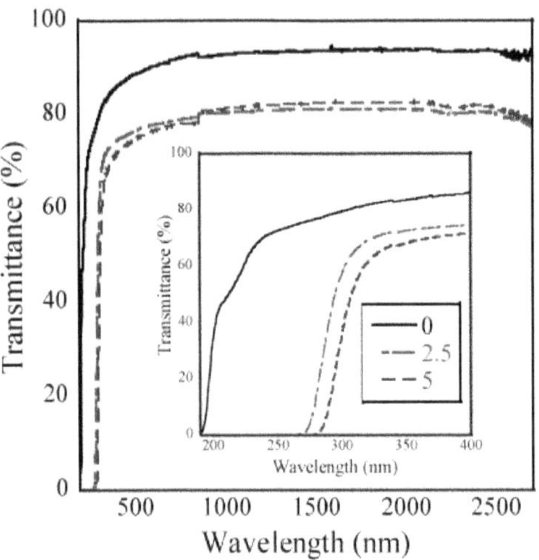

Figure 2. In-line transmittance of xSZP (x=0, 2·5 and 5) glasses

wavelengths longer than 300 nm. Higher transmittance was observed for the Sn-free sample, but the reason is unclear. The difference could be due to the polishing condition and the increase of refractive index with SnO content involving the surface reflectance. In addition, very low content of impurity phases (metal/nanoparticles/nanocrystals) under the detection limit of XRD may contribute to the transmittance.

Figure 3 depicts the PL QY as a function of Sn concentration. The PL QY of the 2·5SZP and 5SZP samples were 95 and 70%, respectively. Very weak host emission was also detected in Sn-free $60ZnO.40P_2O_5$ glass, and the QY was a few %.

PL decay time profiles are shown in Figure 4. The PL decay time profiles of the Sn-doped glasses were well reproduced by a two-component exponential function, and the decay times were approximately 2 and 5·5 μs. As shown in these decay time profiles, the Sn 5% doped sample was slightly faster than the 2·5% doped sample. On the other hand, the PL decay time of the Sn-free sample was fast (1·1 and 7·5 ns). For the

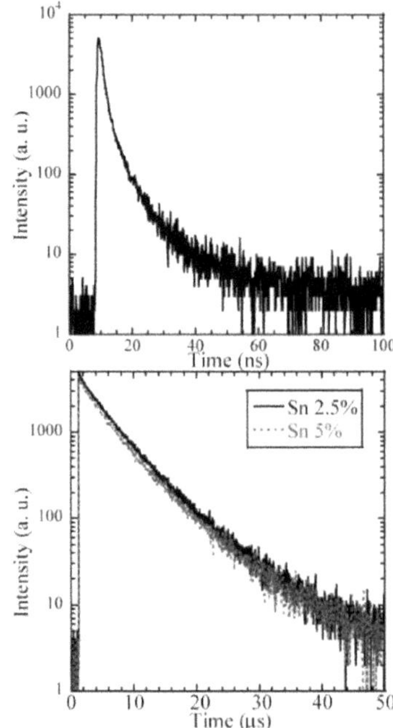

Figure 4. The PL decay time profiles of nondoped (top) and Sn-doped (bottom) samples

Sn-doped glasses, the observed optical properties are consistent with previous reports.[12,13,21]

After optical characterisation, we investigated the ionising radiation induced luminescence properties. Figure 5 shows the x-ray induced radioluminescence (scintillation) spectra. The spectral shapes are similar to those in the PL spectra, and the emission peak of the Sn-doped samples appears from 300 to 700 nm. As for PL, the Sn-free $60ZnO.40P_2O_5$ sample also exhibited a weak emission from 400 to 500 nm. The origin of this emission for the Sn-free sample can be ascribed to some kind of defects, since defects of oxide materials generally show scintillation around 300–500 nm.

The scintillation decay time profiles are shown in Figure 6. The Sn-free glass exhibited a quite fast decay and the primary decay time was around 11 ns. On the other hand, the scintillation decays of the Sn-doped

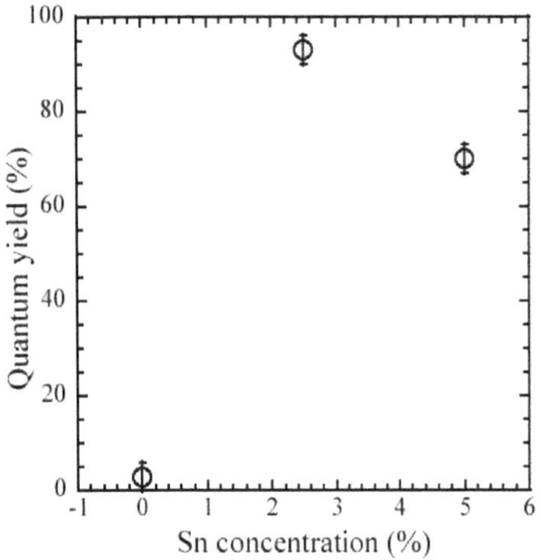

Figure 3. PL QY plotted against Sn concentration

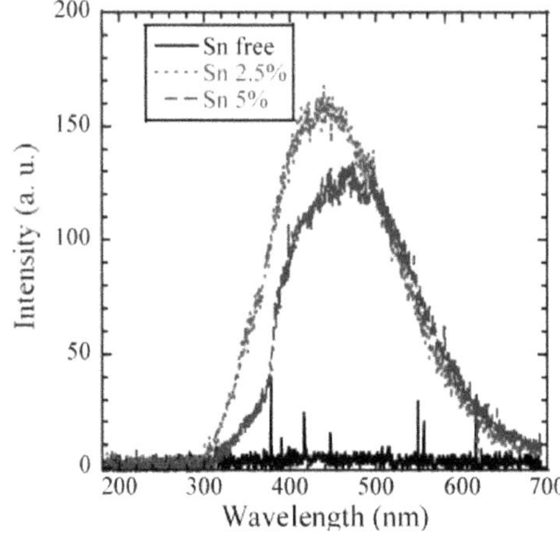

Figure 5. X-ray induced radioluminescence spectra of xSZP glasses

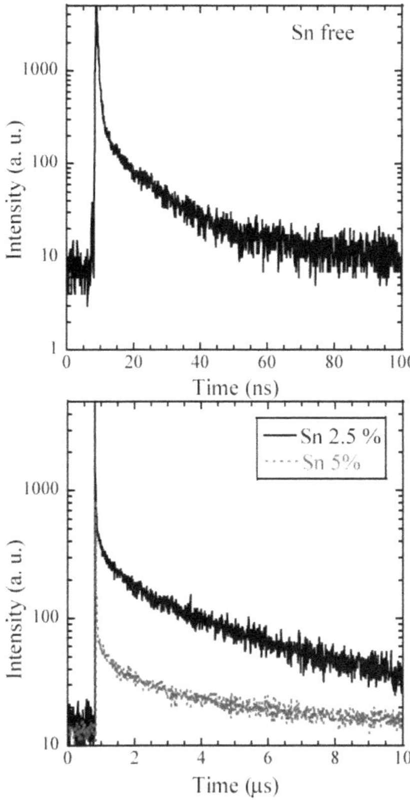

Figure 6. X-ray induced scintillation decay time profiles of xSZP glasses

glasses are characterised by three components: very fast 7–9 ns, intermediate 500 ns, and slow 4 μs. The very fast component coincides well with the primary decay time of the Sn-free glass, and so the origin is clearly the host emission. The slow component is of order μs, and it is ascribed to the Sn^{2+} emission. The origin of the intermediate component is also the host, since the Sn-free sample showed a weak but long decay component (see Figure 6).

The TSL glow curves of xSZP glasses (x=0, 2·5 and 5) are exhibited in Figure 7. In all samples, a glow peak was observed around 80°C. With Sn-doping, the 80°C glow peak became smaller, and instead a new

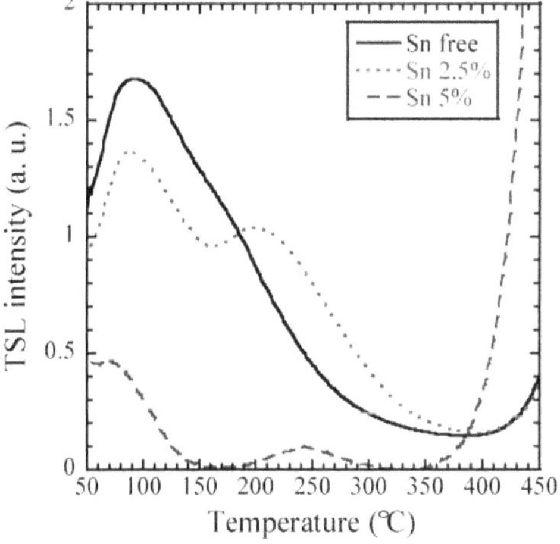

Figure 7. TSL glow curves of xSZP glasses. The heating rate was 1°C/s

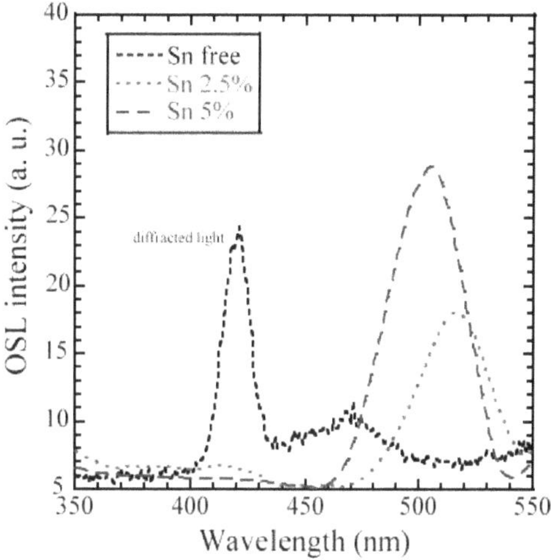

Figure 8. OSL spectra of xSZP (x=0, 2·5 and 5) glasses under 630 nm stimulation

peak around 250°C was observed. TSL around room temperature causes the afterglow and degrades scintillation responses. In this case, Sn-doping had a positive effect for ionising radiation detection. By conducting a conventional initial rise method,(2) trap depths of 80 and 250°C for the Sn-doped samples were evaluated and deduced to be approximately 5 and 10 meV.

The OSL spectra under 630 nm stimulation are shown in Figure 8. For the Sn-doped samples, the OSL appeared round 500 nm, while that for the Sn-free sample appeared at 470 nm. Compared with radioluminescence, the OSL peaks shifted to longer wavelength for the Sn-doped samples. Sn^{2+} emission in radioluminescence showed a broad peak and there would be several emission components. In OSL, the longer wavelength component was stimulated. On the other hand, the OSL of the Sn-free sample was weak and at similar wavelength in radioluminescence (400–500 nm). Finally, taking into account scintillation, TSL, and OSL properties, the energy level diagram relating to ionising radiation induced luminescence is shown in Figure 9.

Conclusions

xSnO.(60−x)ZnO.40P_2O_5 (x=0, 2·5 and 5) glasses were prepared by a conventional melt-quenching method. Investigations of the optical and scintillation

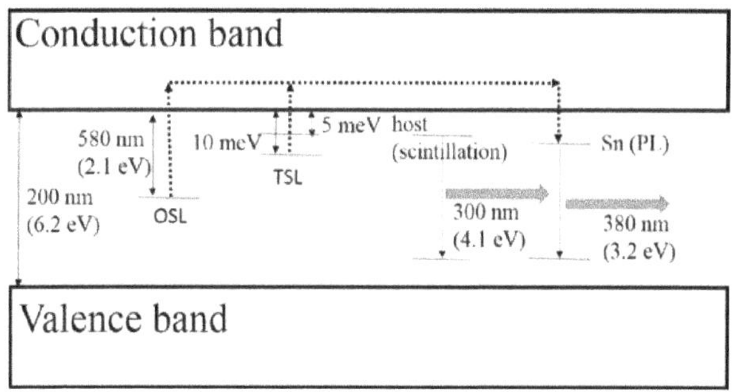

Figure 9. Energy level diagram of ionising radiation induced luminescence in xSZP (x=0, 2·5 and 5) glasses

properties showed emission peaks at around 300–700 nm and 400–500 nm, for the Sn-doped and Sn-free samples, respectively. The photoluminescence and scintillation decay times of Sn-doped and Sn-free samples are a few ms and a few ns, respectively. In the thermally stimulated luminescence glow curves, peaks appeared at 80 and 250°C. The first of these is ascribed to the host, whilst the second is induced by Sn-doping. Under 630 nm stimulation, optically stimulated luminescence appeared at a similar wavelength to radioluminescence. Finally, the energy level diagram for ionising radiation induced luminescence was proposed.

Acknowledgements

This work was mainly supported by a Grant in Aid for Scientific Research (A)-26249147 from the Ministry of Education, Culture, Sports, Science and Technology of the Japanese government (MEXT), and partially by JST A-step. Partial assistance from Nippon Sheet Glass Foundation for Materials Science and Engineering, Tokuyama Science Foundation, Iketani Science and Technology Foundation, Hitachi Metals Materials Science Foundation, Mazda Foundation, JFE 21st Century Foundation, the Asahi Glass Foundation, the Cooperative Research Project of Research Institute of Electronics, Shizuoka University, the Collaborative Research Program of the Institute for Chemical Research, Kyoto University (2014-31), and the Kyoto University SPIRITS Program are also gratefully acknowledged.

References

1. Yanagida, T. Study of rare-earth-doped scintillators. *Opt. Mater.*, 2012, **35**, 1987–1992.
2. McKeever, S. W. S. *Thermoluminescence of Solids*, Cambridge University Press, Cambridge, 1988.
3. Yukihara E. G. & McKeever S. W. S. *Optically Stimulated Luminescence: Fundamentals and Applications*, Wiley, Chichester, 2011.
4. Miyamoto, Y., Takei, Y., Nanto, H., Kurobori, T., Konnai, A., Yanagida, T., Yoshikawa, A., Shimotsuma, Y., Sakakura, M., Miura, K., Hirao, K., Nagashima, Y. & Yamamoto, T. Radiophotoluminescence from silver-doped phosphate glass. *Radiat. Meas.*, 2011, **46**, 1480–1483.
5. Yanagida T., Yoshikawa A., Yokota Y., Kamada K., Usuki Y., Yamamoto S., Miyake M., Baba M., Sasaki K. & Ito M. Development of Pr:LuAG Scintillator Array and Assembly for Positron Emission Mammography. *IEEE. Nucl. Trans. Sci.*, 2010, **57**, 1492–1495.
6. Totsuka, D., Yanagida T., Fukuda K., Kawaguchi N., Fujimoto Y., Yokota Y. & Yoshikawa, A. Performance test of Si PIN photodiode line scanner for thermal neutron detection. *Nucl. Instrum. Meth. A*, 2011, **659**, 399–402.
7. Yanagida T., Fujimoto Y., Kurosawa S., Kamada K., Takahashi H., Fukazawa Y., Nikl M. & Chani V. Temperature dependence of scintillation properties of bright oxide scintillators for well-logging. *Jpn. J. Appl. Phys.*, 2013, **52**, 076401.
8. Yamaoka, K., Ohno, M., Terada, Y., Hong, S., Kotoku, J., Okada, Y., Tsutsui, A., Endo, Y., Abe, K., Fukazawa, Y., Hirakuri, S., Hiruta, T., Itoh, K., Kamae, T., Kawaharada, M., Kawano, N., Kawashima, K., Kishishita, T., Kitaguchi, T., Kokubun, M., Madejski, G. M., Makishima, K., Mitani, T., Miyawaki, R., Murakami, T., Murashima, M., Nakazawa, M., Niko, K., Nomachi, H., Oonuki, M., Sato, K., Suzuki, G., Takahashi, M., Takahashi, H., Takahashi, I., Takeda, , T. Tamura, S., Tanaka, T., Tashiro, T., Watanabe, M., Yanagida, S.& Yonetoku, T. Development of the HXD-II Wideband All-sky Monitor Onboard Astro-E2. *IEEE. Trans. Nucl. Sci.*, 2005, **52**, 2765–2772.
9. Ito T., Yanagida T., Sato M., Kokubun M., Takashima T., Hirakuri S., Miyawaki R., Takahashi H., Makishima K., Tanaka T., Nakazawa K., Takahashi T. & Honda T. A 1-Dimensional Gamma-ray Position Sensor based on GSO:Ce Scintillators Coupled to a Si Strip Detector. *Nucl. Instr. Meth. A*, 2007, **579**, 239–242.
10. http://www.crystals.saint-gobain.com/
11. http://www.c-technol.co.jp/technol_eng/first_pages/02personal_dosi.html
12. Masai, H., Takahashi, Y., Fujiwara, T., Matsumoto, S. & Yoko T. High Photoluminescent Property of Low-Melting Sn-Doped Phosphate Glass. *Appl. Phys. Express*, 2010, **3**, 082102.
13. Masai, H., Fujiwara, T., Matsumoto, S., Takahashi, Y., Iwasaki, K., Tokuda, Y. & Yoko, T. White Light Emission of Mn-Doped $SnO-ZnO-P_2O_5$ Glass Containing No Rare Earth Cation. *Opt. Lett.* 2011, **36**, 2868–2870.
14. Masai, H., Yamada, Y., Suzuki, Y., Teramura, K., Kanemitsu, Y. & Yoko, T. Narrow Energy Gap between Triplet and Singlet Excited States of Sn^{2+} in Borate Glass. *Sci. Rep.* 2013, **3**, 3541.
15. Masai, H., Yanagida, T., Fujimoto, Y., Koshimizu, M. & Yoko, T. Scintillation Property of Rare Earth-free SnO-doped Oxide Glass. *Appl. Phys. Lett.*, 2012, **101**, 191906.
16. Masai, H., Tanimoto, T., Fujiwara, T., Matsumoto, S., Takahashi, Y., Tokuda, Y. & Yoko, T. Fabrication of Sn-doped zinc phosphate glass using a platinum crucible. *J. Non-Cryst. Solids*, 2012, **358**, 265–269.
17. Yanagida, T., Kamada, K., Fujimoto, Y., Yagi, H. & Yanagitani, T. Comparative study of ceramic and single crystal Ce:GAGG scintillators. *Opt. Mater.*, 2013, **35**, 2480–2485.
18. Yanagida, T., Fujimoto, Y., Ito, T., Uchiyama, K. & Mori, K. Development of X-ray induced afterglow characterization system. *Appl. Phys. Exp.*, 2014, **7**, 062401.
19. Yanagida, T., Fujimoto, Y., Kawaguchi, N. & Yanagida, S. Dosimeter properties of AlN. *J. Ceram. Soc. Jpn.*, 2013, **121**, 988–991.
20. Yanagida, T., Fujimoto, Y., Watanabe, K., Fukuda, K., Kawaguchi, N., Miyamoto, Y. & Nanto, H. Scintillation and Optical Stimulated Luminescence of Ce doped CaF_2. *Radiat. Meas.*, 2014, **71**, 162–165.
21. Masai, H., Tanimoto, T., Fujiwara, T., Matsumoto, S., Tokuda, Y. & Yoko, T. Correlation between emission property and concentration of Sn^{2+} centre in the $SnO-ZnO-P_2O_5$ glass. *Opt. Express*, 2012, **20**, 27319–27326.

Phys. Chem. Glasses: Eur. J. Glass Sci. Technol. B, August 2016, **57** (4), 173–182

Structure and high temperature behaviour of sodium aluminophosphate glasses

L. van Wüllen & V. Sabarinathan*

Institute of Physics – Augsburg University, Universitätsstr. 1 - D86159 Augsburg, Germany

Manuscript received 30 September 2014
Revised version received 14 August 2015
Accepted 7 September 2015

A study is presented of the structure of Al-rich sodium aluminophosphate glasses, its dependence on the processing conditions, and its evolution upon heat treatment at high temperatures. Sol-gel and two different melt-quench techniques were employed for the glass synthesis. The influence of the processing conditions and heat treatment on the glass structure was studied by employing in situ and ex situ MAS (magic angle spinning) NMR spectroscopy and advanced dipolar NMR techniques: REAPDOR (Rotational Echo Adiabatic Passage Double Resonance), and DQ-SQ (Double Quantum – Single Quantum) homonuclear correlation spectroscopy. The results indicate a dependence of the crystallisation tendency on the processing conditions: Sol-gel and fast-quenched glasses exhibit a retarded tendency towards crystallisation as compared to conventionally quenched glasses, which may be linked to an increased relative fraction of AlO_5 and AlO_6 units.

Introduction

Phosphate based glasses represent a technologically important class of materials, and find application as laser hosts,[1] fast ion conductors, controlled drug release materials,[2] biomaterials,[3,4] or as host material for radioactive waste.[5,6] The inherent instability towards moisture attack and the poor chemical durability of binary phosphate glasses can be overcome by the addition of glass formers such as B_2O_3, Al_2O_3 or SiO_2. The mixing of the different network formers inhibits the hydrolysis and/or dissolution of phosphate chains or individual phosphate units.

Owing to their technological importance many studies have been directed to an elucidation of the network organisation of these glasses.[7,8] Among the various analytical tools available, especially solid state NMR spectroscopy has been shown to provide very valuable information about the structural motifs on short and intermediate length scales. Structural motifs on a length scale of 1–2 Å can be successfully analysed employing magic angle spinning (MAS) NMR spectroscopy,[9,10] providing information about the nature of the network forming polyhedra, the local building units, i.e. borate, aluminate, silicate or phosphate polyhedra with different degrees of connectivity.[11] Information about the interconnection of these local fragments towards extended structural units on a length scale of 2–8 Å may then be obtained via an exploration of the homo- and heteronuclear dipolar couplings between two nuclei (these scale with the inverse cube of the internuclear distance) under the conditions of fast MAS. Here, a wide variety of multidimensional solid state NMR strategies has been developed in the last two decades to quantitatively determine homo- (2D-exchange-,[12,13] 2D-RFDR-(Radio Frequency Driven Recoupling)[14,15] and 2D-double quantum NMR experiments[16–24]) and heteronuclear (REDOR (Rotational Echo Double Resonance) and related approaches[25–31]) dipolar couplings and to analyse these with respect to the spatial distribution of a given chemical species and connectivity patterns. Accompanied by solid state NMR approaches, which trace the existence of chemical bonds via an evaluation of the scalar homo- and heteronuclear J-couplings (e.g. INADEQUATE (Incredible Natural Abundance DoublE QUAntum Transfer Experiment), HMQC (Heteronuclear Multiple Quantum Coherence), J-resolved 2D-NMR spectroscopy[32–35]), these approaches have contributed enormously to the identification of connectivity patterns in amorphous solids.[36–39] Since the structure of a glass at ambient temperature, and consequently its physical and chemical properties, are a direct consequence of the equilibria within the melt and their corresponding kinetics, a detailed knowledge of the evolution of the glass structure with temperature (i.e. network connectivity, phase separation, crystallisation) is necessary for a controlled optimisation of the material's key properties. Two different approaches are principally possible to follow the structural evolution of glasses with temperature. In *ex situ* studies, the structural changes with temperature are studied at room temperature on samples which have been prepared following different cooling rates (corresponding to different fictive temperatures, T_f). This approach, however, owing

Corresponding author. Email leo.van.wuellen@physik.uni-augsburg.de
Original version presented at Int. Conf. on Phosphate Glasses, Pardubice, Czech Republic, 2–4 July 2014
DOI: 10.13036/17533562.57.4.068

to the limited cooling rates, only opens a narrow range of accessible fictive temperatures. The *in situ* approach, on the other hand, offers the chance to trace the structural changes in the network organisation upon heat treatment as they occur. Again, especially solid state NMR has been identified as one of the most valuable analytical approaches. Stebbins and co-workers have provided important contributions about structural relaxation processes and dynamic species exchange in silicate, borate and borosilicate glasses.(40–46) Contributions from our laboratory include *in situ* studies on aluminophosphate glasses(47,48) and boro- and phosphosilicate glasses.(49,50)

As a consequence of the equilibria in the melt and their temperature dependence, the accessible compositional range within a glass system and the amount of mixing/demixing of the network forming polyhedra may critically depend on the processing conditions. As found by many authors, for aluminophosphates, the compositions with high Al_2O_3 contents are normally not accessible employing melt-quench techniques. According to Zhang & Eckert,(51) especially glasses along the compositional line Na_2O–$AlPO_4$ are only accessible via the sol-gel route. In the present study we attempted to extend the accessible compositional range of melt-quenched (MQ) glasses in the system Na_2O–Al_2O_3–P_2O_5 towards Al_2O_3 contents in excess of 25% by using a quadruple roller quench apparatus (see below) which provides faster quenching rates as compared to conventional melt-quench techniques. The structure of the resulting glasses at ambient temperature was analysed in detail employing advanced solid state NMR techniques, and compared to glasses that were made via sol-gel or conventional MQ techniques. The crystallisation behaviour of the glasses as a function of the synthesis conditions was studied employing *ex situ* NMR on heat treated samples, and with *in situ* NMR techniques at temperatures up to 625°C.

Experimental

The synthesis of the sol-gel (SG) glasses was performed according to the route outlined in Ref. 52 using aluminum lactate, phosphoric acid (H_3PO_4) and sodium acetate as precursor materials. MQ glasses were melted at 1200–1450°C, using Al_2O_3, Na_2CO_3, $NaPO_3$, $H_2Na_2P_2O_7$ and $Al(PO_3)_3$ in appropriate amounts. For the glass samples synthesised in this work, the prefix MQ indicates a normal melt-quenching processing of the melt (quenching of the glass melt onto a copper plate), whereas the prefix QRQ denotes glasses which have been prepared employing a quadruple roller quench apparatus. In this approach, the glass melt is fed through two pairs of counter-rotating steel cylinders. The upper pair is rotating at a maximum speed of 4000 rpm, whereas the maximum speed of the lower pair is 2000 rpm.

Table 1. Nominal compositions of the glasses studied in this work. All glasses were synthesised following a sol-gel process or melt-quenching techniques, i.e. either melt-quenching on a copper plate, or quenching between two pairs of twin rollers

Glass	*Na_2O/%*	*Al_2O_3/%*	*P_2O_5/%*
NAP-1	33	33	33
NAP-2	20	40	40
NAP-3	22	34	44
NAP-4	50	20	30
NAP-5	40	20	40

The spacing between the cylinders may be adjusted between 0–5 mm; for the glasses in this work the spacing was adjusted to 50 μm. The realisation of this apparatus follows a suggestion by Prof. Jörn Schmedt auf der Günne, Siegen, Germany. The nominal compositions of the glasses studied in this work are given in Table 1.

Solid state NMR experiments were performed employing a Bruker Avance III spectrometer operating at 7 T, a Bruker DSX spectrometer operating at 9·4 T, or a Varian DNMR spectrometer operating at 11·7 T. For the experiments performed at room temperature, a range of 4 mm triple resonance probes has been used (Bruker machines). The *in situ* NMR experiments were performed employing a Doty-HT-MAS NMR probe. For the experiments at 11·7 T, a 1·6 mm Varian T^3 MAS probe was employed. For the single pulse acquisition experiments at ambient temperatures, typical pulse lengths used were 4, 1·5 and 2 μs for ^{31}P, ^{27}Al and ^{23}Na, respectively. Recycle delays of 1 s and 60 s were employed for ^{23}Na/^{27}Al and ^{31}P, respectively; spinning speed was 10–14 kHz. For the (partially) crystallised samples, the delay times were increased to 100 – 300 s. For the REDOR and REAPDOR (Rotational Echo Adiabatic Passage Double Resonance) measurements, RF amplitudes of 47·5 and 62·5 kHz were used for ^{27}Al and ^{31}P, respectively. Spinning speeds of 10–14 kHz were applied unless otherwise stated. The 2D-^{27}Al DQ-SQ-NMR experiment was performed at a field of 9·4 T and a spinning speed of 14·0 kHz. The excitation/reconversion times were set to eight rotor periods; the increment in t_1 was set to one rotor period (71·4 μs).

For sample NAP-5, the glass transition temperature, T_g, was determined from DSC measurements (Netsch). The T_g values proved to be identical (445±5°C) for glasses from the three different synthesis routes.

Results and discussion

The ^{27}Al and ^{31}P MAS NMR spectra of the sol-gel glasses of composition NAP-1 are shown in Figure 1. The temperatures given in the figure refer to the heat treatment temperatures, at which the gels were kept for 5 h. The findings correspond well to the results reported by Zhang & Eckert.(51) Whereas Al is

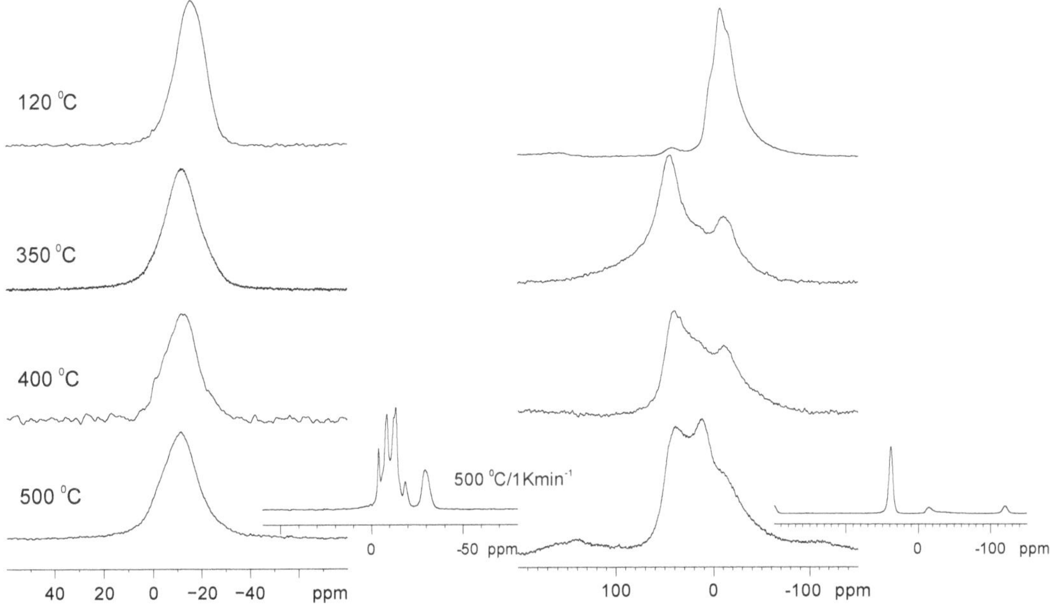

Figure 1. ^{31}P (left) and ^{27}Al MAS NMR spectra (right) for the sol-gel glasses SG-NAP-1 for the indicated heat treatment temperatures. Heating rates of 0·5 $K min^{-1}$ were applied. Spectra recorded at a field of 7·04 T

found predominantly in octahedral coordination in the non-heat treated or samples heat treated at lower temperatures, at higher heat treatment temperatures a change of the coordination to AlO_4 units is observed. With heating ramps >1 $K min^{-1}$, partial crystallisation occurs at temperatures slightly above the gel to glass conversion at 450°C. Here, the signal in the ^{31}P MAS NMR spectrum around −30 ppm and the ^{27}Al MAS NMR signal around 40 ppm indicate the formation of $AlPO_4$[53] (see below). Employing heating ramps of 0·5 $K min^{-1}$, crystallisation is avoided until a temperature of 500°C. In the ^{31}P MAS NMR spectrum for the sample heat treated above 400°C, three signals at −3·0 ppm (23%), −12·5 ppm (70%) and −23·0 ppm (7%) are observed. These were assigned by Zhang & Eckert to Q^0_{2Al}, Q^0_{3Al} and Q^0_{4Al} units, respectively,[51] which is supported by the results of the ^{31}P{^{27}Al} REAPDOR NMR experiment (see below). The ^{27}Al MAS NMR spectrum features the signals for AlO_4, AlO_5 and AlO_6 environments. The individual Al sites exhibit quite a distribution, leading to broad overlapping lines, even at a field of 11·7 T (data not shown). Attempts to simulate the spectra employing the Czjzek option in DMFIT[54] produced non-convergent fits. Thus, in order to obtain reliable results, the spectra were simulated according to the following procedure. First, the isotropic chemical shifts and estimates for the quadrupole parameter (second order quadrupole effect, $C_Q\sqrt{(1+\eta^2/3)}$) for the components were deduced from MQMAS (multiple quantum magic angle spinning) experiments on selected samples (data not shown). These values were then used as input parameters and optimised together with a Gaussian distribution of C_Q and relative areas for the various Al components, employing home-built MATHEMATICA code. Using this approach, the data compiled in Table 2 were obtained, and Figure 2 shows three representative spectra together with the simulations.

Having identified the local network polyhedra, the connectivity motifs between the phosphate and aluminate species was evaluated employing dipolar NMR spectroscopy, ^{27}Al{^{31}P} REDOR NMR and

Table 2. Parameters for the decomposition of the ^{27}Al MAS NMR spectra of samples NAP-1 and NAP-2

		δ_{iso}/ppm	C_Q/MHz (dist C_Q/MHz)	η_Q	rel. area/%
		site 1/AlO_4			
SG	NAP-1	48	3·2 (3·0)	0·5	60
MQ	NAP-1	47	3·2 (2·0)	0·5	72
QRQ	NAP-1	48	3·2(2·0)	0·5	66
SG	NAP-2	47	3·2 (2·5)	0·5	57
MQ	NAP-2	47	3·2 (2·0)	0·5	56
QRQ	NAP-2	47	3·2 (2·0)	0·5	60
QRQ	NAP-3	47	3·2 (2·5)	0·5	85
		site 2/AlO_5			
SG	NAP-1	21	3·0 (3·0)	0·5	18
MQ	NAP-1	15	3·0 (2·0)	0·5	11
QRQ	NAP-1	15	3·0 (3·0)	0·5	14
SG	NAP-2	16	3·0 (2·0)	0·5	8
MQ	NAP-2	14	3·0 (3·0)	0·5	18
QRQ	NAP-2	14	3·0 (3·0)	0·5	20
QRQ	NAP-3	16	3·0 (2·0)	0·5	9
		site 3/AlO_6			
SG	NAP-1	−6	3·0 (3·5)	0·5	22
MQ	NAP-1	−7	3·0 (3·0)	0·5	16
QRQ	NAP-1	−7	3·0 (3·5)	0·5	20
SG	NAP-2	−7	3·0 (2·5)	0·5	35
MQ	NAP-2	−11	3·0 (4·0)	0·5	5
QRQ	NAP-2	−11	3·0 (4·0)	0·5	6
QRQ	NAP-3	−8	3·0 (3·0)	0·5	6
		site 4/Al_2O_3			
SG	NAP-1				
MQ	NAP-1				
QRQ	NAP-1				
SG	NAP-2				
MQ	NAP-2	16	2·2	0·1	21
QRQ	NAP-2	16	2·2	0·1	14
QRQ	NAP-3				

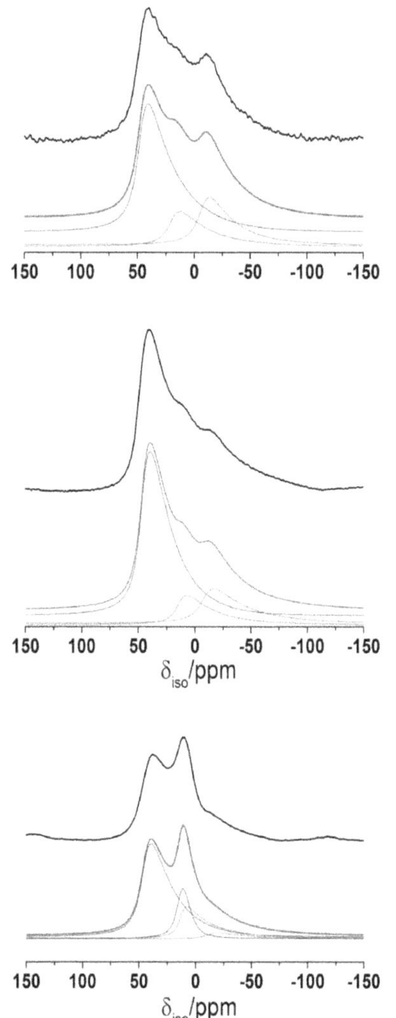

Figure 2. ^{27}Al MAS NMR spectra together with the decomposition into individual lines for glasses SG-NAP-1-400 (top), QRQ-NAP-1 (middle) and QRQ-NAP-2 (bottom). For the parameters of the decompositions see Table 2. Spectra recorded at a field of 7·04 T

^{31}P{^{27}Al} REAPDOR NMR. In REDOR, the results from a rotor-synchronised spin-echo experiment for the observed (S) nuclei, defining the full echo intensity S_0, are compared to spectra resulting from an experiment in which the heteronuclear dipolar coupling between nuclei S and I has been reintroduced by the action of rotor-synchronised π-pulses (I-channel) in addition to the S-spin spin-echo pulses. The difference of the spectra from the two experiments then only contains contributions from S nuclei experiencing a dipolar coupling to I nuclei. The magnitude of the REDOR effect depends on the strength of dipolar coupling (dipole coupling constant $D=\gamma_I\gamma_S\hbar\mu_0/2\pi r^3 4\pi$) and the dipolar evolution time, the latter of which can be controlled by the number of rotor cycles and the MAS frequency. In the case of quadrupolar nuclei as the dephasing I nuclei (as is the case with ^{27}Al), REAPDOR, a variation of REDOR, provides a more efficient dephasing through the application of a long I channel pulse (one third of a full rotation period) in the middle of the pulse sequence.

The ^{27}Al{^{31}P} REDOR NMR and ^{31}P{^{27}Al} REAPDOR NMR data for the sol-gel glass heat treated at 500°C are shown in Figure 3. As observed by Zhang & Eckert,[51] the different Al environments exhibit comparable dephasing behaviour. Simulations for the REDOR evolution curves were performed employing the SIMPSON[55] software package. Since the exact spin geometry and the distribution of the dipolar and quadrupolar interaction parameters are unknown, the data were analysed employing a two-spin approximation. The resulting REDOR evolution curve from such a two-spin approximation (D=700 Hz) is given together with the result for a full simulation assuming an AlP_4 five spin system (with a tetrahedral arrangement of the four ^{31}P spins around the central ^{27}Al spin), with Al–P distances of 3·3 Å (corresponding to individual ^{27}Al–^{31}P dipolar coupling constants

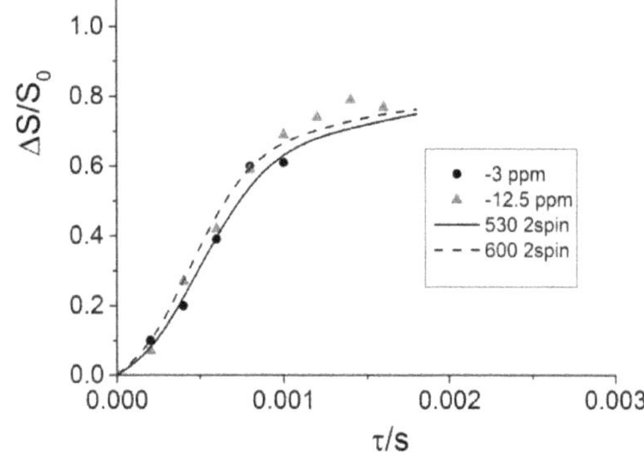

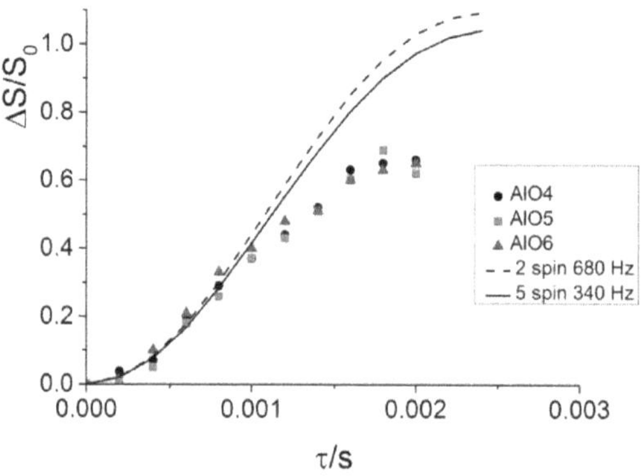

Figure 3. ^{31}P{^{27}Al} REAPDOR (left) and ^{27}Al{^{31}P} REDOR NMR data for the sol-gel glass SG-NAP-1 heat treated at 500°C. The lines in the left figure represent the result of a SIMPSON simulation (two-spin approximation), assuming dipolar coupling strengths of 530 Hz (solid curve) and 600 Hz (dashed curve), respectively. For the dominating AlO_4 signal in the REDOR experiment (right figure), a simulation employing the full AlP_4 spin geometry (solid line) and those using a two-spin approximation (dashed line) produce rather similar results, giving an Al–P distance of 3·3 Å. Data recorded at a field of 7·04 T

of 350 Hz). As shown for REDOR[56,57] and REAPDOR,[58] the analysis has to be restricted to the initial part of the evolution curve. As obvious from these results, the two-spin approximation produces rather similar results to the calculations employing the full spin geometry, thus providing reliable information about the structural motifs present in the glasses. The individual dipolar coupling constants between the involved nuclei, ^{27}Al and ^{31}P, are calculated according to $D_{ind}=\sqrt{(D^2_{2spin}/n)}$, with n denoting the number of ^{31}P nuclei interacting with ^{27}Al, and D_{2spin} the dipolar coupling as obtained from the two-spin approximation. This approach closely resembles the simulations of the REDOR curves employing the second moments approach following Eckert and co-workers,[39,57] but has the advantage of being applicable to the analysis of REAPDOR experiments and experiments in which pulse lengths $\neq\pi$ are used (e.g. constant time (CT)-REDOR). The second moment is then accessible from

$$M_2 = \frac{4}{15} 4\pi^2 I(I+1) \sum_j D^2_{ind}$$

With the help of the ^{31}P{^{27}Al} REAPDOR NMR experiment, the assignment to Q^0_{nAl} units (as given above) can be checked. The simulations employing the two-spin approximation produce dipolar coupling constants of 530 and 600 Hz for the signals at −3·0 ppm and −12·5 ppm, respectively (the intensity of the signal at −24 ppm proved too low for a meaningful analysis of the dipolar evolution curve). Assuming a $P(OAl)_2$ three spin system (−3·2 ppm) and a $P(OAl)_3$ four-spin system (−12·5 ppm), these values translate to individual dipolar coupling constants of 365 Hz (−3·2 ppm) and 350 Hz (−12·5 ppm), respectively, or interatomic P–Al distances of 3·25 and 3·28 Å, in reasonable agreement with published data for crystalline model compounds.[59–61] Thus, the REAPDOR results clearly corroborate the assignment of the signals at −3·0 ppm and −12·5 ppm to Q^0_{2Al} and Q^0_{3Al} units, respectively.

According to Zhang & Eckert,[51] glasses along the line Na_2O–$x$$AlPO_4$ are not accessible using MQ techniques. For NAP-1 (x=2) in our study, use of conventional MQ techniques usually resulted in opaque glasses, and only in rare cases was a clear, transparent glass obtained. However, glasses prepared via fast cooling employing the quadruple roller machine always proved completely transparent and x-ray amorphous. To our knowledge, glasses of this composition have not been successfully prepared before by using a MQ route. We note, however, that, especially for thin roller quenched glasses, the appearance of a clear glass and the absence of any Bragg reflections are not unequivocal criteria for the presence of a homogeneous, completely amorphous material. The MAS NMR spectra (cf. top spectra in Figure 6, see below) exhibit a clear resemblance to those of the sol-gel glass heat treated at ≥400°C, which indicates that both MQ and sol-gel glasses share common local structural motifs. We note, however, that the MQ glass has significantly lower amounts of AlO_5 and AlO_6 units (cf. Table 2) as compared to the sol-gel or QRQ glasses. In the ^{31}P MAS NMR spectra (cf. insert to Figure 4) four signals are found at 5·0 ppm (3%), −2·5 ppm (17%), −12·1 ppm (71%) and −22·7 ppm (8%), which can be assigned to Q^0_{1Al}, Q^0_{2Al}, Q^0_{3Al} and Q^0_{4Al} units.[51] Again, this assignment is corroborated by the ^{31}P{^{27}Al} REAPDOR NMR data, which are shown for the quadruple roller-quenched sample in Figure 4. Here, the simulations employing the two-spin approximation produce dipolar coupling constants of 520 and 620 Hz for the two signals at −3·2 ppm ($P(OAl)_2$ three-spin system, P–Al distances of 3·25 Å) and −12·5 ppm ($P(OAl)_3$ four-spin system, P–Al

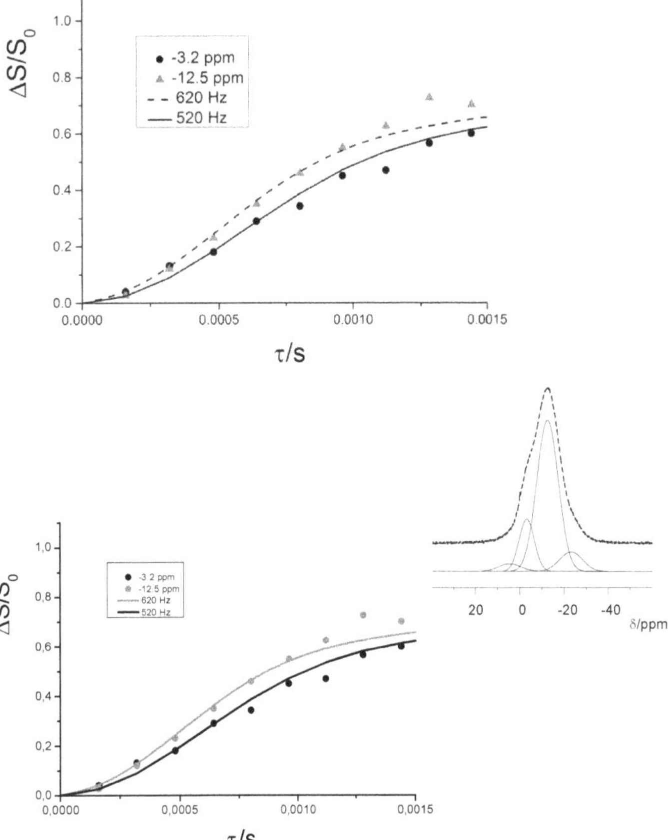

Figure 4. ^{31}P{^{27}Al} REAPDOR NMR data for the quadruple roller quenched glass QRQ-NAP-1. The solid lines represent results of a SIMPSON simulation (two-spin approximation) with dipolar coupling constants of 520 Hz (solid curve) and 620 Hz (dashed curve). The inset shows the ^{31}P MAS NMR spectrum together with its deconvolution into four individual lines. Data recorded at a field of 7·04 T

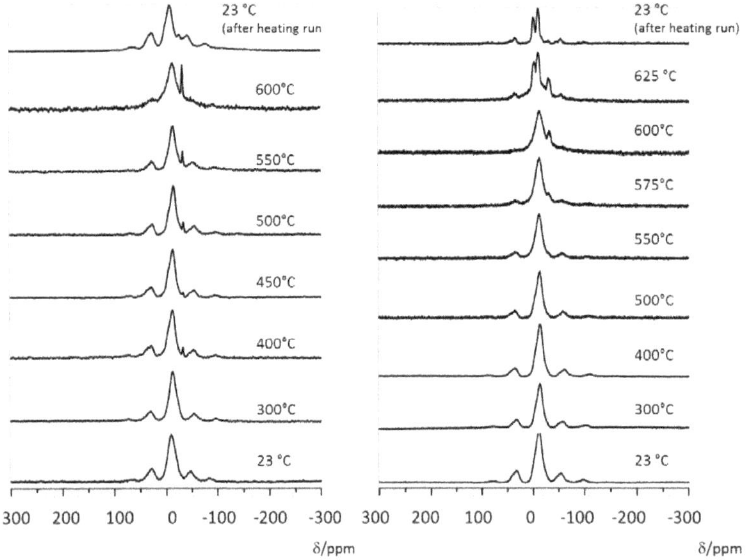

Figure 5. In situ ^{31}P MAS NMR spectra for the sol-gel glass (left) and the melt quenched glass (right) NAP-1 at the indicated temperatures. Spectra recorded at a field of 7·04 T

distances of 3·28 Å). The intensity of the remaining signals proved too low for a meaningful analysis. In addition, the $^{27}Al\{^{31}P\}$ CT-REDOR NMR experiment for the melt quenched glass[62,63] produces an effective dipolar coupling constant of 720 Hz for the dominant AlO_4 environment (corresponding to four ^{31}P neighbours at a distance of 3·27 Å, data not shown), which corroborates the assumption of a full coordination by phosphate groups. Thus, glass synthesis following the sol-gel route, as well as MQ routes, produce glasses in which more or less the same structural motifs on the investigated length scales are found. However, the glasses prepared via the three different routes exhibit a quite different crystallisation behaviour. This was investigated employing *ex situ* and *in situ* NMR approaches. The *in situ* ^{31}P MAS NMR spectra for the sol-gel and MQ glasses are shown in Figure 5. The appearance of a ^{31}P MAS NMR signal around −30 ppm indicates the crystallisation of an $AlPO_4$ phase at around 400°C for the sol-gel sample, and at 550°C for the MQ sample. Up to a temperature of 600°C, no further crystallisation is observed, but the remaining glass matrix becomes mobile, as indicated by the coalescence of the spinning sidebands and the individual signals. However, with a field of 7 T the dispersion of the chemical shift in combination with the maximum MAS frequency of 5 kHz yields ^{31}P MAS NMR spectra in which the spinning sidebands are only barely resolved, rendering an exact analysis of correlation times impossible. However, from the observed coalescence, we can roughly estimate the correlation time at 600°C as $\tau_c \approx 1\times10^{-4}$ s. The crystallisation of the remaining glass matrix is then observed for the MQ glass starting at a temperature of 625°C with two new signals arising at −2 ppm and −12 ppm. The *ex situ* studies were performed on glass samples heat treated at either 580°C for 1 h or 600°C for 24 h, respectively. The results are shown in Figure 6. Although in all cases the same end product of crystallisation is observed (two signals at −2 ppm and −11 ppm in the ^{31}P MAS NMR spectra, together with a signal at −30 ppm, which can be assigned to $AlPO_4$), the glasses definitely show a clear dependence of the crystallisation behaviour on the heat treatment history. Specifically, the MQ glass crystallises significantly faster than the other two glasses; the QRQ and sol-gel glasses exhibit a more similar crystallisation behaviour as the two melt quenched (MQ and QRQ) glasses. Crystallisation commences with the formation of $AlPO_4$ at 580°C, which for the MQ glass is accompanied by the crystallisation of at least one additional crystalline phase (see below). We do not have an explanation for this observation. One might speculate that, since the fast quenched QRQ glass features a higher fictive temperature, corresponding to an equilibrium frozen in at a higher temperature, the higher (configurational) entropy exacerbates the crystallisation of the

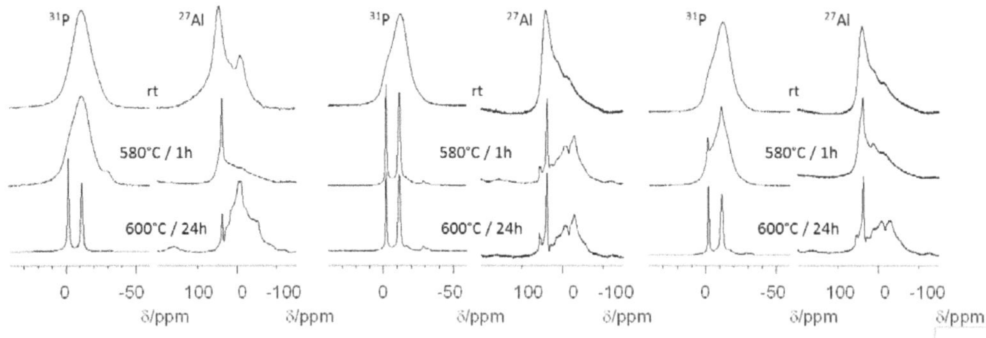

Figure 6. Ex situ ^{31}P and ^{27}Al MAS NMR spectra (^{31}P: left columns; ^{27}Al: right columns) for the three base glasses of composition NAP-1 and the corresponding spectra for the glasses heat treated at 580°C and 600°C; left: sol-gel glass; middle: MQ glass; right: QRQ glass. Spectra recorded at a field of 7·04 T

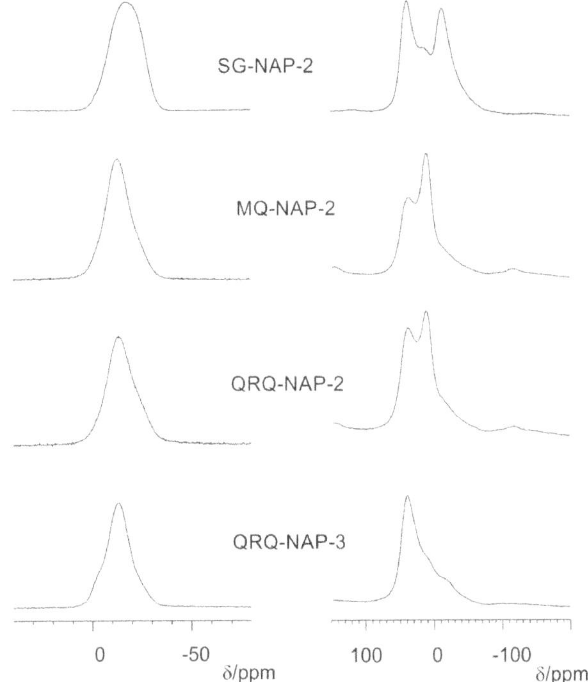

Figure 7. ^{31}P and ^{27}Al MAS NMR spectra for the MQ, QRQ and sol-gel NAP-2 glasses, and glass QRQ-NAP-3. Spectra recorded at a field of 7·04 T

QRQ glass relative to the MQ glass. However, we are aware of the fact that the dominant crystallisation mechanism will involve the surface. Of course, small variations in the composition may have drastic effects on the crystallisation behaviour, and compositional differences between the differently prepared samples cannot be completely ruled out. We note, however, that the glass synthesis from melt (MQ and QRQ) used the same temperature programme for the melting of the batches.

Attempts to prepare glasses with the extremely Al-rich composition NAP-2 ($Na_2O.4AlPO_4$) using a conventional quenching technique produced opaque, non-transparent materials. In the powder diffraction pattern for the MQ sample, reflections are found which can be assigned to α-Al_2O_3. The preparation following the QRQ route (after melting at 1450°C for 1 h) led to a completely transparent glass. However, in the XRD pattern, the α-Al_2O_3 reflections are also visible, but to a much lesser extent. The ^{31}P MAS NMR spectra (see Figure 7) exhibit three signals at −3·5 ppm, −12·5 ppm and −22·5 ppm for all three variants (SG, MQ and QRQ), which can be assigned as above to the structural units Q^0_{2Al} (−3·5 ppm), Q^0_{3Al} (−12·5 ppm) and Q^0_{4Al} (−22·5 ppm). The relative ratio of these signals, however, depends strongly on the synthesis condition (see Table 3). Interestingly, the intensity of the ^{31}P high field signal (−22·5 ppm) decreases with increasing heat treatment temperature for the sol-gel glass, and with decreasing fictive temperature for the melt-quenched glasses. In the ^{27}Al MAS NMR spectra, the presence of α-Al_2O_3 contributes a signal

Table 3. Relative fractions (in %) of the individual local phosphate structural units in glasses of composition NAP-2

	Sol-gel 300°C	Sol-gel 400°C	MQ	QRQ
Q^0_{2Al}	4	6	12	13
Q^0_{3Al}	41	51	64	57
Q^0_{4Al}	55	43	22	29

with δ_{iso}=14 ppm and C_Q=2·38 MHz, and thus would overlap with the signal originating from AlO_5 units. Simulations as described above produced relative fractions of 14% of the Al present in the sample as α-Al_2O_3 in the QRQ glass. The results of the ^{27}Al{^{31}P} REDOR NMR experiment (see Figure 8) reveal a decreasing coordination of the higher coordinated Al species by phosphate units. According to our results only the AlO_4 environments are fully coordinated by phosphate units. All these observations indicate that the homogeneous mixing of Al and P building units for this glass composition is only possible for the sol-gel glass; during the gel-to-glass conversion (at approximately 400°C) even here a partial demixing occurs, as evidenced by the decreasing number of Q^0_{4Al} units (see Table 3). The melt-quenched glasses exhibit an increased mixing with increasing fictive temperature. Again, sol-gel glass and QRQ glass exhibit a similar behaviour. For this extremely Al-rich composition, the quench rates accessible with the quadruple roller quench apparatus obviously are not quite high enough to fully incorporate the Al present in the melt into the glass matrix. Decreasing the Al/P ratio from 1 to 0·75 (NAP-3; again a composition that has not previously been accessible employing MQ techniques; ^{27}Al and ^{31}P MAS NMR spectra are shown in Figure 7), however, produces transparent, x-ray amorphous glasses when prepared via the QRQ route. We hope to be able to circumvent the remaining last traces of α-Al_2O_3 precipitates via a realisation of even higher fictive temperatures/quenching rates.

The obtained relative fractions of Q^0_{nAl}-phosphate

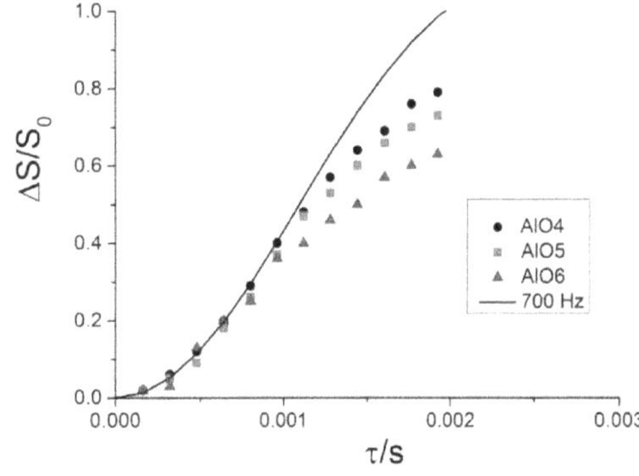

Figure 8. [27]Al{[31]P} REDOR data for the MQ glass of composition NAP-2. The REDOR evolution curve for the signal of the AlO_4 environment can be satisfactorily simulated (two-spin approximation) assuming an effective dipolar coupling constant of 700 Hz (corresponding to an $Al(OP)_4$ unit with Al–P distances of 3·3 Å). Data recorded at a field of 7·04 T

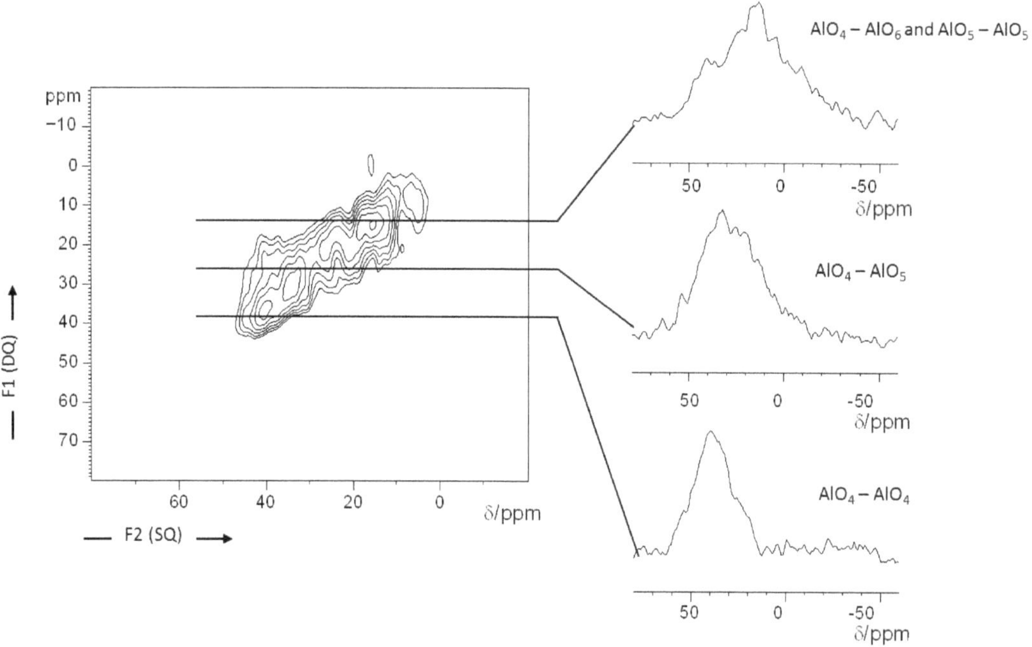

Figure 9. 2D ^{27}Al DQ-SQ NMR spectrum for the MQ glass of composition NAP-2. The value for k was set to 0·25, thus producing a chemical shift scaling in F1 of 0·5.[22,24] *With chemical shifts of 39 ppm (AlO_4), 17 ppm (AlO_5) and −10 ppm (AlO_6), the slices parallel to F2 indicate AlO_4–AlO_4 (39 ppm), AlO_4–AlO_5 (27 ppm) and AlO_4–AlO_6 and AlO_5–AlO_5 (13 ppm) correlations, respectively. Spectrum recorded at a field of 9·4 T*

environments and AlO_x units allows for a rough calculation of the maximum coordination of Al units by phosphate species, taking the amount of α-Al_2O_3 into account (which does not participate in the network formation). According to the results, an average number of three linkages to Al are available per phosphorus atom. These are needed to accommodate 4·4 linkages per Al to P (via oxygen). Thus, a portion of the oxygen atoms of the AlO_x environments remains unsaturated, which has to be accounted for by direct Al–O–Al bonding, even though this entails an incomplete charge balancing on these oxygen atoms.[7,64] This connectivity motif can be directly detected by employing advanced solid state NMR methods that trace the homonuclear dipolar ^{27}Al–^{27}Al couplings via a correlation of double quantum (DQ) and single quantum (SQ) coherences. First implementations are due in particular to Amoureux[18] and Taulelle;[19–21] the applicability to amorphous systems has first been published by Eden.[22–24] In short, the experiment relies on the generation of double quantum coherences between two ^{27}Al central transitions with the help of symmetry-based pulse sequence schemes (here ($BR2_2^1$)[18]). After reconversion into longitudinal polarization the single quantum coherence is then detected after the reading pulse. The main advantage of this two-dimensional DQ–SQ correlation over a SQ–SQ correlation experiment lies in the fact that here connectivity between two equivalent sites may be clearly detected whereas they are obscured by the diagonal signal in the SQ–SQ experiment. Any intensity in the DQ–SQ spectrum indicates connectivity between two Al-sites, thus Al–Al connectivity is clearly indicated in the 2D-DQ–SQ ^{27}Al spectrum depicted in Figure 9. Without attempting a quantitative evaluation, correlations of AlO_4 units to AlO_4, AlO_5 and AlO_6 environments are obvious. Very recently, Ren *et al*[65] established connectivity between all types of Al species present in binary aluminosilicate glasses using the same approach.

The crystallisation products after heat treatment of the MQ glasses of composition NAP-2 are identical to those for glass NAP-1; apart from the precipitation of $AlPO_4$, which commences at 580°C, at least one additional crystalline phase is formed (^{31}P signals at −2 ppm and −11 ppm, with a relative ratio of 1:1). So far it has not been possible to discover the nature of this crystalline phase; the XRD pattern could not be assigned to any known aluminophosphate phase. In the ^{23}Na MAS NMR spectrum, two different sites with a ratio of 1:1 can be identified (see Figure 10. Site 1: δ_{iso}=−2·4 ppm, C_Q=3·3 MHz, η_Q=0·9. Site 2: δ_{iso}=−5·7 ppm, C_Q=3·5 MHz, η_Q=0·9). In the ^{27}Al MAS NMR spectrum, apart from the signal at 38·5 ppm ($AlPO_4$), two signals are present (Site 1: δ_{iso}=22·0 ppm, C_Q=7·5 MHz, η_Q=0·5. Site 2: δ_{iso}=32·0 ppm, C_Q=5·7 MHz, η_Q=0·8), again with a ratio of 1:1. The spectra were recorded at two different field strengths and simulated employing one unique set of parameters. With the help of homo- and heteronuclear dipolar NMR spectroscopy, we checked whether all identified signals belong to a single phase. As evidenced from ^{31}P RFDR (radio frequency driven recoupling) NMR experiments, the two ^{31}P signals at −2 ppm and −11 ppm belong to the same phase; from J-resolved 2D ^{31}P MAS NMR, we can deduce that both phosphate units

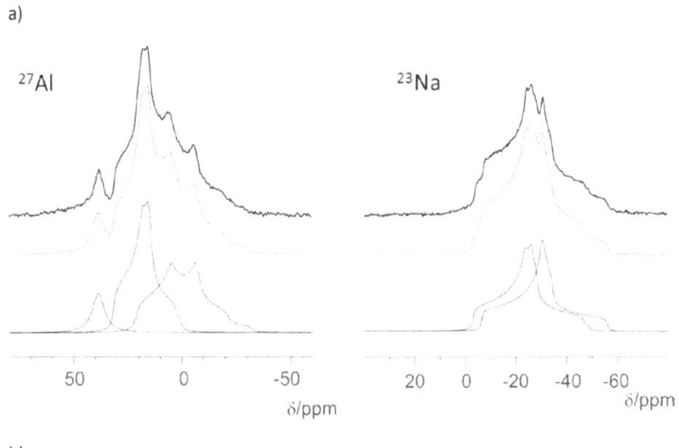

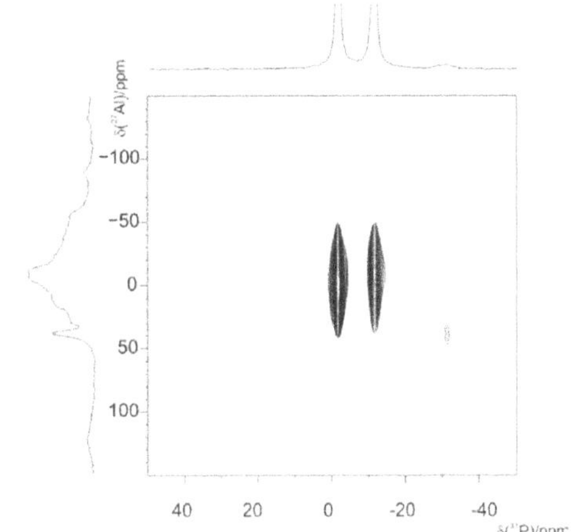

Figure 10. (a) ^{23}Na and ^{27}Al MAS NMR spectra for glass SG-NAP-1, heat treated at 600°C for 24 h. Spectra were acquired at a B_0 field strength of 11·7 T; ν_{MAS}=30 kHz; (b) ^{31}P{^{27}Al} CPMAS HETCOR MAS NMR spectrum. For the CP contact, very low RF powers (^{27}Al: 1 kHz; ^{31}P: ramped at 12 kHz) were used. Spectra recorded at a field of 11·7 T

do not have any connectivity to further phosphate groups. In the ^{27}Al{^{31}P} CPMAS (cross-polarization magic angle spinning)-HETCOR (heteronuclear correlation) spectrum (see Figure 10(b)), we can identify correlations between the ^{31}P signal at −30 ppm and the ^{27}Al signal at 38·5 ppm, corroborating the assignment of these signals to $AlPO_4$. In addition, the correlation between the two ^{31}P signals at −2 ppm and −11 ppm to both ^{27}Al signals suggests that both P species and both Al species belong to one single phase. This is supported by the results of ^{23}Na/^{31}P and ^{27}Al/^{31}P dipolar NMR experiments (REDOR/REAPDOR) (data not shown). Further work to elucidate the nature of this phase is in progress.

Glasses of compositions NAP-4 and NAP-5 have previously been prepared, employing sol-gel and MQ techniques.[51] The ^{31}P MAS NMR spectra of the NAP-5 glasses exhibit three clearly resolved signals at 6·9 ppm, −2·5 ppm and −10·8 ppm. With the help of ^{31}P{^{27}Al} REAPDOR NMR spectroscopy, as shown in Figure 11, these can be assigned to phosphate units with one, two or three AlO_x neighbours. An

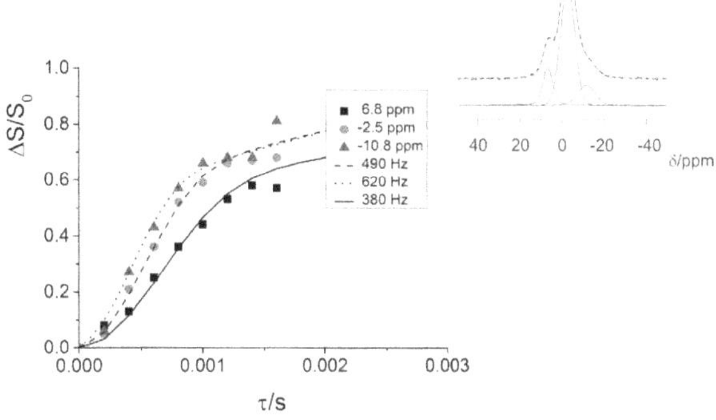

Figure 11. ^{31}P{^{27}Al} REAPDOR NMR data for glass MQ-NAP-5. The solid lines represent results of a SIMPSON-simulation (two-spin approximation) with dipolar coupling constants of 380 Hz (solid curve), 490 Hz (dashed curve) and 620 Hz (dotted curve). The inset shows the ^{31}P MAS NMR spectrum together with its deconvolution into three individual lines. Data recorded at a field of 7·04 T

analysis employing the two-spin approximation leads to coupling constants of 380, 490 and 620 Hz for the signals at 6·9 ppm, −2·5 ppm and −10·8 ppm, respectively, corresponding to Q^0_{1Al} (6·8 ppm, P–Al distance 3·2 Å), Q^0_{2Al} (−2·5 ppm, P–Al distance 3·3 Å) and Q^0_{3Al} (−10·8 ppm, P–Al distance 3·3 Å). For both compositions, the QRQ and sol-gel glasses exhibit a slightly (glass NAP-5) and significantly (glass NAP-4) higher fraction of the higher coordinated AlO_x environments. This is shown for NAP-4 in Figure 12. As for the glass compositions described above, the sol-gel and QRQ glasses seem to exhibit a retarded crystallisation tendency compared to MQ glass. Since it is impossible with our setup to assure constant synthesis parameters for the QRQ glasses (especially with respect to the time elapsed between removing

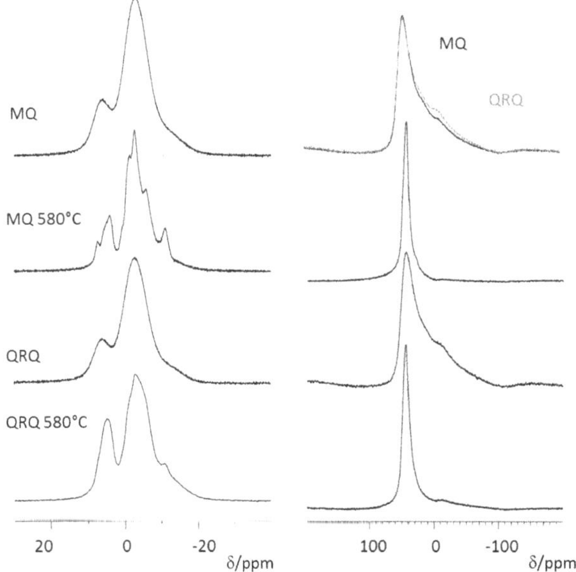

Figure 12. ^{31}P (left) and ^{27}Al MAS NMR spectra (right) for the MQ and QRQ glasses of composition NAP-4. Spectra recorded at a field of 7·04 T

the melt from the oven and pouring it onto the top rollers), these effects vary in magnitude. We note, however, that the increase in the higher coordinated Al species is, again, accompanied by a retarded tendency towards crystallisation. In glass NAP-5 the differences in the relative fractions of the AlO_x species are only marginal, consequently entailing only small differences in the crystallisation behaviour (data not shown).

Conclusions

Employing a quadruple roller quench (QRQ) setup, the preparation of melt-quenched aluminophosphate glasses in a previously inaccessible compositional range has proved successful. Upon heat treatment above the glass transition temperature, all glasses suffer from crystallisation. The sol-gel and QRQ glasses exhibit a rather similar high temperature behaviour, significantly different from that of the MQ glasses. For NAP-2, the retarded tendency towards crystallisation may be linked to a higher degree of mixing between phosphate and aluminate units. The relative fraction of higher coordinated Al species was found to increase with the fictive temperature of the glasses. A corresponding increase in the fraction of AlO_5 species with increasing cooling rate has also been observed in aluminosilicates.[66,67] According to the authors of this study, the AlO_5 environments contribute significantly to the entropy of the glasses. The notion that the sol-gel glasses and the glasses with a high fictive temperature show a close relationship and share a reduced tendency towards crystallisation may thus be explained by the higher entropy of these glasses compared to the conventionally melt-quenched glasses.

Acknowledgments

We are very grateful for valuable assistance and discussions with Prof. Jinjun Ren, Münster/Shanghai. Financial support from the Deutsche Forschungsgemeinschaft is gratefully acknowledged.

References

1. Weber, M. J. *J. Non-Cryst. Solids*, 1990, **123**, 208–222.
2. Pessani, J. P., Al-Ibrahim, N. S., Buzalap, M. A. R. & Toumba, K. J. *J. Appl. Oral Sci.*, 2008, **16**, 238–246.
3. Vogel, W., Holand, W., Naumann, K. & Gummel, J. *J. Non-Cryst. Solids*, 1986, **80**, 34–51.
4. Vogel, J. Wange, P. & Hartmann, P. *Glastech. Ber.-Glass Sci. Technol.*, 1997, **70**, 220–223.
5. Weber, W. J., Ewing, R. C., Angell, C. A., Arnold, G. W., Cormack, A. N., Delaye, J. M., Griscom, D. L., Hobbs, L. W., Navrotsky, A., Price, D. L., Stoneham, A. M. & Weinberg, M. C. *J. Mater. Res.*, 1997, **12**, 1946–1978.
6. Donald, I. W., Metcalfe, B. L. & Taylor, R. N. J. *J. Mater. Sci.*, 1997, **32**, 5851–5887.
7. Brow, R. K. *J. Non-Cryst. Solids*, 2000, **263**, 1–28.
8. Kirkpatrick, R. J. & Brow, R. K. *Solid State Nucl. Magn. Reson.*, 1995, **5**, 9–21.
9. Andrew, E. R., Bradbury, A. & Eades, R. G. *Nature*, 1959, **183**, 1802–1803.
10. Lowe, I. J. *Phys. Rev. Lett.*, 1959, **2**.
11. Eckert, H. *Prog. Nucl. Magn. Reson. Spectrosc.*, 1992, **24**, 159–293.
12. Born, R., Feike, M., Jäger, C. & Spiess, H. W. *Z. Naturforsch. A*, 1995, **50**, 169–176.
13. Hartmann, P. Jana, C. Vogel, J. Jager, C. *Chem. Phys. Lett.*, 1996, **258**, 107–112.
14. Bennett, A. E., Ok, J. H., Griffin, R. G. & Vega, S. *J. Chem. Phys.*, 1992, **96**, 8624.
15. Griffiths, J. M., Lakshmi, K. V., Bennett, A. E., Raap, J., Vanderwielen, C. M., Lugtenburg, J., Herzfeld, J. & Griffin, R. G. *J. Am. Chem. Soc.*, 1994, **116**, 10178–10181.
16. Lee, Y. K., Kurur, N. D., Helmle, M., Johannessen, O. G., Nielsen, N. C. & Levitt, M. H. *Chem. Phys. Lett.*, 1995, **242**, 304–309.
17. Tycko, R. & Dabbagh, G. *Chem. Phys. Lett.*, 1990, **173**, 461–465.
18. Wang, Q., Hu, B., Lafon, O., Trebosc, J., Deng, F. & Amoureux, J. P. *J. Magn. Reson.*, 2009, **200**, 251–260.
19. Mali, G. Fink, G. & Taulelle, F. *J. Chem. Phys.*, 2004, **120**, 2835–2845.
20. Mali, G. & Taulelle, F. *Chem. Commun.*, 2004, 868–869.
21. Mali, G., Kaucic, V. & Taulelle, F. *J. Chem. Phys.*, 2008, **128**.
22. Eden, M. *Solid State Nucl. Magn. Reson.*, 2009, **36**, 1–10.
23. Eden, M. *J. Magn. Reson.*, 2010, **204**, 99–110.
24. Lo, A. Y. H. & Eden, M. *Phys. Chem. Chem. Phys.*, 2008, **10**, 6635–6644.
25. Gullion, T. & Schaefer, J. *J. Magn. Reson.* 1989, **81**.
26. Gullion, T. *Chem. Phys. Lett.*, 1995, **246**, 325–330.
27. Gullion, T. *Concepts Magn. Reson.*, 1998, **10**, 277–289.
28. Grey, C. P. & Veeman, W. S. *Chem. Phys. Lett.*, 1992, **192**, 379–385.
29. Gullion, T. & Vega, A. J. *Prog. Nucl. Magn. Reson. Spectrosc.*, 2005, **47**, 123–136.
30. van Wüllen, L., Muller, U. & Jansen, M. *Angew. Chem.-Int. Ed.*, 2000, **39**, 2519–2521.
31. van Wüllen, L. & Jansen, M. *J. Mater. Chem.*, 2001, **11**, 223–229.
32. Brown, S. P., Perez-Torralba, M., Sanz, D., Claramunt, R. M. & Emsley, L. *Chem. Commun.*, 2002, 1852–1853.
33. Massiot, D. Fayon, F. Alonso, B. Trebosc, J. Amoureux, J. P. *J. Magn. Reson.*, 2003, **164**, 160–164.
34. Fayon, F., Massiot, D., Levitt, M. H., Titman, J. J., Gregory, D. H., Duma, L., Emsley, L. & Brown, S. P. *J. Chem. Phys.*, 2005, **122**.
35. Fayon, F., King, I. J., Harris, R. K., Evans, J. S. O. & Massiot, D. *Comptes Rendus Chim.*, 2004, **7**, 351–361.
36. van Wüllen, L., Tricot, G. & Wegner, S. *Solid State Nucl. Magn. Reson.*, 2007, **32**, 44–52.
37. Schaller, T., Rong, C., Toplis, M. J. & Cho, H. *J. Non-Cryst. Solids*, 1999, **248**, 19–27.
38. van Wüllen, L. & Schwering, G. *Solid State Nucl. Magn. Reson.*, 2002, **21**, 134–144.
39. Eckert, H., Elbers, S., Epping, J. D., Janssen, M., Kalwei, M., Strojek, W. & Voigt, U. In: *New Techniques in Solid-State NMR*, Vol. 246, 2005, pp. 195–233.
40. Farnan, I. & Stebbins, J. F. *J. Am. Chem. Soc.*, 1990, **112**, 32–39.
41. Farnan, I. & Stebbins, J. F. *Science*, 1994, **265**, 1206–1209.
42. Stebbins, J. F., Sen, S. & Farnan, I. *Am. Mineral.*, 1995, **80**, 861–864.
43. Stebbins, J. F. & Ellsworth, S. E. *J. Am. Ceram. Soc.*, 1996, **79**, 2247–2256.
44. Stebbins, J. F. & Sen, S. *J. Non-Cryst. Solids*, 1998, **224**, 80–85.
45. Kanehashi, K. & Stebbins, J. F. *J. Non-Cryst. Solids*, 2007, **353**, 4001–4010.
46. Wu, J., Potuzak, M. & Stebbins, J. F. *J. Non-Cryst. Solids*, 2011, **357**, 3944–3951.
47. van Wüllen, L., Wegner, S. & Tricot, G. *J. Phys. Chem. B*, 2007, **111**, 7529–7534.
48. Wegner, S., van Wüllen, L. & Tricot, G. *J. Phys. Chem. B*, 2009, **113**, 416–425.
49. Wegner, S., van Wüllen, L. & Tricot, G. *Solid State Sci.*, 2010, **12**, 428–439.
50. Venkatachalam, S., Schröder, C., Wegner, S. & van Wüllen, L. *Phys. Chem. Glasses Eur. J. Glass Sci. Technol. B*, 2014, **55**, 280–287.
51. Zhang, L. & Eckert, H. *J. Phys. Chem. B*, 2006, **110**, 8946–8958.
52. Zhang, L. & Eckert, H. *Solid State Nucl. Magn. Reson.*, 2004, **26**, 132–146.
53. Prabakar, S., Rao, K. J. & Rao, C. N. R. *Mater. Res. Bull.*, 1991, **26**, 805–812.
54. Massiot, D., Fayon, F., Capron, M., King, I., Le Calve, S., Alonso, B., Durand, J. O., Bujoli, B., Gan, Z. H. & Hoatson, G. *Magn. Reson. Chem.*, 2002, **40**, 70–76.
55. Bak, M. Rasmussen, J. T. & Nielsen, N. C. *J. Magn. Reson.*, 2000, **147**, 296–330.
56. Chan, J. C. C., Bertmer, M. & Eckert, H. *J. Am. Chem. Soc.*, 1999, **121**, 5238–5248.
57. Bertmer, M. &Eckert, H. *Solid State Nucl. Magn. Reson.*, 1999, **15**, 139–152.
58. van Wüllen, L., Muller, U. 7 Jansen, M. *Chem. Mater.*, 2000, **12**, 2347–2352.
59. Zhao, D. *Acta Cryst. E*, 2011, **67**, I64.
60. Alkemper, J., Paulus, H. & Fuess, H. *Z. Kristallogr.*, 1994, **209**, 616–616.
61. Chudinova, N. N., Palkina, K. K., Karmanovskaia, N. B., Maximova, S. I. & Chibiskova, N. T. *Dokl. Akad. Nauk SSSR*, 1989, **306**, 635-638.
62. Echelmeyer, T., Wegner, S. &van Wüllen, L. *Ann. Rep. NMR Spectrosc.*, 2012, **75**, 1–23.
63. Echelmeyer, T., van Wüllen, L. & Wegner, S. *Solid State Nucl. Magn. Reson.*, 2008, **34**, 14–19.
64. Brow, R. K., Kirkpatrick, R. J. & Turner, G. L. *J. Am. Ceram. Soc.*, 1993, **76**, 919–928.
65. Ren, J., Zhang, L. & Eckert, H. *J. Phys. Chem. C*, 2014, **118**, 4906–4917.
66. Thompson, L. M. & Stebbins, J. F. *Am. Miner.*, 2011, **96**, 841–853.
67. Stebbins, J. F., Dubinsky, E. V., Kanehashi, K. & Kelsey, K. E. *Geochim. Cosmochim. Acta*, 2008, **72**, 910–925.

Phys. Chem. Glasses: Eur. J. Glass Sci. Technol. B, August 2016, 57 (4), 187–192

Spectroscopic studies of zinc manganese metaphosphate glasses

J. Holubová,[1] Z. Černošek,[1] E. Černošková[2] & M. Nádvorník[1]

[1] *Department of General and Inorganic Chemistry, Faculty of Chemical Technology, University of Pardubice, Studentská 573, 532 10 Pardubice, Czech Republic*
[2] *Joint Laboratory of Solid State Chemistry of Institute of Macromolecular Chemistry of Czech Academy of Sciences, v.v.i., and University of Pardubice, Studentská 84, Pardubice, Czech Republic*

Manuscript received 14 October 2014
Revised version received 29 July 2015
Accepted 7 September 2015

Metaphosphate glasses from the xMnO.(50−x)ZnO.50P_2O_5 system were prepared, spanning the whole composition range $0 \leq x \leq 50$. Their structure and properties were studied by Raman spectroscopy, electron paramagnetic resonance, UV-Vis spectroscopy, magnetic susceptibility and DC conductivity. The metaphosphate structure of the glasses was maintained throughout the whole compositional range. Manganese was found to be predominantly in the +II oxidation state, but the colour of the glasses arises from traces of Mn(III). EPR spectra reveal that Mn(II) has slightly axially distorted octahedral coordination, with predominantly covalent Mn–O bonds in the xy-plane, whilst the bonds in the z-direction have a prevalently ionic character. The DC conductivity follows a simple Arrhenius relation, and the activation energy and pre-exponential factor are found to be compositionally independent, with mean values E^*=1·13±0·05 eV and σ_0=4·3±2·1 $\Omega^{-1}cm^{-1}$. The low value of σ_0 indicates band tail conduction. Optical absorption spectra show that localised manganese states in the energy gap are responsible for the DC conductivity of these glasses.

1. Introduction

Phosphate glasses offer an extremely wide range of possibilities for tailoring of physico-chemical properties for specific technological applications. These glasses have many desirable properties in comparison with silicate glasses, such as low melting temperature, high thermal expansion coefficient, and transmission in the ultraviolet region.[1–3] The introduction of a transition metal, such as manganese, molybdenum, etc. usually in multivalent states, into phosphate glasses leads to changes in the structure and results in interesting changes of their electrical, optical or magnetic properties, opening up new opportunities for finding new applications of these materials.[4–6] Recently, glasses containing MnO have frequently been studied, because manganese may be advantageously used as a paramagnetic probe for exploring the structure of phosphate or borophosphate glasses.[7–12]

In terms of P_2O_5 content, and thus the content of basic phosphate structural units, Q^0–Q^3, phosphate glasses can be divided into several groups.[1,13] Our attention is focused on the metaphosphate glasses with PO_3^- as the basic structural unit (i.e. Q^2 units), for which we can expect a wide glass forming area due to the known fact that metaphosphoric acid and its salts produce linear and cyclic polymer structure. The main objective of this work was to study metaphosphate glasses from the xMnO.(50−x)ZnO.50P_2O_5 system, varying the composition between the two end-member stoichiometric compositions, zinc and manganese metaphosphate.

Corresponding author. Email Jana.Holubova@upce.cz
Original version presented at Int. Conf. on Phosphate Glasses, Pardubice, Czech Republic, 2–4 July 2014
DOI: 10.13036/17533562.57.4.075

2. Experimental

Metaphosphate glasses from the ternary system ZnO–MnO–P_2O_5 were prepared with compositions xMnO.(50−x)ZnO.50P_2O_5, x=0, 2, 5, 7, 10, 15, 20, 30, 40, and 50 by a conventional melt-quenching method, using ZnO, MnO (both 99·9%, Sigma-Aldrich) and H_3PO_4 (85%, p.a., Sigma-Aldrich) as ingredients. The glasses were prepared in 20 g batches; the appropriate amounts of oxides and acid were mixed in platinum crucibles and heated slowly up to 600°C for approximately 2 h for the final calcination. Then the mixture was heated up to 1100–1260°C according to the manganese content, and melted for 10 min at the final temperature in an electric furnace. The molten material was quenched at room temperature by pouring into a graphite mould, and the prepared glasses were slowly cooled to room temperature. Homogeneous glasses were prepared over the whole concentration range, and the amorphous character was confirmed by x-ray diffraction. Chemical composition was checked by x-ray fluorescence analysis (μ-XRF analyser EAGLE II).

The density of the glasses, ρ, was determined using the standard Archimedean method with toluene as the immersion liquid (ρ_0=0·8669 g cm^{-3}). The molar volume, V_m, was calculated according to relation $V_m=\bar{M}/\rho$, where $\bar{M}$ is the average molar weight of the glass.

Raman spectra were recorded on bulk samples at room temperature using a Raman confocal microscope LabRam HR (HORIBA Jobin Yvon). The spectra were recorded in back-scattering geometry under excitation with a Nd:YAG laser with wavelength 532 nm and power 5 mW.

The magnetic susceptibility measurements were made using a Variable Temperature Gouy Balance System (Newport Instruments, UK).

Optical absorption spectra of the glasses were recorded in the range 190–1200 nm on a UV-Vis-NIR spectrometer Lambda 1050 (Perkin-Elmer) using glass plates of ~2 mm thickness.

EPR measurements on powder samples were carried out on the X-band (~9·5 GHz) spectrometer ERS 221 (ZWG Berlin) at room temperature. Computer simulation of the spectra was performed using WINEPR SimFonia (Bruker).

The DC conductivity was measured by the method of V–A characteristics (from ±3 V to 0 V) on a computer driven apparatus with a Keithley 6480 picoammeter, and graphite electrodes, over the temperature range 50–330°C, and vacuum ~10 Pa.

3. Results and discussions

The XRF analysis of the final glass compositions is summarised in Table 1. The analysed compositions are very close to the nominal batch values, giving good confidence in the composition of the glasses. It was found that the composition does not change significantly, and that there is no expressive P_2O_5 loss at higher melting temperatures.

3.1. Raman Spectroscopy

The Raman spectra of Zn-Mn metaphosphate glasses, see Figure 1, have a very simple pattern, dominated by two strong bands. For the base glass

Table 1. Nominal and analysed compositions of MnO–ZnO–P_2O_5 glasses, and summary of DC conductivity parameters

Batched/Analyzed/mol% ±0·7 MnO	ZnO	P_2O_5	$\sigma_0/\Omega^{-1}cm^{-1}$	E^*/eV
0/0	50/ 52·7	50/47·3	insulator	
2/2·0	48/ 49·6	50/48·4	2·79	1·12
5/4·7	45/ 46·4	50/48·9	3·65	1·13
7/7·7	43/ 42·6	50/49·7	3·15	1·16
10/11·8	40/ 38·7	50/49·5	17·33	1·22
15/14·6	35/ 35·0	50/50·4	2·11	1·10
20/19·9	30/ 29·8	50/50·3	6·39	1·18
30/29·9	20/ 20·1	50/50·0	4·81	1·12
40/39·6	10/ 9·8	50/50·6	5·46	1·11
50/50·4	0/0	50/49·6	3·96	1·11

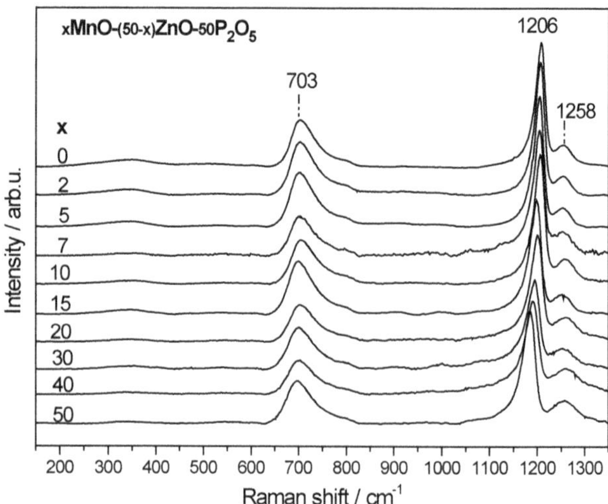

Figure 1. Raman spectra of xMnO.(50–x)ZnO.50P_2O_5 glasses

50ZnO.50P_2O_5, the first band is at 1207 cm^{-1}, and can be assigned to the symmetric stretching mode of the terminal oxygen, $\nu_s(PO_2^-)$ on Q^2 tetrahedra. The second band at 703 cm^{-1} is due to the symmetric stretching vibration of P–O–P bridges, ν_s(P–O–P).[13] The energies of these bands decrease gradually to 1184 cm^{-1} and to 697 cm^{-1}, respectively, as Zn is substituted by Mn. Besides these two bands, a broad band is observed at 1258 cm^{-1}, shifting slightly to 1252 cm^{-1} with increasing Mn content. This band can be assigned to the asymmetric stretching mode, $\nu_{as}(PO_2^-)$, of Q^2 units.[7,15]

The Raman spectra show unambiguously that the basic glass structure consists mainly of Q^2 structural units, forming a metaphosphate network with a chain (or ring) arrangement. The metaphosphate structure of the glasses is maintained throughout the whole compositional range.

3.2. UV-Vis spectroscopy and magnetic susceptibility

The prepared base zinc metaphosphate glass is colourless, while the colour of glasses containing manganese changes from light to very dark purple colour with increasing Mn content. Previous studies of glasses containing Mn have found that manganese can be present in +II and +III oxidation states, and only trivalent manganese ions are responsible for the purple colour of the glasses.[7,12] Mn(II) has a d^5 electronic structure with ground level configuration $^6A_{1g}$, while all excited states are spin quartet states, and thus all electronic transitions are parity and mainly spin forbidden. Hence, the optical absorption bands are very weak. The ground level of d^4 Mn(III), ^5D, is split into four states ($^5B_{1g}$, $^5A_{1g}$, $^5B_{2g}$ and ^5E_g) in an octahedral field, leading to three spin-allowed electronic transitions. One of these transitions ($^5B_{1g}\rightarrow{}^5B_{2g}$, ~530 nm) is responsible for the colouring of the glasses.[7]

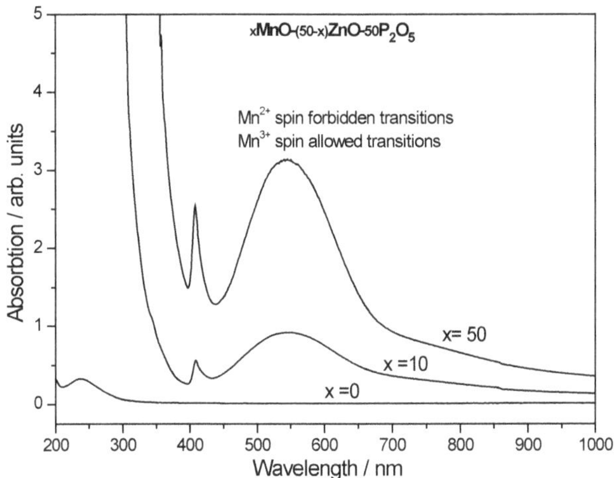

Figure 2. Representative optical absorption spectra of xMnO.(50–x)ZnO.50P$_2$O$_5$ (x=0, 10 and 50) glasses

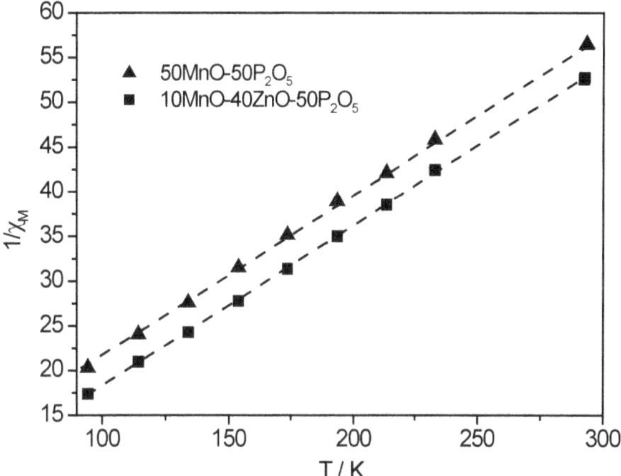

Figure 3. Temperature dependence of reciprocal magnetic susceptibility for xMnO.(50–x)ZnO.50P$_2$O$_5$ glasses with x=10 and 50

We have found that Mn-free glass shows no absorption band in the visible range, while the spectra of glasses containing manganese exhibit a broad band with maximum intensity at ~545 nm, and a narrow band at 409 nm, see Figure 2. The broad absorption band at ~545 nm results from the superposition of d–d spin-forbidden transitions of Mn(II) and spin-allowed transitions of Mn(III) that are much more intense, and so the colouration of glass can be explained by the presence of Mn^{3+}. As will be shown later, this band is also important for an explanation of the DC conductivity of the glasses. The latter band results from the d–d transition of Mn(II) and is not significant for this discussion.

An unambiguous determination of the oxidation state of manganese can be obtained by measurement of the magnetic susceptibility. For example, the temperature dependence of the magnetic susceptibility of glasses with 10 and 50 mol % MnO is presented in Figure 3. It is found that the magnetic susceptibility follows the Curie–Weiss law:

$$\chi = \chi_0 + \frac{c_M}{T - \theta_p} \tag{1}$$

where χ_0 is a temperature independent constant, θ_p is the Curie temperature, $c_M=(\mu_{eff}^2 N_A)/3k_B$ is the molar Curie constant, N_A is Avogadro's number, and k_B is the Boltzmann's constant.

From the values of magnetic susceptibility, the effective magnetic moment of manganese ions was evaluated for the selected glasses, and it was found that its value ($5{\cdot}94\mu_B$) does not change with MnO concentration. The experimentally obtained value is very close to the atomic magnetic moment values of free Mn^{2+} ($\mu_{eff}(Mn^{2+})=5{\cdot}92\mu_B$[8]), which implies that, within the sensitivity of the method (±1%), the manganese in the glasses is in the high-spin d^5 configuration, i.e. Mn(II). The purple colour of the glasses indicates the presence of the Mn^{3+} ion, but at concentrations below the limit of sensitivity of magnetic susceptibility, i.e. less than 1 mol% Mn(III).

Magnetic susceptibility also provides information concerning the nature of the magnetic interactions between manganese ions, based on the experimentally obtained value of θ_p. The obtained negative values of θ_p (−2·8 K for x=10; −22·2 K for x=50) indicate that there is antiferromagnetic interaction between the manganese ions. The absolute value of this temperature increases with increasing MnO content, suggesting that the magnetic interactions become stronger with increasing manganese content.

3.3 Density and molar volume

The density of the glass samples and the molar volume are shown in Figure 4 as a function of composition. The density increases from 2·828 g cm^{-3} for the MnO-free glass, to 2·928 for the 50 mol% MnO glass, while the molar volume decreases from 39·48 to 36·35 cm^3 as the MnO content increases. The density of the glasses increases with increasing MnO concentration despite the fact that a heavier element (Zn) is being replaced by a lighter one (Mn) in the same ratio. Similarly there is a decrease in molar volume. This

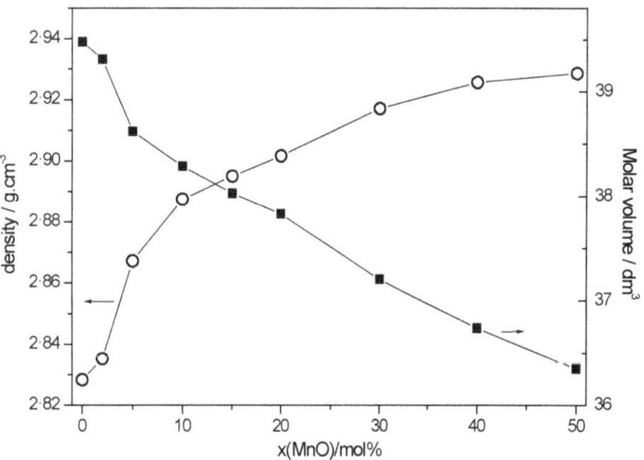

Figure 4. The density and molar volume of xMnO.(50–x) ZnO.50P$_2$O$_5$ glasses

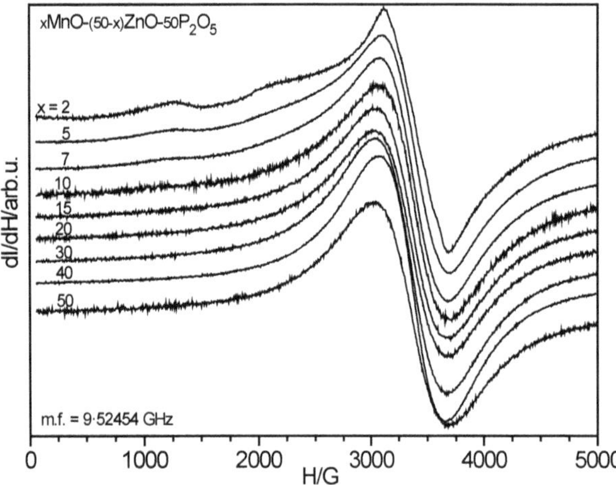

Figure 5. EPR spectra of the glass series xMnO.(50−x) ZnO.50P$_2$O$_5$

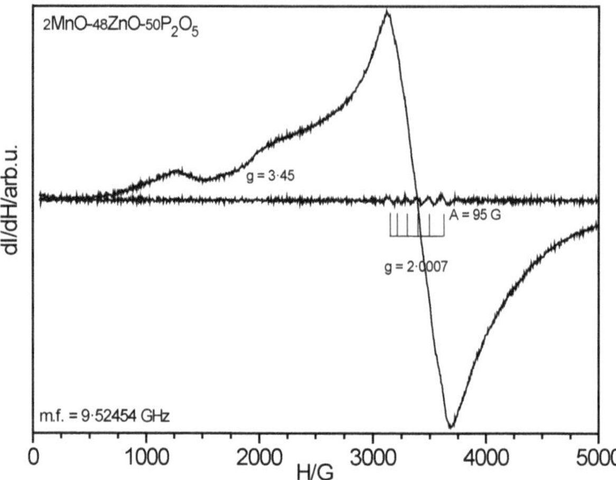

Figure 6. EPR spectrum of the 2MnO.48ZnO.50P$_2$O$_5$ glass

behaviour indicates the increasing compactness of the glassy network with increasing manganese content.

3.4. Electron paramagnetic resonance

No EPR signals were detected for the Mn-free base glass 50ZnO.50P$_2$O$_5$. The glasses containing Mn are paramagnetic, and the EPR spectra of these glasses are shown in Figure 5. Mn(II) with d^5 configuration has a $^6S_{5/2}$ ground state for the free ion and zero orbital angular momentum, and thus the spectra consist mainly of resonance line centred close to the g-factor value of a free electron (g=2·0023).[15] This resonance line arises from the main $M_s=|-1/2\rangle \rightarrow |+1/2\rangle$ transition of Mn^{2+} with approximately octahedral symmetry. The line is broadened by both hyperfine splitting (HFS) arising from the interaction of unpaired electrons and the spin of the ^{55}Mn nucleus (M_I=5/2), as well as by zero field splitting (d^5-system). More information on this EPR prominent ion can be found elsewhere, see for example Ref. 15 and references cited therein.

The compositional dependence of the spectral shape corresponds well to spectra reported in the literature.[7–10] The EPR spectra of glasses with lower Mn concentration (x≤7) contain a second paramagnetic centrum, at g=3·4, see Figure 6. This paramagnetic centrum is associated with magnetically isolated Mn^{2+} ions at tetragonally or rhombically distorted octahedral sites.[7,8,16] This signal is less intense compared to the main signal at g~2, indicating that the majority of Mn^{2+} ions occupy octahedral sites that are only slightly distorted. Six lines of HFS, superimposed on the main resonance line at g~2, can usually be observed for phosphate glasses with manganese concentration less than 1 mol%, because the HFS sextet is a sign of isolated Mn^{2+} ions in a highly ordered environment close to octahedral symmetry.[7–12] The starting concentration of manganese in the glasses used in this study is higher (2 mol%), and therefore the HFS is smeared due to dipole–dipole interaction accompanied by the superexchange magnetic interaction. Only in the case of the lowest manganese content (x=2) can the resolved HFS sextet be isolated by subtracting a numerically filtered experimental spectrum from the original experimental spectrum, see Figure 6. The extracted HFS was found to be A≅95 G, which is in good agreement with Ref. 9.

With increasing Mn content, the resonance with g~3·45 disappears, leaving behind only the broad resonance at g~2, and the shape of the spectra is almost identical. To obtain characteristic parameters, a computer simulation of the spectrum of the glass with the highest Mn content (50MnO.50P$_2$O$_5$) was performed, and the result is shown in Figure 7. The following parameters were obtained: Lorentzian band-shapes of width $l_x=l_y$=500 G, l_z=780 G; principal values of g-tensor $g_x=g_y$=2·0200, g_z=2·0300; zero fine splitting $D=35\times10^{-4}$ cm^{-1}, E=0 cm^{-1} and hyperfine splitting $A_x=A_y$=27 G, A_z=96 G. Generally, the value of the hyperfine splitting, A, is a measure of the delo-

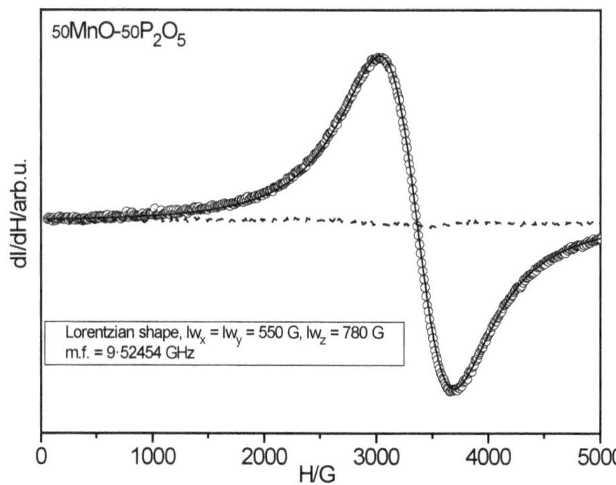

Figure 7. Experimental EPR spectrum of 50MnO.50P$_2$O$_5$ (circles), its computer simulation (solid line) and their diference (dashed)

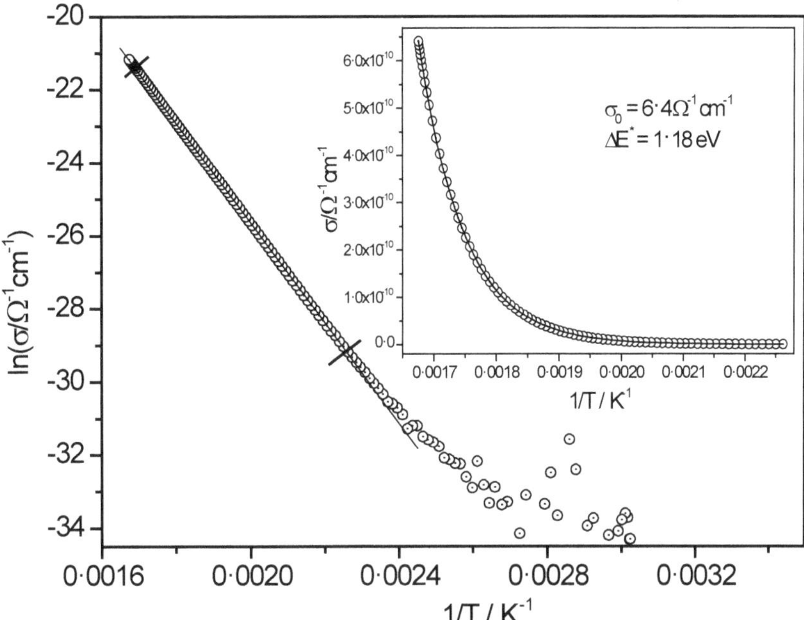

Figure 8. ln(σ) versus inverse temperature (where σ is DC conductivity) for 30MnO.20ZnO.50P$_2$O$_5$ glass (circles), the region of linear dependence is marked out by solid lines. The inset shows the corresponding temperature dependence of σ (circles) and its nonlinear regression (solid line); for details see text

calisation of unpaired electrons, and therefore qualitatively reflects the covalency of bonding between the paramagnetic ion and its ligand, i.e. the smaller the splitting, the higher the delocalisation, and thus the nature of the bonding is more covalent.[17]

Thus, based on computer simulation, one can conclude that Mn(II) has axially distorted octahedral coordination with predominantly covalent Mn–O bonds in the *xy*-plane, whilst bonds in the *z*-direction have a prevalently ionic character. It can therefore be expected that manganese plays a bridging role, covalently crosslinking the Q^2 phosphate chains into *xy*-planes, and subsequently interconnecting these planes by ionic bonds in the *z*-direction. This conclusion is in good agreement with the increasing compactness of the structure as was found above by density measurement.

3.5. DC conductivity

The temperature dependence of the DC electrical conductivity, σ, was measured for all the manganese containing glasses. The base glass 50ZnO.50P$_2$O$_5$ is an insulator; based on the optical spectroscopy measurement its band gap is slightly over 6·5 eV. Since the *I–V* characteristics were linear over the whole range of measurement, the dominant charge carriers in these glasses are electrons.[18]

A commonly used linear transformation (lnσ versus $1/T$) was used for easier visualisation of the experimental data, Figure 8, but to eliminate the influence of the transformation on the results, nonlinear regression of the untransformed data corresponding to the linear part of lnσ versus $1/T$ (see inset to Figure 8) was used to obtain the activation energy and pre-exponential factor according to

$$\sigma=\sigma_o\exp(-E^*/k_BT) \quad (2)$$

where σ_o is the pre-exponential factor, E^* is activation energy for DC conduction, and k_B is the Boltzmann constant. The values of σ_o and E^* are collected in Table 1. It was found that neither activation energy nor pre-exponential factor depend on manganese content, and their mean values were found to be E^*=1·13±0·05 eV and σ_o=4·3±2·1 Ω^{-1}cm^{-1}. The low value of the pre-exponential factor, which is three orders lower that this one for extended state conduction,[19] indicates band tail conduction.

The optical measurements described above reveal deep levels of manganese in the band gap with a maximum at 545 nm, i.e. 2·27 eV (see Figure 2). This energy corresponds closely to twice the value of the DC conductivity activation energy (i.e. to 2·26 eV). From this result and also from the compositional independence of both activation energy and pre-exponential factor, it follows that DC conduction occurs by means of these localised states of manganese in the band gap.

4. Conclusions

Metaphosphate glasses from the *x*MnO.(50–*x*)ZnO.50P$_2$O$_5$ system were prepared over the whole concentration range 0≤*x*≤50. Raman spectroscopy confirmed the preservation of the metaphosphate structure for all the prepared glasses.

Manganese was found to be predominantly (more than 99%) in the +II oxidation state. On the other hand traces of Mn(III) are the reason for the purple colour of the glasses. EPR spectra revealed that the Mn(II) site has axially distorted octahedral coordination, with predominantly covalent Mn–O bonds in the xy-plane, whilst bonds in the z-direction have prevalently ionic character. It can therefore be expected that manganese plays a bridging role, covalently crosslinking Q^2 phosphate chains into xy-planes, and also interconnecting these planes by ionic bonds in the z-direction into a 3D glassy network.

The temperature dependence of the DC conductivity follows a simple Arrhenius relation at higher temperatures, and the activation energy and pre-exponential factor were found to be compositionally independent. Localised levels of manganese in the energy band gap were found by optical spectroscopy at energy twice the DC activation energy, and so DC conduction is carried throughout these states. The pre-exponential factor indicates band tail conduction.

Finally, it can be concluded that manganese is incorporated into the glass network, increasing its dimensionality up to 3D, and thereby increasing its strength, and through its localised states in the band gap making the glasses DC conducting.

Acknowledgment

The authors would like to thank Dr Jiří Schwarz, University of Pardubice, for DC conductivity measurements.

References

1. Brow, R. K. Review: the structure of simple phosphate glasses, *J. Non-Cryst. Solids* 2000, **263&264**, 1–28.
2. van Wazer, J. *Phosphorous and its Compounds*, Vols. 1 and 2, Interscience, New York, 1951.
3. Mackenzie, J. D. *Modern Aspects of the vitreous state*, London: Butterworths, 1964, Vol. 2.
4. Merciera, C., Palavit, G., Montagne, L. & Follet-Houttemane, C. A survey of transition-metal-containing phosphate glasses, *Comptes Rendus Chim.*, 2002, **5**, 693–703.
5. Koudelka, L., Rosslerová, I., Holubová J., Mošner, P., Montagne, L. & Revel, B. Structural study of PbO-MoO_3-P_2O_5 glasses by Raman and NMR spectroscopy, *J. Non-Cryst. Solids*, 2011, **357**, 2816–2821.
6. Boudlich, D., Haddad, M., Nadiri, A., Berger, R. & Kliava, J. Mo^{5+} Ions as EPR Structural Probes in Molybdenum Phosphate Glasses, *J. Non-Cryst. Solids*, 1998, **224**, 135–42.
7. Konidakis, I., Varsamis, C.-P. E., Kamitsos, E. I., Möncke, D. & Ehrt, D. Structure and properties of mixed stroncium-manganese mrtaphosphate glasses, *J. Phys. Chem. C*, 2010, **114**, 9125–9138.
8. Pascuta, P., Bosca, M., Borisi, G. & Culea, E. Thermal, structural and magnetic properties of some zinc phosphate glasses doped with manganese ions, *J. Alloys Compd.*, 2011, **509**, 4314–4319.
9. Toloman, D., Giurgiu, L. M. & Ardelean, I. EPR investigations of calcium phosphate glasses containing manganese ions, *Physica B*, 2009, **404**, 4198–4201.
10. Moguš-Milanković, A., Pavić, L., Srilatha, K., Srinivasa Rao, C., Srikumar, T., Sandhi, Y. & Veeraiah, N. Electrical, dielectric and spectroscopic studies on MnO doped LiI-AgI-B_2O_3 glasses, *J. Appl. Phys.*, 2012, **111**, 013714 (1–11).
11. Kawamo, M., Takebe, H. & Kuwabara, M. Compositionals dependence of the luminiscence properties of Mn^{2+}-doped metaphosphate glasses, *Opt. Mater.*, 2009, **32**, 277.
12. Krishna Mohan, N., Rami Reddy, M., Jayasankar, C. K. & Veeraiah, N. Spectroscopic and dielectric studies on MnO doped PbO–Nb_2O_5–P_2O_5 glassy system, *J. Alloys Compd.*, 2008, **458**, 66–76.
13. Brow, R. K., Alam, M. T., Tallant, D. R. & Kirkpatrick, J. Spectroscopic studies on the structure of phosphate sealing glasses, *MRS Bull.*, 1998, 63–67.
14. Tischendorf, B., Utaigbe, J. U., Wiench, J. W., Pruski, M. & Sales B. C. A study of short and intermediate order in zinc phosphate glasses, *J. Non-Cryst. Solids*, 2001, **282**, 147–158.
15. Griscom, D. L. Electron spin resonance in glasses, *J. Non-Cryst. Solids*, 1980, **40**, 211–272.
16. Bogomolova, L. D., Tepliakov, Yu. G. & Caccavale, F. EPR of some oxide glasses implanted with Mn^+ and Cu^+ ions, *J. Non-Cryst. Solids*, 1996, **194**, 291–296.
17. van Wieringen, J. S. Paramagnetic resonance of divalent manganese incorporated in various lattices, *Discuss. Faraday Soc.*, 1955, **19**, 118–126.
18. Šantic, B., Moguš-Milankovic, A. & Day, D. E. The dc electrical conductivity of iron phosphate glasses, *J. Non-Cryst. Solids*, 2001, **296**, 65–73.
19. *Amorphous Semiconductors*, Ed. M. H. Brodsky, Topics in Applied Physics,, Vol. **36**, Springer Verlag, Berlin-Heidelberg, 1979.

Phys. Chem. Glasses: Eur. J. Glass Sci. Technol. B, October 2016, **57** (5), 206–212

Characterisation of a new NZP material prepared from reactive sintering of a phosphate based glass

S. Chenu, P. Bénard-Rocherullé, R. Lebullenger & J. Rocherullé

Chemical Sciences Institute, Glass and ceramic group, UMR CNRS 6226, University of Rennes, France

Manuscript received 29 September 2014
Revision Received 6 July 2015
Manuscript accepted 7 July 2015

NZP materials are based on the $NaZr_2(PO_4)_3$ type structure, but are difficult to fabricate into monoliths, due to the high temperatures and long sintering times required for the synthesis. We propose reactive sintering of a phosphate based glass ($60NaPO_3.20SnO.20WO_3$) as an alternative route to prepare a new tungsten (IV) and tin (IV) containing crystalline compound with the NZP structure. Using various techniques, we have shown that different parameters influence the achievement of a single crystalline phase with the NZP structure, such as the initial glass composition, glass particle size or curing atmosphere. Compared to the parent glass, the glass-ceramic exhibits an important increase of T_g and of the dilatometric softening point. The elastic properties are also enhanced, while the coefficient of thermal expansion decreases. Leaching tests show that NZP phase is not soluble, in contrast to the glass phase. Indexing of the diffraction pattern was carried out in order to determine the unit cell parameters and the space group. Rietveld refinement of the diffraction pattern, combined with analysis by energy-dispersive x-ray spectroscopy, demonstrates that tungsten is incorporated on a mixed tin and tungsten site, and almost 90% of these octahedral $M^{IV}O_6$ sites are occupied by Sn^{4+} ions, and the remaining 10% by W^{4+} ions.

1. Introduction

Phosphate glasses generally a have low melting temperature, together with a large coefficient of thermal expansion, and a wide range of compositions is available. These glasses have been intensively studied for numerous applications: optical moulding,[1] laser hosts[2] sealing glasses,[3] biomaterials,[4] nuclear waste immobilisation or solid electrolytes.

Among the various glass-ceramics with a high ionic conductivity, the most representative is the NZP family, which is derived from the $NaZr_2(PO_4)_3$ type structure. This structure is a three-dimensional arrangement of ZrO_6 octahedra linked to PO_4 tetrahedra by corner sharing oxygens.[5] The general structural formula is $A_nB_m(PO_4)_3$, where A is generally an alkaline ion or a vacancy[6] and B a transition metal ion or a combination of ions with different valence states (generally +4, such as Zr, Ti, and Sn). The interstitial space in this open framework is occupied by alkali ions, and permits fast ion conduction.[7] Materials with the NZP-type crystal structure are often referred to as 'NASICONs', an acronym for 'Na-Super Ionic CONductors'.

The NZP-type structure allows a large number of ionic substitutions, which leads to compounds with different properties and various possible applications, including anti-thermal shock applications, with low and tunable thermal expansion,[8] or nuclear waste encapsulation and immobilisation of radionuclides,[9] due to their good chemical and thermal stability. The ionic and superionic conductivity is also an interesting property for potential solid electrolytes.[10]

However, NZP is difficult to fabricate into monoliths using conventional solid state reaction, because it requires both high temperatures (higher than 1000°C) and long sintering times (several tens of hours). An alternative, low-temperature route to obtain the NZP crystalline phase is to use reactive sintering of a durable phosphate glass.[11] According to this approach, glass-ceramic bodies from TiO_2-nucleated, zinc and iron phosphate glasses have been prepared. These materials consisted of a fine-grained, internally-nucleated intergrowth of α-$FePO_4$ and a mixed valence Fe/Ti phosphate, namely, $Fe^{2+}Fe^{3+}Ti(PO_4)_3$, the latter crystals having the NZP structure. These glass-ceramics were produced by heat treating precursor glass bodies at temperatures between 625 and 1000°C. When the TiO_2 content was less than 10% by weight and the heat treatment was conducted at temperatures below 800°C, NZP-type crystals constituted the predominant crystal phase. Subsequently, a number of glasses were formulated with the NZP stoichiometry in order to form a glass-ceramic containing NZP as the only crystalline phase and/or to reduce the amount of residual glass. However, the primary aim of this work was to study

Corresponding author. Email jean.rocherulle@univ-rennes1.fr
Original version presented at Int. Conf. on Phosphate Glasses, Pardubice, Czech Republic, 2–4 July 2014
DOI: 10.13036/1753-3562.57.5.063

the thermal expansion of the glass-ceramics without alkaline elements, due to their deleterious effect on the chemical durability. Concerning this last property, it has been shown that it is possible to enhance the durability by different methods: (i) the addition of refractory oxides like Al_2O_3,[12] (ii) the nitridation of glasses[13] (i.e. the nitrogen for oxygen substitution in the vitreous network), and (iii) the addition of transition metal oxides.[14] Tungsten oxide is known to form glasses with $NaPO_3$ over a wide range of composition, and to improve the chemical resistance against atmospheric moisture.[15] SnO can also be added to sodium metaphosphate, $NaPO_3$, and glass formation occurs over a wide range of compositions. An interesting feature of this oxide is that it reduces the glass transition temperature.[16] Therefore, the simultaneous addition of SnO and WO_3 to sodium metaphosphate yields glasses with low characteristic temperatures and good chemical durability.[17] In addition, a composite material including a residual glassy phase and a crystalline phase, with NZP-type structure, containing tungsten(IV) and tin(IV), has been synthesised by reactive sintering of a phosphate glass.[18] The $NaSn_2(PO_4)_3$ crystalline phase was already known,[19] prepared by solid state reaction between the different oxides, and this is the only previous report in the literature of a NZP compound containing both tungsten and tin. In this context, we focused our studies on the $60NaPO_3.20SnO.20WO_3$ glass composition and the corresponding glass-ceramic, with thermal and mechanical characterisation and Rietveld refinement of the crystal structure to precisely determine the chemical composition.

2. Experimental

Glass from the ternary $NaPO_3$–$Sn^{(II)}O$–$W^{(VI)}O_3$ system was prepared using a microwave oven.[17] The microwave route offers huge advantages over conventional methods, with a very short time scale for the preparation, and with homogeneity of the heating. Indeed, heat is generated within the sample by interaction between chemical species and the magnetic field, while in a resistive furnace energy is transmitted to the surface of the sample and then to the whole volume by conduction. A critical requirement is the coupling of at least one of the reactants to the microwave field to initiate and drive the heating.

Typically a batch (~15g) with the necessary quantities of starting materials, $NaPO_3$ (Aldrich), SnO (99·9 % Alfa Aesar) and WO_3 (99+% Aldrich) was placed in a silica crucible and exposed to microwave irradiation for up to 10 min in a small scale oven (Kerwave®, 2·45 GHz, 750 W nominal power) under a nitrogen flow (150 $l h^{-1}$). The crucible was covered by a silica cap that allows the simultaneous transfer of nitrogen and exhaust gas from the melt. This provides a protective atmosphere, avoiding oxidation of tin as mentioned in a previous report.[17] A melting temperature of 1000°C was reached after ~100s of exposure to microwave irradiation,[17] and then the melt was poured between two stainless steel plates, after which the glass was annealed for one hour at a temperature close to T_g to relieve residual stress, before cooling down slowly to room temperature.

The obtained glass WAs finely crushed and the glass powder WAs compressed at room temperature under a pressure of 6 MPa in order to form a green pellet. The diameter of this pellet was about 13 mm and the thickness was 4–5 mm. The pellet was cured for various times and temperatures, and glass reactive sintering occurred to produce a dense glass-ceramic.

The characteristic temperatures were measured using a TA Instruments SDT 2960 calorimeter with an overall accuracy expected to be better than 5°C for this study. The glass samples were contained in a Pt pan, and an empty Pt pan was used as the standard. Non-isothermal experiments were performed under air, using a constant weight of 40 mg. A constant heating rate of $10°C min^{-1}$ was employed. The coefficient of thermal expansion (CTE) was calculated from a thermomechanical experiment (TA Instruments Model TMA 2940), performed with a $5°C min^{-1}$ heating rate and a preload force of 0·1 N. The precision of the CTE determination is $\pm 2\times 10^{-7} K^{-1}$ in the 50–250°C temperature range.

The density of the samples was measured at room temperature using Archimedes' method, with absolute alcohol as the immersion fluid. The error in the determination of the density is $\pm 0{\cdot}02 g cm^{-3}$.

Young's modulus, E, bulk modulus, G, and shear modulus, K, were calculated from experiments using the ultrasonic velocity technique. This technique is based on time-of-flight measurements using the pulse-echo technique.[20] The chemical durability was evaluated from the weight loss of bulk samples immersed in water at 50°C using a static method.[21] A constant ratio of sample surface to solution volume, 0·1 cm^{-1}, was maintained in order to compare the results. The faces of the samples were polished with 1200 grit SiC paper and cleaned in alcohol. After each corrosion test, the samples were dried at 100°C until the weight remained constant. The dissolution rate (DR) of samples was calculated from the measured weight loss, ΔW, sample surface area (SA) and time of immersion, t, using the following equation:

$$DR = \frac{\Delta W(g)}{SA(cm^2)\times t(min)} \quad (1)$$

The error in the DR value reported herein is estimated at 5%. During these durability tests, the pH was measured with a Schott CG819 pH-meter and the precision is ±0·05.

The crystalline phases obtained after the heat treatment were analysed by x-ray powder diffraction (XRD). Data were collected at room temperature with a D8 BRUKER AXS diffractometer (40 kV and 40 mA)) using Cu $K_{\alpha 1}$ radiation ($K_{\alpha 1}$=1·5406 Å), selected with an incident beam curved-crystal Ge monochromator, and using parafocusing Bragg–Brentano geometry and a 1D-LynxEye detector. The patterns were scanned at room temperature, over the angular range from 10 to 130° (2θ), with a step width of 0·008° and a counting time of 250 s. For pattern indexing, the extraction of the angular peak positions was carried out by means of the automatic peak search option in WINPLOTR software. Indexing was performed with the DICVOL04 program,[22] and structural refinement was performed using FULLPROF software.

3. Results and discussion

3.1. Synthesis

In a previous study,[17] we showed that several glass compositions in the $NaPO_3$–SnO–WO_3 ternary phase diagram are able to crystallise and to give a single crystalline phase, isostructural to $NaSn_2(PO_4)_3$, and probably containing tungsten. Indeed, the ionic radii of Sn^{4+} and W^{4+} are relatively close (0·83 and 0·80 Å, respectively[23]), giving credence to a possible substitution of W^{4+} for Sn^{4+}. In a first step, we determined the theoretical composition of the parent glass in order to prepare a NZP glass-ceramic with an equimolar ratio between tin and tungsten. Clearly the glass composition $60NaPO_3.20SnO.20WO_3$ is optimal to obtain the maximum quantity of the $NaSnW(PO_4)_3$ phase from glass reactive sintering. The maximum proportion of this phase is calculated to be 91 wt%, limited to this value because of the amount of sodium present in the glass, which is higher than the requisite quantity to form this phase. This glass reactive sintering has previously been studied by means of DTA and XRD techniques.[18] Various parameters were optimised, and it was shown that three parameters are critical for obtaining a pure NZP phase: (i) a fine powder (<100 μm), because the crystallisation is controlled by a surface dominant mechanism, (ii) the presence of oxygen during sintering (oxygen contained in air is sufficient), and (iii) the presence of tungsten oxide in equimolar ratio with SnO in the initial glass. Moreover, the temperature must be greater than 710°C to obtain a NZP glass-ceramic without any secondary crystalline phases.

In the present study, the phosphate based glass was synthesised in less than 10 min by using a microwave heating device, and then the powdered glass samples were cold pressed, after which they were heated to 750°C for times varying from 1h to 48h, giving glass-ceramics containing a single NZP crystalline phase. These sintered samples are referred to as '750-x', where x corresponds to the heating time expressed in hours. Whatever the sintering time, NZP remains the only crystalline phase to occur.

Table 1. Thermomechanical characterisation of the glass and glass-ceramic samples

	T_g (°C)	α (10^7 K^{-1})	T_d (°C)	E (GPa)	K (GPa)	G (GPa)
Glass	367	164	382	47	34·7	18·4
Glass-ceramic	440	121	460	62·5	40·1	25·2

Glass transition temperature, T_g
Thermal expansion coefficient, α
Dilatometric softening point, T_d
Elastic moduli, E, K and G

3.2. Characterisation

Several physical characteristics of the parent glass and of the 750-1 glass-ceramic are compared in Table 1. They have been measured on bulk samples, except glass transition temperatures. The thermal characteristics (i.e. T_g and T_d) reported for the glass-ceramic are assigned to the residual glassy phase. Nevertheless, the quantity of the crystalline phase is obviously a function of the sintering time, as illustrated by Figure 1. The density of the pellet increases rapidly at first, but after a few hours (~6 h) there is no more increase, and the density reaches a limit close to 3·3 g cm^{-3}. This behaviour is supported by the chemical durability of the glass-ceramic. As shown in Figure 2, the dissolution rate decreases with increasing sintering time, consistent with an increase of the NZP content, assuming that NZP is more durable than the residual glassy phase. Furthermore, for long time durability tests, one can observe that the vitreous phase entirely vanishes, leading to a porous and brittle skeleton of pure NZP crystalline phase, as illustrated by Figure 3. From the leaching tests and the assumption that the NZP phase is not soluble, in accordance with a previous study,[24] its proportion can be estimated by simply measuring the weight loss of the glass-ceramic, which is obviously assigned to the residual and soluble glassy phase. Figure 4 depicts these results for the 750-1 sample, and the weight loss is maximum after 50 h, remaining constant and close to 60% for longer leaching times. From this result, we assume that the NZP weight content is close to 40% when sintering is conducted at 750°C. Furthermore, if one calculates the theoretical quantities (in wt%) of the NZP phase that could be obtained from a $60NaPO_3.20SnO.20WO_3$ glass composition, this value

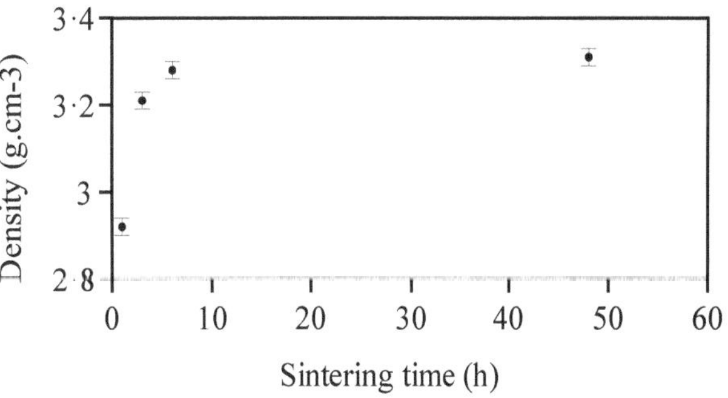

Figure 1. Glass-ceramic densities as a function of the sintering time

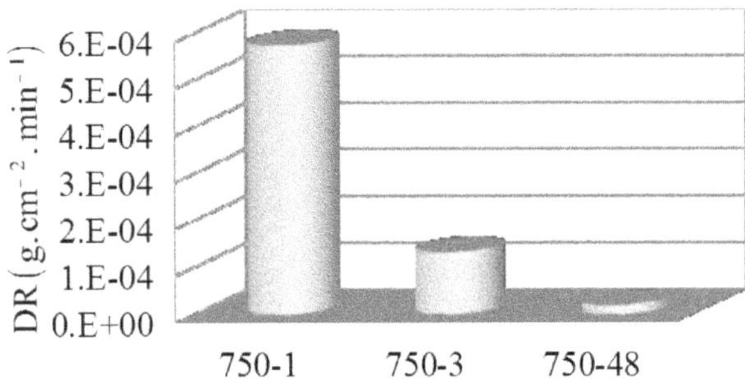

Figure 2. Dissolution rates of the glass-ceramic as a function of sintering time

should be within a range defined from the basis of two hypotheses considering either, a pure tin containing phase (i.e. $NaSn_2(PO_4)_3$) or a tin and tungsten (1:1) containing phase (i.e. $NaSnW(PO_4)_3$).

Clearly, the $60NaPO_3.20SnO.20WO_3$ glass composition is optimal to obtain the maximum quantity of $NaSnW(PO_4)_3$ phase from glass reactive sintering. This value is limited to about 91 wt% because the amount of sodium present in the glass is higher than the requisite quantity to form this phase. On the contrary, considering that only tin is present in the crystalline phase, a lower maximum NZP content is reached, with a value close to 40%. As a consequence, the theoretical quantity of the resulting glassy phase is close to 60%. This value is in close agreement with the result obtained from the leaching experiments giving an indication of the chemical composition of the crystalline phase. Otherwise, the quantity of the crystalline phase can be increased when increasing the sintering temperature up to 850°C.[18] However, the diffraction patterns do not differ from those obtained after sintering at 750°C,[18] indicating the same chemical composition for the crystalline phases whatever the sintering

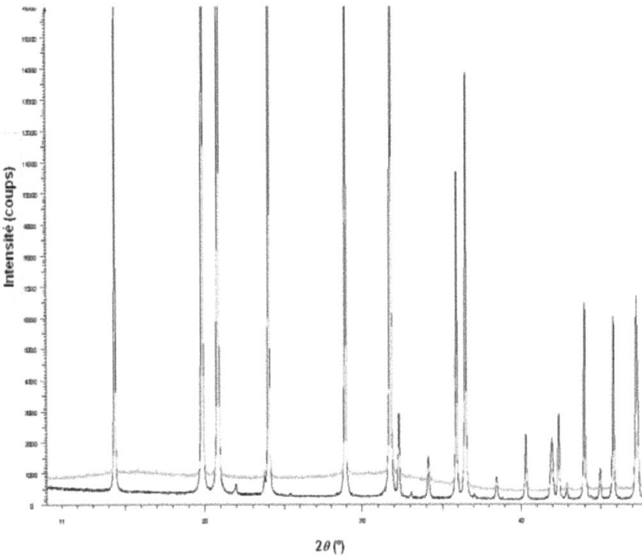

Figure 3. XRD patterns of the 750-1 glass-ceramic before (in green) and after (in black) leaching for 15 days

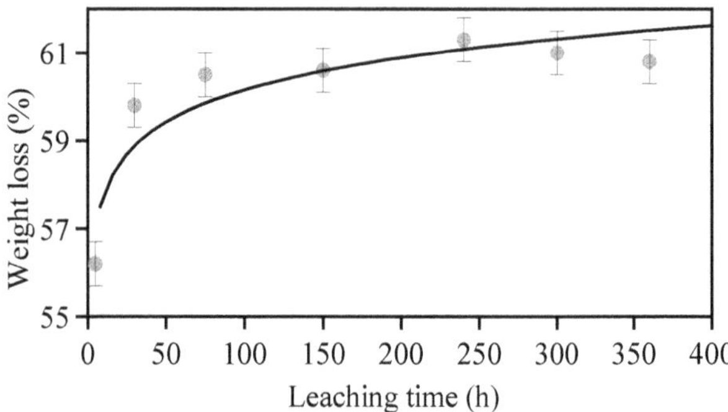

Figure 4. Weight loss of the 750-1 glass-ceramic as a function of leaching time

temperature. Thus the leaching process allows the removal of the glassy phase with the retention of a pure NZP crystalline phase. Thereafter, this pure powder was used for all the x-ray powder diffraction experiments.

3.3. Scanning electron microscopy

SEM micrographs of the glass-ceramic are shown in Figure 5(a) (secondary emission) and Figure 5(b) (backscattered electron emission). As one can see in Figure 5(a), the densification of the green pellet is not achieved because of the presence of pores. Crystals and a residual glassy phase clearly co-exist. The backscattered electron micrograph illustrates a fluctuation in chemical composition due to a difference between the atomic numbers of tin (Z=50) and of tungsten (Z=74). From Figure 5(b), the NZP phase clearly shows a lack in tungsten (dark area), and on the other hand this element is present in the glassy phase (bright area). In addition, SEM micrographs for a sample sintered at 850°C for 3 h are shown in Figure 6(a) (secondary emission) and Figure 6(b) (backscattered electrons). As mentioned above, the 750-1 and 850-3 diffraction patterns do not present any difference. For this last sample, the densification is achieved and the pores have disappeared. As illustrated, the crystal size has substantially increased and the micrograph obtained from the backscattered electrons emission clearly shows a difference in chemical composition between the crystalline and the residual glassy phase. From EDS (energy-dispersive x-ray spectroscopy) analysis, several atomic ratios have been determined and are given in Table 2. The calculated atomic ratios of the parent glass, and for $NaSn_2(PO_4)_3$ and $NaSnW(PO_4)_3$ crystals, are also reported for comparison.

Considering the crystalline phase, the chemical composition is much closer to a pure tin-NZP than it is to a mixed form containing both Sn and W in equal amounts. Thus, for the two glass-ceramics, the Sn/P values are lower than the theoretical one (i.e.

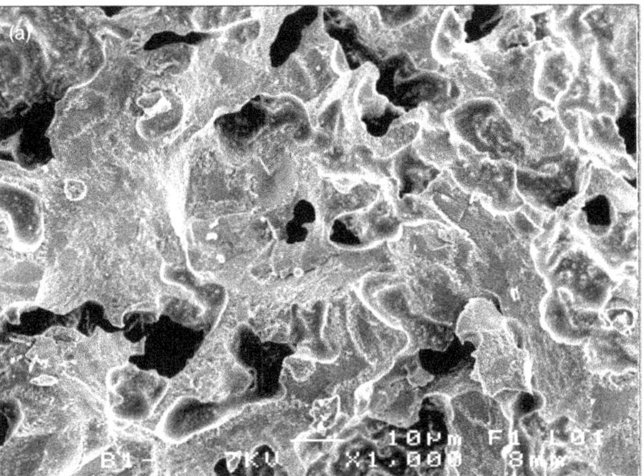
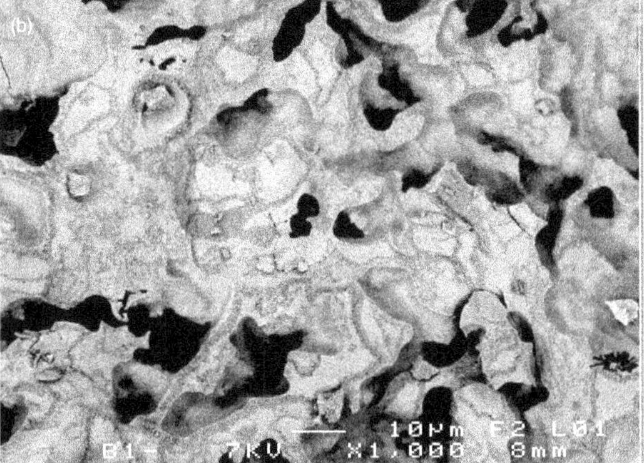

Figure 5. SEM micrographs of the 750-1 glass ceramic. (a) Secondary emission, (b) backscattered electrons

0·67 for a pure tin-NZP) while the W/P values are greater than 0. This indicates a mixed occupation of the tin crystallographic site in the NZP structure. Furthermore, the residual glassy phase is enriched with sodium and tungsten, whereas the tin content is drastically decreased. Nevertheless, all these values suffer from some discrepancy, and it is not possible to claim precisely the chemical composition of the crystalline phase.

3.4. X-ray powder diffraction

X-ray powder diffraction analysis can be carried out

Table 2. Calculated atomic ratios from EDS analysis

	Crystal			Residual glass		
	Na/P	Sn/P	W/P	Na/P	Sn/P	W/P
$NaSn_2(PO_4)_3$	0·33	0·67	0	-	-	-
$NaSnW(PO_4)_3$	0·33	0·33	0·33	-	-	-
750-1	0·31	0·59	0·06	1·19	0·01	0·51
850-3	0·33	0·63	0·04	1·49	0·01	0·60

on single crystalline phase samples like our glass-ceramics. For this purpose, several routines based on at least 20 to 30 reflections, allow to determine the possible unit cells by a complete indexing of all reflections. The main criteria for a good solution are: (i) high figures of merit, (ii) the lowest volume of the unit cell, and (iii) the agreement between calculated and experimental density values, including the number of atoms per unit cell. When different possibilities have been found, the next step is to select a unit cell and to decide on a lattice and possible space group by recalculating the pattern and looking for systematic absences. The contribution of both precision and high resolution powder diffraction has allowed precise determination of the lattice symmetry and space group without any doubt.

The first 20 diffraction lines were indexed on the basis of hexagonal symmetry with the following unit cell dimensions and the corresponding figures of merit:[25]

a=8·5161(5) Å c=22·456(2) Å [V=1410·4 Å^3]

M_{20}=49 F_{20}=73(0·0051; 53)

Starting from this solution, the whole pattern was indexed and refined, giving lattice parameters close to those of the $NaSn_2(PO_4)_3$ phase given in the ICDD-PDF2 database, i.e. 49-1198 (space group: $R\bar{3}c$) and 81-0850 (space group: $R\bar{3}$). The presence of a diffraction line with a significant intensity and located at 38·46° (2θ) corresponds to the presence of the (303) plane, confirming the hypothesis that there is no c-glide plane, and therefore the space group is $R\bar{3}$ rather than $R\bar{3}c$. Due to the slight difference between Sn^{4+} and W^{4+} ionic radii, the results obtained from indexing are very close to those given in the 81-0850 file. However, assuming the presence of tin and tungsten

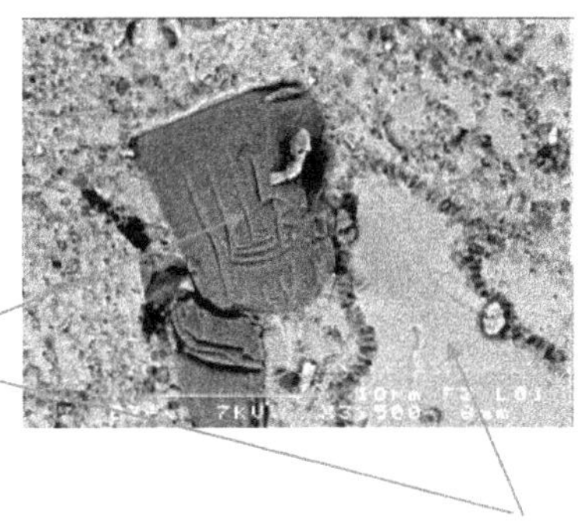

Figure 6. SEM micrographs of the 850-3 glass ceramic. (a) Secondary emission, (b) backscattered electrons

Table 3. Experimental details and results of structural refinement for x-ray powder diffraction on the 750-1 leached glass-ceramic

Experimental	
λ $K_{\alpha 1}$ (Å)	1·5406
Angular range° (2θ)	10–130
Counting time / step	250 s
Step size ° (2θ)	0·008
Unit cell	
a (Å)	8·51532(5)
c (Å)	22·4321(2)
Space group	$R\bar{3}$
Z	6
Refinement	
Number of reflections	565
Number of structural parameters	25
Number of profile parameters	11 (+19 parameters for background)
Number of Wyckoff positions	9
Number of atoms	11
R_P	0·075
R_F	0·026

on the same crystallographic site, and considering the significant difference in the x-ray scattering powder between these elements, Z=50 and Z=74, respectively, a complete structure refinement, should allow the site occupancies to be determined. This is the classical Rietveld method: matching the total experimental pattern by refining both instrumental and structure parameters until the best fit is reached.

The positions reported for $NaSn_2(PO_4)_3$ (ICSD number 072215)[26] were used to start the Rietveld refinement in space group $R\bar{3}$, which was carried out over the complete angular range, assuming the same occupation factor for tin and tungsten. Attempts to refine the cationic distribution were realised, including constraints on site multiplicities, stoichiometry and isotropic thermal factors. Based on 565 reflections, the procedure involves 55 parameters, with 25 structural parameters, 11 profile parameters and 19 parameters for the background. The fitting quality of the experimental data was assessed by computing parameters such as the R factors (R_p=profile factor, and R_F=structure factor). When these parameters reached a minimum value, the best fit to the experimental diffraction data was achieved, and the crystal structure is then regarded as satisfactory. In our case, these reliability factors, R_p=7·5% and R_F=2·6%, are quite good, especially bearing in mind that the crystalline phase has been obtained by a glass reactive sintering conducted at low temperature (750°C) during a short time (1 h). All the characteristics of the Rietveld refinement are summarised in Table 3, and Figure 7 shows the best agreement obtained between observed and calculated profiles. The inset illustrates the quality of the fit in detail over a limited angular range.

The final atomic positions, isotropic thermal displacement parameters and occupation factors of all the elements are given in Table 4. The site occupancies of the M^{IV} cations on the octahedral sites are constrained in order to preserve the stoichiometric composition of the material. The proposed occupancies of the cations are calculated by considering that the heavy metal site has a sum of cationic distribution equal to 0·3333. The site occupancy analysis for sample 750-1 reveals that almost 90% of the octahedral

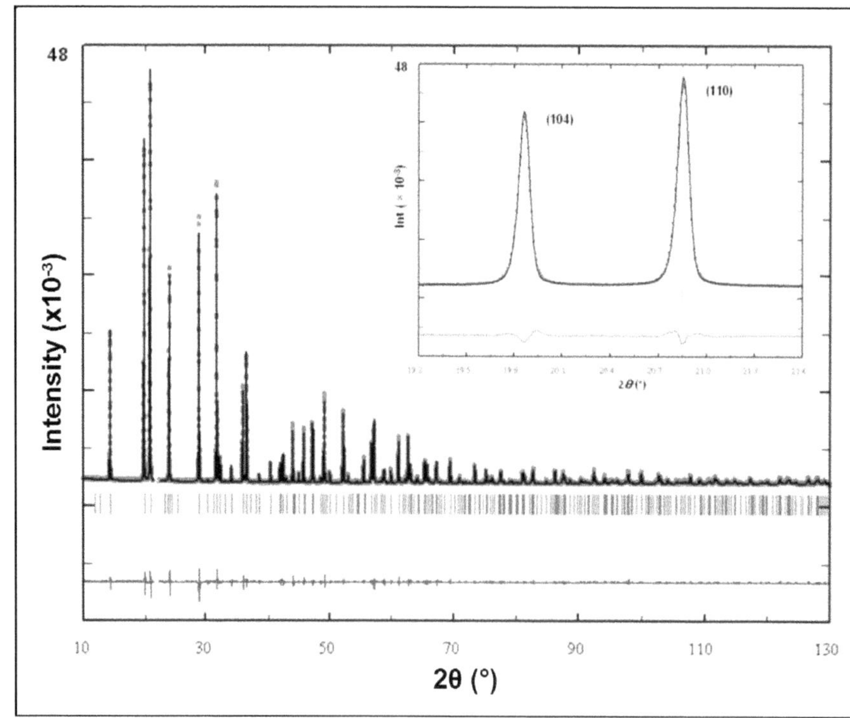

Figure 7. Observed (red points) and calculated (black line) patterns for $NaSn_{1·8}W_{0·2}(PO_4)_3$. The lower curve (blue line) shows the difference between the observed and calculated patterns. Vertical marks (green) indicate the positions of allowed lines. The inset shows (enlarged) the fitting of a part of the pattern

Table 4. Atomic positions, isotropic thermal displacement parameters and occupation factor for the $NaSn_{1.8}W_{0.2}(PO_4)_3$ phase (space group: R)

Ions	Position	x/a	y/b	z/c	$B_{iso}(Å^2)$	Occupation factor
Na1 (Na^+)	3a	0	0	0	4·1(1)	0·1666
Na2 (Na^+)	3b	0	0	0·5	2·8(1)	0·1666
Sn1 (Sn^{4+})	6c	0	0	0·14783(6)	0·49(3)	0·302(2)
W1 (W^{4+})	6c	0	0	0·14783(6)	0·49(3)	0·031(2)
Sn2 (Sn^{4+})	6c	0	0	0·64514(6)	0·49(3)	0·297(2)
W2 (W^{4+})	6c	0	0	0·64514(6)	0·49(3)	0·036(2)
P (P^{5+})	18f	0·2883(5)	−0·0012(8)	0·2503(3)	0·50(2)	1
O1 (O^{2-})	18f	0·155(1)	−0·0604(8)	0·1980(3)	1·16(3)	1
O12 (O^{2-})	18f	0·0162(9)	−0·0187(1)	0·6937(3)	1·16(3)	1
O2 (O^{2-})	18f	0·2065(7)	0·1395(8)	0·0922(3)	1·16(3)	1
O22 (O^{2-})	18f	−0·1785(8)	−0·1783(7)	0·5881(3)	1·16(1)	1

sites are occupied by Sn^{4+} ions and the remaining 10% of the sites by W^{4+} ions. Hence, the Rietveld refinement results suggest that the chemical formula is $NaSn_{1.8}W_{0.2}(PO_4)_3$. The tungsten to tin atomic ratio can also be calculated from the results of EDS analysis performed on the 750-1 sample given in Table 2. This calculation gives a value close to 10%, in accordance with the result of the Rietveld refinement.

4. Conclusions

Various phosphate based ($60NaPO_3.20SnO.20WO_3$) bulk glass-ceramic samples with a single NZP crystalline phase (i.e. $NaZr_2(PO_4)_3$ type) were successfully synthesised by a reactive glass sintering process. Evidence of a tin for tungsten substitution in the crystalline phase has been provided by EDS (energy-dispersive x-ray spectroscopy) analysis. As a consequence, the chemical composition of the residual glassy phase has been modified, leading to changes in thermomechanical characteristics such as glass transition temperature, dilatometric softening point and thermal expansion coefficient. When leached at 50°C in water, the glassy phase was not durable, and thus the crystalline phase was isolated and studied by x-ray powder diffraction. Indexing was performed, allowing the determination of the lattice and of the unit cell parameters. The systematic absences in the diffraction data are consistent with space group $R\bar{3}$. Rietveld refinement was conducted, leading to the determination of the site occupancy in the (Sn,W) $^{IV}O_6$ polyhedra. The result is consistent with the W/Sn atomic ratio determined from EDS analysis. The best fit reveals that almost 10% of the octahedral sites are occupied by W^{4+} ions, giving a NZP phase with a chemical composition $NaSn_{1.8}W_{0.2}(PO_4)_3$.

References

1. Takebe, H., Nonaka, W., Kubo, T., Cha, J., Kuwabara, M. Preparation and properties of transparent $SnO–P_2O_5$ glasses. *J. Phys. Chem. Solids*, 2007, **68**, 983.
2. Campbell, J. H. & Suratwala, T. I., Nd-doped phosphate glasses for high-energy/high-peak-power lasers. *J. Non-Cryst. Solids*, 2000, **263&264**, 318.
3. Donald, I. W. Preparation, properties and chemistry of glass and glass-ceramic-to-metal seals and coatings. *J. Mater. Sci.*, 1993, **28**, 2841.
4. Abou Neel, E. A., Pickup, D. M., Valappil, S. P., Newport, R. J., Knowles, J. C., Bioactive functional materials: a perspective on phosphate-based glasses. *J. Mater. Chem.*, 2009, **19**, 690.
5. Hagman, L. O., Kierkegaard, P., The Crystal Structure of $NaMe_2IV(PO_4)_3$; Me IV=Ge, Ti, Zr. *Acta Chem. Scand.*, 1968, **22**, 1822.
6. Subba Rao, G. V., Varadaraju, U. V., Thomas, K. A. & Sivasankar, B. Metal atom incorporation studies on the phases with NZP structure: $NbTiP_3O_{12}$. *J. Solid State Chem.*, 1987, **70**, 101.
7. Goodenough, J. B., Hong, H. Y-P. & Kafalas, J. A. Fast Na^+ ion transport in skeleton structures. *Mater. Res. Bull.*, 1976, **11**, 203.
8. Breval, E., McKinstry, H. A. & Agrawal, D. K. New [NZP] materials for protection coatings. Tailoring of thermal expansion. *J. Mater. Sci.*, 2000, **35**, 3359.
9. Buvaneswari, G., Varadaraju, U. V., Low leachability phosphate lattices for fixation of select metal ions. *Mater. Res. Bull.*, 2000, **35**, 1313.
10. Zhou, M. & Ahmad, A. Synthesis, processing and characterisation of nasicon solid electrolytes for CO_2 sensing applications. *Sens. Actuators*, 2007, **B122**, 419.
11. Aitken, B. Glass-ceramics containing NZP type-crystals. US Patent n°4784976, 1988.
12. Brow, R. K. Nature of alumina in phosphate glass: I, properties of Sodium aluminophosphate glass. *J. Am. Ceram. Soc.*, 1993, **76**, 913.
13. Le Sauze, A. & Marchand, R. Chemically durable nitrided phosphate glasses resulting from nitrogen/oxygen substitution within PO_4 tetrahedra. *J. Non-Cryst. Solids*, 2000, **263&264**, 285.
14. Araujo, C. C., Strojek, W., Zhang, L., Eckert, H., Poirier, G., Ribeiro, S. J. L. & Messaddeq, Y. Structural studies of $NaPO_3–WO_3$ glasses by solid state NMR and Raman spectroscopy. *J. Mater. Chem.*, 2006, **16**, 3277.
15. Poirier, G., Ottoboni, F. S., Cassanjes, F. C., Remonte, A., Messaddeq, Y. & Ribeiro, S. J. L. Redox behaviour of molybdenum and tungsten in phosphate glasses. *J. Phys. Chem. B*, 2008, **112**, 4481.
16. Bekaert, E., Montagne, L., Delevoye, L., Palavit, G. & Wattiaux, A. NMR and Mössbauer characterisation of tin(II)-tin(IV)-sodium phosphate glasses. *J. Non-Cryst. Solids*, 2004, **345&346**, 70.
17. Chenu, S., Lebullenger, R. & Rocherullé, J. Characterisation of $NaPO_3$–$SnO–WO_3$ glasses prepared by microwave heating. *J. Mater. Sci.*, 2010, **45**, 6505.
18. Chenu, S., Lebullenger, R., Calvez, G., Guillou, G., Rocherullé, Kidari, A., Pomeroy, M. J. & Hampshire, S. Glass reactive sintering as an alternative route for the synthesis of NZP glass-ceramics. *J. Mater. Sci.*, 2012, **47**, 486.
19. Rodrigo, J. L. & Alamo, J. Phase transition in $NaSn_2(PO_4)_3$ and thermal expansion of Na $(PO_4)_3$; (MIV=Ti, Sn, Zr). *Mater. Res. Bull.*, 1991, **26**, 475.
20. Blessing, G. V. The pulsed ultrasonic velocity method for determining material dynamic elastic moduli, In: Dynamic elastic modulus measurements in material. Edited by A Wolfenden, American Society for Testing and Materials, Philadelphia, 1990, **Vol. 1045**, p.47.
21. Takebe, H., Baba, Y. Kuwabara, M. Dissolution behaviour of $ZnO–P_2O_5$ glasses in water. *J. Non-Cryst. Solids*, 2006, **352**, 3088.
22. Boultif, A. & Louër, D. Powder pattern indexing with the dichotomy method. *J. Appl. Cryst.*, 2004, **37**, 724.
23. Shannon, R. D. Revised effective ionic radii and systematic studies of interatomic distances in halides and chalcogenides. *Acta Cryst. A*, 1976, **32**, 751.
24. Breval, E., Harshe, G. & Agrawal, D. K. Synthesis and chemical stability of $NaSn_2P_3O_{12}$. *J. Mater. Sci. Lett.*, 1995, **14**, 728
25. Smith, G. S. & Snyder, R. L. A criterion for rating powder diffraction patterns and evaluating the reliability of powder-pattern indexing. *J. Appl. Cryst.*, 1979, **12**, 60.
26. *ICSD: Inorganic Crystal Structure Database*, Fiz Karlsruhe (D).

Phys. Chem. Glasses: Eur. J. Glass Sci. Technol. B, October 2016, **57** (5), 213–217

Composition optimisation of iron phosphate-based glasses for radioactive sludge

Hiromichi Takebe,[1] Naoto Kitamura,[1] Ippei Amamoto,[2] Hidekazu Kobayashi,[2] Tatsuya Tsuzuki[3] & Naoki Mitamura[3]*

[1] *Graduate School of Science and Engineering, Ehime University, 3 Bunkyo-cho, Matsuyama, Ehime 790-8577, Japan*
[2] *Tokai Reprocessing Technology Development Centre, Japan Atomic Energy Agency, 4-33 Muramatsu, Tokai-mura, Naka-gun, Ibaraki 319-1194, Japan*
[3] *Glass Research Center, Central Glass Co. Ltd., 1510 Ohkuchi-cho, Matsusaka, Mie 515-0001, Japan*

Manuscript received 31 August 2014
Revised version received 4 March 2015
Accepted 22 May 2015

Mixtures of iron phosphate glass granules and $BaSO_4$ powder, a simulated main sludge component, were melted in air using a platinum crucible. The quenched glass samples consisted mainly of oxide constituents, due to the decomposition of $BaSO_4$ during the heating and melting processes. Thermal stability against crystallisation for the glass samples was evaluated using differential thermal analysis, according to the existence of an exothermic crystallisation peak and the temperature difference between the glass transition and onset of crystallisation temperatures. Water durability was determined by the weight loss per unit surface area after immersion in hot water at 120°C for 72 h. Phosphate network species and chemical bonding were characterised by Raman spectroscopy. Glass compositions with the simulated sludge component were optimised in terms of the thermal stability and the water durability, and characterised according to the O/P ratio.

1. Introduction

A catastrophic earthquake and ensuing tsunami occurred on March 11, 2011, in the east part of Japan. A huge amount of water has subsequently been used for cooling the Fukushima Daiichi nuclear power plant. At the initial stage of cooling treatments, a decontamination process based on a co-precipitation with various chemicals was applied for the agglomeration of nuclear matter after the pretreatment of Cs adsorption on zeolite. The decontamination process finally created a sludge including secondary nuclear wastes such as $(Ba,Sr)SO_4$ and potassium caesium nickel ferro cyanide ((K, Cs)ppFeNi).(1)

To immobilise the sludge into a stable glass matrix with minimised volume, iron phosphate (IP) glasses(2) may be favoured because of higher loading capacity and better chemical durability(3) in comparison to standard borosilicate glasses for nuclear waste immobilisation.(4) The purpose of this study is to investigate the effect of $BaSO_4$ addition on the thermal stability and water durability of IP glasses. $BaSO_4$ is one of the main components in the sludge. The properties–structure relationship is also studied. Finally, we propose optimum glass compositions for sludge immobilisation, which have the best thermal stability and water durability. These compositions are characterised according to the O/P ratio.

Corresponding author. Email takebe.hiromichi.mk@ehime-u.ac.jp
Original version presented at Int. Conf. on Phosphate Glasses, Pardubice, Czech Republic, 2–4 July 2014
DOI: 10.13036/17533562.57.5.047

2. Experimental

2.1 Sample preparation

Binary iron phosphate (IP) glasses were prepared from mixtures of Fe_2O_3 and H_3PO_4 as raw materials, melted in a Pt crucible in air at 1200–1300°C. The nominal compositions of IP glasses were $x$$Fe_2O_3$.(100−$x$)$P_2O_5$, where x=30, 35, 40 mol% Fe_2O_3. Melting temperatures required for homogeneous melts, using IP glasses and $BaSO_4$ powders as raw materials, were determined in advance by the direct observation of the sample melting process on a thermocouple filament through a microscope.(5) IP glass granules with diameter less than 300 μm were mixed with $BaSO_4$ commercial powder as a simulated sludge main component, and the mixture was melted in air in a Pt crucible at 1150°C for 1 h. The melt was poured onto a graphite mould, annealed at the glass transition temperature for 1 h, and then cooled to room temperature at a rate of ~1°C/min. X-ray diffraction analysis (PANalytical, X'pert powder diffractometer) was used to confirm the non-crystalline state. The following evaluations were performed only for glass samples without any crystalline phases.

2.2 Evaluation

The residual sulphur content (SO_3 equivalent) was determined by both x-ray fluorescence (XRF) (RIX2100, Rigaku) and SEM-EDX (JSM-6510LA, JEOL) analyses. The ratios of Ba/P and Fe/P were determined by SEM-

Table 1. Nominal compositions and analysed Fe/P, Fe^{2+}/Fe_{total} and O/P ratios of glass samples

Sample	Nominal composition (mole fraction)	Fe^{2+}/Fe_{total} (±0·01)	Fe/P (±0·02)	Ba/P (±0·01)	O/P (±0·04)
IP30	$30Fe_2O_3–70P_2O_5$	0·24	0·44		3·11
10BaIP30	$10BaSO_4–90(0·3Fe_2O_3–0·7P_2O_5)$	0·10	0·43	0·08	3·19
20BaIP30	$20BaSO_4–80(0·3Fe_2O_3–0·7P_2O_5)$	0·11	0·44	0·18	3·32
IP35	$35Fe_2O_3–65P_2O_5$	0·12	0·52		3·25
10BaIP35	$10BaSO_4–90(0·35Fe_2O_3–0·65P_2O_5)$	0·12	0·55	0·08	3·38
20BaIP35	$20BaSO_4–80(0·35Fe_2O_3–0·65P_2O_5)$	0·14	0·52	0·19	3·43
IP40	$40Fe_2O_3–60P_2O_5$	0·14	0·65		3·43
10BaIP40	$10BaSO_4–90(0·4Fe_2O_3–0·6P_2O_5)$	0·12	0·63	0·09	3·50

EDX analysis. The Fe^{2+}/Fe_{total} fraction was determined by a titration method using ~2 mM $KMnO_4$ aqueous solution in the same manner as in a previous study.[6] The O/P ratio was calculated from the evaluated values of the Fe^{2+}/Fe_{total}, Ba/P and Fe/P ratios. Three measurements were made for composition analysis, and the average value was determined for each sample. FactSage, the Integrated Thermodynamic Databank System,[7] was also used to understand the phase decomposition process and retained sulphur concentration after melting.

The glass transition temperature, T_g, and onset of crystallisation temperature, T_X, were determined by differential thermal analysis (DTA), using a Rigaku Thermo-Plus calorimeter, for a ~40 mg glass chunk over the range from room temperature to 1000°C, with a continuous heating rate of 10°C/min. The thermal stability against crystallisation at high temperature was evaluated by the existence of an exothermic crystallisation peak and the temperature difference between T_X and T_g.[8] Water durability was evaluated by the MCC-2 static leaching method at 120°C for 72 h, using a plate sample of size 10×10×3 mm, and ultra-pure water. The weight loss per unit area, $\Delta W/S$, and the change in the pH of solution were evaluated after the immersion tests. Two samples were used for each composition in the evaluations of thermal stability and water durability, and average values were calculated. Raman spectroscopy was performed using a Nihon-bunko Co. NRS-5100 spectrometer to characterise the phosphate anion species Q^n (where n is the number of bridging oxygens in a PO_4 unit) and P–O–P linkages.

3. Results

3.1 Sample composition

Table 1 summarises the nominal compositions and analysed Fe^{2+}/Fe_{total}, Ba/P, Fe/P and O/P ratios for eight glass samples. The sample name expresses the nominal composition, giving the concentration of $BaSO_4$, and the concentration of Fe_2O_3 in the IP glass granules. The fraction Fe^{2+} is in the range of 0·10–0·14 except for sample IP30 (Fe^{2+}/Fe_{total}=0·24). For instance, the fractions of Fe^{2+} are 0·12 and 0·14 for IP35 and IP40, respectively. These values are lower than those for the same nominal compositions in a previous report.[9] This discrepancy may be due to the choice of raw material for phosphate, e.g. H_3PO_4 and $NH_4H_2PO_4$,[9] and the melting conditions, such as crucible type and melting temperature and time. A previous study[10] of $SnO–ZnO–P_2O_5$ glasses also pointed out that ammonium phosphate $(NH_4)_2HPO_4$ as a raw material for P_2O_5 formed reductive source such as H_2 and H radical its thermal decomposition process at high temperature. Further evaluations using Fe Mössbauer spectroscopy are also need for composition analysis in these glass samples.

No retained SO_3 (<0·1 mol%) was detected in all the $BaSO_4$-loaded IP glasses, according to both XRF and SEM-EDX analyses. This is in good agreement with previous results for IP glasses loaded with ≤20 mol% $BaSO_4$.[11] The calculation of thermodynamic equilibrium by FactSage[7] also revealed that in the mixture of 10 mol% $BaSO_4$ and 90 mol% IP35, for example, the vaporisation of SO_x gases occurs at ≥1400 K (1127°C), and the removal of sulphur may be expected whilst the melts are held at a temperature of 1150°C.

3.2 Thermal stability

Figure 1 shows the DTA curves of the three glass series, IP30, IP35 and IP40. T_g increases monotonically with increasing $BaSO_4$ loading content. Table 2 summarises the characteristic temperatures and the thermal stability of IP and $BaSO_4$-loaded glasses. In all series, T_g increases monotonically with increasing $BaSO_4$ loading content, corresponding to BaO concentration in glasses. Glass samples 10BaIP30, 20BaIP30, IP35 and 10BaIP35 have no exothermic crystallisation peaks; this result reveals that these glasses, with O/P ratio in the range 3·19–3·38 (Table 1), have excellent thermal stability against crystallisation at high temperatures.

Table 2. Glass transition temperature, T_g, onset of crystallisation temperature, T_X, temperature difference, T_X–T_g, and the relative thermal stability of the glass samples

Sample	T_g (±3°C)	T_X (±17°C)	T_X–T_g (°C)	Thermal stability
IP30	526	751	223	Good
10BaIP30	571	Not observed	-	Excellent
20BaIP30	590	Not observed	-	Excellent
IP35	512	Not observed	-	Excellent
10BaIP35	552	Not observed	-	Excellent
20BaIP35	585	719	134	Moderate
IP40	497	704	207	Good
10BaIP40	541	731	190	Good

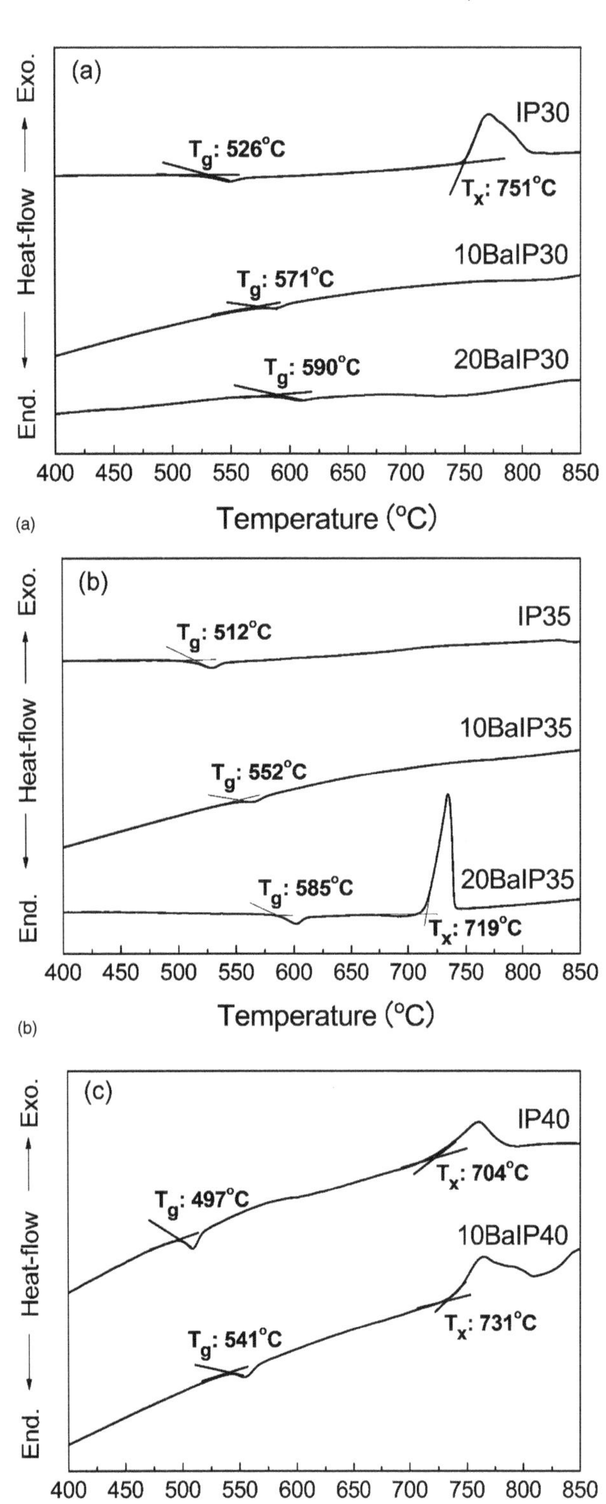

Figure 1. DTA curves of the glass samples. (a) IP30, (b) IP35 and (c) IP40 series

3.3 Water durability

Figure 2 shows the macroscopic appearances after the immersion test. The appearances are classified into four types: fracture with many pieces (type I), reacted with cracks (type II), almost no macroscopic change (type III), and the formation of a reaction layer on the surface (type IV). For type IV, the area attached with Teflon thread exhibits no macroscopic change, and only the exposed area forms the reaction layer. This layer may consist of a hydrated product observed in other phosphate-based glasses.[12,13] Further kinetic study is necessary to understand the dissolution behaviour of iron phosphate-based glasses.

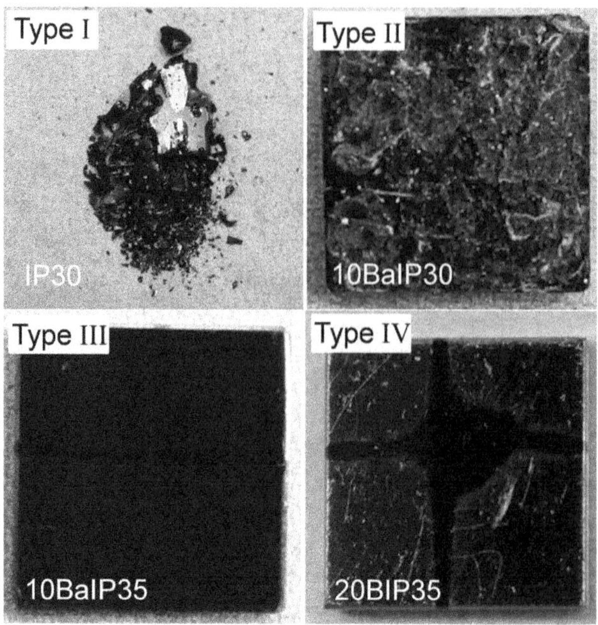

Figure 2. Macroscopic appearances after the immersion test using the MCC-2 static leaching method. Type I: fractured with many pieces, type II: reacted with cracks, type III: no macroscopic change, and type IV: the formation of a protective layer on the surface [Colour available online]

Table 3 summarises the immersion test results: the type of macroscopic appearance, the weight loss per unit surface area, $\Delta W/S$, and the pH of the solution before and after the immersion test. Larger deviation of $\Delta W/S$ in type I samples was caused by the weight loss relating to the morphology of each sample fracture. Comparison of Tables 1 and 3 shows that glass samples with an O/P ratio of 3·32 or more exhibit good water durability (with appearance type III or IV), with relatively small $\Delta W/S$ values less than $1{\cdot}0\times10^{-9}$ kg/mm^2.

Table 3. Results of the immersion test based on the MCC-2 static leaching method: Macroscopic appearance after immersion test, weight loss per unit area, ΔW/S, and the pH change of the solution after immersion

Sample	*Type*	*ΔW/S* ($\times10^{-9}$ kg/mm^2)	*pH* (Before)	(After)
IP30	I	6·1 (±2·9)	6·4	3·5
10BaIP30	II	2·5 (±0·1)	6·4	4·2
20BaIP30	IV	0·8 (±0·2)	6·4	4·8
IP35	I	4·1 (±5·9)	6·5	4·2
10BaIP35	III	<0·3 (±0·1)	6·4	5·2
20BaIP35	IV	0·6 (±0·1)	6·4	5·1
IP40	IV	0·5 (±0·2)	6·4	5·3
10BaIP40	IV	0·6 (±0·1)	6·7	5·2

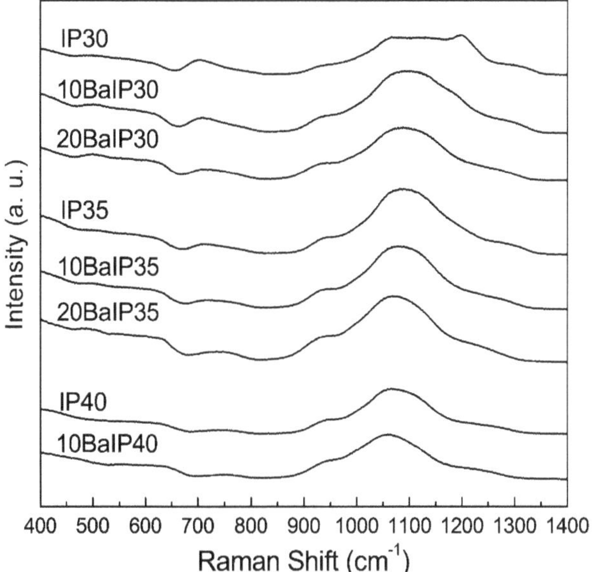

Figure 3. Raman spectra of the IP and $BaSO_4$-loaded IP glass samples

3. 4 Raman spectroscopy

Figure 3 shows the Raman spectra of IP and $BaSO_4$-loaded IP glass samples. The Raman spectra of these iron phosphate-based glasses are qualitatively consistent with the results of previous studies.[6,11] Raman bands in the region of 650–850 cm^{-1} are due to the symmetric stretching modes of bridging oxygen (P–O–P) between Q^2 and/or Q^1 units, whilst those in the region of 850–1350 cm^{-1} are attributed to the symmetric and asymmetric stretching modes of nonbridging oxygen ($P–O_{nb}$) in Q^n units with n=0, 1 or 2.[6] There are overlapping contributions due to the various modes. Systematic variation in Raman spectral shape with composition is observed (Figure 3). The Raman spectrum for each glass sample was decomposed into seven Gausssian peaks at 650–1400 cm^{-1} in the same manner as in a previous study.[14] Peak position and FWHM (full width at half maximum) were fitted to determine the relative intensities of some selected peaks. Spectral decompositions and Raman band assignments for two representative glass samples with and without $BaSO_4$ loading are shown in Figure 4(a) and (b).[6,12,14]

4. Discussion

In this study, glass samples were formed by melting and quenching mixtures of iron phosphate glass granules and $BaSO_4$ powder as the simulated main sludge component. No sulphur component was retained in the glass samples (i.e. <0·1 mol% SO_3 equivalent). Glass compositions optimised for both thermal stability against crystallisation and water durability are described hereinafter.

Figure 5 shows a stability–durability map for the glass samples, related to two parameters obtained from decomposition of the Raman spectra. The values of the vertical and horizontal axes were obtained from the decomposition of the Raman spectra, as shown in Figure 4, for example. The vertical axis is the relative Raman intensity ratio for the symmetric stretching mode of nonbridging oxygen ($P–O_{nb}$) between Q^0 and Q^1 phosphate units. This parameter increases with increasing Q^0 fraction, corresponding to an increase in the O/P ratio (Table 1). The horizontal axis is the relative Raman intensity ratio of the stretching modes between bridging oxygen (P–O–P) for the Q^2 unit and nonbridging oxygen ($P–O_{nb}$) for the Q^1 unit. This parameter expresses the relative fraction of the Q^2 unit to the Q^1 unit, and it decreases with increasing BaO concentration due to the effect of initial $BaSO_4$ loading on the O/P ratio. The thermal stability is related to the degree of polymerisation of the phosphate glass network with P–O–Fe connection by ferric ions (Fe^{3+}).[3] The water durability of phosphate glasses depends mainly on the polarisation state of oxygen atoms around P.[12,13] The minimisation of P–O–P linkages for the Q^2 unit[3,13] is effective for better water durability, due to Q^0 and Q^1 units connecting with other species, e.g. Fe^{3+} polyhedra.

Figure 5 indicates glass samples that have excellent thermal stability without exhibiting crystallisation

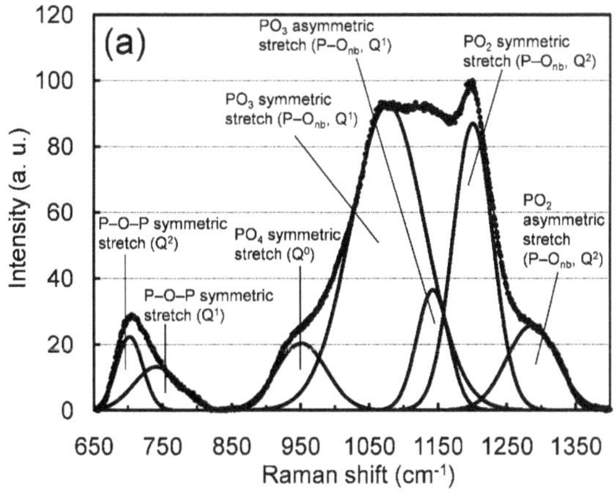

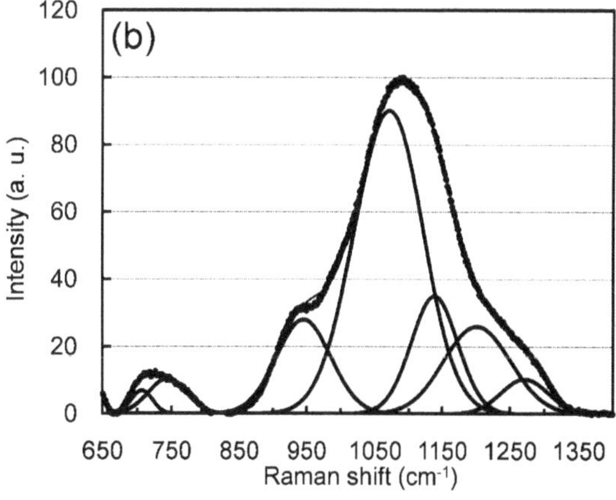

Figure 4. Decompositions of Raman spectra of representative glass samples: (a) IP30 and (b)10BaIP30

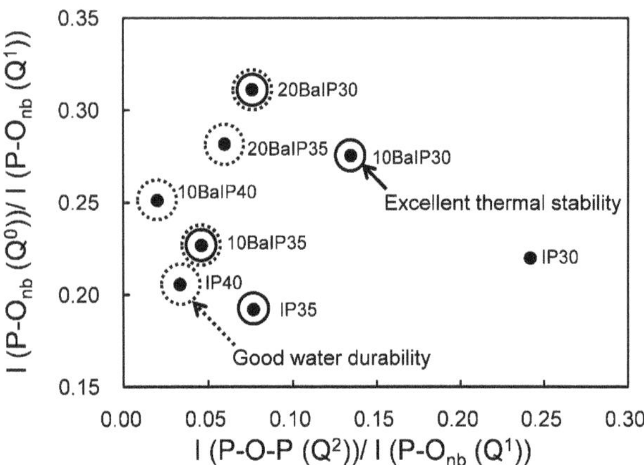

Figure 5. Thermal stability and water durability map expressed by two parameters obtained from the decompositions of Raman spectra of glass samples. The solid circles for 10BaIP30, 20BaIP30, IP35 and 10BaIP35 indicate the glass samples with excellent thermal stability without any crystallisation behaviour in the DTA run (Table 2). The dotted circles for 20BaIP30, 10BaIP35, 20BaIP35, IP40 and 10BaIP40 indicate the glass samples with good water durability without macroscopic shape change and with a relatively small weight loss per unit area (less than 1×10^{-9} kg/mm^2) after the immersion test (Table 3)

behaviour in the DTA run (Table 2) as solid circles, and it indicates those that have good water durability without macroscopic shape change and with relatively small $\Delta W/S$ values (less than $1{\cdot}0\times10^{-9}$ kg/mm^2) in the immersion test (Table 3) as dotted circles. The glass samples with an O/P ratio in the range 3·32–3·38, indicated by both solid and dotted circles in Figure 5, exhibit thermal stability and water durability that are both excellent. These samples are 20BaIP30 and 10BaIP35. Thus, iron phosphate binary glass (IP30) with a relatively low Fe_2O_3 nominal concentration of 30 mol% and with an O/P ratio of 3·11 can be loaded with a high amount of the main sludge component, $BaSO_4$, thus achieving excellent thermal stability and water durability. The location of these glass samples in Figure 5 indicates the following structural features: The optimised glasses have relatively small amounts of Q^2 units, with minimisation of P–O–P linkages, and have moderate amounts of Q^1 and Q^0 units, with the glass network also involving linkages due to Fe^{3+} polyhedra (the fraction of Fe present as ferric ions is >0·8, see Table 1), with the parameter I(P–O–P (Q^2)/I(P–O$_{nb}$ (Q^1)) in the range 0·05–0·08, and with the parameter I(P–O$_{nb}$(Q^0)/I(P–O$_{nb}$(Q^1)) in the range 0·23–0·31 (Figure 5).

5. Conclusions

Glass samples were formed by a conventional melt quenching method, using mixtures of iron phosphate glass granules and $BaSO_4$ powder as a simulated main sludge component. The retained sulphur component in the glasses is <0·1 mol% SO_3 equivalent. The thermal stability against crystallisation at high temperatures was evaluated by a commercial DTA apparatus, and some glass samples exhibit excellent thermal stability without crystallisation behaviour. The water durability, measured by the MCC-2 static leaching method at 120°C for 72 h using a plate sample, indicates that some glass samples have excellent water durability without macroscopic shape change after the immersion test, and also have relatively small weight losses per unit area (less than $1{\cdot}0\times10^{-9}$ kg/mm^2). The optimised glass composition can be designed using the O/P ratio as an analysed parameter, and can be understood in terms of the glass structure evaluated by Raman spectroscopy. A large loading capacity of simulated radioactive sludge component was achieved using iron phosphate binary glass with a relatively low nominal Fe_2O_3 concentration of 30 mol%. The immobilisation of the sludge by solidification of phosphate glass can be optimised by considering the O/P ratio after loading the sludge into the iron phosphate glass.

References

1. Prevost, T., Blasé, M., Paillard, H. & Mizuno, H. Areva's Actiflo™-Rad water treatment system for the Fukushima nuclear power plant. *atw-Int. J. Nucl. Power*, 2012, **57**, 308–313.
2. Amamoto, I., Kobayashi, H., Yokozawa, T., Yamashita, T., Nagai, T., Kitamura, N., Takebe, H., Mitamura, N. & Tsuzuki, T. Applicability of iron phosphate glass medium for loading NaCl originated from seawater used for cooling the stricken power reactors, *Proc. ASME2013 15th Int. Conf. on Environ. Remediation and Radioactive Waste Management*, ICEM2013-96107, pp. 1–8.
3. Yu, X., Day, D. E., Long, J. & Brow, R. K. Properties and structure of sodium-iron phosphate glasses. *J. Non-Cryst. Solids*, 1997, **215**, 21–31.
4. Sengupta, P., A review on immobilisation of phosphate containing high level nuclear wastes within glass matrix – Present status and future challenges. *J. Hazard. Mater.*, 2012, **235&236**, 17–28.
5. Ohta, Y., Morinaga, K. & Yanagase, T. Application of hot-thermocouple method to high temperature chemistry. *Bull. Jpn. Inst. Met.*, 1980, **19**, 239–245.
6. Zhang, L. & Brow, R. K. A Raman study of iron-phosphate crystalline compounds and glasses. *J. Am. Ceram. Soc.*, 2011, **94**, 3123–3130.
7. http://www.factsage.com/
8. Takebe, H., Brady, D. J., Hewak, D. W. & Morinaga, K. Thermal properties of Ga_2S_3-based glass and their consideration during fiber drawing. *J. Non-Cryst. Solids*, 1999, **258**, 239–243.
9. Bingham, P. A. & Hand, R. J. Sulphate incorporation and glass formation in phosphate systems for nuclear and toxic waste immobilisation. *Mater. Res. Bull.*, 2008, **43**, 1679–1693.
10. Masai, H., Tanimoto, T., Fujiwara, T., Matsumoto, S., Takahashi, Y., Tokuda, Y., Yoko, T., Fabrication of Sn-doped zinc phosphate glass using a platinum crucible. *J. Non-Cryst. Solids*, 2012, **358**, 265–269.
11. Bingham, P. A., Hand, R. J., Hannant, O. M., Forder, S. D. & Kilcoyne, S. H. Effects of modifier additions on the thermal properties, chemical durability, oxidation state and structure of iron phosphate glasses. *J. Non-Cryst. Solids*, 2009, **355**, 1526–1538.
12. Takebe, H., Baba, Y. & Kuwabara, M. Dissolution behaviour of ZnO-P_2O_5 glasses in water. *J. Non-Cryst. Solids*, 2006, **352**, 3088–3094.
13. Takebe, H., Kobatake, T. & Saitoh, A. Dissolution behaviour of SnO–P_2O_5 and SnO–P_2O_5–B_2O_3 glasses in water. *Phys. Chem. Glasses: Eur. J. Glass Sci. Technol. B*, 2013, **54**, 182–186.
14. Ma, L., Brow, R. K. & Choudhury, A. Structural study of Na_2O-FeO-Fe_2O_3-P_2O_5 glasses by Raman and Mössbauer spectroscopy. *J. Non-Cryst. Solids*, 2014, **402**, 64–73.

Phys. Chem. Glasses: Eur. J. Glass Sci. Technol. B, December 2016, 57 (6), 245–253

Molecular dynamics modelling of sodium and calcium metaphosphate glasses for biomaterial applications

B. Al Hasni,[1,2] *R. A. Martin,*[3] *C. Storey,*[1] *G. Mountjoy,*[1,*] *D. M. Pickup*[1] *& R. J. Newport*[1]

[1] *School of Physical Sciences, University of Kent, Canterbury, CT2 7NH, UK*
[2] *Engineering Department, College of Applied Sciences, Sohar, Sultanate of Oman*
[3] *Aston Institute of Materials Research & Aston Research Centre for Healthy Ageing, Aston University, Birmingham, B4 7ET, UK*

Manuscript received 9 April 2015
Revised version received 14 June 2015
Accepted 15 September 2016

Many phosphate-based glasses (PBGs) for biomaterial applications have been developed from the Na_2O–CaO–P_2O_5 system. The common base compositions for PBGs are (55–x)Na_2O–xCaO–45P_2O_5 and 20Na_2O.30CaO.50P_2O_5, where the latter corresponds to a metaphosphate, and there is a wealth of experimental data for 50Na_2O.50P_2O_5, 50CaO.50P_2O_5 and 20Na_2O.30CaO.50P_2O_5 metaphosphate glasses. A classical molecular dynamics (MD) method has been used to model Na_2O–CaO–P_2O_5 glass structures, and the results have been closely compared with experimental data for the same glasses. The MD models show the phosphate network to be dominated by Q^2 units, as expected, and to have short range order parameters that are in good agreement with neutron and x-ray diffraction results. Typical coordination numbers of Na and Ca are 5 and 6, respectively. The modifier cation distributions have been examined in detail through the correlation functions $T_{MM}(r)$, with M=Na and/or Ca. The $T_{MM}(r)$ functions show the expected dependence of peak position on modifier cation size, and peak height on modifier cation concentration. The numbers of M–M nearest neighbours are in agreement with a statistical model, in which modifier cations are bonded to nonbridging oxygens, O_{nb}, and $M(O_{nb})_N$ polyhedra are predominantly connected to each other by corner-sharing.

1. Introduction

Historically, phosphate glasses did not have wide applications due to poor durability. In recent decades, however, research into an expanding range of compositions has resulted in several phosphate-based glasses (PBGs) with interesting applications, such as for biomaterials[(1)] and for nuclear waste disposal.[(2)] Many phosphate glasses have been developed from sodium and calcium phosphate compositions, where the addition of Na and/or Ca increases durability. In particular, PBGs for biomaterial applications have been developed based on the Na_2O–CaO–P_2O_5 system.[(1)] (Note that PBGs are different to "Bioglass®" glasses, in which P_2O_5 is a minor component.[(3)]) Previous investigations of glasses in the Na_2O–CaO–P_2O_5 system include studies of properties and glass forming ability,[(4,5)] dissolution in water,[(6)] and dissolution in simulated body fluid.[(7)] Two common base compositions for PBGs are (55–x)Na_2O–$x$$CaO$–45$P_2O_5$ and 20Na_2O.30CaO.50P_2O_5, the latter being metaphosphate. Additional cations have been added to these base compositions to improve functionality.[(1)]

Phosphate glass networks consist of PO_4 tetrahedra, which undergo significant changes with composition (for reviews of phosphate glass structure, see Refs 8 and 9). In pure P_2O_5 glass, all of the PO_4 tetrahedra are connected by three bridging oxygens, and these tetrahedra are denoted Q^3. Na and Ca are archetypal alkali and alkaline earth modifier cations (respectively), and the addition of such modifier cations causes the depolymerisation of the phosphate network, and the introduction of Q^2 tetrahedra. which have two bridging oxygens. The average bond valence of phosphorous bonds to nonbridging oxygens is 2 in Q^3 tetrahedra and decreases to 1·5 in Q^2 tetrahedra, and to 1·33 in Q^1 tetrahedra. Glasses of ultraphosphate composition have a phosphate network with mixed Q^2 and Q^3 tetrahedra and, according to 2D ^{31}P NMR studies,[(10)] there does not appear to be ordering of these units. At the metaphosphate composition, there is an overwhelming dominance of Q^2 units, i.e. rings of Q^2 tetrahedra and/or long chains of Q^2 tetrahedra that are rarely terminated by Q^1 tetrahedra or branched by Q^3 tetrahedra. The modifier cation distribution remains a more challenging topic of investigation in oxide glass structure.[(11)] Recent work on the modifier cation distribution functions in zinc[(12)] and rare earth[(13)] phosphate glasses illustrates the scope for further studies of this topic.

Metaphosphate glasses, and in particular sodium and calcium metaphosphate glasses, are valuable model systems for the study of phosphate glass

Corresponding author. Email g.mountjoy@kent.ac.uk
Original version presented at Int. Conf. on Phosphate Glasses, Pardubice, Czech Republic, 2–4 July 2014
DOI: 10.13036/17533562.57.6.080

structure. In addition, they are relevant to understanding the structures of PBGs derived from the $20Na_2O.30CaO.50P_2O_5$ base composition. The present work addresses this goal using classical molecular dynamics (MD) modelling, which is a widely used technique to study glass structure.[11] While there have been several MD modelling studies of PBGs from the $(55-x)Na_2O-xCaO-45P_2O_5$ base composition (for example, see Ref. 14), there have not been any reports for MD modelling of PBGs from the $20Na_2O.30CaO.50P_2O_5$ base composition. Modelling work always benefits from comparison with experimental data, and neutron and x-ray diffraction data play a special role because they are dependent on the correlation function, $T(r)$,[15] which gives information about nearest neighbour distances, R, and coordination numbers, N. Another reason for modelling sodium and calcium metaphosphate glasses is that multiple diffraction data are available for these glasses.

The literature includes many experimental studies on the structure of sodium and calcium metaphosphate glasses (further details will be given in Section 3). Data have been reported for neutron and x-ray diffraction studies of $50Na_2O.50P_2O_5$,[16–21] $50CaO.50P_2O_5$,[16,22,23] and $20Na_2O.30CaO.50P_2O_5$[24,25] glasses. Several ^{31}P NMR studies have included glasses with compositions equal to or close to the $50Na_2O.50P_2O_5$,[26,27] $50CaO.50P_2O_5$[28,29] and $20Na_2O.30CaO.50P_2O_5$[16] compositions, and confirm the overwhelming dominance of Q^2 units. Results from infrared spectroscopy of $50Na_2O.50P_2O_5$[30] and $20Na_2O.30CaO.50P_2O_5$[24] glasses, and Raman spectroscopy of $50CaO.50P_2O_5$ glass[31] also show the dominance of Q^2 units. ^{23}Na NMR has been used to investigate the Na cation distribution in $50Na_2O.50P_2O_5$ glass,[32] and was interpreted as being consistent with a random hard sphere distribution. X-ray absorption spectroscopy studies of sodium and calcium phosphate glasses are not reported due to the difficulty of using this technique on Na and Ca, arising from their low atomic numbers.

There are limited previous modelling studies of sodium and calcium metaphosphate glasses. Diffraction data have been used to produce RMC models of $50Na_2O.50P_2O_5$,[18] $50CaO.50P_2O_5$,[22] and $20Na_2O.30CaO.50P_2O_5$[33] glasses. Nearly all MD modelling has been done using classical molecular dynamics with empirical interatomic potentials. Work of this kind has previously been reported for $CaO-P_2O_5$,[34] $Na_2O-P_2O_5$,[35] and $Na_2O-CaO-P_2O_5$[14] glasses. *Ab initio* molecular dynamics studies of phosphate glasses are rare due to the computational expense. They have, however, been reported for molecular clusters of $Na_2O-P_2O_5$,[36,37] and for $CaO-P_2O_5$[37] and $(55-x)Na_2O-xCaO-45P_2O_5$[39] glasses with a necessarily small model size of approximately 100 atoms. In this paper, we report classical molecular dynamics modelling of $50Na_2O.50P_2O_5$, $50CaO.50P_2O_5$ and $20Na_2O.30CaO.50P_2O_5$ glasses, using interatomic potentials of the same form as used in recent studies of rare earth phosphate glasses around the metaphosphate composition.[40,41] Our preliminary results for $50CaO.50P_2O_5$ glass were presented recently.[11]

Table 1. Interatomic potential parameters (see Equation (1))

i–j	A_{ij}	ρ_{ij}	C_{ij}	q_i
P–O	27722	0·1819	86·86	3·0
Ca–O	7747	0·2526	93·10	1·2
Na–O	4383	0·2438	30·70	0·6
O–O	1992	0·3436	192·58	−1·2

2. The molecular dynamics method

MD modelling has been carried out using rigid ion potentials, because this allows a large number of time steps, which is important for modelling glasses (shell model potentials are available from Ref. 38, but have the disadvantage of requiring approximately ten times more time steps). P–O, Na–O, Ca–O and O–O interatomic interactions were included, and the potential parameters were taken from those derived by Teter,[42] and were evaluated to establish their suitability (discussed below). Other potential parameters from Teter have proven effective for modelling silicate[43] and aluminate[44] glasses, and the parameters for P–O, Ca–O and O–O interactions have previously been used to model phosphate glasses.[11,40,41] The potentials, $V_{ij}(r)$, have the form

$$V_{ij}(r)=\frac{q_i q_j}{4\pi\varepsilon_0 r}+A_{ij}\exp\left(\frac{-r}{\rho_{ij}}\right)-\frac{C_{ij}}{r^6} \qquad (1)$$

where i and j are element types, r is interatomic distance, q_i is charge, and A_{ij}, ρ_{ij} and C_{ij} are potential parameters given in Table 1 (with $\varepsilon_0=8{\cdot}854\times10^{-12}$ $C^2N^{-1}m^{-2}$). In addition, three body O–P–O and P–O–P bond bending interactions were taken from Ref. 45, and have the form

$$V_{iji}(\theta)=\frac{1}{2}k_{iji}\left(\theta-\Theta_{iji}\right)^2 \qquad (2)$$

Table 2. Details of simulated crystal structures (parentheses show comparison with experimental crystal structures)

	o-P_2O_5	o′-P_2O_5	$NaPO_3$	$Ca(PO_3)_2$
Vol (Å^3)	678·0 (−2·4%)	329·5 (+2·4%)	539·6 (+2·8%)	873·5 (−4·6%)
a (Å)	16·85 (+3·3%)	9·063 (−1·4%)	12·45 (+2·8%)	6·824(−2·5%)
b (Å)	8·457 (+4·2%)	5·128 (+4·9%)	6·172 (−0·4%)	7·591 (−1·6%)
c (Å)	4·772 (−9·4%)	7·091 (−1·0%)	7·019 (+0·4%)	16·86 (−0·5%)
R_{PO} (Å)	1·51 (−0·03 Å)	1·51 (−0·03 Å)	1·51 (−0·03 Å)	1·48 (−0·05 Å)
R_{MO} (Å)	-	-	2·53 (+0·06 Å)	2·48 (+0·01 Å)

Table 3. Details of MD simulations

	Na	Ca	P	O	Box size (Å)	Density (g/cm^3) Model	Experiment
$50Na_2O.50P_2O_5$	250		250	750	25·75	2·48	2·37,[19] 2·47,[16] 2·50[21]
$50CaO.50\ P_2O_5$		125	250	750	25·07	2·61	2·61,[16] 2·69[23]
$20Na_2O.30CaO.50P_2O_5$	300	225	750	2250	36·48	2·59	2·59,[24] 2·61[25]

where j is the element type of the central atom, with parameters k_{iji}=3·5 eV and Θ_{iji}=109·47° for O–P–O, and k_{iji}=3·0 eV and Θ_{iji}=135·5° for P–O–P. The potential parameters have been evaluated by using the GULP program[46] (Version 1.2, 2002) to model the crystal structures of o-P_2O_5, o′-P_2O_5 and the metaphosphates $NaPO_3$ and $Ca(PO_3)_2$ (crystal structures were obtained from CDS database[47]). The results are shown in Table 2. It can be seen there is a tendency for R_{PO} to be slightly too short.

MD simulation was used to obtain models of the atomic structure of $50Na_2O.50P_2O_5$, $50CaO.50P_2O_5$ and $20Na_2O.30CaO.50P_2O_5$ metaphosphate glasses. Random starting configurations were used with a cubic box and periodic boundary conditions. The numbers of atoms, box size and densities are shown in Table 3. One objective of the modelling work was to produce models that agree with the experimental densities. However, use of a fixed pressure NPT algorithm with P=1 atm produced a model with a density ~8% higher than experiment. Hence a fixed volume NVT algorithm was used to obtain a model matching the experimental density. As a consequence, the models obtained have unphysical negative pressures of 1·0–1·5 GPa. This means a greater volume for the structural configuration, but does not noticeably increase the values of R and N, because the latter are dominated by the short-range part of the potentials. (It is often difficult to accurately model both density and pressure of glasses using MD; for example, see Ref. 48.)

The MD modelling used the DLPOLY program,[49] with time steps of 1 fs. A Berensden NVT algorithm was used with a relaxation time of 2 ps. A short range cut-off of 10 Å was used for all except the Coulomb potential, and the Coulomb potential was calculated using the Ewald method with a precision of 10^{-5}. The modelling used six stages. The first three stages were temperature baths of 40 000 time steps (with equilibration) at 6000 K, 3000 K, and 1500 K, and with linear thermal expansion coefficients of 1·03, 1·015, and 1·005, respectively. (A trajectory of 40 000 time steps at 6000 K is sufficient to allow diffusion over the box length.) This was followed by a temperature quench (with equilibration) from 1500 K to 300 K at a quench rate of 10^{13} K s^{-1} (i.e. 60 000 time steps). This quench rate is typical in MD studies of glasses.[45,43,48] Due to constraints on computing time, all MD studies of glasses use quench rates which are many orders of magnitude higher than in experiments (the role of quench rates is under ongoing investigation; for example, see Ref. 50). Despite this, MD studies have

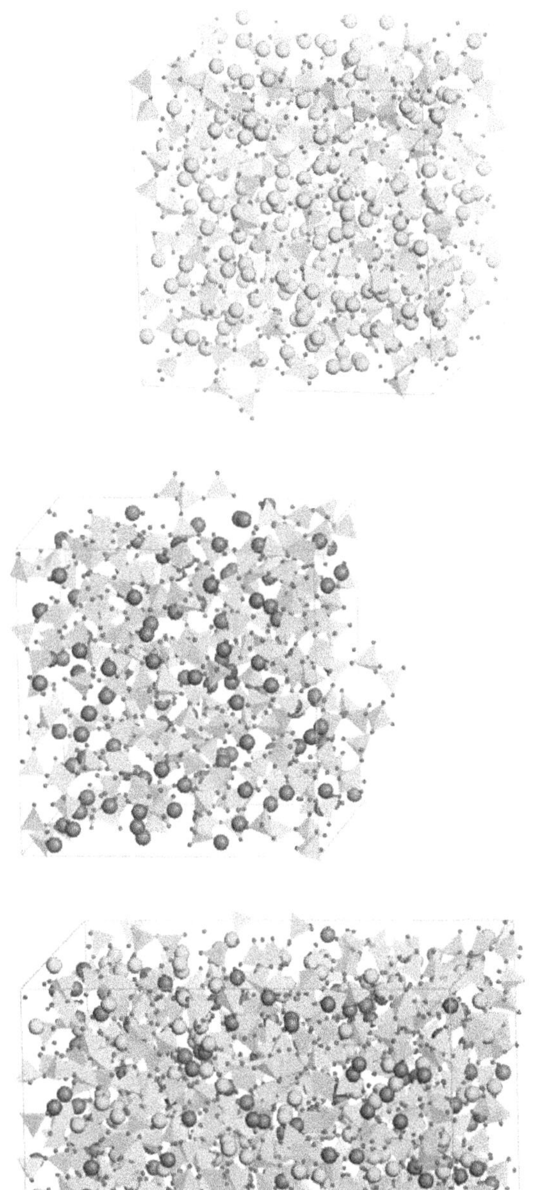

Figure 1. Images of MD models of (a) $50Na_2O.50P_2O_5$, (b) $50CaO.50P_2O_5$ and (c) $20Na_2O.30CaO.50P_2O_5$ metaphosphate glasses (red dots are oxygen, tetrahedra show the PO_4 network, light blue spheres are Na, and dark green spheres are Ca) [Colour available online]

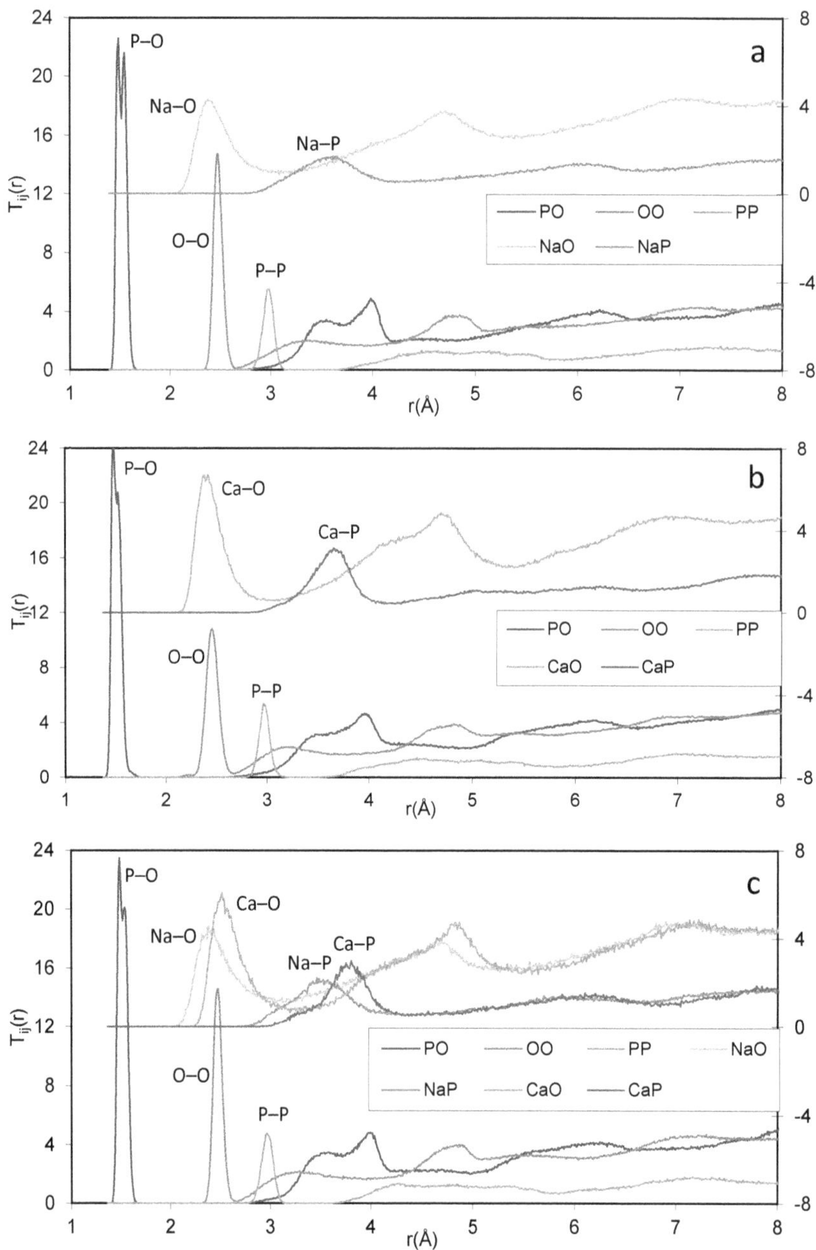

Figure 2. Partial correlation functions, $T_{ij}(r)$, for MD models of (a) $50Na_2O.50P_2O_5$, (b) $50CaO.50P_2O_5$ and (c) $20Na_2O.30CaO.50P_2O_5$ metaphosphate glasses. Note that the left vertical axis is for X–X correlations (X=P, O), the right vertical axis is for M–X correlations (M=Na, Ca), and M–M correlations are shown in Figure 4 [Colour available online]

been able to provide key insights into glass structures. The final two stages were temperature baths of 40 000 time steps at 300 K, the first with equilibration and the second without equilibration. During the final stage, the structural parameters were sampled (every 400 time steps) to account for the disorder caused by thermal vibrations which are present in experimental results.

3. Results

Figure 1 shows images of the MD models of $50Na_2O.50P_2O_5$, $50CaO.50P_2O_5$ and $20Na_2O.30CaO.50P_2O_5$ metaphosphate glasses and the tetrahedral PO_4 network is evident. Figure 2 shows the partial correlation functions, $T_{ij}(r)$, for the models, which are

$$T_{ij}(r)=\frac{1}{r}\left(\frac{1}{N_i}\sum_{\ell=1}^{N_i}\sum_{m\neq\ell}^{N_j}\delta\left(r-R_{\ell m}\right)\right)=4\pi r\rho_j g_{ij}(r) \qquad (3)$$

where $g_{ij}(r)$ is the partial pair distribution function with $g_{ij}(r)\to 0$ as $r\to 0$ and $g_{ij}(r)\to 1$ as $r\to\infty$, ρ_j is the atomic number density, and ℓ and m are indices over atoms of element types i and j. Figure 3 shows experimental x-ray and neutron diffraction structure factors $S(Q)$. The $S(Q)$ functions for the model have been calculated using

$$Q\left(S(Q)-1\right)=\sum_{ij} w_{ij}(Q)\int\left(T_{ij}(r)-4\pi r\rho_j\right)\sin(Qr)\,\mathrm{d}r \qquad (4)$$

where $w_{ij}(Q)$ is the weighting factor for scattering

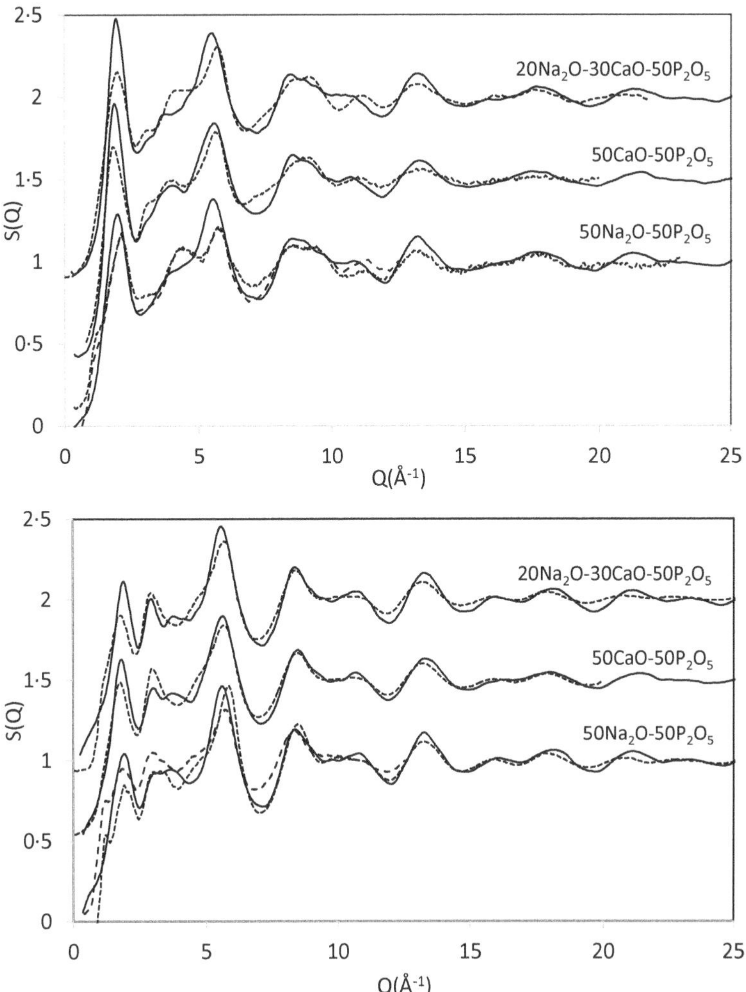

Figure 3. Diffraction structure factors, S(Q), from MD models (solid lines) and experiments (broken lines, see text for references), for $50Na_2O.50P_2O_5$, $50CaO.50P_2O_5$ and $20Na_2O.30CaO.50P_2O_5$ metaphosphate glasses for (upper) x-ray diffraction and (lower) neutron diffraction

from correlations between element types i and j.[51] (The $S(Q)$ functions were calculated using $T_{ij}(r)$ up to r=10 Å, and in this case a window function was found to be unnecessary.) $S(Q)$ for the models are also shown in Figure 3, and are in fair agreement with the experimental $S(Q)$, although there is some mismatch in peak heights and positions. As is usual for molecular dynamics models of glasses, there is less good agreement with $S(Q)$ from x-ray diffraction, because the x-ray structure factor is more influenced by the medium range order associated with the distribution of cations. (In comparison $S(Q)$ from neutron diffraction is more influenced by short range order associated with the distribution of oxygen).

The first peak in $T_{PO}(r)$ at ~1·5 Å (see Figure 2) represents P–O nearest neighbours. As expected, 33% of oxygen atoms are bridging oxygens (denoted O_b) with two bonds to P, and 67% of oxygen are nonbridging oxygen (denoted O_{nb}) with one bond to P. The split first peak in $T_{PO}(r)$ shows that P–O_{nb} bonds are shorter than P–O_b bonds, as expected from neutron diffraction studies.[52] 99–100% of P atoms have 4-fold coordination (the remainder being 5-fold coordinated defects). The first peak in $T_{OO}(r)$ at ~2·5 Å is due to O–P–O configurations, and the average O–$\hat{P}$–O bond angle in the models is tetrahedral, i.e. 109°. The width of the first peak in $T_{OO}(r)$ is approximately 30% larger for $50CaO.50P_2O_5$ glass compared

Table 4. Q^n distributions from ^{31}P NMR and MD modelling (Typical experimental uncertainties are ± a few %)

Glass	Ref.	Q^0 (%)	Q^1 (%)	Q^2 (%)	Q^3 (%)	Q^4 (%)	Average Q^n
$50Na_2O.50P_2O_5$	16, 27			100			
	26		2	98			
	MD	5	18	48	28	1	2·0
$50CaO.50P_2O_5$	16		<5	>95			
	29		3	92	5		
	MD	2	23	50	22	3	2·0
$20Na_2O.30CaO.50P_2O_5$	24		5	95			
	MD	3	23	48	28	1	2·0

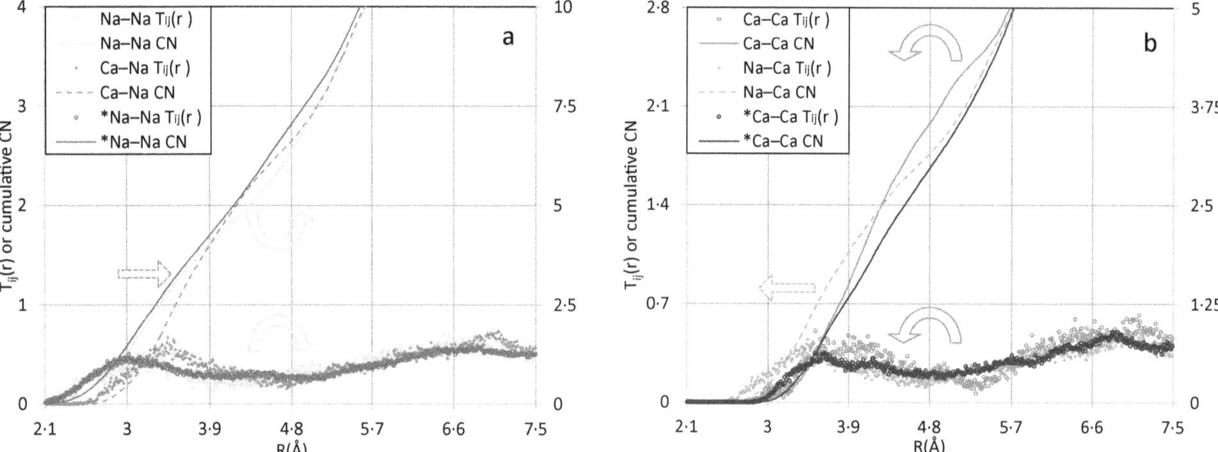

*Figure 4. Modifier cation (M=Na, Ca) partial correlation functions, $T_{MM}(r)$, and cumulative coordination numbers, $CN_{MM}(r)$, for MD models of metaphosphate glasses. (a) M–Na correlations for $20Na_2O.30CaO.50P_2O_5$ and $50Na_2O.50P_2O_5$ glasses (left and right vertical axes, respectively, scaled in proportion to Na content). (b) M–Ca correlations for $20Na_2O.30CaO.50P_2O_5$ and $50CaO.50P_2O_5$ glasses (left and right vertical axes respectively, scaled in proportion to Ca content) *denotes $50Na_2O.50P_2O_5$ and $50CaO.50P_2O_5$ binary metaphosphate glasses [Colour available online]*

to the two glasses containing Na. There is a narrow first peak in $T_{PP}(r)$ at ~3·0 Å representing P–P nearest neighbours in the phosphate network, and the average P–Ô–P bond angle in the models is ~147°.

The connectivity of the phosphate network can be described by the Q^n distribution, and the average as expected for metaphosphates is n=2·0. Table 4 shows that Q^2 species is predominant (~50%) in the models, as expected. However, unlike the results from ^{31}P NMR studies (also shown in Table 4), the models have significant amounts of both Q^1 and Q^3 groups (~20% each), with a slight remainder of Q^0 and Q^4. This broad Q^n distribution is typical of MD models of glasses made using rigid ion potentials with very high quench rates (the use of shell model potentials has been shown to improve this feature of MD models[38]).

The experimental results for nearest neighbour distances, R_{ij}, and coordination numbers, N_{ij}, are reported in Table 5 for P–O, M–O, O–O and P–P correlations. There is good agreement between the modelling and experimental results, considering reported experimental uncertainties. The x-ray results for O–O correlations, and the neutron results for P–P correlations, are somewhat uncertain due to the small weighting of the corresponding $T_{ij}(r)$ (see Equation (4) above). The model values of R_{PO} are ~0·05 Å shorter than the diffraction results of 1·56 Å, and this can be due to the potential parameters, as discussed in Section 2. The model values of R_{OO}=2·46–2·47 Å are somewhat shorter than the diffraction result of 2·52 Å.

The first peak in $T_{MO}(r)$ at ~2·4 Å (see Figure 2) represents M–O nearest neighbours (where M=Na or Ca). In these MD models, made using rigid ion potentials, N_{MO} is overwhelmingly due to O_{nb}, and the contribution of O_b is approximately 0 for N_{CaO}[11] and 0·2 for N_{NaO}). There is a broad second peak in $T_{OO}(r)$ from 3·0 to 3·5 Å, representing O_{nb} coordinated to M, i.e. O–M–O configurations within $M(O_{nb})_N$ polyhedra. The M–P correlations are first prominent in $T_{MP}(r)$ from 3–4 Å. The distances (R_{MP}) and numbers (N_{MP}) of nearest neighbours are shown in Table 5. There is a first peak in $T_{MM}(r)$ around 4 Å. This arises due to the sharing of O_{nb}, i.e. M–O_{nb}–M configurations, due to corner- or edge-sharing of two $M(O_{nb})_N$ polyhe-

Table 5. Nearest neighbour distances, R_{ij}, and coordination numbers, N_{ij}, from neutron and x-ray diffraction experiments (via peak fitting of T(r)), RMC models, and MD models (Typical experimental uncertainties are ±0·02 Å in R, and ±10% in N)

Glass, ref.	R_{PO} (Å)	N_{PO}	R_{NaO} (Å)	N_{NaO}	R_{CaO} (Å)	N_{CaO}	R_{OO} (Å)	N_{OO}	R_{PP} (Å)	N_{PP}
$50Na_2O.50P_2O_5$										
ND,[16] XRD[17]	1·54	3·95	2·39	5·08	-	-	2·52	3·96	2·93	2·08
ND & XRD[21]	1·57	4·00	2·38	5·00	-	-	2·53	4·00	-	-
MD	1·50	4·00	2·38	4·96	-	-	2·48	4·01	2·99	2·00
$50CaO.50P_2O_5$										
ND & XRD[22]	1·56	3·85	-	-	2·48	6·95	2·51	4·00	2·97	2·00
RMC[22]	1·57	4·00	-	-	2·48	5·63	2·51	3·95	2·93	2·00
XRD[23]	1·56	4·00	-	-	2·39	7·0	2·53	3·95	-	-
MD	1·49	4·00	-	-	2·40	5·80	2·46	4·00	2·95	2·00
$20Na_2O.30CaO.50P_2O_5$										
XRD,[24] ND[25]	1·55	3·75	2·46	6·30	2·50	7·10	2·52	4·20	2·94	2·05
RMC[33]	1·57	4·00	2·43	4·20	2·45	5·5	2·49	3·85	2·94	2·00
MD	1·53	3·98	2·40	5·21	2·53	6·00	2·48	4·01	2·99	2·00

dra. The comparison of R_{MO} values from the MD models with those from diffraction studies is problematic for the $20Na_2O.30CaO.50P_2O_5$ glass because the Na–O and Ca–O nearest neighbour peaks are overlapping in diffraction data.

In order to study the M–M correlations more thoroughly in the models, the relevant $T_{ij}(r)$ are further analysed in Figure 4 and Table 6. Figure 4 shows the $T_{NaNa}(r)$ and $T_{CaCa}(r)$ in $50Na_2O.50P_2O_5$ and $50CaO.50P_2O_5$ glasses. As expected, these $T_{ij}(r)$ show a reduction height for the $20Na_2O.30CaO.50P_2O_5$ glass that is in proportion to the modifier content, being 4/7=57% of Na and 3/7=43% of Ca. Figure 4 compensates for these proportions by using different vertical axes for the $20Na_2O.30CaO.50P_2O_5$ glass. In the $20Na_2O.30CaO.50P_2O_5$ glass there are Na–Ca and Ca–Na correlations which are not present in the $50Na_2O.50P_2O_5$ and $50CaO.50P_2O_5$ glasses. As expected, $T_{CaNa}(r)$ primarily differs from $T_{NaNa}(r)$ by having a nearest neighbour peak which is shifted to longer distances (see dashed red arrow in Figure 4), due to the larger atomic radius of Ca compared to Na. Conversely, $T_{NaCa}(r)$ primarily differs from $T_{CaCa}(r)$ by having a nearest neighbour peak that is shifted to shorter distances (see dashed green arrow in Figure 4), due to the smaller atomic radius of Ca compared to Na.

Table 6. Analysis of the number cation nearest neighbours N_{MP} and N_{MM} (where M=Na, Ca) in MD models (using cut offs of 4·20 and 4·80 Å, respectively). The values of N_{MM} estimated using the method described in the text are also shown, and compared with the values of N_{MM} in models as percentages

	$50Na_2O.50P_2O_5$	$50CaO.50P_2O_5$	$20Na_2O.30CaO.50P_2O_5$
M–P			
R_{NaP} (Å)	3·56		3·50
N_{NaP}	4·7		5·6
R_{CaP} (Å)		3·67	3·77
N_{CaP}		5·8	5·7
Na–Na			
MD	7·0		2·6
estimated	5×(2·5−1)=7·5 7·0/7·5=93%		5·55×(1·94−1)×4/7=2·9 2·6/2·9=89%
Ca–Na			
MD			2·6
estimated			=2·9 2·6/2·9=89%
Ca–Ca			
MD		3·1	2·0
estimated		6·0×(1·5−1)=3·0 3·1/3·0=103%	5·55×(1·94−1)×3/7=2·2 2·0/2·2=91%
Na–Ca			
MD			1·95
estimated			=2·2 1·95/2·2=89%

4. Discussion

The MD models obtained have fairly realistic features in terms of short range order and phosphate network connectivity. The coordination numbers, N_{ij}, and nearest neighbour distances, R_{ij}, are in reasonable agreement with those reported in experimental studies, and a tetrahedral phosphate network with dominant Q^2 connectivity was achieved. There is scope for improvement, in that some of the nearest neighbour distances differ from experimental values by up to 0·05 Å, and there is significantly more Q^1 and Q^3 content than observed in experiment. While the former is dependent on the availability of improved potential parameters, the latter is a fundamental limitation of MD modelling of glasses (as discussed in Section 2). Nevertheless, the models enable access to information about modifier cation distributions that is extremely difficult to study experimentally. Therefore it is worthwhile to devote further consideration to interpreting the features in $T_{MM}(r)$ (see Figure 4).

As discussed in Section 3, the $20Na_2O.30CaO.50P_2O_5$ glass has $T_{MM}(r)$ functions that are scaled in proportion to the modifier cation content (M=Na or Ca), and have nearest neighbour peaks shifted in proportion to the atomic radius of the central modifier cation. These effects are consistent with a random or statistical model for the modifier cation distribution. They contrast with more extreme models of chemical ordering, in which (i) Na–Ca nearest neighbours are preferred over Na–Na or Ca–Ca (and the first peaks in $T_{NaNa}(r)$ and $T_{CaCa}(r)$ are much reduced), or (ii) Na–Na and Ca–Ca nearest neighbours are preferred over Na–Ca (and the first peaks in $T_{NaCa}(r)$ and $T_{CaNa}(r)$ are much reduced). A random or statistical model for modifier cation distribution is also physically reasonable, given that an MD model of glass structure is quenched from the melt with a very high quench rate. Although the $T_{MM}(r)$ functions are found to scale with modifier cation content, which must be true at large distances, this is not necessarily the case for the nearest neighbour peaks at short distances.

The numbers, N_{MM}, of modifier cation nearest neighbours are given in Table 6, for a cutoff distance of 4·80 Å. Although the value(s) of the cutoff(s) may be debated, 4·80 Å provides a single criterion that corresponds reasonably with the first minimum after the nearest neighbour peak for all $T_{MM}(r)$ functions. As discussed in Section 3, the nearest neighbour peak in $T_{MM}(r)$ at around 4 Å is due to M–O_{nb}–M configurations, i.e. corner- or edge-sharing between two $M(O_{nb})_N$ polyhedra. Knowledge of M–O_{nb} coordination numbers, N_{MOnb}, (see Table 5) can provide a basis for predicting the number of M–M nearest neighbours, N_{MM}, as follows: For $20Na_2O.30CaO.50P_2O_5$ glass, the average number of O_{nb} nearest neighbours to an M atom is

$$N_{MOnb}=(4/7)\times N_{NaO}+(3/7)\times N_{CaO}=5\cdot 55 \quad (5)$$

(neglecting M–O_b bonding, see Section 3). Expressed as an average number of O_{nb}–M nearest neighbours per O_{nb} this is

$$N_{OnbM}=(7/20)\times N_{MOnb}=1\cdot 94 \quad (6)$$

(note that, for every seven modifier cations, there are 30 oxygens, of which 67% or 20 are O_{nb}). Assuming each O_{nb} which is coordinated to a central M should be coordinated to an additional (N_{OnbM}−1) neighbouring M gives

$$N_{MM}=N_{MOnb}\times(N_{OnbM}-1)=5\cdot 22 \quad (7)$$

(note that we have assumed modifier cations are coordinated to O_{nb}). For a statistical distribution (4/7)=57% of neighbouring M should be Na, or

$$N_{MNa}=(4/7)\times N_{MM}=2\cdot 98 \quad (8)$$

and similarly $N_{MCa}=(3/7)\times N_{MM}=2\cdot 24$. The same analysis of the $50Na_2O.50P_2O_5$ and $50CaO.50P_2O_5$ glass models gives $N_{NaNa}=7\cdot 5$ and $N_{CaCa}=3\cdot 0$, respectively. These estimated values of N_{MM} are shown in Table 4, and it is seen that they are about 110% of the values of N_{MM} in the MD models. The above analysis neglects any M–O bonding to O_b, and MD studies have shown no M–O bonds to O_b in $CaO–P_2O_5$ glass,[11] and 10% of M–O bonds to O_b in $(55-x)Na_2O.xCaO.45P_2O_5$ glasses.[38,39] The above analysis also assumes that $M(O_{nb})_N$ polyhedra are only connected by corners, i.e. corner-sharing. If edge-sharing occurred instead of corner-sharing, then the above analysis would over-estimate N_{MM} by a factor of two, which does not appear to be the case.

Considering the role of M content, the $20Na_2O.30CaO.50P_2O_5$ glass has 60% Ca content relative to the $50CaO.50P_2O_5$ glass, and 60% of $N_{CaCa}=3\cdot 1$ is 1·86, whereas a slightly higher value of $N_{CaCa}=2\cdot 0$ is observed for the $20Na_2O.30CaO.50P_2O_5$ glass. Conversely, the $20Na_2O.30CaO.50P_2O_5$ glass has 40% Na content relative to the $50Na_2O.50P_2O_5$ glass, and 40% of $N_{NaNa}=7\cdot 0$ is 2·8, whereas a slightly lower value of $N_{NaNa}=2\cdot 6$ is observed for the $20Na_2O.30CaO.50P_2O_5$ glass. This subtle difference in N_{MM} appears to correspond to subtle changes in $T_{MM}(r)$ for the mixed glass. Closer inspection of Figure 4 shows that the nearest neighbour peaks in $T_{NaNa}(r)$ for $20Na_2O.30CaO.50P_2O_5$ glass show slightly more nearest neighbours at larger distances compared to the $50Na_2O.50P_2O_5$ glass (see curved light yellow arrows in Figure 4). This can be understood as the mixing of larger Ca ions, which tends to push apart the neighbouring Na ions slightly. Conversely the nearest neighbour peaks in $T_{CaCa}(r)$ for the $20Na_2O.30CaO.50P_2O_5$ glass show that there are slightly more nearest neighbours at shorter distances compared to the $50CaO.50P_2O_5$ glass (see curved light blue arrows in Figure 4). This can be understood as the mixing of smaller Na ions, which tends to allow neighbouring Ca ions to be closer together.

Conclusions

Classical molecular dynamics modelling with rigid ion potentials has been used for the first time to model the $20Na_2O.30CaO.50P_2O_5$ metaphosphate base composition for phosphate bioglasses (PBGs). The model of $20Na_2O.30CaO.50P_2O_5$ glass has been compared with models of $50Na_2O.50P_2O_5$ and $50CaO.50P_2O_5$ metaphosphate glasses to better understand the nature of cation distributions in the mixed glass. The models show that the phosphate network is dominated by Q^2 units, as expected, and has short range order parameters that are in good agreement with neutron and x-ray diffraction data. Typical coordination numbers of Na and Ca are 5 and 6, respectively. The cation distributions have been examined through detailed analysis of the correlation functions $T_{MM}(r)$ with M=Na and/or Ca, for which experimental information is difficult to obtain. The $T_{MM}(r)$ functions show the expected dependence of peak position on modifier cation size, and of peak height on modifier cation concentration. The numbers of nearest neighbours, N_{MM}, are in agreement with a random statistical model, in which modifier cations are bonded to nonbridging oxygen atoms, O_{nb}, and $M(O_{nb})_N$ polyhedra are predominantly connected to each other by corner-sharing. The presence of a random statistical distribution of modifier cations in these MD models is unsurprising, given the very high quench rates which are a feature of MD modelling of glasses.

References

1. Abou Neel, E. A., Pickup, D. M., Valappil, S. P., Newport, R. J. & Knowles, J. C. *J. Mater. Chem.*, 2009, **19**, 690–701.
2. Bingham, P. A., Hand, R. J., Hannant, O. M., Forder, S. D. & Kilcoyne, S. H. *J. Non-Cryst. Solids*, 2009, **355**, 1526–1538.
3. Hench, L. L., Splinter, R. J., Allen, W. C. & Greenlee, T. K. *J. Biomed. Mater. Res. Symp.*, 1971, **5**, 117–141.
4. Uo, M., Mizuno, M., Kuboki, Y., Makishima, A. & Watari, F. *Biomaterials*, 1998, **19**, 2277–2284.
5. Parsons, A. J., Burling, L. D., Scotchford, C. A., Walker, G. S. & Rudd, C. D. *J. Non-Cryst. Solids*, 2006, **352**, 5309.
6. Delahaye, F., Montagne, L., Palavit, G., Touray, J. C. & Baillif, P. *J. Non-Cryst. Solids*, 1998, **242**, 25–32.
7. Ahmed, I., Lewis, M., Olsen, I. & Knowles, J. C. *Biomaterials*, 2004, **25**, 491–499.
8. Brow, R. K. *J. Non-Cryst. Solids*, 2000, **263&264**, 1–28.
9. Hoppe, U., Walter, G., Kranold, R. & Stachel, D. *J. Non-Cryst. Solids*, 2000, **263&264**, 29–47.
10. Alam, T. M. *J. Non-Cryst. Solids*, 2000, **274**, 39.
11. Mountjoy, G., Al-Hasni, B. M. & Storey, C. *J. Non-Cryst. Solids*, 2011, **357**, 2522–2529.
12. Hoppe, U., Walter, G., Carl, G., Neuefeind, J. & Hannon, A. C. *J. Non-Cryst. Solids*, 2005, **351**, 1020–1031.
13. Hoppe, U. *J. Phys.: Condens. Matter*, 2008, **20**, 165206.
14. Di Tommaso, D., Ainsworth, R. I., Tang, E. & de Leeuw, N. H. *J. Mater. Chem. B*, 2013, **1**, 5054–5066.
15. Wright, A. C. In: *Experimental Techniques of Glass Science*, Eds. C. J. Simmons & O. H. El-Bayoumi, The American Ceramic Society, Westerville,

1993, p. 205.
16. Pickup, D. M., Ahmed, I., Guerry, P., Knowles, J. C., Smith, M. E. & Newport, R. J. *J. Phys.: Condens. Matter*, 2007, **19**, 415116.
17. Pickup, D. M. Private communication, 2009.
18. Hall, A., Swenson, J., Adams, S. & Meneghini, C. *Phys. Rev. Lett.*, 2008, **101**, 195901.
19. Zotov, N., Schlenz, H., Brendebach, B., Modrow, H., Hormes, J., Reinauer, F., Glaum, R., Kirfel, A. & Paulmann, C. *Z. Naturforsch.*, 2003, **58a**, 419.
20. Zotov, N., Kirfel, A., Beuneu, B., Delaplane, R., Hohlwein, D., Reinauer, F. & Glaum R. *Physica B*, 2004, **350**, E1071–E1073.
21. Hoppe, U., Stachel, D. & Beyer, D. *Phys. Scripta*, 1995, **T57**, 122–126.
22. Wetherall, K. M., Mountjoy, G., Pickup, D. M. & Newport, R. J. *J. Phys.: Condens. Matter*, 2009, **21**, 035109.
23. Hoppe, U., Walter, G. & Stachel, D. *Phys. Chem. Glasses*, 1992, **33**, 216–221.
24. Pickup, D. M., Guerry, P., Moss, R. M., Knowles, J. C., Smith, M. E. & Newport, R. J. *J. Mater. Chem.*, 2007, **17**, 4777–4784.
25. Moss, R. M., Abou Neel, E. A., Pickup D. M., Twyman, H. L., Martin, R. A., Henson, M. D., Barney, E. R., Hannon, A. C., Knowles, J. C. & Newport, R. J. *J. Non-Cryst. Solids*, 2010, **356**, 1319–1324.
26. Montagne, L., Palavit, G. & Delaval, R. *J. Non-Cryst. Solids*, 1997, **215**, 1–10.
27. Prabakar, S., Wenslow, R. M. & Mueller, K. T. *J. Non-Cryst. Solids*, 2000, **263&264**, 82.
28. Fletcher, J. P., Kirkpatrick, R. J., Howell, D. & Risbud, S. H. *J. Chem. Soc. Faraday Trans.*, 1993, **89**, 3297–3299.
29. Witter, R., Hartmann, P., Vogel, J. & Jager, C. *Solid State Nucl. Mag. Res.*, 1998, **13**, 189–200.
30. Moustafa, Y. M. & El-Egili, K. *J. Non-Cryst. Solids*, 1998, **240**, 144–153.
31. Pemberton, J. E., Latifzadeh, L., Fletcher, J. P. & Risbud, S. H. *Chem. Mater.*, 1991, **3**, 195–200.
32. Alam, T. M., McLaughlin, J., Click, C. C., Conzone, S., Brow, R. K., Boyle, T. J. & Zwanziger, J. W. *J. Phys. Chem. B*, 2000, **104**, 1464–1472.
33. Moss, R. M. *PhD Thesis*, 2009, University of Kent, Canterbury, UK.
34. Belashchenko, D. K. & Ostrovskii, O. I. *Inorg. Mater.*, 2002, **38**, 146–153.
35. Speghini, A., Sourial, E., Peres, T., Pinna, G., Bettinelli, M. & Capobianco, J. A. *Phys. Chem. Chem. Phys.*, 1999, **1**, 173–177.
36. Uchino, T. & Yoko, T. *J. Non-Cryst. Solids*, 2000, **263&264**, 180–188.
37. Parsons, A. J., Ahmed, I., Rudd, C. D., Cuello, G. J., Pellegrini, E., Richard, D. & Johnson, M. R. *J. Phys.: Condens. Matter*, 2010, **22**, 485403.
38. Ainsworth, R. I., Di Tommaso, D., Christie, J. K. & de Leeuw, N. H. *J. Chem. Phys.*, 2012, **137**, 234502.
39. Tang, E., Di Tommaso, D. & de Leeuw, N. H. *Adv. Eng. Mater.*, 2010, **12**, B331–B338.
40. Clark, E. B., Mead, R. N. & Mountjoy, G. *J. Phys.: Condens. Matter*, 2006, **18**, 6815–6826.
41. Martin, R. A., Mountjoy, G. & Newport, R. J. *J. Phys.: Condens. Matter*, 2009, **21**, 075102.
42. Teter, D. Private communication, 2004.
43. Cormack, A. N. & Du, J. *J. Non-Cryst. Solids*, 2004, **349**, 66–79.
44. Thomas, B. W. M., Mead, R. N. & Mountjoy, G. *J. Phys.: Condens. Matter*, 2006, **18**, 4697–4708.
45. Liang, J. J., Cygan, R. T. & Alam, T. M. *J. Non-Cryst. Solids*, 2000, **263&264**, 167.
46. Gale, J. D. *J. Chem. Soc. Faraday Trans.*, 1997, **93**, 629–637.
47. Fletcher, D. A., McMeeking, R. F. & Parkin, D. *J. Chem. Inf. Comput. Sci.*, 1996 **36**, 746–749 (United Kingdom Chemical Database Service).
48. Smith, W., Forester, T. R., Greaves, G. N., Haytera, S. & Gillan, M. J. *J. Mater. Chem.*, 1997, **7**, 331.
49. Smith, W. & Forester, T. *J. Molec. Graphics*, 1996, **14**, 136–141.
50. Corrales, L. R. & Du, J. *Phys. Chem. Glasses*, 2005, **46**, 420–424.
51. Elliott, S. R. *Physics of Amorphous Materials*, 1990, Longman, Harlow.
52. Hoppe, U., Kranold, R., Stachel, D., Barz, A. & Hannon, A. C. *Z. Naturforsch.*, 2000, **55a**, 369.

Phys. Chem. Glasses: Eur. J. Glass Sci. Technol. B, December 2016, 57 (6), 254–266

Model considerations of changes in the average M–O coordination numbers of ternary alkali mixed glass former phosphate glasses, $A_2O–M_yO_z–P_2O_5$

*U. Hoppe**

Institut für Physik, Universität Rostock, Universitätsplatz 3, 18051 Rostock, Germany

Manuscript received 28 August 2014
Revised version received 3 February 2015
Accepted 10 March 2015

Several mixed glass former $M_yO_z–P_2O_5$ glasses show changes of the oxygen coordination numbers of the M atoms in definite intervals of chemical composition. A third component, such as an alkali oxide, A_2O, complicates the description of the behaviour. Recent diffraction results for sodium germanophosphate glasses suggest assuming the Ge–O coordination numbers to depend on the amount of the available O atoms that is reduced by a definite fraction of nonbridging O atoms. This latter fraction supplies the counter charge for the alkali ions. The model is corroborated by the tendencies of the changes of the oxygen coordination numbers that are reported for glass former atoms of ternary phosphate glasses of similar compositions. Some of the obvious differences that exist with the predictions of the simple model are due to the poor ability of glass formers to form a five-fold coordinated network unit.

1. Introduction

Phosphate glasses show a couple of peculiarities[1–3] that are relatively well understood for binary phosphate systems, i.e. for additions of one further oxide to P_2O_5. On the other hand, a great variety of properties are only accessible for practical use with multicomponent systems. The addition of even just one more oxide increases the glass forming range and modifies the properties considerably. Appropriate extension of the structural model becomes more complicated if the increased variety of possible atomic environments is taken into account. The present work is aimed at adapting the characteristics of the model for binary phosphate glasses to ternary systems. Here, the focus is on systems with two additional oxides that behave quite differently; one is a glass former or an intermediate oxide (M_yO_z), and the other is an alkali oxide (A_2O). Such glasses have received much interest due to their non-linear changes of properties – a phenomenon known as the 'mixed glass former effect' (MGFE).[4] The present work is focused on the compositional behaviour of the average M–O coordination number, N_{MO}, of such glasses.

The structural features of binary glasses, $M_yO_z–P_2O_5$ or $A_2O–P_2O_5$, are briefly reviewed as follows: The addition of the second oxide is accompanied by the maximum rupture of P–O–P bridges, which disassembles the tetrahedral phosphate network step by step, from the three-connected PO_4 tetrahedra (Q^3) of vitreous P_2O_5, via chain (Q^2) and end groups (Q^1), to isolated PO_4 (Q^0).[1] Here Q^n indicates a PO_4 unit with n bridging oxygens. This effect can be reduced if the second oxide is itself a glass former and its content is high. All terminal O atoms, including the P=O double bond of the Q^3, participate in the coordination of the M or A atoms as much as possible.[2] The character of the one P=O bond of the Q^3 becomes shared in the Q^2, Q^1 and Q^0 groups, and thus the fifth valence of the P atom is then delocalised over several P–O bonds. Accordingly, the negative charge that is deposited on the nonbridging oxygen (NBO) and used for charge compensation of the A^+ ions is less than one electron unit. The bond order of the M–O bonds of a bridging oxygen (BO) in M–O–P bridges is less than unity. The fifth phosphorus valence is shared on two, three or four bonds in the Q^2, Q^1 or Q^0 units, respectively. Hence, the charges on the NBOs or the bond orders in the M–O bonds change with the type of Q^n group. These variations, together with the changes of the ratio of available O atoms per M site, can influence the M–O coordination numbers. Several examples for changes of the coordination numbers in binary phosphate glasses are known, for example, with the oxides M_yO_z=ZnO,[5] Al_2O_3,[6] La_2O_3,[7] Ga_2O_3,[8] TeO_2,[9] or GeO_2.[10,11] The corresponding average N_{MO} values follow the number of terminal O sites in PO_4 units that are available for each M atom.[3] This change is only possible in a definite interval of chemical composition provided that the corresponding N_{MO} values belong to the range of possible oxygen environments of the given M atom.

Corresponding author. Email uwe.hoppe@uni-rostock.de
Original version presented at Int. Conf. on Phosphate Glasses, Pardubice, Czech Republic, 2–4 July 2014
DOI: 10.13036/17533562.57.6.041

Ternary M_yO_z–A_2O–P_2O_5 glasses can be formed for chemical composition ranges that considerably exceed those for binary systems. The extrapolation of known N_{MO} values for binary systems to distant ternary glass compositions is less certain. If the M atoms were to capture all O atoms to increase the value of N_{MO}, at the expense of the NBOs of the A^+ fraction, then that would clearly increase the range of large N_{MO} values. Such behaviour is known for many related crystal structures. Moreover, the networks of the ternary glasses may possess considerable numbers of M–O–M bridges in the presence of P–O–P bridges, as is found, for example, for sodium germanophosphate glasses.[12] The reduced tendency for the maximum rupture of the phosphate network could indicate a reduced tendency for large values of N_{MO}. However, comparison of the experimental values of N_{GeO} for similar glasses[13] suggests a model with a maximum value of N_{GeO}, and a mixture of germanophosphate and sodium phosphate regions. It is assumed that a moderate preference of Ge for octahedral environments produces this behaviour. The details of the model[13] are given in Section 2. The success of the model suggests examining the behaviour of the N_{MO} values for other but similar ternary phosphate systems. The considerations are based on the M–O coordination numbers that are available from the literature. Most N_{MO} values were determined by magic angle spinning (MAS) nuclear magnetic resonance (NMR) spectroscopy, and the small error bars for these values justify a detailed analysis.

2. Modelling changes in the M–O coordination number

Firstly, N_{MO} values for binary M_yO_z–P_2O_5 glasses are calculated. For definite glass compositions the M atoms can form oxygen environments with all O atoms in bridging positions M–O–P.[3] All O atoms that are located outside of the P–O–P bonds are involved in these environments, which means the terminal O of all existing Q^n species. This behaviour of O sites can be maintained in a definite range of composition if the M–O coordination number changes for compensation. The increase of the bond order of the M–O bonds, and the decrease of the number of available O atoms with increasing M_yO_z content, can be balanced by decreasing the values of N_{MO}.[5–11] The appropriate change of the average value of N_{MO} is given by

$$N_{MO}=2[z/y+n(P_2O_5)/n(MO_{z/y})] \quad (1)$$

where $n(\ldots)$ denotes the mole fraction of the given oxide. ZnO–P_2O_5 glasses, for example, exhibit N_{ZnO}=6 and 4 for ZnO contents of 33 and 50 mol%, respectively.[5] The changes of N_{MO} are assumed to be driven by a general preference of the O atoms for bridging sites between the glass-forming units or highly charged

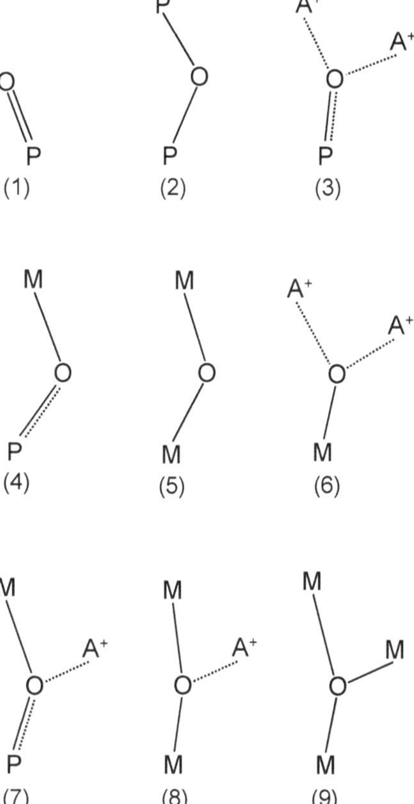

Figure 1. Suite of the expected oxygen sites in A_2O–M_yO_z–P_2O_5 glasses,[12,13] and a few other sites that are expected for glassy samples outside the range of the considered compositions

ions, which also means the avoidance of terminal P=O bonds. The atomic radii of the particular M atoms limit the N_{MO} range to one or other oxygen environment. Other effects may dominate the structural evolution, and can suppress any change of the value of N_{MO}. For example, binary CuO–P_2O_5 glasses do not show any change of the oxygen environment of the Cu(II) sites;[14] strongly distorted CuO_6 octahedra are stabilised by ligand field effects throughout all glass compositions.

The addition of A_2O as a third oxide may increase the flexibility of the structure, and this could allow the value of N_{MO} to persist close to its most preferred value. However, the occupation of the different oxygen sites should be considered also. Nine possible O sites for A_2O–M_yO_z–P_2O_5 glasses are shown in Figure 1. The O sites (3) or (6) are shown with two A^+ neighbours, but other numbers of A^+ neighbours such as one, three or four can exist, as well. If possible, the O site (1) is avoided in phosphate glasses, and this site should be absent in the structures of the Na_2O–GeO_2–P_2O_5 glasses studied.[13] Also, site (9) should not exist in these glasses as long as other sites could be formed. Site (9) could result from high pressure processing.

Two possibilities are seen as the extreme scenarios, which are explained for the case of Na_2O–GeO_2–P_2O_5

glasses as follows:[12,13] (i) All oxygen added by Na_2O or GeO_2 is used for the rupture of P–O–P bridges, and the created terminal O atoms coordinate the Ge to maintain GeO_6 units as much as possible. In this case, Equation (1) has to be changed to

$$N_{MO}=2[z/y+(n(P_2O_5)+n(A_2O))/n(MO_{z/y})] \quad (2)$$

This mechanism is valid for the $KGeOPO_4$ crystal[15] and some other related compounds where O sites (3) and (6) do not exist. The charge compensation of the GeO_6 octahedra and K^+ ions is balanced by means of the O sites (7) and (8). (ii) The GeO_2 component is accommodated simply as neutral GeO_4 units. Formally, this behaviour seems possible with a constant value for N_{MO}. However, the consequent P–O–Ge bridges to such units do not contribute to smooth distributions of the valences on the four P–O bonds of the PO_4 units and that is not expected.

The N_{GeO} values of sodium and potassium germanophosphate glasses are found between four and six,[13,16] i.e. between those of the scenarios (ii) and (i). It is assumed that the A^+ ions consume part of the O atoms as NBO neighbours solely for their own charge compensation. A new relation between average N_{MO} and the glass composition is derived from Equation (2) by creating a fraction of NBOs. This fraction reduces the amount of O atoms that is available for M–O bonds. A free factor is added that takes into account the existence of the NBOs. By introducing this factor as a ratio $n(NBO)/n(A)$ the amount of NBO is related to the fraction of its potential A^+ neighbours and the equation is

$$N_{MO}=2\left\{z/y+\left[\frac{n(P_2O_5)+n(A_2O)\left(1-\frac{n(NBO)}{n(A)}\right)}{n(MO_{z/y})}\right]\right\} \quad (3)$$

The composition dependence of the N_{GeO} values obtained for the GeO_2–$NaPO_3$ system[13] shows that a special ratio $n(NBO)/n(A)$ close to ~1·5 is favoured. This value is equivalent to a model with only BOs between the P and Ge atoms (O sites 2, 4, 5), NBOs that coordinate the Na^+ ions (O sites 3, 6), and a maximised value for N_{GeO}. The O sites (7) and (8) do not occur. The O site (6) is less probable for reasons of charge balance,[12] but that is not essential here. If the sole existence of O sites (2–6) is assumed, then the counter charge of the Na^+ ions must be supplied by NBOs only.

The determination of the ratio $n(NBO)/n(A)$ that supplies this full charge compensation is a little uncertain. The average negative charge that is deposited on an NBO of a PO_4 depends on the type of Q^n group, with values of 0·5, 0·66, and 0·75 e^- for Q^2, Q^1 and Q^0, respectively. Furthermore, the negative charge deposited on the NBO of a MO_m unit (where $m=2z/y$) is $1e^-$. The distributions of Q^n are sometimes unknown, and moreover the electron charge of a given Q^n could be distributed on its NBOs in slightly different portions. The charge on a NBO of a Q^n depends also on possible linkages with neutral MO_m units, which have a similar effect as a link with a PO_4 unit. Different Q^n sites may be preferred as neighbours of either A^+ or M sites if mixtures of various Q^n groups exist. It could be argued that the preferred neighbour of an A^+ ion is the Q^n with the larger electron charges on its NBOs. On the other hand, a randomly balanced distribution or the opposite preference could exist. These problems were discussed for Na_2O–GeO_2–P_2O_5 glasses, concerning the coexistence of Q^1 and Q^0 groups.[13] The ratios $n(NBO)/n(A)$ for Q^1 and Q^0 with 1·5 and 1·33, respectively, do not differ much. For the comparisons in Ref. 13 and given below, it is found that the uncertainties in the experimental values of N_{MO} and the ratios $n(NBO)/n(A)$ possess a similar magnitude. Hence, it is possible to estimate the behaviour of N_{MO} for similar ternary phosphate glasses on the basis of only their chemical compositions. The possible range of oxygen environments of the M atoms is taken into account. The validity of Equation (3), according to the scenario of exclusive formation of the O sites (2–6), is examined on the basis of reported N_{MO} values. Since the alkali germanophosphate system is the prototype of the model, it is considered again in detail before starting the other comparisons.

Lastly, recall a special case of the ratio $n(NBO)/n(A)$ which, for example, has already been used in the interpretation of the Al–O coordination numbers of ternary Na_2O–Al_2O_3–P_2O_5 glasses.[17] The evolution of N_{AlO} was related to the ratio $n(O)/n(P)$, implying a continuous change of the ratio $n(NBO)/n(A)$, depending on $n(O)/n(P)$. The ratio $n(NBO)/n(A)$ is related to the average n of that Q^n distribution which corresponds to the maximum rupture of P–O–P bridges. Only the O sites (2–6) should exist. The model implies, on average, that the oxygen valences in M–O bonds of M–O–P bridges given in valence units (vu) are equal to the electron charges on the NBOs as neighbours of the A^+ ions. This balance seems reasonable and is expressed by

$$n(NBO)/n(A)=N_{MO}y/2z \quad (4)$$

The average value of N_{MO} can take on any necessary value. The M atoms do not have any preference for a definite Q^n in competition with the A^+ ions and vice versa. Then, Equation (3) is changed to

$$N_{MO}=2\frac{\left\{\frac{z}{y}+\frac{\left[n(P_2O_5)+n(A_2O)\right]}{n(MO_{z/y})}\right\}}{\left[1+\frac{y}{z}\frac{n(A_2O)}{n(MO_{z/y})}\right]} \quad (5)$$

This simple model implies continuous changes of the

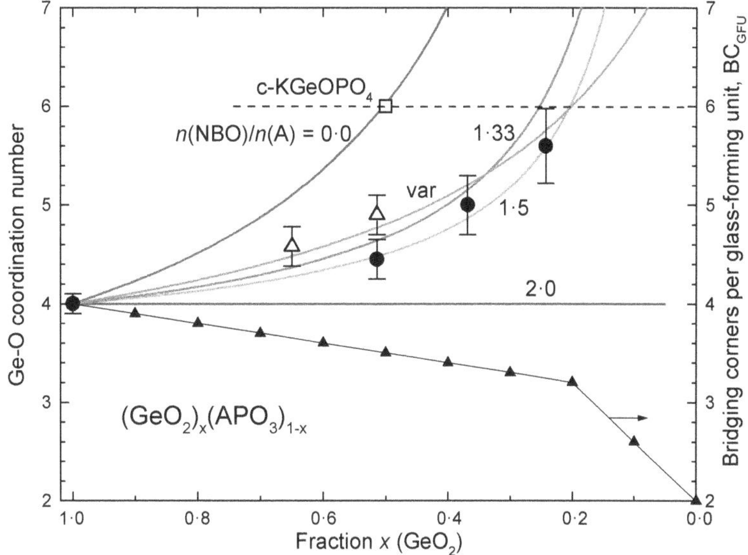

Figure 2: Comparison of the experimental values of N_{GeO} for $(GeO_2)_x(APO_3)_{1-x}$ glasses (A=Na – solid circles[13]; K – open triangles[16]) with the model N_{GeO} behaviour according to Equations (3) or (5), given by coloured solid lines. The lines for different ratios n(NBO)/n(A) of 1·33, 1·5, and 2·0 are related to the NBOs of Q^0, Q^1, and Q^2 neighbours, respectively, and 0·0 means the absence of NBO. The label 'var' means variable n(NBO)/n(A), in balance with N_{MO} according to Equation (4). The value of N_{GeO} for crystalline $KGeOPO_4$[15] is indicated by an open square. The line with solid triangles shows the average number of bridging corners per glass-forming unit, BC_{GFU}, as calculated by Equation (6) [Colour available online]

structural parameters. It could be valid up to the limit of binary $A_2O–M_yO_z$ glasses, where it would finish with $n(NBO)/n(A)=1$ in the case of $MO_{2z/y}$ structural units. But here, in this approach, the model is developed for the P_2O_5-rich region.

3. M–O coordination numbers for various ternary phosphate glasses

3.1 Alkali germanophosphate glasses

The average value of N_{MO} depends on the mole fractions of the three components according to Equations (3) or (5), and one fraction is fixed by the two others. Commonly, the compositional range of glass formation for a ternary system is illustrated in an equilateral triangle (a ternary phase diagram), in which the single components mark the three corners. The graphical presentation of the overall behaviour of N_{MO} is difficult, but most experimental studies of N_{MO} are made along cross-sections of this triangle, and thus the plots follow these lines.

The first example shows a comparison of model predictions with the values of N_{GeO} that result from diffraction experiments on $(GeO_2)_x(APO_3)_{1-x}$ glasses (cf. Figure 2).[13,16] Coloured solid lines indicate the model N_{GeO} behaviour according to Equations (3) or (5) and calculated for different values of the ratio $n(NBO)/n(A)$. The ratios 2·0, 1·5, and 1·33 as indicated in the plot represent the NBOs of Q^2, Q^1, and Q^0 groups, respectively. A ratio $n(NBO)/n(A)=0$ corresponds to the absence of NBOs. The label 'var' indicates the use of variable $n(NBO)/n(A)\geq 1$, with balance between N_{MO} and the charges on the NBOs according to Equation (4). The model functions with $n(NBO)/n(A)=1·33$ or 1·5 (for Q^0 and Q^1) are close to each other, and also close to that of the model 'var'. The N_{GeO} values of the series of $(GeO_2)_x(NaPO_3)_{1-x}$ glasses[13] are well approximated by these three functions, but the best agreement is obtained with the ratio according to Q^1 groups (1·5). One of the N_{GeO} values (x=0·5, A=Na) differs a little from the N_{MO} model with variable $n(NBO)/n(A)$ (var). Recently, a series of $(Na_2O)_{0·33}[(GeO_2)_{2x}(P_2O_5)_{1-x}]_{0·67}$ glasses has been analysed by ^{31}P MAS NMR.[18] Though only two of our samples[13] have suitable compositions for comparison with this series, the two N_{GeO} values also follow the model N_{MO} according to Q^1 NBOs.

The behaviour of these N_{GeO} values was the starting point for our model.[13] Thus, only the O sites (2–5) should exist, and the O site (6) can be excluded on the phosphate-rich side of the glasses studied. Hence, the corners of the PO_4 units interact separately with either the Ge atoms via bridging O sites, or with the Na^+ ions via NBOs in non-directional bonds. Since the NBOs are shared between two or more Na^+ neighbours, the sodium-rich phosphate regions form interconnected sub-structures among the Ge-rich phosphate nanodomains. The corresponding ordering on the scale of intermediate lengths (~1 nm) was related to the first diffraction peak that was found at very small magnitudes of scattering vector, Q, of ~8 nm^{-1}.[13]

Figure 2 also shows two values N_{GeO} from an earlier study of $K_2O–GeO_2–P_2O_5$ glasses.[16] The values are located in a range between the lines labelled 1·33, 'var' and 0, and correspond to $n(NBO)/n(K)\approx 1·0$. Accordingly, part of the countercharge of the K^+ ions is supplied from the BOs in P–O–Ge or Ge–O–Ge bridges, i.e. the O sites (7) or (8), as shown in Figure 1. Nevertheless, most of the countercharge of the K^+ ions is supplied by the NBOs. First diffraction peaks at very small Q-values were detected and interpreted in terms of the ordering of Ge- and K-rich phosphate nanodomains.[19] In the extreme case of the full absence of NBOs, the N_{GeO} values would approach the blue line [n(NBO)/n(A)=0·0], and the special first diffraction peak should vanish. That behaviour is found for the structure of the crystalline $KGeOPO_4$.[15]

The validity of a model can be tested by consideration of the behaviour of glass properties. The changes of N_{MO} correspond to changes of the fractions of BOs, f_{BO}, and NBOs, f_{NBO}. The values of N_{GeO} for $(GeO_2)_x(NaPO_3)_{1-x}$ glasses should show a sharp change in slope for $x≈0·2$ because N_{GeO} greater than 6 is not expected. The average N_{GeO} values approximately follow the model function given for n(NBO)/n(Na)=1·5 (Q^1) for x>0·2. It is assumed that all Ge are in octahedral sites for glasses with x<0·2. The behaviour of the glass transition temperature, T_g, has been found to show a change in slope close to this composition.[12] The increase of T_g for x<0·25 is followed by a plateau for x>0·25. GeO_6 units with six bridging corners strengthen the phosphate network with an increased fraction of BOs, and this is related to the T_g increase. The subsequent change to GeO_4 units reduces this effect. The network stiffness, as expressed by T_g, is influenced by the average number of bridging corners per glass-forming unit, BC_{GFU}. This value is calculated as

$$BC_{GFU}=2c_Of_{BO}/(c_P+c_{Ge})=2(N_{PO}c_P+N_{GeO}c_{Ge}-c_O)/(c_P+c_{Ge}) \quad (6)$$

where c_i are the atomic fractions of the elements. The resulting curve is also given in Figure 2. The factor BC_{GFU} provides a possible explanation for the behaviour of T_g. A similar interpretation in terms of BOs has been given for the behaviour of T_g for alkali borophosphate glasses.[4] The O sites (7) or (8) contribute to the BO fraction if they are present.

3.2 Alkali silicophosphate glasses

The existence at ambient pressure of six-coordinated Si in oxide glasses prepared by melt quenching was first shown for the $Na_2O–SiO_2–P_2O_5$ system.[20] In contrast to diffraction results on the average value of N_{GeO}, the fractions of the various oxygen environments of Si are resolved by ^{29}Si MAS NMR. Crystalline SiP_2O_7,[21] of a composition in the vicinity of those of the glasses,[20] shows only octahedral Si sites. Though Equation (3) suggests values between five and six, the total N_{SiO} values of the glassy samples were found to be between four and five.[20] These smaller N_{SiO} values indicate the existence of PO_4 (Q^3) with terminal P=O bonds, even though these units could be largely avoided by forming more SiO_6 octahedra. Other series of ternary silicophosphate glasses with 25 mol% A_2O (A=Na, K) have been measured by x-ray absorption spectroscopy (XAS),[22] and values for N_{SiO} were obtained with larger uncertainty. The average values of N_{SiO} for these glasses are a little less than six, though six could be possible for most samples to avoid the terminal P=O bonds. In crystalline $Rb_2SiP_4O_{13}$,[23] which is in line with this compositional series, silicon is only in SiO_6 octahedra. This crystal has no P=O bonds, but only BOs and NBOs on O sites (2–4), see Figure 1. Unfortunately, the ternary Ge- and Si-containing glasses under discussion do not have exactly analogous compositions.

Though the experimental values for N_{SiO}[20,22] do not follow the predicted model behaviour, the total N_{SiO} values are clearly larger than four, as is assumed according to scenario (ii). The N_{SiO} values are just not six for some of the compositions. Hence, similar to the behaviour of N_{GeO},[13] the increase of N_{SiO} can be interpreted as a result of overbonded P–O in P–O–Si bridges with the P_2O_5 component. However, the ternary phosphate glasses with Si show significantly less tendency to octahedral units than for Ge.

3.3 Alkali borophosphate glasses

The structural behaviour of alkali borate glasses is characterised by the occurrence of three- and four-coordinated B sites, indicating the limiting values for the average coordination number, N_{BO}. An excellent probe of B–O coordination numbers is given by ^{11}B MAS NMR spectroscopy. Several series of ternary

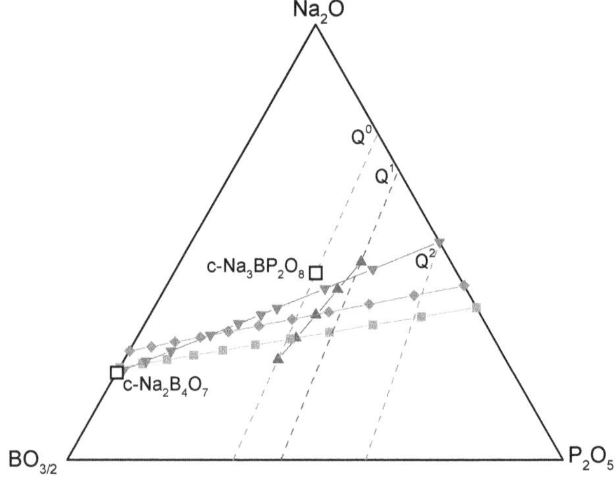

Figure 3. Compositions of $Na_2O–B_2O_3–P_2O_5$ glasses and crystals. The lines with symbols indicate the chemical compositions of the samples used in the ^{11}B MAS NMR measurements of different authors – down triangles,[24] diamonds,[25] squares,[26] up triangles.[27] The dashed lines marked with Q^2, Q^1, and Q^0 connect the binary compounds with the equivalent type of Q^n group. Open squares indicate crystalline compounds: Only BO_4 units exist in $Na_3BP_2O_8$[28] while a BO_3/BO_4 mixture (50%/50%) occurs in $Na_2B_4O_7$[29] [Colour available online]

alkali borophosphate glasses have been investigated in the last few years. Here, the focus is on investigations of the sodium borophosphate system, and the glass compositions that have been studied[24–27] are given in Figure 3. In addition, three straight lines connect those binary phosphate compositions that are characterised by the same Q^n groups.

The N_{BO} values of the four studied glass series[24–27] have essentially similar behaviour: N_{BO} is four in the triangle between the P_2O_5 corner and the Q^0 line. A decrease of N_{BO} occurs immediately beyond this line for all glasses of smaller P_2O_5 content. Formally, the Q^0 line connects the Na_3PO_4 and BPO_4 compositions, which are known as Q^0 structures. The obvious N_{BO} change at the Q^0 line does not mean that the corresponding borophosphate glasses are formed of only Q^0 groups. ^{31}P MAS NMR studies on such glasses show the co-existence of Q^1 and Q^0, and possibly a few Q^2, at these compositions.[4,25,26] The Q^0 groups could be assumed as a mixture of BPO_4 and Na_3PO_4 structures. However, it is unlikely that the corresponding Q^0 groups have either four BO_4 or four Na^+ neighbours. Mixed PO_4 environments are expected, as occur in crystalline $Na_3BP_2O_8$.[28] Evidence for phase separation in BPO_4 structures is not known.[26] The scenario according to Equation (3), which is valid on the Q^0 line, suggests the absence of O sites (7) or (8), see Figure 1; here they are not needed for charge balance. Finally, N_{BO} behaves according to the well known borate anomaly[30] for compositions approaching binary alkali borate compositions. Figure 4 shows the behaviour of N_{BO} measured by ^{11}B MAS NMR.[26] The values are constant at four for small B_2O_3 contents, but on crossing the Q^0-line (1·33), N_{BO} starts to decrease. The influence of the borate anomaly leads to elevated values of N_{BO} above the model function for variable n(NBO)/n(A). The borate anomaly is associated with O site (8), and possibly (7) – see Figure 1, instead of NBOs. The decrease of N_{BO} ends close to the binary sodium borate glass of maximum N_{BO}~3·5. This value is also known for crystalline $Na_2B_4O_7$,[29] of similar composition, which does not contain any NBO.

As known for BPO_4, the underbonded B–O bonds of a BO_4 unit are well balanced with four Q^0 neighbours (4×0·75vu=3vu). Also, the BO_4 tetrahedron must be able to co-exist with Q^1 or Q^2 neighbours for the P_2O_5-rich compositions. Nominally, the four Q^1 corners supply insufficient valence to balance B–O bonds; 4×0·67vu=2·67vu. An O site (7) does not compensate for this deficit. The missing valence must be supplied from either a B–O–B bridge, or links with special Q^1 or Q^2 units. These units must possess varying valence on those three or two PO_4 corners that do not belong to the P–O–P bridges. This varying valence could be balanced by a NBO corner with A^+ neighbours. Some examples of such BO_4 environments are suggested in Ref. 4. The fractions of PO_4 with varying numbers of PO_4 and BO_4 neighbours in $(K_2O)_{1/3}[(B_2O_3)_x(P_2O_5)_{1-x}]_{2/3}$ glasses have been determined by ^{31}P MAS NMR.[4] In this paper the net charges on the NBOs per K^+ ion and the number of B–O–P bridges per B site are calculated from these reported results. A charge of ~1e$^-$ per K^+ and ~4 B–O–P per B site are determined for glasses with x≤0·3. Thus, the full counter charge of K^+ is supplied from NBOs and the valences of the BO_4 are stabilised from PO_4 neighbours only.

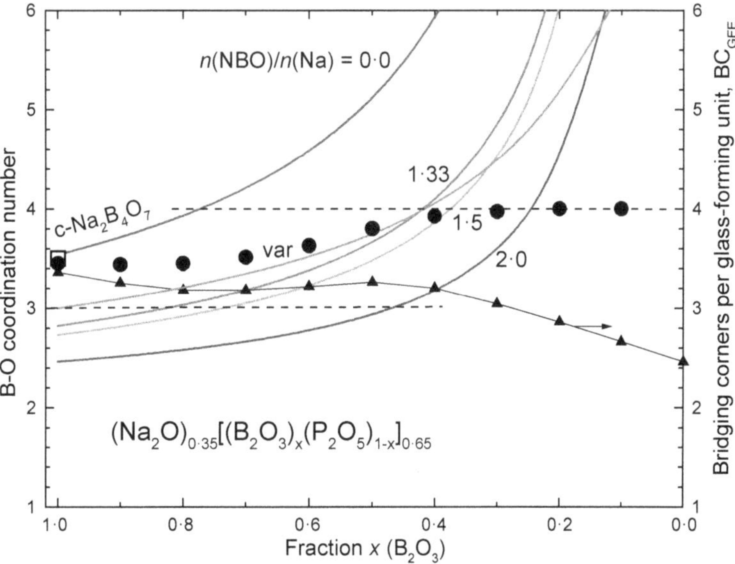

Figure 4. Comparison of the experimental (^{11}B MAS NMR) N_{BO} values for $(Na_2O)_{0·35}[(B_2O_3)_x(P_2O_5)_{1-x}]_{0·65}$ glasses[26] (solid circles) with the model N_{BO} behaviours according to Equations (3) and (5), given by coloured solid lines. The ratios n(NBO)/n(A) are related to three Q^n groups, variable Q^n ('var') and the absence of NBOs. The values of N_{BO} for crystalline $Na_2B_4O_7$,[29] with a BO_4/BO_3 ratio of 50:50, is marked with an open square. The line with solid triangles shows the average number of bridging corners per glass-forming unit, BC_{GFU} [Colour available online]

From these characteristics of P_2O_5-rich borophosphate glasses, it is concluded that the Na^+ ions do not coordinate any corners of the BO_4 units. Since the Na^+ ions share their NBO neighbours the sodium phosphate regions form interconnected sub-structures. As outlined above,[4] many PO_4 units have B and Na^+ neighbours on different corners at the same time. Thus, alternating orders of sodium phosphate and borophosphate moieties should exist on the scale of intermediate lengths, with nanodomains similar to those used to explain the diffraction pre-peaks for the alkali germanophosphate glasses.[13,19] This decoupling of the Na^+ sites from the BO_4 units for glasses with $x \leq 0{\cdot}4$ is thought to be related to the maximum in the Na^+ ion conductivity.[31]

The average numbers of bridging corners per glass-forming unit for the $(Na_2O)_{0{\cdot}35}[(B_2O_3)_x (P_2O_5)_{1-x}]_{0{\cdot}65}$ series[26] is calculated according to Equation (6) and shown in Figure 4. The increase of BC_{GFU} starts from sodium phosphate (x=0) due to the addition of four-connected BO_4, and this correlates well with the increase of the experimental T_g.[31] A plateau of T_g ensues for $x>0{\cdot}4$ and BC_{GFU} also has a plateau in this region. Thus, the change in slope of N_{BO} versus x at the Q^0 line marks the change of the T_g behaviour.

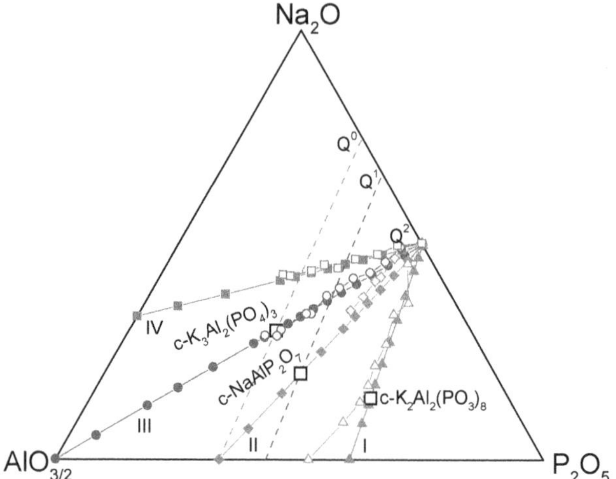

Figure 5. Compositions of Na_2O–Al_2O_3–P_2O_5 glasses and crystals. The straight lines with solid symbols follow the directions of the compositional series I, II, III, and IV.[17,32] *The lines with open symbols indicate the analysed sample compositions. Dashed lines marked with Q^2, Q^1, Q^0 connect binary compounds with the same type of Q^n group. The larger open squares indicate related crystalline compounds: AlO_4 tetrahedra occur in $K_3Al_2(PO_4)_3$.*[33] *AlO_6 octahedra are found in $NaAlP_2O_7$*[34] *and $K_2Al_2(PO_3)_8$*[35]*. [Colour available online]*

3.4 Alkali alumino- and gallophosphate glasses

The structure and properties of ternary sodium aluminophosphate glasses have been studied thoroughly.[17,32] In phosphates, Al atoms occupy sites with four, five and six oxygen neighbours, as shown by ^{27}Al MAS NMR. Brow has studied glassy samples close to the sodium-phosphate rich side of the ternary phase diagram (Figure 5).[32] The compositions of the glasses were analysed,[32] but systematic differences from the batch compositions did not occur, except for the Al-rich end of series I. This series follows the line of the mixture of Q^2 structures. Structural motifs of $Al(PO_3)_3$ and $NaPO_3$ glasses could co-exist, which means AlO_6 octahedra in accordance with Equation (3), and this behaviour is observed. A small decrease of N_{AlO} at the Al-rich end of series I is interpreted as arising from the loss of P_2O_5 during melting.[17] Separate work on binary Al_2O_3–P_2O_5 glasses has shown that N_{AlO} decreases continuously with the addition of Al_2O_3.[6] Deviations from a mixture of $Al(PO_3)_3$ and $NaPO_3$ motifs with O sites (7) or (8) would result in $N_{AlO}>6$. Any $N_{AlO}<6$ would require terminal P=O bonds (O site (1)). Since the other series (II, III, IV) show a strong tendency for AlO_6 octahedra, other AlO_n polyhedra are also not expected for series I.

The samples of series II, III and IV of Na_2O–Al_2O_3–P_2O_5 glasses[17] between the P_2O_5 corner and the Q^1 line (Figure 5) have Al–O coordination numbers close to six. A decrease to N_{AlO} values close to ~4 ensues if the glass compositions of series III and IV cross the Q^1 line and approach the Q^0 line. A value N_{AlO}=4 in the case of Q^0 groups is reminiscent of N_{BO}=4 for the corresponding borophosphate glasses. (The N_{AlO} values of series III are shown in Figure 6.) Thus, the N_{AlO} values for ternary Na_2O–Al_2O_3–P_2O_5 and binary Al_2O_3–P_2O_5 glasses differ clearly for compositions between the Q^1 and Q^2 lines. Though only a small range between the Q^2 and Q^1 lines is accessible, N_{AlO} for the binary glasses decreases continuously, with $5{\cdot}3 \geq N_{AlO} \geq 4{\cdot}8$.[6] The additional Na_2O fraction of the ternary glasses helps Al to maintain AlO_6 units in this range.[17] Possibly, this behaviour is accompanied by the O sites (7) or (8), as is found in the structure of crystalline $NaAlP_2O_7$.[33] With the assistance of such oxygens, AlO_6 octahedra can be formed until N_{AlO} reaches the line with n(NBO)/n(Na)=0 (Figure 6). But this behaviour is not realised, and N_{AlO} roughly follows the model function for n(NBO)/n(Na)=1·33 (Q^0 units). Similar to the $K_3Al_2(PO_4)_3$ crystal[35] and following Equation (3), AlO_4 units dominate the Al_2O_3 component when approaching the Q^0 line (see Figure 5). A further decrease of N_{AlO} is not observed, and the few samples beyond the Q^0 line have constant N_{AlO} values.

The factor BC_{GFU} increases clearly for $0<x<0{\cdot}2$ because the addition of AlO_6 shows a stronger effect compared with that of BO_4 units (cf. Figures 4 and 6). The subsequent decrease of N_{AlO} is compensated with greater numbers of Al centred polyhedra and BC_{GFU} shows a plateau for $x>0{\cdot}2$. Here also, the change in slope for BC_{GFU} is well related to a change in slope for the reported T_g values.[32]

The structures of gallophosphate glasses should

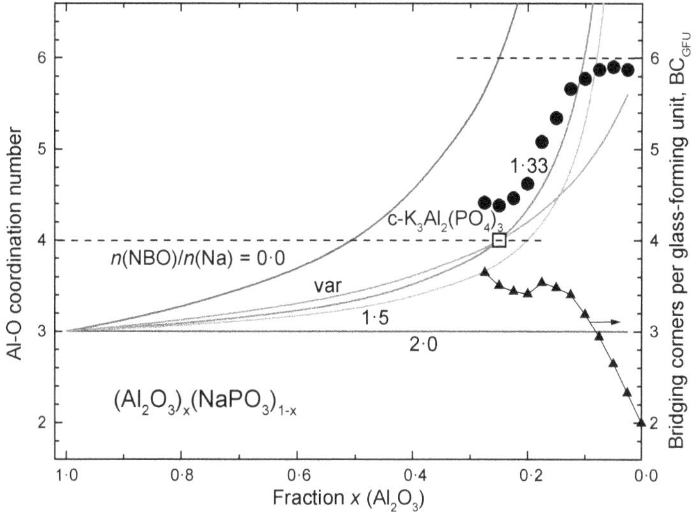

Figure 6. Comparison of the experimental values of N_{AlO} for $(Al_2O_3)_x(NaPO_3)_{1-x}$ glasses of series III in Figure 5 (solid circles[17]) with the model predictions of N_{AlO} according to Equations (3) and (5) (coloured solid lines). The ratios n(NBO)/n(A) are related to three Q^n species, variable Q^n ('var'), and the absence of NBO. The open square indicates the value of N_{AlO} for crystalline $K_3Al_2(PO_4)_3$.[33] The line with solid triangles shows the average number of bridging corners per glass-forming unit, BC_{GFU} [Colour available online]

behave similarly to aluminophosphate glasses due to the similarity in Ga–O and Al–O bonding. Ternary gallophosphate glasses are advantageous for a systematic structural study due to the large compositional range of glass formation. For example it is helpful to consider the series of pyrophosphate glasses $(Ga_{4/3}P_2O_7)_x(Na_4P_2O_7)_{1-x}$.[36] This series with n(NBO)/n(A)=1·5 for Q^1 units should show constant N_{GaO}=4·5 according to Equations (3) or (5). Contrary to the model and analogous to the ternary aluminophosphate glasses,[17] a strong tendency for GaO_6 octahedra exists at pyrophosphate compositions. The published values of N_{GaO} from ^{71}Ga MAS NMR and EXAFS experiments[36] are shown in Figure 7. The NMR results have smaller uncertainty, and also GaO_4, GaO_5 and GaO_6 units are identified. A clear dominance of GaO_6 units was observed. The N_{GaO} value for the sample with x~0·6 shows a visible decrease of the GaO_6 fraction, and an increase of the GaO_4 fraction. Presumably, the small Na_2O content of this sample is already not sufficient to stabilise GaO_6 octahedra with the needed amount of O sites (7) or (8). These changes end with the full absence of NBOs. That is reached either for crystalline $KGaP_2O_7$,[37] still with some K^+ ions, or the binary $Ga_{4/3}P_2O_7$ glass that is free of A^+ ions. The latter structure should show

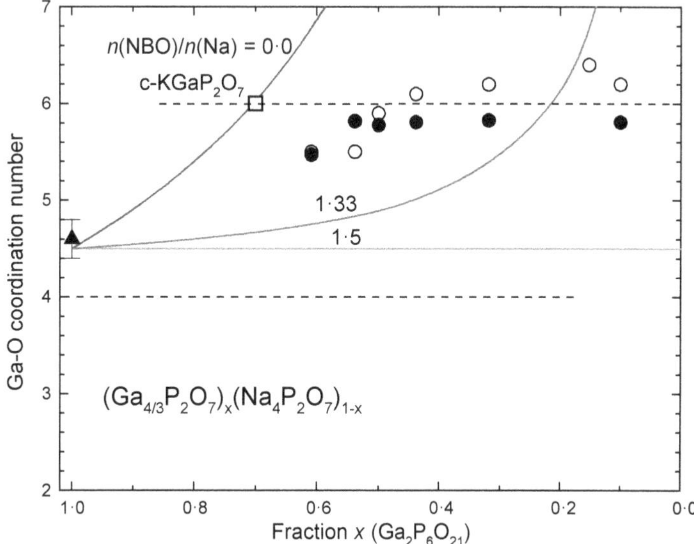

Figure 7. Comparison of the experimental values of N_{GaO} for $(Ga_{4/3}P_2O_7)_x(Na_4P_2O_7)_{1-x}$ glasses obtained by ^{71}Ga MAS NMR (solid circles) and EXAFS (open circles)[36] with the model predictions of N_{GaO} from Equation (3) (coloured solid lines). n(NBO)/n(Na)=0 means the absence of NBO. Constant N_{GaO}=4·5 results for Q^1 groups (1·5) as expected for diphosphate glasses, and n(NBO)/n(Na)=1·33 for Q^0 groups. In addition, the values of N_{GaO} for a binary glass of diphosphate composition[8] (solid triangle) and for crystalline $KGaP_2O_7$[37] (open square) are shown [Colour available online]

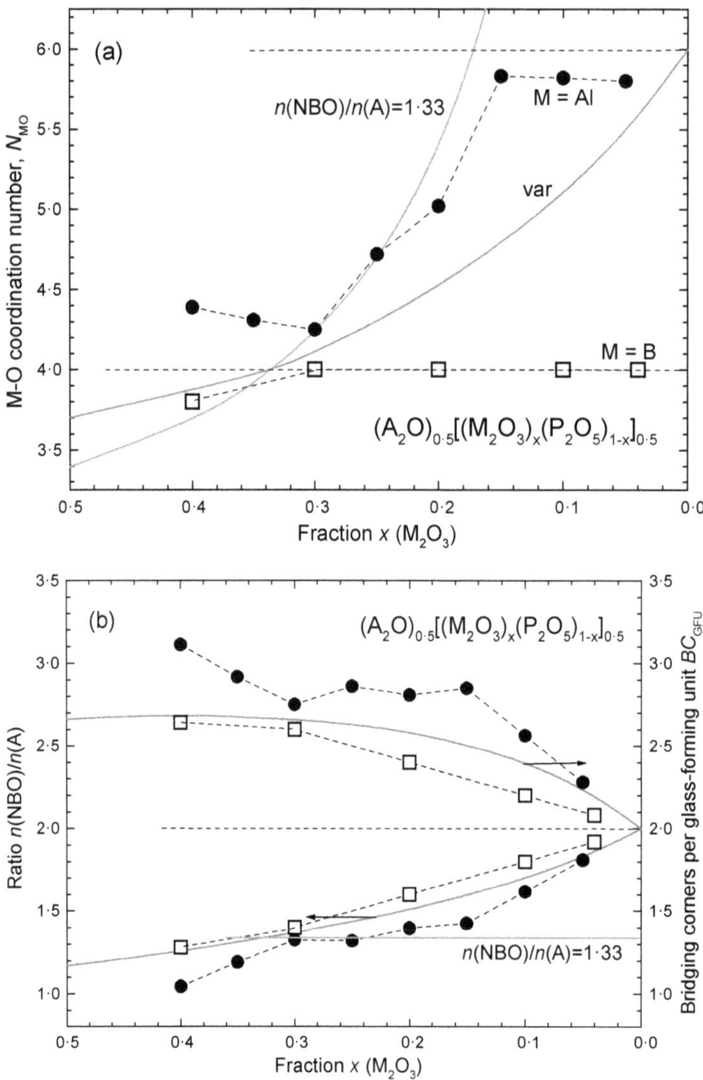

Figure 8. Comparison of the experimental M–O coordination numbers (a) and the corresponding values of bridging corners per glass-forming unit and ratios n(NBO)/n(A) (b) for the glasses $(A_2O)_{0·5}[(M_2O_3)_x(P_2O_5)_{1-x}]_{0·5}$ with M=Al, A=Na[17] (solid circles); M=B, A=Li[39] (open squares) with model functions calculated for variable n(NBO)/n(A) ratio (red solid lines) and fixed n(NBO)/n(A)=1·33 according to Q^0 groups (orange solid lines) [Colour available online]

only O sites (2), (4) and (5), and a Ga–O coordination number of ~4·5 is expected according to Equation (1); this is what was found by x-ray diffraction, giving N_{GaO}=4·6±0·2.[8]

Recently, a series of $(Ga_2O_3)_x(NaPO_3)_{1-x}$ glasses was measured by various ^{71}Ga, ^{31}P and ^{23}Na NMR techniques, and a comprehensive description of the structure was presented.[38] The distribution of GaO_4, GaO_5 and GaO_6 sites was determined from MAS and static NMR spectra. The authors emphasised the poor resolution of the ^{71}Ga NMR signals for the three Ga sites,[38] and that is the reason for differences between the two datasets. Nevertheless, the values of N_{GaO} behave similarly to the values of N_{AlO} shown for the Al series III[17] (cf. Figure 6). The fractions of nine differently linked PO_4 species (P and Ga neighbours) were analysed.[38] The change of the Q^n distribution makes use of all oxygen from the added Ga_2O_3 and Na_2O for rupturing of P–O–P bridges. The three groups Q^2, Q^1 and Q^0 co-exist in the pyrophosphate sample. The stabilisation of GaO_6 units at this composition is explained with elevated negative charge on the NBOs in the Q^1 groups and not with O sites (7) as in Ref. 17. A preference of GaO_6 for the Q^2 groups was not detected.

3.5 Comparison of boro- and aluminophosphate glasses

Reliable comparisons of structural parameters for M=B or Al can be done using samples of analogous compositions; this is possible for series IV of the sodium aluminophosphate system,[17] by comparing with the results for a series of $(Li_2O)_{0·5}[(B_2O_3)_x(P_2O_5)_{1-x}]_{0·5}$ glasses.[39] Figure 8(a) shows the evolution of the M–O coordination numbers versus M_2O_3 fraction, x. The model N_{MO} values are calculated according to Equation (3) with fixed n(NBO)/n(A)=1·33 (Q^0), and Equation (5) with variable $n(NBO)/n(A)$ as described in Section 2. The model N_{MO} values decrease continuously with increasing M_2O_3 content, and the latter model N_{MO} values (var) end with three for x=1. However, the experimental values of N_{AlO} and N_{BO} differ visibly from the model behaviour, and show agreement only for selected compositions. N_{AlO} starts at ~6 for small x, as predicted by the model (var). However, N_{AlO} is constant for low x. The value of N_{AlO} starts to decrease only when the ratio $n(NBO)/n(A)$ approaches 1·33 (Q^0 groups). Finally, N_{AlO} approaches ~4 for a glass composition x~0·3, again in agreement with the model. Further decrease of N_{AlO} is not observed. The maximum value N_{BO}=4 is observed for all samples rich in P_2O_5. Only when crossing the $n(NBO)/n(A)$ ratio of Q^0 units for x=0·33 do the values of N_{BO} start to decrease, here similar to the model N_{MO}.

The evolution of the values of $n(NBO)/n(A)$ (Figure 8(b), lower part), and the corresponding BC_{GFU} according to Equation (6) (Figure 8(b), upper part), are compared with experimental values for ternary boro- and aluminophosphate glasses. The deviation of the actual ratio $n(NBO)/n(A)$ from the corresponding model function (var) in the range $0<x<0·3$ can be interpreted as follows ('var' means that A^+ and M interact with the same Q^n distribution): The Li^+ ions of the borophosphate glasses (for which $n(NBO)/n(A)$ is greater than the model function) prefer Q^2 or Q^1 groups, whilst the Na^+ ions of the aluminophosphate glasses (for which $n(NBO)/n(A)$ is less than the model function) prefer Q^1 and Q^0 groups, see the constant line of 1·33 (Q^0 groups). Consequently, the BO_4 units interact with Q^1 or Q^0 groups, whereas the AlO_6 interact with Q^2 groups. This yields a consistent interpretation if one takes into account the valence in the M–O bonds with the different Q^n groups. It is difficult to estimate the contribution of the O sites (7) and (8) for stabilisation of the AlO_6 octahedra. The strong change in the behaviour of BC_{GFU} of the aluminophosphate glasses at x=0·15 can be related to the change found in the T_g values,[32] whereas a change of BC_{GFU} or T_g[39] for the borophosphate glasses does not occur at this composition.

4. Discussion

4.1 General features and exceptions to the model

The increase of M–O coordination numbers of MO_k units in links with PO_4 units is a unique characteristic of phosphate compounds and glasses. The fifth valence of the phosphorous influences the distribution of the bond valences in M–O–P bridges, with underbonded M–O and overbonded P–O. The corresponding changes of average N_{MO} values of binary M_yO_z–P_2O_5 glasses are simply related to the glass composition as given by Equation (1) but, of course, only up to the limits of the MO_k environments that are accepted by the M atoms. On the other hand, various scenarios for the evolution of N_{MO} are possible for multicomponent phosphate glasses, and quantitative predictions for N_{MO} seem to be difficult. Here, the focus is on 'simple' A_2O–M_yO_z–P_2O_5 glasses, with a glass-forming oxide and an alkali oxide. The simplest model for N_{MO} is given by Equation (5), which is described at the end of Section 2. Equation (4), which expresses equal balance of the electron charges on NBO for the A^+ ions or valence units in M–O bonds, is the critical point. The corners of the different Q^n are used by the A^+ ions and M atoms in portions proportional to their charges and mole fractions. The oxygen in M–O–P bridges is not shared with A^+ ions. The validity of this reasonable assumption could allow generalisation of this simple N_{MO} approach to many multicomponent phosphate glasses.

Phosphate glasses with an alkali oxide with non-directional ionic bonds of the A^+ to the NBOs, and a glass-forming oxide with M–O–P bridges, produce a comparably simple situation. However, the model N_{MO} values approximate the experimental N_{MO} values satisfactorily over only part of the glass formation range. Note, however, that the approach is very simple and agreement in distinct points can already be seen as a success. The limitations of acceptable N_{MO} values for a given M atom (which is due to the radius ratio of M to O atoms) are not seen as a deficit of the model.

Nevertheless, only the rough trends of N_{SiO} values for Na_2O–SiO_2–P_2O_5 glasses[20,22] could be reproduced. The value N_{BO}=4 for Na_2O–B_2O_3–P_2O_5 glasses[24–27] is correctly predicted for the Q^0 line (Figure 3). The values N_{AlO}=4 and 6 for Na_2O–Al_2O_3–P_2O_5 glasses[17] are well predicted for the Q^0 and Q^2 lines, respectively (Figure 5). The N_{GeO} values for Na_2O–GeO_2–P_2O_5 glasses[13] are correctly predicted, but with a constant ratio $n(NBO)/n(A)$=1·5 and Equation (3), rather than with Equation (5) (cf. Figure 2). Despite these deficits, the simple model reveals an applicable basis for predictions. Obviously, the O atoms prefer the bridging sites (2, 4, 5 in Figure 1) between the glass-forming units or highly charged ions, and avoid the sites (7) and (8). That relates the N_{MO} values to definite chemical compositions. The tendency to balanced distributions of charges seems to be favoured, which is expressed by the relation between the M and A sites given by Equation (4).

4.2 Continuous changes of the M–O coordination number

The model implies a continuous change of the average N_{MO} values with the glass composition, which requires flexible reconstructions of the atomic structure. Such flexibility is given by the interaction of the A^+ ions with the NBOs. The A^+ ions have flexible environments and share NBO neighbours of different charge with appropriate numbers and distances of A^+

ions. In principle, continuous changes of the average N_{MO} value could be possible from side of the M atoms, as well, with appropriate mixtures of the various MO_3, MO_4, MO_5, MO_6 or other units. However, the MO_k polyhedra need numbers k of O neighbours with definite valence in the M–O bonds in bridges with definite Q^n groups. Ideally, GeO_6 units need Q^1 neighbours (6×0·67vu=4vu), AlO_6 units need Q^2 (6×0·5vu=3vu) and AlO_4 units need Q^0 (4×0·75vu=3 vu). AlO_5 trigonal bipyramids are possible with three Q^1 and two Q^2 (3×0·67vu+2×0·5vu=3vu), and so are GeO_5 square pyramids with four Q^0 and a single bond (4×0·75vu+1vu=4vu). Thus MO_5 units need a suite of definite Q^n neighbours in the right positions. Some shift of electron charges from or to an adjacent NBO of the PO_4 could smooth the environment of the MO_5 unit.

In addition, the MO_5 unit could also be less acceptable due to its lower internal symmetry, in comparison with MO_4 and MO_6 polyhedra. This difference could produce a comparably small gain in bond energy. The measured distributions of AlO_k units for series III or IV of Na_2O–Al_2O_3–P_2O_5 glasses show that only minor fractions of AlO_5 units are present,[17] with analogous behaviour also for Na_2O–Ga_2O_3–P_2O_5 glasses,[36,38] and only a tiny fraction of SiO_5 is present in Na_2O–SiO_2–P_2O_5 glasses.[20] As a consequence, the mixtures of MO_4 and MO_6 do not allow the continuous behaviour of average N_{MO} which is described by the simple model. For ternary Na_2O–M_2O_3–P_2O_5 glasses, the MO_5 group is almost missing that could interact with Q^1 units. The alternative is the observed excess of MO_6 (M=Al,Ga) in the vicinity of the pyrophosphate composition that is accompanied by O sites (7) of the Q^1.[17,36] Otherwise, Q^1 units are the ideal neighbours for SiO_6 or GeO_6 octahedra. Meanwhile MO_5 pyramids should coordinate Q^0 units, and their numerical shortage complicates the transition to N_{SiO}=6 for silicophosphate glasses.[20,22] The small value of N_{GeO} for $(GeO_2)_x(NaPO_3)_{1-x}$ glasses with x~0·5,[13] in comparison with that of the model 'var' (Figure 2), can also be attributed to the poor contribution of GeO_5 polyhedra. Support for these interpretations comes from the observed distributions of AlO_k polyhedra in binary Al_2O_3–P_2O_5 glasses.[6] In these glasses, A^+ ions (and O sites (7)) do not exist. They would be needed for charge compensation of the AlO_6 octahedra with Q^1 units. Also the charge compensation that was suggested for the GaO_6–Q^1 interaction in ternary gallophosphate glass[38] cannot be effective in binary glasses. Therefore, binary M_2O_3–P_2O_5 glasses must contain considerable fractions of MO_5 units when approaching the pyrophosphate composition. Actually, a considerable fraction of AlO_5 units is formed,[6] though it never reaches 50%, and the N_{AlO} values obey Equation (1). Thus it is concluded that the simple N_{MO} model for ternary A_2O–M_yO_z–P_2O_5 glasses fails in definite compositional ranges due to insufficient participation of MO_5 units. The real behaviour is controlled by a preference for MO_6 or MO_4 units, whereby a weak preference is sufficient for the observed differences. These shortcomings of the model illustrate the problems for a general description of N_{MO} in phosphate glasses. If the third component were also a glass-forming or intermediate oxide, with a valence that differs from that of M, then even more complex interactions could be relevant.

4.3 Effects of Q^n distributions and longer range considerations

In the considerations of the evolution of the average value of N_{MO}, the linkages of the glass-forming units, PO_4 and MO_k, were not analysed in detail. The model with a variable ratio $n(NBO)/n(A)$ is simply related to the average n of the Q^n distribution with the maximum rupture of P–O–P bridges according to the $n(O)/n(P)$ ratio. A difference from binary Q^n distributions that was detected by ^{31}P MAS NMR for Na_2O–Ga_2O_3–P_2O_5 glasses[38] is described as $2Q^1 \rightarrow Q^2+Q^0$, and that has no effect on the average n of Q^n. In the case of Na_2O–GeO_2–P_2O_5[12,18] and A_2O–B_2O_3–P_2O_5 glasses,[4,25,26,39] it was found that the rupture of the P–O–P bridges is significantly reduced compared with the available oxygen fraction. Does that behaviour influence the predicted N_{MO} values? Surprisingly, the experimental N_{GeO} values were found to follow the model values[13] very well, even though the Q^n distribution[12] differs clearly from that of the maximum rupture according to the ratio $n(O)/n(P)$. But if one considers the relation that creates additional P–O–P bridges with

$$2M\text{–}O\text{–}P \leftrightarrow M\text{–}O\text{–}M + P\text{–}O\text{–}P \quad (7)$$

so the number of M–O bonds is not changed, and thus N_{MO} can remain unchanged. Several authors report M–O–M bridges in such glasses, even for samples of relatively low M_yO_z contents.[18,25] But this detailed information on the Q^n distributions is not needed for the model considerations on the M–O coordination numbers.

Knowledge of the P–O and M–O coordination numbers and the glass composition is sufficient for the determination of the fractions of BO and NBO. And the average numbers of bridging corners per glass-forming unit (BC_{GFU}) were calculated by Equation (6). Despite the lack of knowledge of the relation of BC_{GFU} to the absolute values of T_g, the obvious changes in slope of BC_{GFU}, and hence of the N_{MO} values of several A_2O–M_yO_z–P_2O_5 glasses, could be well related to the changes in slope in the compositional behaviour of T_g. These considerations assume all the BOs are simply crosslinking the network entities. Internal bridges of superstructural units would interfere with the topological behaviour. Such effects should be corrected as was shown for the fragility of borate glasses.[40]

An ordering on the scale of intermediate lengths (~1·0 nm) has been detected by means of pre-peaks in the diffraction data of alkali germanophosphate glasses.[13,19] These peaks were attributed to the ordering of separate alkali- and germanophosphate nanodomains. Corners of the GeO_k polyhedra do not interact with the A^+ ions in P_2O_5-rich glasses, and the networks are constructed of the O sites (2–5). A similar ordering could exist for alkali borophosphate glasses, and an analogous structural scenario has been discussed,[26,31] though BO_4 tetrahedra are smaller than GeO_6 octahedra. This ordering is not expected for alkali alumino- or gallophosphate glasses, where the AlO_6 or GaO_6 octahedra are stabilised by A^+ ions. Only glasses of compositions on the Q^0 line with $N_{AlO}(N_{GaO})$=4 could exhibit a similar ordering.

5. Conclusions

The observed changes of N_{MO} for binary M_yO_z–P_2O_5 glasses are well related to the ratio $n(O)/n(P)$, i.e. to the available terminal oxygen fraction. Such clear behaviour is not expected for ternary A_2O–M_yO_z–P_2O_5 glasses. However, if A^+ and M sites interact separately with the phosphate component of P_2O_5-rich glasses, A^+ with NBOs and M in M–O–P bridges, a balance between the counter charge for A^+ and the valence M–O bonds could occur. This scenario would cause similarly simple relations between the average value of N_{MO} and the glass composition, i.e. according to Equations (3) or (5).

In the case of alkali borophosphate glasses, the equations of the simple model are only valid on the line between the A_3PO_4 and BPO_4 compositions with N_{BO}=4. Beyond this line, N_{BO} starts to decrease as the P_2O_5 content decreases.

For alkali alumino- and gallophosphate glasses, the predicted N_{MO} values (M=Al, Ga) of the simple model approximate the experimental values for glasses on the lines between the A_3PO_4 and MPO_4 compositions with $N_{MO}\approx 4$, and between the APO_3 and $M(PO_3)_3$ compositions with $N_{MO}\approx 6$. Glasses between these two compositional lines show an excess of MO_6 octahedra. The reason for this alternative is attributed to the poor stability of MO_5 pyramids, and that mixtures of MO_4 and MO_6 cannot compensate for the lack of MO_5 polyhedra. MO_6 units are improper neighbours for Q^1 groups, but the surplus charge is compensated by A^+ ions. With further decrease of the P_2O_5 content, N_{MO} follows the model behaviour according to Equation (3) with a ratio $n(NBO)/n(A)$=1·33 (Q^0 groups), until the Q^0 compositions with N_{MO}=4 are reached.

Unlike Al or Ga, Q^1 groups are the optimal neighbours for GeO_6 octahedra in alkali germanophosphate glasses, and N_{GeO}=5·6 was found close to the Q^1 line of compositions. With further decrease of the P_2O_5 content, N_{GeO} follows a behaviour according to Equation (3) in the case of A=Na, best approximated with $n(NBO)/n(A)$=1·5 (Q^1 groups). The analogous alkali silicophosphate glasses also show a fraction of Si in SiO_6 octahedra, which do not usually exist in glasses prepared at ambient pressure. However, these fractions are smaller than predicted by the model, and thus a fraction of PO_4 units must exist with corners formed of P=O bonds.

The value of N_{MO} has an effect on the fraction of bridging oxygens. Hence changes in slope of the composition-dependence of N_{MO} are also visible in the behaviour of the glass transition temperature.

References

1. Van Wazer, J. R. Phosphorus and Its Compounds, Vol. 1, Interscience, New York, 1958, p. 717 ff.
2. Brow, R. K. Review: the Structure of simple phosphate glasses. *J. Non-Cryst. Solids*, 2000, **263&264**, 1–28.
3. Hoppe, U., Walter, G., Kranold, R. & Stachel, D. Structural specifics of phosphate glasses probed by diffraction methods: a review. *J. Non-Cryst. Solids*, 2000, **263&264**, 29–47.
4. Larink, D., Eckert, H., Reichert, M. & Martin, S. W. Mixed Network Former Effect in Ion-Conducting Alkali Borophosphate Glasses: Structure/Property Correlations in the System $[M_2O]_{1/3}[(B_2O_3)_x(P_2O_5)_{1-x}]_{2/3}$ (M = Li, K, Cs). *J. Phys. Chem. C*, 2012, **116**, 26162–26176.
5. Hoppe, U., Walter, G., Kranold, R., Stachel, D. & Barz, A. The dependence of structural peculiarities in binary phosphate glasses on their network modifier content. *J. Non-Cryst. Solids*, 1995, **192&193**, 28–31.
6. Brow, R. K., Click, C. A. & Alam, T. M. Modifier coordination and phosphate glass networks. *J. Non-Cryst. Solids*, 2000, **274**, 9–16.
7. Hoppe, U., Metwalli, E., Brow, R. K. & Neuefeind, J. High-energy X-ray diffraction study of La co-ordination in lanthanum phosphate glasses. *J. Non-Cryst. Solids*, 2002, **297**, 263–274.
8. Hoppe, U., Ilieva, D. & Neuefeind, J. The structure of gallium phosphate glasses by high-energy X-ray diffraction. *Z. Naturforsch.*, 2002, **57a**, 709–715.
9. Hoppe, U., Gugov, I., Bürger, H., Jóvári, P. & Hannon, A. C. Structure of tellurite glasses – effects of K_2O or P_2O_5 additions studied by diffraction. *J. Phys.: Condens. Matter*, 2005, **17**, 2365–2386.
10. Zwanziger, J. W., Shaw, J. L., Werner-Zwanziger, U. & Aitken, B. G. A neutron scattering and nuclear magnetic resonance study of the structure of GeO_2-P_2O_5 glasses. *J. Phys. Chem. B*, 2006, **110**, 20123–20128.
11. Hoppe, U., Brow, R. K., Tischendorf, B. C., Jóvári, P. & Hannon A. C. Structure of GeO_2-P_2O_5 glasses studied by x-ray and neutron diffraction. *J. Phys.: Condens. Matter*, 2006, **18**, 1847–1860.
12. Ren, J. & Eckert, H. Quantification of Short and Medium Range Order in Mixed Network Former Glasses of the System GeO_2-$NaPO_3$: A Combined NMR and X-ray Photoelectron Spectroscopy Study. *J. Phys. Chem. C*, 2012, **116**, 12747–12763.
13. Hoppe, U., Wyckoff, N. P., Brow, R. K., von Zimmermann, M. & Hannon, A. C. Structure of Na_2O-GeO_2-P_2O_5 glasses by X-ray and neutron diffraction. *J. Non-Cryst. Solids*, 2014, **390**, 59–69.
14. Hoppe, U., Kranold, R., Barz, A., Stachel, D., Schöps, A. & Hannon, A. C. X-ray and neutron scattering studies of the structure of copper phosphate glasses. *Phys. Chem. Glasses*, 2007, **48**, 188–194.
15. Voronkova, V. I., Yanovskii, V. K., Sorokina, N. J., Verin, I. A. & Simonov, V. I. Ferroelectric phase-transition and atomic-structure of $KGeOPO_4$ crystals. *Kristallografiya*, 1993, **38**, 147–151.
16. Hoppe, U., Brow, R. K., Wyckoff, N. P., Schöps, A. & Hannon, A. C. Structure of potassium germanophosphate glasses by X-ray and neutron diffraction. Part 1: Short-range order. *J. Non-Cryst. Solids*, 2008, **354**, 3572–3579.
17. Brow, R. K., Kirkpatrick, R. J. & Turner, G. L. Nature of Alumina in Phosphate Glass: II, Structure of Sodium Aluminophosphate Glass. *J. Am. Ceram. Soc.*, 1993, **76**, 919–928.
18. Behrends, F. & Eckert, H. Mixed Network Former Effects in Oxide Glasses: Spectroscopic Studies in the System $(M_2O)_{1/3}[(Ge_2O_4)_x(P_2O_5)_{1-x}]_{2/3}$. *J. Phys. Chem. C*, 2014, **118**, 10271–10283.
19. Hoppe, U., Walter, G., Brow, R. K. & Wyckoff, N. P. Structure of potassium germanophosphate glasses by X-ray and neutron diffraction: 2. Medium-range order. *J. Non-Cryst. Solids*, 2008, **354**, 3400–3407.

20. Dupree, R., Holland, D. & Mortuza, M. G. Six-coordinated silicon in glasses. *Nature*, 1987, **328**, 416–417.
21. Hesse, K. F. Refinement of the crystal-structure of silicon diphosphate, SiP_2O_7 A_{IV}-phase with 6-coordinated silicon. *Acta Crystallogr. B*, 1979, **35**, 724–725.
22. Ide, J., Ozutsumi, K., Kageyama, H., Handa, K. & Umesaki, N. XAFS study of six-coordinated silicon in R_2O-SiO_2-P_2O_5 (R = Li, Na, K) glasses. *J. Non-Cryst. Solids*, 2007, **353**, 1966–1969.
23. Königstein, K. & Jansen, M. A simple route to silicon in octahedral oxygen coordination. *Chem. Ber.*, 1994, **127**, 1213–1218.
24. Videau, J. J., Ducel, J. F. & Couzi, M. A structural approach to (1−x) $NaPO_3 \cdot xNa_2B_4O_7$ glasses. The correlation of structure with physical and chemical properties. *Phys. Chem. Glasses*, 1994, **35**, 253–257.
25. Zielnok, D., Cramer, C. & Eckert, H. Structure/Property Correlations in Ion-Conducting Mixed-Network Former Glasses: Solid-State NMR Studies of the System Na_2O-B_2O_3-P_2O_5. *Chem. Mater.*, 2007, **19**, 3162–3170.
26. Christensen, R., Olson, G. & Martin, S. W. Structural Studies of Mixed Glass Former $0.35Na_2O + 0.65[xB_2O_3 + (1 - x)P_2O_5]$ Glasses by Raman and ^{11}B and ^{31}P Magic Angle Spinning Nuclear Magnetic Resonance Spectroscopies. *J. Phys. Chem. B*, 2013, **117**, 2169–2179.
27. Carta, D., Qiu, D., Guerry, P., Ahmed, I., Abou Neel, E. A., Knowles, J. C., Smith, M. E. & Newport, R. J. The effect of composition on the structure of sodium borophosphate glasses. *J. Non-Cryst. Solids*, 2008, **354**, 3671–3677.
28. Xiong, D. B., Chen, H. H., Yang, X. X. & Zhao, J. T. Low-temperature flux syntheses and characterizations of two 1-D anhydrous borophosphates: $Na_3B_6PO_{13}$ and $Na_3BP_2O_8$. *J. Solid State Chem.*, 2007, **180**, 233–239.
29. Neumair, S. C., Sohr, G., Vanicek, S., Wurst, K., Kaindl, R. & Huppertz, H. The new high-pressure sodium tetraborate HP-$Na_2B_4O_7$. *Z. Anorg. Allg. Chemie*, 2012, **638**, 81–87.
30. Vogel, W. *Glass Chemistry* (Second Edition – Translation of the German third edition, translated by N. Kreidl & M. Lopes Barreto), Springer-Verlag, Berlin – Heidelberg, 1994, p. 144.
31. Christensen, R., Olson, G. & Martin, S. W. Ionic Conductivity of Mixed Glass Former $0.35Na_2O + 0.65[xB_2O_3 + (1\text{-}x)P_2O_5]$ Glasses. *J. Phys. Chem. B*, 2013, **117**, 16577–16586.
32. Brow, R. K. Nature of Alumina in Phosphate Glass: I, Properties of Sodium Aluminophosphate Glass. *J. Am. Ceram. Soc.*, 1993, **76**, 913–918.
33. Oudahmane, A., Avignant, D. & Zambon, D. Dipotassium dialuminium cyclooctaphosphate. *Acta Crystallogr. E*, 2010, **66**, i49-u114.
34. Alkemper, J., Paulus, H. & Fuess, H. Crystal-structure of aluminum sodium pyrophosphate, $NaAlP_2O_7$. *Z. Kristallogr.*, 1994, **209**, 616–616.
35. Devi, R. N. & Vidyasagar, K. Solid-state synthesis and characterization of novel aluminophosphates, $A_3Al_2P_3O_{12}$ (A = Na, K, Rb, Tl): Influence of A^+ ions on the coordination of aluminum. *Inorg. Chem.*, 2000, **39**, 2391–2396.
36. Belkébir, A., Rocha, J., Esculcas, A. P., Berthet, P., Poisson, S., Gilbert, B., Gabelica, Z., Llabres, G., Wijzen, F. & Rulmont, A. Structural characterization of glassy phases in the system Na_2O-Ga_2O_3-P_2O_5 by MAS NMR, EXAFS and vibrational spectroscopy. I. Cations coordination. *Spectrochimica Acta A*, 2000, **56**, 423–434.
37. Genkina, E. A. & Timofeeva, V. A. Synthesis and structure of double pyrophosphates $KFeP_2O_7$ and $KGaP_2O_7$. *J. Struct. Chem.*, 1989, **30**, 149–151.
38. Ren, J. & Eckert, H. Intermediate Role of Gallium in Oxide Glasses: Solid State NMR Structural Studies of the Ga_2O_3-$NaPO_3$ System. *J. Phys. Chem. C*, 2014, **118**, 15386–15403.
39. Muñoz, F., Montagne, L., Pascual, L. & Durán, A. Composition and structure dependence of the properties of lithium borophosphate glasses showing boron anomaly. *J. Non-Cryst. Solids*, 2009, **355**, 2571–2577.
40. Sidebottom, D. L. & Schnell, S. E. Role of intermediate-range order in predicting the fragility of network-forming liquids near the rigidity transition. *Phys. Rev. B.*, 2013, **87**, 054202(1–6).

CONFERENCE PARTICIPANTS

Prof. Mario Affatigato
Coe College, USA
maffatig@coe.edu

Dr Ifty Ahmed
University of Nottingham, United Kingdom
ifty.ahmed@nottingham.ac.uk

Dr Lyubomir Alexandrov
Institute of General and Inorganic Chemistry, Bulgaria
l_lubo79@abv.bg

Prof. Bojja Appa Rao
Osmania University, India
apparao.bojja@gmail.com

Dr Salah Arafa
The American University, Egypt
smarafa@aucegypt.edu

Prof. Valerii Arbuzov
Research and Technological Institute of Optical Material Science, Russia
arbuzov@goi.ru

Dr Vladimir Aseev
ITMO University, Russia
aseev@oi.ifmo.ru

Prof. Ryszard Barczynski
Technical University of Gdansk, Poland
jasiu@mif.pg.gda.pl

Dr Emma Barney
University of Nottingham, United Kingdom
emma.barney@nottingham.ac.uk

Dr Axelle Baroni
Laboratoire de Physique Théorique de la Metiere Condensée, France
axelle.baroni@impmc.jussieu.fr

Prof. Ganghishetti Bhikshamaiah
Osmania University, India
gbhyd08@gmail.com

Dr Yaroslav Biryukov
St. Petersburg State University, Russia
y.p.biryukov@gmail.com

Dr Lucica Boroica
National Institute for Laser, Plasma and Radiaton Physics, Romania
boroica_lucica@yahoo.com

Dr David Bouttes
PMMH - EPCI, France
david.bouttes@espci.fr

Dr Olivier Bouty
CEA Marcoule, France
olivier.bouty@cea.fr

Ing. Pavlína Bozděchová
Preciosa a.s., Czech Republic
pavlina.bozdechova@preciosa.com

Dr Caio Barca Bragatto
Universidade Federal de Sao Carlos, Brazil
caio.bragatto@gmail.com

Prof. Delia S. Brauer
Friedrich Schiller University Jena, Germany
delia.brauer@uni-jena.de

Prof. Vadim Brazhkin
Institute for High Pressure Physics, Russia
brazhkin@hppi.troitsk.ru

Prof. Richard Brow
Missouri S&T, USA
brow@mst.edu

Prof. Rimma Bubnova
Institute of Silicate Chemistry, Russia
rimma_bubnova@mail.ru

Ing. Daniel Bustin
Prayon S.A., Belgium
dbustin@prayon.be

Dr Thierry Cardinal
ICMCB-CNRS, France
cardinal@icmcb-bordeaux.cnrs.fr

Dr Thibault Charpentier
CEA Saclay, France
thibault.charpentier@cea.fr

Mr Hsin-Yin Chiang
National United University, Taiwan
ever.win@msa.hinet.net

Ing. Mária Chromčíková
A.Dubček University of Trenčín, Vila Glass Center, Slovakia
maria.chromcikova@tnuni.sk

Dr Laurent Cormier
CNRS - UPMC, France
cormier@impmc.upmc.fr

Prof. Giovanna D´Angelo
University of Messina, Italy
gdangelo@unime.it

Dr Giuseppe Dalba
University of Trento, Italy
dalba@science.unitn.it

Dr Leire del Campo
CNRS/CEMHTI, France
leire.del-campo@cnrs-orleans.fr

Prof. Boris Denker
A.M. Prokhorov General Physics Institute, Russia
denker@lst.gpi.ru

Ing. Elena Derkacheva
Institute of Silicate Chemistry, Russia
derkachevael@gmail.com

Mr Gunhter Desinger
Heraeus Noblelight GmbH, Germany
gunther.desinger@heraeus.com

Dr Manfred Dubiel
Martin Luther University of Halle-Wittenberg, Germany
manfred.dubiel@physik.uni-halle.de

Prof. Doris Ehrt
University Jena, Germany
doris.ehrt@uni-jena.de

Dr Roland Ehrt
IGK Roland Ehrt Ingenieurleistung Glas-Keramik-QM, Germany
roland.ehrt@glaskeram.de

Prof. Viacheslav Eremyashev
South Ural State University, Russia
vee-zlat@mineralogy.ru

Dr Hua Fan
CNRS/CEMHTI, France
hua.fan@cnrs-orleans.fr

Prof. Evelyne Fargin
ICMCB-CNRS, France
fargin@icmcb-bordeaux.cnrs.fr

Dr Franck Fayon
CNRS/CEMHTI, France
franck.fayon@cnrs-orleans.fr

Prof. Steve Feller
Coe College, USA
sfeller@coe.edu

Dr Guillaume Ferlat
UPMC/IMPMC, France
ferlat@impmc.upmc.fr

Prof. Stanislav Filatov
St. Petersburg State Univ, Russia
filatov.stanislav@gmail.com

Dr Shingo Fuchi
Aoyama Gakuin University, Japan
fuchi@ee.aoyama.ac.jp

Dr Yutaka Fujimoto
Tohoku University, Japan
fuji-you@qpc.che.tohoku.ac.jp

Prof. Jerzy Garbarczyk
Warsaw University of Technology, Poland
garbar@if.pw.edu.pl

Ing. Tadeáš Gavenda
Institute of Chemical Technology, Czech Republic
tadeas.gavenda@vscht.cz

Mrs. Radka Gegova
Institute of General and Inorganic Chemistry, Bulgaria
r.gegova@svr.igic.bas.bg

Ing. Alain Germeau
Prayon S.A., Belgium
agermeau@prayon.be

Mr Matthias Glätzle
University of Innsbruck, Austria
matthias.glaetzle@uibk.ac.at

Dr Liudmila Gorelova
St.Petersburg State University, Russia
gorelova.ljudmila@gmail.com

Dr Alex C. Hannon
ISIS, Rutherford Appleton Lab, United Kingdom
alex.hannon@stfc.ac.uk

Dr Ruzha Harizanova
University of Chemical Technology and Metallurgy, Bulgaria
ruza_harizanova@yahoo.com

Dr Dongbing He
Shanghai Institute of Oprics and Fine Mechanics, China
hdb798123@163.com

Dr Christian Hermansen
Aalborg University, Denmark
chh@bio.aau.dk

Dr Petr Hockicko
University Žilina, Slovakia

Dr Jana Holubová
Univeristy of Pardubice, Czech Republic
jana.holubova@upce.cz

Prof. Tsuyoshi Honma
Nagaoka University of Technol, Japan
honma@mst.nagaokaut.ac.jp

Dr Uwe Hoppe
Rostock University, Germany
uwe.hoppe@uni-rostock.de

Prof. Shinya Hosokawa
Kumamoto University, Japan
hosokawa@sci.kumamoto-u.ac.jp

Ing. Branislav Hruška
A.Dubček University of Trenčín, Vila Glass Center, Slovakia
branislav.hruska@tnuni.sk

Prof Hubert Huppertz
University of Innsbruck, Austria
hubert.huppertz@uibk.ac.at

Dr Seiji Inaba
Asahi Glass Co., Ltd., Japan
seiji.inaba@agc.com

Dr Raluca Iordanescu
National Institute R&D for Optoelectronics INOE 2000, Romania
iorda85@yahoo.com

Mr Yoshiki Ishii
Niigata University, Japan
f12a030g@alumni.niigata-u.ac.jp

Ing. Petr Kalenda
University of Pardubice, Czech Republic
st26865@student.upce.cz

Dr Efstratios Kamitsos
National Hellenic Research Foundation, Greece
eikam@eie.gr

Dr Naoyuki Kitamura
National Institute of Advanced Industrial Science and Technology, Japan
naoyuki.kitamura@aist.go.jp

Mr Ian Kleman
Coe College/Washington High School Cedar Rapids, USA
sfeller@coe.edu

Prof. Jonathan Knowles
University College London, United Kingdom
j.knowles@ucl.ac.uk

Prof. Elena Kolobkova
University ITMO, Russia
kolobok106@rambler.ru

Prof. Takayuki Komatsu
Nagaoka University of Technology, Japan
komatsu@mst.nagaokaut.ac.jp

Prof. Ladislav Koudelka
University of Pardubice, Czech Republic
Ladislav.Koudelka@upce.cz

Prof. Sergey Krivovichev
St. Petersburg State University, Russia
s.krivovichev@spbu.ru

Prof. Scott Kroeker
Univeristy of Manitoba, Canada
scott.kroeker@umanitoba.ca

Prof. Denise Krol
University of California, USA
dmkrol@ucdavis.edu

Dr Maria Krzhizhanovskaya
St.Petersburg State University, Russia
krzhizhanovskaya@mail.ru

ing. Piotr Kupracz
Gdansk University of Technology, Poland
pkupracz@mif.pg.gda.pl

Dr Mijai Lee
KICET, Korea
im1004@kicet.re.kr

Ing. Philippe Lehuede
Corning SAS CETC, France
lehuedep@corning.com

Dr Michael Leister
Chemische Fabrik Budenheim KG, Germany
michael.leister@budenheim.com

Dr Gérald Lelong
UPMC/IMPMC, France
gerald.lelong@impmc.upmc.fr

Ing. Florian Lindner
Leibniz Institute of Photonic Technology, Germany
florian.lindner@ipht-jena.de

Prof. Marek Liška
A.Dubček University of Trenčín, Slovakia
marek.liska@tnuni.sk

Dr Nadja Lönnroth
Corning Incorporated, USA
lonnrothnt@corning.com

Prof. Steve Martin
Iowa State University, USA
swmartin@iastate.edu

Ing. Emmanuel Martins
Prayon S.A., Belgium
emartins@prayon.be

Prof. Matthieu Micolaut
Paris Sorbonne Universites, France
mmi@lptl.jussieu.fr

Prof. Andrea Moguš-Milankovič
Ruder Boškovič Institute, Croatia
mogus@irb.hr

Dr Doris Möncke
Otto-Schott Institute for Material Research, Germany
doris.moencke@uni-jena.de

Dr Lionel Montagne
ENS Chimie de Lille, France
lionel.montagne@univ-lille1.fr

Dr Petr Mošner
University of Pardubice, Czech Republic
Petr.Mosner@upce.cz

Dr Gavin Mountjoy
University of Kent, United Kingdom
G.Mountjoy@kent.ac.uk

Dr Francisco Muñoz
Ceramics and Glass Institute, CSIC, Spain
fmunoz@icv.csic.es

Mrs Laura Muñoz-Senovilla
Ceramics and Glass Institute, CSIC, Spain
lmsenovilla@icv.csic.es

Ing. Petr Neugebauer
Preciosa a.s., Czech Republic
petr.neugebauer@preciosa.com

Prof. Takahiro Ohkubo
Chiba University, Japan
ohkubo.takahiro@faculty.chiba-u.jp

Dr Yohei Ohsasa
Chiba University, Japan
adfa3645@chiba-u.jp

Dr Nadege Ollier
CEA, France
nadege.ollier@polytechnique.edu

Dr Armenak Osipov
Institute of Mineralogy of UB RAS, Russia
armik@mineralogy.ru

Dr Dimitrios Palles
National Hellenic Research Foundation, Greece
dpalles@eie.gr

Dr Andrew Parsons
University of Nottingham, United Kingdom
andrew.parsons@nottingham.ac.uk

Ing. Luka Pavic
Ruder Boškovič Institute, Croatia
lpavic@irb.hr

Mr Benjamin Perez
Coe College, USA
sfeller@coe.edu

Dr Tomasz Pietrzak
Warsaw University of Technology, Poland
topie@if.pw.edu.pl

Dr Joanna Pisarska
University of Silesia, Poland
joanna.pisarska@us.edu.pl

Prof. Wojciech Pisarski
University of Silesia, Poland
wojciech.pisarski@us.edu.pl

Dr Marcel Potuzak
Corning Incorporated, USA
potuzakm@corning.com

Dr Alex C. Priven
SciGlass, USA
priven@itcwin.com

Mr Antonín Račický
University Pardubice, Czech Republic
antonin.racicky@seznam.cz

Mr Paul Rasmussen
Coe College, USA
sfeller@coe.edu

Ing. Radoslav Raykov
University of Chemical Technology and Metallurgy, Bulgaria
rectorat@uctm.edu

Mrs Rachel Rice
Coe College/Simpson College, USA
sfeller@coe.edu

Prof. Jesús Ma. Rincón
CSIC-IET cc, Spain
jrincon@ietcc.csic.es

Dr Maxime Rioux
Université Laval/COPL, Canada
maxime.rioux.2@ulaval.ca

Ing. Christian Ritzberger
Ivoclar Vivadent AG, Liechtenstein
ch.ritzberger@ivoclarvivadent.com

Prof. Jean Rocherullé
Université de Rennes, France
jean.rocherulle@univ-rennes1.fr

Prof. Ana Candida M. Rodrigues
Universidade Fedral de Sao Carlos, Brazil
acmr@ufscar.br

Prof. Antonella Rossi
ETH Zurich/University of Cagliari, Switzerland/Italy
antonella.rossi@mat.ethz.ch, rossi@unica.it

Dr Ivana Rösslerová
University Pardubice, Czech Republic
rosslerova.ivana@seznam.cz

Dr Ali Saberi
University of Bayreuth, Germany
ali.saberi@uni-bayreuth.de

Dr Gnanamuthu Sahaya Baskaran
Adrhra Loyola College, India
sbalc@rediffmail.com

Dr Akira Saitoh
Ehime University, Japan
asaito@ehime-u.ac.jp

Prof. Philip Salmon
University of Bath, United Kingdom
p.s.salmon@bath.ac.uk

Dr Meike Schneider
Schott AG, Germany
Meike.Schneider@schott.com

Mr Michael Schwall
Schott AG, Germany
michael.schwall@schott.com

Ing. Georgy Shakhgildyan
D.I. Mendeleev University of Chemical Technology, Russia
georgiy.shahgildyan@gmail.com

Dr Kenji Shinozaki
Nagaoka University of Technology, Japan
kshinozaki@mst.nagaokaut.ac.jp

Dr Pavel Shirshnev
NRU ITMO, Russia
pavel.shirshnev@gmail.com

Prof. Yung Shi-Wen
National United University, Taiwan
hwyang@nuu.edu.tw

Prof. David Sidebottom
Creighton University, USA
sidebottom@creighton.edu

Prof. Steve Singleton
Coe College, USA
ssinglet@coe.edu

Dr Christoph Soeller
Heraeus Noblelight GmbH, Germany
christoph.soeller@heraeus.com

Mr Gerhard Sohr
University of Innsbruck, Austria
gerhard.sohr@uibk.ac.at

Mrs Marta Soltys
University of Silesia, Poland
martasoltys@interia.pl

Dr Fabiana Spadaro
Swiss Federal Institute of Technology, Switzerland
fabiana.spadaro@mat.ethz.ch

Dr Masanori Suzuki
Osaka University, Japan
suzuki@mat.eng.osaka-u.ac.jp

Ing. Natalia Anna Szreder
Gdansk University of Technology, Poland
nszreder@mif.pg.gda.pl

Prof. Akira Takada
Asahi Glass Company, Japan
akira-takada@agc.com

Prof Hiromichi Takebe
Ehime University, Japan
takebe.hiromichi.mk@ehime-u.ac.jp

Prof. Yoshikazu Takeda
Aichi Science & Technology Foundation, Japan
takeda@astf.or.jp

Ing. Helene Trégouet
Chimie ParisTech, France
helene.tregouet@chimie-paristech.fr

Dr Grégory Tricot
LASIR/UMR CNRS, France
gregory.tricot@univ-lille1.fr

Mr Jeremy Trimble
Austin Peay State University, USA
jtrimble1@my.apsu.edu

Prof. Leo van Wüllen
Augsburg University, Germany
leo.van.wuellen@physik.uni-augsburg.de

Dr Nataliia Vedishcheva
Institute of Silicate Chemistry, Russia
ionatali386@gmail.com

Prof. Nalluri Veeraiah
Acharya Nagarjuna University, India
nvr8@rediffmail.com

Dr Maryna Vorokhta
University Pardubice, Czech Republic
marina.vorohta@gmail.com

Dr Keita Watanabe
Aoyama Gakuin University, Japan
watanabemonchi@gmail.com

Ing. Anja Winterstein-Beckmann
Otto-Schott Institute for Material Research, Germany
winterstein.anja@uni-jena.de

Prof. Adrian Wright
Univeristy of Reading, United Kingdom
a.c.wright@reading.ac.uk

Prof. Takayuki Yanagida
Kyushu Institute of Technology, Japan
yanagida@lsse.kyutech.ac.jp

Dr Gandhi Yerramreddy
K.V.R. College, India
gandhi_y@rediffmail.com

Dr Randall Youngman
Corning Incorporated, USA
youngmanre@corning.com

Prof. Yuanzheng Yue
Aalborg University, Denmark
yy@bio.aau.dk

Mrs Mariiam Ziiatdinova
D.I.Mendeleev University of Chemical Technology, Russia
m.z.ziyatdinova@gmail.com

Mrs Lidia Žur
University of Silesia, Poland
lidia.zur@us.edu.pl

Prof. Josef Zwanziger
Dalhousie University, Canada
jzwanzig@dal.ca

Author Index

Subject Index

2017
24–28 July

The ninth international conference on
Borate Glasses, Crystals and Melts

The second international conference on
Phosphate Materials

St Anne's College
Oxford, UK

2nd Announcement

The 9th International Conference on
BORATE GLASSES, CRYSTALS AND MELTS
July 24 - 26, 2017

The ninth conference follows the previous meetings held since 1977. The Borate Conference will embrace all current scientific research on borate materials, containing B_2O_3, B_2S_3 or B_2Se_3, including investigations of structure, properties and applications.

Invited speakers:

Honoree, Steve Feller (Coe College, Iowa, USA)
How do the fundamental physical properties of borate glass reveal its underlying atomic structure?

Mario Affatigato (Coe College, Iowa, USA)
44 years of borates: Steve Feller's legacy

Diane Holland (University of Warwick, UK)
NMR studies of alkali borogermanate glasses and comparison with borosilicates

Simona Ispas (University of Montpellier, France)
Ab initio modelling of multicomponent borosilicate glasses.

Mohamed Rahaman (Missouri University of Science and Technology, USA)
Bioactive borate glasses: Compositional effects on structure, reactivity and cell response

Oliver Alderman (Materials Development Inc., USA)
Borate melt structure: Temperature dependent B-O bond lengths and coordination numbers from x-ray diffraction

Jun Matsuoka (University of Shiga Prefecture, Japan)
Thermal properties and viscosity of isotope substituted borate glasses: Concept of vibrational connectivity

Frank Hawthorne (University of Manitoba, Canada)
Bond valence and Lewis-acid-base relations: Their effects on bond topology and chemical composition of borate minerals

The 2nd International Conference on
PHOSPHATE GLASSES, CRYSTALS AND MELTS
July 26 - 28, 2017

This conference follows the successful first conference held in Pardubice, Czech Republic (2014). It will cover the structures, properties and applications of phosphate- and phosphate-containing glasses and crystals, including those based on P-O, P-S, P-Se, and P-O-N networks.

Invited speakers:

Honoree, Uwe Hoppe (Rostock University, Germany)
A career in diffraction:
Probing the structure of phosphate glasses

Alex Hannon (ISIS Neutron and Muon Source, UK)
X-rays, neutrons and phosphates

Tsuyoshi Honma (Nagaoka University of Technology, Japan)
Phosphate glass and glass-ceramics for rechargeable batteries.

Delia Brauer (Friedrich Schiller University Jena, Germany)
Following phosphate glass dissolution by P-31 NMR.

Morten Smedskjaer (Aalborg University, Denmark)
Structure-property transformations in hot compressed phosphate glasses.

Jincheng Du (University of North Texas, USA)
Structures of rare earth containing phosphate and phosphosilicate glasses, insight from molecular dynamics simulations.

Leo Van Wullen (University of Augsberg Germany)
Glass structure and its evolution at high temperatures:
Phase separation, crystallization and species exchange:
Lessons from solid state NMR spectroscopy

Francisco Munoz (Spanish National Research Council)
Viscosity of phosphate melts

Important dates

Abstract Submission deadline: *24th February 2017*

Confirmation of talk acceptance: *10th March 2017*

Early-bird registration deadline: *31st March 2017*

To register for more information, please visit the conference website:

WWW.BORATE-PHOSPHATE.SGT.ORG

Conference Themes

Major conference themes include:

- **Energy** including nuclear waste storage and electrolytes, super ionic systems and ionic conductivity.
- **Healthcare** including bone repair, bioactivity, dental and medical applications, glass ceramics and crystallisation.
- **Optical and data storage materials** including optical properties, glass fibres, luminescence, laser hosts.
- **Property-structure relationships** including mixed network glasses, structural modelling and simulations, mechanical properties, thermal properties, glasses in extreme conditions, glass ceramics, phase separation and inhomogeneities, thermodynamics, corrosion.
- **Fundamentals of the glassy state** including glass structure, short and intermediate range order, the glass transition, the boson peak, glass relaxation, crystallisation.
- **Novel glasses, crystals and melts** including synthesis and characterisation.
- **Technology and Industry** including current applications and future developments.

Conference Site – St Anne's College, Oxford, UK.

St Anne's College is one of the largest constituent colleges of the University of Oxford. It is located 15 minutes walk from the centre of Oxford and provides outstanding conference facilities. **Oxford** is home to the oldest University in the English speaking world and it is visited by leading academics from all over the world. It is also famous for its historic buildings, museums and shops. The architectural style of central Oxford has led to it being known as the "city of dreaming spires" – a term first used by poet Matthew Arnold. The conference venue is easily accessible from all major UK airports, with regular bus and train services running from Gatwick and Heathrow.

Committees

Local Organising Committee:

E.R. Barney (Chair: emma.barney@nottingham.ac.uk), A.C. Hannon, A.C. Wright, G. Mountjoy, I. Ahmed, J.K. Christie, J.V. Hanna, M. Marshall, N. Karpukhina, N.S. Barrow, P.A. Bingham, P.S. Salmon, D. Holland, R.A. Martin, ***C. Brown (Secretary: christine@sgt.org)***

Borate International Organising Committee:

A.C. Hannon (UK; chair: alex.hannon@stfc.ac.uk), A.C. Wright (UK), A.G. Clare (USA), A. Takada (Japan), L. Cormier (France), S. Kroeker (Canada), N.M. Vedishcheva (Russia), J.W. Zwanziger (Canada), S.A. Feller (USA), H. Huppertz (Austria), N. Umesaki (Japan), R.E. Youngman (USA), E.I. Kamitsos (Greece), D. Möncke (Germany), E.R. Barney (UK), R. Iordanova (Bulgaria), F. Rocca (Italy), A.A. Osipov (Russia), L. Koudelka (Czech Republic), R.K. Brow (USA).

Phosphate International Organising Committee:

R.K. Brow (USA; chair: brow@mst.edu), L. Koudelka (Czech Republic), S.W. Martin (USA), L. Montagne (France), U. Hoppe (Germany), D. Ehrt (Germany), I. Ahmed (UK), F. Munoz (Spain), H. Takebe (Japan), H. Yang (Taiwan), A. Ghosh (India), J.E. Garbarczyk (Poland), Y. Messaddeq (Canada), H. Eckert (Brazil), A.C. Hannon (UK).

The conference is organised by the Society of Glass Technology and the conference proceedings will be published in Glass Technology: European Journal of Glass Science and Technology Part A and Part B

Images: 1; https://en.wikipedia.org/wiki/Oxford#/media/File:Oxford_Montage_2012.png.
2; http://www.wphna.org/Oxford2014/about-oxford/, 3; http://www.st-annes.ox.ac.uk/home.
4; https://commons.wikimedia.org/wiki/File:Punting,_Cambridge,_Summer.jpg. 5; http://www.blenheimpalace.com/
6; https://en.wikipedia.org/wiki/Oxford#/media/File:High_Street,_Oxford,_England,_1890s.jpg.

WWW.BORATE-PHOSPHATE.SGT.ORG

www.ingramcontent.com/pod-product-compliance
Ingram Content Group UK Ltd.
Pitfield, Milton Keynes, MK11 3LW, UK
UKHW050615260726
13967UKWH00008B/2880